En vertu de l'article 1er de la Convention signée le 14 décembre 1960, à Paris, et entrée en vigueur le 30 septembre 1961, l'Organisation de Coopération et de Développement Économiques (OCDE) a pour objectif de promouvoir des politiques visant :

— à réaliser la plus forte expansion de l'économie et de l'emploi et une progression du niveau de vie dans les pays Membres, tout en maintenant la stabilité financière, et à contribuer ainsi au développement de l'économie mondiale ;
— à contribuer à une saine expansion économique dans les pays Membres, ainsi que les pays non-membres, en voie de développement économique ;
— à contribuer à l'expansion du commerce mondial sur une base multilatérale et non discriminatoire conformément aux obligations internationales.

Les pays Membres originaires de l'OCDE sont : l'Allemagne, l'Autriche, la Belgique, le Canada, le Danemark, l'Espagne, les Etats-Unis, la France, la Grèce, l'Irlande, l'Islande, l'Italie, le Luxembourg, la Norvège, les Pays-Bas, le Portugal, le Royaume-Uni, la Suède, la Suisse et la Turquie. Les pays suivants sont ultérieurement devenus Membres par adhésion aux dates indiquées ci-après : le Japon (28 avril 1964), la Finlande (28 janvier 1969), l'Australie (7 juin 1971) et la Nouvelle-Zélande (29 mai 1973). La Commission des Communautés Européennes participe aux travaux de l'OCDE (article 13 de la Convention de l'OCDE). La Yougoslavie prend part à certains travaux de l'OCDE (accord du 28 octobre 1961).

L'Agence de l'OCDE pour l'Energie Nucléaire (AEN) a été créée le 1er février 1958 sous le nom d'Agence Européenne pour l'Énergie Nucléaire de l'OECE. Elle a pris sa dénomination actuelle le 20 avril 1972, lorsque le Japon est devenu son premier pays Membre de plein exercice non européen. L'Agence groupe aujourd'hui tous les pays Membres européens de l'OCDE, ainsi que l'Australie, le Canada, les États-Unis et le Japon. La Commission des Communautés européennes participe à ses travaux.

L'AEN a pour principal objectif de promouvoir la coopération entre les gouvernements de ses pays participants pour le développement de l'énergie nucléaire en tant que source d'énergie sûre, acceptable du point de vue de l'environnement, et économique.

Pour atteindre cet objectif, l'AEN :

- *encourage l'harmonisation des politiques et pratiques réglementaires notamment en ce qui concerne la sûreté des installations nucléaires, la protection de l'homme contre les rayonnements ionisants et la préservation de l'environnement, la gestion des déchets radioactifs, ainsi que la responsabilité civile et l'assurance en matière nucléaire ;*
- *évalue la contribution de l'électronucléaire aux approvisionnements en énergie, en examinant régulièrement les aspects économiques et techniques de la croissance de l'énergie nucléaire et en établissant des prévisions concernant l'offre et la demande de services pour les différentes phases du cycle du combustible nucléaire ;*
- *développe les échanges d'informations scientifiques et techniques notamment par l'intermédiaire de services communs ;*
- *met sur pied des programmes internationaux de recherche et développement, et des entreprises communes.*

Pour ces activités, ainsi que pour d'autres travaux connexes, l'AEN collabore étroitement avec l'Agence Internationale de l'Energie Atomique de Vienne, avec laquelle elle a conclu un Accord de coopération, ainsi qu'avec d'autres organisations internationales opérant dans le domaine nucléaire.

FOREWORD

Safety evaluations of repositories for the final disposal of spent nuclear fuel and high-level radioactive waste use models to predict the long-term performance of the disposal system. It is important that these models can be used with a satisfactory level of confidence. Efforts are therefore directed at the validation of models, particularly when they are to be applied to specific repositories and sites.

The first symposium on the validation of geosphere models (GEOVAL) was organised by the Swedish Nuclear Power Inspectorate (SKI) and held in Stockholm in 1987. The increasing importance of validation studies at the international level and, more specifically, within the programme of the NEA Radioactive Waste Management Committee, led to a joint NEA/SKI sponsorship of the GEOVAL-90 Symposium.

The objective of GEOVAL-90 was to review the progress made during the last three years in the validation of geosphere models, to review plans for further work, and to discuss validation strategies in the context of nuclear waste disposal programmes, regulatory requirements and scientific progress. GOEVAL-90 focussed on flow and transport in the geosphere and the disturbed zone around excavations. Attention was given to the validation of coupled geochemical, thermo-hydromechanical and fracture-flow models.

These proceedings reproduce the papers contributed to the symposium and a summary of the panel discussions. The opinions, conclusions and recommendations presented in these proceedings are those of the authors only, and do not necessarily express the views of any Member country or international organisation. They are published on the responsibility of the Secretary-General.

Programme Committee

Kjell ANDERSSON, SKI, Sweden (Chairman)
Johan ANDERSSON, SKI, Sweden (Scientific Secretary)
Arnold BONNE, SCK/CEN, Belgium
Daniel GALSON, OECD/NEA
Claes THEGERSTRÖM, OECD/NEA
Chin-Fu TSANG, LBL, United States

AVANT-PROPOS

Les évaluations de sûreté des dépôts pour l'évacuation définitive des combustibles irradiés et des déchets de haute activité font appel à des modèles pour prévoir les performances à long terme du système d'évacuation. Il est essentiel que ces modèles puissent être utilisés avec un niveau de confiance satisfaisant. Des efforts sont donc consacrés à la validation des modèles, notamment lorsqu'on applique ceux-ci à des dépôts et des sites particuliers.

La première conférence sur la validation des modèles relatifs à la géosphère (GEOVAL) a été organisée par le Service d'inspection nucléaire suédois (SKI) à Stockholm, en mai 1987. L'importance croissante des études de validation au niveau international, et plus spécialement au sein du Programme du Comité de la gestion des déchets radioactifs de l'AEN, est à l'origine du copatronage par l'AEN et le SKI de la Conférence GEOVAL-90.

L'objectif de GEOVAL-90 était de passer en revue les progrès accomplis au cours des trois dernières années sur la validation des modèles relatifs à la géosphère, d'examiner les projets d'activités futures et de débattre des stratégies de validation dans le contexte des programmes d'évacuation des déchets nucléaires, des exigences réglementaires et des progrès scientifiques. L'accent de GEOVAL-90 portait sur les écoulements d'eau et le transport dans la géosphère et la zone perturbée autour des excavations. La validation des modèles couplant les phénomènes géochimiques et thermo-hydro-mécaniques ainsi que les écoulements dans les fractures, a fait l'objet d'une attention particulière.

Ce compte rendu reproduit les exposés présentés à la Conférence ainsi qu'un résumé des discussions de la Table ronde. Les opinions, conclusions et recommandations présentées dans ce compte rendu sont celles des différents auteurs et n'expriment pas nécessairement les points de vue d'un pays Membre ou d'une organisation internationale quelconque. Il est publié sous la responsabilité du Secrétaire général.

Comité du Programme

Kjell ANDERSSON, SKI, Suède (Président)
Johan ANDERSSON, SKI, Suède (Secrétaire scientifique)
Arnold BONNE, SCK/CEN, Belgique
Daniel GALSON, OCDE/AEN
Claes THEGERSTRÖM, OCDE/AEN
Chin-Fu TSANG, LBL, Etats-Unis

TABLE OF CONTENTS
TABLE DES MATIERES

OPENING OF THE SYMPOSIUM

L. Högberg, Director General, SKI 21
J-P. Olivier, OECD/NEA 23

Session 1 - Séance 1

OPENING SESSION

SEANCE D'OUVERTURE

Chairman - Président

D.H. ALEXANDER (United States)

IN SEARCH OF TRUTH: THE REGULATORY NECESSITY OF VALIDATION

U. Niederer (HSK, Switzerland) 29

VALIDATION OF CONCEPTUAL MODELS OF FLOW AND TRANSPORT IN POROUS FRACTURED MEDIA

G. de Marsily (ENSM, France) 36

VALIDATION OF GEOCHEMICAL MODELLING AND DATA FOR ASSESSING THE PERFORMANCE OF GEOLOGICAL DISPOSAL OF NUCLEAR WASTES

F.P. Sargent (AECL, Canada) 51

VALIDATION STRATEGY AND STRATEGIC VALIDATION. AS SEEN BY A SAFETY ASSESSOR

T. Papp (SKB, Sweden) 62

Session 2 - Séance 2

PROGRESS IN VALIDATION OF FLOW AND TRANSPORT MODELS

ETAT D'AVANCEMENT DES TRAVAUX DE VALIDATION SUR LES MODELES D'ECOULEMENT ET DE TRANSPORT

Chairmen - Présidents

E. PELTONEN (Finland)
N. CHAPMAN (United Kingdom)
C.-F. TSANG (United States)
T. NICHOLSON (United States)

2.1. Introduction

THE STUDY OF RELEVANT AND ESSENTIAL FLOWPATHS

O. Brotzen (F. Brotzen Consultants, Sweden) 73

RECENT ACCOMPLISHMENTS IN THE INTRAVAL PROJECT - A STATUS REPORT ON VALIDATION EFFORTS

T.J. Nicholson (NRC, United States) 81

HYDROCOIN LEVEL 3 - TESTING METHODS FOR SENSITIVITY/UNCERTAINTY ANALYSIS

B. Grundfelt, B. Lindbom, A. Larsson (KEMAKTA), K. Andersson (SKI), Sweden 93

IAEA CO-ORDINATED RESERACH PROGRAMME ON "THE SAFETY ASSESSMENT OF NEAR-SURFACE RADIOACTIVE WASTE DISPOSAL FACILITIES"

S. Hossain (IAEA) 104

2.2. Flow and Transport in Crystalline Rock

Ecoulement et transport dans les roches cristallines

DUAL-POROSITY MODELLING OF INFILTRATION EXPERIMENTS ON FRACTURED GRANITE

P.A. Smith, J. Hadermann, K. Bischoff (PSI, Switzerland) 110

WATER FLOW AND SOLUTE TRANSPORT THROUGH FRACTURED ROCK

J.E. Bolt, D.M. Pascoe, V.M.B. Watkins (Elcon Western Ltd.),
P.J. Bourke, R.D. Kingdon (AEA Technology), United Kingdom 119

MODELLING OF THE LARGE SCALE REDOX FRONT EVOLUTION IN AN OPEN PIT URANIUM MINE IN POÇOS DE CALDAS, BRAZIL

L. Romero, L. Moreno, I. Neretnieks (KTH, Sweden) 131

ANALYZING TRANSPORT IN LOW PERMEABLE FRACTURED ROCK USING THE DISCRETE FRACTURE NETWORK CONCEPT

B. Dverstorp (KTH), J. Andersson (SKI), Sweden 139

MODELLING ACTIVITIES FOR THE GROUNDWATER TRACER TEST PROGRAM AT THE WHITESHELL NUCLEAR RESEARCH ESTABLISHMENT BOREHOLE SITE

N.W. Scheier, L.H. Frost, C.C. Davison (AECL, Canada) 148

RESULTS OF A CHANNELLING EXPERIMENT IN STRIPA

H. Abelin, L. Birgersson, T. Ågren (CHEMFLOW AB), I. Neretnieks,
L. Moreno (KTH), Sweden 157

MODELLING TRACER TESTS IN A HIGHLY CONDUCTIVE FRACTURE ZONE IN THE FINNSJÖN AREA IN CENTRAL SWEDEN

R. Nordqvist, P. Andersson, E. Gustafsson (SGAB), P. Wikberg (SKB),
Sweden 165

EVALUATION OF EARTH TIDE MATRIX DISPERSION

U. Kuhlmann (ETH), S. Vomvoris, P. Hufschmied (NAGRA), Switzerland 173

VALIDATION OF FRACTURE FLOW MODELS IN THE STRIPA PROJECT

A. Herbert (AEA Technology), D. Hodgkinson (INTERA-ECL), United Kingdom,
W. Dershowitz (Golder Associates), J. Long (LBL), United States 182

PRESENT STATUS OF DEVELOPMENT FOR SAFETY ASSESSMENT ON GEOLOGICAL
DISPOSAL OF HIGH-LEVEL RADIOACTIVE WASTES

M. Kawanishi, Y. Tanaka, H. Komada (CRIEPI, Japan) 197

THE DISPERSAL OF COLLOIDS IN FRACTURED ROCK

P. Grindrod (INTERA-ECL, United Kingdom) 209

COMPLEXITY IN THE VALIDATION OF GROUND-WATER TRAVEL TIME IN
FRACTURED FLOW AND TRANSPORT SYSTEMS

P.B. Davies, R.L. Hunter (SNL), J.F. Pickens (INTERA Inc.),
United States 218

SOME PROPERTIES OF A CHANNEL NETWORK MODEL

H.S. Lee (KAERI, Korea), L. Moreno, I. Neretnieks (KTH, Sweden) 226

VALIDATION OF A CONCEPTUAL MODEL FOR FLOW AND TRANSPORT IN FRACTURES

A. Hautojärvi, V. Taivassalo, S. Vuori (VTT, Finland) 234

SITE CHARACTERIZATION FOR THE SWEDISH HARD ROCK LABORATORY

G. Bäckblom, P. Wikberg (SKB), G. Gustafson (Chalmers University of
Technology), R. Stanfors (IDEON), Sweden 242

VALIDATION OF TRANSPORT MODELS FOR USE IN REPOSITORY PERFORMANCE
ASSESSMENTS: A VIEW

C.P. Jackson, D.A. Lever, P.J. Sumner (AEA Technology, United Kingdom) .. 250

2.3. Flow and Transport in Unsaturated Media

Ecoulement et transport dans les milieux non saturés

A STOCHASTIC APPROACH FOR VALIDATING MODELS OF UNSATURATED FLOW

D. McLaughlin, S. Luis (MIT, United States) 258

HYDROLOGIC MODELING AND FIELD TESTING AT YUCCA MOUNTAIN, NEVADA

D.T. Hoxie (USGS, United States) 266

LABORATORY RESEARCH PROGRAM TO AID IN DEVELOPING AND TESTING
THE VALIDITY OF CONCEPTUAL MODELS FOR FLOW AND TRANSPORT
THROUGH UNSATURATED POROUS MEDIA

R.J. Glass (SNL, United States) 275

VALIDATION EFFORTS IN MODELING FLOW AND TRANSPORT THROUGH
PARTIALLY SATURATED FRACTURED ROCK - THE APACHE LEAP TUFF
STUDIES

T.C. Rasmussen, D.D. Evans (UAZ), T.J. Nicholson (NRC), United States ... 284

VALIDATION EFFORTS IN MODELING PARTIALLY SATURATED FLOW AND
TRANSPORT IN HETEROGENEOUS POROUS MEDIA - THE LAS CRUCES
TRENCH STUDIES

T.J. Nicholson (NRC), R.G. Hills (N.M. University),
P.J. Wierenga (UAZ), United States 292

THE WATERTIGHTNESS TO UNSATURATED SEEPAGE OF UNDERGROUND CAVITIES
AND TUNNELS

J.R. Philip (CSIRO, Australia) 301

2.4. Flow and Transport in media with high Salinity

Ecoulement et transport dans les milieux à forte salinité

APPROACHES TO THE VALIDATION OF MODELLING OF GROUNDWATER FLOW RELATED TO THE PERFORMANCE ASSESSMENT OF RADIOACTIVE WASTE DISPOSAL SITES IN ROCK-SALT

P. Glasbergen (RIVM, Netherlands), R. Storck (GSF, Federal Republic of Germany) 309

VERIFICATION AND VALIDATION OF COUPLED FLOW AND TRANSPORT MODELS

S.M. Hassanizadeh (RIVM, Netherlands) 317

MODELLING OF VARIABLE-DENSITY GROUNDWATER FLOW WITH RESPECT TO PLANNED RADIOACTIVE WASTE DISPOSAL SITES IN WEST GERMANY - VALIDATION ACTIVITIES AND FIRST RESULTS

K. Schelkes, P. Vogel, H. Klinge, R.-M. Knoop (BGR, Federal Republic of Germany) 328

FLOW AND TRANSPORT AT HIGH SALINITY

G. Arens, E. Fein (GSF, Federal Republic of Germany) 336

2.5. Flow and Transport in Clay

Ecoulement et transport dans l'argile

MIGRATION EXPERIMENTS IN THE UNDERGROUND FACILITY AT MOL TO VALIDATE SAFETY ASSESSMENT MODEL

M. Monsecour, M. Put, A. Fonteyne (SCK/CEN, Belgium), H. Yoshida (PNC, Japan) 344

2.6. General Papers on Flow and Transport

Articles généraux sur l'écoulement et le transport

THE JAERI PROGRAM FOR DEVELOPMENT OF SAFETY ASSESSMENT MODELS AND ACQUISITION OF DATA NEEDED FOR ASSESSMENT OF GEOLOGICAL DISPOSAL OF HIGH-LEVEL RADIOACTIVE WASTES

H. Matsuzuru (JAERI, Japan) 352

SENSITIVITY ANALYSIS OF GROUNDWATER FLOW MODEL USING THE DIFFERENTIAL ALGEBRA METHOD

H. Kimura (JAERI), A. Isono (Information and Mathematical Science Lab.), Japan 360

USE OF LARGE-SCALE TRANSIENT STRESSES AND A COUPLED ADJOINT-SENTIVITY/KRIGING APPROACH TO CALIBRATE A GROUNDWATER-FLOW MODEL AT THE WIPP SITE

R.L. Beauheim (SNL), A. Marsh LaVenue (INTERA), United States 369

UNCERTAINTIES IN GROUNDWATER TRANSPORT MODELLING - A COMPONENT OF UNCERTAINTY IN THE PERFORMANCE ASSESSMENT OF LOW-LEVEL RADIOACTIVE WASTE DISPOSAL SITES

C.Y. Hung (EPA, United States) 377

A COMPARISON OF RESULTS FROM GROUNDWATER FLOW MODELLING FOR TWO CONCEPTUAL HYDROGEOLOGICAL MODELS FOR THE KONRAD SITE

G. Arens, E. Fein, R. Storck (GSF, Federal Republic of Germany) 389

ARE PARTICLE-TRACKING ALGORITHMS THE ADEQUATE METHOD TO DETERMINE MIGRATION PATHS FOR PERFORMANCE ASSESSMENT?

P. Bogorinski (GRS, Federal Republic of Germany), A. Boghammar (Kemakta, Sweden) 396

SKB - SWEDISH HARD ROCK LABORATORY. PREDICTIVE GROUNDWATER FLOW MODELLING OF A LONG TIME PUMPING TEST AT ÄSPÖ

B. Lindbom, B. Grundfelt, A. Boghammar (Kemakta), M. Liedholm, I. Rhén (VIAK AB), Sweden 405

Session 3 - Séance 3

PROGRESS IN VALIDATION OF GEOCHEMICAL MODELS

ETAT D'AVANCEMENT DES TRAVAUX DE VALIDATION DES MODELES GEOCHIMIQUES

Chairman - Président

T. NICHOLSON (United States)

THE CHEMVAL PROJECT - AN INTERNATIONAL STUDY AIMED AT THE VERIFICATION AND VALIDATION OF EQUILIBRIUM SPECIATION AND CHEMICAL TRANSPORT MODELS

D. Read, T.W. Broyd (W.S. Atkins, United Kingdom), B. Côme (CEC) 417

HYDROGEOCHEMICAL MODELLING OF THE NEEDLE'S EYE NATURAL ANALOGUE (SCOTLAND)

P.J. Hooker (BGS, United Kingdom), Ph. Jamet, P. Lachassagne, E. Ledoux (ENSM), P. Escalier des Orres (CEA), France 425

THE RELAY SUBSTANCE METHOD: A NEW CONCEPT FOR PREDICTING POLLUTANT TRANSPORT IN SOILS

M. Jauzein, C. André, R. Margrita (CENG), M. Sardin, D. Schweich (CNRS), France 433

LESSONS LEARNED FROM MODEL VALIDATION - A REGULATORY PERSPECTIVE

P. Flavelle, S. Nguyen, W. Napier (AECB), D. Lafleur (Intera), Canada ... 441

RELATIONS BETWEEN GROUNDWATER COMPOSITION AND FRACTURE FILLING MINERALOGY IN SWEDISH DEEP ROCKS

K. Andersson (Lindgren o Andersson HB, Sweden) 449

ALTERATION OF CHLORITE AND ITS RELEVANCE TO THE URANIUM REDISTRIBUTION IN THE VICINITY OF THE ORE DEPOSIT AT KOONGARRA, NORTHERN TERRITORY, AUSTRALIA

T. Murakami, H. Isobe (JAERI, Japan) 458

VERIFICATION OF THE STELE CODE WITHIN THE FRAMEWORK OF THE CHEMVAL PROJECT

Ph. Jamet (ENSM), Ph. Jacquier (CEA), France 466

IMPACT COMPUTER CODE: A HELP IN DESIGN AND INTERPRETATION OF LABORATORY EXPERIMENTS

M. Jauzein, C. André, R. Margrita (CENG), M. Sardin, D. Schweich (CNRS), France 474

MIGRATION OF CHROMIUM IN THE SATURATED ZONE. SIMULATION OF COLUMN EXPERIMENTS

E. Tevissen, M. Jauzein, C. André, R. Margrita (CENG), M. Sardin, D. Schweich (CNRS), France 482

APPLICATION OF THE ALLIGATOR RIVERS ANALOGUE FOR VALIDATION OF SAFETY ASSESSMENT METHODOLOGIES

K. Skagius, K. Pers, F. Brandberg (KEMAKTA), S. Wingefors (SKI), Sweden, P. Duerden (ANSTO, Australia) 489

Session 4 - Séance 4

PROGRESS IN VALIDATION OF COUPLED THERMO-HYDRO-MECHANICAL EFFECTS

ETAT D'AVANCEMENT DES TRAVAUX DE VALIDATION DES EFFETS COUPLES THERMO-HYDRO-MECANIQUES

Chairman - Président

B. CÔME (CEC)

VALIDATION OF TWO ROCK MECHANICS CODES AGAINST COLORADO SCHOOL OF MINES BLOCK TEST DATA

O. Stephansson, T. Savilahti (Luleå University, Sweden), N. Barton, P. Chryssanthakis (NGI, Norway) 503

MODELLING THE ROOM 209 EXCAVATION RESPONSE TEST IN THE CANADIAN UNDERGROUND RESEARCH LABORATORY

T. Chan, E. Kozak, B.W. Nakka (AECL, Canada),
A. Winberg (SGAB, Sweden).. 511

PROGRESS IN VALIDATION OF STRUCTURAL CODES FOR RADIOACTIVE WASTE REPOSITORY APPLICATIONS IN BEDDED SALT

D.E. Munson (SNL), K.L. DeVries (RE/SPEC Inc.), United States 522

MODELLING THE EFFECT OF GLACIATION, ICE FLOW AND DEGLACIATION ON LARGE FAULTED ROCK MASSES

P. Chryssanthakis, N. Barton (NGI, Norway), J. Lanru, O. Stephansson
(Luleå University, Sweden) 530

STOCHASTIC THREE DIMENSIONAL JOINT GEOMETRY MODELLING INCLUDING A VERIFICATION TO AN AREA IN STRIPA MINE

P.H.S.W. Kulatilake, D.N. Wathugala (UAZ, United States),
O. Stephansson (Luleå University, Sweden) 542

ROCK MASS RESPONSE TO GLACIATION AND THERMAL LOADING FROM NUCLEAR WASTE

B. Shen, O. Stephansson (Luleå University, Sweden) 550

THREE DIMENSIONAL COUPLED THERMO-HYDRAULIC-MECHANICAL ANALYSIS CODE WITH PCG METHOD

Y. Ohnishi, S. Akiyama (Kyoto University), M. Nishigaki (Okayama
University), A. Kobayashi (Hazama Corporation), Japan 559

Session 5 - Séance 5

VALIDATION STRATEGY

STRATEGIE DE VALIDATION

Chairman - Président

P.E. AHLSTRÖM (Sweden)

VALIDATION OF MATHEMATICAL MODELS

R.G. Sargent (Syracuse University, United States) 571

A PROPOSED STRATEGY FOR THE VALIDATION OF GROUND-WATER FLOW AND SOLUTE TRANSPORT MODELS

P.A. Davis (SNL), M.T. Goodrich (GRAM Inc.), United States 580

NATURAL ANALOGUE STUDIES USEFUL IN VALIDATING REGULATORY COMPLIANCE ANALYSES

D.H. Alexander (DOE), A.E. Van Luik (Battelle), United States 589

SUFFICIENT VALIDATION: THE VALUE OF ROBUSTNESS IN PERFORMANCE ASSESSMENT AND SYSTEM DESIGN

C. McCombie, P. Zuidema, I.G. McKinley (NAGRA, Switzerland) 598

A FRAMEWORK FOR VALIDATION AND ITS APPLICATION ON THE SITE CHARACTERIZATION FOR THE SWEDISH HARD ROCK LABORATORY

G. Bäckblom, G. Gustafson, R. Stanford, P. Wikberg (SKB, Sweden) 611

SOME CONSIDERATIONS FOR VALIDATION OF REPOSITORY PERFORMANCE ASSESSMENT MODELS

N. Eisenberg (NRC, United States) 619

Session 6 - Séance 6

PANEL DISCUSSION

TABLE RONDE

Chairman - Président

S.P. NEUMAN (United States)

PANEL DISCUSSION

A. Larsson, Editor (KEMAKTA, Sweden) 633

CONCLUDING REMARKS

K. Andersson (SKI, Sweden) 657

LIST OF PARTICIPANTS - LISTE DES PARTICIPANTS 659

OPENING OF THE SYMPOSIUM

OUVERTURE DU SYMPOSIUM

INTRODUCTION AND WELCOME

Lars HÖGBERG
Director General
Swedish Nuclear Power Inspectorate

I extend a very hearty welcome to this Geoval 90 Symposium and a special welcome to Mr. Olivier the representative of our co-organising body the OECD/NEA.

In my opinion, the future of nuclear power heavily depends on the successful handling of two issues. The first issue concerns reactor safety and can be expressed as success in avoidance of reactor accidents with significant radioactive releases to the environment, in particular of nuclides causing long term ground contamination.

The second issue concerns success in gaining public acceptance for methods and actual installations for final disposal of spent fuel and high level waste. This symposium is one important step among the many, many remaining to reach success on the second issue.

The present Swedish nuclear waste management policy is largely based on the conviction that the same generation that reaps the benefits of nuclear power also has a moral obligation to carry the financial and technical burden to develop and build final repositories for the nuclear waste created in the process; repositories that are acceptable with respect to safety and radiation protection. Therefore, in Sweden, we have a legislation setting aside about 0.02 SEK for each nuclear kWh produced to pay for, waste management, now and in the future. Therefore, we also have legislation requiring the nuclear utilities through their joint company SKB to pursue a vigourous research and development program with the objective to have a final repository for spent fuel ready for commissioning in the first decades of the next century. And therefore, finally, the Swedish regulators have to pursue vigourous efforts to match the SKB program in the development of regulatory guidelines, for example, for the performance assessments to be submitted as a part of the licensing process.

Fulfillment of this moral obligation of the present generation is, in my opinion, an important prerequisite for gaining public confidence and acceptance with regard to national nuclear programmes. Some might argue that scientific and technical progress will be so enormous in the next 100 years that we should adopt a policy of "wait and see". To me, this is counter-productive to creating confidence and public acceptance. The repositories should be designed in such a way that the future generations have some options left to monitor and improve on the technical solutions, if they so wish, but we should not leave it to them to develop and build the repositories for the waste we created. Also I have more faith in the stability of some rock formations in the 10 000 to

100 000 year time frame than I have in the stability of society in the 100 year time frame.

Incidently, if the same approach discussed above had been chosen in the past for many types of chemical and other types of waste, we should certainly have less environmental problems to cope with today.

As performance assessments of high level waste repositories have to be based on calculations using a complex set of models, the validation of these models is extremely important both in order to create a firm basis for technical design work and as a basis for public confidence and acceptance.

The process of validation of models going all the way to both scientific and public acceptance, is a complex on which is discussed more in detail in the invited paper by Dr. Niederer. Let me just underline the importance of this process. Based on more than 15 years of experience of decision making on nuclear issues in Sweden, I have found that key decision makers in the political arena may, of course, not develop an in-depth scientific knowledge in the areas involved. However, they have in many cases developed a shrewd understanding of the process of scientific quality assurance through an open and critical discussion involving experts from many countries and institutions. This is especially important for small countries like Sweden where the number of domestic experts in each area by necessity is limited. This demonstrates the importance of symposia such as this Geoval 90 also with respect to public acceptance.

In this context, I would like to express my appreciation of the efforts made by the OECD Nuclear Energy Agency in promoting open and critical international scientific discussions on important nuclear safety issues. The former chairman of the NEA radiation waste management Committee, Dr. Rometsch, recently said that, in his opinion, one of the principal tasks of the NEA is and should remain "the assurance of quality of thought". I fully agree with Dr. Rometsch and I gratefully acknowledge the NEA involvment in Geoval 90 as a part of the fulfillment of this task.

So, in summary, I hope and expect that your discussions during this week will be both open, critical, and productive taking one more step towards validated models that can provide a firm basis for both hardware design and public acceptance. With this I leave the floor to Mr. Jean-Pierre Olivier - for many years responsible for NEA activities in the areas of waste management and radiation protection. Thank you.

INTRODUCTORY REMARKS

J.-P. Olivier
Head, Radiation Protection and Waste Management Division
OECD Nuclear Energy Agency

Ladies and Gentlemen,

I am particularly happy to welcome you today to GEOVAL 90, on behalf of the OECD Nuclear Energy Agency, who is co-sponroring the meeting with SKI.

As both SKI and NEA were interested in the issue of validation of models used in assessing the safety of radioactive waste repository systems and had their own plans for meetings in this field, the logical approach was to join together for the organisation of GEOVAL 90, and I sincerely hope that our joint efforts have provided a good frame for your discussions.

Before expanding further on validation, I would like to mention briefly NEA's main concerns and objectives in the area of radioactive waste management. Being the nuclear department of the OECD, the Organisation for Economic Co-operation and Development, well known as a reliable source of information on economic data and statistics and as a forum among industrially developed countries, the NEA has similar objectives in its more limited area of work. NEA's main priorities are in nuclear safety, radioactive waste management and radiation protection, in line with member governments interests for safey and regulatory matters.

In radioactive waste management, we already began to focus on disposal issues 15 years ago when our Radioactive Waste Management Comittee was set up. Treatment and conditioning of waste were deliberately given a low priority as these topics were covered by other international organisations and also by the nuclear industry itself. In 1977, we published the so-called "Polvani Report" on Objectives, Concepts and Strategies for the Management of Radioactive Waste, presenting the disposal issue in a broad perspective and emphasizing geological disposal research and safety assessment needs. Then we sponsored the International Stripa Project in this country.

In 1984, before the ICRP, we published a report on Long-Term Radiation Protection Objectives for Waste Dsiposal, introducing explicitly the concept of risk. I have read in the proceedings of the GEOVAL 87 meeting that the practical application of this concept raises difficulties for some of you. I can tell you that we were fully aware of this problem when the risk concept was proposed, but we felt that we could not ignore the probabilistic nature of future events that might affect the integrity of repositories. Logically, therefore, we gave a high priority to the work of our PSAC Group, the Probabilistic

Systems Assessment Code User Group, which was set up some five years ago. Today this Group is very active, and puts more and more emhasis on the modelling of realistic cases, including verification and quality assurance aspects.

In order to keep safety-related issues in perspective, and to address the right priorities on safety assessment matters, the NEA Performance Assessment Advisory Group (PAAG) fulfils an overview function and advises the RWMC on the NEA programme of work in this area. The main objective of PAAG is to assist in the development of methods and tools of high quality for safety assessment and to promote a balanced and coherent use of these methodologies within radioactive waste disposal programmes. As a parallel to PAAG, a new NEA Co-ordinating Group on Site Evaluation and Design of Experiments (SEDE) was set up recently, focussing on methodologies for data collection at specific sites. The common denominator for modellers, experimentalists, performance assessors and regulators is necessarily the scientific basis behind safety assessments, and many of you already know that NEA efforts in radioactive waste management are indeed concentrated in this area. In this respect, I cannot overstress the importance of validation.

I personally found the papers and panel discussions of GEOVAL 87 extremely useful, illustrating the sense of responsibility which characterise the management of radioactive waste, and the systematic search for some kind of validation for the models used, whatever is meant by validation. Of course, it is inherent to long-term assessments that we will never reach full proof and full validation, but is is also a rule of the game that we have to rely on all possible supportive material or evidence to comfort our final judgments about the safety of waste repositories. It is another rule of the game that at a given point in time, differing from one country to the next, a licence application will be made for a specific site and will have to be judged on the basis of available knowledge. This will automatically put a limit on our validation efforts, which will never be as deep as we would like them to be in an ideal situation.

This being said, we at NEA believe that the state-of-the-art for performance assessment of radioactive waste repositories is already well advanced, as discussed at the October 1989 joint CEC/IAEA/NEA Symposium in Paris. We are convinced, however, that many of the tools and models which have been or are being developed need to be verifed and checked whenever possible against observations from real situations and, therefore, be "validated".

We also recognise the need for future and new research, aiming at the reduction of uncertainties in performance assessment, the detailed characterisation of potential disposal sites and the ultimate confirmation of the safety of disposal systems. Given present plans, there are at least two decades of work ahead of us before the first high-level waste repositories become operational. GEOVAL 90 should be seen in this perspective.

Before closing I would like to express the appreciation of the NEA at being associated with SKI in this particular circumstance. NEA was involved in HYDROCOIN, and is now participating in the INTRAVAL Project secretariat. On each of those activities, as well as during NEA regular meetings, we have been able to enjoy fully co-operation with Lars Högberg and his staff, Alf Larsson who has now left the Inspectorate, Kjell Andersson and Johan Andersson notably for the preparation of this meeting, and Soren Norrby who is one of the vice-chairmen of the NEA RWMC. I would like on this occasion to pay a tribute to their competence and valuable initiatives in the area of waste management and international co-operation.

Finally, let me wish you a successful meeting, with fruitful discussion helping each of us to bridge the gap between educated guesses, theoretical models, scientific doubts and reasonable assurance.

Thank you.

Session 1

OPENING SESSION

Séance 1

SEANCE D'OUVERTURE

Chairman - Président

D.H. ALEXANDER
(United States)

IN SEARCH OF TRUTH: THE REGULATORY NECESSITY OF VALIDATION

U. Niederer
Swiss Nuclear Safety Inspectorate (HSK)
Wuerenlingen, Switzerland

ABSTRACT

Quid est veritas?
(Io. 18, 33)

A look at modern ideas of how scientific truth is achieved shows that theories are not really proved but accepted by a consensus of the experts, borne out by often repeated experience showing a theory to work well. In the same sense acceptability of models in waste disposal is mostly based on consensus. To obtain consensus of the relevant experts, including regulators, all models which considerably influence the results of a safety assessment have to be validated. This is particularly important for the models of geospheric migration because scientific experience with the deep underground is scarce. Validation plays a special role in public acceptance where regulators and other groups, which act as intermediaries between the public and the project manager, have to be convinced that all the relevant models are correct.

LA NOTION DE VERITE SCIENTIFIQUE : NECESSITE REGLEMENTAIRE DE LA VALIDATION

RESUME

L'analyse qui est faite aujourd'hui du processus d'élaboration de la vérité scientifique montre que les théories ne sont pas vraiment prouvées mais acceptées sur la base d'un consensus des experts, qui s'appuie sur la constatation répétée que l'expérience reflète correctement la théorie. Il en va de même pour l'acceptabilité des modèles relatifs à l'évacuation des déchets. Pour arriver à une convergence d'opinions des experts concernés, y compris les autorités de sûreté, il est nécessaire que soient validés tous les modèles ayant une influence décisive sur les résultats de l'analyse de sûreté. Vis-à-vis de l'acceptation par le public, la validation joue un rôle particulier car des groupes d'experts, qui agissent en tant qu'intermédiaires entre le public et le responsable de l'élimination des déchets, doivent être convaincus de la validité de chacun des modèles pertinents.

1. INTRODUCTION

In the context of waste management, validation has become a very popular term in recent years. Assessing the safety of a system as complicated as a waste repository is only possible if the system is first conceptually separated into a number of components, for each of which a model is then constructed to describe its performance. Modelling thus became a primary part of safety assessment. However, experience with performing and reviewing comprehensive safety assessments has shown that credibility given to the results of the assessments heavily depends on the validity of the models used in safety analysis. The safety to be assessed is the safety of the real system, not of the model. Hence validation, in the sense of testing a model in the real world, receives ever more attention. The goal of validation is to demonstrate that one really understands the relevant processes. Our models we understand by definition because we made them; what we have to demonstrate is that we understand the nature behind the models as well.

Validation in a broader sense has a long tradition in science and goes back at least to the beginnings of modern science around 1600 where scientists tried to test their theories by experiments. What has changed since is the jargon: In waste management rather than of theories we speak of models, and instead of testing theories we validate our models. But the essential concept and the ultimate goal have always been the same: To find truth. Hence a look at philosophy of science, which is concerned with the question of scientific truth, might perhaps help to understand some aspects of validation. This approach is followed in section 2, while section 3 adresses validation in waste management in particular, and section 4 discusses the role validation can play in public acceptance.

The present paper partly draws from discussions within a Swedish-Swiss working group on regulatory guidance of waste disposal [1] where similar issues were adressed.

2. THE "PROOF" OF A THEORY

To prove a theory is to show that the theory is true. Yet the concepts of truth and proof in an absolute sense are only meaningful in mathematics, where a theorem is proved, i.e. is shown to be true, if it can be logically deduced from a set of axioms. However, mathematics is a closed science, that is to say its objects are themselves defined by the axioms and do not, strictly speaking, exist outside the mathematicians mind. All other sciences, or rather empirical sciences, such as physics are concerned with objects existing in the real world which the scientist is not free to axiomatize at will. The axioms are themselves subjected to the question of truth. What then is truth in science, or more precisely, how can a scientific theory be proved?

The discipline called philosophy of science has given several answers to this question which partly contradict, partly supplement each other. One tentative answer is that scientific progress is achieved by <u>falsification</u> of wrong theories. It should be clear however that this method, although useful for a first screening of competing theories, cannot positively prove a theory. To have survived all falsification attempts is of course a necessary but not a sufficient condition for truth.

The traditional opinion among scientists is that a theory has to rest on <u>positive proof</u>, that is a theory must be able to explain the pertinent observations and experimental data. Again this is a necessary but not a sufficient condition. The totality of empirical knowledge is necessarily incomplete, hence a theory is true only with respect to the given thesaurus of common scientific experience. The broader the empirical knowledge which corroborates a theory, the better is the theory . Yet, this statement needs qualification: The broadening of our knowledge has to be independent of the theory it is to support. Increased experimentation in an area where the theory has already been successfully proven will not do; rather the experience ought to be pushed to the limits where either the theory breaks down or else new proofs are found. There is a lesson to be learned for validation in waste disposal: The broader our knowledge and understanding of the disposal system, the more credible is the successfully validated model. But the broadening of our knowledge must not be governed by the model itself, hence we have to push our investigation of the disposal system farther than is originally required by the model.

The role of falsification and positive evidence is generally accepted but neither is considered as sufficient to prove a theory. What then is the additional ingredient that establishes a theory as true? The most recent and - in our context - most interesting answer was given by Thomas Kuhn [2], and the answer is surprising and perhaps not very flattering for the make-believe strictness of the supposedely exact sciences. After analyzing many so-called scientific revolutions and taking into account psychological aspect - which cannot be neglected because sience, after all, is made by human beings - Kuhn concludes that the proof of a scientific theory largely rests on <u>consensus</u>. In other words, and to put it simply and brutally, a scientific theory by definition is true if it has gained broad consensus among the experts of that particular science. Of course, a scientist will - conciously or not - use a number of rational acceptance criteria before he consents to a theory; such criteria could be mutual consistency of the basic assumptions or positive evidence in as wide an area of application as possible. However, acceptance of a theory is ultimately based on the feeling that "it works", a feeling that has grown from repeated successful use of the theory. The Copernican system of the world was eventually accepted not on the basis of a formal proof but because more and more astronomers became familiar with it and used it as a convenient tool in their daily work. Thus a theory is not accepted as true by way of a mathematical proof but by way of consensus; and consensus is achieved when a majority of experts feel sufficiently convinced that the theory works. It is perhaps the greatest merit of Kuhn to have identified and analyzed this irrational but all too human component of science.

To translate these conclusions into the context of validation in waste disposal: It does not make sense to demand strict proof that a model is correct; it makes a lot of sense, however, to promote consensus by providing ample positive evidence for the correctness of the model. In this sense, validation is primarily a means to achieve consensus. That the consensus has to include the regulatory authorities is obvious.

3. VALIDATION IN WASTE DISPOSAL

There must be no doubt that validation has to play a crucial role in waste disposal. Our safety assessments rest on modelling or, to be more precise, on predictive modelling. Safety will be present - or absent, if we are not careful - in the real world, but the safety assessment itself takes place in a model world. The modeller as such is never quite sure whether the models adequately reflect all the relevant features of the real disposal systems. As we cannot wait till the effects of a waste repository are experienced, validation is the only way to connect the model world to the real world. Models, of course, are an indispensable tool without which calculations cannot even start. However, somewhat more caution in the use of models and more concern about the questions of model applicability and transferability are advisable. There is still a gap between the "scientist" who wants to understand nature but takes his time, and the "modeller" who wants to produce numbers and presses ahead with fragmentary knowledge. The time has come to bridge this gap, and the way to do it is by validation. All models which have a considerable influence upon the results of a safety assessment are to be validated.

The particular goal of validation in waste disposal is to remove the conceptual uncertainties of the models. Validation of the models, in conjunction with thorough investigation of the systems described by the models, has to provide answers to questions like these: Does a process really function the way we modelled it? Did we perhaps overlook a process which may drastically alter the outcome? Do we have a sufficient understanding of the system? Is the model applicable to the particular situation where we want to use it? The range of applicability of a model, as touched upon by the last question, is an important issue. Ideally, validation should allow to delineate the range of applicability because knowing the limits and knowing why a model breaks down at the limits gives a deep insight into the features and capabilities of the model.

Not all of the models customarily used in safety assessments are equally in need of validation. Two criteria may be used to judge the necessity and the extent of validation. The first is relevance for safety: Validation is certainly more necessary for models which have a large (positive or negative) impact upon the results of the safety analysis. The second criterion is connected with the consensus-theory of scientific proof. The ultimate aim of validation is to create widespread consensus, which in turn is based to a large extent on "gut feeling" or some kind of inner conviction that the model is correct. Hence validation is more important in areas where a large body of scientific and technical experience is yet missing. A few examples may clarify this point. Physicists have a very good feeling that the model describing radioactive decay is correct; therefore, the model used to predict the evolution of the <u>inventory</u> of a repository may be considered as already validated. There is less consensus but still a moderate conviction, based on historical experience and everyday life, that the models describing the performance of <u>engineered barriers</u> are probably correct, provided predictions do not reach beyond a few tens or hundreds of years; some validation, particularly concerning long-term behaviour, is still needed in this area. The situation changes when models of <u>geospheric migration</u> are considered. There is very little experience with processes going on in the deep underground, and what has been learned so far is not sufficient to create the necessary consensus on the relevant models. At least in the case of high-level waste disposal, where the long-term safety mostly rests on geological barriers, our understanding of the geological situation (generic and site-specific) has yet to improve, and geospheric processes will be the primary target of validaton efforts. The last example is provided by the models describing the future behaviour of <u>man and society</u>. Here the usefulness of validation has reached its limit; there is very little hope that consensus about the development of our society can ever be achieved. We have to accept this limit and admit in modesty that not all models can be validated.

Finally, a particular aspect of validation has to be adressed which might be called "balance of validation". The models used in safety assessments necessarily give a simplified description of the real system and validation will tell whether or not additional features or processes have to be incorporated into the model. Now a process which was overlooked in the first stage of modelling may influence the safety of a repository both ways: It can increase or it can decrease the real safety as compared to the model safety. While it is understandable that implementing organizations concentrate their modelling and validation efforts on those processes that are beneficial to safety, other processes with potentially negative impact on safety may be as important and are not to be neglected in modelling and validation. To avoid biased end results there must be a balance of validation in the sense that validation has to cover all important processes, irrespective of whether they affect safety in a positive or in a negative way. This could be achieved if validation is applied not only to single, isolated models but, although more difficult, to chains or complexes of models that describe more comprehensive systems.

4. VALIDATION IN PUBLIC ACCEPTANCE

The motion of railroad trains and automobiles is governed by Newton's laws of motion, a theory that was already accepted around 1700. Nevertheless, when this new kind of locomotion was introduced in the last century many people, including learned and knowledgeable people, protested and denounced the very idea as ridiculous because the human body would never survive velocities of, say, 60 km/h. The argument, of course, went against the relativity of motion, but the deeper reason for bringing it up was that, although the theory of motion did have the necessary consensus to be accepted, the consensus only included physicists and engineers but not the layman, for which high speeds were outside everyday experience. Nowadays everybody accepts the theory because everybody experiences motion daily; everybody feels confident that the theory works.

The situation is less comfortable in waste disposal because disposal systems are far outside the usual experience of most scientists, let alone the public. Consensus, therefore, is restricted to the few that have sufficient knowledge and experience to understand the specific problems of waste disposal. How, then, is the public to be convinced that a given waste repository is acceptably safe? How can the public find truth among the various statements and allegations of experts, counter-experts and false experts? A large part of the answer is that consensus has to be created among those groups which are in a particular position to act as intermediary between the public and the organization concerned with the implementation of waste disposal projects. Two such groups may be mentioned.

The first group is composed of scientists which by their scientific background, although not directly involved in waste disposal, are well able to judge some particular aspect of a disposal project. Thus, in the context of geospheric migration or, more generally, whenever geologic features of a site are concerned, it is the scientific community of geologists that plays an important role in public acceptance. Scientifically sound and transparent site selection and characterization and successful validation of the geological barrier models are necessary conditions to gain consensus of this expert group.

The second group of experts which has to accept the role of a mediator between the public and the implementing organization is the regulators. For right or wrong, the regulator is given more credit by the public than the implementor. This may be so because the regulator has the official task of critically reviewing the implementors work, while the implementor, particularly by people living at a proposed repository site, is invariably seen as someone who tries to talk them into accepting an unwanted burden. The implementor cannot fully convience the public but he can and must at least convince the regulator, because the regulator will have to decide, on behalf of the government, but ultimately on behalf of the people, whether or not a proposed waste disposal facility is acceptable. Here again we find the crucial role of validation to promote conviction and consensus: An indispensable condition for the regulators acceptance of a facility is his conviction that the models used in demonstrating safety are correct; he has to be part to the consensus, otherwise he is not able to credibly present the case to the public.

Consensus is but one aspect of scientific truth. However, as far as public acceptance is concerned, it is the only one that really counts. Without consensus among the relevant experts there is no confidence, and without confidence there will be no chance of public acceptance. The best way to achieve consensus is careful validation of all the essential models.

[1] Swedish Nuclear Power Inspectorate, Swiss Nuclear Safety Inspectorate, National Institute of Radiation Protection (Sweden): "Regulatory Guidance for Waste Disposal", an advisory document (to be published).

[2] Kuhn Th.: "The Structure of Scientific Revolutions", 2nd edition, Chicago/London 1970.

VALIDATION OF CONCEPTUAL MODELS
OF FLOW AND TRANSPORT IN POROUS OR FRACTURED MEDIA

G. de MARSILY
University Paris VI and Paris School of Mines
Paris, France

ABSTRACT

In low permeability media, flow and transport conditions may more often than not be in a state of non- equilibrium because of relatively recent geologic changes to which these media have been subjected (glaciations, sea level and temperature changes, etc.) and which may still be influencing the behaviour of the system. Our predictions of its future behaviour could be very much in error if what we observe today were interpreted as a stable situation, thus ignoring the very long transient response of such systems. However, such transient behaviours can be of great interest for the validation of our models, if we can use them to reconstruct the present observed state of the system, as the calculated result of the succession of events that have occurred in the past. It is clear, however, that using models to try to reconstruct the past is somewhat difficult and this area is currently being developed, e.g. in the petroleum industry, where the goal is to reconstruct the thermal and hydraulic history of sedimentary basins in order to estimate the potential of oil formation. For nuclear waste disposal the reconstruction of the past is also an excellent means of increasing our understanding of what may happen in the future. In particular, if the transport models used to predict movements of radionuclides in the geosphere were also used to reconstruct the transient movements of those chemical species or environmental tracers, that are active on the sites considered as potential repositories, this would constitute a great step forward in the validation and confidence build-up of these models and the resulting performance assessments.

Examples are given of transient behaviour and use of chemical species migration in the Paris Basin and in the Belgian Boom clay.

VALIDATION DES MODELES CONCEPTUELS D'ECOULEMENT
ET DE TRANSPORT DANS LES MILIEUX POREUX OU FISSURES

RESUME

Dans les milieux faiblement perméables, les conditions d'écoulement et de transport sont rarement à l'équilibre du fait des changements "récents" subis (glaciation, changement du niveau de la mer, de la température...). Nos prédictions seraient erronées si nous interprétions les états actuellement observés comme des régimes permanents, en ignorant les très longs transitoires que connaissent ces systèmes. Ces transitoires peuvent cependant être utiles pour valider les modèles si l'on peut reconstruire l'état actuel observé comme la conséquence calculée des évènements antérieurs. De telles reconstructions, difficiles, sont actuellement tentées par l'industrie pétrolière, pour reconstruire l'histoire hydraulique et thermique des sédiments afin d'estimer leur potentiel pétrolier. Dans le problème du stockage des déchets nucléaires, reconstruire les évènements passés est aussi une excellente façon de mieux comprendre ce qui peut se passer dans le futur. En particulier, si les modèles de transport qui seront utilisés pour prédire le mouvement des radionucléides, sont aussi utilisés pour reconstruire la migration des traceurs environnementaux, ceci fournirait une validation de ce modèle et donc renforcerait la crédibilité des analyses de sûreté.

Des exemples de comportement transitoire de migration de traceurs environnementaux sont choisis dans le Bassin de Paris et l'argile de Boom en Belgique.

1. INTRODUCTION

It has recently been reported that in low permeability media and/or large aquifer systems non-equilibrium conditions could exist even if present observations show constant measurements with time. This is due to the slow transmission of perturbations in such systems and to the relatively short duration of our observations. In the Pierre Shale formation, in North Dakota, USA, for instance, NEUZIL (1989) showed that the pressure profile within the shale is not in steady state but probably reflects the loading of the formation by an ice cap during the last glaciations. If the migration of elements had been predicted from an assumed steady state, it would have resulted in a flow direction opposite to the actual one. Given the duration of our observations, the transient state of the system is imperceptible. Similarly, DIENG et al (1990) reported that the famous piezometric depression of the FERLO in the Senegal Basin, which is a closed cone 40 m below sea level, can be explained by the change in the sea level that occurred about 12,000 years ago as a result of the melting of the ice. The relatively large distance to the sea and the relatively low hydraulic conductivity of the medium explain why the depression is still filling up today but this process is so slow that the rate of filling cannot be detected at the historic scale.
As a first guess, it is relatively easy to determine the time needed for a hydraulic system to reach piezometric equilibrium after a sudden perturbation has been propagated through it in one direction ; one can show (see e.g. MARSILY, 1986) that equilibrium is approximately reached when :

$$\frac{t}{x^2}\,\frac{K}{S'} > 1$$

where t is the time elapsed since the start of the perturbation, x is the size of the dimension in which the perturbation propagates, K is the hydraulic conductivity and S' the specific storage coefficient. Alternatively, (K/S') can be replaced by (T/S), where T is the transmissivity of the aquifer, and S the storage coefficient (or effective porosity in a phreatic aquifer). As an example, let us consider a plastic clay layer of 100 m thickness, 10^{-10} ms^{-1} hydraulic conductivity, and 10^{-2} m^{-1} specific storage (corresponding to a 10^{-6} Pa^{-1} compressibility). A vertical perturbation such as a change in pressure on one side of the clay layer will propagate through it and reach approximate equilibrium in 31,700 years. Since these values are not outside the range of those of a potential host rock, it is clear that the "recent" climatic variations which have occurred since the last glaciation can still have a significant effect on the behaviour of such systems. We shall give a hypothetical example of such behaviour in the Boom clay in Belgium.
Let us now turn to tracer transport. In a medium of 10^{10} ms^{-1} hydraulic conductivity with a 1% hydraulic gradient, a conservative tracer will move a few meters by advection and diffusion in 10,000 years, if the kinematic porosity is that of a clay, e.g. 20-30% ; it will move 3 km in a fractured medium with a 10^{-4} fracture porosity or hundreds of meters if matrix diffusion makes the matrix porosity available instead of just the fracture porosity. Many "signals" can be assumed to exist during such a period : those given by radioactive tracers but also the change in $^{18}O/^{16}O$ in recharge water which occurs when the temperature changes at the surface after a glaciation. Our low- permeability media are thus in the process of recording the signals, that we should try to detect and analyse in order to validate our models. We will show another example of such validation on the aquifers of the Paris Basin.
It is worth noting here that the attempts at reconstructing the present, observable state of nature, that we generally consider as "steady-state" because we cannot perceive its rate of change, are not particular to the nuclear waste problem. The petroleum industry has

developed models such as TEMIS (DOLIGEZ et al, 1987) which represent the evolution of a sedimentary basin from the time of the depositing of its first layer to the present day ; the mechanisms that are represented include sedimentation, compaction, tectonic uplift or subsidence, faulting and hydraulic fracturing, groundwater flow, thermal flow and even geochemical transformations, particularly the transformation of organic matter (kerogene) into oil or gas, when the sediments are brought within the "oil window", i.e. a well-defined range of temperature and pressure where petroleum products are formed. The challenging task of reconstructing the complete history of a geologic basin over hundreds of millions of years is considered feasible by the industry and would have the general objective of helping petroleum exploration by predicting the amounts of oil and gas that have been produced and their potential subsequent migration in the carrier beds. Examples can also be found in BETHKE et al (1988).

2. HYPOTHETICAL TRANSIENT BEHAVIOUR OF THE BOOM CLAY, BELGIUM

The following example is hypothetical since no detailed information is yet available on the pressure distribution within the Boom clay, which is considered in Belgium as a potential host rock formation for nuclear waste disposal. The modelling of the hydrology of the system has been done jointly by the CEN-SCK in Belgium and the Paris School of Mines (PATYN, 1985, PATYN et al, 1990). Briefly, the approximately 100 m-thick Boom clay is surrounded by two aquifers, the Neogene sands above and the Rupelian sands below. Piezometric surveys in each aquifer show that, in the central part of the system, the head in the upper aquifer is higher than the one in the lower aquifer by approximately 2 to 5 m ; under steady state assumptions, this would mean that the leakage flux from one aquifer to the next through the Boom clay must be oriented downwards in that area. The multilayered aquifer model which was fitted on the observed piezometric data in the aquifers is a steady-state model ; in order to fit the model, a leakage flux had to be prescribed flowing into the underlying aquifer, the Rupelian sands ; without this leakage flux the observed piezometric surface in the Rupelain sands could not be reproduced. This leakage flux was actually imposed by using trial and error to fit the hydraulic conductivity of the Boom clay ; given e.g. 2 m of head difference between the two aquifers, an order of magnitude of 10^{-10} ms^{-1} for the hydraulic conductivity of the Boom clay was necessary to obtain the needed leakage flux to the Rupelian sands in the steady-state assumption. The leakage flux is thus on the order of 2.10^{-12} ms^{-1} with the vertical hydraulic gradient on the order of 2%.
This relatively high value for the hydraulic conductivity of the Boom clay has always been regarded as controversial in Belgium, since most laboratory or even in- situ measurements of the hydraulic conductivity of the Boom clay give values on the order of 10^{-12} ms^{-1}. One possible explanation of this discrepancy could be that it is a "scale effect", i.e. the laboratory measurements are on the scale of small cores, and even the in-situ measurements are on the scale of a few decimeters, whereas the estimates made by the model represent average values for meshes of several kilometers. Alternatively, is the regional hydrologic model wrong ?
Three considerations are worth noting here. Firstly, there are unfortunately no undisturbed piezometric measurements within the Boom clay to help determine if the vertical piezometric profile is indeed in steady state, as the model assumes. Secondly, even if the mass balance in the lower aquifer clearly indicates that there is an incoming leakage flux from the Boom clay, the mass balance in the upper phreatic aquifer is not sensitive enough to evaluate the outgoing leakage flux towards the Boom clay ; in other words, the hydrologic model cannot tell if the flux through the Boom clay is in steady state and, consequently, if the flux entering the Boom clay from above is equal to that leaving it from below. Thirdly, recent studies in the Netherlands concerning the hydrologic regime during a glaciation have produced the

following plausible scenario (GLASBERGEN, 1990) : when the permafrost develops in front of and around the edge of an ice cap during a glaciation, the upper aquifers of the system become frozen ; however, further upstream, beneath the ice cap, the insulating quality of the ice prevents the ground from freezing and high water pressures develop, in equilibrium with the thickness of the ice (Figure 1). The water is then forced to flow deep into the ground, in the lower aquifers, below the permafrost (arrow in Figure 1). Evidence of the existence of this type of flow is said to come from the deep aquifers in the Netherlands, where the very young age of the waters cannot be explained by the present observed flow conditions.

Let us now build a hypothetical scenario for the Boom clay situation. Suppose that, due to the permafrost in the upper Neogene aquifer, the flow of water was much higher in the lower Rupelian sands 100,000 years ago. Because of the known thickness of the ice cap in Belgium at this time, we can assume that the head in the lower Rupelian sands was 100 m above its present value and that it remained identical for tens of thousands of years, so that a true steady state was reached which had, for example, a more or less constant head profile in the Boom clay (Figure 2, profile 1). Suppose then that 10,000 years ago, the ice melted and that the head in the Rupelian aquifer fell almost instantly to its present value. This "perturbation" will slowly propagate itself through the Boom clay, and the present head profile in the clay could be the second one shown in Figure 2. This is a transient profile, the final steady-state value has not yet been reached. Assuming uniform material properties in the clay, one can easily calculate the transient flux from the Boom clay to the lower Rupelian aquifer ; one finds (see e.g. MARSILY, 1986) :

$$\phi = \frac{2\Delta hK}{\sqrt{\Pi}}\sqrt{\frac{S'}{4Kt}}\mathrm{Exp}\left(-\frac{\rho^2 S'}{4Kt}\right)$$

where Δh is the head drop at time $t = 0$ (100 m here), e is the thickness of the clay (100 m here), and K and S' are the hydraulic conductivity and specific storage coefficient. If we take $K = 10^{-12}$ ms^{-1} and $S' = 10^{-3}$, the present leakage flux, 10,000 years after the ice melt, is equal to 3.10^{-12} ms^{-1}. The numbers have been chosen to make this transient leakage flux almost identical to the one calibrated by trial and error on the steady-state hydrologic model (2.10^{-12} ms^{-1}) while using the laboratory-measured value of the hydraulic conductivity.

This hypothetical example does not by any means prove that the Boom clay is in a transient state of flow at present ; it only shows that the apparent discrepancy between two values for the hydraulic conductivity of the clay, determined by two different methods, could be made consistent if a transient behaviour could be established.

The validating of a groundwater flow model at such sites, when particular interest is being paid to what happens in low permeability media, must therefore take into consideration that the present, observed state may be the result of a long transient history ; it is thus necessary to study and to try to reconstruct the flow conditions during glacial periods and during ice melts. If environmental tracers were used (e.g. ^{14}C or ^{18}O or ^{2}H, or simply the soluble species) and considering the even longer time required for the system to reach equilibrium after a perturbation, it would be virtually impossible to make any sense of the data if these transient conditions were not understood. In the next section we shall see an example of how such data are used to validate a model.

3. USE OF ENVIRONMENTAL TRACERS TO VALIDATE A FLOW MODEL

3.1 The flow model

In recent years, the hydrology of the deep aquifers of the Paris Basin has attracted continuous interest. One reason for this is the development of the geothermal resources of the Dogger and Triassic aquifers for space heating in the Paris area, another is the over-exploitation of the shallower Albian aquifer and, more recently, the prospection of the Oxford clay and the Liassic clay in the north-eastern part of the basin as potential host rock formations for nuclear waste disposal. Data are also available from petroleum exploration in the Dogger.

With the help of these data, WEI (1990), WEI et al (1990) developed a multilayered aquifer model of the entire Paris Basin, representing six successive aquifers : Triassic, Dogger, Lusitanian, Portlandian, Neocomian and Albian. Between each of these aquifers, low permeability layers form aquitards, through which vertical leakage fluxes can occur, as in the case of the Boom clay. The hydrologic model was calibrated by trial and error, based on the following information :

- piezometric measurements in each aquifer ;
- transmissivity data from pumping tests in each aquifer ;
- estimates of the recharge rates on the outcrops of each aquifer.

However, most of the data came from the Dogger aquifer. Concerning the low-permeability layers between the aquifers, no hard data were available, only the thickness and the lithologic description from which the vertical permeability was "guestimated".

One of the major features of the aquifers of the Paris Basin is that the deeper the aquifer, the higher the hydraulic head ; this can be explained by the fact that the Paris Basin has a synclinal shape (Figure 3) ; therefore the deeper aquifers have their outcrops in the periphery of the basin, at a higher elevation than the shallower ones ; their general hydraulic head is thus higher than that of the shallower aquifers and the general leakage flux in the basin is directed upwards. The multilayered model did not assume steady-state conditions in the system. However, the dominant perturbation was assumed to be the heavy exploitation of the Albian aquifer, which started more than 100 years ago, and is believed to have influenced most of the aquifers of the basin. The trial and error calibration was thus made in transient conditions, over the last 100 years, with the calibration criterion that the observed and the calculated head distribution in each aquifer be as close to each other as possible. The initial conditions, 100 years ago, were assumed to be in pseudo-steady state.

The uncertainty in the true parameter values after calibration was, however, considered important, given the paucity of hydrologic data and the large number of unknowns : since the transmissivity and storage coefficient of each aquifer and the vertical hydraulic conductivity and specific storage coefficient of each aquitard are all potentially variable in space. Three environmental tracers were used to try to validate the flow model : Sodium Chloride, Helium and ^{14}C.

3.2 Validation with salinity

The origin of Sodium Chloride in the basin is believed to be concentrated mainly in the deep Triassic layers. Rock salt is indeed found (and mined) in the eastern part of the basin, and it is known to extend westwards almost as far as the city of Reims. The salt stratum is protected above and below by confining layers, but in some areas these layers disappear and the salt can come into direct contact with the Triassic aquifers. Furthermore, vertical faults are known to exist, through which leakage can take place from one layer to the next, and salt can again be leached (Figure 4.). As a result, the water of the Triassic aquifers is generally a brine (100 to 200 g/l). In the overlying Dogger aquifer, the salinity is between

30 and 6 g/l ; in the other aquifers, the salinity generally decreases with the elevation from a maximum of 10 g/l in some parts of the Lusitanian down to negligible amounts in the Albian. The general explanation of this feature is that the salt is transported upwards from the Triassic to the upper aquifers with the leakage flux ; in each aquifer, this salt flux is diluted by the incoming fresh water flux from the recharge on the outcrops. Hence, the reduction in the salinity. A first validation of the flow model can thus be obtained if the model can reproduce this salinity without further calibration. Convective salt transport in the aquifers and aquitards was calculated, assuming conservative transport and steady state (the Paris Basin was not covered by ice during the last glaciation, and the present general flow pattern is believed to have been in place for millions of years). The observed salt concentration in the lower unit, the Triassic, was prescribed, and the concentration in all other aquifers was calculated. As an example, the calculated salt concentrations in the Dogger and the Portlandian are presented in Figure 5 together with locally measured values ; the agreement is excellent which suggests that the general balance between the vertical leakage flux and the recharge on the outcrop of each aquifer is correctly estimated. This partly validates the ratio between the transmissivities in the aquifers and the vertical hydraulic conductivity in the aquitards, which were almost totally unknown.

3.3 Validation with Helium transport

A second validation was obtained using Helium concentrations measured in the Triassic and Dogger aquifers (no He measurements were yet available in the other aquifers). In the Paris Basin these concentrations are abnormally high, on the order of 1 to 5 10^{-5} mol/l. The origin of He in aquifers is twofold : (i) internal production in the aquifers by radioactive decay of U and Th ; (ii) flux of He dissolved in the water coming into the aquifer by leakage (there is almost no He in the recharge water). This leakage water itself contains He of two origins : the He of the aquifer from which the leakage comes, plus the He produced within the aquitards, "leached" by the leakage flux flowing through them. For instance, the He concentration in the Dogger is the balance between the internal production within the Dogger, plus the total production of the underlying Liassic aquitards, which is carried upwards by the ascending leakage flux, plus the initial He amount contained in this water when it leaves the Triassic to permeate the Liassic aquitard. However, this statement is made on the assumption that the flow system is in steady state, so that the He produced in the Liassic aquitard is "leached" away at the same rate as that at which it is formed ; the same leakage velocity must therefore have been active for the whole time of convective transfer through the Liassic aquitard, which is on the order of 20 million years.
The production of He in aquifers or aquitards is easily related to the amount of radioactive U and Th and to a "production efficiency coefficient" defined by TORGENSEN (1980), which is the ratio of the Helium entering into the fluid to the Helium actually produced within the rock ; this coefficient is in general close to 1 (TORGENSEN and IVEY, 1985). The amounts of U and Th in the sediments were known in the Dogger and taken from tables for similar formations for the Lias. The concentration of He in the Triassic aquifer was prescribed as observed, and the concentration of He in the Dogger was calculated with the above assumptions. Figure 6 is a map of the calculated He concentration in the Dogger with the observed values in a dozen wells. Again, the agreement is quite good, thus providing a second validation of the water flowing into the Dogger either through the ascending vertical leakage from the Triassic or through the horizontal recharge on the outcrops. Note that the relatively high concentration of He in the Triassic, which was precribed as observed in the model, must have part of its origin in deep crustal Helium. Preliminary estimates indicate that in order to explain the present He concentration in the lower aquifer there must be, below the 700 m of Triassic sediments, another 600 m of metamorphic crust in

the process of degassing and producing Helium. This figure is not unreasonable, given the work of TORGENSEN and CLARKE (1985) in the Great Artesian Basin in Australia. The presence of mantellic Helium could also be invoked, and determined by the $^{3}He/^{4}He$ ratio. As an environmental tracer Helium is thus of great importance for the study of leakage fluxes in low permeability media.

3.4 Validation with ^{14}C

A third attempt at validation was made with a dozen measurments of ^{14}C in the water of the Dogger aquifer. Given the half-life of ^{14}C (5,570 years), the only possible origin of ^{14}C in the Dogger water is the recharge on the outcrops : the transit time of the leakage water through the Liassic aquitard is too long to contain any ^{14}C. If the flow model is able to reproduce the observed ^{14}C content of the Dogger, it will bring a partial validation of the recharge rate and water velocity in this formation. However, a new parameter has to be introduced into the calculations (contrary to what has been done so far for the other tracers) : the product of the effective thickness of the Dogger aquifer times the kinematic porosity. The groundwater flow model is indeed fitted with the value of the transmissivity (hydraulic conductivity times thickness of aquifer) and to calculate the pore water velocity, we must divide it by the thickness times the porosity and introduce this product into the model. As these two quantities are not very well known, there is room for uncertainty and the validation will not be very strong.
Two additional problems must be mentioned : we know that in a limestone aquifer, such as the Dogger, there is an exchange between the ^{14}C in the dissolved carbonates and the solid carbonates ; correction methods have been developed using ^{13}C concentrations to account for this exchange, but the validity of such corrections is sometimes questioned (see e.g. FONTES and GARNIER, 1979), adding a further uncertainty. The second problem is that the measured ^{14}C concentrations are very small, giving groundwater ages between 22,000 and more than 40,000 years, which is the detection limit (IMRG, 1988). In some cases, the validity of the measured values have been questioned by the operator, because of the possibility of counting failure or external contamination. However, a consistent picture from a few repeated measurements shows that there is a measurable gradient in ^{14}C ages in the Dogger towards the center of the basin.
The convective transport of decaying 14C was calculated with the model, and the product thickness * porosity was adjusted so that the calculated ^{14}C concentrations matched the observed ones. Figure 7 a and b show the calculated and measured ^{14}C concentrations and the adjusted effective thickness of the aquifer, assuming a constant porosity value of 15% (value measured on cores in the center of the basin). This adjusted effective thickness is in good agreement with the measured value at the center of the basin, and its reduction towards the periphery is consistent with what is known of the aquifer, although no real measurements are available. Note that the total thickness of the Dogger is much greater (up to 200 m) but that only a few thin layers towards the top of the formation are pervious and form the Dogger aquifer.
Although this validation is perhaps more questionable than the two previous ones, the unquestionable evidence of ^{14}C deep in the center of the Paris Basin, in the Dogger, which initially came as a great surprise, forced us to admit that the Dogger was indeed a flowing system and not a confined and immobile one, as had been assumed when the first geothermal wells were drilled. This information is of no great help in validating the leakage flux in the aquitards but it would be of interest for the estimation of groudwater travel times in a near-by aquifer, if a waste disposal site were to be considered in the Liassic formation.

4. CONCLUSION

We are nearing the time when site-specific models will have to be fitted to observations and used to predict radionuclide migration in the geosphere. One posssible approach is to estimate a plausible range of uncertainty of the parameters representing the site from a few measurements or expert opinions and then, to use Monte Carlo simulations to find distributions of possible outcomes. Although such an approach can be useful for comparing the roles of the successive barriers of disposal, it is very hard to imagine how it can be validated and made convincing enough to prove that the dominant natural features of a site have indeed been understood and incorporated into the predictions.

In this paper, we have given two examples of site- specific models ; in the second one, we have shown that, even with limited information, it is possible to use a number of environmental tracers to support various elements of the assumptions used in the model and to partly validate the parameter values which have been used. Much more could be done if the complex geochemistry of the water had been understood and simulated with a coupled flow, transport and chemical reaction model (this type of work is now in progress in the Dogger aquifer, COUDRAIN-RIBSTEIN et al, 1991). In the first example, which is more hypothetical, we have shown that site-specific attempts at validation must take into account the fact that low permeability systems will, in many cases, not be in steady state and that, as a consequence, the "recent" geologic history of the site over e.g. the last million years must be understood and reconstructed in order to build a representative model of the site which can be validated.

5. REFERENCES

Bethke, C.M., Harrisson, W.J., Upson, C., Altaner, S.P. : "Supercomputer analysis of sedimentary basin", Science, vol. 239, 1988, 261-267.

Coudrain-Ribstein, A., Gouze, P., Michard, G. : "Quantitative study of geochemical reactions, fluid flow and heat transfer in the Dogger aquifer of the Paris Basin", Int. Geology Conf. Basin modelling, advances and applications, Stavanger, Norway, 13-15/3/1991.

Doligez, B., Ungerer, P., Chenet, P.Y., Burrus, J., Bessis, F., Bessereau, G. : "Numerical modelling of sedimentation, heat transfer, hydrocarbon formation and fluid migration in the viking Graben, Nort Sea", In Petroleum Geology of North West Europe, J. Brooks, K. Glennie, eds, Graham and Trotman, 1033-1048, 1988.

Dieng, B., Ledoux, E., Marsily, G. de : "Paleohydrogeology of the Senegal sedimentary basin. A tentative explanation of the piezometric depressions", Journal of Hydrology, 118, 357-371, 1990.

Fontes, J.C., Garnier, J.M. : "Détermination of the initial ^{14}C activity of the total dissolved carbon : a review of the existing models and a new approach", Water Resour. Res., 15-2, 1979, 399-413.

Glasbergen, P. : "The interaction between diapirism, caprock growth and salt solution in The Netherlands as an example of a sedimentary basin" To appear in : Proc. Symp. on Diapirism, Teheran, Dec. 1990.

IMRG (Institut Mixte de Recherches Géothermiques) : BRGM-AFME, Rapport annuel d'activité 1988, BRGM, Orléans, France, 56 p.

Marsily, G. de : "Quantitative Hydrogeology", groundwater hydrology for engineers, Academic Press, New-York, 1986, 440 p.

Neuzil, C.E. : "erosion, rebound, dilatation and fluid pressure in low permeability settings", 28th Int. Geol. Congr., Abstract, Vol. 2, 2-509-510.

Patyn, J. : "Contribution à la recherche hydrogéologique liée à l'évacuation de déchets radioactifs dans une formation argileuse", Doctoral Thesis, Paris School of Mines, Fontainebleau, 1985.

Patyn, J., Ledoux, E., Bonne, A. : "Geohydrological research in relation to radioactive waste disposal in an argilaceous formation", Journal of Hydrology, 102, 1990, 267-285.

Torgensen, T. : "Control of pore fluid concentration of He-4 and Rn-222 and the calculation of He-4/Rn-222 ages", J. Geochem. Explor. 13, 1980, p. 57-75.

Torgensen, T., Clarke, W.B. : "Helium accumulation in groundwater. I : An evaluation of sources and the continental flux of crustal 4He in the Great Artesian Basin, Australia", Geochemica Cosmochimica Acta. 49, 1985, 1211-1218.

Torgensen, T., Ivey, G.N. : "Helium accumulation in groundwater. II: A model for the accumulation of crustal 4He degassing", Geochemica Cosmochemica Acta. 49, 1985, p. 2445-2452.

Wei, H. F. : "Modélisation tridimensionnelle du transfert d'eau, de chaleur et de masse dans l'aquifère géothermique du Dogger dans le Bassin de Paris", Doctoral Thesis, Paris School of Mines, Fontainebleau.

Wei, H.F., Ledoux, E., Marsily, G. de : "Regional modelling of groudwater and salt and environmental tracer transport in deep aquifers in the Paris Basin". To appear, J. of Hydrology, 1990.

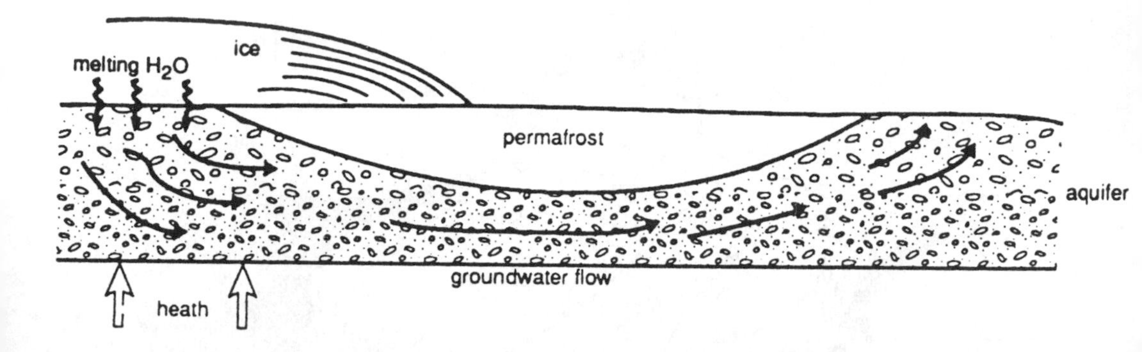

Fig. 1. Groundwater flow under (semi) artic conditions
(after Glasbergen, 1990)

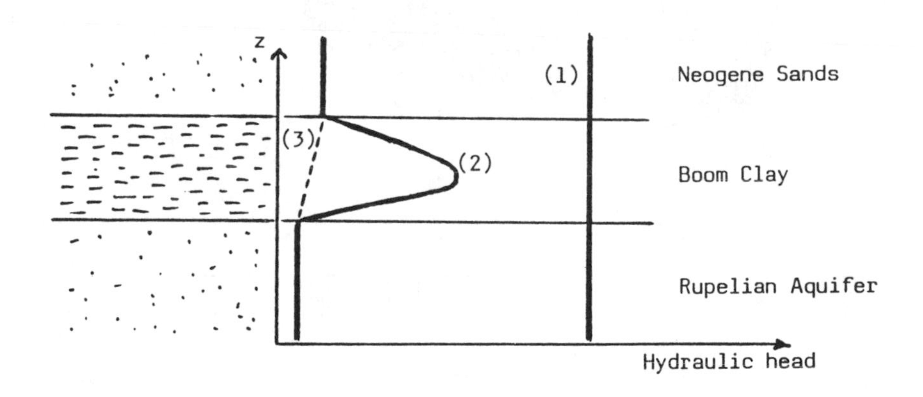

Fig. 2. Possible piezometric profiles in the Boom Clay :

(1) Initial steady state profile during glaciation

(2) Possible present transient profile

(3) Final steady state profile.

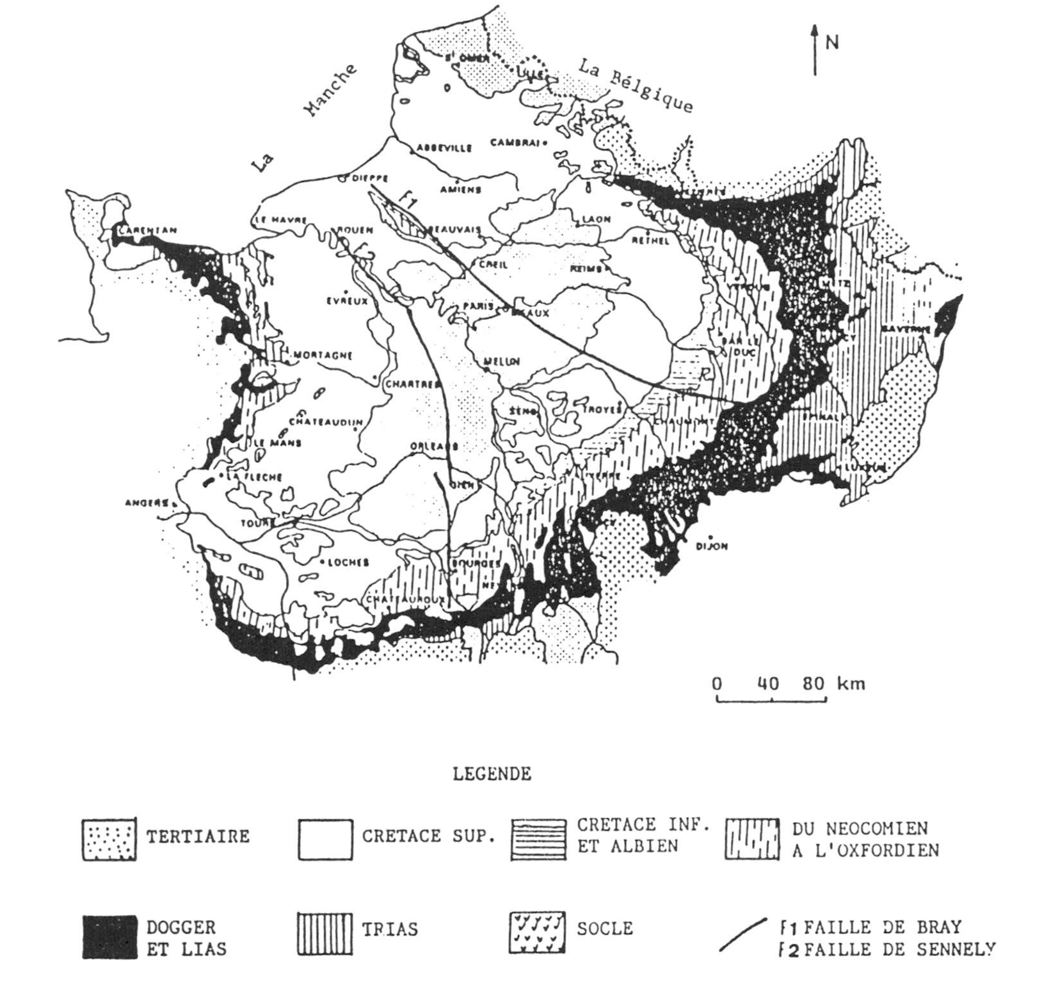

Fig. 3. Geologic map of the Paris Basin
(after BRGM, 1980, in WEI, 1990)

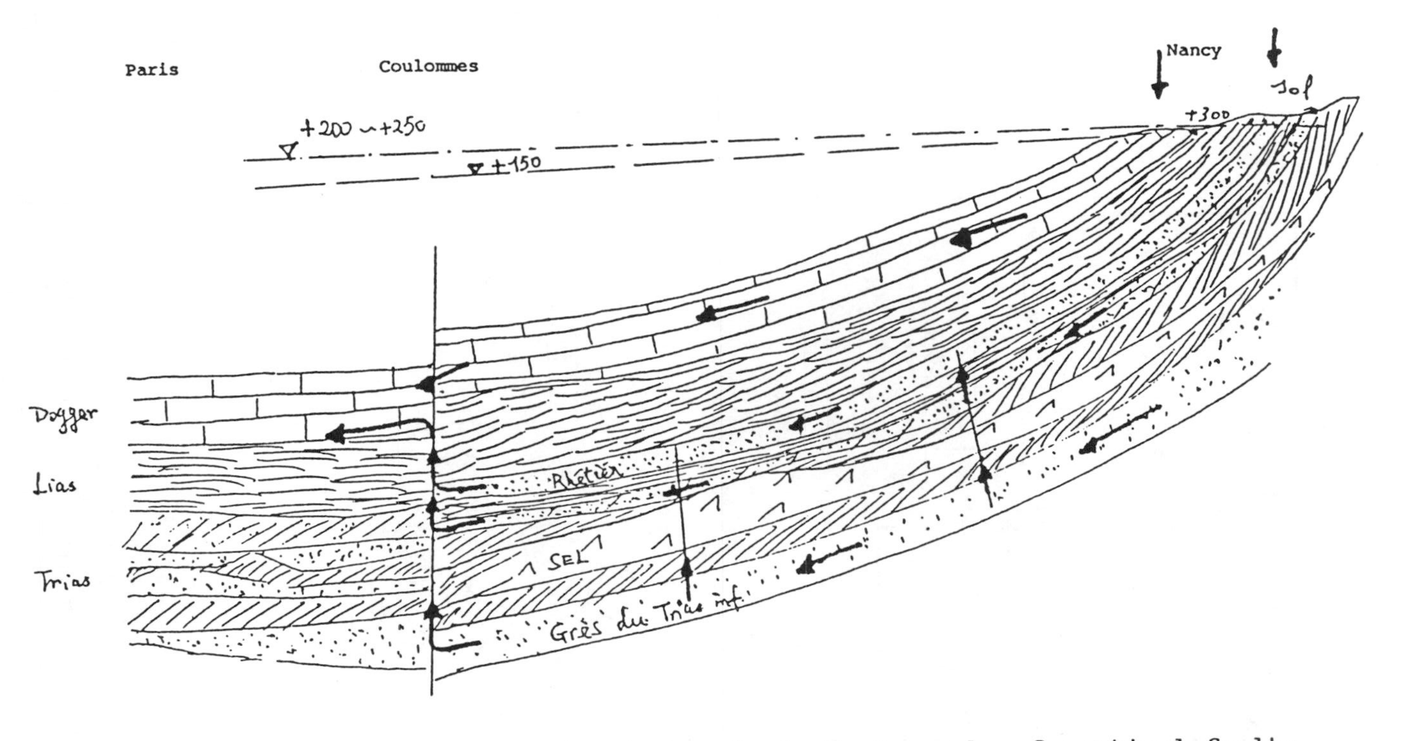

Fig. 4. E-W cross-section in the Paris Basin, role of vertical faults through the Triasic salt (after WEI, 1990)

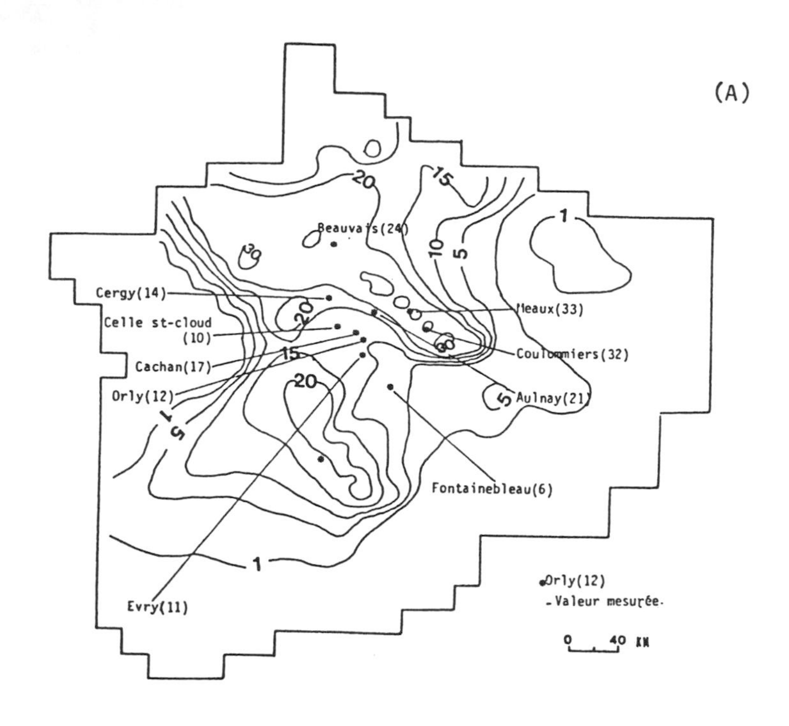

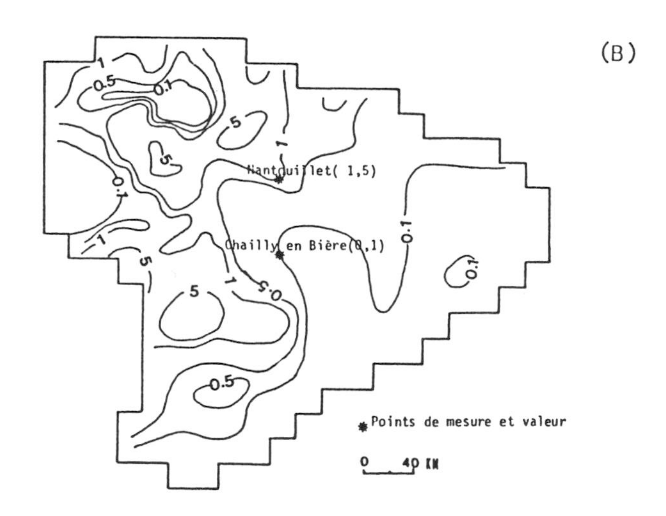

Fig. 5. (after WEI, 1990)

A : Calculated and measured salinity in the Dogger, in g/l

B : Calculated and measured salinity in the Portlandian, in g/l

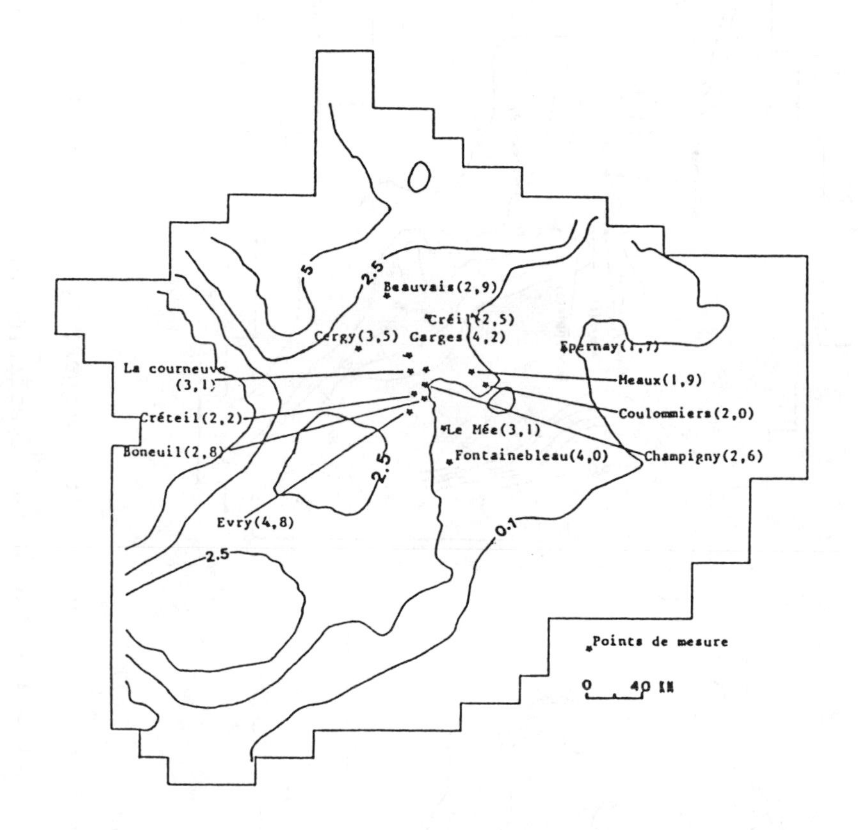

Fig. 6. Calculated and measured concentration of Helium in the Dogger, in 10^{-5} mol/l (after WEI, 1990)

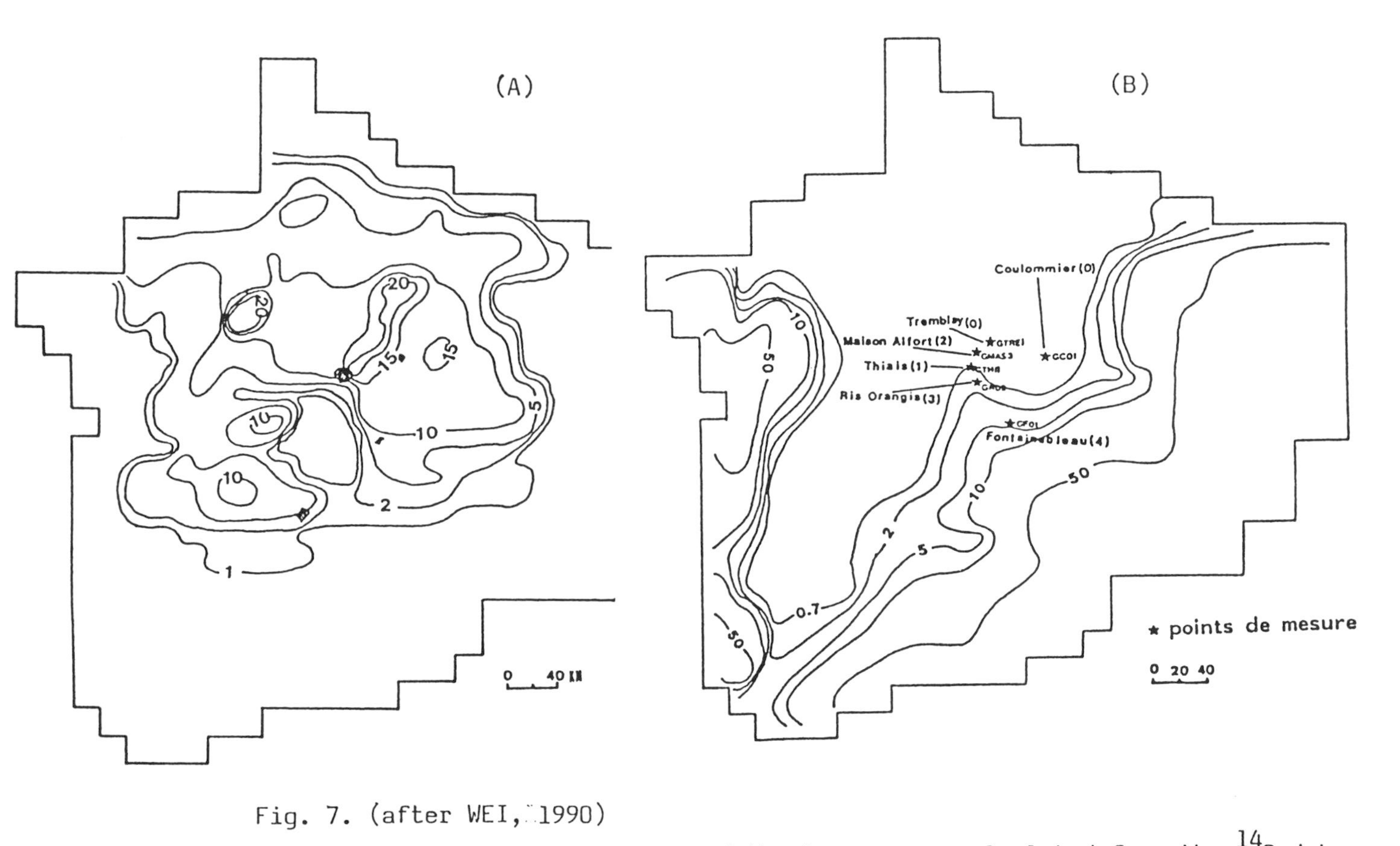

Fig. 7. (after WEI, 1990)

A : effective thickness of the Dogger, m, calculated from the ^{14}C data

B : measured and calculated ^{14}C activity in the Dogger, in % modern ^{14}C

VALIDATION OF GEOCHEMICAL MODELLING AND DATA FOR ASSESSING THE PERFORMANCE OF GEOLOGICAL DISPOSAL OF NUCLEAR WASTES

F.P. Sargent
Atomic Energy of Canada Limited
Pinawa, Manitoba, Canada, R0E 1L0

ABSTRACT

Geochemical modelling and data for predicting radionuclide transport in the geosphere are reviewed. The major conclusion reached is the need to consider validation of the use of such codes and data for specific geological sites, disposal concepts, and designs.

VALIDATION DE LA MODELISATION ET DES DONNEES GEOCHIMIQUES PERMETTANT D'EVALUER L'EFFICACITE DE L'EVACUATION DES DECHETS NUCLEAIRES DANS DES FORMATIONS EN MILIEU GEOLOGIQUES

RESUME

L'auteur examine les méthodes de modélisation et les données géochimiques utilisées dans la prévision du transport des radionucléides dans la géosphère. La principale conclusion est qu'il faut valider l'utilisation de ces programmes et de ces données pour des conditions bien définies qu'il s'agisse des sites géologiques, des principes d'évacuation et de la conception des ouvrages.

1. INTRODUCTION

The purpose of this invited paper is to review progress and the scientific needs of geochemical models and data validation for performance assessment of geological concepts for nuclear waste disposal. It will be shown that geochemical models and data for radionuclide transport in the geosphere should be reviewed in the broader context of the overall disposal concept. The paper draws extensively on an earlier review of geochemical modelling needs by Sargent and McKinley [1].

The definition of validation and the level required for acceptance of waste disposal concepts and licensing of sites has been discussed at length at GEOVAL 1987, and no doubt we will go over many of the elements again at GEOVAL 1990. For the purpose of this paper, I am adopting the following definitions paraphrased from the OECD/NEA Stripa Project:

- Validation is defined as ensuring that the computer models and data are an appropriate representation of the geochemical reality of radionuclide behaviour in the geosphere surrounding a disposal vault.

- A model is considered to be validated for use in a given application when it has been shown to provide a representation of the process or system involved that is acceptable to a group of experts.

Validation of geochemical models for radionuclide transport in the geosphere cannot be treated in isolation. Consideration has to be given to the major elements of the overall disposal system, i.e., disposal vault, geosphere, and biosphere, and how they interact or couple. Since the only credible way for radionuclide release from a vault to occur is by transport in groundwater through the geosphere to the biosphere, the modelling of radionuclide behaviour in groundwater and the evolution of the geosphere are key areas for data and model validation. They are the whole purpose of GEOVAL 1990.

Transport of radionuclides in the geosphere is governed by two major factors: solubility (including speciation) and radionuclide partition from groundwater onto the minerals present (sorption). Hence it is important to have a good understanding and a full description of the geosphere minerals and groundwaters.

It is important also to understand how the presence of the vault constituents can alter geochemical conditions in the geosphere and the biosphere and the degree to which the alterations are time-dependent and reversible. This cannot be done without considering the geochemical conditions in all components of the system and developing a realistic, self-consistent database. In practice, it is only possible to demonstrate both realism and self-consistency within the context of a specific design (vault) for a specific site (geosphere and biosphere).

It is possible to validate the representation of particular geochemical processes in models independently, and it is possible to validate geochemical data on the behaviour of minerals, ions and complexes independently. However, the possible range of geochemical behaviour in a natural system is constrained by the actual distribution of mineral constituents and geochemical conditions, and the actual nature of the alteration in conditions that will be caused by the disposal vault. Validation of a representation of the geochemical behaviour of a disposal system should demonstrate both that the processes included in the models and the data on geochemical behaviour used in the models are appropriate for the conditions in each component of the disposal system and for the interactions between components. Where appropriate data do not exist, or the conditions expected are outside the range for which data are available, the basis for the data selection made should be presented.

2. DESCRIPTION OF THE GEOSPHERE

Earlier performance assessments in the 1970s used generic one dimensional models of the geosphere. At the time, data from actual sites were limited and these models were useful for scoping calculations and sensitivity analyses. However, non-site-specific models are of limited value in a licensing process precisely because they cannot be validated as representative of the conditions at the site being licensed. During the 1980s a number of national radioactive waste management programs have progressed to detailed site characterization. AECL's experience from field investigations at several research areas on the Canadian Shield has been that patterns of faulting and fracturing, the fracture-filling mineralogy and other geohydrological and geochemical conditions controlling groundwater flow and radionuclide migration vary significantly as a result of site-specific conditions. Consequently, the geometry of the pathways and the hydrogeological and geochemical conditions along the pathways that provide the basis for geochemical modelling can be quite different even for sites in rock with nominally similar lithology and physical properties. We have been unable to devise a defensible generic model for the geosphere for use in environmental and safety assessments and instead rely upon models developed from site-specific data.

The exercise of incorporating geochemical data consistent with conditions at a specific site quickly identifies gaps in the geochemical databases. The detailed characterization investigations carried out at the site of AECL's Underground Research Laboratory have shown that the principal pathways of groundwater transport can be identified and that the mineralogy and geochemistry along the pathways can be characterized. However, there is a shortage of sorption data as a function of varying groundwater salinity and redox condition for key radionuclides on minerals along the pathways. Furthermore, much of the available data has not been validated by expert review. Most of the remainder of this paper reviews progress in addressing these and other issues.

3. GEOCHEMICAL MODELLING

Geochemical models and the associated databases are commonly divided into two subsets: models dealing with the major rock and mineral-forming elements, and models dealing with trace radionuclides. The models are used to predict the stability of solids, the solubility of radionuclides and the chemical speciation of dissolved species in groundwaters. The 1979 American Chemical Society Symposium Series, 93: The 1979 publication "Chemical Modeling in Aqueous Systems," edited by E.A. Jenne [2], is an excellent comprehensive review of the models and data available at that time. In particular, the paper by Nordstrom et al. entitled "Comparison of Computerized Chemical Models for Equilibrium Calculations in Aqueous Systems" is an authoritative review of the computer models and chemical data available in 1978 [3]. In his preface to ACS-93, Jenne writes the following comments about the paper: "the authors concluded that the differences in the thermodynamic data selected and treatment of redox are the predominant causes of the discordant results obtained." Similar conclusions were reached in the more recent CHEMVAL exercise conducted as part of the CEC MIRAGE project [4]. The codes compared gave similar predictions if the same database was used. The major discrepancies between the databases were due to incorrect data, omission of important aqueous or solid species, and incorrect procedures in deriving parameter values from experimentally determined data.

Some more recent reviews with narrower scope have been published for codes currently in use. For example, the EPRI series of reports released in 1984 [5] were commissioned to assess the safety of disposal of wastes from coal-fired power plants in the U.S.A. Another series of reviews is part of the CEC MIRAGE Project and CHEMVAL [6]. The geochemistry and performance assessment modelling needs for high-level waste disposal in granitic rocks were reviewed in 1983 at a workshop held at Minster Lovell in the United Kingdom and co-sponsored by Atomic Energy of Canada Limited and the Commission of European Communities (Euratom) [7]. Recommendations for research on integrated system analysis, sorption of key nuclides, and thermodynamic data on iron-containing solids were highlighted and are still valid.

The assumption of local thermodynamic chemical equilibrium in the geosphere for the speciation and solubility of the migrating radionuclide is reasonable. Hence, the primary validation requirement is for peer reviewed data.

4. SOLUBILITY LIMITATION

The release of radionuclides from a disposal vault and their transport in the surrounding geosphere is limited by the solubility of the migrating radionuclide. These solubilities can be predicted from thermodynamic data and a good description of the geochemical conditions in the groundwater. In particular, redox, E_H, salinity, pH, and groundwater composition need to be specified. A comprehensive report on the solubility model and database for U, Np, Pu, Th, and Tc has recently been published by Lemire and Garisto [8]. However, there is a general need to validate such a database by extensive peer review.

5. THERMODYNAMIC DATA

A number of different databases are being used throughout the world in various geochemical modelling codes. There is a need to coordinate the continual changes being made to these databases by individual users to ensure internal consistency. The databases for radionuclides are the most critical since errors could lead to greater errors in predicting the radiological impact of waste disposal than errors in the databases for rock-forming minerals. Currently, there is no single accepted database for key radionuclides. Reviews of existing data are currently being coordinated by the NEA and IAEA. They are briefly summarized below. The CHEMVAL exercise coordinated by the CEC, is also making important contributions and is reviewed elsewhere in this symposium.

The OECD/NEA has coordinated development of CODATA-compatible chemical thermodynamic databases for a number of key radionuclides since 1984. Critical reviews are currently in progress for six elements: uranium, plutonium, neptunium, americium, technetium, and iodine. Critical review of all the existing data in the scientific literature is being performed by groups of experts using procedures that will allow subsequent peer reviews to follow and verify the evaluation path. The uranium database will be released for review in 1990. In addition to the recommended database for each element, this review process will also identify the areas where the existing data need to be strengthened by means of direct measurement. It is possible that the number of experiments required will be sufficiently large to necessitate a coordinated international program.

The IAEA has been and is coordinating a number of programs contributing to the supply of thermodynamic data. The agency is publishing a handbook on the chemical thermodynamics of actinide elements and compounds as a series. The published volumes on actinide elements, aqueous ions and a draft of the volume on actinide complex aqueous ions are very useful sources, and have been used by the NEA review teams with the full cooperation of the IAEA. Unfortunately, the volume on aqueous organic complexes apparently is no longer planned to be completed. It is recommended that work on this document be resumed, possibly through an internationally coordinated project, similar to that previously described for the actinide elements.

The activities referred to above are essentially reviews of existing and often quite old data. The number of laboratories producing new data is quite small, and hence the publication of the results of new work can have a major impact. For example, the Parks and Pohl determination of the solubility of UO_2 under alkaline conditions indicated that the previously estimated value was too high by several orders of magnitude [9]. For some radionuclides in certain groundwaters, the available data are uncertain or nonexistent because they are very difficult to measure directly. The recently developed unified theory of metal ion complex formation can be used to remove some of the gaps or, at least, indicate the possible importance of missing data [10].

The CEC continues to coordinate and fund a number of programs to generate thermodynamic data through the MIRAGE Project. The annual plenary sessions, to which non-CEC participants are also invited, provide a valuable annual review of the status of these programs.

6. SORPTION/RETARDATION DATA

The measurement of sorption coefficients has been a major activity in waste management R&D throughout the world. These measurements have been conducted for a wide range of radionuclides, minerals, groundwaters, and experimental conditions. Recently, the OECD/NEA has compiled a Sorption Database (SDB) consisting primarily of data generated from static batch radionuclide sorption studies carried out in laboratories worldwide. The database serves as a useful bibliography of sorption data; however, the direct use of this data in performance assessments should be approached with caution. It is essential to determine if the data were measured under conditions appropriate for the disposal system being assessed. For example, the majority of the k_d values were determined under aerated conditions and are not necessarily appropriate for the reducing conditions found in many groundwaters.

The movement from assessments of generic geologies to those for specific sites has had a major impact on the sorption data required. For example, the geosphere model currently being used by AECL in SYVAC for the Lac du Bonnet batholith requires sorption data for sixteen minerals under two redox conditions as a function of ionic strength and radionuclide concentration. These large data requirements have led to the need to focus on measuring sorption for a restricted number of key elements in detail under conditions closely approximating those of the disposal system. Experimental work generally has to be carried out under very tightly controlled atmosphere conditions. Simple batch sorption methods may be useful in some cases, but dynamic methods may be required to confirm their applicability and may be essential for microporous rock systems. The experimental measurements need to address the expected heterogeneity of the geological system and possible variations in important hydrochemical parameters (e.g., pH, redox, nuclide concentration) with distance from the repository. Finally, the data analysis and reduction must be capable of synthesizing global sorption parameters for performance assessment that reasonably reflect an appropriate degree of the complexity of the natural system. One way to approach the acquisition of both the required sorption data and a description of the geochemical heterogeneity of the geosphere pathway is the close integration of laboratory and field studies of radionuclide migration [11a,b,;12].

Currently, there is no internationally accepted peer-reviewed sorption database. Apart from the NEA sorption database, there have been only quite restricted databases published in recent years. However, as performance assessment studies become more site-specific, there will be a greater need to publish and peer review the data. McKinley and Grogan [13] have recently reviewed sorption databases for recent Swiss Repository

Safety Assessments and Vandergraaf [14] has prepared a similar review for the Canadian program. The absence of a peer-reviewed sorption database is a major weakness in code validation. Fortunately, the NEA plans to hold a Sorption Workshop in 1990 or 1991 that will begin to address this weakness.

7. COLLOIDS, ORGANICS, AND BACTERIA

The standard approach to geochemical modelling has knowingly tended to overlook some areas that are increasingly being recognized as important, in particular, the impact of colloids, organics, and bacteria. International working groups have been established to consider these topics:

- The CEC Colloids and Complexants Working Group is examining various aspects of the characterization of natural colloids and dissolved organic carbon. This group, known informally as the COCO Club, has conducted two field sampling comparison exercises for colloids in the U.K. and Switzerland, and two laboratory intercomparison exercises for characterization of organics in groundwater [15].
- An ad hoc group set up to study microbiology in nuclear waste disposal (MIND) held its first meeting in Baden in 1987, hosted by NAGRA [16], and the second meeting was held in Stockholm in 1988, hosted by SKB [17].

The role of natural organics as complexants is a particularly important area where thermodynamic data are lacking. An extensive review of the complexation of radionuclides by natural organics has recently been carried out and indicated that the general behaviour might be modelled by a few defined ligand groups [18]. In addition to natural organics present in groundwater, there is the possible generation of new organics by processes occurring in the waste disposal system. This is a particularly acute problem for LLW containing large amounts of organic materials, such as cellulose, which, at least temporarily, generate a complexing agent that greatly enhances the solubility of transuranics such as plutonium [19]. There are no reports of solubility enhancements due to organics in high-level-waste systems, although it seems to be an area worthy of reexamination. However, once again the absence of a peer-reviewed internationally accepted database for organics is a major weakness in code validation for some disposal concepts.

8. SALINITY AND HIGH-IONIC-STRENGTH EFFECTS

The discovery of high-salinity groundwaters in crystalline rock formations in Canada, Europe, and Scandinavia [20] has had a major impact on the geochemical modelling and data requirements for performance assessment and the design of engineered barriers. Lemire recently summarized the effects on calculated equilibrium concentrations and speciation in the uranium-water system [21]. It was concluded that the effects of increasing ionic strength on UO_2 solubility under reducing conditions would not be great at 25°C. However, there are very large uncertainties in extrapolating to higher temperature, primarily because of the lack of good data at temperatures other than 25°C and a lack of knowledge of changes in speciation. For this reason, the calculated solubility of radionuclides at the higher temperatures expected in the near-field could quite possibly be inaccurate. If better estimates are required, they may be obtained by measurements on solutions at higher temperatures. Such experiments however, are very difficult and are best postponed until groundwater data are available from specific sites investigated for waste disposal. These measurements could play a major role in code validation.

High salinities could also have a major impact on the rate of migration of radionuclides if the sorption values are significantly reduced.

9. EFFECTS OF DISPOSAL VAULT ON GEOSPHERE

Most performance assessments assume an invariant geosphere and neglect any effects of the disposal vault. Considering the duration of the thermal transient and the large amounts of engineering materials in the vault, the impact on the geosphere needs examination starting at the boundary. Changes in the nature of the fracture-filling minerals is an obvious place to start.

10. VALIDATION EXERCISES AND NATURAL ANALOGUES

A number of validation exercises and natural analogue studies are being conducted throughout the world. Many of these have also been co-ordinated internationally and will be discussed at GEOVAL 1990. It is clear that validation and acceptance of geochemical codes and data are considerably strengthened by such exercises. The validation of the use of simple sorption models for predicting the retardation of radionuclide migration is a key area. Hence, large-scale laboratory and field migration studies need to continue.

CONCLUSIONS

1. Geochemical models and data in performance assessment need to be validated in the broader context of the overall disposal concept and in detail within the overall geosphere model.

2. The geochemical models used in performance assessment must be designed for incorporating site- and design-specific data to be relevant in the licensing process for a disposal facility.

3. Solubilities and speciation of key radionuclides can be predicted by means of available thermodynamic codes and databases. The amount of data available for some radionuclides could be insufficient. There is a need to validate these predictions by means of direct observation of solubility and speciation.

4. The lack of a fundamental theoretical basis for modelling the sorption/retardation process in terms of the underlying thermodynamics and kinetics of the processes remains a weakness. However, significant progress is being made in laboratory and field investigations of radionuclide migration to bound the uncertainties in the use of empirically determined sorption data and simple sorption isotherms.

5. The absence of a peer-reviewed internationally accepted database for sorption coefficients is a major weakness.

6. The universal assumption of an invariant geosphere, unaffected by time or the presence of a vault and contents is a major difficulty in validation.

ACKNOWLEDGEMENTS

This paper drew heavily on a previous review co-authored with I.G. McKinley. S.H. Whitaker is acknowledged for significant contributions to the current paper and B.A. St Denis for typing the manuscript. "The Canadian Nuclear Fuel Waste Management Program is jointly funded by AECL and Ontario Hydro under the auspices of the CANDU Owners Group."

REFERENCES

(1) Sargent, F.P. and I.G. McKinley : "Geochemical and Chemical Data and Modelling for the Safety Assessment of Geological Disposal of High-Level Nuclear Fuel," Presented at the International Symposium on Safety Assessment of Radioactive Waste Repositories, Paris, France, 1989 October.

(2) E.A. Jenne (Editor): "ACS Symposium Series, 93: Chemical Modeling in Aqueous Systems," American Chemical Society, Washington, DC, 1979.

(3) E.A. Jenne (Editor): "ACS Symposium Series, 93: Chemical Modeling in Aqueous Systems," American Chemical Society, Washington, DC, 1979, pp. 857-894.

(4) Côme, B. and T.W. Boyd : "CEC Activities for Verification and Validation of Geosphere Models Within the MIRAGE Project," In GEOVAl 1987, Proceedings of a Symposium on Verification and Validation of Geosphere Performance Assessment Models, Stockholm, Sweden, 1987, Volume 1, pp. 85-105.

(5) Kincaid, C.T., J.R. Morrey and J.E. Rogers : "Geohydrochemical Models for Solution Migration," Volumes 1 (1984), 2 (1984), and 3 (1986), Electric Power Research Institute Report, EPRI EA-3417.

(6) Elsewhere in these proceedings.

(7) Chapman, N.A. and F.P. Sargent (Editors): "The Geochemistry of High-Level Waste Disposal in Granitic Rocks," Proceedings of an AECL/CEC (Euratom) Workshop, Commission of the European Communities Report, EUR-9162. Also available as Atomic Energy of Canada Limited Report, AECL-8361, 1984.

(8) Lemire, R.J. and F. Garisto : "The Solubility of U, Np, Pu, Th, and Tc in a Geological Disposal Vault for Used Nuclear Fuel," Atomic Energy of Canada Limited Report, AECL-10009, 1989.

(9) Parks, G.A. and D.C. Pohl : "Hydrothermal Solubility of Uraninite," Geoch. Cosmochim. Acta, 52, 1988, pp. 863-875.

(10) Brown, P.L. and H. Wanner : "Predicted Formation Constants Using Unified Theory of Metal Ion Complexations," OECD/NEA Paris Publication 1987. (See also references cited in this report).

(11a) McKinley, I.G., W.R. Alexander, C. Bajo, V. Frick, J. Hadermann, F.A. Herzog, and E. Höhn : "The Radionuclide Migration Experiment at the Grimsel Rock Laboratory, Switzerland," Materials Research Symposium Proceedings, 112, (Scientific Basis for Nuclear Waste Management XI) 1988, pp. 179-187.

(11b) Bradbury, M. (Editor): "The Laboratory Investigations in Support of the Migration Experiment at the Grimsel Test Site," Nationale Genossenschaft für die Lagerung Radioaktiver Abfaelle Report, NAGRA-NTB-88-23, Nagra, Baden, Switzerland, 1988.

(12) Sargent, F.P. and T.T. Vandergraaf : "Radionuclide Migration R&D in the Canadian Nuclear Fuel Waste Management Program," Rad. Waste Manage. Nucl. Fuel Cycle, 10(1-3), 1988, pp. 21-40.

(13) McKinley, I.G. and H.A. Grogan : "Radionuclide Sorption Database for Swiss Repository Safety Assessments," presented at Migration 1989, 1989, and to be published in Radiochemica Acta.

(14) Vandergraaf, T.T. and K.V. Ticknor : "A Compilation and Evaluation of Radionuclide Sorption Coefficients Used in the GEONET Submodel of SYVAC for the Whiteshell Research Area," in preparation, to be published as an Atomic Energy of Canada Limited Report.

(15) Longworth, G., C.A.M. Ross, C. Degueldre, and M. Ivanovich : "Interlaboratory Study of Sampling and Characterization Techniques for Groundwater Colloids," Atomic Energy Research Establishment Report, AERE-R-13393, 1989.

(16) Grogan, H.A. and J.M. West (Editors): NAGRA Report, NTB 88-54, Nagra, Baden, Switzerland, 1988.

(17) McCabe, A.M. (Editor): "MIND Group - Formulation of a Collaborative Technical Programme," Proceedings of the 2nd Meeting of the Microbiology in Nuclear Waste Disposal (MIND) Working Group, Stockholm, Sweden, May 1988. Available from CEGB, Berkeley Nuclear Laboratories, Berkeley, U.K., as CEGB RD/B/6181/R89, March 1989.

(18) Grauer, R. : "Zur Koordinationschemie der Huminstoffe," Paul Scherrer Institute Report, PSI-24, 1989.

(19) Cross, J.E., F.T. Ewart, and B.F. Greenfield : "Modelling the Behaviour of Organic Degradation Products," Materials Research Society Symposium Proceedings 127, (Scientific Basis for Nuclear Waste Management XII) 1989, pp. 715-722.

(20) Fritz, P. and S.K. Frape (Editors): "Saline Water and Gases in Crystalline Rocks," Geological Association of Canada Special Paper 33, 1987.

(21) Lemire, R.J. : "Effects of High Ionic Strength Groundwaters on Calculated Equilibrium Concentration in the Uranium-Water System," Atomic Energy of Canada Limited Report, AECL-9549, 1988.

VALIDATION STRATEGY AND STRATEGIC VALIDATION.
As seen by a Safety Assessor

Tõnis Papp
SKB, Stockholm, Sweden

ABSTRACT

To have a final repository accepted and approved, a high safety level is required. It must be shown that the risk posed by the repository is low, and that the long term repository performance can confidently be evaluated. The confidence issue seems today to be the most difficult item to address when licensing a disposal facility. Validation is a part of achieving sufficient confidence in our predictive capabilities, and it is necessary to develop a strategy for the priorities.

Aspects on validity and validation, from safety assessment and repository development point of view, are discussed based on the KBS-3 concept and the SKB time schedule for site investigations.

At best, validation of predictive models provides uncertainty bands around the results within which the real outcome will fall. The late availability of local and concept specific data will postpone quantification until late in the repository development process. For a long time, we have to use validity in an unquantified, generic way. When addressing our capacity in general to validate models in settings similar to the finally selected one, it is important that local and generic factors are addressed separately.

Only few radionuclides dominate the potential hazard of spent fuel after the first 1000 years. A high quality of data for these nuclides is essential for the confidence in the results. Large validation efforts should be given to barriers early in the release sequence and to processes causeing release of radionuclides around 10 000 years. Due to the reduction of radiotoxicity with time there seems to be litle need for modelling change in the geosphere, after about 100 000 years.

Since there are many ways by which to achieve a safe repository, a close cooperation must exist between designers of the repository and those trying to validating the assessment models. It might well be that a repository built for high confidence is more acceptable than a repository built for maximum safety.

STRATEGIE DE VALIDATION ET VALIDATION STRATEGIQUE. Point de vue d'un responsable en matière de sûreté

Tönis Papp
SKB, Stockholm, Suède

RESUME

Pour qu'un dépôt définitif soit accepté et approuvé, il faut qu'il présente un degré de sûreté élevé. Il faut montrer qu'il ne présente que peu de risques et que l'on peut avoir confiance dans l'évaluation de son comportement à long terme. Il semble qu'aujourd'hui la confiance soit le problème le plus délicat à traiter quand il s'agit d'octroyer une autorisation pour une installation de stockage. La validation est l'un des moyens de parvenir à un degré de confiance suffisant dans nos capacités de prévision et il est nécessaire d'élaborer une stratégie pour définir les priorités.

Se basant sur le projet KBS-3 et le calendrier d'étude des sites du SKB, l'auteur examine les aspects liés à la validité et à la validation, du point de vue de l'évaluation de la sûreté et de la mise au point des dépôts.

Au mieux, les modèles de prévision de la validation permettent de déterminer, autour des résultats, des zones d'incertitude dans lesquelles viendra se placer la réalité des faits. On ne pourra faire des prévisions chiffrées qu'à un stade avancé du processus de mise au point du dépôt car on ne disposera que tardivement de données précises concernant les conditions locales et le projet. Pendant longtemps le concept de validité ne peut être utilisé que de façon générique et qualitative. Dans l'étude de notre aptitude générale à valider des modèles dans des conditions comparables à celles du site qui sera finalement retenu, il importe de traiter séparément les facteurs locaux et les facteurs génériques.

Au bout de 1 000 ans le danger potentiel du combustible irradié ne tient plus, pour l'essentiel, qu'à quelques radionucléides. Il est donc capital pour la confiance que l'on peut avoir dans les résultats, que les données relatives à ces nucléides soient d'excellente qualité. Il faudra déployer des efforts de validation importants en ce qui concerne les barrières dans les premières phases de la séquence de rejet et les processus provoquant le rejet des radionucléides après une dizaine de millénaires. Compte tenu de la décroissance de la radiotoxicité avec le temps, il ne semble pas très utile de modéliser les changements pouvant intervenir dans la géosphère au-delà de 100 000 ans.

Comme il existe de nombreux moyens de réaliser un dépôt sûr, les responsables de la conception des dépôts, d'une part, et ceux de la validation des modèles d'évaluation, d'autre part, doivent coopérer étroitement. Il se pourrait qu'un dépôt construit dans l'optique d'une confiance renforcée soit plus acceptable qu'un dépôt construit dans l'optique d'une sûreté maximale.

INTRODUCTION

To have a final repository accepted and approved, a high safety level is required. It must be shown that the risk posed by the repository is low, and that the long term repository performance can confidently be evaluated. The confidence issue seems today to be the most difficult item to address when licensing a final disposal facility for long lived waste.

An important part of the confidence is the validity of the models used to predict future repository performance.

Since the word "valid" is used in many different fields ranging from statistical sociology to legal matters and science it has different connotations in different communities. Here validity is used in a rather generic way as an indicator of the capacity of a model to describe nature or processes therein. It is not seen as something absolute, the validity can be high or low and its acceptability depends on the specific situation where the model is used.

Some aspects on validity and validation, seen as important from the safety assessment and repository development point of view, will be discussed based on the repository concept in KBS-3, the Swedish time schedule and planning for how the site specific information will be collected, and Swedish geology.

GENERAL FRAMEWORK FOR VALIDATION IN SWEDEN

At certain times in a repository development programme decisions have to be taken to focus the efforts in an appropriate way. For the Swedish programme (as given in SKB R&D-program 89), the time table is based on a sequence of decisions focusing the efforts towards a so called Preliminary Safety Assessment for the siting in the early years of the next decade:

- 3 candidate sites for the repository are appointed	1992
- After surface based investigations 2 sites are selected for detailed characterization	1994
- Based on the studies of alternative repository designs a conceptual design will be selected	1995
- Detailed investigations incl shaft or tunnel down to repository depth at site 1, starting	1996
- Investigations at site 2 are planned to start about 2 years later	1998
- Optimization of the repository design starts	1998
- Preparation of Preliminary Safety Report starts	2001
- Siting application will be submitted	2003

During this sequence safety evaluations will have to provide an information basis for the decisions. Investigations, predictions and control measurements will have to establish the local validity of the used models as the siting and construction of the disposal facility is progressing. Models validated to an appropriate level must be available at certain times. Since the validity level is coupled to the availability of relevant information, the order in which the investigations will be performed at the candidate sites is important.

At an early phase of the investigations only limited data are available from repository depth. The characteristics qualifying a site as a candidate site in 1992 are quite general. It must be an uncomplicated, stable site with large enough volumes and high potential for the geosphere to provide a good safety barrier (low ground water flow and high nuclide retention). At this stage only the potential of the site must be shown. Questions to be answered are: Can we model the site? Are the rock volumes between the major zones large enough? Could an entrance shaft or tunnel be located without impairing the suitability of the site?

The pre-investigations, 1992-1994, will lead to a gradual increase in the amount and quality of data available for assessing the suitability of the candidate sites and for positioning the repository within a site. At this stage only a few drilled holes will be available for characterization of the rock at repository depth. A preliminary evaluation of the safety must be made, and also a programme for the further characterization.

At the next stage starting around 1996, shafts or tunnels to repository depth will provide more detailed information for the safety assessment. When the investigations have reached appropriate depth, the repository level(s) has to be decided, and the area that will be used for disposal must be appointed.

Some years later alternative systems for the tunnel layout will be tested with boreholes in the available deposition volume. Only late in the investigations, or during construction, there will be good data available to decide the actual positions for the canisters, and what tunnel sections not to use due to less good quality of the rock mass.

After each step the influence of new data on lay-out and safety must be evaluated and it must be controlled if factors prohibiting the use of the site have emerged. It must also be evaluated how the disturbances caused by the investigations will affect a possible future repository.

THOUGHTS ON VALIDATION STRATEGY

At best, validation of predictive models will provide uncertainty bands around the results within which we believe that the real outcome will fall. The availability of local and concept specific data will, however, delay the possibilities for quantification until late in the repository development process. We have thus, for a long time, to use the validity in a much less quantified way.

Some of the models for predicting the long term performance of a repository are highly site or system specific, other are generic. Models for radioactive decay, for instance, are considered fully generic, and can be utilized irrespective of site or repository concept. Most generic models need some adaption to repository specific circumstances by selection of input variables, like for instance equilibrium models for chemical speciation. Some of these models are well tested and are regarded as well validated in the normal parameter space. Uncertainties are mainly introduced by the data set or are due to applying the models in an unusual parameter space. The selection of certain materials in the repository, like copper or bentonite will, of course, require selection of models addressing specific phenomena, like pitting corrosion in copper or the degradation of bentonite to illite.

Probably the models most sensitive to repository specific data are the models for ground water movements and radionuclide transport. To build up a continuous description of the repository and its geologic setting, models of statistical or of inter-polative/extrapolative nature are often used to extend limited amounts of measured data. And the validity of the geosphere transport models will strongly depend on the validity of models used for the structural description of the repository and site.

For the focusing process, as indicated above, there is already at an early stage a need to show in a generic way that we understand and can evaluate the barrier capacity of the geos-phere. So, even if present attempts to validate models for ground water flow or nuclide transport can be seen as only large scale testing of validation methodology, their value as indicators of our ability for site and concept specific valida-tion is very important.

For the "generic" validity, addressing our capacity to build valid models for geotransport in settings similar to the finally selected site, it is important that the different phenomena involved in a model are addressed individually. Uncertainties or lack of validity in models caused by local or conceptual phenomena should, if possible, be kept apart from those caused by generic processes.

This principle should also be used on models for other barriers in the repository. If the barrier structure (geometric form, homogeneity of properties, etc) is important for the performance, the local or concept specific factors affecting the predictions should be kept apart from the generic factors and discussed separately.

It seems that a substantial international cooperation has been established both in validation of models for geosphere transport and chemistry, and that there is an international interest also in the quality of the corresponding data bases. I hope that an international cooperation will be established also for the validity of other models in the assessment sequence, like canister corrosion, release of radionuclides from the waste matrix, etc.

A difficult area with regard to methodology is the validity we can obtain by natural analogues. It might be difficult to establish a quantified level of validity provided by a natural analogue. But not finding unexpected phenomena in a system where the external parameters are defined by nature is of primary importance when we want to show our understanding of the systems involved .

Another important issue needing more discussions is the difference in validation methodology between the mathematical models used in research and models used in assessing the total repository performance. The process of transferring the validity of comprehensive research models to simplified assessment models must be done carefully and in a transparent way. The same problem is even more pronounced for the simplified models often used in probabilistic assessments.

WHICH PROCESSES ARE IMPORTANT?

Another aspect beside the question how to do the validation is "what to validate and when to validate". Obviously the models most important to validate are those that are used to evaluate the barriers most important to the safety of the repository. In a multibarrier concept there should, however, be such a redundancy between the barriers, that the failure of any single barrier shall not create a totally unacceptable situation. Which of the mutually redundant barriers should then be regarded as the most important? One way of ranking them is to give priority to barriers that have their role early in the release sequence. E.g.

- isolation by the canister,
- solubility limitations for species in the near-field,
- retardation in the geosphere.

In quantifying the performance certain models play a large role for many barriers and should consequently be given priority:

- Ground water movements and their time dependence.
- Chemical speciation in the liquid phase.
- Diffusion in near-field.
- Radioactive decay and radiolysis.
- Sorption, mineralization and co-precipitation.

Other ways to rank models is according to the potential safety effect a barrier can provide, or according to how easy it is to establish an acceptable proof for the intended performance of the barrier. It seems that a ranking on potential performance is most relevant in an early design phase of the repository, and the provability becomes the dominating issue during the licensing.

Looking at KBS-3, the two barriers/processes having greatest potential for reducing the environmental effects are the long lived cannister and the diffusion/sorption of radionuclides in the micro-fissures of the rock matrix. Barriers having the most provable longterm performance seem to be the thermodynamically stable copper canister and the solubility limitations for the fuel inside the bentonite.

Whatever principles are used to decide the validation efforts it is important that there is an information exchange between the repository designers and the validators. Based on the belief that we can arrange effective and safe repositories in many ways by giving different emphasis to different barriers, the "best" barrier mix will to a large extent be based on what potential there is to obtain a high quality validation of the models used.

SAFETY ROLE OF THE REPOSITORY

Since the activity in the waste is decaying, the required level of validation might be reduced for predictions far into the future.

Ideally, the required protection level of a repository should be defined by evaluating the consequences of "not taking care" of the waste and compare those with the acceptance criteria given by the society. In practice there are, however, some problems in defining the case of no protection.

How then can the required protection level of the repository be quantified?

Attempts have been made to calculate something that could be called a "volume for needed dilution". It is defined as the water volume necessary to dilute the radionuclides in e.g. one tonne of spent fuel to a concentration equal to the maximum concentration allowed in drinking water. It can be seen as an index for the potential toxicity for the waste if the pathway to man is by drinking water. The toxicity can be given in absolute terms or, for instance, relative to the toxicity of natural radioactive substances in the bedrock.

The problem with using the needed dilution volume as an index for required safety is that there might be other paths through the biosphere for human exposure than intake by water.

In figure 1 a relative toxicity index calculated in a somewhat different way is presented. The radioactivity in one tonne of spent fuel is multiplied with nuclide specific dose-factors indicating what individual dose one Bq of an isotope would give rise to if released to the environment.

The dose-factors are calculated by assuming that the radionuclides will reach the biosphere in a lake with a nearby small family farm. A well is situated in the groundwater inflow area of the lake and used for drinking and for watering the kitchen garden and farm animals. All food is produced on the farm, or taken from the lake. The cereals are produced in fields on old lake sediments.

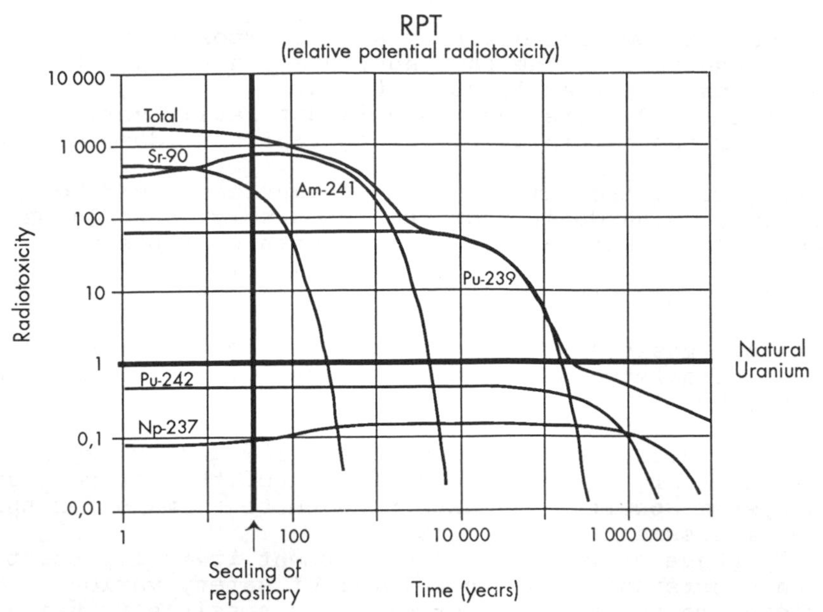

Fig. 1 The potential radiotoxicity of spent fuel compared to natural uranium

The toxicities are summarized for the nuclides in one ton of fuel and divided by a toxicity calculated in the same way for one tonne of naturally occurring uranium in equilibrium with its daughters. This gives a relative potential toxicity index for spent fuel compared with that for natural uranium.

Recognizing:

- that the Swedish bedrock in general contains the same amount of natural uranium in the 500 m rock above the repository as in the repository,
- that the spent fuel to 95 % consists of the same natural uranium that once was mined for the fuel fabrication,
- that there exists natural uranium ores of the same uranium concentrations and to the same amounts as in such a repository,
- that the natural radionuclides, chemically very similar to the radionuclides in the fuel, will have to be transferred to man through the same biosphere as the one used for the critical group above,
- and, finally, that the dose from the natural radionuclides in bedrock gives a dose of the same order as the design limits for nuclear activities in Sweden,

the curve can be regarded as a reasonable measure of the protection level which must be provided by the repository at different times.

Some conclusions can be drawn from this diagram:

- There are only a few nuclides that are of importance for the toxicity of the fuel if the radionuclides can be isolated from the biosphere for at least 1000 year.
- After about 100 000 years the waste is not really very different from the naturally occurring radionuclides in nature.
- After that period, the same type of precautions would be needed to avoid high doses from the repository as would be exercised around a naturally occurring uranium deposit.

THOUGHTS ON STRATEGIC VALIDATION

Based on the discussions above a strategic selection, from the point of view of safety assessments, of what to validate and when to do it would be:

- The "generic" validation of groundwater flow and transport models is necessary to show the potential of the methodology. Site and system specific validations have to be made for the quantified assessments for licensing.
- At an early phase of repository development it is important to have an understanding of what level of safety various barriers can provide and of whether it is possible to get a reasonable validation of the models needed to quantify the barrier performance.
- When coming close to the licensing, the focus in the validations should be on trying to establish a site specific quantification of the uncertainties.
- The largest efforts to get a high validity in barrier modelling should be given to barriers early in the release sequence.
- The highest need for proven validity would be for processes which could cause a release of radionuclides to the biosphere in a time span of the order of 10 000 years.
- There seems to be less need for prediction of the behaviour of the geosphere after about 100 000 years.
- There are only a few radionuclides dominating the potential hazard of the spent fuel after the first 1000 years. A high quality of the database for these nuclides is essential for the confidence in the assessment results.

All the validation efforts must be seen as parts of the general issue of achieving sufficient confidence in our predictive capabilities, and it is necessary to develop a strategy for where to put the priorities. A priority list suitable for one repository or adapted to the time schedule of one group, is not necessarily good for another concept or programme. An open international exchange of both validation strategies and validation results is good and should be extended beyond the geo-sciences.

Since there are many ways by which to achieve a safe repository, it is essential that there exists a close cooperation between those designing the repository and those studying the possibility to validate the models for assessments of the performance of the system. It might well be that a repository built for high confidence is more acceptable than a repository built for maximum safety!

Session 2

PROGRESS IN VALIDATION OF FLOW AND TRANSPORT MODELS

Séance 2

ETAT D'AVANCEMENT DES TRAVAUX DE VALIDATION SUR LES MODELES D'ECOULEMENT ET DE TRANSPORT

Chairmen - Présidents

E. PELTONEN (Finland)
N. CHAPMAN (United Kingdom)
C.-F. TSANG (United States)
T. NICHOLSON (United States)

THE STUDY OF RELEVANT AND ESSENTIAL FLOWPATHS

O. Brotzen
F. Brotzen AB Consultants
Djursholm, Sweden

ABSTRACT

At a KBS-type repository, flowpaths relevant to performance interact with the waste packages. Their proximal parts provide essential portions of the flow resistance and transport time. They can be studied by local investigations. The farfield, in contrast, remains inaccessible for detailed study. Validation regarding long-term groundwater transport of nuclides should focus on (1) nearfield mass transfer and retardation under natural gradients, and (2) sub-surface dilution at discharge. Tentative in-situ experiments are briefly considered. Validation concerning other groundwater effects appears more urgent. Inflow to tunnels and shafts is irrelevant to backfilled repositories.

ETUDE DE LA CIRCULATION DES FLUIDES : VOIES D'ECOULEMENT ESSENTIELLES ET IMPORTANTES

RESUME

Dans un dépôt de type KBS, les voies d'écoulement qui revêtent de l'intérêt pour le comportement du dépôt sont celles qui réagissent avec les colis de déchets. Les plus proches interviennent de façon essentielle dans la résistance à l'écoulement et le temps de transport. On peut étudier ces voies d'écoulement par des recherches in situ. En revanche, le champ lointain ne peut pas faire l'objet d'études détaillées. S'agissant du transport à long terme des radionucléides par les eaux souterraines, la validation devrait porter essentiellement sur (1) le transfert de masse en champ proche et le ralentissement sous l'effet de gradients naturels, puis (2) la dilution souterraine au moment du rejet. L'auteur analyse brièvement quelques expériences préliminaires in situ. Il apparaît plus urgent de procéder à la validation d'autres effets de l'eau souterraine. L'entrée d'eau dans des galeries et des puits est un phénomène qui n'intervient pas dans le cas des dépôts comblés.

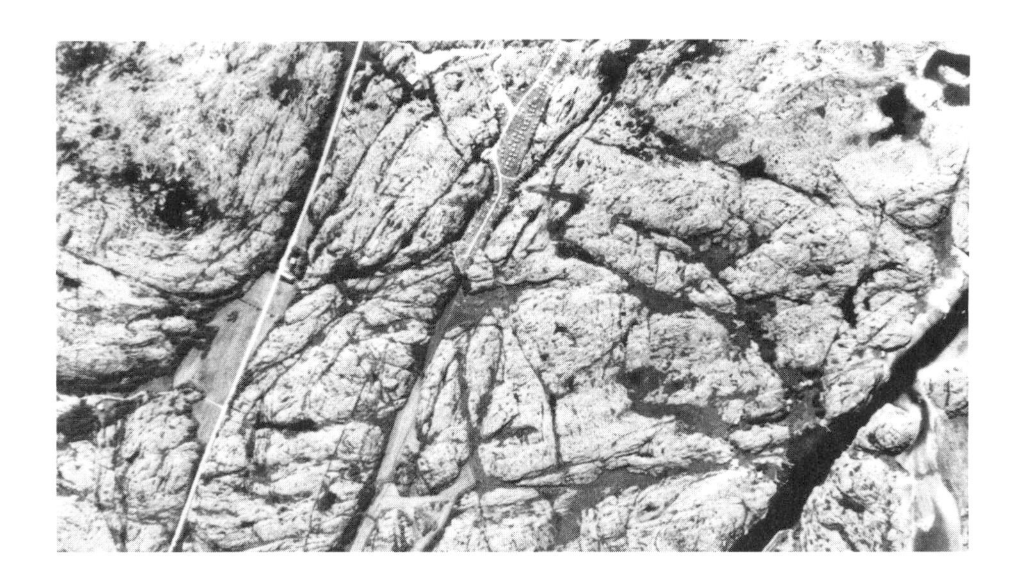

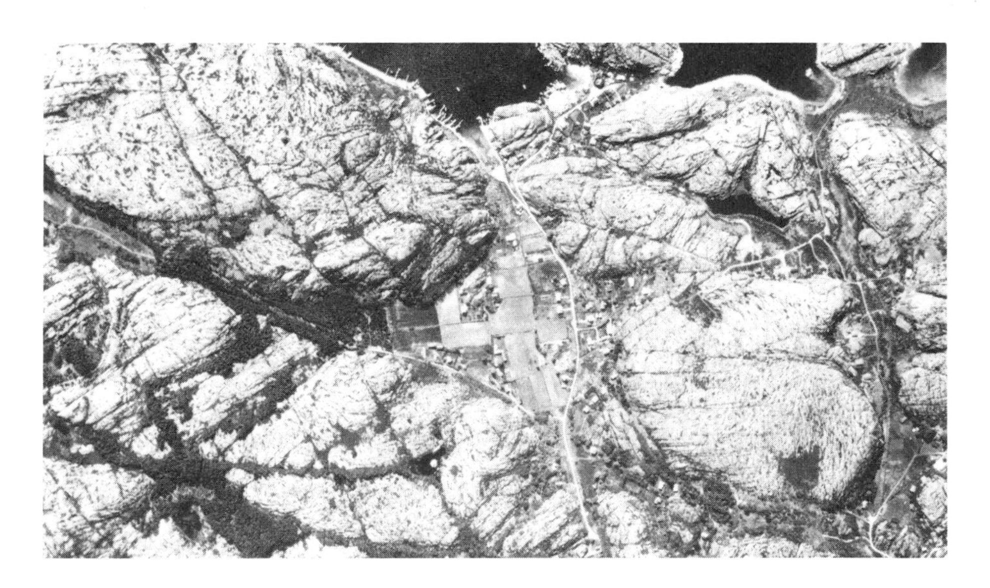

Figure 1. Vertical views of common fracture patterns in granite. Top: Ordinary fracture network. Bottom: Fracture network with marked larger bedrock blocks of differing interior fracturing. Length of strips about 2.0 and 2.5 km, respectively. LMV photo. These pictures suggest that mass transfer within blocks, mass transfer in the dominant fracture network, subsurface dilution in areas of groundwater discharge, and inflow to underground facilities, must be evaluated separately.

1. INTRODUCTION

Radionuclide transport by groundwater is often seen as a crucial issue in nuclear waste management. Such transport would take place in large and sometimes complex flow systems. Transport calculations therefore are thought to require extensive and detailed information on the entire flow system. Results for sites in fractured granite, or similar heterogeneous aquifers, are consequently judged to be uncertain concerning the validity of available site-specific data and fundamental models. More acute analysis rather suggests that local conditions, controlling groundwater interaction with the individual waste packages, are key factors, whereas the impact of farfield transport appears conjectural. Specific flowpaths, relevant to performance, deserve experimental study. Local observations could then replace more generalized assumptions regarding the surrounding heterogeneous aquifers, and observed standard deviations or similar estimates could be used to evaluate uncertainty. This might call for changes in current routines for site investigation and validation work.

2. REPOSITORY CONDITIONS

At a KBS-type repository in fractured rock, such as granite, HLW will be emplaced in selected, favorable rock-volumes and be contained in individual packages. These comprise a longlived canister surrounded by a buffer mass. The canister prevents and delays groundwater contact with the waste. The buffer reduces mechanical strain and groundwater interaction with the canister. Chemical conditioners and getters may be added to the buffer, and the transmissivity of the surrounding rock may be reduced by pre-excavation grouting. This arrangement should increase safety and put off uncertainties regarding the farfield [1, p 119, 126]. After closure, groundwater conditions will recover from operational disturbances. Thermal perturbations will decay after some thousand years, well within the lifetime of the proposed canisters. Topographically induced hydraulic gradients will then generally, in platform areas of moderate relief,resemble those before construction. Long-term performance, after recovery, is a major concern. Groundwater effects unaffected by repository drainage therefore constitute a primary target for validation. The present, pre-operational, conditions must be mastered before future changes, if at all important, regarding hydraulic gradients, temperature effects, fracture minerals, groundwater chemistry and rock stresses, can be credibly addressed.

3. GROUNDWATER EFFECTS AND EXPLORATORY ASPECTS

Groundwater may affect performance by:

- Interaction of the water, and substances carried by it, with the manmade containment, eventually causing its deterioration.
- Interaction with the fuel, *if* the container should fail.
- Transport of radionuclides through the containment, *if* they

are released from the fuel.
- Transport of radionuclides through the farfield.
- Subsurface dilution, and sometimes precipitation, of nuclides before surfacing, if they survive decay during transport.

The central issue here is local mass transfer towards the waste packages and away from them, to be studied by site- and pre-emplacement investigations. Similar study of the farfield is difficult. The structures and fractures here mainly remain unknown. This also applies to the spatial distribution of fracture minerals and their effects on retardation. Retardation of sorbed nuclides is an important factor in safety calculations, but field data are very few. These practical limitations create a sampling problem. It concerns groundwater flow and transport, undisturbed by inflow to a repository, in their bearing on a single, discrete and non-random, distribution of waste packages in a specific, non-uniform fracture network, most of which is inaccessible for sampling.

4. FLOWPATH RELEVANCE RELATED TO REPOSITORY PERFORMANCE

Relevant flowpaths are those which interact with a waste package. Common fracture patterns, cf Figure 1, show a rather continuous fracture network and intervening, less fractured rock blocks. Such low-conductive blocks, possibly further sealed by pre-excavation grouting, are the most likely canister sites. They thus would contain the proximal parts of the relevant flowpaths. These carry a negligible fraction of the total groundwater flow.

Most of the groundwater flows in the surrounding, more conductive network, which essentially controls hydraulic head. For these reasons investigations of the network around a repository or underground laboratory cannot characterize the proximal parts of relevant flowpaths. Borehole measurements indicate that the hydraulic conductivity in the network is several (2-7) orders of magnitude higher than in the intervening blocks, and that head variations at depth are generally rather continuous. This suggests that the proximal parts of relevant flowpaths are essential, whereas their distal - network - parts might contribute only slightly to the total flow resistance and residence time relating to the waste packages. In that case even the dominant fractures and fracture zones at greater distance from the canister sites would be relatively unimportant for repository performance, unless the hydraulic gradients are found, by direct measurements, to be drastically reduced by them. Otherwise their detailed hydraulic characterization is cumbersome and rather useless. At times they may, however, have some influence on the layout of the underground facility.

Transport in individual flowpaths, not involved in mass transfer regarding the canister sites, appears irrelevant to repository performance. Their combined flow, however, issuing together with that of the relevant flowpaths, affects performance by causing subsurface dilution. The total flowrate is best estimated by local drawdown experiments at discharge. Dilution may also be determined by appropriate tracer tests, but can not be calculated from any amount of data on inflow to a repository and its hydraulic and statistical analysis. This actually reduces the impact of exploratory limitations, which normally leave dominant portions of the distal network inaccessible for detailed work.

5. EARLIER FIELD EXPERIMENTS

Much earlier work considers the inflow of groundwater and solute transport to underground facilities, i e flow in the outer conductive network under centripetal gradients, and associated model development and statistical work regarding the farfield. This concerns conventional underground storage and the recovery of groundwater, petroleum and natural gas, but has little bearing on barrier performance at a backfilled repository. A different approach, with externally controlled tracer tests, was pioneered at Savannah River [2]. On that basis the writer designed a comprehensive orientation study regarding sorbed and unsorbed tracers, and the effects of grouting on transport [3]. Other early tracer tests, in less conductive rock, were made at Stripa [4].

These experiments demonstrated retardation of sorbed nuclides (i e Cs, Sr) in the ungrouted rock, and the effective containment of unsorbed (Br) and sorbed (Sr) tracers alike, by conventional bentonite grouting. Apparent transport porosities (p) around 0.0004 and 0.002 are indicated for the ungrouted rock, with "transmissivities" around 5E-11 and 3E-6 m^2/s, respectively. Later Swedish tracer tests [5,6,7,8,9] yield similar data. Together they suggest that log p may, as an approximation, be a linear function of log K, where K is the measured hydraulic conductivity of the most conductive one m thick section of the tested flowpath:

$$\log p = 0.17 \log K - 1.7 \pm 0.3.$$

This may be used for approximate predictions, based on Darcy's law, of transport times in similar rocks, provided the pertinent distances, gradients, transmissivities and retardation factors are known. It may also help to quantify the impact of the nearfield by considering a schematic, but rather realistic case of linear, laminar flow under a uniform gradient along a flowpath composed of three different parts of the following nature:

	Nearfield	Transition	Farfield
K, conductivity m/s	10^{-10}	10^{-8}	10^{-6}
Length m	10	20	1000
Calculated:			
Porosity	4.0×10^{-4}	8.7×10^{-4}	1.9×10^{-3}
Relative Flowrate/m^2	1:	100:	10000
-"- Residence Time	23:	1:	1;

Here the nearfield is seen to be the essential part in limiting the flowrate and to provide the essential portion (about 90 %) of the time for radioactive decay.

6. VALIDATION CONCERNING NUCLIDE TRANSPORT

Much validation considers groundwater transport of radioactive substances from a repository to the surface. Here in-situ tracer experiments would seem most valuable. Other approaches, by extraction or excavation of representative fractures and complex flowpaths, for estimates of void volume and sorptive area, appear generally inapplicable to significant rock volumes under realistic repository conditions. On the other hand tracer experiments under natural gradients cannot be expected to cover significant distances in the essential parts of relevant flowpaths within realistic periods of time. It thus appears necessary to perform such tests under gradients made stronger by pumping. Truly representative tests require that these artifical gradients coincide in direction with the natural ones, otherwise irrelevant flowpaths may be activated. The results may be evaluated for lower gradients if Darcy's law is applicable.

Such validation therefore requires experimental evidence of:
- Our ability to identify and characterize at least the essential parts of relevant flowpaths
- The applicability of Darcy's law
- The validity of estimates of dilution.

Characterization would comprise determinations of hydraulic gradients and conductivities, together with apparent transport porosities and retardation factors. Information on dispersion also is of interest. It seems likely, however, that dispersion in a potential release would be dominated by differences between flowpaths from separate canister locations rather than internal dispersion of individual flowpaths. Furthermore only dispersion in the rising part of breakthrough appears significant, if residence time and radioactive decay are at all important. For the same reason times to first arrival of nuclides in significant amounts appear more important than average travel time. In many cases dispersion in individual flowpaths would be covered by the standard deviation indicated above for the transport porosity p. Regarding tracer tests it may be added that forceful injection of tracers may activate irrelevant flowpaths and affect travel distances. Tracers therefore should be introduced passively into relevant flowpaths. For reliable estimates of dilution no significant fraction of the tracers should travel to unknown sinks.

7. APPLICATION TO A REPOSITORY

Validation work will differ during the development of a repository. At first, tracer tests may be performed in boreholes from the surface, and directed towards areas of discharge by pumping at downstream locations. For passive introduction of tracers, and direct measurement of the associated flowrates, special equipment, such as the point dilution cell, [10] may be used. The retardation of strongly sorbed tracers may be studied by comparison with that of weakly and unsorbed ones, upon sealing and overcoring the borehole-sections used for their introduction.

Such experiments may be used to test Darcy's law and to estimate apparent transport porosity, retardation, dispersion and subsurface dilution. Actual field data here should be more representative than laboratory tests on isolated samples and calculations based on much simplified geometric assumptions. Retardation data from tracer tests are probably the best estimates of the sorptive area in the flowpaths. The transport of unsorbed and sorbed tracers alike may be affected by diffusion into the rock matrix. The measured delay and relative retardation of sorbed tracers therefore would include these effects. Here supplementary laboratory studies may be required. Special interest also pertains to the comparison of tracer tests with single-hole-and interference tests of hydraulic conductivity in the same test sections. Probably simple injection and withdrawal tests yield minimum estimates of flow resistance due to spherical activation of flowpaths. If so, they may be satisfactory for setting upper limits to mass transfer.

Borehole studies are limited in their spatial coverage. The relative ease with which forced gradients are set up in the general direction of natural flow is a great advantage.Coverage of the entire flowpath to the surface in single tracer experiments, and the exact location of the pumping hole may not be critical in this work because the proximal flowpaths join a pervasive conductive network. - In later underground work numerous locations are accessible for study, but large distortions occur in the field of hydraulic po-tentials. The exploratory shafts and drifts will promote groundwater inflow and be surrounded by extensively disturbed hydraulic gradients. Considerable skin effects will reside in their walls. Transport experiments on long term performance therefore must be located outside their influence. Its extent may be strongly redu-ced by pre-excavation grouting. Such grouting, combined with backfilling, probably is the best way to prevent hydraulic shortcir-cuiting by a repository and its disturbed walls. Grouting of the canister locations also may greatly reduce the impact and varia-tion of the local hydraulic conditions, and hence overall uncer-tainty. It would form the outermost engineered barrier. Field tests of its radial extent, durability and hydraulic effects in itself should be an important task of hydrogeological validation and demonstration.

In summary, a host rock may be likened to the proverbial haystack, and the relevant flowpaths to the sought-after needles. They are to be found and studied at the canister sites. Here the first three steps of the conditional sequence of groundwater effects, listed in section 3, primarily control safety. Therefore the hydrogeological aspects and field performance of local barriers, chemical reactions and nuclide retention should be investigated and validated, to see if multiple failure and farfield transport can at all take place. It may be added that even major flowpaths probably would serve as barriers against actinide release by chemical constraints [11]. They thus may be studied as barriers, and perhaps as time-scale models of low-conductive flowpaths. - In contrast, disregard of post-closure gradients, local barriers and exploratory limitations, as in the study of inflow and "haystack" models for the farfield, in the disturbed walls of tunnels and shafts, appears like barking up the wrong tree. It delays much-needed validation of vital performance aspects and interferes with relevant work by creating extensive adverse hydraulic gradients. In the end it may cast doubt on our professional judgement and the reliability of safe disposal.

ACKNOWLEDGEMENTS

My warm thanks to SKB, Swedish Nuclear Fuel and Waste Management Co., for funding much of this work, and to Prof. G de Marsily, Laboratory of Applied Geology, Paris, and Messrs K Ahlbom, E Gustafsson and A Winberg of Swedish Geological Co, who kindly read and commented on earlier versions of this report.

REFERENCES

1. Aka-utredningen: "Använt kärnbränsle och radioaktivt avfall", II, SOU 1976:31, Industridepartementet, Stockholm 1976, 1-222.
2. Webster D S, Proctor J F, Marine I W: "Two-well tracer tests in fractured crystalline rock", US Geol Survey, Water Supply Paper 1544-1, 1970.
3. Landström O, Klockars C-E, Holmberg K-E: "In-situ experiments on nuclide migration in crystalline rocks", KBS TR 110, Stockholm, 1978.
4. Lundström L, Stille H: "Large scale permeability test of the granite in the Stripa mine and thermal conductivity test", Swed. Amer. Coop. Progr., Techn. Proj. Rep. 2, Lawrence Berkeley Lab.-7052, Berkeley, 1978.
5. Gustafsson E, Klockars C-E: "Studies of groundwater transport in fractured crystalline rock under controlled conditions using non-radioactive tracers", KBS TR 81-07, Stockholm 1981.
6. Klockars -E, Persson O: "The hydraulic properties of fracture zones and tracer tests with non-reactive elements in Studsvik", KBS TR 82-10, Stockholm 1982.
7. Gustafsson E, Klockars C-E: "Study of strontium and cesium migration in fractured crystalline rock", KBS TR 84-07, Stockholm 1984.
8. Andersson P, Klockars C-E: "Hydrogeological investigations and tracer tests in a well-defined rock mass in the Stripa mine", KBS TR 85-12, Stockholm 1985.
9. Abelin H, Birgersson L, Gidlund J, Moreno L, Neretnieks I, Wide'n H, Ågren T: "3-D migration experiment - Report 3 Part 1, Performed experiments, results and evaluation", Stripa Project TR 87-21, SKB Stockholm 1987.
10 Gustafsson E, Ericsson C-O: Characterization of fracture zones in the Brändan area, Finnsjön study site: Determination of groundwater flow by the point dilution method in packed-off section in the boreholes BFi 1 and HFi 1. SKB Progress report, Stockholm, in prep.
11. Brotzen O: "On the environmental impact of a repository for spent nuclear fuel", KBS TR 83-29, Stockholm 1983.

RECENT ACCOMPLISHMENTS IN THE INTRAVAL PROJECT - A STATUS REPORT ON VALIDATION EFFORTS

Thomas J. Nicholson
Waste Management Branch
U.S. Nuclear Regulatory Commission
Washington, D.C. 20555, USA

ABSTRACT

The INTRAVAL Project is an integrated international effort dealing with validation of geosphere transport models which began in October 1987. Its purpose and scope developed from two earlier projects, INTRACOIN and HYDRCOIN, which focused on assessment of transport and ground-water flow models, respectively. The unique aspect to INTRAVAL is the active interaction between the experimentalists and modelers simulating the selected test cases for examining model validation issues. The test cases selected consist of laboratory and field transport experiments and natural analogue studies that incorporate hydrogeologic and geochemical processes relevant to safety assessments of radioactive waste disposal. These test cases cover a range of spatial and temporal scales, hydrogeologic conditions and media for various radionuclide transport phenomena. The success to date has centered on the selection, documentation, simulation and analysis of these 17 test cases. The sharing of ideas on development and testing of conceptual models employed by the various 25 project teams in their simulations of specific test cases has begun the evolution of a validation strategy. The conceptualization of ground-water flow and radionuclide transport through various media is being actively tested using these specially selected, and in certain cases specifically designed, data sets. A second set of test cases are under development for an additional three-year Phase II effort to build on the successes of the Phase I work.

PROGRES REALISES RECEMMENT PAR LE PROJET INTRAVAL – RAPPORT SUR L'ETAT D'AVANCEMENT DES TRAVAUX DE VALIDATION

Thomas J. Nicholson
Waste Management Branch
U.S. Nuclear Regulatory Commission
Washington, D.C. 20555, USA

RESUME

Le projet INTRAVAL, qui a débuté en octobre 1987, est une entreprise intégrée à l'échelon international en vue de valider des modèles de transport dans la géosphère. Son objectif et sa portée dérivent de deux projets antérieurs, les projets INTRACOIN et HYDROCOIN, qui portaient respectivement, sur l'évaluation des modèles de transport et celles des modèles d'écoulement de l'eau souterraine. Ce qui distingue le projet INTRAVAL c'est la collaboration entre les expérimentateurs et les concepteurs des modèles qui effectuent les simulations des cas témoins retenus, pour étudier les questions posées par la validation des modèles. Les cas témoins mettent en jeu des expériences de transport en laboratoire et in situ et des études d'analogues naturels qui prennent en compte des processus hydrogéologiques et géochimiques revêtant de l'importance pour l'évaluation de la sûreté de l'évacuation des déchets radioactifs. Ces cas témoins permettent d'étudier, dans une gamme d'échelles spatiales et temporelles, de conditions hydrogéologiques et de milieux, divers phénomènes intervenant dans le transport des radionucléides. Les progrès les plus notables ont été accomplis dans le choix, la documentation, la simulation et l'analyse de ces 17 cas témoins. Grâce à la mise en commun des idées sur la conception et l'essai des modèles théoriques employés par les 25 équipes de projets dans leurs simulations de cas témoins concrets, une stratégie de la validation a commencé à se dessiner. La représentation théorique de l'écoulement de l'eau souterraine et du transport des radionucléides à travers divers milieux fait l'objet d'essais intensifs à partir de ces séries de données spécialement choisies et parfois élaborées pour répondre à des besoins particuliers. Une deuxième série de cas témoins est en cours d'élaboration en vue d'une phase II d'une durée de trois ans pour approfondir les résultats de la phase I.

1. INTRODUCTION

The performance assessment of a repository for radioactive waste involves evaluation of the long-term behavior of the entire repository system. The repository system includes the engineered surface and subsurface facilities, the excavated cavities and engineered barriers, and the disturbed and undisturbed natural settings. This evaluation will involve quantitative modeling of both the natural and engineered subsystems and their interaction. A common aspect to such assessments is the evaluation of the potential for radionuclide transport from the repository through the hydrogeologic system to the accessible environment. Therefore, a major technical issue to be resolved is the development of procedures for validating the geosphere transport models to be used.

Based upon lessons learned from two earlier international cooperative projects dealing with analysis of transport models, INTRACOIN, and hydrogeologic flow models, HYDROCOIN [1,2,3,4,5,6], international consensus developed prior to and during the GEOVAL Symposium in Stockholm in April 1987 to begin a new project dealing with validation of geosphere transport models. This new international cooperative project named INTRAVAL began in October 1987. As with the preceding projects, INTRAVAL is organized and managed by the Swedish Nuclear Power Inspectorate, SKI. The project proposal was based upon a technical proposal developed by an international ad-hoc group from eight selected nuclear waste programs and institutes [7,8].

2. PROJECT OVERVIEW

The INTRAVAL Project was thus initiated as an integrated international effort towards the validation of geosphere transport models. The purpose of INTRAVAL is to increase the understanding of how various geophysical, hydrogeologic and geochemical processes of importance to radionuclide transport can be described by specially formulated numerical models. This objective is being realized through simulations and analysis of laboratory and field experiments and natural analogue studies chosen as validation test cases. The INTRAVAL test cases represent a range of spatial and temporal scales, hydrogeologic conditions and media as well as transport phenomena.

INTRAVAL has been structured to enhance active interaction between the laboratory and field experimentalists on one hand and the modelers on the other hand. To aid in development of a unified validation strategy and to integrate lessons learned from the various test cases, a Validation Overview and Integration Committee, VOIC, was formed to assist the Project Secretariat and the Coordinating Group. INTRAVAL consists of project teams from 33 technical and scientific organizations set up by 21 funding organizations from 12 countries [9]. In addition, the OECD/NEA, and HMIP in the United Kingdom participate in the Project Secretariat. KEMAKTA Consultants Company serves as the Principal Investigator to SKI.

INTRAVAL has been set up as a three year project (Phase I), to be completed by October 1990, with an optional additional three year period (Phase II). Recently, the Project Secretariat has recommended that the Phase II effort be approved by the INTRAVAL parties and is scheduled to begin in April 1991.

2.1 Flow and Transport Phenomena

The following relevant processes and geometric factors affecting radionuclide transport are being studied in the INTRAVAL project;

- spatial and temporal variability of hydraulic and transport parameters,
- issues of temporal and spatial scaling,
- dispersion created by variability in the velocities between and within the pores, fractures, and channels,
- fracture flow along preferential pathways,
- complexation of solutes with particulates,
- mineralization and coprecipitation of contaminants,
- channel networks of fractures,
- unsaturated flow through coupled matrix-fracture systems,
- matrix diffusion including diffusion in pores and on surfaces,
- fracture diffusion including diffusion in variable fracture apertures and fracture surfaces,
- anion exclusion,
- variable geochemical factors such as Eh, pH, Kd through various pathways,
- density-driven flow and density-effects on transport.

2.2 Selection Criteria for Phase 1 Test Cases

Criteria were developed by the ad-hoc group to assist in screening and selecting test cases for Phase 1 of INTRAVAL [8].

- The phenomena involved in the experiment should be relevant to the process of transport of radionuclides in the geosphere surrounding a repository.

- The experiments should be well documented in order to make the modeling teams acquainted with the experiment and to make it possible to achieve as comprehensive a database as possible.

  It should be possible to acquire additional information on the experiment. Therefore, a person or a group of persons with detailed knowledge of the experiment should be available for advice to the project.

- The experiment should include as many independently determined data as possible so as to allow meaningful model selection and validation.

2.3 Phase 1 Test Cases

Initially seven test cases were proposed and incorporated into the project [8]. Additional test cases have been added and one project was withdrawn. Presently, there are 17 test cases based upon laboratory, field and numerical experiments in the project.

Table 1 lists the current test cases, and the experimentalists responsible for providing the data sets to the modelers. These experiments are testing conceptual models and examining groundwater flow and transport processes that are relevant to disposal of radioactive waste in the subsurface. Separate data sets for model calibration and subsequent model validation are being developed for many of the test cases.

For example, active interaction between modelers and experimentalists who are developing and implementing a series of tracer tests in saturated fractured granite is ongoing for the Finnsjön test case. In the Apache Leap Tuff test case, a series of laboratory core (centimeter scale), laboratory block (meter scale), and borehole (tens of meters scales) experiments are ongoing for liquid and vapor transport in unsaturated fractured rock [10,11]. Various international cooperative projects such as the OECD/NEA projects at Stripa and Alligator Rivers as well as numerous national projects cooperate with the INTRAVAL Project [9].

3. INTEGRATION

The integration of the various Phase I test cases and the future Phase II test cases into a validation strategy is being pursued by the various project teams during the numerous scheduled INTRAVAL Workshops (4th workshop recently held in Las Vegas, Nevada, USA, February 1990). The following series of questions developed during the formulation of the format for the individual INTRAVAL test case descriptions assist in this effort;

1. What are the relevant processes of the experiment and how does the model consider them?

2. What are the geometric and spatial framework of the hydrogeologic system and do the modeling assumptions conform to them?

3. Are the simplifying assumptions inherent in the model compatible to the hydrogeologic, hydraulic, and geochemical components of the system being modeled and consistent with the model used?

4. Are the model inputs representative of the system?

5. Are the field and laboratory experiments detailed enough to provide unique sets of databases that characterize the governing processes?

6. Is the measurement scale compatible to the scale of the relevant processes?

7. Is there a coherent validation strategy that allows for additional data collection and determination of "goodness-of-fit" criteria for comparison of the simulation results versus the independent experimental databases?

Table 1: INTRAVAL Test Cases*

case 1a: Radionuclide migration in intact rock and clay samples by diffusion and advection based on laboratory experiments performed at Harwell, U.K.

case 1b: Uranium migration in crystalline bore cores based on experiments performed at PSI, Switzerland

case 2: Radionuclide migration in single natural fissures in granite, based on laboratory experiments performed at KTH, Sweden

case 3: Tracer tests in a deep basalt flow top performed at the Hanford reservation, Washington, USA case 4: Flow and tracer experiments in crystalline rock based on the Stripa 3-D experiment performed within the International Stripa Project

case 5: Tracer experiments in a fracture zone at the Finnsjön research area, Sweden

case 6: Synthetic database, based on single fracture migration experiments in Grimsel Rock Laboratory in Switzerland, by NRC/NAGRA/PNL

case 7a: Redox-front and radionuclide movements in an open pit uranium mine. Natural analogue studies at Poços de Caldas, Minas Gerais, Brazil, by SKB/NAGRA/U.K. DoE/U.S. DOE

case 7b: Morro do Ferro colloid migration studies. Natural analogue studies at Poços de Caldas, Minas Gerais, Brazil, by SKB/NAGRA/U.K. DoE/U.S. DOE

case 8: Natural analogue studies at the Koongarra site in the Alligator Rivers area of the Northern Territory, Australia, by the international ARAP Project

case 9: Radionuclide migration in a block of crystalline rock based on laboratory experiments performed at AECL, Whiteshell, Canada, by AECL/U.S. DOE

case 10: Evaluation of unsaturated flow and transport in porous media using an experiment with migration of a wetting front in a superficial desert soil performed within a U.S. NRC trench study in Las Cruces, New Mexico, U.S.A., by University of Arizona

case 11: Evaluation of flow and transport in unsaturated fractured rock using studies at the U.S. NRC Apache Leap Tuff Site near Superior, Arizona, U.S.A., by University of Arizona.

case 12: Experiments with changing near-field hydrologic conditions in partially saturated tuffaceous rocks performed in the G-Tunnel Underground Facility at the Nevada Test Site performed by the Nevada Nuclear Waste Storage Investigation Project of the U.S. DOE

case 13: Experimental study of brine transport in porous media performed at RIVM, Netherlands

case 14a: Pumping test in highly saline groundwater performed at the Gorleben site, FRG

case 14b: Saline groundwater movement in an erosional channel crossing the salt dome at the Gorleben site, FRG

case 15 Salt test case, to be decided.

---

*Note: The subsequent test case descriptions and their documentation in the INTRAVAL reports address these questions (see INTRAVAL [9]).

The Pilot Groups sponsoring the test cases and the various project teams modeling them are actively developing technical reports that illustrate validation strategies using the Phase I results. In particular, the testing of various transport models in which numerous processes and parameter variations exist has been possible through alternative modeling approaches by the project teams. The simulation results reported at the recent Las Vegas workshop are an example of this and will be synthesized into an integration summary and discussion of model validation analyses.

4. PROGRESS AND ACCOMPLISHMENTS

4.1 Laboratory experiments (Test cases 1a, 1b, 2 and 9)

A series of laboratory studies on radionuclide transport through clay and granite cores have been simulated by several of the project teams and reported at the INTRAVAL workshops. Several modeling approaches focusing on different transport processes (e.g. advection-dispersion, chemical retardation, matrix diffusion, and channeling) as well as stochastic-derived parameter values have been pursued. In the comparison of results, a number of statistical methods have been used to evaluate the significance of variations from the laboratory data. Both the selection of transport models used, and the analysis of independent and dependent parameters that significantly affect the derived curve fits represent the state-of-the art in transport modeling. For example, attempts have been made to determine if there is any significant difference between matrix diffusion model and channeling model using statistical analysis.

To date, most teams have only been able to analyze a subset of the available database, yet the progress demonstrated in their simulation results and active discussions at workshops, indicates an evolution in their analysis strategy and potential closure on which models best represent the transport processes. This work has already provided technical suggestions to the experimentalists on how to revise and rerun the laboratory studies.

4.2 Field Experiments (Test cases 3, 4, 5, 10, 11 and 12)

The three unsaturated test cases are designed to examine spatial variability in heterogenous porous and fractured media. Coupled processes such as thermal-hydrologic effects on simultaneous vapor-liquid flow and transport are being simulated. Due to both limited data and extremely large computer storage requirements to simulate the unsaturated experiments, most of the simulations were on the core studies and short-duration (order of days) field studies with relatively little progress accomplished on the long-term (weeks to months) field studies. With the development of fast iterative solvers [12, 13], and the collection of tritium, bromide, and chloride samples for the second field study at the Las Cruces Trench, significant progress has recently been achieved [14]. The Apache Leap Tuff test case similarly is progressing with the recent experimental design modeling by the University of Arizona pilot group [15].

As the first major tracer experiment to study spatial variability for long time periods and large distances (order of tens of meters) in saturated fractured rock, the Stripa 3D experiment has stimulated development of different modeling approaches that have been presented at the INTRAVAL workshops [16].

These simulation results and ensuing discussion will benefit the experimental design of future tracer experiments in fractured rock.

For example, the Finnsjön test case from the SKB Fracture Zone Project, has also been modeled by several project teams interested in designing and simulating tracer tests in fractured media. Limited datasets were given to the modelers so that they could model the experiment prior to the actual tests. The various project teams made predictive modeling assessments of the radially convergent tracer tests prior to the field experiment. Similarly, predictive modeling of the dipole test was performed. The subsequent field test data is substantial and shows a complex hydrogeologic structure at the site which the modelers hope to simulate and interpret. The excellent interaction between the experimentalists and modelers demonstrates the value in modeling for experimental design of field tests and interpretation of complex hydrogeologic conditions.

4.3 Salt Studies 13, 14 and 15

The three salt test cases are designed to examine density-driven flow and density effects on transport due to brine migration in porous media in both laboratory experiments (test case 13) and at the Gorleben site (test cases 14 a&b). The two field experiments at this site examine pumping test results for saline groundwater movement through an inhomogeneous porous aquifer, and through an erosional channel transecting the salt dome structure. Additional data sets are to be made available during the first half of 1990.

Test case 15 is still in the developmental stage and is anticipated to involve modeling of flow and transport through the complex hydrogeologic environment associated with salt dome structures (e.g., faults and caprock), and the effect of brine on radionuclide transport.

4.4 Natural Analogues (Test cases 7 and 8)

The INTRAVAL Project includes the international natural analogue projects Poços de Caldas and Alligator Rivers as test cases. The detailed analyses of these cases come comparatively late in the INTRAVAL process due to their complexity. At the Helsinki workshop some modelling results were presented on the redox front in Poços de Caldas and the hydrology in Alligator Rivers.

4.5 Synthetic Experiment (Test case 6)

A synthetic experiment has been designed for testing flow and transport models which utilize a limited database for simulating tracer migration experiments. The "synthetic system" from which partial data sets are created and provided to the modeling teams is patterned on a large-scale tracer test at the Grimsel Rock Laboratory in Switzerland. The objective of this case is to examine the "identifiability problem" which arises due to inherent variability in natural hydrogeologic systems, their complex boundary conditions and structures, and competing processes in the sense that one or more, and in various combinations, can produce similar effects on the observed transport data.

5. FUTURE DEVELOPMENTS

5.1 Selection Criteria for Phase II Test Cases

At the recent Las Vegas meeting, the Coordinating Group reviewed a set of test cases for inclusion into Phase II. Criteria for reviewing the nominated test cases were developed by VOIC to assist the INTRAVAL Secretariat and Coordinating Group in this effort. In particular, ongoing Phase I experiments and others outside of INTRAVAL were presented and examined for their suitability in examining validation issues relevant to transport affected by coupled processes, and incorporating an integrated series of spatial and temporal experimental scales .

The following list of favorable criteria were used in reviewing the nominated Phase II test cases by the Coordinating Group Meeting in Las Vegas:

1. The test cases should be ongoing experiments that would continue through the Phase II period.

2. The test cases should have an experimental design oriented toward a problem that has both scientific and performance assessment relevance.

3. The test cases should be interactive, allowing for input from the INTRAVAL modeling groups on experimental design and types of data collected.

4. The test cases should already contain sufficient datasets to allow initial modeling efforts for evaluating theories and/or concepts.

5. The test cases should be comprehensive enough to cover relevant flow and transport issues presently missing from the Phase I efforts. These may include:

    a. colloidal transport,
    b. vapor-gas transport,
    c. coupled processes which include thermo-hydrochemical, hydro-mechanical, and thermo-hydromechanical processes,
    d. microbial-induced transport,
    e. large salinity variations,
    f. density dependent flow (both temperature and chemical effects),
    g. density dependent transport,
    h. effect of large ionic strengths on transport.

Furthermore, it was considered essential that the test cases should cover various geometric, temporal and process ranges, especially:

range of geologic scales	cm, m, 100's m, to kilometer
range of time scales	less than a week, weeks to year, much greater than a year
range of variabilities	homogeneous properties, isotropic to anisotropic variations, variable properties as function of other parameters or processes, systematic vs. non-systematic variations
coupled effects	thermo-hydromechanical, etc.
multiple discontinuities	fracture zones, networks, dykes, faults, etc.

The preliminary list of candidate test cases to be incorporated into the Secretariat's Phase II proposal includes:

1. Three-dimensional flow and tracer experiments through fracture channels at the International Stripa Project site,
2. Tracer experiments in a complex fractured zone at the Finnsjon research area.
3. Partially saturated flow and transport through heterogeneous soils field studies at the Las Cruces Trench site.
4. Unsaturated flow and transport through fractured rock field studies at the Apache Leap Tuff site.
5. Field studies of radionuclide migration in the weathered zone at the Koongarra uranium deposit at the international ARAP site,
6. Flow and brine transport studies at the Gorleben salt dome site,
7. Field studies of tritium migration in the Boon clay at the Mol, Belgium field site.
8. Brine flow through bedded evaporites at the WIPP site, Carlsbad, New Mexico, USA.

The project teams and VOIC members are developing a Phase I technical report. The final Phase I report will outline the test case descriptions, modeling results, and will integrate the results into a systematic hierarchy of processes being studied, geologic complexities, and spatial and temporal scales involved. The subsequent groupings and intercomparisons will help to draw generalizations and possibly to develop unifying theories on flow and transport.

5.2 Further Validation Needs

The INTRAVAL project is an important step in the development of validation strategies, planning and executing specially-designed radionuclide transport experiments, and testing of models. Because of the large financial commitments necessary to accomplish these tasks and the urgent need to resolve radioactive waste disposal issues, international cooperation can greatly enhance the technical and timely resolution of the attendant uncertainties. Future work developing from the INTRAVAL exercises involves not only the Phase II considerations, but also the integration of this work into other ongoing validation studies for other performance assessment issues (e.g., engineered system performance and biosphere processes).

6. CONCLUSIONS

INTRAVAL has quickly developed into a major international effort studying validation issues associated with geosphere transport models as demonstrated by the large number of active participants from throughout the world, and the great variety of laboratory and field experiments, natural analogue studies, and numerical experiments involved.

It should be remembered that real progress in validation can only be achieved when the work on the individual test cases focus on specific relevant issues associated with radionuclide transport experiments and simulation studies. Considerable progress has been achieved along these lines. It should also be remembered that validation is a learning process. Both the experimental work and the modeling of specific databases help to increase the knowledge of crucial performance issues, and advance the state-of-the-art in transport modeling.

It is also necessary to constantly integrate and evaluate the results in the context of performance assessment, and to develop strategies for further validation work considering the needs of the various disposal programs. This important aspect to INTRAVAL will be the cornerstone in the discussions on the definition of the Phase II effort.

The active involvement of the various national and international cooperative programs has contributed greatly to INTRAVAL's success. Similarly, the positive and collegial interaction between the various experimentalists and modelers as demonstrated in the numerous formal workshops and special meetings convened by the test case participants, has contributed to INTRAVAL's progress and achievements.

REFERENCES

1. INTRACOIN Final Report Level 1, February 1984, Swedish Nuclear Power Inspectorate

2. INTRACOIN Final Report Levels 2 and 3, May 1986, Swedish Nuclear Power Inspectorate

3. Cole, C.R. and T.J. Nicholson, Co-Chairman, 1986, Proceedings of the Symposium on Ground-Water Flow and Transport Modeling for Performance Assessment of Deep Geologic Disposal of Radioactive Waste: A Critical Evaluation of the State-of-the Art, May 20-21, 1985. Albuquerque, New Mexico, NUREG/CP-0079 (PNL-SA-13796, CONF-8505180). U.S. Nuclear Regulatory Commission, Washington, DC

4. The International HYDROCOIN Project, Level 1: Code Verification, NEA-SKI, Paris 1988

5. The International HYDROCOIN Project, Level 2: Model Validation, Draft Stockholm, 1989

6. Cole, C.R., T.J. Nicholson, P.A. Davis, and T.J. McCartin, 1987, "Lessons Learned from the HYDROCOIN Experience", Proceedings of the GEOVAL 87 Symposium on Verification and Validation of Geosphere Performance Assessment Models, Swedish Nuclear Power Inspectorate, Stockholm, Sweden, April, 1987

7. INTRAVAL Ad Hoc Group, June 1987, "Geosphere Transport Model Validation - A Status Report", SKI 87:4, Swedish Nuclear Power Inspectorate, Stockholm, Sweden

8. INTRAVAL Ad Hoc Group, July 1987, "INTRAVAL Project Proposal", SKI 87:3, Swedish Nuclear Power Inspectorate, Stockholm, Sweden

9. INTRAVAL Progress Reports No. 1-4, Swedish Nuclear Power Inspectorate, Stockholm

10. Evans, D.D. and T.J. Nicholson (editors), 1987, Flow and Transport Through Unsaturated Fractured Rock, AGU Geophysical Monograph 42, American Geophysical Union, Washington, DC, 187 pp.

11. Evans, D.D., T.C. Rasmussen, and T.J. Nicholson, 1987, "Fracture System Characterization for Unsaturated Rock", in Proceedings of the Waste Management 87 at Tucson, Arizona, March, 1987, Edited by Roy G. Post, University of Arizona, Tucson, AZ

12. Bouloutas, E.T., and M.A. Celia, "Fast Iterative Solvers for Linear Systems Arising in the Numerical Solution of Unsaturated Flow Problems", submitted for publication to Advances in Water Resources, January 1990.

13. Huyakorn, P.S., J.B. Kool, and J.B. Robertson, "WAM2D - Variably Saturated Analysis Model in Two Dimensions", NUREG/CR-5352 (HGL/89-01), U.S. Nuclear Regulatory Commission, Washington, DC, May 1989

14. Wierenga, P.J. and others, "Soil Physical Properties at the Las Cruces Trench Site", Soil and Water Science Report No. 89-002, Department of Soil and Water Science, University of Arizona, Tucson, AZ, 1989

15. Yeh, T.C., T.C. Rasmussen, and D.D. Evans, "Simulation of Liquid and Vapor Movement in Unsaturated Fractured Rock at the Apache Leap Tuff Site - Models and Strategies", NUREG/CR-5097, U.S. Nuclear Regulatory Commission, Washington, DC, March 1988

16. Dverstorp, B. and J. Andersson, "Application of the Discrete Fracture Network Concept With Field Data: Possibilities of Model Calibration and Validation", Water Resources Research, Vol. 25, No 3, pp 540-550, March 1989

HYDROCOIN LEVEL 3 - TESTING METHODS FOR SENSITIVITY/UNCERTAINTY ANALYSIS

Bertil Grundfelt, Björn Lindbom, Alf Larsson
KEMAKTA Consultants Co., Stockholm, Sweden

Kjell Andersson
Swedish Nuclear Power Inspectorate, Stockholm, Sweden

ABSTRACT

The HYDROCOIN study is an international cooperative project for testing groundwater hydrology modelling strategies for performance assessment of nuclear waste disposal. The study was initiated in 1984 by the Swedish Nuclear Power Inspectorate and the technical work was finalised in 1987. The participating organisations are regulatory authorities as well as implementing organisations in 10 countries. The study has been performed at three levels aimed at studying computer code verification, model validation and sensitivity/uncertainty analysis respectively. The results from the first two levels, code verification and model validation, have been published in reports in 1988 and 1990 respectively. This paper focusses on some aspects of the results from Level 3, sensitivity/uncertainty analysis, for which a final report is planned to be published during 1990.

For Level 3, seven test cases were defined. Some of these aimed at exploring the uncertainty associated with the modelling results by simply varying parameter values and conceptual assumptions. In other test cases statistical sampling methods were applied. One of the test cases dealt with particle tracking and the uncertainty introduced by this type of post processing. The amount of results available is substantial although unevenly spread over the test cases. It has not been possible to cover all aspects of the results in this paper. Instead, the different methods applied will be illustrated by some typical analyses.

HYDROCOIN STADE 3 - ESSAI DE METHODES APPLICABLES A L'ANALYSE DE SENSIBILITE/INCERTITUDE

Bertil Grundfelt, Björn Lindbom, Alf Larsson
KEMAKTA Consultants Co., Stockholm, Suède

Kjell Andersson
Swedish Nuclear Power Inspectorate, Stockholm, Suède

RESUME

L'étude HYDROCOIN est un projet international en coopération dont le but est de mettre à l'épreuve des stratégies de modélisation hydrogéologiques permettant d'évaluer le comportement des dispositifs d'évacuation des déchets nucléaires. L'étude a été lancée en 1984 par le Service suédois d'inspection de l'énergie nucléaire (SKI) et les travaux techniques ont été achevés en 1987. Les autorités réglementaires ainsi que les organismes d'exécution de 10 pays participent au projet. L'étude a été réalisée en trois stades successifs : vérification des programmes de calcul ; validation des modèles et analyse de sensibilité/incertitude. Les résultats des deux premiers stades, vérification des programmes de calcul et validation des modèles, ont fait l'objet de rapports publiés, respectivement, en 1988 et 1990. Le présent document porte essentiellement sur certains aspects des résultats du stade 3, analyse de sensibilité/incertitude, dont le rapport final devrait être publié dans le courant de l'année 1990.

En ce qui concerne le stade 3, sept cas témoins ont été définis. Certains d'entre eux visent à étudier l'incertitude liée aux résultats de la modélisation en faisant simplement varier les valeurs des paramètres et les hypothèses théoriques. Dans d'autres cas témoins, on a appliqué des méthodes statistiques d'échantillonnage. L'un des cas témoins porte sur le traçage des particules et sur l'incertitude qu'introduit ce type de traitement a posteriori. La quantité de résultats disponibles est considérable, encore qu'inégalement répartie entre les divers cas témoins. Il n'a pas été possible dans le présent document de traiter tous les aspects des résultats. On a préféré illustrer les diverses méthodes mises en oeuvre par quelques analyses représentatives.

INTRODUCTION

Mathematical modelling of groundwater flow and radionuclide transport is a central part of the performance assessment of repositories for radioactive waste. The acceptability of a disposal concept to the public, the scientific community and the safety authorities is a function of the credibility of the safety analysis and thus of the mathematical models used in the performance assessment. The Swedish Nuclear Power Inspectorate, SKI, has initiated three international cooperation projects aimed at increasing the understanding and credibility of models describing groundwater flow and radionuclide transport. The HYDROCOIN study is the second of these studies. It is devoted to testing of groundwater flow models and is performed at three levels aimed computer code verification, model validation and sensitivity/uncertainty analysis. These three issues form an integral part in the building of confidence in the models. It is obvious that a model's ability to properly represent the processes that govern the groundwater flow should be demonstrated,i.e. that the model is valid. It is also obvious that it should be shown that the computer coding of the model is done correctly, i.e. that the code is verified. Code verification and model validation have been the subject of HYDROCOIN Levels 1 and 2 [1,2]. This paper focusses on the third level in which different methods for sensitivity/uncertainty analysis have been tested.

The sensitivity/uncertainty analysis can be used in the performance assessment to explore the sensitivity of conclusions drawn to the uncertainties in the model parameters. It can, however, also be used as a tool in the model validation to investigate whether the agreement between the model results and experimental data is consistent with the uncertainty in the experimental data and the model parameters and to identify the model parameters to which the results are most sensitive. First a brief overview of methods available for sensitivity/uncertainty analyses is given. This is then followed by some example applications of some of the described methods.

METHODS FOR SENSITIVITY/UNCERTAINTY ANALYSIS

This section is intended to give a very brief overview of the methods available for sensitivity/uncertainty analysis. Due to space restrictions it is far from comprehensive. Subsequent sections of this paper will illustrate some HYDROCOIN applications of some of the methods described.

The summary of methods focusses on methods for the quantification of sensitivities and uncertainties related to the model parameters as this has been the emphasis of the work within the HYDROCOIN study. The more complicated issues of uncertainties induced by the spatial variability of the model parameters and uncertainties in the conceptual model used are mentioned only very briefly.

Uncertainty Analysis

The term uncertainty shall in this context mean lack of knowledge or ignorance of the exact state of the physical world. Traditionally, uncertainties in conjunction with modelling are classified as either parameter uncertainties or uncertainties in the conceptual model chosen. There are methods available to quantify parameter uncertainties whereas conceptual uncertainties generally have to be addressed by testing several alternative models. In order to be able to choose between such alternative models, it is important to have a quantitative view of whether the agreement between the model result and the "real" results are consistent with the uncertainties in the model parameters, or in other words, if the hard data available quantify the model parameters well enough.

The methods used to quantify the parameter uncertainties start from the assumption that probability density functions (pdf) can be defined for the model parameters. These pdfs are measures of one's ignorance of the actual parameter values and are hence subjective and dependent on expert opinion. It is obvious that it is much more difficult to assign pdfs to parameters that vary in space and time than to those that are expected to be constant.

The uncertainty analysis aims to determine the uncertainty in a consequence of the model, i.e. a model result or performance measure, resulting from the uncertainty in the model including the model parameters. The entity calculated is generally the cumulative distribution function of the performance measure:

$$G(C^*) = \int_{\underline{\alpha}} f(\underline{\alpha})\, d\alpha_1 d\alpha_2 \ldots d\alpha_N \qquad (1)$$

with $\underline{\alpha}$ chosen such that $C(\underline{\alpha}) < C^*$

where: $G(C^*)$ is the probability of the performance measure C being less than C^* and $f(\underline{\alpha})$ the joint pdf for the model parameters $\underline{\alpha} \equiv \{\alpha_1, \alpha_2, \ldots \alpha_N\}$

The integral in (1) must, except for very simple cases, be evaluated numerically using discrete approximations or, more commonly, sampling schemes. The sampling schemes employed include *i)* expert choice, *ii)* point estimate, *iii)* Monte Carlo, *iv)* Latin Hypercube, and *v)* importance sampling. These methods all have their advantages and disadvantages. The expert choice methods cannot be readily automated in computer programmes as can the point estimate method. This latter method, however, gives rise to a large number of samples (2^N if the number of model parameters is N). Monte Carlo, MC, is a simple and straightforward, random-number based sampling scheme. Its main drawback is that it often requires a large number of samples in order to give a converged solution to the integral in (1). One technique to reduce the number of samples required is the Latin Hypercube Sampling, LHS. For favourable conditions, in particular for monotonous performance measures, the LHS technique both reduces the number of samples needed and ensures that the performance measure is evaluated also for extremes in the parameter ranges. The importance sampling often reduces the number of samples necessary to obtain a converged solution to the integral in (1) compared to the Monte Carlo method. It is, however, difficult to handle parameter correlations in the importance sampling method. The LHS method appears to be the most commonly used sampling scheme for uncertainty analysis of groundwater flow models.

Sensitivity Analysis

The sensitivity analysis provides answers to questions like "What happens if the value of parameter α_i is changed by x %?" or "Which are the most important parameters?". The most straightforward way to answer these questions is to calculate the local sensitivities, i.e. the partial derivatives of the performance measure with respect to the model parameters:

$$S = \partial C / \partial \alpha_i \qquad (2)$$

The local sensitivities do not, however, account for the expected variability of a given parameter. This can be accounted for if the local sensitivities are normalised by multiplication by a measure of the parameter's uncertainty, e.g. the standard deviation of the pdf of the parameter:

$$S_{norm} = \sigma_i (\partial C / \partial \alpha_i) \qquad (3)$$

The evaluation of S_{norm} thus requires that the parameter uncertainties are estimated. The relative importance of the various parameters can then be analysed by comparing such normalised sensitivities.

The evaluation of the sensitivities according to (2) or (3) can be made e.g. by discrete approximations of the derivatives, by differentiating a response-surface function which has been fitted to the model results, or by the adjoint technique. The choice of method depends on the number of model parameters, the number performance measures which are of interest etc. The adjoint technique results in significant computational savings if the number of model parameters is significantly greater than the number of performance measures studied. However, in the normal case this method requires significant coding efforts for the evaluation of the derivatives needed.

As an alternative to the evaluation of S or S_{norm}, correlations or regressions between performance measures and model parameters can be used to evaluate the relative importance of parameters. The correlation coefficients express to what extent the performance measure is a linear function of the model parameters. This means that in the common situation when the model response to a certain parameter is strongly non linear, the calculated correlation coefficient between that parameter and the chosen performance measure might show a low correlation despite that the model's sensitivity to that parameter is high. One way of tackling this problem is to calculate rank correlations instead of simple correlations. In this case the parameter values and their associated values of the performance measure are ranked and the correlation coefficients are calculated from the ranks instead of the actual values. A high value of the rank correlation coefficient (i.e. a value close to +1 or -1) indicates a monotonous relation between the parameter and the performance measure rather than a linear one.

In regression analysis, a response surface in the form of a polynomial is fitted to the model. The degree of the polynomial used is often restricted to two, i.e. the polynomial is assumed to be a linear combination of terms of the form 1, α_i and $\alpha_i\alpha_j$. Similarly to the correlation analysis, the regression analysis can be based on ranks rather than actual parameter values.

In the general case both the calculation of correlation coefficients and the regression analysis require sampling techniques since the integrals involved cannot be solved analytically. In fact, it is very common to combine an uncertainty analysis by sampling with a regression analysis for the identification of the most important parameters.

Spatial variability

The spatial variability of the permeability is a major source of uncertainty in groundwater flow modelling. The approaches used assume that the permeability is a random field. The permeability values may be spatially correlated, i.e. the parameter values at two points are correlated to a degree which decreases with the separation between the points. In the case when the flow equation is solved with the finite-difference or finite-element method, permeability values for the individual grid blocks could be interpolated from measured data using the kriging technique which can take spatial correlations into account.

SOME EXAMPLES FROM HYDROCOIN LEVEL 3

This section is devoted to three examples of analyses performed within HYDROCOIN Level 3. The examples represent three different approaches to sensitivity/uncertainty analysis. Since they have been performed for different test cases, a direct comparison between the three approaches is difficult.

The first analysis is an application of parameter variations using the expert choice method [3]. This analysis is probably a typical example of a method used by many modellers. In the second analysis [4], LHS is used in conjunction with regression analysis for uncertainty analysis and sensitivity analysis. The pdfs of the parameters have in this analysis been assumed to be a priori known. In the third analysis [5], the uncertainty/sensitivity analysis is performed subsequent to a model calibration/validation exercise. In this approach probabilities are assigned to the different parameter values based on a posteriori pdfs derived from the goodness-of-fit-function used in the model calibration phase.

Parameter Variations

The example chosen to illustrate the application of parameter variations as a method to explore the sensitivity of modelling results to the model parameters is based on an analysis of Level 3 Test Case 5A by the Swedish Geological Co and KEMAKTA Consultants Co [3]. Test case 5A is defined as the study of the sensitivity of the groundwater flow rate at a given point and the groundwater travel time along a flow line starting at the same point at the Fjällveden study site in south central Sweden to variations in the permeability distribution and the position of the groundwater table [6]. The analysis described here is restricted to varying the permeability distribution. The groundwater flow model used is a three-dimensional, steady-state, finite-element model.

The Fjällveden study site was first modelled during the KBS 3 study [7]. The approach used at that time to assign permeability values to the different parts of the model was to first identify the data points that belonged to the fracture zones. These amounted to 14 data points out of the total number of data points that was a little more than 200. In a second step the geometric mean of the permeabilities at different depth intervals were calculated and an equation for the depth dependance of the permeability was fitted to the means. The form of this equation was:

$$K = a \cdot z^{-b} \quad (4)$$

where: z is the depth below the ground surface and
a and b are fitting parameters.

In a third step an attempt was made to narrow down the confidence interval of the fit by separating out data points connected to a gneissic granite which was identified as being more permeable than the predominant gneiss. An anisotropic permeability tensor with principal directions lying parallel to the strike of the gneissic granite was evaluated. Subsequent to the KBS 3 study, additional calculations were made mainly to elucidate the impact of the depth dependance of the permeability in the fracture zones [8].

In the present analysis totally 9 runs are presented including those performed in the previous studies. The differences between the various runs are different averaging techniques for the permeability (geometric and arithmetic), different depth dependance of the permeability (power function or stepwise constant), isotropic versus anisotropic permeability, and differences in permeability between different fracture zones depending on their orientations relative to the direction of the principal rock stresses. Figure 1 displays the permeabilities used in the different runs.

In the project team report [3], the 9 runs have been divided into three main groups describing variations of *i)* the rock mass permeability, *ii)* the hydraulic contrast between the rock mass and the fracture zones, and *iii)* the permeability in the fracture zones. The project team has thus been able to draw some conclusions about how the overall results are influenced by the permeability field. One such conclusion is that the runs with anisotropic permeability have yielded somewhat higher flow rates than the isotropic runs. It is concluded that the main reason is that the permeability tensor is oriented essentially parallel to the general direction of the hydraulic gradient and that the effective permeability therefore increases when the anisotropy is introduced.

The analysis of Fjällveden has included a number of parameter variations selected by a team of people who have been deeply involved in the analysis of the field data. Despite that the parameter variations presented are a compilation of calculations made at three different occasions with partly different objectives, and that the selection of parameter variations therefore has not been made systematically, it has been possible to gain some understanding of the sensitivity of the system. It is, however, difficult from the available analysis to identify the most important parameters and to get a quantitative measure of the sensitivity to these parameters. This would have required a more systematic selection of parameter variations and e.g calculation of local sensitivity coefficients or performing a regression analysis.

The current analysis is probably representative for many sensitivity analyses performed in the performance assessment context. It has the advantage that it is easily performed as it does not require any software in addition to the groundwater flow model. A careful planning of the parameter variations is, however, needed to optimise the utility of the analysis. A good knowledge of the site conditions facilitates such a planning. The analysis presented here is in some sense an analysis of uncertainties in the conceptual model. In this case the model has not been calibrated to any data from the field site. It is therefore not possible to apply quantitative criteria to select one out the alternative models as being most representative of the field situation.

Application of Latin Hypercube Sampling and Regression Analysis

A project team from U.S Nuclear Regulatory Commission performed as part of their HYDROCOIN effort [4] an uncertainty analysis using the LHS method followed by the application of regression analysis to identify the most important parameters. The analysis was performed for Test Case 1 [9] which is based on a vertical two-dimensional cross-section through a 25 m deep clay formation. The clay is horizontally stratified with four layers having different permeabilities and porosities. Two concrete blocks are located in the second layer from the top. The driving force for the groundwater flow is created by the infiltration through the top surface and a difference in water level in two pits at either end of the domain. Figure 2 shows the geometry of the system along with a base-case set of parameter values.

The uncertainty analysis was performed for two conceptual models differing form each other in the water level in the left pit which was set to 18 and 25 m respectively. This influences the vertical extent of the unsaturated zone which was thickest in the 18 m model. Three performance measures were defined namely *i)* the shortest travel time from a concrete block to the model boundary, *ii)* the travel time from top right corner to the boundary, and *iii)* the volumetric flux through the concrete blocks calculated separately for each block.

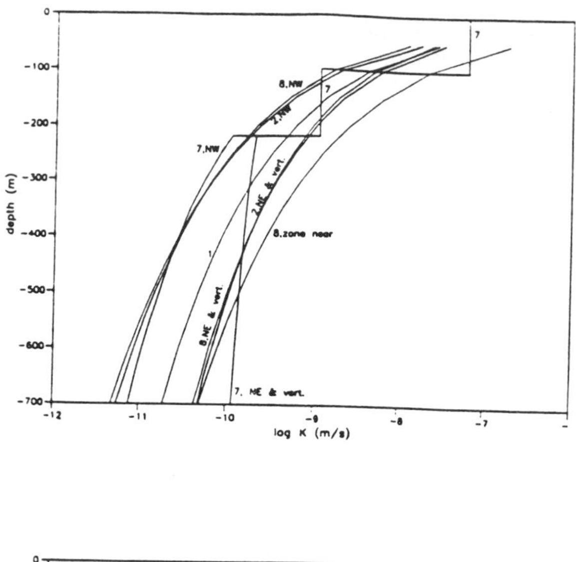

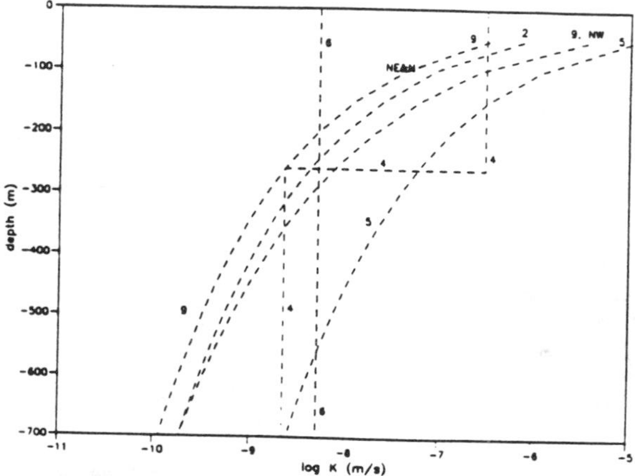

Figure 1 Depth dependance of the permeability of the rock mass (left) and the fracture zones (right).

The model has 11 parameters; 5 permeabilities, 5 porosities, and 1 infiltration rate. The permeabilities were assumed to follow lognormal pdfs with the base-case values denoted in Figure 2 as mean values and a range of two orders of magnitude. The porosities were assumed to follow normal distributions with the mean equal to the base-case value in Figure 2 and a range corresponding to ± 25 % of the mean. The infiltration rate was taken to be uniformly distributed with a mean corresponding to the base case value and a range of ± 100 % of the base-case value.

From these parameter distributions, 20 sets of parameter values were sampled. The performance measures were then calculated for the two conceptual models and the 20 parameter sets. The results were compiled as complementary probability distributions. Following this, multiple regression analysis was performed in order to determine which parameters that are the most important. The regression analysis was performed both for the calculated values of the performance measures with respect to the sampled values and for the ranks of the calculated performance measures with respect to the ranks of the sampled parameter values. Figure 3 shows the calculated complementary probability distributions for the volumetric flux through the concrete blocks and the results of the corresponding regression analysis. It is evident from the figure that the permeability of the concrete blocks is the most important parameter for the flux in the block. This conclusion is of course logical.

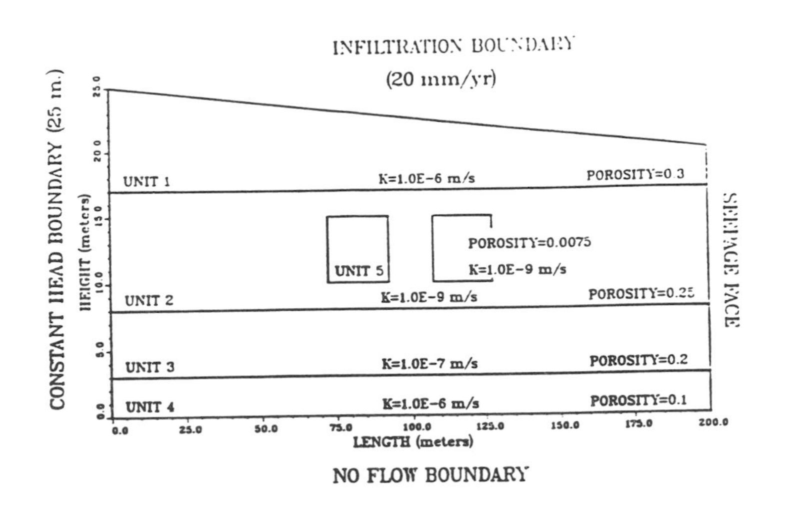

Figure 2 *Geometry of the modelled domain in Test Case 1. Shown are also the base-case set of parameter values.*

Statistical sampling and regression analysis is a powerful tool for uncertainty and sensitivity analysis. One advantage is that it covers the entire parameter space. It requires that pdfs can be assigned to the model parameters prior the modelling. This step is by necessity subjective and may in some instances significantly influence the results. If the performance measure shows a strongly non linear functional dependance of the model parameters, the regression analysis may give rise to problems. However, virtually all known sensitivity analysis methods have difficulties to cope with such situations. The LHS sampling and regression analysis scheme requires that software for sampling and multiple regression analysis is available.

Sensitivity/Uncertainty Analysis Based on Expert Choice Parameter Sets and A Posteriori Estimates of Parameter Pdfs

A project team from Colenco in Switzerland has presented an analysis of Test Case 5B [5] which is based on geohydrological investigations at a site at Chalk River Nuclear Laboratories, Ontario, Canada [6]. The analysis is an integrated model calibration, validation and uncertainty analysis approach. The model has first been calibrated to measured hydraulic heads using a goodness-of-fit-function accounting both for the differences between calculated and measured heads and the differences between the used values of model parameters and a priori information about these parameters (e.g. hydraulic conductivities evaluated from injection tests in packed-off sections of bore holes). The calibration step involved the evaluation of the goodness-of-fit function in totally 29 model runs with different parameter sets. The parameter sets each comprising 33 parameters were chosen by expert choice and the runs were ranked in the order of decreasing values of the goodness-of-fit function.

The 29 calibration runs were then used to perform a sensitivity and uncertainty analysis. As performance measures the groundwater fluxes at two points and the groundwater travel times along flow lines starting from these points were used. The sensitivity analysis was performed as a regression analysis assuming linear regression between the performance measures and the 33 model parameters. Correlation coefficients were evaluated allowing for the linearity assumption to be tested. Each parameter was normalised by dividing the parameter value by the standard deviation of the 29 sample values. The performance measures were

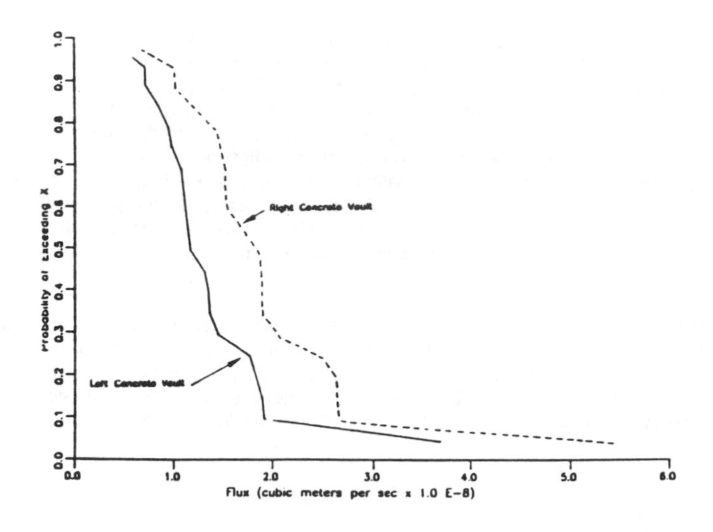

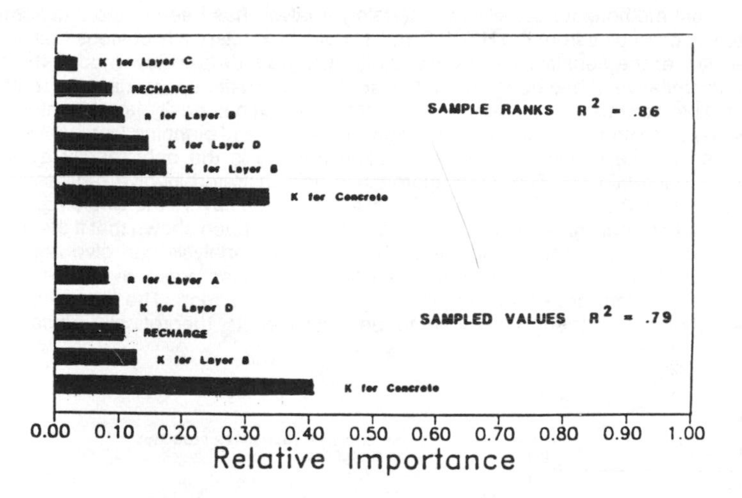

Figure 3 Calculated complementary probability distribution (left) and relative parameter importance (right) for Test case 1 using the volumetric flux through the concrete blocks as performance measure.

normalised in the same way.

The conclusions that could be drawn regarding the parameter sensitivities were judged to be logical for all parameters but one. The slope of the regression line indicated that an increase of a hydraulic conductivity in one part of the domain would tend to lower the flow rate in a way which would not a priori expected. A possible explanation might be that the number of parameter sets available for the regression is limited and that they have not been selected randomly. In general the slopes of the regression line is small. The project team has shown by applying rank regression that this is not caused by non linearities in the functional dependance between the performance measure and the model parameters. The conclusion is therefore that the influence of the individual parameters is limited.

The uncertainty analysis requires that probabilities can be assigned to the individual parameter values. In the present analysis the parameter sets were generated in order to minimise the goodness-of-fit function. The parameter sets chosen were not a priori given any probabilities. These therefore had to be evaluated a posteriori taking account of the information available from head measurements and from model calibration. This was done by using Bayes' theorem to establish a relationship between the conditional pdf of a parameter set given the head measurements and the elements of the goodness-of-fit function expressed as a log-likelihood function, and making use of the fact that the log-likelihood function is χ^2-distributed in the vicinity of its minimum. By doing this it was shown that there was a rather sharp separation between probable or rather likely parameter sets and highly improbable parameter sets. Figure 4 shows the resulting complementary probability distribution for the groundwater travel time along a flow line. The travel time exceeds 21 days with a confidence level of 99.7 %.

The approach demonstrated in the presented analysis is a comprehensive method for model calibration, validation and sensitivity/uncertainty analysis accounting for prior information on the model parameter values. The analysis is approximative. In particular it is difficult to assess parameter pdfs from a very limited number of parameter sets which are not statistically distributed in the parameter space. The statistical analysis involved in the evaluation of the goodness-of-fit function and the a posteriori estimate of the parameter pdfs therefore needs further discussion.

CONCLUSIONS

An overview of different methods for sensitivity/uncertainty analysis has been made and some methods have been demonstrated by examples from the HYDROCOIN Level 3 exercise. It is obvious that, in order to provide a comprehensive view of the sensitivity and uncertainty of a groundwater flow model, the analysis method applied should be quantitative. It has been shown that several conclusions regarding the model sensitivity can be drawn from a simple parameter variation. This is a method which is easily implemented since it does not require any software in addition to the groundwater flow model. Careful planning is, however, needed to make the analysis systematic. The simple parameter variation method is the only readily available method for analysing conceptual uncertainties. The other approaches demonstrated involve regression analysis for the sensitivity analysis. This method has the advantage that it in principle covers the whole parameter space while simple parameter variations only give the local sensitivity. It has also been shown that if the sampled parameter sets are not distributed over the parameter space, the regression analysis can give regressions which are physically unexpected. Two methods of uncertainty analysis have been presented. These differ in the sense that the pdfs of the parameters are estimated either a priori or a posteriori. The latter approach involves the application Bayes' theorem. Although deemed to be possible, its theoretical foundation needs further discussion.

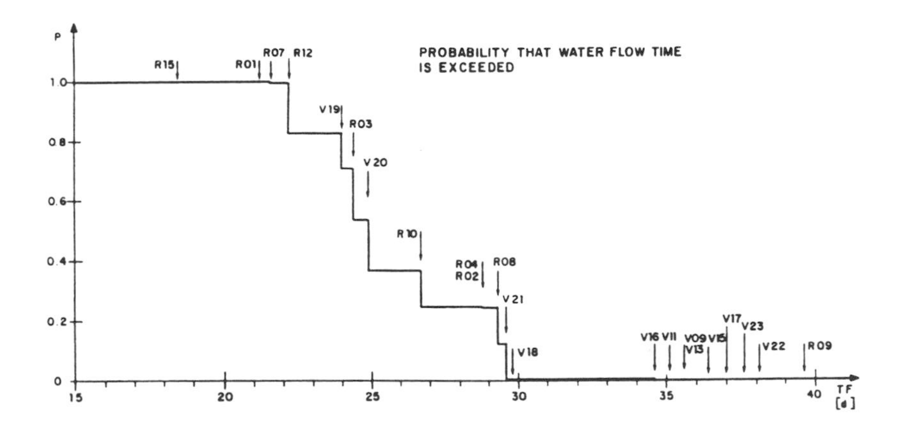

Figure 4 *Evaluated complementary probability distribution for the groundwater travel time along a flow line. The travel times of the individual calibration runs have been indicated.*

REFERENCES

[1] The international HYDROCOIN Project Level 1: Code verification, OECD/NEA Swedish Nuclear Power Inspectorate, 1988.

[2] The international HYDROCOIN Project Level 2: Model Validation, OECD/NEA Swedish Nuclear Power Inspectorate, 1990.

[3] Larsson N-Å., A. Markström, *Groundwater Numerical Modelling of the Fjällveden Study Site -Evaluation of Parameter Variations, A HYDROCOIN Study - Level 3, Case 5A*, SKB Technical Report 88-11, Swedish Nuclear Fuel and Waste Management Co., Stockholm, Sweden, 1987.

[4] McCartin T.J., P.A. Davis, M.T. Goodrich, *NRC Model Simulations in Support of the Hydrologic Code Intercomparison Study (HYDROCOIN), Level 2 - Model Validation, Level 3 - Sensitivity and uncertainty analysis*, Report NUREG-1249, Volume 2, U.S. Nuclear Regulatory Commission, Washington D.C., U.S.A., 1989 (DRAFT).

[5] Job D., G. Resele, C. Wacker, *Hydrogeologic Modelling of the Chalk River Block - Calibration, Validation, and* Sensitivity and Uncertainty Analysis of a Numerical Groundwater Model, National Cooperative for the Storage of Radioactive Waste (NAGRA), Baden, Switzerland, 1988 (DRAFT).

[6] Grundfelt B., *Definition of HYDROCOIN Level 3 Case 5 - Crystalline Rock*, KEMAKTA Consultants Co., 1986.

[7] *Final Storage of Spent Nuclear Fuel - KBS 3*, Swedish Nuclear Fuel and Waste Management Co, Stockholm, Sweden, 1983.

[8] Carlsson L., A. Winberg, B. Grundfelt, *Hydraulic Properties and Modelling of Potential Repository Sites in Swedish Crystalline Rock*, IAEA-SR-104/15, IAEA Seminar on Site Investigation Techniques and Assessment Methods for Underground Disposal of Radioactive Wastes, Sofia, Bulgaria, 1984.

[9] Hodgkinson D.P., T.W. Broyd, D.J. Noy, G.M. Williams, R.W. Paige, A.W. Herbert, C.P. Jackson, *Specification of a Test Problem for HYDROCOIN Level 3 Case 1: Sensitivity Analysis for Near-Surface Disposal in Argillaceous Media*, AERE-R. 11987, Harwell Laboratory, Harwell, U.K., 1985.

IAEA CO-ORDINATED RESEARCH PROGRAMME ON "THE SAFETY ASSESSMENT OF NEAR-SURFACE RADIOACTIVE WASTE DISPOSAL FACILITIES"

S. Hossain
International Atomic Energy Agency
Vienna, Austria

ABSTRACT

The International Atomic energy Agency is starting a new Co-ordinated Research Programme on "The Safety Assessment of Near-Surface Radioactive Waste Disposal Facilities" with the acronym INSARS (International Near-Surface Radioactive Waste Disposal Safety Assessment Reliability Study). This is programme is aimed at improving confidence which can be attached to the results of safety assessments. A Consultants' Meeting was held recently in the Agency Headquarters at Vienna to define the objectives, scope and preferred approach for conducting the programme and to formulate potential intercomparison/validation test cases for the programme. This paper describes the main conclusions and recommendations of the meeting.

PROGRAMME DE RECHERCHE COORDONNE DE L'AIEA SUR "L'EVALUATION DE LA SURETE DES INSTALLATIONS DE STOCKAGE DE SURFACE DE DECHETS RADIOACTIFS"

RESUME

L'Agence Internationale de l'Energie Atomique met en place un nouveau programme de recherche coordonné sur "l'évaluation de la sûreté des installations de stockage de surface de déchets radioactifs" ayant pour sigle INSARS (International Near-Surface Radioactive Waste Disposal Safety Assessment Reliability Study). Ce programme a pour objectif d'accroître la confiance que l'on peut accorder aux résultats des évaluations de sûreté. Une réunion de consultants s'est tenue récemment au siège de l'Agence de Vienne pour définir les objectifs, l'approche la plus indiquée pour conduire le programme et pour formuler les cas tests potentiels d'intercomparaison/validation. L'auteur expose les conclusions et recommandations les plus importantes de cette réunion.

1. INTRODUCTION

The safe management and disposal of radioactive wastes from the nuclear fuel cycle is increasingly being seen as a necessary condition for the future development of nuclear energy. In many countries near-surface disposal (1) is the preferred option for the comparatively large volumes of low- and intermediate-level wastes which arise during nuclear power plant operations, and nuclear fuel reprocessing and also for the wastes arising from radionuclide applications in hospitals and research establishments.

It is obviously necessary to show that waste disposal methods are safe and that both man and the environment will be adequately protected. With the increased public awareness of the waste management issue the topic has become very important. The overall objective of safety assessment is the quantitative prediction of the performance of the whole disposal system over the time for which the waste remain hazardous. National regulatory authorities require that safety assessment are performed in order to show that the prescribed radiological and more specific performance criteria can be complied with. Safety assessments are important in every phase of repository development:

- system selection
- site selection
- repository design, construction, operation, shutdown and sealing
- licensing processes relevant to these phases

The modelling techniques needed for use in these safety assessments have been developed over the last decade to a fairly sophisticated level in several industrialized countries. However there remains a need to provide assurance of the reliability of the predictions of these models. Although the developing countries are generally at a preliminary stage in developing such methodologies, it can be envisaged that there will be an increasing need for them in these countries.

The Agency sponsored a Co-ordinated Research Programme (CRP) on a similar topic from 1985 to 1989 with the title "The Migration and Biological Transfer of Radionuclides from Shallow Land Burial". This CRP has proved to be very popular providing a useful forum for exchange of information, both for scientists from industrialized and developing countries. However it was observed that there is a difference of interest between, on the one hand, the industrialized and more developed countries with established disposal programmes, and developing countries which are still at the planning stage. It was suggested that a future CRP should be more focussed, aiming at the resolution of a particular issue or towards the improvement of knowledge in a particular area of importance and in which international co-operation would be useful. Thus the current CRP is focussed on safety assessment of near-surface radioactive waste disposal facilities and specifically on studies aimed at improving confidence in the results of safety assessments.

---

(1) In near-surface disposal (also called shallow-land or shallow-ground burial) radioactive wastes are normally emplaced at or near the ground surface, generally not below 50 m in depth.

A Consultants' Meeting (CM) was held in the IAEA Headquarters, Vienna during 12-16 March 1990 for the planning of this new CRP. The CM was attended by experts on safety assessment methodology from Czechoslovakia, France, India, UK and USA, and representatives of the secretariats of the ongoing related international studies INTRAVAL and BIOMOVS. The main objectives of the meeting were:

- to define the scope and objectives, and preferred approach for conducting the programme

- to formulate potential intercomparison/validation test cases for the programme

The acronym INSARS (International Near-Surface Radioactive Waste Disposal Safety Assessment Reliability Study) was proposed as name for the CRP. In the following Sections of this paper the conclusions and recommendations based on the discussions and deliberations of the CM are described under separate headings.

2. OBJECTIVES, SCOPE AND EXPECTED RESULTS

The presentations and discussions in the CM confirmed that the main objectives of the CRP should be

- to improve the confidence which can be attached to the results of safety assessments for near-surface radioactive waste disposal facilities by means of model intercomparison and validation

- to help in establishing international consensus on the approach to safety assessment

- to facilitate exchange of information on safety assessment, its documentation and wider dissemination

The CM was aware of the need not to overlap with ongoing international programmes (INTRAVAL, BIOMOVS and PSACOIN). The emphasis in this CRP is therefore on the practical application of assessment models. Participants will be encouraged from a range of backgrounds with the objective of attracting those who have responsibilities in performing assessments of actual disposal systems. It is intended that the CRP should provide the possibility for the testing of models of a simple and robust type as well as those of a more complex and sophisticated nature. The objective therefore is that the CRP should have a wide potential scope and a practical orientation attracting participants from both industrialized and developing Member States.

For intercomparison exercises, the programme should emphasize a deterministic approach considering pathways of the whole system from source to man. This is in recognition of the most commonly used approach currently being applied for near-surface disposal safety assessments in Member States. Both an undisturbed and an intruder disturbance scenario should be included. Recognizing the contents of previous and ongoing international studies in this area it is considered that this CRP should focus on near-field aspects, on intrusion scenarios and on the overall safety assessment of near-surface disposal facilities. A range of end points will be considered including release terms and dose to man. As the CRP progresses, based on the views expressed by participants during the Research Co-ordination Meetings (RCMs), the variety of pathways may be augmented.

An important aim of the study will be the production of handbooks or documents which contain digests of all of the information developed on methods and data and on the results of intercomparison/validation exercises relevant to the safety assessment of disposal facilities of this type. Such documents should be valuable information sources for newcomers to the field of safety assessment.

3. POTENTIAL INTERCOMPARISON/VALIDATION TEST CASES

The experts attending the meeting reviewed known data bases for existing sites in their countries. Based on this review and a survey of available safety assessment methodologies, the group concluded that the first phase of the CRP should be an intercomparison exercise emphasizing a limited number of important pathways. This is intended as a base test case in order to initiate the CRP and provide a basis for discussion at the first RCM. Based on that discussion and the interest of RCM participants, further test cases will be developed. The following two main scenarios will be considered in this first test case:

- drinking well water outside the site boundary (Figure 1)
- house building intrusion (Figure 2)

Another conclusion of the planning meeting was that validation data bases were not readily available for near-surface disposal facilities. A major problem, true even of the best characterized sites, is a lack of good source term information. Records kept in the early days of waste disposal are not specific as to the radionuclide content or physical and chemical nature of the wastes. Information from existing near-surface disposal sites may be adequate to validate some aspects of pathways and other data bases may be available to achieve the same results. It was decided that the participants in the CRP would be used as resources for identification of potential data sets and in formulating validation exercises in the later stages of the CRP.

4. APPROACH FOR THE CONDUCT OF THE CRP

4.1 Structure of the CRP

Based on the experiences from previous and ongoing international projects of this type it is proposed to have

- A Secretariat, provided by IAEA, to provide technical support and administrative assistance
- A Co-ordinating Group, with Chairman, to provide planning and guidance

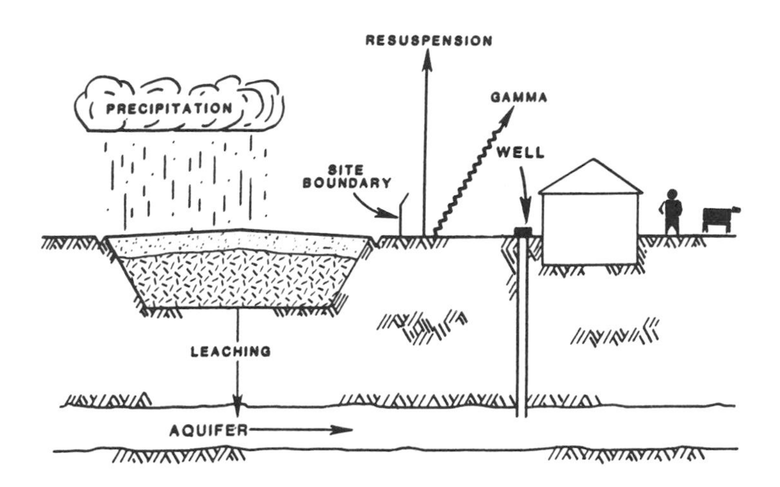

Figure 1. Sketch of the scenario of drinking well water outside the site boundary.

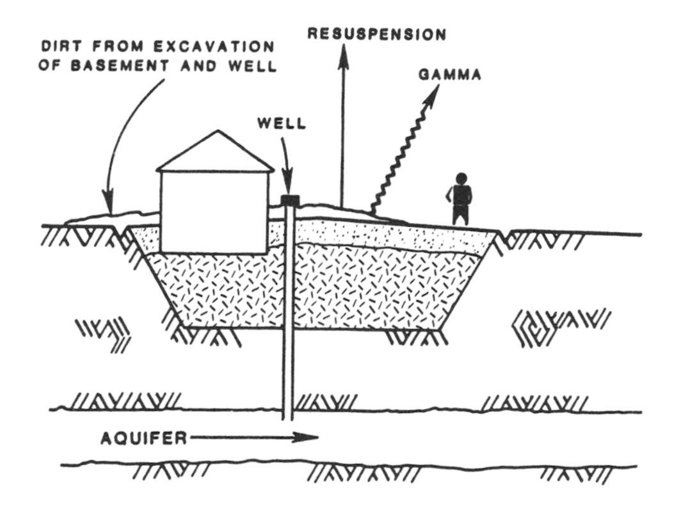

Figure 2. Sketch of the house building intruder scenario.

- Pilot Groups, with Chairmen, to establish and perform trial evaluations of the test cases prior to issue to participants and guide in assessing the intercomparison of results

- CRP Participants who are those actively involved in carrying out the test cases.

4.2 Participation

The following Member States, though not exhaustive, are identified as potential participants of the CRP:

Argentina	India
Australia	Japan
Austria	Korea
Belgium	Mexico
Brazil	Spain
Canada	Sweden
China	Switzerland
Czechoslovakia	UK
Finland	USA
France	USSR
Hungary	Yugoslavia

4.3 Work Plan and Time Schedule

The CRP is expected to last in its present phase for 4 years with a total of 4 RCMs, the first being in March 1991 in IAEA, Vienna. A preliminary programme was proposed as follows:

- Preliminary version of test case 1	May 1990
- Pilot runs of test case 1	Summer 1990
- Invitation to potential participants	June 1990
- 1st Pilot Group Meeting to finalize test case 1	October 1990
- Issue of test case 1 to participants	November 1990
- 1st RCM	March 1991

The intention of the first RCM will be to discuss the initial results of test case 1, to explore participants for future test cases and also any potential sites or experimental results for validation.

DUAL-POROSITY MODELLING OF INFILTRATION EXPERIMENTS ON FRACTURED GRANITE

P.A.Smith, J.Hadermann, K.Bischoff
Paul Scherrer Institut
CH–5252 Villigen PSI, Switzerland.

ABSTRACT

The results of dynamic infiltration experiments on core samples of fractured granite from the NAGRA Grimsel Test Site (GTS) are analysed in terms of single and dual porosity models, in which the radionuclides are transported by advection-dispersion through a planar fracture. In the dual porosity model, diffusion into the adjacent altered rock zones (matrix diffusion) is included. In the experiments, a spike of tracer containing non-sorbing (^{82}Br) and slightly sorbing (^{24}Na) radionuclides is introduced into the infiltration fluid. The transit time of fluid through a sample gives the advection velocity for the models and is estimated from the breakthrough of ^{82}Br. The remaining parameters, determined by fitting a solution of the governing equations to the ^{24}Na breakthrough curve, are the longitudinal dispersion length, the sorption retardation factor for the fracture and, in the case of the dual porosity model, the retardation factor and diffusion coefficient for the matrix. They are compared with independent experimental measurements, such as sodium batch sorption experiments on fracture infill material.

RESUMÉ

Les résultats des expériences d'infiltration dynamique effectuées sur des échantillons de carotte du granite, provenant du Laboratoire Souterrain du Grimsel (LSG) de la CEDRA, ont été analysé en terme de modèle de porosité simple et double, dans lesquelles les radionuclides sont transportés par advection-dispersion dans des fractures planes .Dans le modèle de porosité double la diffusion dans une zone limitée de la roche adjacente et altérée (diffusion dans la matrice) a été considérée. Dans les expériences, des traceurs non-sorbant (^{82}Br) et faiblement sorbant (^{24}Na) ont été introduits dans la solution d'infiltration. Le temps transit de ce fluide à travers l'échantillon donne la vitesse d'advection pour les modèles et est estimé à l'aide des courbes de transport du (^{82}Br). Les autres paramètres, determinés par une adaptation de la solution des équations utilisées pour la courbe de transport du ^{24}Na sont: la longeur de dispersion longitudinale, le facteur de retardation pour la fracture, et, dans le cas du modèle du porosité double, le facteur de retardation et le coefficient de diffusion pour la matrice. Ces résultats sont comparés avec des mèsures experimentales indépendantes comme la détermination des coéfficients de sorption du sodium sur le materiel de diaclase du Grimsel.

1. INTRODUCTION

The modelling of radionuclide transport through the geosphere is necessary in the safety assessment of repositories for high-level nuclear waste in deep-lying geological formations. Confidence in a model may be gained from its ability to fit dynamic laboratory and field experiments, which can differ in scale from a few centimetres to tens of metres. Parameters describing the interaction of radionuclides with rock can also be compared with results from static batch sorption tests. Such experiments are presently being performed with sorbing and non-sorbing tracers at the GTS migration site [1] and in the laboratory [2].

In the dual porosity model of nuclide transport in inhomogeneous rocks, Hadermann and Roesel [3] took into account the processes of advection and dispersion in water-conducting structures (idealised fractures or veins) and diffusion out of these structures into a spatially-limited altered zone (the matrix). Sorption of nuclides onto the surfaces of the structures and within the matrix was also incorporated. A computer code RANCHMD has been developed, which solves the coupled partial differential equations describing the dual porosity model [4]. As test of the dual porosity model, comparison was made with high-pressure bore core infiltration experiments reported in [5]. ^{233}U was used as a tracer and the crystalline samples were taken from NAGRA boreholes in northern Switzerland [6].

In this paper, the results of further bore core infiltration experiments on samples of granite from the GTS migration site are presented, in which non-sorbing and slightly sorbing tracers have been used: ^{82}Br and ^{24}Na respectively. The water-conducting structures of each sample are modelled as a single, parallel-walled fracture. The ^{82}Br breakthrough curve is used to determine the advection velocity in the model fracture and to calculate upper and lower limits for its width. Best fits are then obtained for the ^{24}Na breakthrough curve based on single porosity (without matrix diffusion) and dual porosity models. The purpose is to determine whether it is necessary to invoke the process of matrix diffusion in order to obtain a good fit and, if so, whether sorption and diffusion parameters for the matrix are consistent with dynamic field experiments at GTS and with static batch sorption tests.

2. HIGH-PRESSURE APPARATUS

A description of the essential parts of the high-pressure infiltration apparatus is given in [5]. Briefly, groundwater is pressed by a HPLC pump through the rock sample, which itself is confined in a pressure cell within a pressure vessel. At the high-pressure side, a valve allows for in-line injection of a measured pulse of the radionuclides of interest. At the low-pressure side, the eluting solution emerges as a succession of drops. The number of drops is recorded on-line by a counter, which triggers a fraction sampler. The mean drop volume is measured periodically by weighing. The radioactivity in the collected samples, on the other hand, is measured off-line and batchwise. The core sample is mounted between two stainless steel end pieces, held in position by a rubber sleeve.

3. TRANSPORT MODELS

3.1 Single-Porosity Model

In the single-porosity model, we distinguish two regions within the fractured granite of the infiltration experiments: intact rock, in which transport of radionuclides is neglected and fractures, which are amalgamated into a single parallel-walled, fluid-conducting zone of half-width (aperture) b, in which transport of radionuclides is by advection and dispersion. We adopt a Cartesian coordinate system (x, z): the z axis is parallel to the direction of fluid flow and the x axis is perpendicular to the plane of the conducting zone. Advection is accounted for by a fluid velocity u, dispersion by a longitudinal dispersion length a_L and sorption by a retardation factor R_f. The concentration of sorbing radionuclide (^{24}Na) in the infiltration experiments is well below the natural sodium concentration in the infiltration fluid. Sorption is therefore determined by isotopic exchange, and R_f is independent of concentration. Transport along the conducting zone is described by the advection-dispersion equation

$$\frac{\partial C}{\partial t} = S\frac{\partial^2 C}{\partial z^2} + T\frac{\partial C}{\partial z} + Q_f. \tag{1}$$

$C(z,t)$ is the concentration of a radionuclide in the liquid phase in the conducting zone and Q_f is a sink term denoting its decay. The coefficients of equation (1) are defined by

$$S = \frac{a_L u}{R_f} \tag{2}$$

$$T = -\frac{u}{R_f}. \tag{3}$$

3.2 Dual-Porosity Model

In the dual-porosity model, a third region within the fractured granite is distinguished, namely rock matrix with connected pore spaces. This represents, for example, altered rock adjacent to the fractures and is modelled as zones of thickness x_{max}, porosity ϵ_p and density ρ_p. Advection in the rock matrix is neglected and transport is by diffusion perpendicular to the conducting zone, with an effective diffusion coefficient $\epsilon_p D_p$. Sorption in the matrix is accounted for by a retardation factor R_p. R_p is again independent of radionuclide concentration, since sorption is determined by isotopic exchange of the tracer and, if $\epsilon_p \ll 1$ and $\epsilon_p \ll \rho K_D$, is related to the volume-based sorption constant K_D by

$$\epsilon_p R_p \sim \rho_p K_D. \tag{4}$$

The assumptions and approximations of the dual-porosity model are discussed in [3], and the governing equations presented. Transport along the conducting zone is described by

$$\frac{\partial C}{\partial t} = S\frac{\partial^2 C}{\partial z^2} + T\frac{\partial C}{\partial z} + U\frac{\partial P}{\partial x'}\mid_{x'=0} + Q_f. \tag{5}$$

Equation (5) contains a term in addition to those in equation (1), which represents the flux of radionuclide between the conducting zone and the matrix. Diffusion through the matrix is described by

$$\frac{\partial P}{\partial t} = V\frac{\partial^2 P}{\partial x'^2} + Q_p\ ;\ \mid x' \mid \geq 0. \tag{6}$$

$P(x',t)$ is the concentration of a radionuclide in the liquid phase within the matrix and Q_p is a sink term to account for its decay. Initial conditions and boundary conditions for the transport models are given in [4]. We have introduced a coordinate transformation in the x-direction

$$x' = \frac{x-b}{x_{max}-b} \tag{7}$$

This transformation enables solutions to be obtained without specifying the unknown quantity x_{max}. x_{max} appears in neither the governing equations nor the boundary conditions; it is incorporated into the two new coefficients U and V, defined by

$$U = \frac{1}{b}.\frac{\epsilon_p D_p}{R_f}.\frac{1}{(x_{max}-b)} \tag{8}$$

$$V = \frac{D_p}{R_p} \cdot \frac{1}{(x_{max} - b)^2}. \quad (9)$$

3.3 Method of Solution

The governing equations for both models are solved using code RANCHMD [4]. A set of time dependent, ordinary differential equations is obtained using the Lagrange interpolation technique and integrated by Gear's variable order predictor-corrector method.

Equation (1), the governing equations equation of the single-porosity model, has two independent coefficients S and T, which are determined by fitting the model to the time-history of ^{24}Na concentration with S and T as regression parameters. The Levenberg-Marquardt method is used to fit the model to the experimental data by the minimisation of the χ^2 merit function. The minimisation of χ^2 yields the regression parameters themselves and also their standard errors. The coefficients S and T are defined in terms of the three unknowns a_L, R_f and u. From equations (2) and (3)

$$a_L = -\frac{S}{T}, \quad (10)$$

and

$$R_f = -\frac{u}{T}. \quad (11)$$

We take u as equal to the ratio of the sample length Z to the transit time of (non-sorbing) ^{82}Br through the sample.

Equation (5) and equation (6), the governing equations of the dual-porosity model, have two further independent coefficients U and V. They are determined by fitting the dual-porosity model to the time-history of ^{24}Na concentration with S, T, $log_{10}U$ and $log_{10}V$ as regression parameters. $log_{10}U$ and $log_{10}V$ are used, rather than the coefficients themselves, in order to obtain meaningful error estimates for the regression parameters, since solutions to the governing equations are insensitive to large changes in U and V. The coefficients U and V are defined in terms of the five additional unknowns ρ_p, b, K_D, $\epsilon_p D_p$ and x_{max}. From experiments on mylonite [7], the matrix was assumed to have a density $\rho_p = 2600 m^3 kg^{-1}$. A lower bound for b can be obtained from the pressure gradient along the conducting zone $\Delta P/Z$ and an upper bound from the volume flow rate of the infiltration fluid q (see section 4), both of which are recorded during the infiltration experiments. The remaining three unknowns cannot be determined uniquely from the coefficients. We therefore eliminate x_{max}, which is likely to be the most sample-dependent quantity, and determine the product of K_D and $\epsilon_p D_p$. From equations (3), (4), (8) and (9)

$$\epsilon_p D_p K_D = -\frac{u^2 b^2 U^2}{\rho_p V T^2} \quad (12)$$

4. FRACTURE APERTURE

The fracture aperture can be estimated using two different approaches. The first estimate is based on an expression for hydraulic conductivity K for porous and equivalent porous media, which takes into account the constrictivity and shape of the pore spaces [8]

$$K = \frac{\rho g m^2 \epsilon^{2r-1}}{\mu k_0}. \quad (13)$$

Here, ρ and μ are the density and viscosity of the fluid, g is the gravitational acceleration, k_0 is a dimensionless shape factor which can vary between 2 and 3, ϵ is the porosity, m is the hydraulic radius (the ratio of the volume of interconnected pores to their surface area) and r is an empirical constant. Equation(13) reduces to the expression for laminar flow through a planar fracture (eg.[9]) for $r = 1$ and $k_0 = 3$. In [8], it was found that equation(13) reproduced measured values for rocks (including granite with natural fractures) and other materials with various porosities and pore shapes with the constant r equal to 2. Using Darcy's law, the hydraulic conductivity may be obtained from the specific discharge resulting from a known pressure gradient. Equating m in equation(13) to b in the transport models, and the specific discharge to ϵu, gives

$$b = \epsilon^{1-r} \left(-\frac{k_0 \mu u Z}{\Delta P} \right)^{\frac{1}{2}}. \tag{14}$$

The principal problem here is that a sample with one single fracture is clearly neither a porous nor an equivalent porous medium and the dynamic porosity ϵ is, in addition, related to the fracture half-width b. For such a sample, ϵ is dependent on the sample size (specifically the diameter in the present case). In the absence of a better relationship, we use equation(14) to give an estimated lower limit (b_l) for the half-width b, taking for the value of ϵ the measured dynamic porosity and setting k_0 to 2.5.

The second possibility is to use the relation

$$b = \frac{q}{2uD}, \tag{15}$$

where we have assumed that the fracture has a constant cross sectional area $2bD$, where D is the sample diameter and q the volume flow rate. The problem here is that, although a fracture zone may appear to follow a near-straight path across the centres of a sample, the fractures themselves are not straight and may branch many times, so that D may be an unreliable estimate of their total length. Again in the absence of a better relationship, we use equation(15) to provide an estimated upper limit (b_u) for the half-width b.

5. RESULTS

Table I gives details of the conditions during each infiltration experiment. The three experiments are identified by the names BOMI57, BOMI58 and BOMI60. BOMI57 and BOMI58 were carried out on the same sample, but with different rates of flow of the infiltration fluid. BOMI60 was carried out using the same flow rate as BOMI58, but with a second sample of greater length. Each of the samples had a diameter D of 4.6cm.

Table II gives the parameters derived from the breakthrough of ^{82}Br. The dynamic porosity, the ratio of the conducting-zone volume to the sample volume, is obtained from the equation

$$\epsilon = \frac{4q}{\pi D^2 u}. \tag{16}$$

b_l and b_u, the lower and upper bounds to the half-width of the fracture aperture are obtained from equation(14) and equation(15) respectively. ϵ and b_u vary little between BOMI57 and BOMI58, which indicates that the geometry of the fracture did not change significantly during the course of the two experiments. Unfortuately, the model parameters as given in sections 3.1 and 3.2 could not be extracted from the experimental ^{82}Br break-through, since the spreading of the curves is mainly generated by the experimental apparatus and is not amenable to modelling.

The fitted transport model results and the experimental ^{24}Na outlet concentrations are plotted in figure 1. The parameters obtained from the fitting procedure are given in table III. The uncertainties given are the standard errors in the regression parameters. χ^2_{min}, the minimum value of the χ^2 merit function for the fitted models, is also given. For BOMI57 and BOMI58, it can be seen that a marked reduction in χ^2_{min} can be obtained by the incorporation of matrix diffusion. A smaller reduction was obtained for BOMI60 (but note that BOMI60 was ended while the outlet concentration was relatively high). In general, allowing U and V to vary results in an increase in the standard errors in the other parameters. To investigate the possibility of several χ^2 minima, various initial guesses were used for the regression parameters; the minima found for each experiment coincide (at least to within the standard error in each parameter).

a_L, R_f are determined from the regression parameters S and T using equation (10) and equation (11), and are shown in table IV with the corresponding uncertainties. For the dual-porosity model incorporating matrix diffusion, upper and lower limits for $\epsilon_p D_p K_D$ are determined from equation (12) using regression parameters $log_{10}U$ and $log_{10}V$, with both b_l and b_u as the aperture half-width, and are also shown in table IV. Note that the product of the effective diffusion coefficient and matrix sorption constant is smallest for the lower bound of the aperture width. The reason is that a small aperture increases the flow of tracer into the matrix and hence less sorption and diffusion reproduce the experimental results.

5. DISCUSSION AND CONCLUSIONS

For two out of the three sets of experimental results (BOMI57 and BOMI58), the dual-porosity model gave a much closer fit to ^{24}Na breakthrough (as measured by χ^2) than the single-porosity model. For the third set of results (BOMI60), a relatively small improvement in the fit was obtained by the incorporation of matrix diffusion, although it was still possible to extract sorption and diffusion parameters. The most significant difference between the experiments in this respect is in the proportion of the tracer pulse recovered at the outlet. The final concentration of ^{24}Na in BOMI57 and BOMI58 was 20% or less of the peak concentration. In BOMI60, the final concentration was about 60% of the peak concentration.

The major ions in groundwater have diffusion coefficients in free water D_0 in the order of $10^{-9} m^2 s^{-1}$ at $25°C$ [10]. The effective diffusion coefficient is given by

$$\epsilon_p D_p = \epsilon_p D_0 \frac{\delta}{\xi^2}, \tag{17}$$

where δ is the constrictivity and ξ is the tortuosity. Values for the ratio δ/ξ^2 between about 0.5 and 0.01 are commonly observed in diffusion experiments of non-sorbing ions in porous geological materials [10]. A matrix porosity of 0.01 [7] gives a likely range for the effective diffusion coefficient of 10^{-13} to $5 \times 10^{-12} m^2 s^{-1}$. Laboratory batch sorption experiments give values of K_D of between 10^{-3} and $4 \times 10^{-3} m^3 kg^{-1}$ [2], on the basis of which the product of the effective diffusion coefficient and the sorption constant would lie in the range $10^{-16} < \epsilon_p D_p K_D < 2 \times 10^{-14} m^5 kg^{-1} s^{-1}$. From table IV, this range is contained between the bounds derived from BOMI60, and contains the upper bound derived from BOMI57 and BOMI58. However, Baeyens *et al.* [11], by interpreting CEC measurements in terms of an ion exchange model, obtained smaller values of K_D: between 3×10^{-4} and $4 \times 10^{-4} m^3 kg^{-1}$ for groundwater of the composition found at the GTS. This gives a range $3 \times 10^{-17} < \epsilon_p D_p K_D < 2 \times 10^{-15} m^5 kg^{-1} s^{-1}$, which is contained between the bounds derived from BOMI58 and BOMI60, and contains the upper bound derived from BOMI57.

In conclusion, a clear improvement in the fit to the dynamic experiments is obtained by incorporating matrix diffusion, using values for the product of the effective diffusion coefficient and the volume based sorption coefficients which are consistent with the results of static batch-sorption experiments. It was shown that there is insufficient information in the present breakthrough curves to obtain the sorption and diffusion parameters separately. This is principally because, in the dual-porosity model, x_{max}, the thickness of the rock-matrix region with connected pore spaces, is unknown. We have therefore been limited to obtaining upper and lower bounds for the product $\epsilon_p D_p K_D$. This limitation could be overcome if it were possible to fit the dual-porosity model to the breakthrough curve of ^{82}Br as well as that of ^{24}Na. However, with the present apparatus, we are unable to resolve the spreading of the ^{82}Br tracer pulse due to longitudinal dispersion and matrix diffusion. Work is currently being carried out to overcome this problem.

REFERENCES

[1] McKinley,I.G., Alexander,W.R., Bajo,C., Frick,U., Hadermann,J., Herzog,F.A. and Höhn,E. : "The radionuclide migration experiment at the Grimsel Rock Laboratory, Switzerland", Scientific Basis for Nuclear Waste Management XI, MRS Pittsburgh, 1988, 179.

[2] Aksoyoglu,S., Bajo,C., Baeyens,B., Bradbury,M. and Mantovani,M. : "Batch sorption experiments with iodine, bromine, strontium and cesium on Grimsel mylonite", PSI-Report and NAGRA technical report (in preparation) 1989.

[3] Hadermann,J. and Roesel,F. : "Radionuclide chain transport in inhomogeneous crystalline rocks. Limited matrix diffusion and effective surface sorption", EIR-Bericht Nr. 551 and NTB 85–40, February 1985.

[4] Jakob,A., Hadermann,J. and Roesel,F. : "Radionuclide chain transport with matrix diffusion and non-linear sorption", PSI-Report Nr.54 and NTB 90–05, November 1989.

[5] Bischoff,K., Wolf,K. and Heimgartner,B. : "Hydraulische Leitfaehigkeit, Porositaet und Uranrueckhaltung von Kristallin und Mergel: Bohrkern-Infiltrationsversuche", EIR-Bericht Nr. 628 and NTB 87–07, April 1987.

[6] Hadermann,J. and Jakob,A. : "Modelling of small scale infiltration experiments into bore cores of crystalline rock and break-through curves", EIR-Bericht Nr. 622 and NTB 87–07, April 1987.

[7] Meyer,J., Mazurek,M. and Alexander,W.R. : Ch.2 in Laboratory Investigations in Support of the Migration Experiments at the Grimsel Test Site. M.Bradbury (Ed.) NTB 88–23, PSI-Report Nr. 28, 1989.

[8] Brace,W.F. : "Permeability from resistivity and pore shape", Journal of Geophysical Research, 82, 23, (1977) 3343-3349.

[9] Snow,T.D. : "Rock fracture spacings, openings and porosities", J.Soil Mech. Found. Div. Amer. Soc. Civil. Eng. 94(SM1), (1968), 73-91.

[10] Freeze,R.A. and Cherry,J.A : "Groundwater", Prentice-Hall Inc., Eaglewood Cliffs, N.J., 1979.

[11] Baeyens,B. and Bradbury,M. : Ch.6 in Laboratory Investigations in Support of the Migration Experiments at the Grimsel Test Site, M.Bradbury (Ed.) NTB-88-23, PSI-Report Nr. 28, 1989.

ACKNOWLEDGEMENTS

We would like to thank A.Jakob and W.R.Alexander for their interest in this work and for fruitful discussions. We would also like to thank B.Wernli and K.Bleidissel for carefully executing all activity measurements and for experimental assistance. Partial financial support by NAGRA is gratefully acknowledged.

Table I : Experimental conditions: Z is the length of the sample, ΔP the difference in pressure across the sample, q the volume flow rate and t_L the duration of the tracer pulse.

	Z (cm)	ΔP ($\times 10^5 Nm^{-2}$)	q ($\times 10^{-10} m^3 s^{-1}$)	t_L (s)
BOMI57	3.05	20	2.333	427.9
BOMI58	3.05	21	1.050	952.4
BOMI60	4.38	14	1.050	952.4

Table II : Parameters derived from the breakthrough of ^{82}Br. Δt is the transit time through the dead volume, t_t the initial breakthrough time of ^{82}Br, u the fluid velocity, ϵ the dynamic porosity and b_l and b_u are lower and upper bounds to the aperture half-width.

	Δt (s)	t_t (s)	u ($\times 10^{-6} ms^{-1}$)	ϵ	b_l ($\times 10^{-6} m$)	b_u ($\times 10^{-4} m$)
BOMI57	210.0	2653.	12.48	0.011	1.77	2.0
BOMI58	466.7	5762.	5.76	0.011	1.18	2.0
BOMI60	466.7	6098.	7.78	0.008	2.76	1.5

Table III : Fitted parameters for the infiltration experiments with ^{24}Na. χ^2_{min} is the merit function for each fit. † single-porosity model calculations.

	S ($\times 10^{-9} m^2 s^{-1}$)	$-T$ ($\times 10^{-8} ms^{-1}$)	$-log_{10}\frac{U}{1.ms^{-1}}$	$-log_{10}\frac{V}{1.m^2s^{-1}}$	χ^2_{min} ($\times 10^{-4}$)
BOMI57 †	11.2 ± 0.3	14.8 ± 2.2	–	–	5.15
BOMI57	15.1 ± 0.8	15.1 ± 0.8	9.6 ± 0.3	14.03 ± 0.07	1.94
BOMI58 †	7.3 ± 0.1	$17. \pm 2.$	–	–	4.78
BOMI58	8.0 ± 0.3	$17. \pm 5.$	10.8 ± 0.2	16.0 ± 0.2	1.41
BOMI60 †	9.26 ± 0.06	5.9 ± 0.9	–	–	0.145
BOMI60	8.6 ± 0.2	20.8 ± 1.6	9.3 ± 0.5	12.3 ± 0.9	0.129

Table IV : The longitudinal dispersion length a_L, retardation factor for the conducting zone R_f and the product of the effective diffusion coefficient and volume-based sorption constant $\epsilon_p D_p K_D$, calculated from the regression parameters in table III. † single-porosity model calculations.

	a_L ($\times 10^{-2} m$)	R_f (–)	$\epsilon_p D_p K_D$ ($m^5 kg^{-1} s^{-1}$)	
			$b = b_l$	$b = b_u$
BOMI57 †	7.6 ± 1.3	$84. \pm 12.$	–	–
BOMI57	13.7 ± 4.4	$113. \pm 31.$	1.05×10^{-16}	1.34×10^{-12}
BOMI58 †	4.3 ± 0.6	$34. \pm 4.$	–	–
BOMI58	4.7 ± 1.6	$34. \pm 10.$	1.54×10^{-18}	4.44×10^{-14}
BOMI60 †	15.7 ± 2.4	$132. \pm 19.$	–	–
BOMI60	4.1 ± 0.4	$37. \pm 3.$	2.05×10^{-18}	6.07×10^{-15}

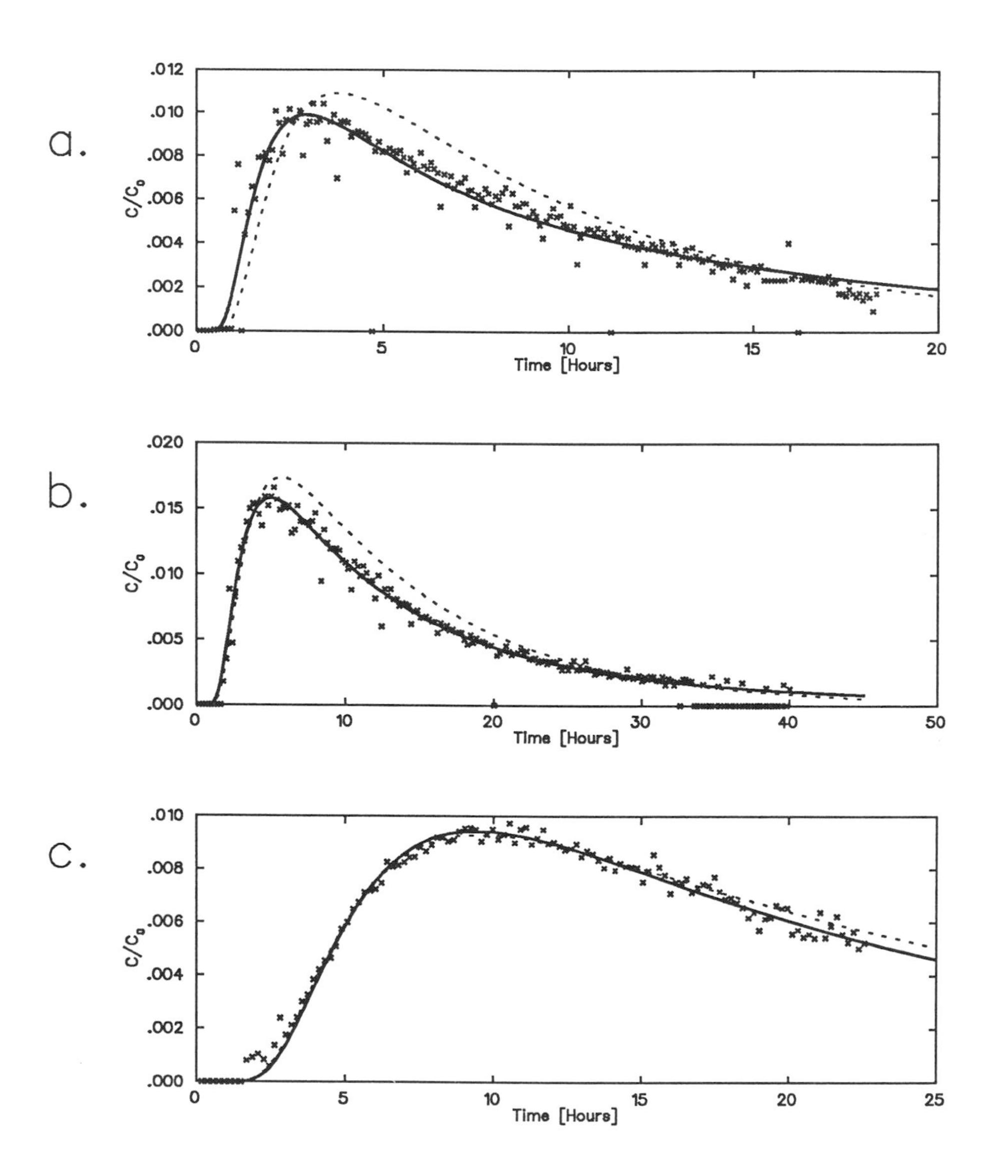

Figure 1. Time history of the concentration of ^{24}Na at the outlet. C_0 is the initial concentration at injection. a: BOMI57, b: BOMI58, c: BOMI60
× × × × experimental data, - - - - single-porosity model, —— dual-porosity model

WATER FLOW AND SOLUTE TRANSPORT THROUGH FRACTURED ROCK

J.E. Bolt, D.M. Pascoe, V.M.B. Watkins
Elcon Western Ltd., Redruth, Cornwall
P.J. Bourke
Chemistry Division, Harwell laboratory, Oxon
R.D. Kingdon
Theoretical Physics Division, Harwell Laboratory, Oxon

ABSTRACT

In densely fractured slate at the Nirex research site in Cornwall, the positions, orientations and hydraulic conductivities of the 380 fractures intersecting a drill hole between 9 and 50 m depths have been individually measured. These data have been used:

- to determine the dimensions of statistically representative volumes of the sheetwork of fractures;

- to predict; using discrete flowpath modelling and the NAPSAC code; the total flows into the fractures when large numbers are simultaneously pressurised along various lengths of the hole;

Corresponding measurements, which proved the modelling and validated the code to factor of two accuracy, are reported.

Possibilities accounting for this factor are noted for experimental investigation, and continuing, more extensive inter-hole flow and transport measurements are outlined.

The application of this experimental and theoretical approach for calculating radionuclide transport in less densely fractured rock suitable for waste disposal is discussed.

ECOULEMENT DE L'EAU ET TRANSPORT DE SOLUTES A TRAVERS UNE ROCHE FISSUREE

J.E. Bolt, D.M. Pascoe, V.M.B. Watkins
Elcon Western Ltd., Redruth, Cornwall
P.J. Bourke
Chemistry Division, Harwell laboratory, Oxon
R.D. Kingdon
Theoretical Physics Division, Harwell Laboratory, Oxon

RESUME

Dans une ardoise fortement fissurée au Centre de recherches de Nirex en Cornouailles, on a déterminé la position, l'orientation et la conductivité hydraulique de chacune des 380 fissures recoupant un trou de forage entre 9 et 50 m de profondeur. Ces données ont été utilisées :

- pour déterminer les dimensions de volumes statistiquement représentatifs du réseau de fissures ;

- pour prévoir - en utilisant un modèle à variables discrètes de la circulation de l'eau et le programme de calcul NAPSAC - l'écoulement total dans les fissures lorsque l'on applique simultanément à un grand nombre d'entre elles une pression le long de tronçons plus ou moins longs du trou de forage ;

Les auteurs présentent les mesures obtenues qui démontrent la justesse du modèle et valident le programme de calcul à un facteur deux près.

En vue d'études expérimentales, les auteurs donnent un certain nombre de raisons pouvant expliquer cet écart et esquissent un plan de mesures plus poussées de l'écoulement et du transport entre les trous de forage.

Enfin, les auteurs examinent comment cette approche expérimentale et théorique pourrait s'appliquer au calcul du transport des radionucléides dans des roches moins intensément fissurées se prêtant à l'évacuation de déchets.

Introduction

In the near field with relatively high, dissolved radionuclide concentration gradients round deeply buried waste, diffusion will be the main means of leakage out of the waste packages. Farther out where these gradients will have decayed, convection with any water movement will be the only likely means of transport back to the surface. Knowledge of the far field water flow is therefore necessary for the assessment of the safety of burial.

In hard fractured rock most of the flow occurs through the fractures. Such flows and their convection of solutes are being studied with relevance to disposal of wastes in several countries[1-4]. A programme in the U.K. was begun in Cornish granite and produced results and fracture data which helped the development of modelling of flow and convection through many fractures over long distances.

Because of the low density of fractures and consequently large dimensions of representative volumes of the rock, validation experiments to prove the modelling over such distances were not practical at this site. The field work was therefore transferred to a much more densely fractured slate, with smaller representative volumes at another site in Cornwall. This paper reports the progress that is now being made there to validate the modelling.

Previous Field Results

The previous work[5] in Cornish granite gave the positions and orientations of all fractures observed in core and geophysical logs from drill holes from 30 to 700 m depths. These fractures were flow tested one at a time to determine their hydraulic conductivities or effective apertures. The results showed that most (97+%) of the flow from the holes entered only a small (~1/10) fraction of the fractures and that the intact rock between them was substantially impermeable.

Measurements[6] of flows between holes lying along a single fracture proved that these flows were channelled in the fracture plane by areas of contact between its faces. This accounts for at least some fractures' taking no flow because they must, by chance, have been intersected by the holes through these contact areas. The fraction of the total fracture facial area wetted by the flows may therefore be taken as the fraction of all fractures found to be conductive.

In inter-hole radioactive tracer tests, a detector was "yo-yoed" in other holes, while tracer was pumped from source holes into fractures one at a time. The numerous positions of tracer arrivals found showed that there are many intersections between the fractures.

During the flow testing, two faults in the rock were found with conductivities at least a hundred times bigger than those of the many fractures tested. Similar occurrences of a few large faults have been noted in the other investigations mentioned. If such faults are found in the vicinity of disposal sites, the waste will presumably be put as far as possible from them and the rock between it and them may be the main barrier to leakage.

Flow and Convection Envisaged

The above results suggest that the natural movement of water from waste back to the surface or nearest large fault will occur through a random, three dimensional sheetwork of channelled, planar flow paths. Any radionuclides being convected with this flow will beneficially be dispersed to lower concentrations in the water by mixing in fractures and at intersections. They will also beneficially be retarded by diffusion and sorption through the wetted areas of the fracture faces into the rock between them.

Modelling Needed

Because of the very large number of flow paths involved, long distance modelling to assess safety of waste burial will probably have to treat fractured rock as a continuum with equivalent permeability, dispersivity and flow-to-rock mass transfer coefficient for quantifying the flow, mixing and retardation respectively(7). In deep, less permeable, hard rock suitable for burial, the average fracture separation is likely to be dekametres and hence, statistically representative volumes of the fracture sheetwork have hectometre dimensions. Direct measurement of the equivalent properties are therefore both:

- impractical because of the difficulty of flow and convection measurements over such long distances at depth;
- unreliable because flows from holes (and tunnels) of diameters less than the fracture separation depend mainly on the apertures locally round the holes and are insensitive to the interconnectivity which controls the permeability and dispersivity.

To avoid these problems, the following approach is being tried in the present programme. The above phenomena will be controlled by the discrete nature of the many flowpaths and Monte Carlo modelling to take this into account will need the distributions of the fracture variables listed here.

Phenomenon	Fracture Data Needed
Flow	Separations (or density) Orientations Conductivity (or apertures)
Dispersion	Intersections = f (separations, orientations) Channelling
Retardation	Wetted areas Separations

A suitable code, NAPSAC(8), for this modelling has been developed for the Stripa Project. Work at the new, densely fractured slate site is providing the data needed by NAPSAC to predict flows through representative volumes and, because of such volumes have small dimensions, direct measurement of these flows are being made to validate the code.

If this approach is successful, it is suggested that, at disposal sites with flows as envisaged here, the above fracture data be measured and used in a validated NAPSAC code to derive the equivalent properties needed for long distance continuum modelling.

Site and Experimental Methods

The new site is a disused slate quarry mainly of siltstone. This is extensively broken by conductive fractures with decimetre separations.

A number of vertical, 46 mm diameter, holes have been drilled and cored to 50 m depths. Rubber packers were wrapped in soft plastic tape, lowered by rigid rods into the holes and inflated to obtain relief impressions of the fracture apertures in the hole walls. Comparison of logs of these impressions and the cores gave the depths and orientations of all fractures to accuracies of better than ± 1 cm and ± 10° respectively.

All fractures below 9 m were then flow tested using double packers. With axial lengths of 5, 10 and 20 cm between 5 long packers, about 90% of the fractures could be individually isolated for testing one at a time. The other 10% overlapped or intersected within the hole and had to be tested mostly in twos, but at some places in threes and fours together.

All flow tests were made with constant pumped over-pressures of 10^5 Pa above the pressure between the packers for no flow. Tests were continued until steady flows had been obtained, usually after about 1 hour, but with several checks that no measurable (± 3%) changes occurred during further periods up to 10 hours. These steady flows were found to be proportional to over-pressures from 0.3 to 2 x 10^5 Pa at the three depths of 11, 20 and 25 m where this was checked. The reproducibility of the results is shown by the 12% standard deviation of seven repeat tests at one position over a five month period. The 5 m long packers were needed because with 1 m and lesser lengths some increases in the flows were found, presumably due to flow back to the hole above and below short packers.

Fracture Data

No flow tests were made above 9 m depths because of possible easy flow to the surface. In the first hole to be studied in detail, 380 fractures were found between this and 50 m depths. The positions, orientations and hydraulic conductivities (i.e. steady flows per unit over-pressure) of all of these were measured. Between 9 and about 25 m depth, the density of fractures was fairly homogeneous, but below 25 m there was some decrease in density.

The distributions of these variables for the about 25% of the fractures which took measurable flows between 10 and 25 m are given in Figs. 1, 2 and 3. In calculating these distributions it was assumed that, in each of the 8 groups of fractures which were tested in twos, threes and fours and found to take flow, only one arbitrarily chosen fracture took all the flow - this assumption is statistically consistent with only 25% of the individually tested fractures' being conductive. Flows into the other 75% of the fractures and any permeation into the rock between them must have been less than 3% of the total flow from the hole.

Minimum Statistically Representative Volume of Fracture Sheetwork

The flow and position data were used as follows to assess this volume in the homogeneously fractured rock. Starting arbitrarily at a depth of 11.36 m the flows into the individual conductive fractures were cumulatively added for increasing depth and divided by the lengths of hole from 11.36 cm. The results are plotted as average flows per unit length versus length in Fig. 4, where each point represents the accumulation of one extra fracture.

For short lengths below 11.36 m, the points show a wide scatter depending on the occurrence or not of highly conductive fractures. This scatter, however, decreases with increasing length and the approach of the graph to a nearly constant value at about 5 m after adding the flows into 25 fractures shows that such lengths may encompass statistically representative volumes of the sheetwork. Measurement of flow and solute transport over such distances should be practical to validate modelling based on individual fracture data. This modelling with individual fracture data may then be used to assess safety of disposal, in less densely fractured rock.

Modelling Without Fracture Lengths

The specified data requirements of NAPSAC are the distributions of the separations, orientations, effective hydraulic apertures and lengths of representative numbers of fractures. The first three of these have been measured at both sites as described above, but no method of determining lengths through the rock has been found. This lack of fracture length data may, however, not be serious in modelling over representative volumes of the sheetwork for the following reason. The flow data determine the density of conductive fractures along the drill hole. Because these took steady flows, they must all be connected to the sheetwork through which percolation occurs. The remaining uncertainty between these fractures':

- being much longer than their separations and intersecting many other fractures

- being shorter and intersecting fewer other fractures down to a minimum of two others to maintain connectivity

is illustrated in Figs. 5A and B. Comparison of these suggests that the effect of this uncertainty will not be large because for every fracture termination there must, on average, be one local fracture initiation to maintain the homogeneous density and whether the flow passes through more "X" and less "T" intersections or vice versa will not greatly change the above phenomena.

This qualitative view was investigated quantitatively using NAPSAC and the data from Figs. 1, 2 and 3 as follows. The hydraulic apertures needed for NAPSAC were calculated from the conductivity data in Fig. 3 using the Hodgkinson model[9]. This essentially takes into account that the individually measured flows from the drill hole radiate two dimensionally in the fractures until reaching the first intersections with other fractures, and that the pressures must fall toward zero at distances from the hole comparable with the distance between intersections.

With these apertures, the flows from a 4 m length of hole into fractures of various lengths and corresponding numbers of intersections were calculated. The results in Fig. 6 show that, provided the average number of intersections per fracture is a few above the effective percolation threshold of two, the flow is not very sensitive to the exact number, as suggested above.

Initial Validation of Flow Modelling

An initial validation of flow modelling has been carried out as follows. The total simultaneous flows, into all fractures along various pressurised lengths of the drill hole, were predicted using NAPSAC, the data from Figs. 1, 2 and 3 and the assumption that average number of intersections per fracture is a few above two. These simultaneous totals must be less than the totals of the flows into the fractures tested one at a time because when the flow into any fracture reaches its first intersection there will already be a rise in pressure there due to flows into adjacent fractures. These simultaneous totals were then divided by the pressurised lengths and plotted as average flows per unit length versus length in Fig. 7. This graph must decrease asymptotically to zero as the flow changes from three to two dimensional with increase towards an infinitely long source.

Corresponding experimental measurements of these average flows per unit length were made independently of the calculations for eight lengths of 1.25 m, four of 2.5 m, three 33 m, two of 5 m and one of 10 m from 11.36 m downward and are also plotted in Fig. 7. These results, as expected, show large scatter for lengths less than the 5 m length of the representative volume (see above), and reducing scatter as the length increases. Also, the means of these averaged simultaneous flows show a tendency to decrease with increasing length, as explained in the previous paragraph.

The agreement between the NAPSAC predictions and the experimental measurements is to within a factor of about two.

This factor of two may be explained by several possibilities. Firstly, and perhaps most likely, the distributions in Figs. 1, 2 and 3 are not sufficiently well defined to expect better agreement. Secondly, because the holes are vertical, there is some bias against sampling nearly vertical fractures which may distort the orientation data. Thirdly, no account has yet been taken in the modelling of channelling of the flow in fractures. These possibilities are now being looked into as described below.

Conclusions

1. The minimum statistically representative volume of fractured rock has been shown to have dimensions comparable with a few tens of fracture separation lengths.

2. Because of the large dimensions of such volumes of deep rock with low fracture density suitable for waste disposal, direct measurements of equivalent permeability, dispersivity and flow-to-rock mass transfer coefficients needed to quantify flow, mixing and retardation in continuum assessments of waste burial safety, are likely to be impractical and unreliable.

3. An initial validation of NAPSAC flow predictions has been achieved to an accuracy of about two.

4. Depending on continuing success in further validation experiments (below), NAPSAC with practical measurements of the separations, orientations and conductivities of representative numbers of fractures at disposal sites offers the prospect of reliable derivation of the equivalent continuum properties needed for safety assessments.

Continuing Work

To improve the statistical accuracy of the experimental distributions, another vertical hole has been drilled and is being tested similarly to the previous one. To reduce bias in the orientation data, three orthogonal holes, each at 45° to the vertical, are being drilled and will be tested.

In the foregoing analysis, channelling was neglected and it was assumed that flow was confined to the small fraction of fractures found to take flow from the drill hole and that these have uniform apertures. NAPSAC will be modified to take into account that, depending on geological examination, all or most fractures are likely to have some conductive areas which are fractions of their total areas equal to the fractions of all fractures found to take flow, as suggested above. Channelling may not greatly change long distance flow calculations because its effects of producing flow in smaller areas of more fractures tend to cancel each other. It will, however, have considerable effects on dispersion by changing mixing in fractures and at intersections and on retardation by altering the intimacy of contact between the flow and the bulk of the rock.

In continuing validation trials more extensive measurements of pressure and flow will be made at various distances from source holes. Solute transport experiments will be made to determine dispersion through the fractures which have been studied for the flow work. Finally, gas flow tests will be made to obtain some understanding and quantification of two-phase flow through fractures.

Acknowledgements

The authors wish to thank Dr. P.L. Hancock of the Department of Geology in the University of Bristol for frequent advice and guidance about the fracturing of the slate site. Dr. C.P. Jackson and Mr. A.W. Herbert of Theoretical Physics Division have helpfully discussed the modelling and NAPSAC code. Support by Dr. D.R. Woodwark of Chemistry Division for the programme was much appreciated. Funding of the work by UK Nirex Ltd. is gratefully acknowledged.

References

1. Stripa Project: "Programme for the Stripa Project Phase 3, 1986-91", TR 87-09, SKB Stockholm, 1987.

2. M.C. Cacas, E. Ledoux, G.de Marsily, B. Tillie, A. Barbreau, E. Durand, B. Feuga and P. Peaudecerf. Modelling Fracture Flow with a Stoichastic Discrete Fracture Network: Calibration and Validation Water Resources Research, to be published.

3. L. Liedtke. Fracture system flow tests - Grimsel test site, Final Report. BMFT-KWA 53045, 1988.

4. C.C. Davison, V. Guvanesen. Hydrogeological characterisation, modelling and monitoring of Canada's Underground Research Laboratory. DoE/ER 60152/1/Pt.2, 1985.

5. P.J.Bourke, E.M. Durrance, M.J. Heath, D.P. Hodgkinson. "Fracture hydrogeology relevant to radionuclide transport", AERE-R 11414, 1985.

6. P.J. Bourke. "Channelling of flow through fractures in rock", EUR 11455, 1988.

7. C.P. Jackson. Theoretical Physics Division, Harwell Laboratory. Private Communication.

8. A.W. Herbert, "NAPSAC Stochastic network modelling code: Technical summary (Release 1A)". AERE Report in preparation.

9. D.P. Hodgkinson. "Analysis of steady state hydraulic tests in fractured rock", AERE-R 11287, 1984.

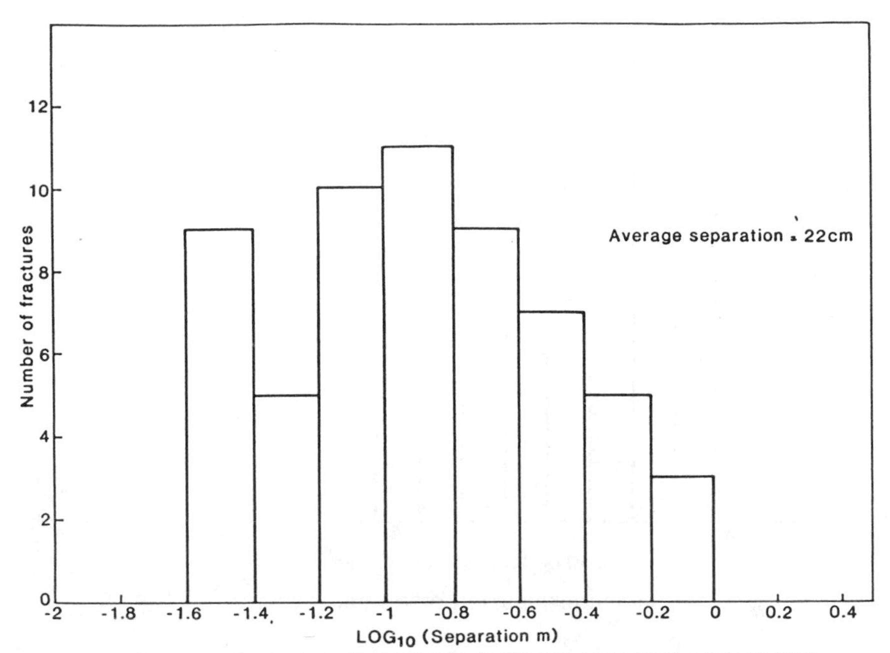

FIG.1. DISTRIBUTION OF SEPARATIONS BETWEEN CONDUCTIVE FRACTURES

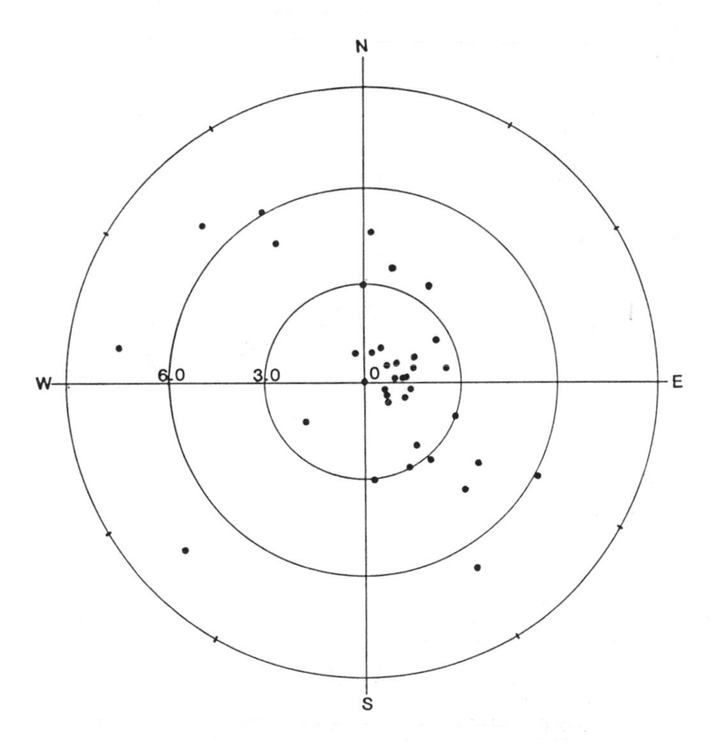

FIG.2. STEREO-PLOT OF NORMALS TO FRACTURE PLANES

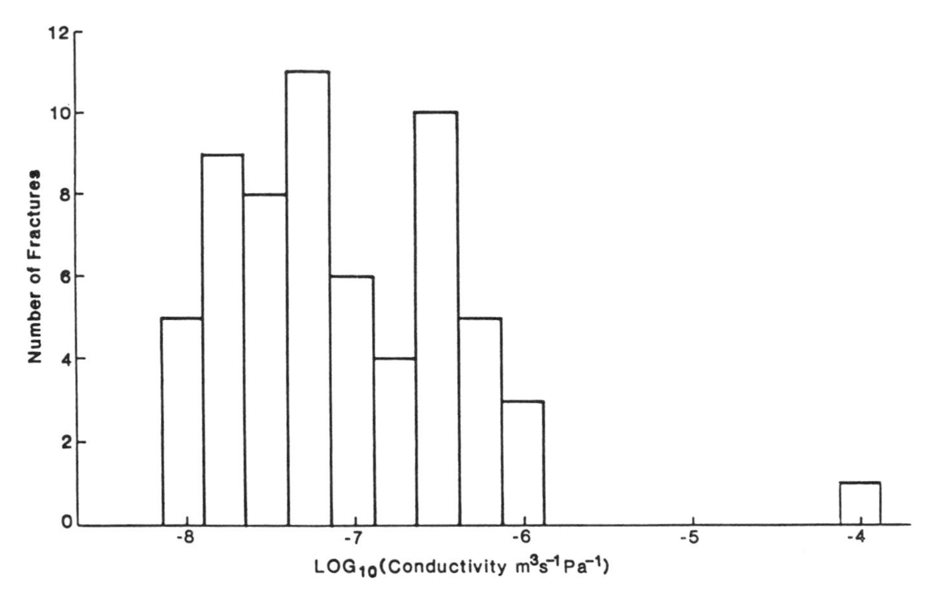

FIG.3. DISTRIBUTION OF CONDUCTIVITIES

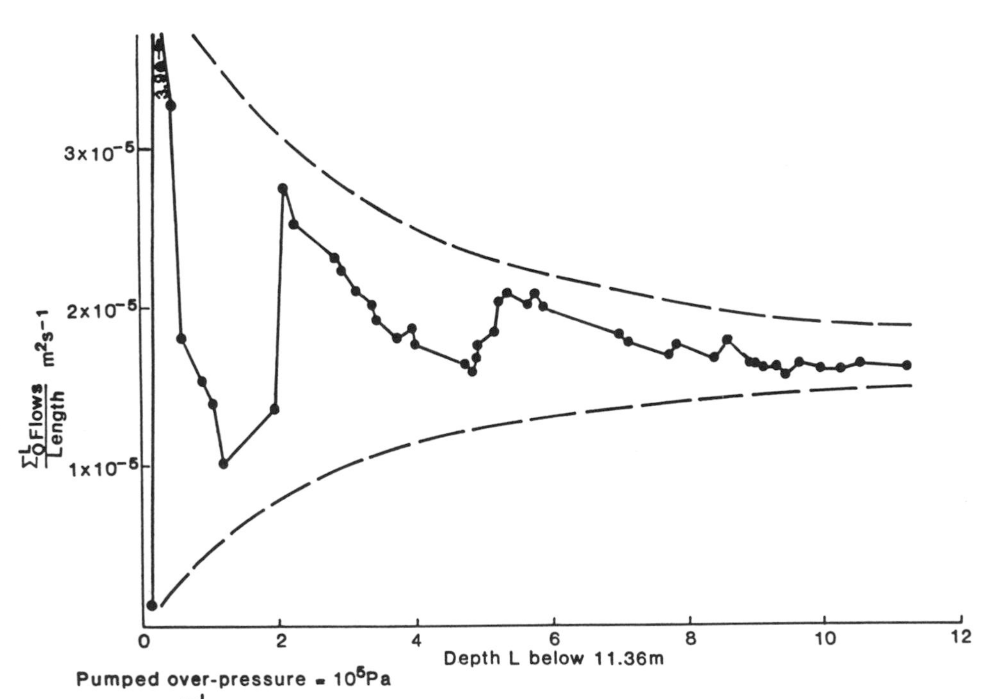

Pumped over-pressure = 10^5Pa

FIGT.4. $\frac{\Sigma_0^L \text{ FLOWS INTO INDIVIDUALLY PRESSURISED FRACTURES}}{\text{LENGTH L}}$ v. LENGTH L

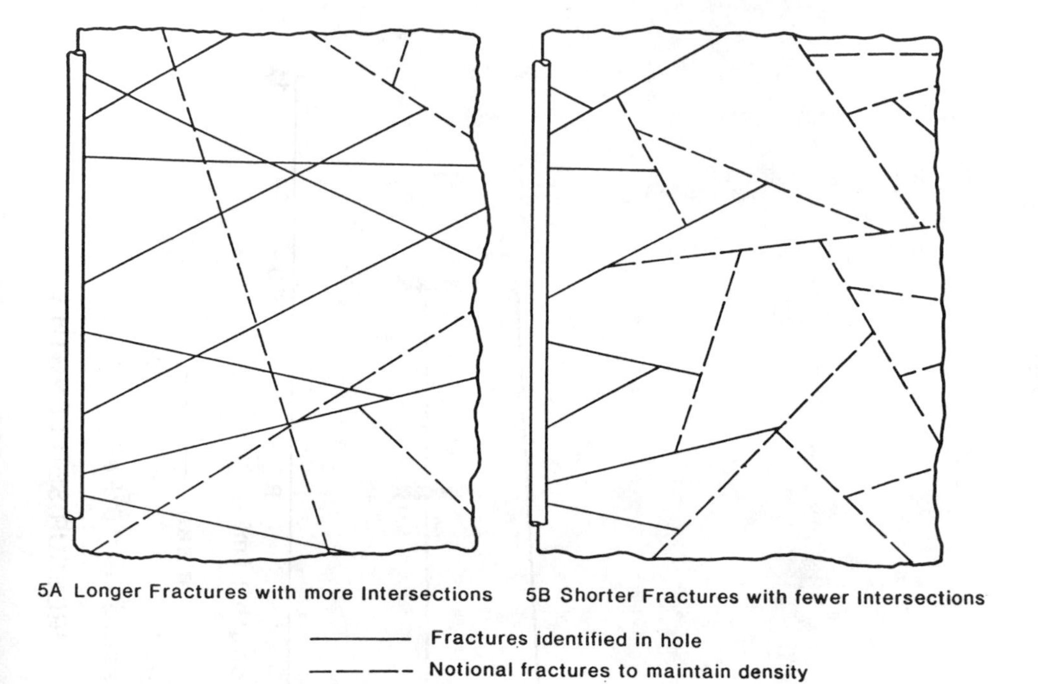

FIG.5. NOTIONAL FRACTURE SHEETWORKS

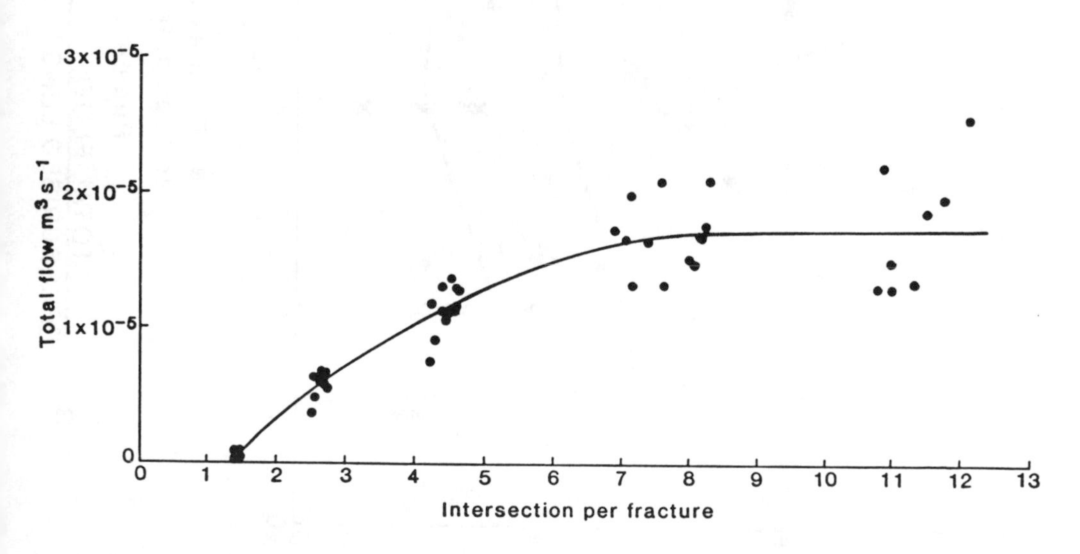

Pressurised Length = 4M
Pumped Over-pressure = 10^5Pa

FIG.6. TOTAL FLOW VERSUS INTERSECTIONS PER FRACTURE

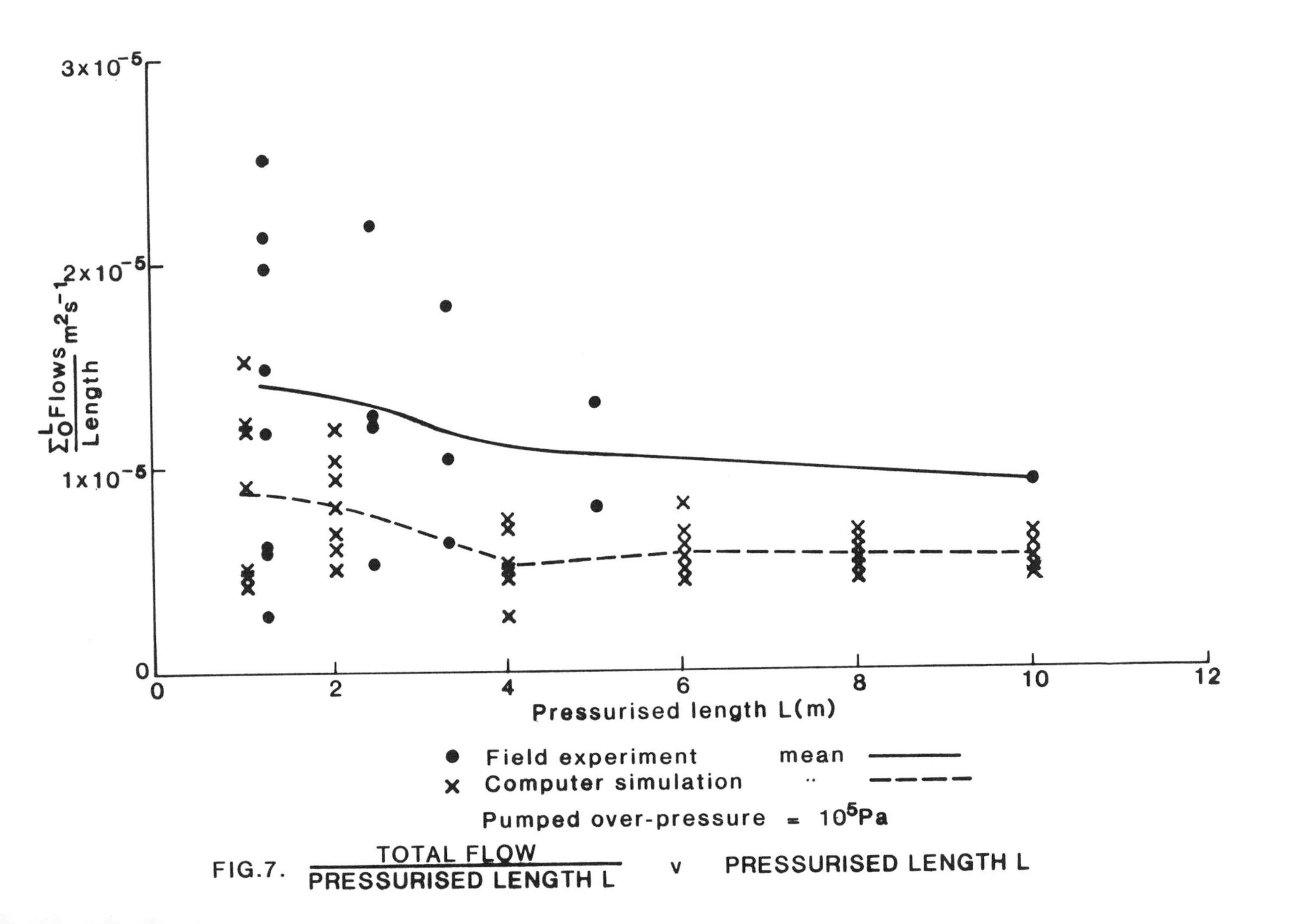

FIG.7. $\frac{\text{TOTAL FLOW}}{\text{PRESSURISED LENGTH L}}$ v PRESSURISED LENGTH L

MODELLING OF THE LARGE SCALE REDOX FRONT EVOLUTION IN AN OPEN PIT URANIUM MINE IN POÇOS DE CALDAS, BRAZIL

L. Romero, L. Moreno, I. Neretnieks
Department of Chemical Engineering
Royal Institute of Technology, Stockholm, Sweden

ABSTRACT

In an open pit uranium mine at Poços de Caldas in Brazil, the upper portions of the rock have been oxidized by infiltrating oxidizing groundwater. The redox front is very uneven and "fingering" is in evidence to depths ranging down to several hundred meters. The redox "fingers" are found in fractures and fractures zones. An attempt has been made to model the development of such redox fingerings along flow channels and to relate the structure to independent observations of flow channels in other crystalline rocks.

MODELISATION DE L'EVOLUTION A GRANDE ECHELLE D'UN FRONT D'OXYDO-REDUCTION DANS UNE MINE D'URANIUM A CIEL OUVERT A POÇOS DE CALDAS, BRESIL

RESUME

Dans une mine d'uranium à ciel ouvert à Poços de Caldas au Brésil, les parties supérieures de la roche ont été oxydées par des infiltrations d'eau souterraine oxydante. Le front d'oxydo-réduction est très irrégulier et on observe des indentations jusqu'à plusieurs centaines de mètres de profondeur. Ces indentations du front d'oxydo-réduction se trouvent dans des fissures et des zones fissurées. Les auteurs ont essayé de modéliser la formation de ce type d'indentations le long de canaux d'écoulement et d'établir un lien entre la structure obtenue à des observations indépendantes de canaux d'écoulement dans d'autres roches cristallines.

INTRODUCTION

Natural analogue studies of migration of radionuclides through geological sites, have been accepted as a means to provide useful information on processes and mechanisms considered to be significant to the radioactive waste disposal concepts. One of these analogues is an open pit Uranium mine, which has been operated by the Nuclebras Company at Poços de Caldas, PDC, in the state of Minas Gerais in Brazil, since 1975. The international Poços de Caldas project is a study of many processes including redox front movement, matrix diffusion, colloids and microorganisms as well as radionuclide migration.

The aim of this paper is to model the movement of the redox front, taking into account the oxidation of pyrite as the principal reaction. Of special interest is also the formation of large scale "fingering" of the redox front along paths of higher hydraulic conductivity. The present day redox front has evolved over long times and erosion must probably be accounted for in the modelling of the formation and movement of the front. The topography, hydraulic properties of the rock, and rates of infiltration of water are certain to have changed over time so that the present day shape and location of the front is the result of processes which have varied over time. We will attempt to account for the influence of erosion as well as the stochastic nature of the flow processes involved.

DESCRIPTION OF THE MINE

The open pit mine lies along a valley bottom. It is several hundreds of meters wide, nearly 1 km long and more than hundred meters deep in some places. The mine has near vertical terraced walls where the rock is accessible and visible. The bedrock is crystalline and consists mainly of phonolites and nepheline-syenites. The deeper portions of rock contain about 2 per cent by weight of pyrite (FeS_2) and are strongly reducing. However, the upper portions have became oxidized by infiltration of rainwater. The hydraulic conductivity is around 10^{-6} m/s [1], with a trend to decrease with depth. The bedrock porosity ranges around 4-20 %, the higher porosity is found in the oxidized and the lower one in the reduced region. Samples taken from boreholes in the reduced rock, show a characteristic reducing groundwater of low pH, around 5-6 with values of Eh-measurements ranging around -450 mV and relatively low ionic strength [2].

The redox front movement and the deposition of Uranium are clearly seen on the walls of the mine and in the boreholes from the floor. Figure 1 shows a cross-section of the rock below the present day mine level. There is a very sharp redox front delineating the upper oxidized rock from the deeper reduced rock. The reducing zone is rich in pyrite and the oxidized zone is practically devoid of pyrite. There are also many fingers of oxidized rock extending much further downward than the average depth of the front. The fingering is often associated with fractures and these fractures are sloping at various angles so that the fingers are often not vertical but some low angle. Uraninite nodules are found in many places just below the redox front in the reduced rock.

Erosion of the rock over the last 90 million years, has been estimated to be 3 - 9 km in depth from erosion data [3].

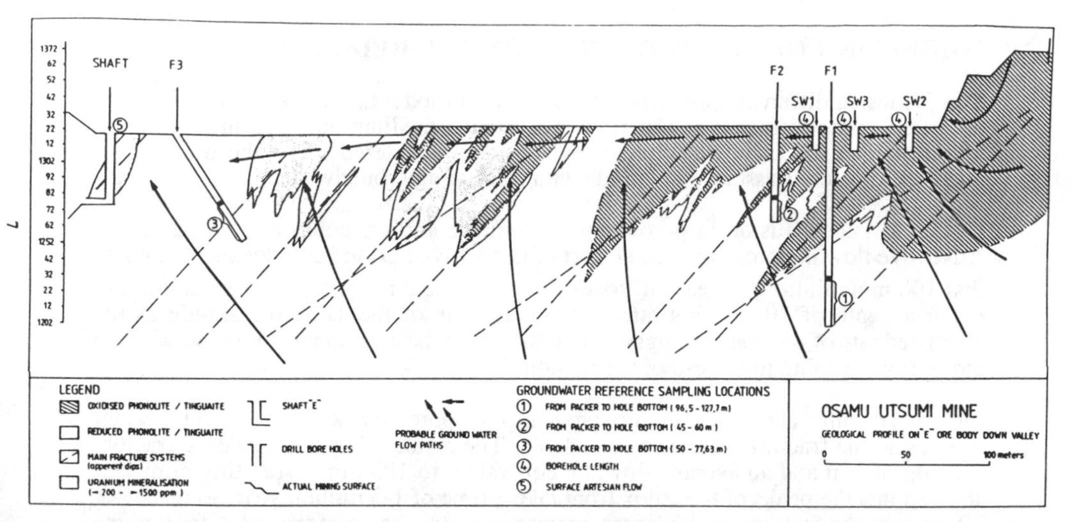

Figure 1. A cross-section of the uranium mine showing the redox front "fingers".

LOCATION OF THE REDOX FRONT

The mine has been mapped in great detail during the excavation. A set of maps provided by the mine geologists, which have been based on geochemical analytical data, shows the location of the redox front for any given section. From these data a 3-D computer model has been devised showing the location of the redox front located between 1200 - 1400 m above sea level. The area covered is about 216.000 m^2 (600 by 360 m). Figure 2 illustrates the 3-D distribution of the redox front. The picture clearly shows the existence of several downward penetrating "fingers" of oxidized rock, along and adjacent to fractures or fracture zones of greater permeability than the rock matrix itself. Some 200 individual "fingers" were identified in the mapped area.

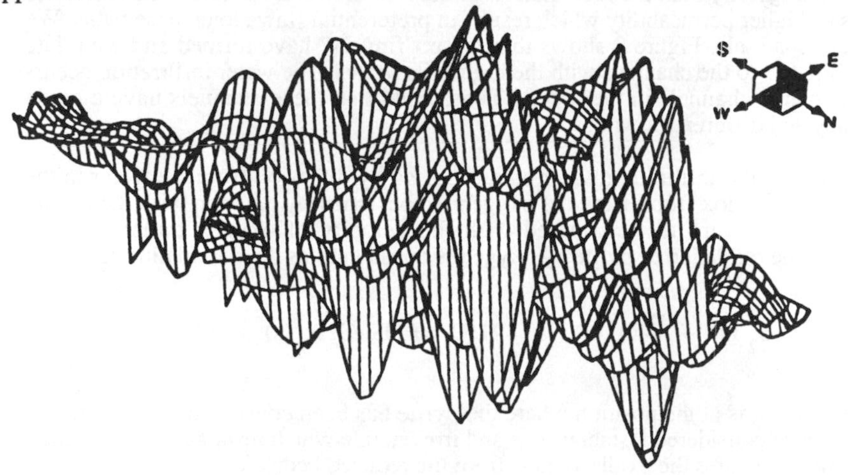

Figure 2. A three dimensional view of the redox "fingers" in an area of 600 x 360 m. Slanting view from below.

MODELLING OF THE MOVEMENT OF THE REDOX FRONT

Zhu [4] tested different approaches in an effort to model the redox front movement, considering the diffusion and/or flow in a porous medium and also in a fractured medium. The oxidation reaction of pyrite was considered to be the dominating reaction. Oxygen is assumed to intrude with infiltrating groundwater.

He found that diffusion in porous rock, from the surface could not penetrate far. Advective flow in porous rock could carry the dissolved O_2 to large depths. Assuming that 100 mm rainfall per year infiltrates it would take 3 million years for the front to reach a depth of 70 m. This rate of movement is of the same magnitude as the estimated rate of erosion and suggests that there may be a stationary situation with the redox front keeping just ahead of the erosion.

Flow in fractures and diffusion of oxygen into the porous rock matrix from the water flowing in the fracture were also considered. The results obtained for dense fracture spacing of 1 m and an average flowrate equivalent to 100 mm infiltrating rainwater showed that the peaks of the redox front take a time of 1-3 million years to reach 40-70 m along the fractures. With larger fracture spacings the tip of the redox front in the fracture can move very much faster. It was found that the spacing of conductive fractures and thus the water flowrate per fracture was one of the most important entities which determine the depth of redox front in the fractures.

CONCEPTUAL MODEL

In the present model we assume that there are isolated channels in the fractures in the rock and that different channels carry different flowrates.

A part of the rainwater infiltrates into the rock. The water is saturated with oxygen and has a concentration of approximately 10 ppm. The rock contains pyrite which is the main reducing component. Some of the water flows into the porous rock matrix directly but some of the water flows into channels, fractures, and fracture zones with higher permeability than the rock matrix. In the fractures and fracture zones there are regions of higher permeability which results in preferential flow along these paths. We call them channels. Figure 2 shows that redox "fingers" have formed and we relate these "fingers" to the channels with the higher flowrates. The water infiltration occurs through sparse channels in the rock. It is assumed that these channels have a given frequency and different flowrates.

The oxygen in the infiltrating water is transported by diffusion from the water in the channel to the redox front. Here the oxygen is consumed by reaction with the pyrite. The reaction is very fast compared with the velocity of oxygen transported by diffusion. The oxidation of pyrite by oxygen is considered as the dominating reaction.

$$FeS_2\ (s) + \frac{15}{4} O_2\ (aq) + \frac{7}{2} H_2O \Rightarrow Fe(OH)_3\ (s) + 4\,H^+ + 2\,SO_4^{2-}$$

The redox front is at the position where the pyrite has been consumed. For simplicity the reaction is considered instantaneous and irreversible which results in a sharp redox front which separates the oxidized zone from the reduced bedrock.

The bedrock is modelled as a fractured porous medium containing sparse channels in the fracture planes. At early times, the advance lines of the redox front perpendicular to the channel into the matrix may be represented by a linear model. The oxidized zone is nearby rectangular initially, but with time this zone becomes oval and finally circular. This is shown in Figure 3, for a horizontal cross-section of the rock. For modelling purposes we assume that propagation of the front is cylindrical already from the start.

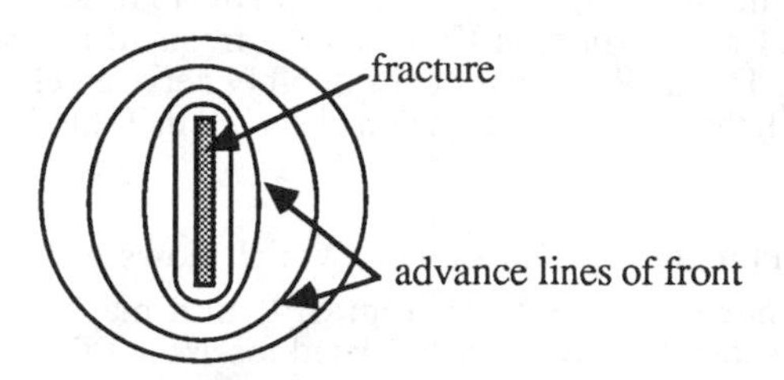

Figure 3. Redox front develops outward from channel by diffusion with time

The rock is assumed to contain a large number channels. The channels may have different flowrates and widths. Figure 4 shows a cross-section of rock with independent channels. Every channel has on the average a cross-section of rock which may be oxidized by oxygen diffusing from that channel. The channels are independent for some distance but otherwise are part of a channel network.

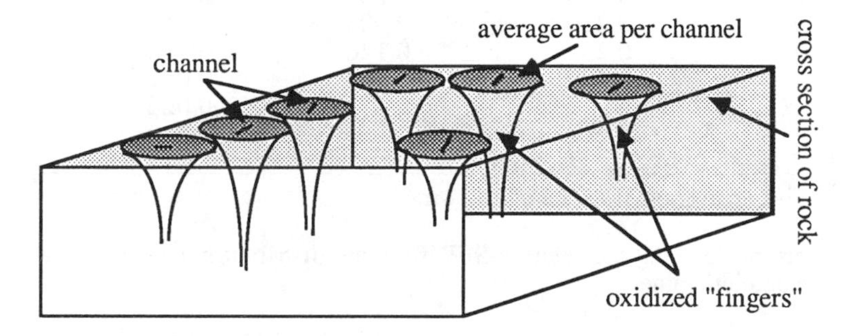

Figure 4. Cross-section of rock with independent channels.

THE MATHEMATICAL MODEL

The mathematical model was based on the assumption of a fast reaction and a cylindrical geometry for the spreading of the redox front from the channel. The solution to the advective transport and diffusion equations was developed using the same method as Cooper and Liberman [5]. For a constant flowrate a semianalytical solution was obtained. When erosion was included and the water flowrate was allowed to vary in time, a numerical scheme based on implicit techniques was used.

CALCULATIONS OF THE MOVEMENT OF THE REDOX FRONT ALONG A CHANNEL

There is no information on widths, frequency and flowrate distribution of channels in the rock in the uranium mine. There are, however, observations in several tunnels in crystalline rocks in Sweden which show that channel widths range from a few centimeter to tens of centimeters and up to one meter. The frequency of channels range from 1 per 20 m^2 in Stripa [6] to about 1 per 100 m^2 in SFR [7] and Kymmen [8] in the good rock. In fracture zones in Kymmen the frequency was nearly one order of magnitude larger. The flowrates vary considerably between channels. In the tunnels charted in SFR [7], the flowrate distribution is shown in Table 1.

Table 1. Fraction of the total flowrate which flows in different categories of channels. The data are from SFR in a mapped area of 14000 m^2. Category 7 and 8 are extrapolated for use at PDC.

Channel category	Relative flowrate	Fraction of spots	Fraction of flowrate	Flowrate per channel in PDC.* $m^3/s*10^6$
1	32	0.012	0.131	2.758
2	16	0.024	0.148	1.550
3	8	0.073	0.207	0.723
4	4	0.250	0.305	0.312
5	2	0.232	0.126	0.138
6	1	0.409	0.084	0.052
7	1/2			0.026
8	1/4			0.013

* These values give an average flowrate of 100 l/m^2 a.

Calculations were made using the relative SFR flowrate distribution applied to the conditions at the uranium mine.

If there were no erosion and the channels extended "forever" with the same flowrate, the rate of movement of the tip of the redox front decreases inversely to the square root of time. This means that for a constant rate of erosion there will be a time at which the front moves as fast as erosion takes place. This is shown in Figure 5.

Figure 5 shows the oxidized length along a channel for channels with little flow, categories 4-8. For example, the stationary length is 0.2 km for channels in category 6. The fastest channels would extend many kilometers if they were isolated. Even assuming that these channels are horizontal for long distances, the distance is too long to be reasonable even accounting for erosion.

The above calculations were based on the assumption that a channel extends forever with the same flowrate all along and for all times. This is not a reasonable assumption for long distances, because of the "network" structure of channels.

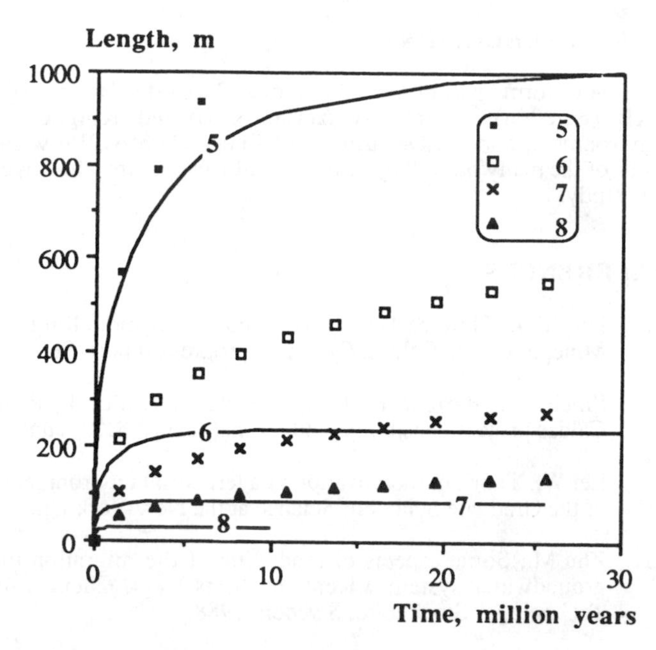

Figure 5. Length of oxidized "fingers" as a function of time considering no erosion and considering erosion (full lines) for independent channels. The number are flow categories of Table 1.

DISCUSSION AND CONCLUSIONS

It is seen from Figure 5 that for channels with little flow, categories 6-8, the redox front in independent channels would stabilize at 20-200 m below the constantly eroding ground surface. For the channels with larger flowrate the length of the oxidized channels would become very large. It is conceivable that channels would keep their identity for tens to hundreds of meters because the channels are sparse. For larger distances the channels are bound to intersect other channels and form a network. This would lead the water in different channels to mix their waters and the channels would loose their identity. The fronts would not penetrate as far as the individual channel concept would indicate.

Furthermore, the flowrate at larger depths will decrease and also become more horizontal before finally turning upward. Very long channels will thus come up to the surface.

The frequency of channels found in the Swedish crystalline rocks was on the order of 1/20 m^2 to 1/100 m^2. The frequency of redox fingers at PDC was on the order of 1/1000 m^2. At PDC the fingers in the mined away rock could not be reconstructed. The figure for PDC is thus truncated and gives too few channels. The resolution of the PDC mapping is very coarse and finer channels will not be found. Visual observations in the vertical walls show many more fine channels but these have not been quantified.

ACKNOWLEDGMENTS

This work forms part of the international Poços de Caldas project, jointly funded by SKB (Sweden), Nagra (Switzerland), United Kingdom Department of the Environment, and the Department of Energy U.S.A. We wish to acknowledge the work of the many other Project Principal Investigators who have provided the data for this study.

REFERENCES

1. Noy D.J., Holmes D.C., Hydrogeological modelling of the Osamu Utsumi Mine. Poços de Caldas, Quarterly progress report, Sept. 1987 - Nov. 1987.

2. Smellie J., Barroso L., Chapman N., McKinley I., Penna E., The Poços de Caldas project Feasibility study, Final report, SKB, June1986-1987.

3. Lei W., Thorium mobilization in a terrestrial environment, Thesis at the faculty of the Graduate School of Science at the New York University, 1984.

4. Zhu M., Some aspects of modelling of the migration of chemical species in groundwater system, Licentiate thesis Dept. Chem. Eng. Royal Institute of Technology, Stockholm, Sweden, 1988.

5. Cooper R.S., Liberman D.A., Fixed-bed adsorption kinetics with pore diffusion control, Ind. Eng. Chem. Fundam., 9, 4, 620, 1970.

6. Abelin H., Neretnieks I., Tunbrant S., Moreno L., Final Report of the Migration in a single fracture-Experimental results and evaluation. Stripa Project Report 85-03. OECD/NEA, SKB 1985.

7. Bolvede P., Christianson R., SKB Forsmarksarbetena SFR. Vattenförande sprickor inom lagerområdet. VIAK, Stockholm (in Swedish), Water bearing fractures in the repository area, 1987.

8 Palmqvist K., Stanfors R., The Kymmen power station TBM tunnel. Hydrogeological mapping and analysis. SKB Technical report, 87-26, 1987.

ANALYZING TRANSPORT IN LOW PERMEABLE FRACTURED ROCK USING THE DISCRETE FRACTURE NETWORK CONCEPT

B. Dverstorp
The Royal Institute of Technology, Stockholm, Sweden
J. Andersson
The Nuclear Power Inspectorate, Stockholm, Sweden

ABSTRACT

Tracer migration experiments in an experimental drift of the Stripa research mine, Sweden, are analyzed with a discrete fracture network model. Synthetic experiments with the discrete model explain the seemingly unstructured and erratic transport behaviour observed in the field experiment. A high variability of the flow channel transmissivity results in non-Fickian dispersion and channeling effects which may severly reduce the possibilities of estimating effective transport parameters (e.g. Peclet number and porosity) for predictive use. Consequently, transport predictions with an averaging continuum model will be hazardous in this type of rock.

ANALYSE DU TRANSPORT DANS UNE ROCHE FISSUREE DE FAIBLE PERMEABILITE AU MOYEN DU MODELE A VARIABLES DISCRETES SIMULANT UN RESEAU DE FISSURES

RESUME

Au moyen d'un modèle à variables discrètes simulant un réseau de fissures, les auteurs analysent des expériences de migration de traceurs réalisées dans une galerie expérimentale de la mine de recherche de Stripa (Suède). Des expériences de synthèse avec le modèle à variables discrètes permettent de comprendre pourquoi les caractéristiques du transport observées lors de l'expérience in situ semblaient non structurées et erratiques. Une forte variabilité de la transmissivité des canaux d'écoulement entraîne une dispersion non fickienne et des effets de canalisation qui peuvent compromettre sérieusement l'estimation de certains paramètres qui interviennent effectivement dans la prévision du transport (par exemple, nombre de Peclet et porosité). C'est pourquoi il serait imprudent, dans ce type de roche, de faire des prévisions concernant le transport au moyen d'un modèle à variables continues donnant une valeur moyenne.

INTRODUCTION

The quality of predictions of transport in heterogeneous fractured rock is due to the adequacy of the conceptual and mathematical model used but also to the possibility of estimating relevant transport parameters. The choice of conceptual model for this purpose basically stands between a discrete fracture model and some kind of equivalent continuum description of the rock. The latter approach relies on the possibilities of averaging flow and transport in terms of effective transport parameters like Peclet number and flow porosity whereas discrete models aim to reproduce the actual structure of the individual flow channels in the rock.

Spatial variations of permeability in a heterogeneous medium, hence varying channel aperture in case of fractured rock, result in channeling effects. Flow and transport concentrates to a few preferential high permeability flow paths which may significantly deviate from the average flow and transport behaviour in terms of porosity, dispersion and capacity for chemical surface reactions like sorption and matrix diffusion. Field experiments (e.g. Abelin et al., 1987 and Cacas et al. 1990) support this channeling phenomenon on large scale in fractured rock. Tsang and Tsang (1989) illustrate how flow channeling occur in a two-dimensional heterogeneous medium with spatially varying aperture. Moreno et al. (1989) extend this analysis to include solute transport. They illustrate the difficulties of making point measurements in a spatially heterogeneous medium and suggest spatial averaging to overcome this problem. However, due to the effects of flow channeling together with other potential network effects averaging of transport in fractured rock is a complex matter which has to be fully understood to be able to make meaningful model predictions. Still, in large scale practical applications, limited computational resources and lack of detailed data on individual flow and transport paths often restrict the model choice to simplified continuum type models which use effective transport parameters.

Hydraulic and tracer tests in fractured rock provide data for estimation of effective transport parameters. However, experimental evidence (e.g. Abelin et al., 1987) demonstrates that flow and transport properties may vary even on large scales. If the scale of this variability is in the same order or larger than the scale of the experiment the estimated transport parameters may be so sensitive to statistical variations that predictions will be questionable. Therefore it is essential to develop experimental designs that reveal the scale and degree of variability of the effective transport parameters.

The present study uses synthetic experiments with a discrete fracture network model to address the above aspects of transport in fractured rock. Dverstorp and Andersson (1989) provide realistic model parameters obtained by calibration of their discrete fracture model for flow on the trace geometry and inflow to an experimental drift in the Stripa research mine, Sweden. The validity of modeling transport in fractured rock with effective transport parameters is explored by fitting a one-dimensional advection-dispersion model to breakthrough curves from the synthetic experiments.

SYNTHETIC EXPERIMENTS

Transport simulations

The discrete fracture network model, DISCFRAC, originally developed by Andersson and Dverstorp (1987) generates realizations of fracture networks in three dimensions and solves for steady state flow and transport. The conceptual model for flow and transport within the fracture is similar to the one proposed by Cacas et al. (1990). Flow takes place in rectangular channels, with a constant width of 0.2 m, leading from the center of one fracture through the intersection with another fracture and further on to the center of the next fracture. DISCFRAC uses a particle following algorithm for the transport calculations.

This analysis focus on the effects of the fracture network and therefore neglect local dispersion in the flow channels of the fractures. Fracture channel transmissivities follow a lognormal distribution with log mean, μ_{lnT}, and log standard deviation, σ_{lnT}. The field data analysis indicates a value between 1 and 2 of σ_{lnT}. However, the estimate of σ_{lnT} is associated with a high uncertainty. Therefore, this analysis includes a range of values, σ_{lnT} equal to 0, 2, 4 and 6, to highlight the effects of varying channel transmissivities.

The simulation domain is a rectangular box with sides X_L, Y_L and Z_L. A hydraulic head difference between the prescribed head faces perpendicular to the x-direction drives the flow through the domain. The faces along the x-direction are no-flow boundaries. In each simulation case a hydraulic head gradient equal to 0.1 is applied on the domain. The length of the domain, X_L, varies from 10 m to 30 m whereas the cross section area (Y_L*Z_L = 30*30 m^2) is held constant. The length of the domain is increased by equal expansion in positive and negative x-direction about the mid point of the domain. All particles are released from the fracture intersection that carries most of the flux within a central 10*10 m^2 square on the upstream face. Particle sampling on the entire outflow face provides an areally averaged breakthrough curve. Each transport simulation uses 40 000 particles, and for each simulation case 30 Monte-Carlo realizations provide ensemble mean and variance of the flow and transport parameters obtained from individual realizations.

Continuum model

The continuum model is the equation for advective one-dimensional transport along a channel including dispersion in the direction of flow and sorption onto the fracture surface (see e.g. Abelin et al., 1987). Fitting this transport equation to the breakthrough curves from the discrete model, by the method of least squares, provides estimates of Peclet number, Pe, and water residence time, t_w. For later use we define a Peclet number

$$Pe = \frac{U_f \; x}{D_L} \qquad (1)$$

and a dispersion length

$$\alpha = \frac{x}{Pe} \tag{2}$$

where U_f is the global fluid velocity and D_L the longitudinal dispersion coefficient. x is a characteristic length of the transport problem considered. In this analysis x equals the length of the simulation domain, X_L. Thus for a constant dispersion coefficient, D_L, the Peclet number will increase linearly with increasing domain length. The dispersion length, α, is a measure of the scale of the discontinuities of the fractured medium.

VALIDITY OF CONTINUUM MODEL

Model fit in individual realizations

Figure 1 illustrates simulated breakthrough curves (solid) for two individual fracture network realizations in a 20 m long domain. The standard deviation of the channel transmissivity σ_{lnT} takes on the value of 0 and 4 in Figures 1a and 1b, respectively. Dashed curves represent the fitted AD-model.

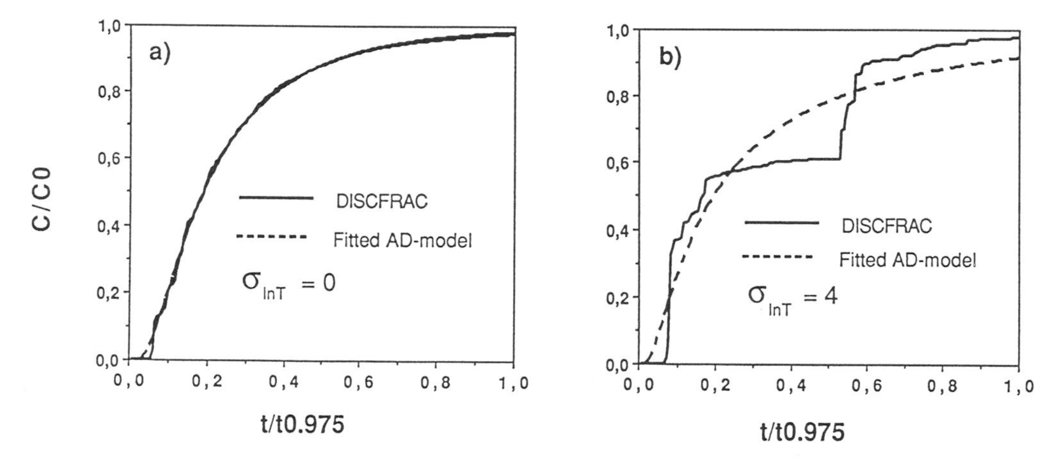

FIGURE 1. Breakthrough curves (solid) from simulations with a) σ_{lnT} equal to 0 and b) σ_{lnT} equal to 4. Dashed curves represent fitted AD-model.

The variability of the channel transmissivity has several effects on a breakthrough curve. For the cases with zero variability the particles are evenly distributed in the fracture network which results in a relatively smooth breakthrough curve (see Figure 1a). The tailing of the breakthrough curve is exclusively due to varying lengths of the transport paths through the network. The fitted AD-model is in general a good representation of the simulated breakthrough curve, except for small discrepancies, usually at early breakthrough times.

Increasing the value of σ_{lnT} results in a redistribution of the transport paths. Transport concentrates to fewer high flux channels in the network. This is reflected in the breakthrough curve as fewer and more pronounced peaks. Furthermore, large transmissivity contrasts produce significant dispersion which results in tailing and longer median transport time, $t_{0.50}$. The fit of the AD-model gets worse with increased variability of the channel transmissivity. Occasional realizations, with σ_{lnT} equal to 4 or 6, result in anomalously irregular breakthrough curves with one or only a few pronounced peaks. Figure 1b illustrates the cumulative breakthrough curve from one of these cases. The smooth curve of the fitted continuum model (dashed) clearly fails to represent such irregular breakthrough.

Scale dependences

Estimation of effective transport parameters and model predictions are often made on different scales. The length scale of a tracer migration tests providing effective transport parameters is typically less than a few tenths of meters whereas transport predictions may be desired over several hundreds of meters of rock. Therefore, knowledge of potential scale effects on the effective transport parameters is crucial for reliable transport predictions.

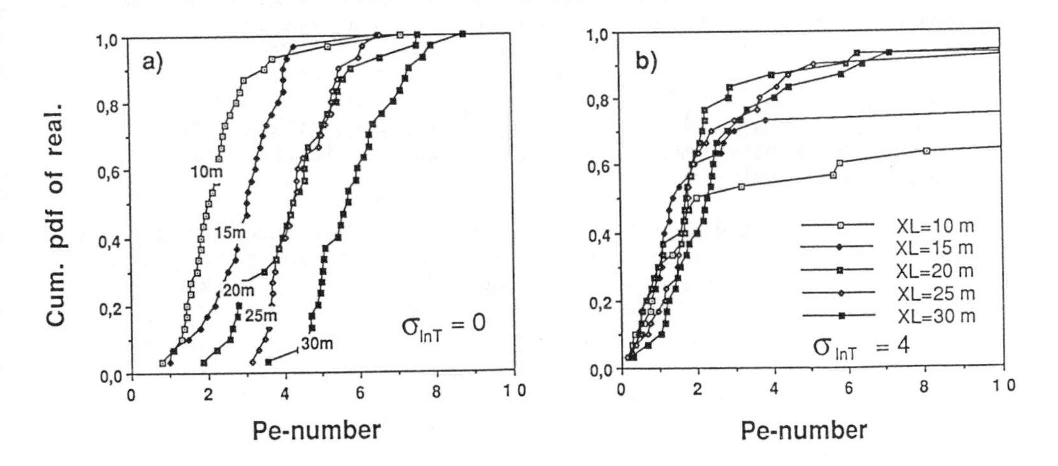

Figure 2. Peclet number in individual network realizations. One curve of 30 realizations for each transport distance. The value of σ_{lnT} equals 0 and 4 in Figures 2a and 2b, respectively.

Figure 2a illustrates the evolution of Peclet number with increasing transport distance when the variability of the channel transmissivity is zero. It contains a cumulative distribution curve of Peclet numbers for each of the five transport distances. Each curve represents the values from 30 different fracture network realizations. The ensemble mean Peclet number increases linearly with increasing transport distance. Hence, the dispersion coefficient, D_L, is constant and specific for the studied medium which implies a Fickian type of dispersion behaviour. The standard deviation of the Peclet number is roughly constant which means that the relative variability decreases with increasing transport distance. These results imply that the possibilities of estimating an effective Peclet number for predictive use are quite good. Furthermore, the smallest scale of the synthetic experiments, X_L equal to 10 m, is not too small to provide reliable estimates of an effective Peclet number. The dispersion length α, equation 2, is constant and approximately equal to 5 m.

For non-zero variability of the channel transmissivity the regular trends observed above becomes less obvious while channeling effects become more important. Figure 2b illustrate the evolution of dispersion when σ_{lnT} equals 4. As in Figure 2a the Peclet numbers are presented in cumulative distribution curves, one curve of 30 realizations for each transport distance. The curves in Figure 2b show that the trend towards higher Peclet numbers with increasing transport distance, disregarding realizations with anomalously high Peclet number, becomes less pronounced. In fact, there is no increase at all when σ_{lnT} equals 4 and 6 which implies a non-Fickian dispersion behaviour. The value of the dispersion length α is in the same order or exceeding the largest scale of the studied transport problem. Thus an increase in domain length, X_L, also results in an increased dispersion length; hence constant Peclet number. A constant Peclet number also indicates that mixing between flow channels is negligible. This is due to flow channeling which concentrates flow and transport to preferential, poorly interconnected, flow paths with low resistance to flow. An interesting extension of this analysis would be to establish if the concept of a constant dispersion length becomes valid on larger scales. However, this would require further simulations in a significantly larger domain, say above 100*100*100 m^3. At any rate, the above results demonstrate that the use of effective transport parameters, such as Peclet number, for predictive use may become hazardous if the variability of the channel transmissivity is high.

Another important feature of the curves in Figure 2b is the occurrence of occasional anomalously low-dispersive realizations with very high Peclet numbers, hence extreme channeling. The frequency of these anomalous realizations increases with increasing value of σ_{lnT}. More than 80% of all realizations in Figure 2b have Peclet numbers below 5 whereas occasional channeling realizations results in Peclet number exceeding 100. Obviously, the concept of ensemble average becomes meaningless in the presence of such anomalous values. Instead, extreme channeling must be explicitly accounted for when making transport predictions.

Comparison with field data

The Stripa 3D experiment comprises migration experiments on different scales. Different non-sorbing tracers were injected at different levels in three vertical boreholes above the experimental drift. Water and tracers emerging into the drift were continuously recorded in plastic sampling sheets covering the ceiling and walls of the drift. Each sheet cover $2m^2$ of rock.

Abelin et al., 1987 evaluate the migration experiments by fitting different continuum models to experimental breakthrough curves. The following comparison uses the results for the one-dimensional advection-dispersion (AD-) model.

The experimental results give the impression of a very erratic and unstructured transport behaviour; the tracers are very unevenly distributed in the drift, the Peclet numbers are consistently very low (typically below 4) except for a few anomalous values and the water residence times show no correlation with the distance between the drift and the injection point. However, the synthetic experiments with the discrete model indicate that this transport behaviour is to be expected in fractured rock with non-zero variability of the flow channel transmissivity. In fact, the simulations reveal a structure in the transport behaviour that can be recognized in the seemingly erratic experimental results.

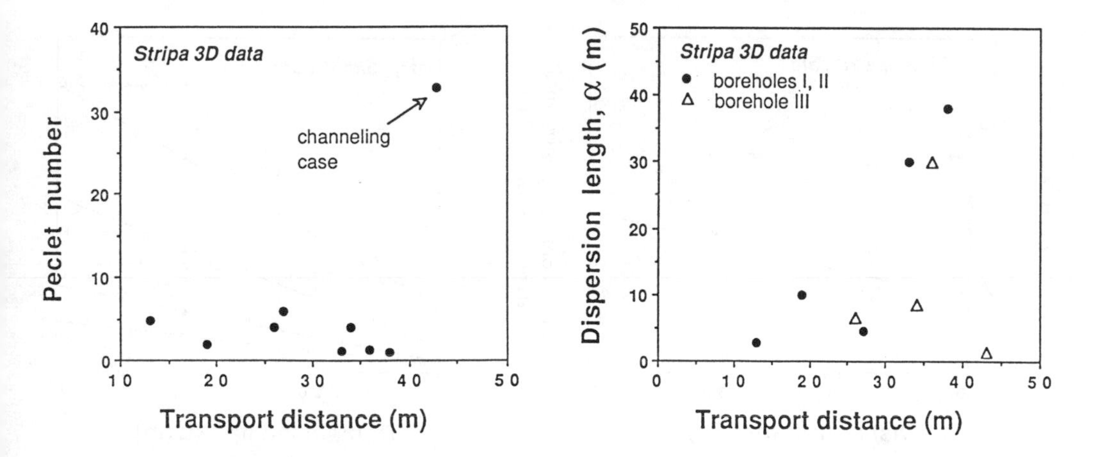

Figure 3. a) Peclet number and b) dispersion length as a function of distance between injection point and sampling area in the Stripa 3D experiment.

Figure 3a is a plot of Peclet number versus transport distance for the field experiment. Apart from one anomalously high value, the Peclet number is essentially constant with increasing transport distance which implies a non-Fickian dispersion behaviour in the investigated rock. These results are consistent with the synthetic experiments with σ_{lnT} equal to 4 (see Figure 2b). Figure 3b displays the evolution of the dispersion length, α, with transport distance. The tracers injected in boreholes I and II appear to give a linear increase of α with increasing transport distance whereas the tracers injected in borehole III result in a more erratic behaviour with some very small values of α at large transport distances. However, a high value of the variability of the channel transmissivity would explain the deviating results for borehole III. Simulations with σ_{lnT} equal to, or exceeding, 4 in the 30 m long domain demonstrates that the dispersion length in individual realizations with pronounced channeling effects may be less than 2 m although the typical value is an order of magnitude larger. Borehole III also deviates from the other two boreholes in terms of a significantly larger water inflow rate. Dverstorp and Andersson (1989) show that the calibrated discrete model for flow, with σ_{lnT} equal to 2, cannot explain this anomaly. However, increasing the value of σ_{lnT} to e.g. 4 might sufficiently increase the variability of the model results to cover the hydraulic properties of borehole III.

The water residence times of the different tracer experiments show no correlation with the distance between the injection point and sampling area. Again, this is consistent with the simulation results because the spread of water residence times, t_w, between realizations is very large compared to the differences in ensemble mean t_w between the different transport distances when σ_{lnT} is large. The high variability thus conceals the trend towards larger t_w with increasing transport distance

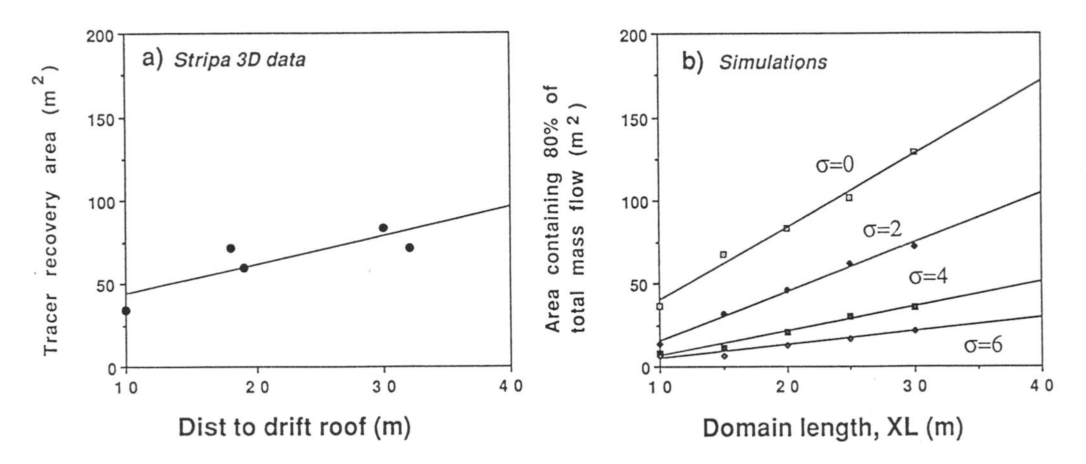

FIGURE 4. a) Tracer recovery area in the ceiling of the experimental drift as a function of distance between injection point and drift roof b) Simulated tracer recovery area as a function of domain length X_L and variability of channel transmissivity, σ_{lnT}.

The areal distribution of tracers is very uneven both in the field experiment and in the synthetic experiments (with σ_{lnT} equal to 4 or 6) which makes it difficult to quantify the lateral dispersion. However, comparing the tracer recovery area in the field experiment with the simulations gives valuable information on σ_{lnT}. Figure 4a shows the recovery area in the ceiling of the drift as a function of the distance between the injection point and the ceiling of the drift. The recovery area was calculated by summing the areas of the sampling sheets containing measurable amounts of tracers. Figure 4b is a corresponding plot of the area for 80% mass recovery on the out flow face in the simulations. The simulations with σ_{lnT} equal to 4 closely match the slope of the experimental curve in Figure 4a and thus explain the lateral dispersion observed. The shift along the ordinate is due factors like tracer detection limit, the chosen mass recovery percentage in Figure 4b and exaggerated mass recovery due to the no-flow boundary conditions in the synthetic experiments.

CONCLUSIONS

Synthetic transport experiments, performed with a discrete fracture network model, explain the seemingly erratic results from the Stripa 3D field migrations experiment. The need to identify the variability of the channel transmissivity, σ_{lnT}, is one of the most important conclusions from this analysis. High values on σ_{lnT}, say above 2, results in non-Fickian dispersion behaviour and extreme channeling effects which severly reduce the possibilities of predicting transport with the continuum model.

The synthetic experiments demonstrate that the Peclet number may be identical in two different realizations despite different values on σ_{lnT}. However, the Peclet number evolves differently with increasing scale of the transport problem depending on the value of σ_{lnT}. Thus, transport predictions based on parameter estimates from a single migration experiment will be most uncertain unless σ_{lnT} otherwise is known. Furthermore, the variability of effective transport parameters like Peclet number and water residence time increases rapidly with increasing value on σ_{lnT}. Thus parameter estimates from a single migration experiment may become extremely sensitive to statistical variations. These results strongly motivate multiple migration experiments performed on different scales (e.g. different distances between injection and sampling area) when estimating effective transport parameters in fractured rock.

REFERENCES

Abelin H., L. Birgersson, J Gidlund, L. Moreno, I. Neretnieks, H. Widen and T Ågren, 3-D Migration experiment, Report 3, Performed experiments, Results and evaluation, *Stripa Tech. rep. TR 87-21*, Svensk Kärnbränsleförsörjning, 1987.

Andersson J. and B. Dverstorp, Conditional simulations in three-dimensional networks of discrete fractures, Water Resour. Res., 23(10), 1876-1886, 1987.

Cacas M. C., E. Ledoux, G. de Marsily, A. Barbreau, P. Calmels, B. Gaillard and R. Margritta, Modeling fracture flow with a Stochastic discrete fracture network: Calibration and validation, 2. The transport model, Water Resour. Res. 26(3), 491-500, 1990.

Dverstorp B. and Andersson J., Application of the discrete fracture network concept with field data: possibilities of model calibration and Validation, Water Resour. Res., 25(3), 540-550, 1989.

Moreno L., C. F. Tsang, Y. Tsang and I. Neretnieks, Some anomalous features of flow and transport arising from fracture aperture variability, submitted to Water Resour. Res., September, 1989.

Tsang Y. W., C. F. Tsang, Flow channeling through strongly heterogeneous permeable media, submitted to Water Resour. Res., January, 1989.

MODELLING ACTIVITIES FOR THE GROUNDWATER TRACER TEST PROGRAM AT THE WHITESHELL NUCLEAR RESEARCH ESTABLISHMENT BOREHOLE SITE

N.W. Scheier, L.H. Frost, C.C. Davison
Atomic Energy of Canada Limited
Whiteshell Nuclear Research Establishment
Pinawa, Manitoba, Canada R0E 1L0

ABSTRACT

This paper describes a study of the large-scale groundwater flow and solute transport properties of a major low-dipping fracture zone in a granitic rock mass. A porous media conceptual model of the fracture zone was constructed and the MOTIF finite-element computer code was used to solve the equations describing flow and transport. The hydraulic properties of the model could be reasonably calibrated to simulate a long-term pumping test. The calibrated flow model gave fair predictions of hydraulic conditions during a subsequent conservative tracer test. The model solute transport properties were then calibrated using tracer breakthrough data. The study shows strong evidence that distinct channelling and high porosities exist within the fracture zone.

ACTIVITES DE MODELISATION LIEES AU PROGRAMME D'ESSAI DE TRACAGE DE L'EAU SOUTERRAINE AU SITE DE FORAGE DE L'ETABLISSEMENT DE RECHERCHES NUCLEAIRES DE WHITESHELL

RESUME

Dans la présente communication, on décrit une étude de l'écoulement d'eaux souterraines à grande échelle ainsi que les propriétés de migration des solutés d'une vaste zone de fissures à faible pendage dans un massif granitique. On a simulé la zone fissurée au moyen d'un modèle mathématique d'un milieu poreux et on s'est servi du programme de calcul à éléments finis, MOTIF, pour résoudre les équations décrivant l'écoulement et la migration. On pourrait étalonner avec assez de précision les propriétés hydrauliques du modèle pour simuler un essai de pompage de longue durée. Le modèle d'écoulement étalonné a prédit correctement les conditions hydrauliques au cours d'un essai classique postérieur aux traceurs. On a ensuite étalonné les propriétés de migration des solutés du modèle à l'aide de données sur le passage des traceurs. L'étude fournit la preuve solide qu'il y a des cheminements (voies) individuels et une forte porosité dans la zone de fissures.

INTRODUCTION

The Canadian Nuclear Fuel Waste Management program is currently assessing the concept of disposing of nuclear fuel waste in a vault deep in plutonic rock of the Canadian Shield. Transport by groundwater moving through fractures in the rock is the most likely pathway for radionuclides to reach the surface environment. As part of its investigations, Atomic Energy of Canada Limited (AECL) in cooperation with the Power Reactor and Nuclear Fuel Development Corporation Japan, has been studying the large-scale groundwater flow and solute transport properties of a major low-dipping fracture zone in a granitic rock mass at the Whiteshell Nuclear Research Establishment (WNRE) borehole site.

SITE CHARACTERISTICS

The WNRE borehole site is situated on the southern margin of the Lac du Bonnet Batholith, a large granitic pluton in southeastern Manitoba. An existing set of boreholes defining an east-west section was used in this study (Figure 1). No direct geological or hydrogeological information is available for the region perpendicular to the section.

A major, eastward dipping, subhorizontal fracture zone of variable thickness and fracture intensity was identified as a hydraulic connection between boreholes WN-1, WN-4, WN-8 and WN-10 (Figure 1). The natural groundwater flow within the zone appears to be westward with a small component of vertical leakage to the fracture zone from above and below.

Short-term single-hole permeability and interborehole pulse tests estimated the near-field transmissivity of the fracture zone surrounding boreholes WN-1 and WN-4 to be 2×10^{-6} m^2/s, whereas the average transmissivity for the zone between boreholes WN-8 and WN-1 or WN-4 was about 3×10^{-4} m^2/s. In view of this inhomogeneity, a numerical modelling approach was adopted for proper analyses of long-term pumping and tracer tests within this zone.

Further information on the site investigations is presented in Frost, Scheier and Davison [1] and the references made therein.

MATHEMATICAL MODEL

Throughout this analysis it was assumed that the fracture zone can be modelled as an equivalent porous medium with respect to fluid flow and solute transport. This approach has been validated for fluid flow in similar fracture zones at the nearby Underground Research Laboratory [2]. It was also assumed that vertical leakage from the adjacent rock mass is insignificant. Therefore a two-dimensional, depth-integrated representation of the fracture plane was adopted.

The equation describing fluid flow is, in indicial notation:

$$\frac{\partial}{\partial x_i} T_{ij} \frac{\partial h}{\partial x_j} = S \frac{\partial h}{\partial t} \quad (1)$$

where x_i = Cartesian coordinates, T_{ij} = transmissivity tensor, h = piezometric head, S = storativity, and t = time.

The transmissivity can be expressed as:

$$T_{ij} = \frac{k_{ij} \rho g b}{\mu} \quad (2)$$

where k_{ij} = permeability tensor, ρ = fluid density, g = gravitational acceleration, b = fracture zone thickness, and μ = dynamic viscosity.

The storativity can be expressed as:

$$S = \rho g \, [(1-\Theta)\alpha + \Theta\beta]b \qquad (3)$$

where Θ = porosity, α = compressibility of rock within the fracture zone, and β = fluid compressibility.

Assuming the flow porosity and transport porosity are equal then the interstitial velocity components are:

$$v_i = \frac{T_{ij}}{\Theta b} \frac{\partial h}{\partial x_j} \qquad (4)$$

The equation describing conservative solute transport is:

$$\frac{\partial}{\partial x_i} (D_{ij} \frac{\partial C}{\partial x_j}) - \frac{\partial}{\partial x_i} (v_i C) = \frac{\partial C}{\partial t} \qquad (5)$$

where D_{ij} = dispersion coefficient tensor and C = solute concentration.

The dispersion coefficients are:

$$D_{ij} = a_T v \delta_{ij} + (a_L - a_T) \frac{v_i \; v_j}{v} + D_o \tau \delta_{ij} \qquad (6)$$

where a_T = transverse dispersivity, v = interstitial velocity magnitude, δ_{ij} = Kronecker delta, a_L = longitudinal dispersivity, D_o = molecular diffusion coefficient, and τ = tortuosity.

The equations have been solved using the finite-element computer code MOTIF developed by AECL. The ability of MOTIF to correctly model fluid flow in a fractured rock mass has been validated by predicting the hydrogeological impact of excavating the access shaft for the Underground Research Laboratory [2].

The region of the fracture zone included in the model was a rectangular area measuring 2549 m from east to west and 2336 m from north to south. As there is little information available on boundary conditions, the model boundaries were located at sufficient distances to ensure that boundary conditions did not significantly influence predictions near the tracer test site. The natural groundwater flow appears to be from east to west towards the Winnipeg River; therefore a no-flow condition was assumed for the north and south boundaries. The east and west boundaries were assumed to have constant heads estimated by linear extrapolation from the measured values at boreholes WN-1 and WN-10.

Meshes and time-stepping sequences of differing refinement were tested to check the numerical convergence of predictions. Figure 2 illustrates the central portion of the mesh selected for most calculations.

LONG-TERM PUMPING TEST

In order to better evaluate the hydraulic characteristics of the major fracture zone, a 9 day pumping test was performed. Groundwater was pumped from the fracture zone through borehole WN-8 at a near-constant rate of 30 L/min. A maximum piezometric head drawdown of 3.53 m was observed in the fracture zone at the pumping well compared to 2.45 m, 2.81 m and 2.70 m in observation wells WN-1, WN-4 and WN-10, respectively (Figure 3).

The results of the pumping test were used to calibrate the numerical model. Initially the transmissivity of the major fracture zone was assumed to be isotropic with a value of 4×10^{-4} m^2/s, except for the area within a 15 m radius of boreholes WN-1 and WN-4 that was assigned a value of 2×10^{-6} m^2/s. The storativity was assumed to be a uniform 9×10^{-4}. The initial model predicted the drawdown at the pumping well, borehole WN-8, reasonably well but underpredicted the drawdowns at WN-1, WN-4 and WN-10 by an order of magnitude.

The relatively high drawdown at the distant boreholes in comparison to that at the pumping well (between 69% and 80%) suggested that the boreholes were within, or close to, an east-west trending, high permeability channel in the fracture zone. Similar large scale permeability channelling has been observed elsewhere in major fracture zones within the Lac du Bonnet Batholith [3]. Therefore, a high permeability channel was added to the model. After considerable calibration by adjusting transmissivity, storativity and channel geometry, a good match of the maximum drawdown was obtained for all boreholes (Figure 3). Although, both the drawdown and recovery were somewhat underpredicted by the model at other times, further model calibration was not considered warranted at this stage due to the lack of sufficient geological and hydrogeological data to establish a unique parameter set.

The final configuration of the high permeability channel is shown in Figure 2. Its transmissivity was estimated to be 5×10^{-3} m^2/s. The transmissivity of the surrounding region of the fracture zone was estimated to be 2×10^{-6} m^2/s, the value initially estimated for the area surrounding boreholes WN-1 and WN-4. A uniform storativity of 9.7×10^{-4} was estimated for the entire zone.

TRACER TEST

The calibrated model was used to help plan a conservative groundwater tracer test within the fracture zone between boreholes WN-4 and WN-8. Transient groundwater flow during the test was simulated to estimate piezometric drawdowns and groundwater fluxes. Tracer transport was then simulated to predict possible breakthrough times and concentrations at boreholes.

During the tracer test, groundwater was pumped from the fracture zone through borehole WN-8 at a near constant rate of 10 L/min, and a pulse of 3850 L of 1900 mg/L iodide tracer was injected into the fracture zone at borehole WN-4 285 m away.

The predicted drawdown in piezometric head at boreholes WN-8 and WN-1 is shown in Figure 4. The injection of the pulse of tracer at WN-4 had a negligible impact on these predictions except in the short term at WN-4.

Two tracer transport simulations were examined: one using an assumed longitudinal dispersivity value of 50 m and the other a value of 100 m. The ratio of longitudinal to transverse dispersivity was assumed to be 10:1. Molecular diffusion was neglected due to the relatively high velocities predicted. The "effective thickness", the product of porosity and fracture zone thickness, was assumed to be 0.1 m. For the case of a longitudinal dispersivity of 50 m, the model predicted the arrival of 0.1 mg/L of iodide approximately 157 days into the test and a peak of 1.1 mg/L at 425 days. For the case of a longitudinal dispersivity of 100 m, the arrival time of 0.1 mg/L was reduced to 83 days and the peak concentration was reduced to 0.6 mg/L at 200 days.

The piezometric heads measured in the fracture zone during the tracer test gradually declined towards a steady state condition. The responses at boreholes WN-8 and WN-1 are shown in Figure 4. The model slightly underpredicted drawdown at early times. After 10 days for the pumping well (WN-8) and 15 days for the observation boreholes, the predicted drawdowns exceeded those measured and a steady divergence occurred.

The tracer test was terminated after 338 days with no detectable breakthrough of iodide tracer at the withdrawal borehole (WN-8). This suggested that the "effective thickness" of the fracture zone was greater than 0.1 m and/or the longitudinal dispersivity was less than 50 m.

The drawdowns measured during the tracer test were then used to recalibrate the model transmissivity, storativity and channel geometry. In the final recalibrated model the channel geometry was unchanged. The transmissivity of the high permeability channel was reduced to 1×10^{-3} m^2/s from 5×10^{-3} m^2/s; the transmissivity of the surrounding rock mass was increased to 8×10^{-6} m^2/s from 2×10^{-6} m^2/s; and the storativity was decreased slightly from 9.7×10^{-4} to 9.0×10^{-4}. The revised predictions of drawdown at boreholes WN-8 and WN-1 showed close agreement with the measured values (Figure 4). The small differences that still existed between the measured drawdowns and those predicted with the recalibrated model, particularly in the far-field, were likely caused by spatial variations in the hydrogeological character of the major fracture zone and/or groundwater leakage from the rock mass adjacent to the major fracture zone.

TRACER RECOVERY

Nineteen days after the pumping from borehole WN-8 was stopped, withdrawal from the tracer injection zone of borehole WN-4 was started to obtain more information on the nature of the iodide transport in the vicinity of WN-4. The pumping rate was 0.4 L/min for the first 15 days and 1.3 L/min for the next 62 days.

The recalibrated model was used to predict the groundwater flow field following the end of pumping from borehole WN-8. Figure 5 shows the predicted and measured recovery in piezometric heads in the fracture zone at borehole WN-8. The predictions and measurements are reasonably close.

The model transport parameters were then recalibrated using iodide concentrations measured in the groundwater recovered from borehole WN-4. Measured and predicted concentrations are presented in Figure 6. The predictions were based on the revised flow field, an "effective thickness" of 1.0 m, longitudinal dispersivities ranging from 1 to 50 m and a ratio of longitudinal to transverse dispersivity of 10:1. The longitudinal dispersivity within the major fracture zone in the vicinity of borehole WN-4 appeared to be about 5 m.

CONCLUSIONS

The hydraulic and transport parameters of a two-dimensional model of the plane of the fracture zone were reasonably calibrated to simulate experimental results. The basic parameters permeability, porosity and rock compressibility were calculated using the calibrated transmissivity, "effective thickness" and storativity, together with an estimated fracture zone thickness. Based on the geological logs of boreholes WN-1, WN-4 and WN-8, a fracture zone thickness of 10 m was assumed. Therefore, the estimated permeabilities were 1×10^{-11} m^2 for the highly conductive channel and 8×10^{-14} m^2 for the remainder of the fracture zone. These values fall within the range 10^{-11} to 10^{-15} m^2 calculated for similar fracture zones in the Lac du Bonnet granite batholith at the Underground Research Laboratory [4]. The estimated porosity was 0.1. This value is higher than some presented in the literature, (e.g., Anderson et al. [5] report values between 3.8×10^{-3} and 5.9×10^{-2} for a similar fracture zone in granitic rock in Sweden); however, it is consistent with other observations in the Lac du Bonnet Batholith. The estimated rock compressibility of 10^{-8} Pa^{-1} is at the high end of the range for jointed rock reported by Freeze and Cherry [6]. The calibrated longitudinal dispersivity of 5 m is within the range reported by Anderson et al. [5] for similar experiments conducted over distances of 155 to 189 m. Because of the limited amount of data available from the tracer experiment, it was not possible to make an estimate of the transverse dispersivity value.

The calibrated flow model was moderately successful in predicting different pumping scenarios. It should be also noted that it may be possible to construct other models that simulate the field measurements equally well. Insufficient tracer data made it impossible to attempt to validate the solute transport model.

In general, this study has improved our knowledge of groundwater flow and solute transport in fractured granite. There is strong evidence that distinct channelling and high porosities can exist within major fracture zones. A porous medium representation may be appropriate for modelling both flow and transport on the scale of hundreds of metres in such media; however, there is a need to obtain good tracer transport data from field tests performed at these large scales to validate these models.

REFERENCES

[1] Frost, L.H., N.W. Scheier and C.C. Davison: "Summary of the Groundwater Tracer Test Program at the WNRE Borehole Site - 1987 to 1989", to be published as Atomic Energy of Canada Limited Technical Record*.

[2] Guvanasen, V., J.A.K. Reid and B.W. Nakka: "Predictions of Hydrogeological Perturbations due to the Construction of the Underground Research Laboratory," Atomic Energy of Canada Limited Technical Record TR-344*, 1985.

[3] Davison, C.C. and E.T. Kozak: "Hydrogeological Characteristics of Major Fracture Zones in a Large Granite Batholith of the Canadian Shield", Proc. Fourth Canadian/American Conference on Hydrogeology, National Water Well Association, Dublin, Ohio, 1988, p. 52-59.

[4] Davison, C.C.: "Monitoring Hydrological Conditions in Fractured Rock at the Site of Canada's Underground Research Lab", Ground Water Monitoring Review, 4 (4), 1984, 95-102.

[5] Andersson, J-E, L. Ekman, E. Gustafsson, R. Nordqvist, and S. Tiren: "Hydraulic Interference Tests and Tracer Tests within the Brändan Area, Finnsjön Study Site. The Fracture Zone Project-Phase 3", SKB Technical Report 89-12, Swedish Nuclear Fuel and Waste Management Co., Stockholm, 1989.

[6] Freeze, R.A. and J.A. Cherry: Groundwater, Prentice-Hall Inc., Toronto, 1979.

---

* Unrestricted, unpublished report available from SDDO, Atomic Energy of Canada Limited Research Company, Chalk River, Ontario, Canada KOJ 1JO.

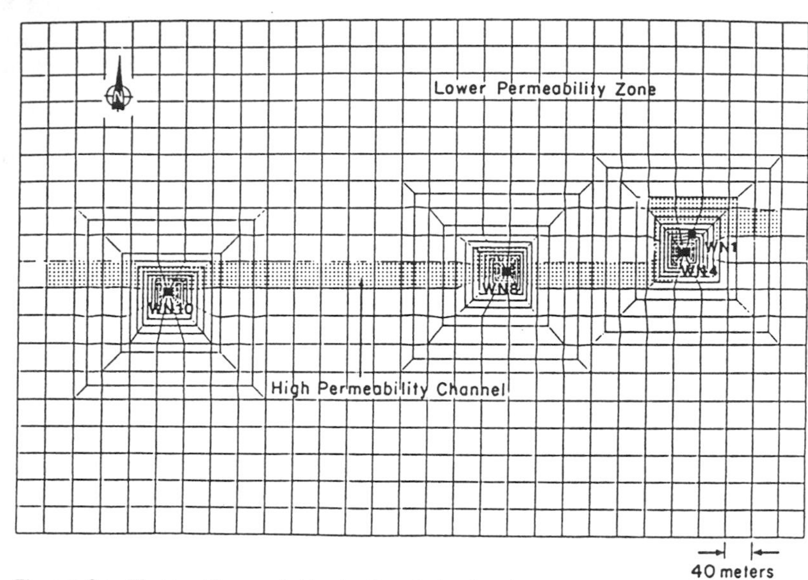

Figure 2: Finite Element Mesh-Borehole Region

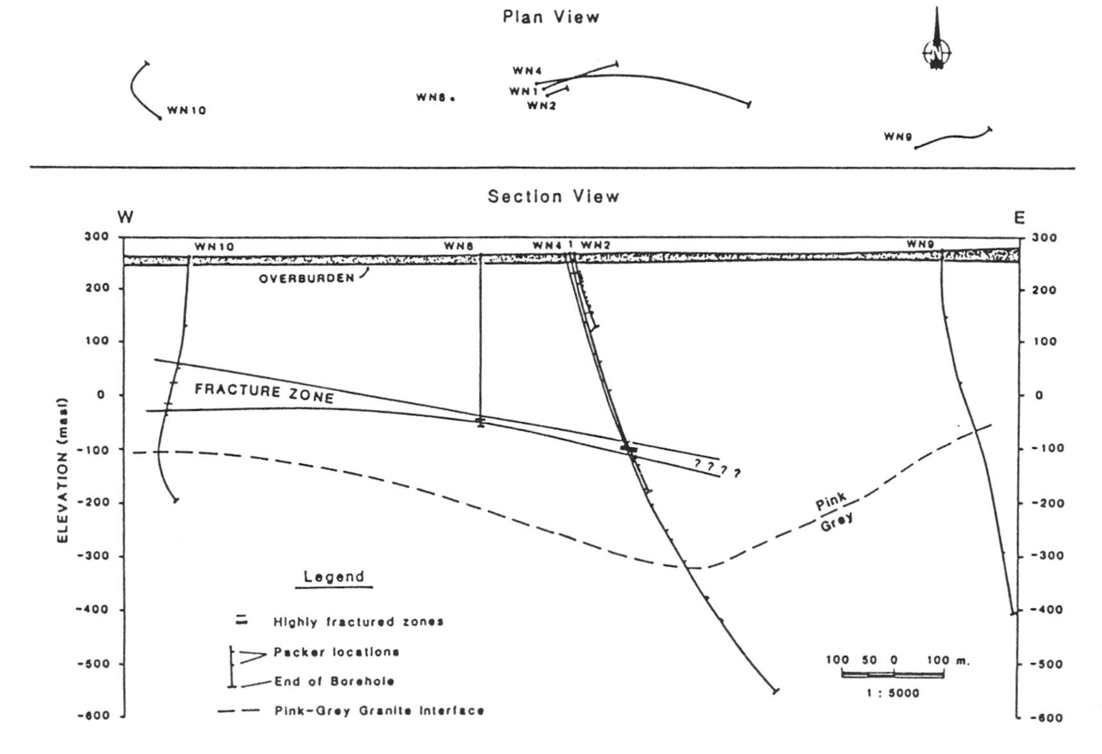

Figure I: WNRE Borehole Site

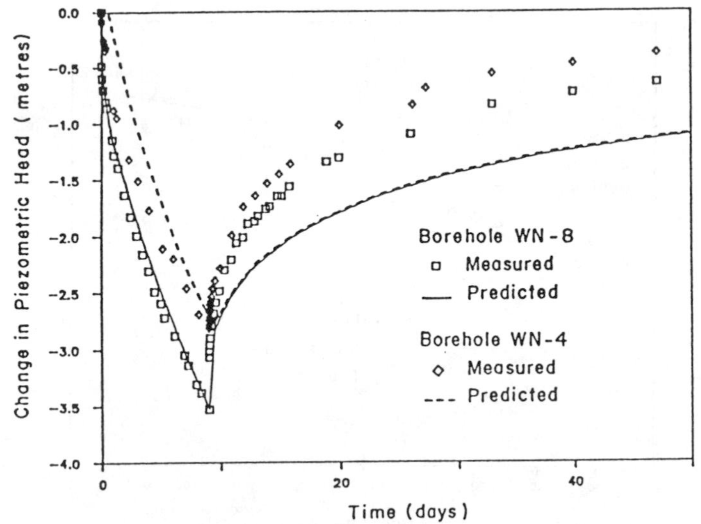

Figure 3: Long - Term Pumping Test

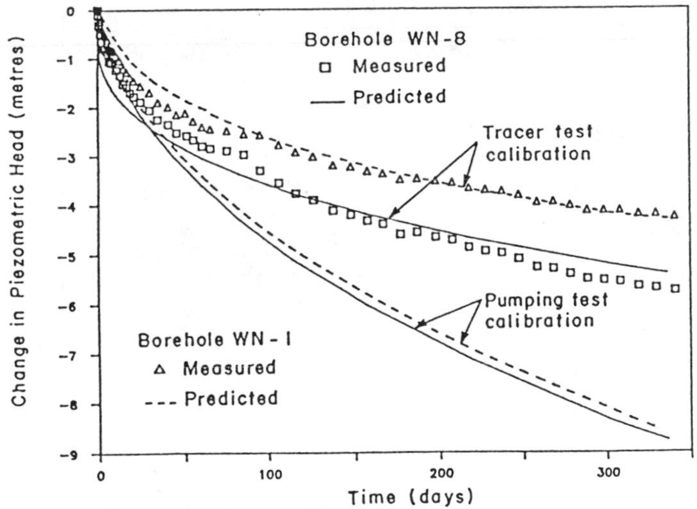

Figure 4: Tracer Test

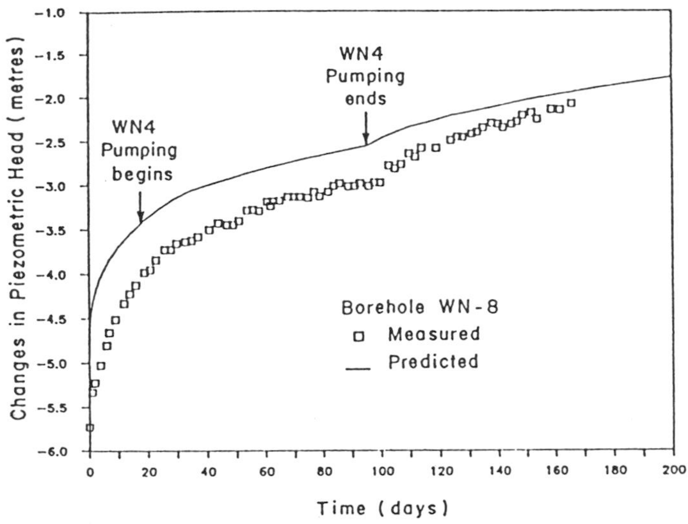

Figure 5: Piezometric Head Recovery

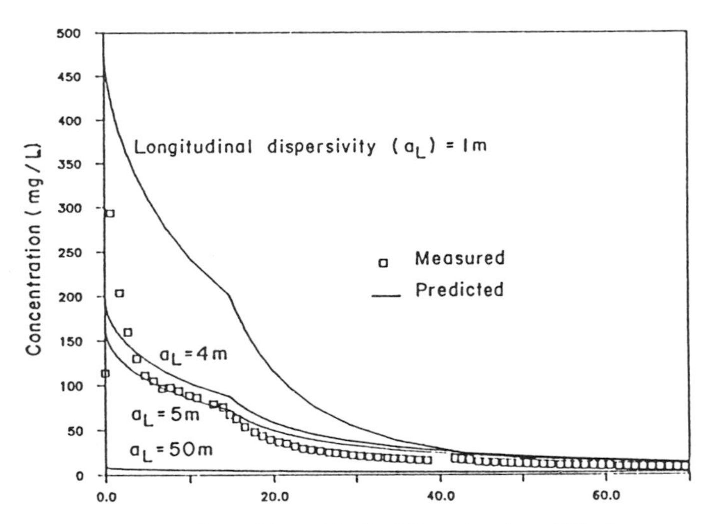

Figure 6: Iodide Recovery at Borehole WN-4

RESULTS OF A CHANNELLING EXPERIMENT IN STRIPA

Harald Abelin, Lars Birgersson, Thomas Ågren
CHEMFLOW AB, Stockholm, Sweden

Ivars Neretnieks, Luis Moreno
Department of Chemical Engineering
Royal Institute of Technology, Stockholm, Sweden

ABSTRACT

A set of experiments has been performed in the Stripa experimental mine to study channeling in individual natural fractures on a scale of 2 m. Two different techniques were used.

The so called single hole experiments were made by drilling 2 m long holes in the plane of some fractures and measuring the injected flowrate into the fracture in some detail. By using a specially designed packer the fracture running along the hole could be pressurized in 5 cm long sections. The flowrate into the fracture was measured in about 40 sections on each side of the hole. The results clearly show that the water flowrate varies considerably along the fracture.

In the so called double hole experiment two holes were drilled in the same fracture plane and between-hole-tests were made using pressure pulses by pressurizing the whole or parts of one hole and monitoring the pressure response in the other hole. In addition tracer tests were made between the holes using 5 different tracers.

RESULTATS D'UNE EXPERIENCE DE CANALISATION MENEE A STRIPA

RESUME

Une série d'expériences a été réalisée dans la mine expérimentale de Stripa pour étudier les phénomènes de canalisation dans des fissures naturelles individuelles sur une longueur de 2 m. Deux techniques différentes ont été employées.

Les expériences dans un forage unique ont consisté à creuser des forages d'une longueur de 2 m dans le plan de quelques fissures et à mesurer avec précision le débit du fluide injecté dans la fissure. Grâce à un obturateur spécialement conçu, il a été possible de mettre sous pression des tronçons de 5 cm de long de la fissure parallèle au forage. Le débit dans la fissure a été mesuré dans environ 40 tronçons de chaque côté du forage. Les résultats font clairement apparaître des variations considérables du débit le long de la fissure.

Dans l'expérience de mesures entre deux forages, deux forages ont été exécutés dans le même plan de fissure et des essais sont effectués dans l'espace qui les sépare, le principe consistant à mettre par intermittence sous pression l'ensemble ou certaines parties de l'un des forages et de surveiller les variations de pression dans l'autre. De plus, des essais de traçage utilisant cinq traceurs différents ont été réalisés entre les forages.

BACKGROUND AND INTRODUCTION

There has been an increasing interest in describing flow and transport in low permeability fractured rock recently because many countries are considering to build repositories for high level nuclear waste at large depths in the ground in different rock types including crystalline rock. Crystalline rock is fractured and most of the water flow takes place in the fractures. Lately is has been recognized that most of the water flows only in a small part of every fracture and that this may have a strong impact on the transport of escaping radionuclides [1,2,3,4]. The water flowpaths in the fractures may connect to form a network of pathways, some of which may be faster than others. The surface area of the fractures which is in contact with the mobile water will determine how much surface area is available for sorption and retardation of the nuclides.

The channeling experiments were designed to study the transmissivity and aperture variations in fractures at depth in crystalline rock. Two types of experiments were designed. In the single hole experiments a hole was drilled more than 2 m into the plane of the fracture and the flowrates were measured in 50 mm sections using a specially designed injection packer. Photographs were also taken inside the hole along the fracture to determine the visible fracture aperture and to obtain other information. In the double hole experiment two parallel holes were drilled in the plane of a fracture at a center distance of nearly 2 m. Hydraulic tests and tracer tests were made between the two holes to obtain information on connections in the plane of the fracture and to obtain information on residence time distributions in different paths (channels) [5].

EXPERIMENTAL DESIGN

To be able to investigate the fracture characteristics along a line in the fracture plane, a large diameter (200 mm) hole was drilled along the fracture plane to a depth of about 2.5 m from the face of the drift, see Figure 1.

Water was injected with a constant overpressure all along the intersection of the fracture. The equipment allowed for individual flowrate monitoring over 50 mm sections, 20 at a time, with a resolution of 1/100 ml per hour. The "left" and "right" side of the hole was tested separately. The injection tests gave information on flow characteristics and their distribution in the fracture plane along two lines, 200 mm apart. These tests are called the single hole experiments and have been performed in 5 different fractures.

One of the fractures used in the single hole tests has been investigated in more detail, with measurements using a second hole, parallel to the first, at a distance of 2 m. Pressure pulse tests and tracer tests were made between the holes. Different tracers were injected at different sections along the fracture.

Prior to any test, the boreholes were photographed all along the fracture visible in the hole. These enlarged photos were used, i.e. to measure the visual aperture of the fracture along the two sides (both sides of the borehole) that later were subjected to water injection.

The single hole experiments

The single hole experiments gave information on flowrate variations over 50 mm sections on both sides of the fracture and in addition photographs made on totally 12 fractures gave information on fracture aperture variations and number of fracture intersections.

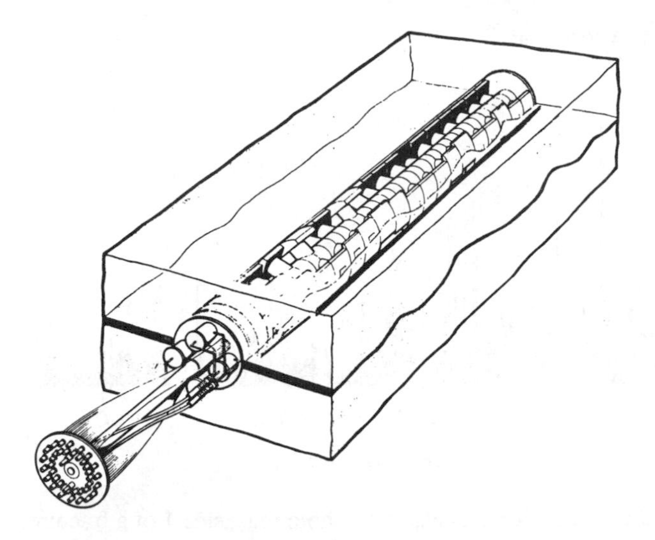

Figure 1. A fracture with an investigation borehole and the Multipede packer.

These data were analyzed to obtain means and variances of the visible aperture and variograms. In addition the flowrate measurements were used to determine means, variances and variograms.

Double hole experiment

Pressure pulse tests between different sections in the two holes in the same fracture were obtained and connections identified. The injection hole was pressurized either in one 50*50 mm window, the rest of the hole being closed off by an inflatable packer, or in other tests by pressurizing the whole hole. The rubber "feet" of the Multipede packer were used to monitor the pressure in the receiving hole. The rest of the monitoring hole was kept at zero head. The fracture was sealed at the face of the drift. The pressure connection information formed the basis for the design of the tracer tests from one hole to the other. 5 different tracers were injected at different locations in one of the holes and monitored for in the other hole.

RESULTS AND INTERPRETATION

The single hole experiments gave information on visual apertures and Figure 2 below shows an example of these data for one side of a fracture extending nearly 2.5 m into the rock. The measurements are made as averages over 10 mm length along the fracture. Figure 3 shows the variation of injected flowrate in 50 mm sections along both sides of the same fracture. Figure 4 shows the number of fracture intersections on side 1 of the same hole.

The mean flowrates in the 50 mm sections for the 5 holes with each 2 sides were between 0.04 and 1 ml/h. The logarithmic (base 10) standard deviation of the flowrate distribution varied between 0.32 and 1.02. No correlation between visual apertures or number of fracture intersections on one hand and the injected flowrates on the other hand were found. Visual apertures were order(s) of magnitude larger than hydraulic apertures.

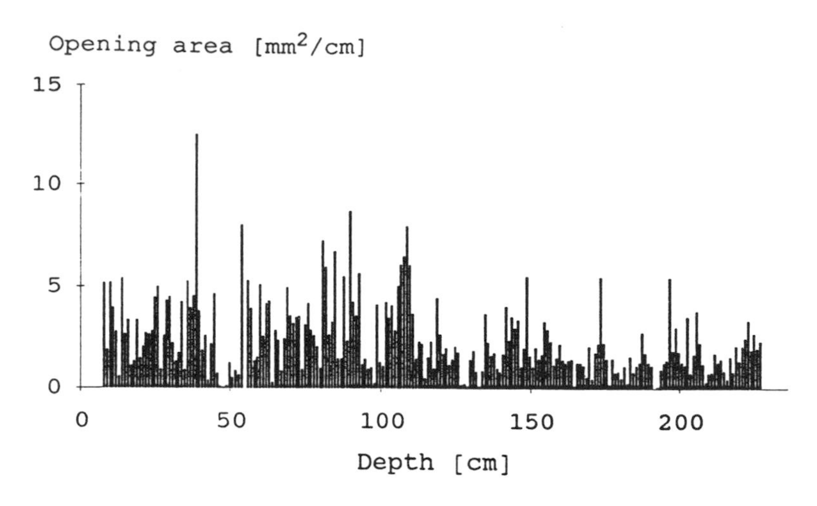

Figure 2. Example of fracture apertures based on photo logs, side 1 of a fracture.

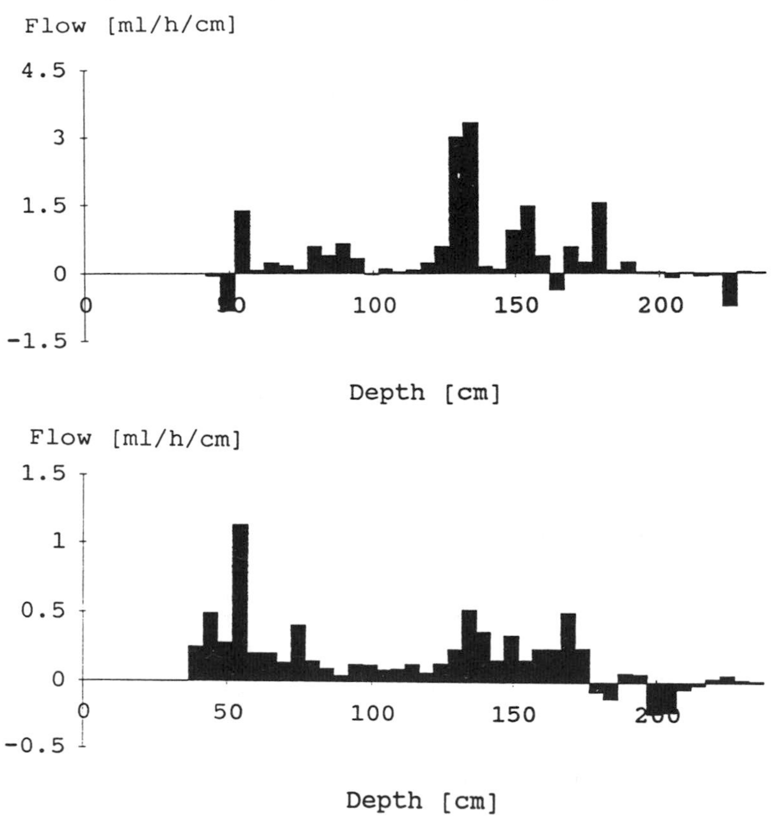

Figure 3. Injected flowrate in both sides of a fracture in 50 mm sections along the hole.

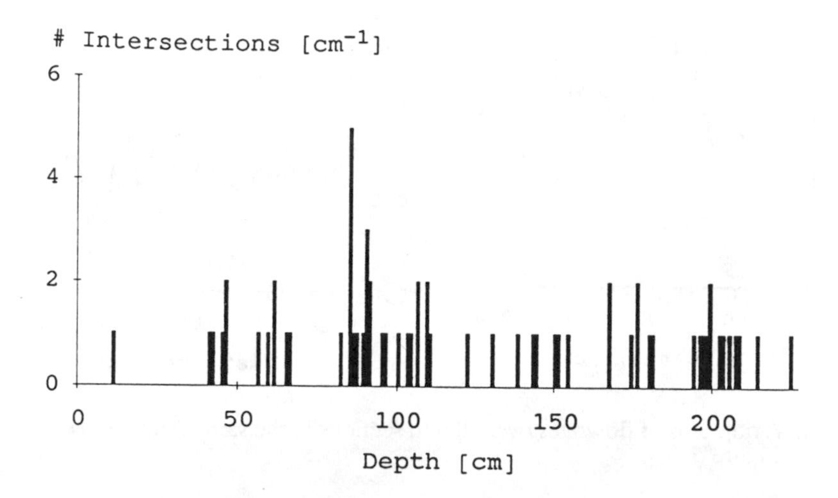

Figure 4. Number of fracture intersections seen on the photographs taken from within the hole.

Kriging was used to study how far out in the fracture from the observation points along the hole the aperture could be assessed with any degree of confidence. The variograms obtained from both visual apertures and the detailed flow distribution measurements were used. It was found that the prediction of apertures could not be made for more than 10 cm and often less. The flowrate distribution can be estimated somewhat further but still only for so short a distance that it is not very useful. Figure 5 below shows variograms of visual apertures from two different fractures. Figure 6 shows variograms of flowrates into 50 mm sections of the same fractures.

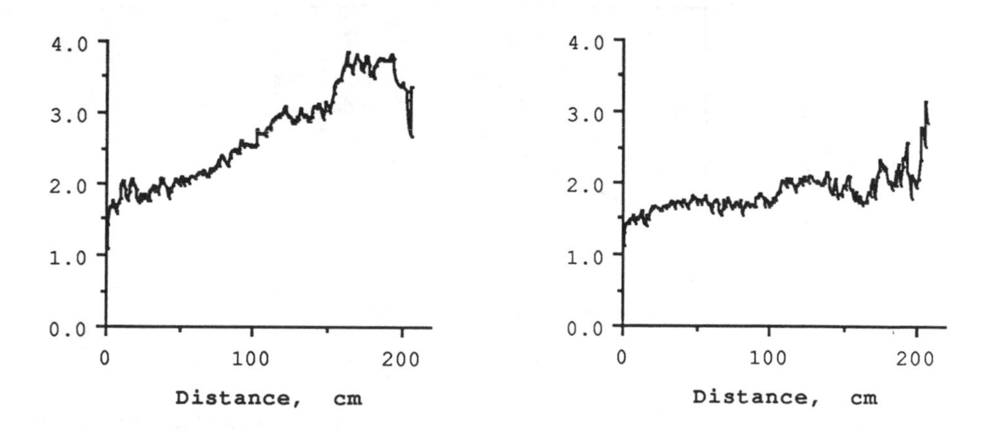

Figure 5. Variograms of visual apertures for two fractures.

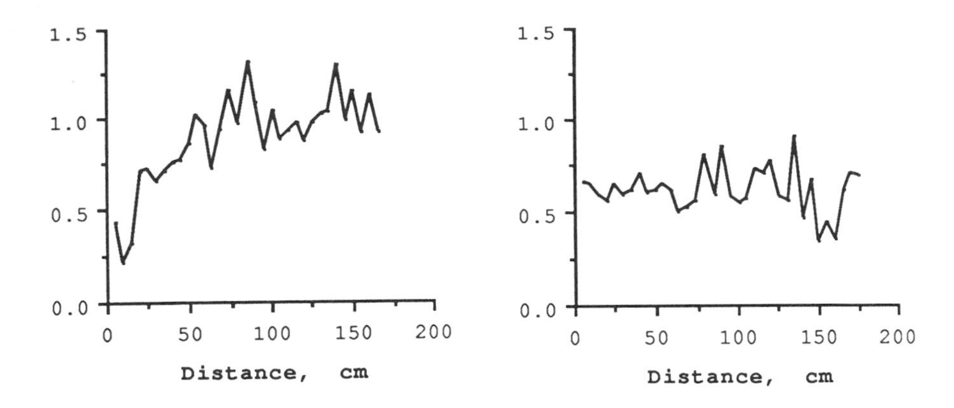

Figure 6. Variograms of flowrates over 50 mm sections in the same fractures as in Figure 5.

One of the questions we addressed was how to translate the flowrate distribution to a transmissivity distribution. The following method was used. A large number of stochastic fractures were generated using the information on the spatial correlation of the variograms based on the flowrate distribution. A log normal distribution of local transmissivities was assumed. The standard deviation of the transmissivities was varied. The flowrate distribution into the fracture in the stochastic fractures was calculated for every generated fracture for boundary conditions similar to those in the experimental fracture. The standard deviation of the flowrate distribution into the stochastic fractures was obtained. Figure 7 below shows the relation between the standard deviation of the flowrates and that of the local transmissivities.

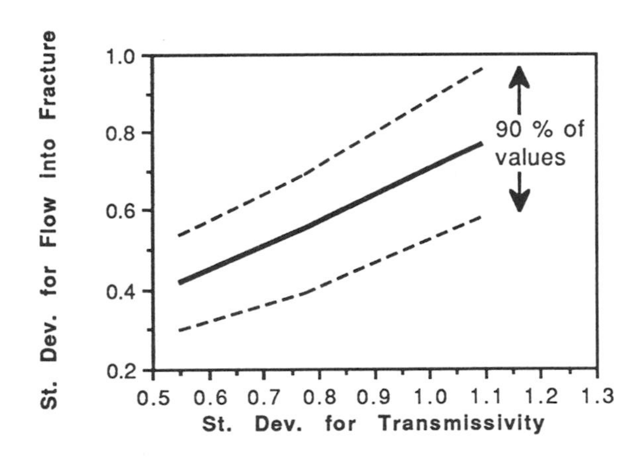

Figure 7. Relation between the standard deviation of the flowrates and that of the local transmissivities.

The standard deviations of the flowrates ranged from 0.32 to 1.02 for the 5 times 2 sides of the fractures measured. The mean was 0.58. It is seen from figure 7 that the standard deviation of transmissivities is expected to be about 30 % larger.

Double hole experiments

The pressure tests in the double hole experiments show that there are pressure connections between the two holes in the fracture. The pressure responses are seen already after a few tens of minutes and in many monitoring sections the pressure has levelled off after 6 to 10 hours. The level reached varies between much less than one percent and up to 5 percent of that in the injection hole (20 m water head). There is a large pressure dissipation somewhere. This may be due to dissipation into other fractures intersecting "our" fracture or due to flow into the monitoring hole by some paths bypassing the monitoring "feet" near the collection hole which is kept at zero head. Where dissipation takes place could not be determined from the pressure tests only. No increased water seepage on the walls of the drift could be noted during the time the injection hole was pressurized.

To study the transport paths and residence times five different tracers were injected at different conductive points in one of the holes and monitored for in the other hole. Figure 8 shows where the tracers were injected and where tracers were found.

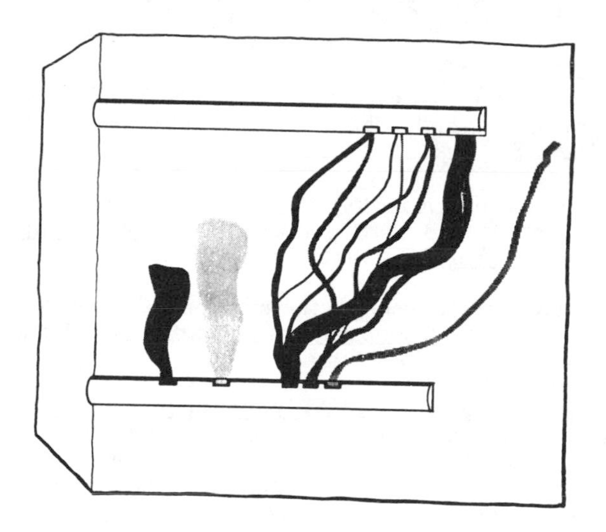

Figure 8. Location of tracer injection points and where tracers where found.

Only two tracers were found in the monitoring sections. The other tracers were later found on the wall of the drift in three separate areas where the proportion of the tracer concentrations were quite different. One of the tracers, Uranin was found with a steady state recovery of about 10 % but in a concentration of 25-50 % in the different sections. Duasyn which was injected only 10 cm from the injection point of Uranin, was found in a steady state recovery of 1 % but in nearly the same concentrations as Uranin.

DISCUSSION AND CONCLUSIONS

The results, regarding channelling, from the single hole experiments are not contradictory to what has been observed in other experiments [4,6,7]. Some parts of the fracture conduct water but a large part of the fracture is almost tight. It is impossible to tell from the photo logs if a fracture will conduct water or not. Fractures with visible large openings will in some cases not take any water. Even if the fractures selected had a similar look at the face of the drift a quarter of them were very tight and could not be used for measurements.

There is a bias in the fracture selection because the fractures that have been investigated are those that from the drift seem to be isolated and planar. However, these fractures might not always participate in the major flow in the rock.

The pressure tests as well as the tracer tests in the double hole experiment show that there are pathways between the two holes which divert the water from "our" fracture plane into other paths. A very detailed survey of the photographs from the two holes used in the double hole experiment did not show any clear location of any conductive intersecting fracture. The diversion pathways probably intersect the fracture somewhere between the holes. The tracers found on the wall of the drift were located at distances from the fracture of at least one meter and cannot be clearly related to fractures visible on the wall of the drift.

LITERATURE

1 Moreno L., Neretnieks I. : "Channeling and its potential consequences for radionuclide transport." Scientific basis for nuclear waste management XI, MRS, Boston, 1987, 169-178.

2 Neretnieks I. : "The Swedish repository for low and intermediate reactor waste - SFR. Radionuclide release and transport calculations." Symposium on Scientific basis for nuclear waste management, Berlin Oct 10-13, 1988.

3 Moreno L., Neretnieks I. : "Channeling in fractured zones and its potential impact on transport of radionuclides." Symposium on Scientific basis for nuclear waste management, Berlin Oct 10-13, 1988.

4 Bourke P. : Paper presented at GEOVAL symposium, Stockholm, April 1987. Proceedings, Swedish Nuclear Power Inspectorate, Stockholm 1988, 167-177.

5 Abelin H., Birgersson L., Ågren T., Neretnieks I. A channeling experiment to study flow and transport in natural fractures. Symposium on Scientific basis for nuclear waste management, Berlin Oct 10-13, p 661-668, 1988.

6 Neretnieks I. : "Channeling effects in flow and transport in fractured rocks - Some recent observations and models." Paper presented at GEOVAL symposium, Stockholm, April 1987. Proceedings, Swedish Nuclear Power Inspectorate, Stockholm 1988, 315-335.

7 Abelin H., Neretnieks I., Tunbrant S., Moreno L.: "Final report of the migration in a single fracture - Experimental results and evaluation." Stripa Project Report 85-03, OECD/NEA, SKB 1985.

MODELLING TRACER TESTS IN A HIGHLY CONDUCTIVE FRACTURE ZONE IN THE FINNSJÖN AREA IN CENTRAL SWEDEN

R. Nordqvist, P. Andersson, E. Gustafsson
SGAB, Uppsala, Sweden
P. Wikberg
Swedish Nuclear Fuel and Waste Management Co., Stockholm, Sweden

ABSTRACT

This paper presents some results from modelling performed in connection with a series of hydraulic and tracer tests in a sub-horizontal fracture zone in central Sweden, aiming at characterizing radionuclide transport in large fracture zones.The modelling was performed with a porous media approach with advection and dispersion as major transport processes, predicting each test based on available information, and with subsequent evaluation after each test. Some tracer breakthrough curves can to a large extent be explained by the presence of more than one major flow path, which is supported by a variety of experimental results. Such results are very useful, as effects of the flow geometry on observed data during a test can be distinguished from effects of other transport processes.

MODELISATION D'ESSAIS DE TRAÇAGE DANS UNE ZONE FISSUREE FORTEMENT CONDUCTRICE DE LA REGION DE FINNSJÖN DANS LE CENTRE DE LA SUEDE

RESUME

Les auteurs exposent quelques résultats de travaux de modélisation réalisés en liaison avec une série d'essais hydrauliques et d'essais de traçage dans une zone fissurée sub-horizontale dans le centre de la Suède, l'objectif étant d'étudier les mécanismes régissant le transport des radionucléides dans des zones fissurées de grande dimension. Le modèle a été élaboré sur la base d'un milieu poreux dans lequel l'advection et la dispersion constituent les mécanismes de transport principaux. Chaque essai a été précédé d'une prévision établie à partir des informations disponibles puis suivi d'une évaluation. Quelques courbes de passage de traceurs peuvent s'expliquer dans une large mesure par la présence d'au moins deux voies d'écoulement principale, hypothèse étayée par divers résultats expérimentaux. Les résultats obtenus sont très utiles car sur les données observées pendant un essai on peut distinguer les effets de la répartition spatiale des écoulements des effets dus aux autres mécanismes de transport.

INTRODUCTION

In crystalline rocks the groundwater flow is concentrated to zones with an increased frequency of connected hydraulically conductive fractures. Any leakage of radionuclides from an underground repository to the biosphere will preferrably occur through such fracture zones. In previous safety analysis (KBS-3), consideration was only given to the retention of radionuclides in the low conductivity rock between the repository and the nearest fracture zone, where the nuclides were conservatively assumed to reach the biosphere without delay.

The fracture zone project in Finnsjön, Sweden, was started 1984 with the aim of defining the characteristics of importance with respect to radionuclide migration in large fracture zones. The 100 m thick sub-horizontal fracture zone has been extensively characterized by means of geological, geophysical, geohydrological, and hydrochemical investigations carried out from the ground surface and from a large number of boreholes [1,2].

The purpose of this paper is to describe the flow and transport modelling that has been carried out in connection with a series of three field experiments conducted within the fracture zone. These experiments consists of hydraulic interference tests, a radially converging tracer experiment and a dipole tracer test. In general, the outcome of each experiment was predicted based on available information, the predictions were subsequently evaluated following an experiment, and the model was updated in order to predict the next experiment. Table I contains a summary of the field experiments discussed in this paper.

Table I.

Summary of field experiments subjected to flow and transport modelling at Finnsjön.

Activity	no of Observation boreholes (no of sections)	no of Injection boreholes (no of sections)	Distance (m)
Hydraulic interference tests	13 (43)	-	155 - 1540
Radially converging tracer test	1 (1)	3 (9)	155 - 189
Dipole tracer test	3 (3)	1 (1)	155 - 189

The dipole test was conducted only in the uppermost, highly conductive, 0.5 m layer of the fracture zone, while the other tests also included other parts of the 100 m thick fracture zone. 11 different tracers were used for the radially converging test, and 16 different tracers for the dipole test (including radioactive tracers).

The work presented in this paper is based on deterministic models. The fracture zone is treated as a porous medium, where dominant transport processes are advection and dispersion. Two slightly different angles of approach are taken. Firstly, flow and transport are modelled in two dimensions, attempting to include geometrical features and physical processes that govern flow and transport in two dimensions. Secondly, transport is modelled in one dimension, using relatively simple models aiming at: capturing the essential features of tracer transport, model validation using all previously collected information, and also to some extent model discrimination. These two approaches will be discussed below.

The Finnsjön tracer tests are one of the test cases within the INTRAVAL project, and are being modelled by several other participants using different conceptual models. Other participants at GEOVAL presenting modelling (posters) of the fracture zone include VTT (Finland) and E.N.S.M. (France).

2-D FLOW AND TRANSPORT MODELLING

The general approach here is to perform a series of different hydraulic and tracer test, integrated with modelling of both flow and transport processes. Validation is accomplished by comparing model predictions with experimental results. In order to validate the 2-D model, it is considered important that both flow (hydraulic heads) as well as transport (concentrations) can be simulated and predicted. Thus, if a consistency of model performance for a series of different tests can be obtained, it is seen as one way to validate a model. However, it should be pointed out that in this case only one basic conceptual model is validated. That is, no attempt is made to discriminate between models based on other approaches and assumptions than those described in the introduction. Referring to Table I, the different modelling steps will be outlined briefly below.

The hydraulic interference tests were predicted based on limited information, where transmissivity values were primarily obtained from single hole measurements and the boundary conditions were simplified. The outcome of the tests showed that the prediction model did not describe the transient flow field satisfactorily. Thus, boundary conditions were revised by adding more recent geological interpretations, and inverse modelling on transient data from eleven boreholes from the interference test was carried out. The inverse modelling generally verified the updated boundary conditions, and the interference tests could be simulated relatively well [3].

Using the updated model from the interference tests, the flow field during the radially converging tracer experiment was predicted. By adding transport parameters (porosity and dispersivities), tracer breakthrough curves were predicted. Generally, the evaluation after the radially converging experiment showed that tracer residence times were underestimated significantly, as rather low flow porosity values were used. Further, it was evident that there were heterogeneities and/or anisotropies in hydraulic parameters that the model did not account for. Nontheless, tracer breakthrough could be simulated relatively well after adjustment of primarily the flow porosity, see Figure 1. A detailed description of the radially converging experiment is given by Gustafsson et al, 1989 [4].

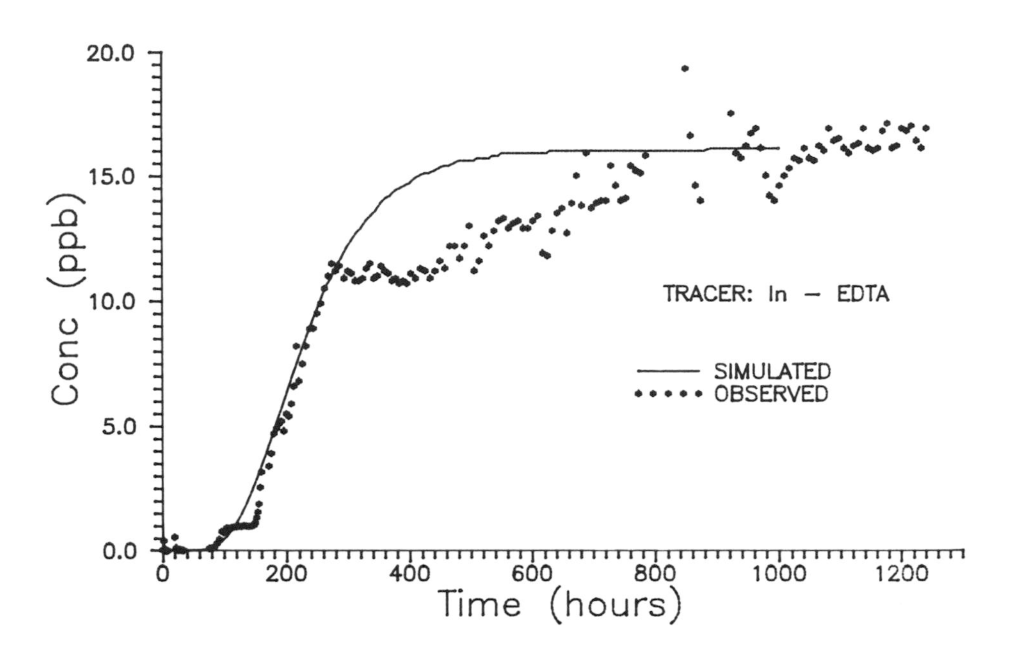

Figure 1. Example of simulated (2-D) versus observed breakthrough data from the radially converging tracer test (conservative tracer).

Following the radially converging test, the dipole tracer test was predicted [5]. The dipole flow field for the fracture plane and tracer breakthrough in the pumped borehole was predicted for conservative as well as sorbing tracers, assuming re-circulation of tracer. As an effect of the measured natural gradient, the predictive modelling indicated that approximately 10 percent of the injected mass would be lost. As an example, Figure 2 shows predicted and observed breakthrough curves for a conservative tracer (preliminary evaluation). Average travel time and peak level is predicted relatively well, but it is evident that more than the predicted 10 percent of the injected tracer mass is lost. Based on measurements of salinity during the experiment, it is indicated that some of the discharged water originates from below where the salinity increases sharply. Thus, the dipole field, which is assumed to be two-dimensional, may have a three-dimensional component. A simple mass balance gives a preliminary estimate of the contribution of high-salinity water to the discharged water, indicating that possibly as much as 30 percent of the injected tracer mass would be lost (or delayed) due to this effect. As far as predicting the flow field, it can be mentioned that the observed head difference between the injection hole and the discharge hole was almost identical to the predicted.

In summary, from a model validation point of view, it has been shown that tracer transport in a general sense can be predicted based on results from field experiments. In this case, hydraulic tests enabled prediction of flow, while the radially converging test was necessary in order to obtain parameters for predicting tracer transport. Although the outcome of the dipole test was fairly well predicted for the discharge well, more data from two other observations wells remain to be evaluated. This will give further indications of the variability of flow and transport properties within the fracture zone. Results from the interference tests, the radially converging experiment, and the dipole test can then be integrated to give a relatively complete picture of the transport properties of the fracture zone, especially with respect to heterogeneities and anisotropies.

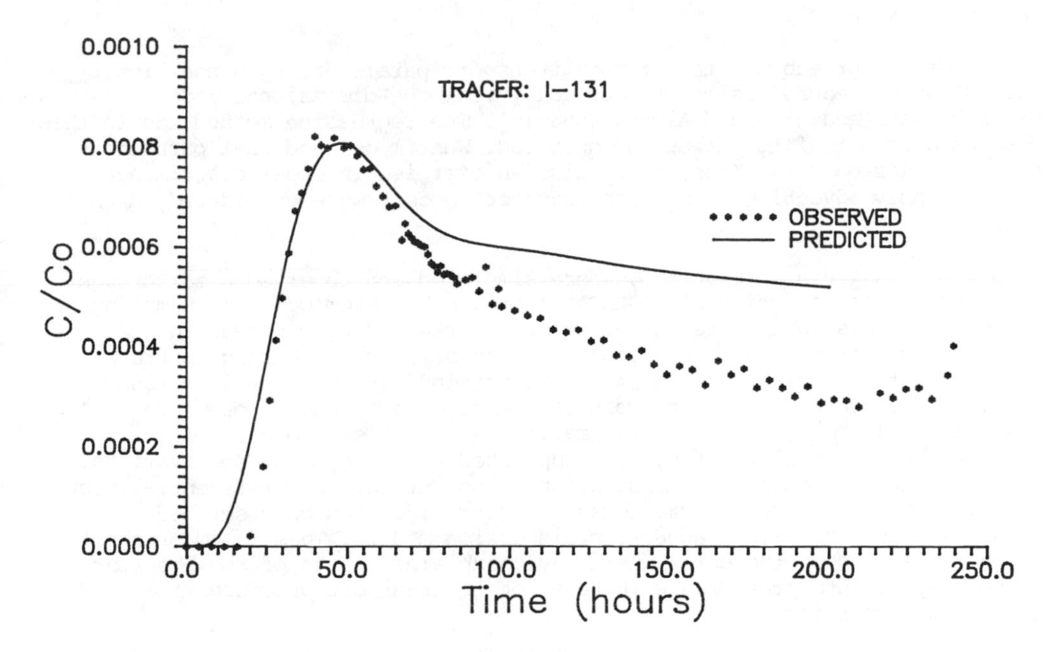

Figure 2. Comparison of model predictions (2-D) and experimantal results from the dipole tracer test (conservative tracer).

1-D TRANSPORT MODELLING

The one-dimensional transport modelling was carried out with a slightly different purpose than the two-dimensional flow and transport modelling. In this case the purpose was to evaluate the breakthrough curves from the radially converging test, in order to be able to make some interpretations primarily about the transport connectivity between the different injection sections and the sampled borehole. It is emphasised that supplementary information from all the experiments provided a basis for these interpretations. The analysis generally assumed steady-state flow, but accounted for variable injection schemes.

An important difference, compared to the 2-D modelling, is that mixing in the sampled borehole of tracers travelling through several different major flowpaths is considered. The concentration in the sampled section is assumed to be a volume-averaged concentration:

$$C = \sum_i f_i C_i$$

where C = tracer concentration in borehole
f_i = fractional volume parameter
C_i = tracer concentration from flowpath i

As independent measurements (other than the breakthrough curves) indicated the existence of such multiple-path transport, this approach proved particularly useful.

The general approach was to estimate transport parameters by a non-linear regression procedure, using various analytical one-dimensional models as given by Van Genuchten and Alves, 1984 [6]. The regression method was in this case used as a tool to answer the question: What model and what parameters explain observed data? Thus, it is also an exercise in model discrimination, although only advection-dispersion transport models were considered in this case.

An example is given in Figure 3, where simulated and observed breakthrough data for a step injection of In-EDTA is presented. From detailed sampling at different levels in the fracture zone during the radially converging test, it was evident that at least two major flow paths was contributing to tracer arrival in the sampled section [4]. Similiar indications were also found from the interference tests, by analysis of primary and secondary pressure responses [3]. Thus, parameter estimation was in this case carried out assuming two major flow paths, well supported by independent observations. Figure 3 can be seen as a demonstration of how effects of preferential flow paths on the breakthrough curves can be identified, and be described relatively well by simple advection/dispersion models. This is important for further analyses of the data, where any unexplained parts of the breakthrough curves may be interpreted as effects of other transport processes than advection/dispersion.

Future work using relatively simple models will include an evaluation of the dipole tracer tests, where the interpretations from the radially converging experiment will be checked and verified. Basically, the models and parameters obtained from one experiment should also explain data from a second experiment with different flow field. One possibility that will be considered is parameter estimation from both experiments simultaneousely.

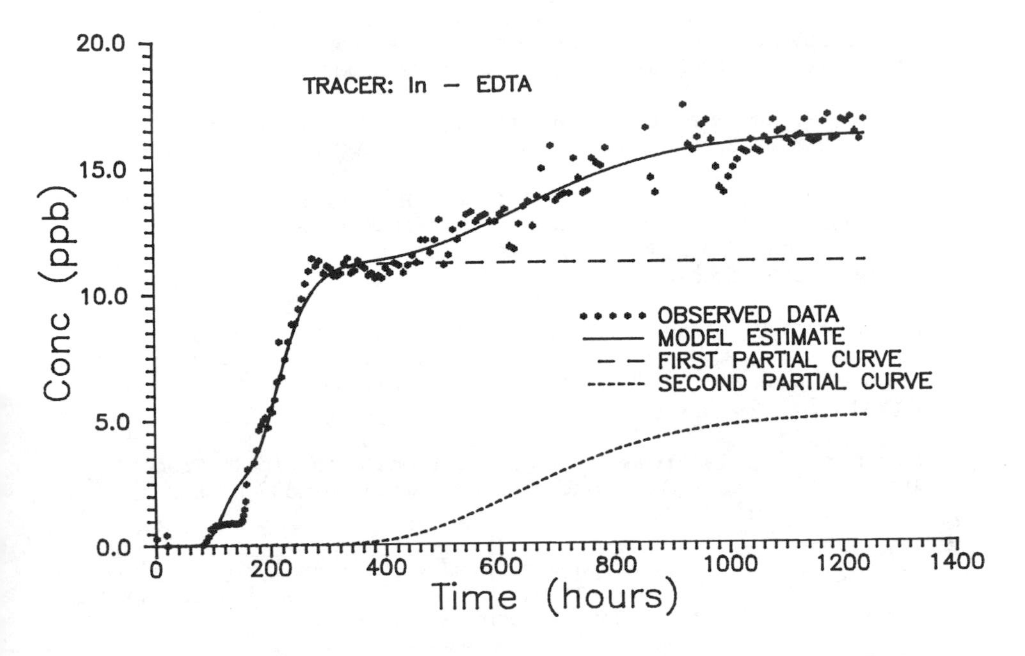

Figure 3. Example of evaluation with a multiple-path model (1-D) of the radially converging tracer experiment.

CONCLUSIONS

For the two tracer experiments discussed, both under induced and different flow fields, it has been shown that tracer transport can be modelled satisfactorily with relatively simple advection/dispersion models. This is the case both for the 2-D as well as for the 1-D modelling.

Simultaneous simulation/prediction of both flow and transport in the fracture zone using the 2-D model seemed to be a workable approach. The fact that both flow and transport can be modelled simultaneousely increases confidence in a model.

From the 1-D modelling of the radially converging experiment, it was evident that data could be explained at least as well as with the 2-D model. The physical meaning of some of the estimated parameters from the 1-D model may be questioned, but the important aspect in this case is that explanation of breakthrough curves with multiple-path models can be supported by other independent experimental results. Any remaining unexplained parts of the breakthrough curves may then be interpreted as effects of other transport processes then advection/dispersion.

REFERENCES

[1] Ahlbom, K., Andersson, P., Ekman, L., Tiren, S., 1987: Characterization of Fracture Zones in the Brändan Area Finnsjön Study Site, Central Sweden. SKB Progress Report 88-09.

[2] Ahlbom, K. and Smellie, J. (editors), 1989: Characterization of Fracture Zone 2, Finnsjön Study Site. SKB Technical Report 89-19.

[3] Andersson, J-E, Ekman, L., Gustafsson, E., Nordqvist, R., Tiren, S., 1988: Hydraulic Interference Tests and tracer Tests within the Brändan Area, Finnsjön Study Site. The Fracture Zone Project - Phase 3. SKB Technical Report 89-12.

[4] Gustafsson, E., Andersson, P., Eriksson, C-O, Nordqvist, R., 1989: Radially Converging Tracer Experiment in a Low-angle Fracture Zone at the Finnsjön Site, Central Sweden. SKB Technical Report, in prep.

[5] Nordqvist, R., 1989: Numerical Predictions of a Dipole Tracer Test in a Fracture Zone in the Brändan Area, Finnsjön. SKB Progress Report 89-34.

[6] Van Genuchten, M. T., and Alves, W. J., 1982: Analytical Solutions of the One-dimensional Convective-dispersive Transport Equation. U.S. Dep. of Agric. Tech. Bull. 1661.

EVALUATION OF EARTH TIDE MATRIX DISPERSION

U. Kuhlmann
Laboratory of Hydraulics, Hydrology and Glaciology
Federal Institute of Technology, Zürich, Switzerland

S. Vomvoris, P. Hufschmied
National Cooperative for the Storage of Radioactive Waste (NAGRA)
Baden, Switzerland

ABSTRACT

The impact of earth tide effects on species retardation migrating in fractured rock formations is investigated. One-dimensional oscillatory flow and solute transport between a fracture and the rock matrix is simulated for some realistic combinations of hydrogeologic parameters. While the time-varying flow field is modeled analytically the advective-disversive equation is solved by use of the finite element method. The results are compared to solutions of the pure diffusion equation to discuss the importance of the earth tide induced matrix dispersion on the basis of an equivalent matrix diffusion coefficient. The calculations show that the mass retained by the matrix is directly proportional to: i) the earth tide amplitude difference between the matrix and the fracture and ii) the dispersivity of the matrix. A comparison with values reported in the literature shows that the additional 'tidal' transport coefficient may be as high as the conventional matrix diffusion coefficient.

EVALUATION DE LA DISPERSION DANS LA MATRICE INDUITE PAR LA MAREE TERRESTRE

RESUME

Les auteurs examinent les effets de la marée terrestre sur le ralentissement de la migration des espèces chimiques dans les formations rocheuses fissurées. Partant d'un certain nombre de combinaisons réalistes de paramètres hydrogéologiques, les auteurs procèdent à une simulation unidimensionnelle du flux oscillatoire et du transport de solutés entre une fissure et la matrice rocheuse. Comme il varie en fonction du temps, le champ d'écoulement fait l'objet d'une modélisation analytique alors que l'équation d'advection-dispersion est résolue par la méthode des éléments finis. On compare les résultats obtenus aux solutions de l'équation de diffusion pure pour examiner, sur la base d'un coefficient équivalent de diffusion dans la matrice, l'importance de la dispersion dans la matrice induite par la marée terrestre. Il ressort des calculs que la masse retenue par la matrice est directement proportionnelle à : i) la différence d'amplitude de la marée terrestre entre la matrice et la fissure et ii) la capacité de dispersion de la matrice. Une comparaison avec les valeurs que l'on trouve dans les communications scientifiques pertinentes montre que le coefficient de transport additionnel imputable à la marée peut égaler le coefficient classiquement retenu pour la diffusion dans la matrice.

Introduction

The earth tide effect refers to the phenomenon under which the small changes of the gravitational field due to the celestial movements cause pressure variations in the groundwater flow systems. Earth tide effects have been observed in various deep boreholes and numerous publications are dedicated to its analysis especially in an inverse mode where large scale aquifer parameters can be inferred (e.g. Bredehoeft, 1967; Van der Kamp and Gale, 1983). Recently, earth tide effects were also monitored in a planar fracture at the Grimsel Test Site - Nagra's Rock Laboratory situated in the Alps and in a Nagra deep borehole in Northern Switzerland (Fig. 1, s. Gemperle, 1989).

With respect to solute transport in fractured rock, tidal pressure variations may have two effects. Firstly, because the tidal pressure variations depend on rock stiffness it is expected that a pulsating exchange of water is induced between fractures and the rock matrix. A substance transported by the fracture flow system would be sequentially "pumped" into and flushed out of the matrix during a tidal cycle. Due to the irreversibility of hydrodynamic dispersion, however, some tracer would be retained in the matrix. Thus, a retarding mechanism similar to matrix diffusion would arise. Secondly, tidal pressure variations may also occur within the plane of the fracture due to lateral variations in fracture stiffness. The corresponding pressure gradients may enhance especially lateral dispersion in a way similar to what has recently been attributed to temporal fluctuations in hydraulic head in heterogeneous aquifer (e.g. Naff et al., 1989).

The present investigation aims at evaluating the magnitude of the earth tide induced matrix dispersion (ETIMD) related to other processes affecting migration of species in fissured rock. We will investigate matrix dispersion in an 'experimental' scale considering flow and transport from a single fracture into an infinite rock matrix based on an one-dimensional approach. To evaluate separately the advective-dispersive component induced by the tidal pressure fluctuations mechanisms such as molecular diffusion, sorption or radioactive decay are disregarded. A number of numerical experiments has been carried out to exemplify the sensitivity of the phenomenon on hydraulic and geologic parameters. These results are compared with solutions of the pure diffusion equation. A 'rule of thumb' is provided which allows to discuss the potential importance of the ETIMD on the basis of an equivalent rock matrix diffusion coefficient.

Basic equations

The process of solute transport from a fracture into the rock matrix is illustrated in Fig. 2. The concentration within the matrix is described by the advective-dispersive equation whose one-dimensional form can be written as

$$\frac{\partial c}{\partial t} + u \cdot \frac{\partial c}{\partial x} - \frac{\partial}{\partial x} \left(D_h \cdot \frac{\partial c}{\partial x}\right) = 0 \qquad (1)$$

where C is the concentration of the substance [M/L³], $D_h = \alpha \cdot |u|$ is the dispersion coefficient [L² /T], α is longitudinal dispersion length [L] and u represents the groundwater flow velocity perpendicular to the plane of fracture [L/T]. The total mass flux into the matrix, at $x=0$, is given by

$$m = u \cdot C_0 - D_h \cdot \frac{\partial C}{\partial x}\Big|_{x=0} \qquad (2)$$

the first and second terms on the right hand side being the advective and dispersive portion, respectively. The total mass retained by the matrix at any time t is obtained by integrating over time

$$M = \int_0^t m \; dt = \int_0^t u \cdot C_0 \; dt - \int_0^t D_h \cdot \frac{\partial C}{\partial x} \; dt \qquad (3)$$

The flow velocity in the matrix which varies with time and space

$$u(x,t) = \frac{K}{n} \cdot \frac{\partial h}{\partial x} \qquad (4)$$

where K is the hydraulic conductivity [L/T] and n is the porosity of the matrix, is obtained solving the one-dimensional groundwater flow equation

$$\frac{\partial h}{\partial t} = \frac{K}{S} \cdot \frac{\partial^2 h}{\partial x^2}$$ where S is the specific storage [L^{-1}].

For periodic fluctuations of the potential head at the fracture interface

$$h(0,t) = A \cdot \cos(\omega t - \varepsilon)$$

where A is the amplitude of the oscillation [L], $\omega = 2\pi / T$ [1/T], T is the time period of the tidal cycle [T], ε is a phase angle and initial condition $h(x,0)=0$ the steady oscillatory solution is (Carslaw and Jaeger, 1959)

$$h(x,t) = A \cdot \cos(t' - x' - \varepsilon) \qquad (5)$$

where $x' = x \cdot (\omega S / 2K)^{1/2}$ and $t' = \omega t$. The velocity is given by

$$u(x,t) = u_0 \cdot e^{-x'} \cdot \cos(t' - x' - \varepsilon - \pi/4) \qquad (6)$$

where $u_0 = A/n \cdot (\omega S K)^{1/2}$ is the maximum velocity at the fracture/matrix interface.

Evaluation of the skin depth (i.e. the depth at which the amplitude of the velocity oscillation falls at e^{-1}) for hydrogeologic conditions of crystalline rock (RUN-2.3, s. Table I) yields a value of 0.37 m, approximately.

Numerical experiments

Due to the spatial and time dependencies of velocity field and parameters an analytical expression solving the advective-dispersive equation (1) for the problem at hand is not readily available. Hence, the following investigations are based on numerical experiments. A number of examples were selected to illustrate the process expected and to provide a data base for a subsequent empirical analysis.

While the flow field is modeled analytically using equation (6) a finite element code is applied to simulate advective-dispersive transport (Atkinson et al., 1988). The rock matrix is represented by one-dimensional, 3-node line elements using quadratic interpolation. The total lengths of the meshes depend on hydrogeologic parameters and vary between 0.4 and 2 m. The degree of discretization is chosen appropriately fine that numerical dispersion is virtually excluded. Nodal spacings increase with the smallest element situated at the fracture boundary to account for the decreasing hydraulic gradient with depth of penetration. The Crank-Nicholson time marching scheme is applied with step length increasing from a quater of an hour up to 2 hours. The total time of the simulations vary between 1 and 4 years.
Table I summarizes the hydrogeologic parameters used for the various data runs.

RUN		1	2.1	2.2	2.3	3
K	[m/s]	10^{-9}	10^{-12}	10^{-12}	10^{-12}	10^{-12}
S	[m^{-1}]	10^{-5}	10^{-7}	10^{-7}	10^{-7}	10^{-7}
n	[-]	0.001	0.001	0.001	0.001	0.001
α	[m]	0.1	0.01	1.0	0.001	0.01
T	[h]	12	12	12	12	12
A	[cm]	7.5	7.5	7.5	7.5	37.5

Tab. I: Parameter values used for the simulations

RUN-2.3 which is assumed to represent hydrogeologic conditions of crystalline rock is selected to illustrate the computed results. The individual terms of the flux through the interfacing plane between fracture and matrix as indicated in equation (2) is examined in Fig. 3. The advective portion is oscillatory constant while the dispersive and hence the total flux decrease with time due to the flattening gradient of the inflowing mass. The movement of the tracer front through the matrix is depicted in Fig. 4. Integration of these profiles along the flow direction at each time step of computation yields the total mass retained in the matrix. The complete results are presented in Fig. 5 and discussed in the following section.

Equivalent diffusivity

For the reasons outlined above it is difficult to consider tidal effects efficiently by use of flow and transport models in regional/site scale. Alternatively, the following empirical approach attempts to lump the complex mechanism of matrix dispersion together in a more simple process of matrix diffusion. To estimate an equivalent diffusivity the numerical results are compared with analytical solutions of the pure diffusion equation which is given by

$$\frac{\partial c}{\partial t} = \frac{\partial}{\partial x}(D_M \cdot \frac{\partial c}{\partial x}) \tag{7}$$

where D_M is the diffusivity of a species in a porous rock matrix. For the case of step input of concentration C_0 at location $x=0$ and time $t=0$ as well as constant diffusivity,

$$\frac{\partial c}{\partial x}(x,t) = C_0 / (\pi D_M t)^{1/2} \; e^{-x^2 / (4 D_M t)} \tag{8}$$

$$q_M = -D_M \cdot \frac{\partial c}{\partial x} = C_0 \cdot (D_M / \pi t)^{1/2} \tag{9}$$

$$\int_0^t q_M \, dt = 2 \cdot C_0 \, (D_M t / \pi)^{1/2} \tag{10}$$

is obtained for the concentration gradient, the mass flux into the matrix, both at $x=0$, and the amount of solute retained by the rock matrix, respectively.

The intention is to compute an equivalent diffusivity D'_M in such a way that the experimental data fit solution (10). Parameter estimation is performed by the method of least squares. The resulting coefficients indicated and plotted in Fig. 5 (solid lines) are in excellent agreement with the 'measured' data. Inspection of the values found from runs 2.i whose parameters have been kept the same exept for the longitudinal dispersivity shows

$$D'_{M,2.1} / D'_{M,2.2} / D'_{M,2.3} \cong \alpha_{2.1} / \alpha_{2.2} / \alpha_{2.3}$$

i.e. a linear dependence of the estimations on α. Further dimensional and physical considerations suggest an approach of $D'_M \cong f(\alpha, u_0)$. In fact, a final analysis supports even an approximation of

$$D'_M = \beta \cdot \alpha \cdot u_0 \tag{11}$$

where $\beta=0.6$ is the mean value for the five data runs. For comparison purposes Fig. 5 contains, additionally, 10% error bounds of the proposed relation related to retained mass ($\beta=0.5/0.7$).

Alternatively, the equivalent diffusivity can be interpreted as the time averaged dispersion coefficient given by

$$\bar{D}_M = 1/t' \cdot \int_0^{t'} \alpha \cdot |u(x,\tau)| \; d\tau \qquad (12)$$

which, in contrary to (11), exhibits a dependency on x. Substituting (6) for $\varepsilon=\pi/4$ yields

$$\bar{D}_M = \alpha u_0 \, e^{-x'} \cdot 1/t' \cdot \int_0^{t'} |\cos(\tau - x')| \, d\tau \qquad (13)$$

For $x=0$ and large t, (13) reduces to $\bar{D}_M \equiv \alpha u_0 \cdot 2/\pi = 0.637 \cdot \alpha u_0$ and is consistent with (11). However, note that (11) holds only for 'early times' when equivalent diffusivities near the fracture/matrix interface dominate the transport process. Numerical test computations for pure matrix diffusion comparing constant with x-varying diffusivities for the parameters of RUN-2.3 indicate that approach (11) can be applied for 10000 years, approximately.

Discussion and conclusion

The performed calculations show that the mass retained by the matrix due to the oscillatory flow in the matrix caused by the earth tides is directly proportional to: i) the earth tide amplitude difference between the matrix and the fracture and, ii) the dispersivity of the matrix. An empirically derived relationship indicates that the ETIMD is proportional to the product of the maximum velocity at the fracture/matrix and the matrix dispersivity. It is expected that the oscillatory motion will enhance the mixing in the matrix in the vicinity of the fracture, contributing thus to sorption and hence increased retardation.

For some realistic combinations of parameters, described in Table I, the value of the equivalent ETIMD coefficient is of the order of 10^{-12} to 10^{-11} m^2/s (RUN-2.2 and RUN-3.0, respectively). A comparison with values reported in the literature (e.g. Lever et al, 1983) which are of the order of 10^{-11} to 10^{-10} m^2/s shows that the additional 'tidal' transport coefficient may be as high as the conventional matrix diffusion coefficient. However, additional sensitivity studies should be performed and particularly with case specific data to substantiate such a conclusion. Even for the parameter combinations treated herein, if one considers that: i) the head fluctuations observed in the borehole are significantly lower than the ones occuring in the fracture due to the compressibility of the borehole measuring volume and, ii) there may exist a zone of 'disturbed' matrix adjacent to the fracture with increased dispersivity, the values estimated in the previous analysis could easily increase by a factor of 5 to 10. An increase of K or S would also increase the estimated ETIMD coefficient.

Additional sensitivity studies are required to investigate the effect of hydraulic diffusivity, porosity, sorption and determine the long-term behavior. Finally, the effect of earth tides on cross-fracture transport is currently being independently investigated. For the performance assessment of a repository one may attempt to account for all these effects through sensitivity analyses of the various lumped parameters. However, the careful consideration of these phenomena is quite important for the interpretation of controlled field scale migration experiments. They may be neglected only if justified after careful consideration of the case specific parameters.

REFERENCES

Atkinson, R., A.W.Herbert, C.P.Jackson, P.C.Robinson and M.G.Williams, NAMMU user guide (rel. 4), Theoretical Physics Division, AERE Harwell, 1988

Bredehoeft, J.D., Response of well-aquifer systems to earth tides, J. Geophys. Res., 72, 3075-3087, 1967

Carslaw and Jaeger, Conduction of heat in solids, Second edition, Oxford Press University, 1959

Gemperle, R., Nagra Grimsel Field Laboratory - Long term (35 day) constant Q test in BOMI 87.009, Contractors Report to Nagra, Solexperts AG, Feb. 1989

Hadermann J. and F.Roesel, Radionuclide chain transport in inhomogeneous crystalline rocks: Limited matrix diffusion and effective surface sorption, Nagra, Technical Report 85-40, Baden Switzerland

Lever, D.A., M.H.Bradbury and S.J.Hemingway, Modelling the effect of diffusion into the rock matrix on radionuclide transport, 1983

Naff, R.L., J.-C. Jim Yeh and M.W. Kemblowski, Reply on a comment by G. Dagan, Water Resour. Res., 25(12), 2523-2525, 1989

Neretnieks I., Diffusion in the rock matrix: An important factor in radionuclide retardation?, J. Geophys. Res., 85(10), 1980

Van der Kamp, G. and J.E.Gale, Theory of earth tide and barometric effects in porous formations with compressible grains, Water Resour. Res., 19(2), 1983

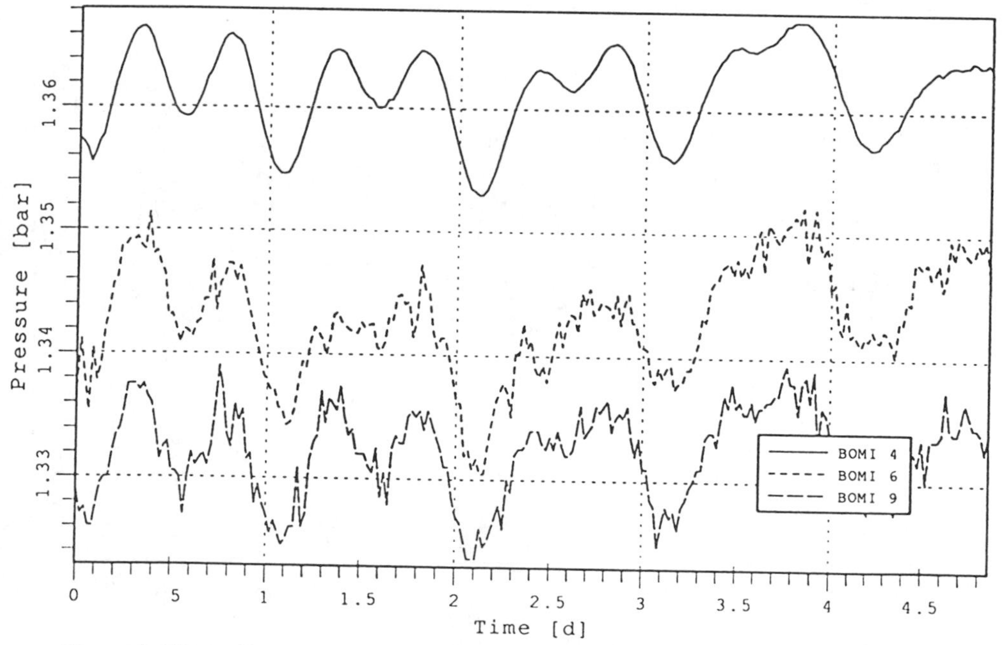

Figure 1: Tidal effects monitored at the Grimsel Test Site during a long term pumping test

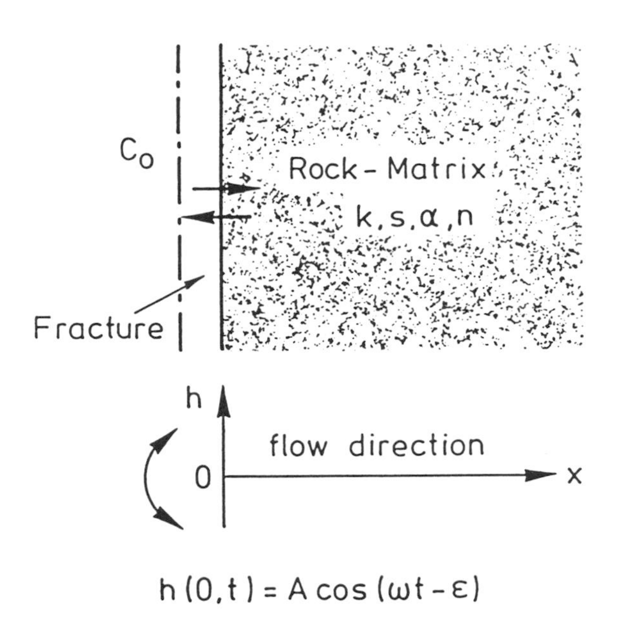

Figure 2: Conceptual model

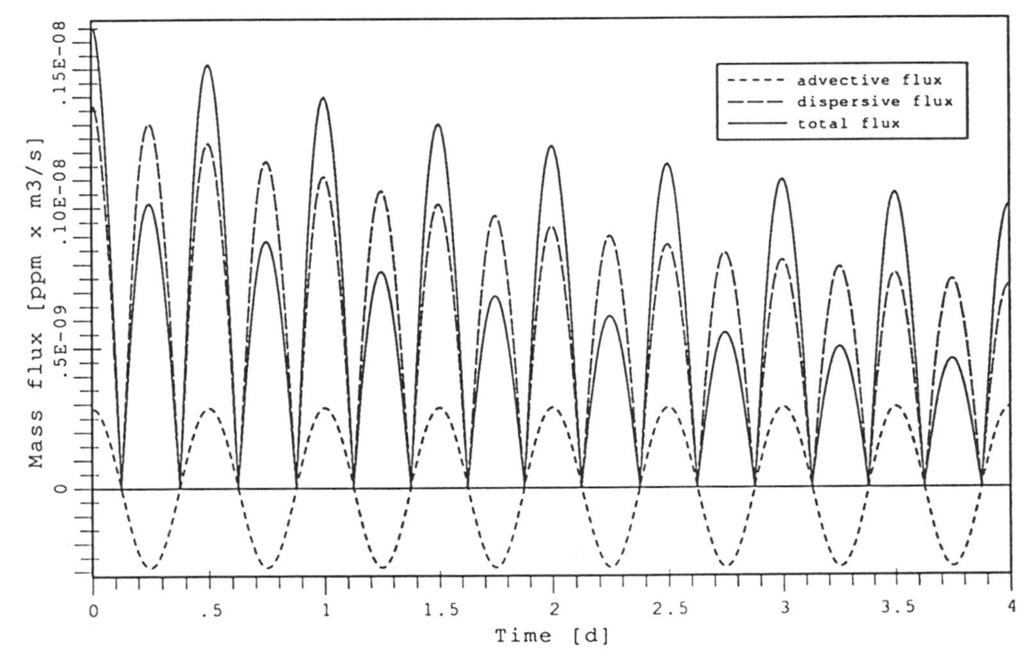

Figure 3: Mass flux through the fracture/matrix interface (RUN-2.3)

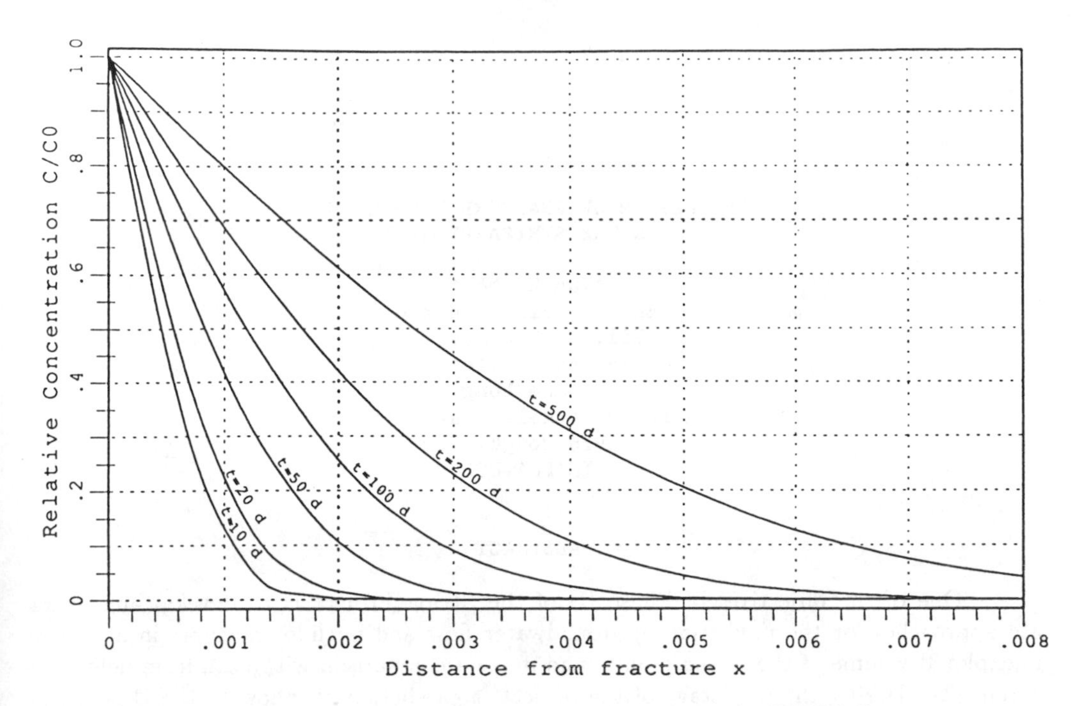

Figure 4: Proceeding of the tracer front through the rock matrix (RUN-2.3)

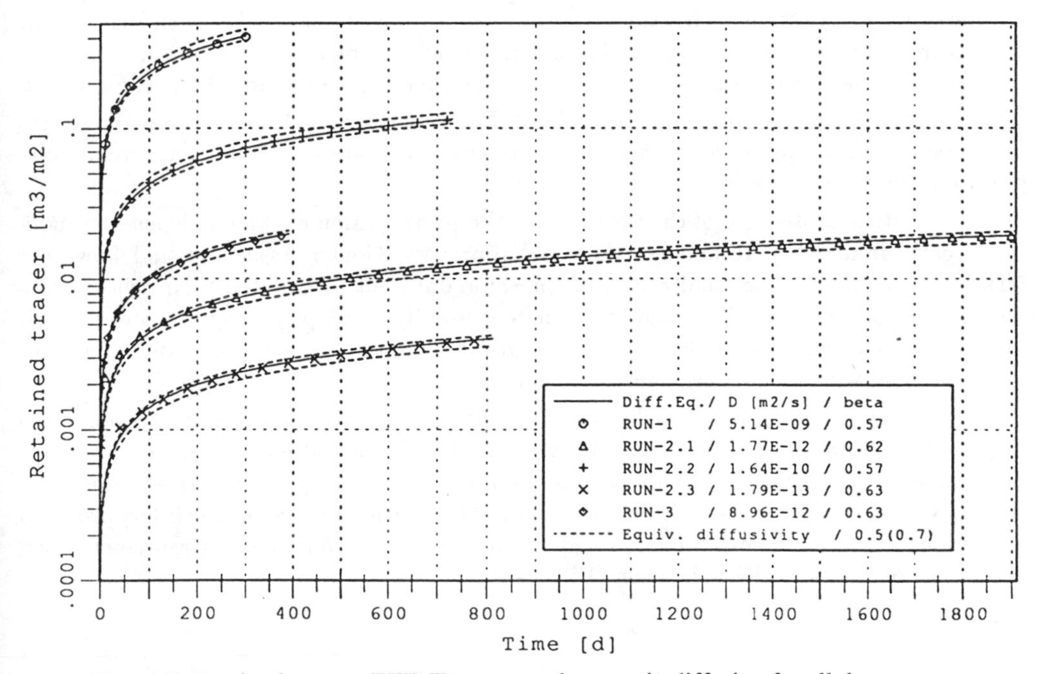

Figure 5: Retained tracer - ETIMD compared to matrix diffusion for all data runs

VALIDATION OF FRACTURE FLOW MODELS IN THE STRIPA PROJECT

Alan Herbert
AEA Technology, Harwell
William Dershowitz
Golder Associates Inc.
Jane Long
Lawrence Berkeley Laboratory
David Hodgkinson
INTERA-ECL

ABSTRACT

One of the objectives of Phase III of the Stripa Project is to develop and evaluate approaches for the prediction of groundwater flow and nuclide transport in a specific unexplored volume of the Stripa granite and make a comparison with data from field measurements. During the first stage of the project, a prediction of inflow to the D-holes, an array of six parallel closely spaced 100m boreholes, was made based on data from six other boreholes. This data included fracture geometry, stress, single borehole geophysical logging, crosshole and reflection radar and seismic tomogram, head monitoring and single hole packer test measurements. Maps of fracture traces on the drift walls have also been made. The D-holes are located along a future Validation Drift which will be excavated. The water inflow to the D-holes has been measured in an experiment called the Simulated Drift Experiment.

The paper reviews the Simulated Drift Experiment validation exercise. Following a discussion of the approach to validation, the characterization data and its preliminary interpretation are summarised and commented upon.

A particular strength of the exercise was the participation of three independent modelling teams from AEA Technology Harwell Laboratory, Golder Associates and Lawrence Berkeley Laboratory. They had access to the same data but formulated their own conceptual models and interpreted parameters for independently developed computer codes. Their work is reviewed, and the predictions of groundwater inflow to D-holes are compared with each other and with independent measurements.

The major achievement of this work is that it has proved feasible to carry through all the complex and interconnected tasks associated with the gathering and interpretation of characterization data, the development and application of complex models, and the comparison with measured inflows. It has been a valuable learning exercise which has provided detailed feed-back to the experimental and theoretical work required for measurements and predictions of flow into the Validation Drift.

PROJET DE STRIPA :
VALIDATION DE MODÈLES D'ÉCOULEMENT DANS LES FISSURES

Alan Herbert
AEA Technology, Harwell
William Dershowitz
Golder Associates Inc.
Jane Long
Lawrence Berkeley Laboratory
David Hodgkinson
INTERA-ECL

RESUME

La Phase III du Projet de Stripa a notamment pour objectif de mettre au point et d'évaluer des méthodes permettant de prévoir l'écoulement de l'eau souterraine et le transport des nucléides dans un volume déterminé non étudié du granite de Stripa, puis de comparer les résultats calculés aux mesures effectuées sur le terrain. Pendant la première phase du projet, on a fait une prévision relative à l'infiltration dans les forages-D, réseau de six forages parallèles rapprochés de 100 m de profondeur, en se fondant sur les résultats obtenus dans six autres trous de forage. Les données prises en compte étaient les suivantes : caractéristiques géométriques des fissures ; contraintes ; diagraphie individuelle des forages ; résultats des investigations par radar de la zone entre les forages et des phénomènes de réflexion et tomogrammes sismiques ; mesures de contrôle de la hauteur d'eau et résultats d'essais avec obturateur dans des forages individuels. Des cartes des traces de fissures sur les parois de la galerie ont été également dressées. Les forages-D sont situés le long du tracé d'une galerie de validation qui sera excavée ultérieurement. L'infiltration d'eau dans les forages-D a été mesurée à l'occasion de l'Expérience effectuée dans une galerie simulée.

Les auteurs analysent les résultats de l'Expérience effectuée dans une galerie simulée. Après avoir exposé la méthode de validation, les auteurs résument et commentent les données obtenues pour la caractérisation et la première interprétation qu'ils en font.

L'un des aspects les plus enrichissants de ces travaux de modélisation est qu'ils ont été menés par trois équipes de recherche indépendantes (AEA Technology Harwell Laboratory, Golder Associates et Lawrence Berkeley Laboratory). Toutes trois ont reçu les mêmes données mais elles ont formulé leurs propres modèles théoriques et interprété les paramètres compte tenu de programmes de calcul élaborés de façon indépendante. Les auteurs analysent le travail accompli et comparent entre elles ainsi qu'avec des mesures effectuées par d'autres équipes de recherche les prévisions relatives à l'infiltration de l'eau souterraine dans les forages-D.

Le mérite principal de ces travaux est d'avoir prouvé qu'il était possible de réaliser toutes les tâches complexes et interdépendantes liées à la collecte et à l'interprétation des données nécessaires pour définir les phénomènes, à l'élaboration et à la mise en oeuvre de modèles complexes et à la comparaison avec les infiltrations mesurées. Ces activités ont été riches d'enseignements et ont fourni en retour des informations détaillées qui pourront être utilisées dans les travaux expérimentaux et théoriques nécessaires pour mesurer et prévoir l'écoulement dans la galerie de validation.

1 Introduction

Safety assessments of the geological disposal of radioactive wastes require reliable predictions of the likely water-borne transport of radionuclides. Low permeability hard rocks, in which the water flow is predominantly through fractures, are being investigated as possible host formations in a number of countries.

The techniques for characterizing and modelling fluid movement and solute transport through low permeability heterogeneous fractured rocks are less developed than those used for higher permeability water supply aquifers and oil reservoirs. Also, the stringent safety requirements for the release of radioactive substances to our environment, and the need to make predictions over long time scales, place exacting requirements on proving the validity of the experimental and theoretical methods.

In response to this challenge, the international Stripa Project has pioneered the development of new techniques at an underground research laboratory in fractured granitic rock in central Sweden [1,2]. In particular, the feasibility and validity of fracture flow modelling, including data collection and interpretation, are being examined as part of Phase 3 of the Stripa Project [3]. This work is focussed on a previously unexplored volume of rock, known as the Site Characterisation and Validation (SCV) block, and consists of two major cycles of data gathering, prediction and validation.

This paper summarises a validation exercise based on the Simulated Drift Experiment (SDE) in the SCV block. The objective was to predict the inflow of water into six pilot boreholes (D-holes) drilled along the line of a future Validation Drift, based on a wide ranging database of fracture characteristics. Subsequently, the flows were measured and compared with the predictions.

The plan of the paper is as follows. The approach to validation is presented in Section 2, followed by an overview of the characterization data and their preliminary interpretation in Section 3. Summaries of the work of the three independent modelling groups are given in Section 4 (AEA Technology Harwell Laboratory), Section 5 (Golder Associates) and Section 6 (Lawrence Berkeley Laboratory). Section 7 compares the predictions with each other and with the measured inflows, and Section 8 presents some conclusions and the outlook for the future.

2 Approach to Validation

In general, there are two major aspects of groundwater transport models which require validation [4]. The first aspect relates to the physical and chemical processes such as advection, diffusion and sorption. The second concerns the structures within which the processes operate.

In the SDE validation exercise, the only process considered was groundwater flow. The physical law describing this process is taken to be that groundwater flux is proportional to the hydraulic head gradient. This has been extensively validated for porous media (Darcy's law) and for slow flow along tubes (Poiseuille flow) and thus does not need to be considered further in the present study.

The key questions are through what structures do the flows take place, and how are they characterized and modelled? Thus this study is purely concerned with the validation of models for the distribution of hydraulic transmissivity through the rock mass.

The evolving approach used for the present validation exercise is discussed below.

2.1 Relevance and Purpose

The first consideration is the relevance of fracture flow models and the role that they are intended to fulfil, since their validity can only be judged with respect to their intended purpose. Clearly some type of fracture flow modelling is relevant since fractured rocks are being considered as hosts for radioactive waste repositories, and water-borne migration of radionuclides is a central concern. The purposes of fracture flow modelling within this exercise are to interpret geometrical and hydraulic measurements and to predict the pattern and approximate magnitude of flows within the SCV block. The project is concerned both with the fracture zones, which carry most of the water through the site, and with the averagely fractured rock, which would be the most favoured location for waste canisters.

2.2 Performance Measures

A set of performance measures are being developed which can be used to assess the utility of the characterization, interpretation and modelling approaches. These address the order-of-magnitude of the total flow and its spatial distribution in the fracture zones and averagely fractured rock.

A key element of the Stripa validation stategy is that predictions should be made in the absence of any knowledge of the corresponding measurements, in order to obtain an objective measure of the degree of validation.

2.3 Peer Review

Fracture flow modelling work for the Stripa Project is overseen by the Stripa Fracture Flow Modelling Task Force which is a peer review group consisting of delegates from each of the participating countries. It guides and reviews the work and in particular is charged with recommending criteria for the verification and validation of the fracture flow models.

2.4 Iteration and Interaction

The level of knowledge of fracture flow characterisation and modelling is developing throughout the course of the Stripa Project. Thus it was not possible to specify all details at the outset. In view of this, the SCV project has made a virtue of necessity and has formulated a project plan which explicitly includes iteration and feedback from earlier phases [3]. In particular, the validation of fracture flow models is proceeding in two phases, of which the SDE exercise is the first.

The SCV project has been considerably strengthened by cross-fertilisation of ideas resulting from close interactions between experimentalists, from a wide variety of disciplines, and modellers.

2.5 Multiple Approaches

Confidence in our ability to predict fluxes of groundwater through fractured rocks is enhanced by pursuing a number of independent modelling approaches. This strength through diversity is achieved within the SCV project by involving three modelling groups from AEA Technology Harwell Laboratory [5,6], Golder Associates [7] and Lawrence Berkeley Laboratory [8,9]. The groups all had access to the same set of characterization data including a preliminary interpretation [10,11,12] and were free to form their own conceptual model of the flow paths through the rock mass, and to interpret parameter values from the measured data. Also, they used independently developed computer codes.

This resulted in three models all of which are consistent with the available data. The spread of results from the three groups gives some indication of the uncertainties and biases arising from conceptual model formulation and data interpretation [13].

2.6 Verification

The numerical accuracies of the computer codes used in the SDE validation exercise have been verified for a range of problems as part of their development. In addition, a cross-verification exercise is underway within the Stripa project.

3 Characterization Data and Preliminary Interpretation

A perspective view of the SCV block is shown in figure 1.

Three complementary techniques have been used, to locate five extensive planar geological features within the SCV block, namely geophysical logging, single-hole and cross-hole radar, and cross-hole microseismic. All features are irregular and appear in the tomograms as a series of connected patches rather than as well defined planar zones.

Single borehole hydraulic measurements have been carried out in the N and W boreholes using packer spacings of between 1m and 7m, and have been interpreted using a radial flow model to give transmissivities of the local rock-mass [10,11]. The correlation between transmissivity and the locations of the geophysical zones is not particularly high, indicating that transmissivity is distributed unevenly throughout the plane.

A considerable effort has been devoted to characterizing the statistical properties of the averagely fractured rock between the fracture zones. The flow in this part of the rock mass is primarily confined to a system of interconnected fractures with sub-millimetre apertures and extents of the order of metres. These are far too numerous to characterise and model deterministically, and thus they are considered in a statistical framework.

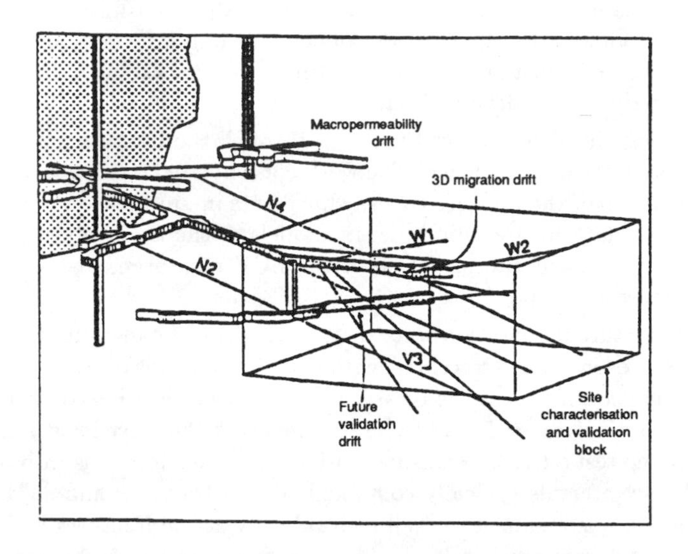

Figure 1: Perspective view of the Site Characterisation and Validation (SCV) block. The x and y axes are mine East and North respectively.

Data on fracture orientations have been obtained from observations in drifts, from oriented cores and from televiewer measurements in boreholes. This raw data has been corrected to take account of biases due to the orientation of the drifts and boreholes, and interpreted into sets with similar orientations. This analysis yielded two well determined clusters and one poorly defined sub-horizontal set.

A fracture trace length distribution was obtained by mapping more than 150m of scanlines in tunnels. In order to deduce a distribution for fracture lengths, this data needs to be corrected for truncation and censoring. Statistics on the spacings between fractures in each set have been compiled from observed intersections with the drill core. Fracture transmissivity distributions have been determined from the single-hole hydraulic test data.

4 AEA Technology Harwell Laboratory Modelling

AEA Technology Harwell Laboratory has developed and applied the NAPSAC discrete fracture code within the Stripa Project [5,6].

The NAPSAC code is able to include known fracture zones explicitly and to generate a network of fractures stochastically in parts of the rock-mass where only statistical properties of the fracture system are known. A number of realisations of the statistical fractures are generated which conform to the observed statistics. This allows the uncertainties associated with unobservable details of the fracture system to be quantified in addition to mean flow rates.

NAPSAC is capable of including variability of the transmissivity within fracture planes. However, in view of the lack of relevant channelling information at that time, the SDE simulations assumed constant fracture apertures. For consistency, this assumption was also made in the analysis of hydraulic data.

The modelling aimed to predict the fluxes through the fracture network from observations of individual fracture properties. The work concentrated on simulating the averagely fractured rock rather than the fracture zones, which were in any case only very simply characterized. The input data for fracture network models of the averagely fractured rock was derived from the preliminary interpretation of the SCV measurements made by the other principal investigators at Stripa.

Since only fracture intersection trace distributions can be measured, an analytic relationship between trace length and fracture length distributions was derived. This relationship accounted for orientation biases in the measurements. However, to account for censoring of the measured intersection traces, it had to be assumed that the trace length distribution was lognormal. Hydraulic tests derived transmissivities for packer intervals in boreholes around the SCV site. These intervals typically contained several fractures and so, to derive single fracture properties, it was assumed that all hydraulically active fractures had effective apertures sampled from the same distribution. All these fractures had to be accounted for, and an iterative computer program calculated maximum likelihood estimates for a lognormal fit to their transmissivity distribution.

Computational constraints made it unfeasible to construct a discrete fracture model of the whole SCV site and so representative cubes of the fracture network were simulated on a variety of scales. These simulations predicted the effective network permeability to be largely independent of details of the realisation of the fracture statistics, and of the size of the region modelled, for simulations on a scale larger than 8m. This surprising result was not inconsistent with relatively few fractures carrying most flow, and it was decided that the most appropriate means to predict fluxes across distances of 100m or more was to use an equivalent-porous-medium model characterized by fracture network simulations. Boundary conditions were interpolated from conventional porous medium models of the mine region. Fracture zones were included explicitly in this model and the resulting inflow to the D-hole array is shown in figure 2. No prediction is made as to the distribution of flow within a given fracture zone, only of the total flux from each zone and flux outside the zones averaged over 10m intervals. The predicted inflow is dominated by the GH,GB and GI zones.

The most important test of the approach is the prediction of the flux from the averagely fractured rock. This directly tests the interpretation of fracture network properties, since there is no explicit calibration of the model against similar inflows elsewhere. In order to investigate this in more detail and predict the distribution of flux to the 0.5m intervals measured, discrete fracture network simulations were used. On this scale, the details of the network are significant and predictions are stochastic: the fracture network approach is most appropriate. Boundary conditions for an 8m cube around the D-holes were inferred from the SCV site model, and several realisations of the network model were calculated. These models predicted that a small proportion of the intersections would carry a large part of the flow, and that all but one or two intersections would carry flows smaller than the measurement limit. This is a significant prediction given the total flux from the averagely fractured rock, but a better test of the model prediction will come when the flows are remeasured with greater accuracy.

5 Golder Associates Modelling

As shown in the AEA Technology Harwell modelling, the limitations of current computer technology make it impossible to model the estimated 10^8 fractures within the SCV site. However, the patterns of head and hydraulic connection seen within the SCV project clearly demonstrate the desirability of such a programme. The approach taken by Golder Associates is to divide the SCV block into two zones: i) a detailed model region, a 20m diameter cylinder around the D-holes where fractures are modelled at their full intensity, and ii) a boundary region, a 200m cube around the D-holes, where fracture intensity and conductivity are reduced to facilitate simulation, but other fracture properties are preserved.

This approach allows preservation of the complex boundary conditions arising from the network of drifts and stopes at the site, and avoids the necessity of applying strong constant head boundary conditions too close to the D-holes (figure 1). The disadvantage of the approach is that the properties of the boundary region depend strongly upon the calibration carried out, and determine the magnitude of the fluxes within the D-holes. As shown below, the coarse calibration carried out for the boundary region resulted in significant errors in the SDE prediction carried out by this approach. The calculations were performed primarily with the FracMan [14] and MAFIC [15] discrete fracture codes developed by Golder Associates.

The discrete fracture approach depends upon the derivation of realistic fracture properties. New techniques were developed for derivation of all of the parameters necessary for discrete fracture modelling.

Orientation was derived from a combination of oriented core and drift mapping statistics by a maximum likelihood clustering technique [7] which determines Fisher distribution parameters for derived fracture sets.

Transmissivity and intensity were derived from fixed-interval-length well tests on 7-metre intervals by a method adapted from Osnes et al. [16]. The method recognizes that there is very poor correlation between fracture intensity and location as seen in boreholes, and apparent packer interval transmissivity. This observation is consistent with a situation in which fractures exhibit a lognormal distribution of transmissivity, such that a large number of fractures exhibit no measurable transmissivity, while a few fractures may support very large fluxes. The method assumes that the net transmissivity of test zone is equal to the sum of the at-borehole transmissivities of the fractures that intersect the test zone. The best estimate for conductive fracture intensity can then be derived from the percentage of packer intervals in which there was no measurable flow, and the best estimate for the distribution of transmissivity is found by deconvolution of the distribution of packer interval transmissivities as the distribution of the sum of conductive fracture transmissivities.

Fracture size was derived from the distribution of fracture tracelengths observed in trace maps. No analytical relationship is available which relates censored, biased fracture tracelengths to fracture size. This transformation was therefore carried out utilizing simulated sampling, in which all bias and censoring processes were preserved.

Fracture location was derived by utilizing the War-Zone fracture geometric conceptual model [7]. In this model, fracture location is defined by a Poisson point process of fracture centres, with increased intensity in fracture zones defined by geologic or geophysical evidence. In this case, geophysical evidence was used to define six fracture zones. Since no evidence was available at the time to justify the use of different fracture geometric or hydrologic properties within fracture zones, a 60% elevation in fracture intensity was used for all fracture zones, based upon fracture statistics in the N- and W-holes.

Due to time constraints, only limited model calibration was carried out to derive the properties of the boundary region. It was originally planned to calibrate the properties of the boundary region against head and flux measurements throughout the SCV site. This would include the effects of drifts and stopes within the SCV block, and would reflect documented hydrologic connections through fractures and fracture zones within the block. However, only one preliminary calibration based upon flux into a single hole was completed. This calibration was made against the flux at the 3-D Drift migration borehole (74 litres/day). Later, following the comparison of the preliminary predictions with measured inflows, the boundary region model was recalibrated against the fluxes in the N- and W-holes. In both cases, fracture properties within the detail model region were not changed; only the boundary region conductive intensity, fracture radius cut-off for simulation, and mean transmissivity were changed.

Only six Monte Carlo simulations were carried out with the model calibrated for the SDE prediction. Five hundred Monte Carlo simulations were carried out with the model calibrated against the N- and W-borehole fluxes. The change in the boundary intensity and transmissivity between the first and second models had a dramatic effect upon the magnitude of flux into the drift, increasing the maximum likelihood estimate of total flux from 2 litres/day to 2 litres/hour. The shape of the SDE inflow response, which is controlled by the full intensity detail model region, was not changed between the models. Figure 2 shows the predicted inflow for the model calibrated to the N- and W-borehole fluxes.

6 Lawrence Berkeley Laboratory Modelling

Lawrence Berkeley Laboratory (LBL) is sponsored by the US Department of Energy to participate in the hydrological modelling of the SCV block. Over the past few years LBL have developed a suite of numerical codes for modelling flow and transport through fractured rock masses. In this work, the channel network generation code CHANGE [17] was used to define a regular grid of conductors within each fracture zone. Also, the three-dimensional finite-element code TRINET [18] was used to model the response of the zones to hydrological perturbations.

The LBL approach [8,9] was to focus on flow through fracture zones, and thus the averagely fractured rock was taken to be impermeable. This concentration of effort on the fracture zones allowed their properties to be considered in greater detail than in the AEA Technology Harwell or Golder Associates work. In particular, the number and positions of zones were re-examined. Also, the heterogeneous substructure within the zones was examined by discretising the zones with a regular grid of equally conductive channels with a fraction of missing links.

The hydrological and geophysical data has been reinterpreted with an eye to allocating all of the high transmissivity regions in the N and W boreholes to fracture zones. This has been achieved by introducing a new zone, GB′, between the GB and the GC zones and parallel to them.

Cross-hole hydraulic data is needed to calibrate the equivalent discontinuum zone model. Unfortunately, no controlled cross-hole tests were performed in advance of this exercise, although such information will be available for the Validation Drift predictions. As a learning exercise, synthetic results of a steady-state cross-hole test were constructed based on some ad hoc measurements performed by the British Geological Survey.

Equivalent patterns of one-dimensional conductors representing the inhomogeneity of the zones have been determined using a simulated annealing algorithm. Several configurations of conductors were found to match the synthetic head data extremely well.

The predicted heads are independent of the conductance of the channels. This was calibrated in order to give the observed flow from W2. The calibrated model was then used to predict flow into the open D-holes, which were treated as 'one big hole' since their separations ($\sim$ 1.4m) are less than the resolution of the grid (10m). The predicted inflow was 8.9 ± 0.1 litres/minute where the variability arises from using widely different network configurations. Thus with the present level of cross-hole information and assumption about overall heterogeneity, the annealing has rather little effect.

The above prediction is directly proportional to the measured flow in W2, which is somewhat anomalous since the transmissivity of W2 is much higher than the other holes. Consequently, the calibration has been repeated using ad hoc measurements of outflows from the other N and W holes. The preliminary prediction is that the mean and standard deviation of the sum of the D-hole flows is 3.1 and 3.1 litres/minute respectively. The distribution of inflow along the D-hole array is shown in figure 2. By calibrating the model sequentially leaving out one measurement of inflow each time, a series of models was used to predict the inflow measurement that was left out. This resulted in an average estimate of prediction error of 4.6 litres/minute.

7 Comparison of Measurements and Predictions

The first, but not necessarily the most important, comparison is of the total steady-state influx into the D-hole array. This is shown in table 1. The numbers given in this and the following tables are preliminary. The radial flow estimate is based on average permeabilities for the fracture zones and average rock [19].

	Mean Inflow (litres/min)	Range (litres/min)
Measurement	1.71	1.67-1.75
Harwell	1.45	0.36-5.80
Golder - 1	0.055	0.001-0.156
Golder - 2	1.5	0.5-95
LBL	3.1	0.0-7.7
Radial Flow	2.3	0.4-7.0

Table 1. Comparison of measured and predicted total inflows to the D-hole array

The immediate comment on table 1 is that, based on the mean values and ranges, all the predictions are reasonably accurate except for the first Golder prediction which suffered from calibration problems. Whilst this is interesting, such a gross measure of performance should not be taken to validate or invalidate any of the modelling approaches at this stage. It is necessary to look in more detail at the pattern of distribution along the holes. This is shown in figure 2, where the predictions from all three groups are compared with the measured results for inflow to 0.5m sections of the combined holes. All modelling groups agree that the fracture zones dominate the inflow, as found experimentally. However, there is less than perfect agreement between the predictions and measurements for the locations of major inflows. This arises, in part, due to the poorly characterised inhomogeneities within fracture zones.

A disappointing feature of the SDE validation exercise is the relatively minor extent to which it has been possible to confront predictions and measurements of inflow from the averagely fractured rock. This part of the rock-mass is likely to be favoured by repository designers as locations for waste canisters, and predictions of the flux are important for assessing near-field corrosion and radionuclide release rates.

The only measured quantity relating to the averagely fractured rock is the difference between the total inflow from all the D-holes and the total inflows from the 25m and 90m fracture zones. Table 2 shows that there is order-of-magnitude agreement between predictions and the measurement for the integrated flow from the average rock, except for the original Golder calculation.

	Integrated inflow from averagely fractured rock (litres/min)	
	Best Estimate	Range
Measurement	0.2	0.0-0.3
Harwell	0.09	0.009-0.9
Golder - 1	0.014	0.0-0.028
Golder - 2	0.4	0.05-50
Radial Flow	0.05	0.005-0.5

Table 2. Comparison of theory and experiment for the integrated inflow to the D-holes from averagely fractured rock.

8 Conclusions and Outlook

The major achievement of the SDE validation exercise is that it has proved feasible to carry through all the complex and interconnected tasks associated with the gathering and interpretation of characterization data, the development and application of complex models, and the comparison with measured inflows. It has been a difficult task even with the limited ambition of predicting groundwater flux rather than solute transport.

A major area of difficulty, which has necessitated considerable development during the project, has been the interpretation of geometrical and hydraulic measurements to give the input parameters required by the models. The conceptual model for the transmissivity distribution through the rock mass, and the interpretation of input parameters, is not unique as witnessed by the differences between the models from the three participating groups. Techniques for the interpretation of characterization data and for calibrating models are the most important area for future research and development. In particular, conditional simulation approaches, where known data is incorporated exactly and inferred information is incorporated statistically, should be investigated.

Lack of appropriate data has meant that little account has been taken of inhomogeneity within fractures and fracture zones in the present study. It is hoped that channelling and cross-hole hydraulic test data from ongoing projects at Stripa will allow greater consideration of these effects in future predictions. In particular the question of whether only a fraction of fractures are conductive, or whether they are all channelled, needs to be resolved.

The single-hole hydraulic tests which provided input to the modelling used packer separations which were too long to allow single fractures to be isolated. This resulted in significant ambiguities in the interpretation of fracture transmissivity distributions. It would be interesting to try some tests with shorter packer intervals to investigate whether these would help resolve these problems.

In the hydraulic characterization and modelling aspects of this exercise, the majority of the effort has been focussed on the averagely fractured rock, rather than on the fracture zones. In contrast, the D-hole inflow measurements concentrated on flows from the fracture zones. It is hoped that the inflow measurements with an improved measurement limit will be performed in future, in order to discriminate between alternative interpretations.

The outlook for the future is that the valuable experience gained from this exercise will be used to improve the predictions of flow into the Validation Drift excavated along the line of D-holes. These predictions will benefit from further characterization data, especially from regions in the vicinity of the drift. The flows will be measured by fixing plastic sheets to the walls.

It is vital that the flow measurements to be made in the Validation Drift should be precise enough to discriminate between alternative interpretations. This should be specified in conjunction with the determination of precise validation criteria.

A number of further phenomena will need to be assessed in connection with inflows to the drift, including the effect of excavation damage and stresses, and two-phase flow near the walls. It is not clear that the benefits of direct access via the drift outweigh the disadvantages of having to account for these extra phenomena.

In addition to the above groundwater flux validation exercise, it is proposed to carry out a tracer test close to the Validation Drift. The results will be compared with predictions made using further features of the models described here. It should be emphasised that the characterization and modelling of transport phenomena is a considerably more difficult undertaking than the flow modelling discussed in this paper. However, doubtless the Stripa Project team will rise to this challenge.

Acknowledgements

The help and encouragement from all those involved with the Stripa Project, especially the members of the Fracture Flow Modelling Task Force, is gratefully acknowledged. The work at Lawrence Berkely Laboratory was funded by the Manager, Chicago Operations, Repository Technology Program, REpository Technology and Transportation Division, U.S. Dept. of Energy under contract number DE-AC03-76SF00098.

References

[1] Stripa Project, Executive Summary of Phase 1, Stripa Project Technical Report 86-04, SKB, Stockholm, 1986.

[2] Stripa Project, Executive Summary of Phase 2, Stripa Project Technical Report 89-01, SKB, Stockholm, 1989.

[3] Stripa Project, Program for the Stripa Project Phase 3: 1986-1991, Stripa Project Technical Report 87-09, SKB, Stockholm, 1987.

[4] Chin-Fu Tsang, Comments on Model Validation, Transport in Porous Media, 2, 623-629, 1987.

[5] A.W. Herbert and B.A. Splawski, A Prediction of Flows to be Measured in the Stripa D-hole Experiment: An Application of the Fracture Network Approach, Draft, September 1989.

[6] A.W. Herbert and J.E. Gale, Fracture Flow Modelling of the Site Characterisation and Validation Area in the Stripa Mine, Harwell Laboratory Report TP.1336, 1989.

[7] J. Geier, W. Dershowitz and G. Sharp, Stripa Simulated Drift Experiment Prediction Report, Golder Associates Report, August 1989.

[8] J.C.S. Long, K. Karasaki, A. Davey, J. Peterson, M. Landsfeld, J. Kemeny and S. Martel, Preliminary Prediction and Inflow into the D-holes at the Stripa Mine, Lawrence Berkeley Laboratory Report, April 1989.

[9] J.C.S. Long, K. Karasaki, A. Davey, J. Peterson, M. Landsfeld and S. Martel, Preliminary Calculations of Inflow to the D-holes at the Stripa Mine.

[10] D. Holmes, Site Characterisation and Validation - Single Borehole Hydraulic Testing Stage 1, Stripa Project Technical Report 89-04, SKB, Stockholm.

[11] O. Olsson, J. Black, J. Gale and D. Holmes, Site Characterisation and Validation Stage 2 - Preliminary Predictions, Stripa Project Technical Report 89-03, SKB, Stockholm, 1989.

[12] P. Andersson, Site Characterisation and Validation - Geophysical Single Hole Logging Stage 3, Stripa Project Technical Report 89-07 SKB, Stockholm, 1989.

[13] M.C. Thorne and J-M. Laurens, The Development of an Overall Assessment Procedure Incorporating an Uncertainty and Bias Audit, Proceedings of the International Symposium on the Safety Assessment of Radioactive Waste Repositories, Paris, October 1989.

[14] Golder Associates Inc., FracMan Version 2.0 Interactive Rock Fracture Geometric Model: User Documentation, prepared for Battelle Office of Waste Technology Development, Willowbrook, Illinois, USA, 1988.

[15] Golder Associates Inc., MAFIC/T Version 1.0 Matrix/Fracture Interaction Code with Solute Transport: User Documentation, prepared for Battelle Office of Waste Technology Development, Willowbrook, Illinois, USA, 1988.

[16] J.D. Osnes, A. Winberg and J. Andersson, Analysis of Well Test Data: Application of Probabilistic Models to Infer Hydraulic Properties of Fractures, Topical Report RSI-0338, RE/SPEC Inc., Rapid City, South Dakota, USA, 1988.

[17] D. Billaux, J.P. Chiles, K. Hestir and J. Long, Three-Dimensional Statistical Modeling of a Fractured Rock Mass - An Example for the Fanay-Augères Mine, International Journal of Rock Mechanics and Mining Science, special issue on Forced Fluid Flow through Fractured Rock Masses, in press.

[18] K. Karasaki, A New Advection-Dispersion Code for Calculating Transport in Fracture Networks, LBL Report, in press, 1988.

[19] Veikko Taivassalo, private communication, 1990.

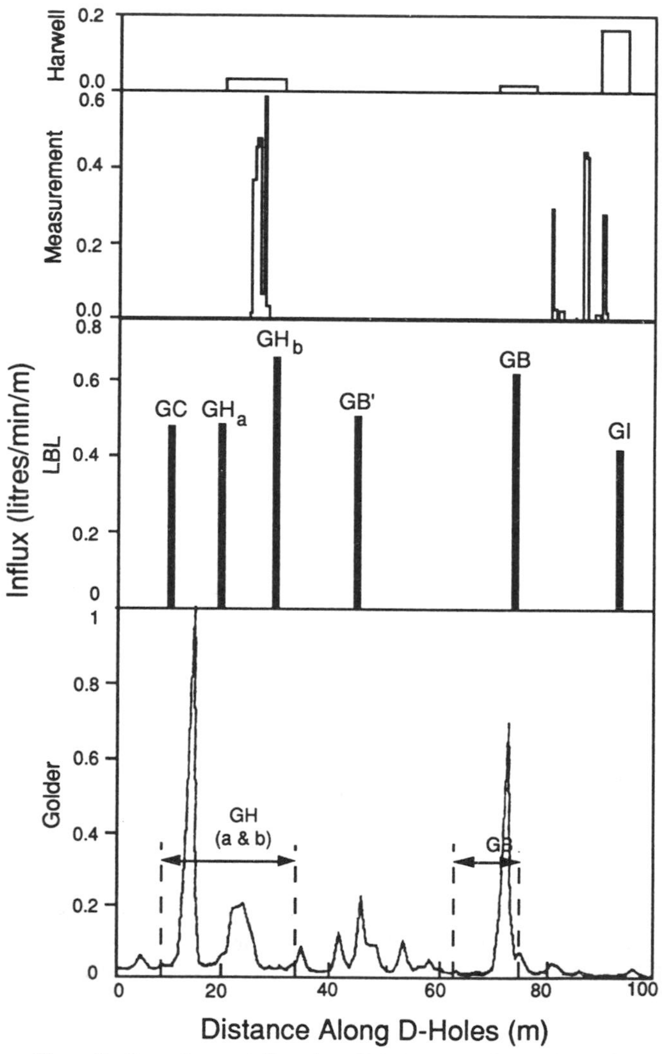

Figure 2: Groundwater influx along D-hole array. Comparison of predictions and experiment.

PRESENT STATUS OF DEVELOPMENT FOR SAFETY ASSESSMENT ON GEOLOGICAL DISPOSAL OF HIGH-LEVEL RADIOACTIVE WASTES

Motoi Kawanishi, Yasuharu Tanaka, Hiroya Komada
Abiko Research Laboratory, CRIEPI, Japan

ABSTRACT

The objectives of this study are to develop the computer codes analyzing the groundwater flow and radionuclide migration in the fractured rocks and to provide the reliable results needed for the safety assessment on the geological disposal of high-level radioactive wastes (HLW). In this paper, we introduce the outline of the whole safety assessment methods for HLW disposal, especially the computer codes to analyze the groundwater flow, heat transfer and radionuclide migration in the fractured rocks developed by CRIEPI (Central Research institute of Electric Power Industry) for the performance evaluation of natural barriers. Furthermore, we present a few applied results using these codes to the experimental results and a case study for the preliminary safety assessment.

EVALUATION DE LA SURETE DE L'EVACUATION DES DECHETS DE HAUTE ACTIVITE DANS DES FORMATIONS GEOLOGIQUES : ETAT D'AVANCEMENT DES TRAVAUX DE MISE AU POINT

RESUME

Cette étude a pour objectifs d'élaborer des programmes de calcul permettant d'analyser l'écoulement de l'eau souterraine et la migration des radionucléides dans les roches fissurées et de fournir les résultats fiables dont on a besoin pour évaluer la sûreté de l'évacuation des déchets de haute activité dans des formations géologiques. Dans le présent document, les auteurs exposent les grandes lignes des méthodes d'évaluation globale de la sûreté de l'évacuation des déchets de haute activité mises au point par le CRIEPI (Institut central de recherche de l'industrie de l'énergie électrique) pour évaluer les barrières naturelles. Ils mettent particulièrement l'accent sur les programmes de calcul élaborés pour analyser l'écoulement de l'eau souterraine, les transferts thermiques et la migration des radionucléides dans les roches fissurées. En outre, les auteurs présentent une comparaison entre quelques résultats obtenus au moyen de ces programmes de calcul et les résultats expérimentaux ainsi qu'une étude de cas en vue de l'évaluation préliminaire de la sûreté.

1. INTRODUCTION

In Japan, the "Long-Term Plan of Nuclear Power Development and Utilization"[1] was published in 1987 by the Japan Atomic Energy Commission. This plan pointed out the importance of the commercialization of the activities required for the establishment of nuclear fuel cycle and indicated the basic strategy for the realization of the geological disposal of high-level radioactive wastes(HLW) and that HLW should be disposed in the underground rocks which are layered at over a few hundred meters deep after the interim storage during 30-50 years in Japan. Therefore, it is necessary to establish the safety assessment method for HLW disposal.

CRIEPI has developed a set of safety assessment methods that is composed of the mathematical models to estimate the performance of the engineering and natural barriers, the food chain/ radiation dose. In this paper, we introduce the present status for development of these safety assessment methods, especially the computer codes to analyze the groundwater flow, heat transfer and radionuclide migration in the fractured rocks for the performance of natural barriers. Furthermore, we describe a few applied results to the experimental results and a case study for the preliminary safety assessment and present the good applicability of these codes.

2. COMPOSITION OF SAFETY ASSESSMENT CODES

Table I shows the composition of the major safety assessment codes developed by CRIEPI for the geological disposal of HLW. In these codes, the performance evaluation models for the engineering barrier have been developed by Komae Research Laboratory of CRIEPI[2]. On the other hand, in this paper, we describe the performance evaluation models for the natural barrier as follows.

2.1 Groundwater flow model

2.1.1 Shallow groundwater flow model (FEGM)

FEGM is a computer model to analyze the grounwater flow in saturated-unsaturated media by the finite element method(FEM) and is compatible to the radionuclide migration model FERM mentioned below[3]. This model has been modified from FEMWATER developed by ORNL[4] to be able to analyze the quasi-three and three dimenional problems. Furthermore, a sub-model for the salt intrusion analysis has been added to this model.

2.1.2 Deep groundwater flow model (GMF)

GMF is a computer model to analyze the saturated groundwater flow in the fractured rocks by the FEM[5]. In this model, the interaction of flow between the fracture and matrix of rocks can be represented by the discrete fracture model as well as by the dual porosity model. In the former model, the fracture and the rock matrix are respectively represented as a line element and an unisotropic porous element. In the latter model, the rectangular or circular block model for the rock matrix combined with a fracture can be used[6]. Table II shows the model list for these elements used in GMF. Furthermore, GMF is able to use the isoparametric elements with four or eight nodes for FEM analysis and is compatible to computer code RMF for the radionuclide migration analysis mentioned below.

2.2 Coupling model for groundwater flow and heat transfer (CHGR)

CHGR is a coupling analysis model for the groundwater flow and heat transfer in fractured rocks by using FEM[5]. In this model, it is considered that the density and viscosity of fluid are depend on the temperature and that the consolidation of rock matrix and the compressibility of fluid exist.

2.3 Radionuclide migration model

2.3.1 Mass transfer model in shallow groundwater (FERM)

FERM is a computer model to analyze the radionuclide migration in the saturated and unsaturated porous media as the two or three dimensional advective-dispersion process by using FEM and is compatible to FEGM code[3]. In this model, the linear and nonlinear distribution equilibrium models or the chemical reaction rate model can be used for the adsorption of radionuclides.

2.3.2 Mass transfer model in deep groundwater (RMF)

RMF is a computer model to analyze the radionuclide migration in the fractured rocks as the two dimensional advective-dispersion process by using FEM and is compatible to GMF code[5]. In this model, the linear and nonlinear distribution equilibrium models for the adsorption and the decay chain of radionuclides are considered. Especially, in the dual porosity element model, the advective-dispersive transport of nuclides in a flow direction are considered for the fracture part and the diffusion in a perpendicular direction to the flow is considered for the matrix block[7].

2.4 Food chain/radiation dose model (FORADO)

FORADO is a computer model to estimate the radiation dose for the safety assessment on the underground disposal of the low level and high level radioactive wastes. This model in which the migration path-way of radionuclides was predetermined as shown in Fig. 1 considering its generality and the own condition in Japan, was developed by CRIEPI based on the NRC model [3,8]. Furthermore, FORADO has recently been modified by adding some version-ups on an engineering workstation system by using the 32 bit computer for serviceable uses on the siting and safety assessment of radioactive waste management[9]. FORADO is composed of the internal radiation dose model(FORADO-I) and the external radiation dose model(FORADO-E) in consideration of Pub.2 and Pub.30 of ICRP.

3. APPLIED RESULTS

3.1 Validation for applicability of models

Fig. 2 shows an example of comparison between the calculated results by Huyakorn[7] and the calculated result by RMF of the present study on a problem for the advective-dispersive transport along the fracture and diffusion into porous matrix of a parallel-fractured porous medium system[7]. In this figure, The calculated values of breakthrough curves for the first component of a three-member chain are compared with each other. The overall agreement between the analytical solution and calculated results by Huyakorn and the present study using the refined grid for FEM elements is quite acceptable.

Fig. 3 shows an applied result by the CHGR code to a test problem for the case 1 of HYDROCOIN level 2[10]. This test is based on a field heat transfer experiment that is potentially useful for validating models of coupled groundwater flow and heat transfer. The optimal parameters for the hydraulic conductivity and thermal conductivity used in this calculation were identified by a try and error method in which the summation of absolute deviation of calculated temperature values from observed ones should be minimized. On those summation values of temperature deviations for a set of hydraulic and thermal conductivity values, as shown in Fig. 4, we can draw the isothermal curves so that the optimal values of parameters can be rationally estimated from relatively few experimental data.

From these results, it was shown that both simulation codes, RMF and CHGR have respectively good applicability. In the same way mentioned above, on the other hand, we have verified that the another simulation codes also have sufficient accuracy for their predictions.

3.2 Preliminary safety assessment for HLW disposal

We tried to apply the simulation codes, GMF, RMF and FORADO to the preliminary safety assessment for the geological disposal of HLW. In this study, we modelized roughly the geohydrological conditions on a generic site for HLW disposal and predetermined the other conditions needed for assessment.

Fig. 5 shows a calculated example of the groundwater flow vector pattern in the fractured rocks. Especially, the relatively high speed flow of groundwater occurs in the fracture of which the hydraulic conductivity was supposed to be greater than that of the matrix blocks. Based on these results, we simulated the radionuclide migration from the release point of a disposal facility and evaluated the radiation dose at an estimation point, which is the ground surface supposed to be 1000 meter distant from the facility, as shown in Fig. 6.

From those results estimated roughly, it was shown that the internal radiation dose to an individual person is sufficiently less than 10 μSv/y (1 mrem/y) that is an objective permitted dose value for LLW disposal in Japan.

4. CONCLUSION REMARKS

In this paper, the outline of the safety assessment method for HLW disposal, especially the performance evaluation models for natural barriers were introduced and a few applied results of those computer codes to the experimental results and the case study on a preliminary safety assessment for HLW disposal were described. The principal results of this paper are summarized as follows.

1) We developed three computer codes using FEM, that are GMF/RMF for the groundwater flow and radionuclide migration in fractured rocks and CHGR for the coupled groundwater flow and heat transfer.

2) From the comparison between experiments and calculated results, it was shown that those computer codes have good applicability.
3) Based on the simulated results for coupling analysis on heat and groundwater flow, we clarified that the optimal parameter values for hydraulic conductivity and thermal conductivity can be rationally estimated from relatively few data by making up the isothermal curves for the temperature deviations between observed and calculated values.
4) From a preliminary safety assessment for HLW disposal on the generic site under predetermined conditions, it was roughly shown that there is a sufficient possibility for the geological disposal of HLW in Japan.

REFERENCES

1. Japan Atomic Energy Commission (1987), "Long-Term Plan of Nuclear Power Development and Utilization", JAEC Report. (in Japanese)

2. Ohe, T., Mayuzumi, M. and Tsukamoto, M. (1989), "Analysis Codes for Engineering Barrier System Performance of High-Level Radioactive Waste Disposal", CRIEPI, Report No. T88049. (in Japanese)

3. Kawanishi, M. et al. (1987), "Computer Models for Safety Assement on Land Disposal of Low-Level Wastes", Proc. of the Symposium on Waste Management '87, Univ. of Arizona, Tucson, Vol. 3, pp. 175-180.

4. Yeh, G. T. and Ward, D. S.(1979), "FEMWATER: A Finite-Element Model of Water Flow Through Saturated-Unsaturated Porous Media", ORNL-5567, ORNL, Oak Ridge, Tennessee.

5. Kawanishi, M., Tsukamoto, M. and Kumai, N. (1988), "FEM Analsis on Transfer of Heat, Groundwater and Radionuclide in Rocks", Proc. of 32th Japanese Conf. on Hydraulics, Tokyo, Japan Soc. Civil Engineers, pp. 631-636. (in Japanese)

6. Huyakorn, P. S., Lester, B.H. and Faust, C.R. (1983), "Finite Element Techniques for Modeling Groundwater Flow in Fractured Aquifers", Water Resour. Res., 19-4, pp.1019-1035.

7. Huyakorn, P. S., Lester, B. H. and Mercer, J. W. (1983), "An Efficient Finite Element Technique for modeling Transport in Fractured Porous Media 1. Single Species Transport and 2. Nuclide Decay Chain Transport", Water Resour. Res., 19-3 and 19-5, pp. 841-854 and pp. 1286-1296.

8. U. S. NRC (1981), "Draft Environmental Impact Statement on 10 CFR Part 61 Licensing Requirements for Land Disposal of Radioactive Waste", NUREG-0782.

9. Kawanishi, M. et al. (1989), "Development of Synthetic Safety Assessment System "SYSA" for Land Disposal of Low-Level Wastes", Proc. of 1989 Joint international Wastemanagement Conf., vol. 1, ASME, JSME and AESJ, Kyoto, Japan, pp.407-414.

10. Hodgkinson, D. and Herbert, A. (1985), "Specification of a Test Problem for HYDROCOIN Level 2 Case 1: Thermal Convection and Conduction around a Field Heat Transfer Experiment", AERE -R. 11627.

Table I Safety Assessment Code for HLW Disposal (by CRIEPI)

Code	Contents of model	Solution technique	Boundary condition	Dimension	Remarks
GMF	①steady and unsteady groundwater flow ②combined flow model in fracture and matrix block of rocks ③4 type of elements	FEM	hydraulic pressure or flux	2D	This model is compatible to RMF.
RMF	①advective dispersion of solutes in fracture and matrix block of rocks ②decay chain of radionuclides ③linear and nonlinear sorption model ④4 type of elements	FEM	concentration or mass flux	2D	This code is compatible to GMF.
CHGR	①steady and unsteady groundwater flow and heat transfer in porous media ②coupling analysis between flow and heat ③linear isoparametric element	FEM	hydraulic pressure or flux;temperature or heat flux	2D, 3D axisym-metry	
FEGM	①steady and unsteady groundwater flow ②in unisotropic porous media ③linear isoparametric element (triangle,square,hexahedron)	FEM	hydraulic pressure or flux;rainfall or seepage flow	2D,3D, quasi-3D	This code is a model modi-fied from FEMWATER and compatible to FERM.
FERM	①advective dispersion of solutes in saturated and unsaturated media ②linear and nonlinear sorption models ③linear isoparametric element (triangle,square,hexahedron)	FEM	concentration or mass flux	2D,3D	This code is a model modi-fied from FEMWASTE and compatible to FEGM.

Table II List of Element Models for GMF and RMF

Name of Element	Charactaristics of Element	Applicicable Computer Code of the Element
① PORO	Two dimensional Anisotropic Porous Media(Isoparametric Element)	A, B, C
② FRAC	Model of Discrete Fracture	A, B, C
③ DBLP	Rectangular Block Model of Matrix in Dual Porosity Element	A, B
④ DBLC	Circular Block Model of Matrix in Dual Porosity Element	A, B

A : Groundwater Flow Analysis (GMF)
B : Radionuclide Migration Analysis(RMF)
C : Coupling Analysis between Heat Transfer and Groundwater Flow(CHGR)

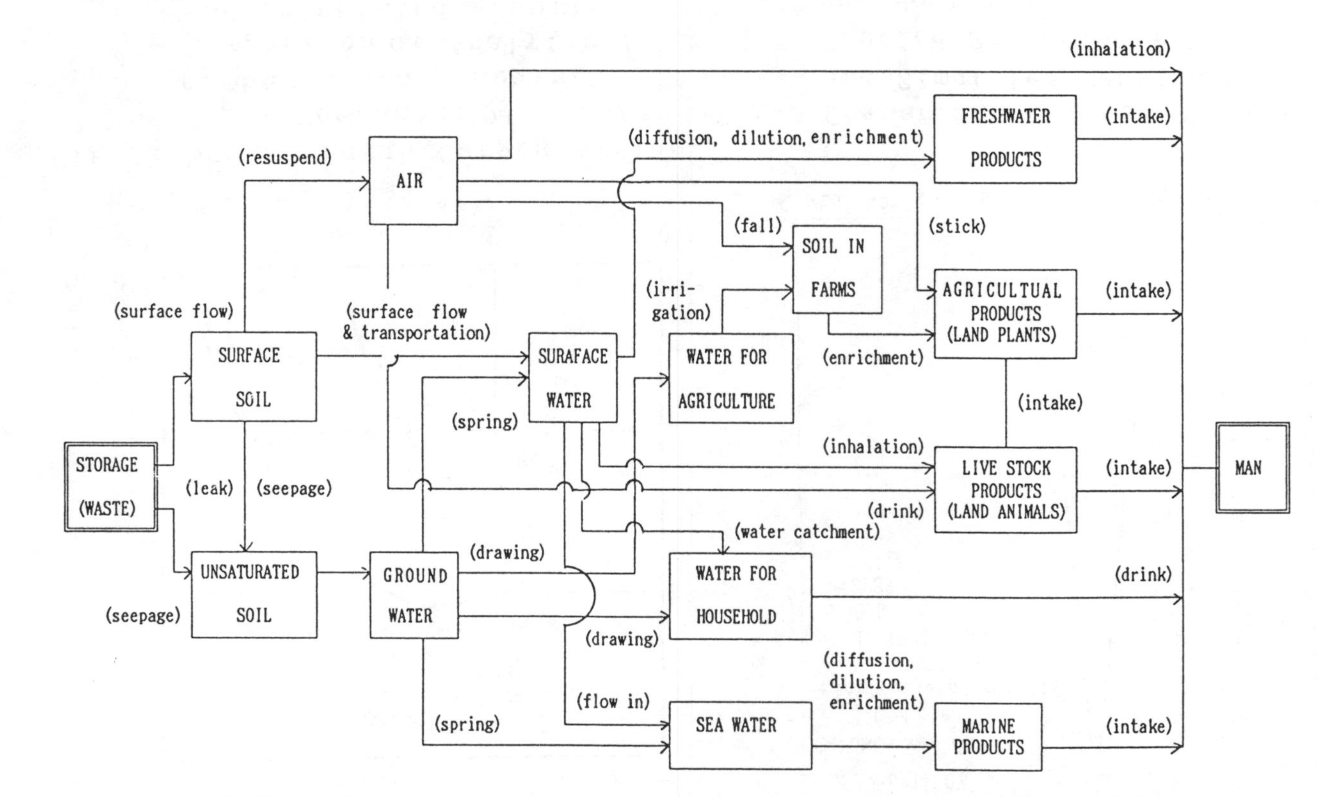

Fig. 1 Predetermined Migration Path-way of Radionuclides in Environments for Safety Assessment (FORADO)[3]

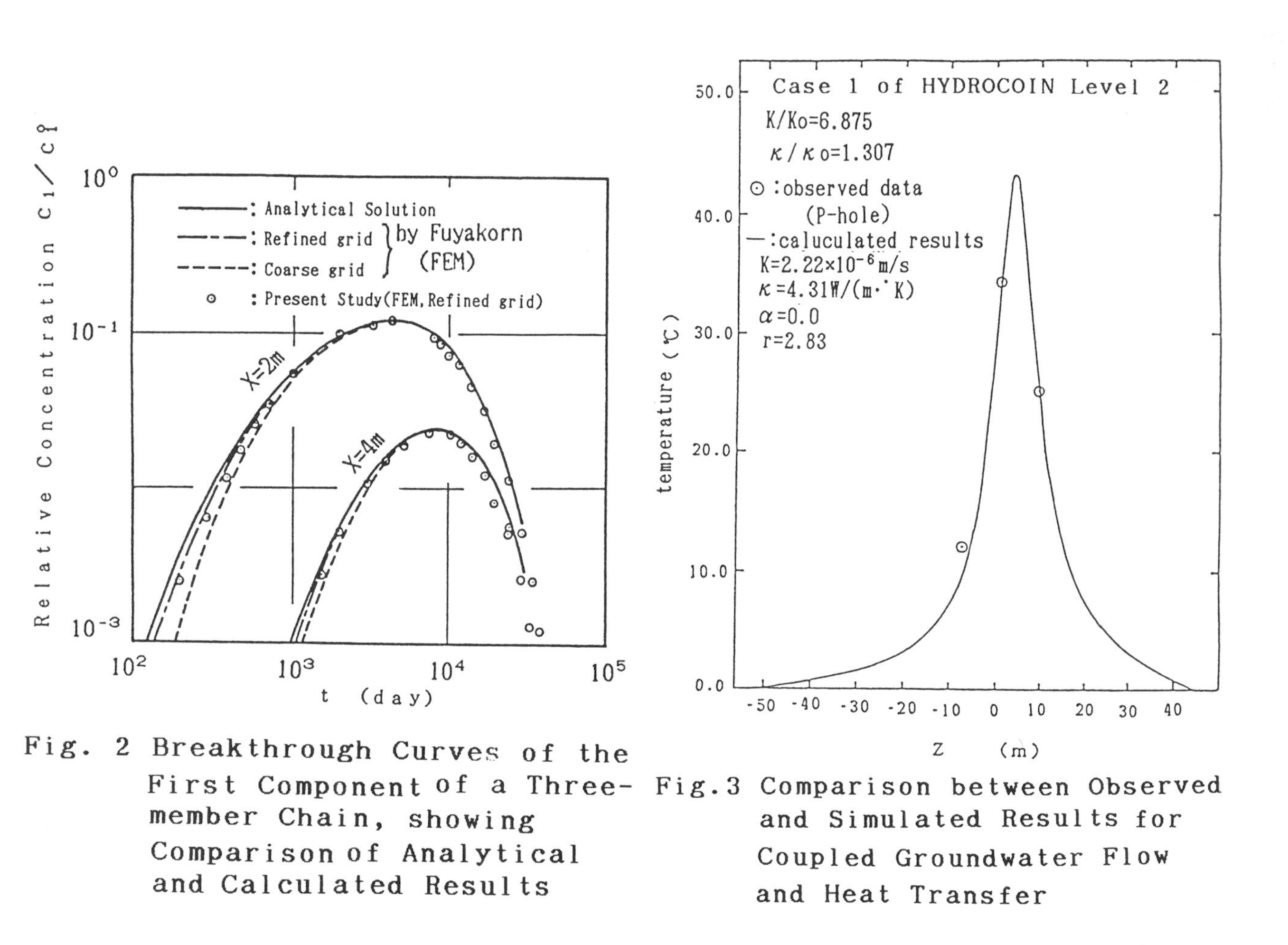

Fig. 2 Breakthrough Curves of the First Component of a Three-member Chain, showing Comparison of Analytical and Calculated Results

Fig.3 Comparison between Observed and Simulated Results for Coupled Groundwater Flow and Heat Transfer

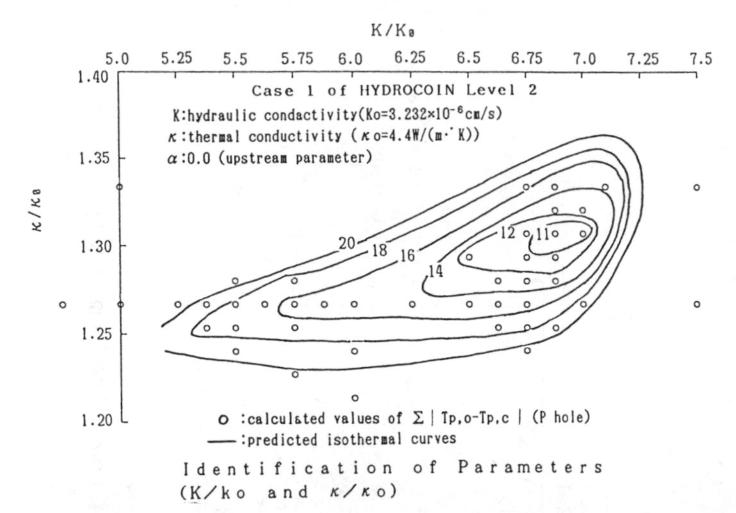

Fig. 4 Isothermal Curves Contoured for Temperature Deviations between Observed and Calculated Values for Identification of Hydraulic and Thermal Conductivities

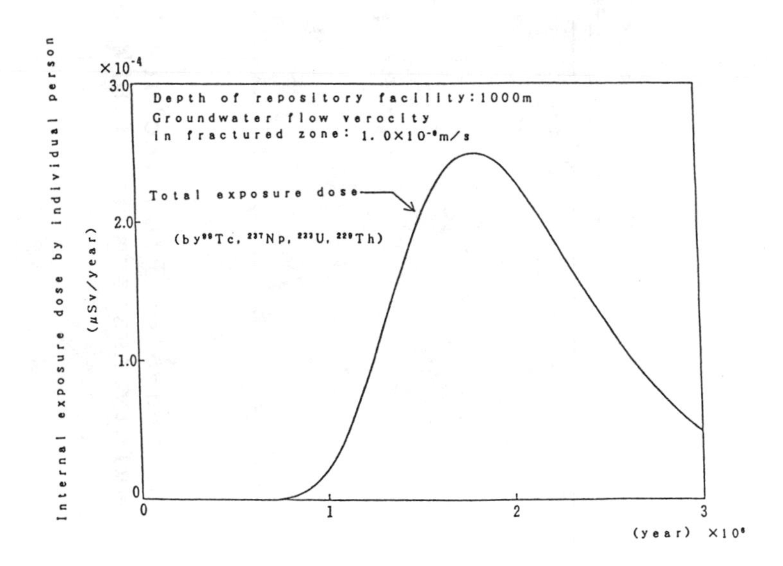

Fig. 6 Estimated Result of Radiation Dose for Preliminary Safety Assessment of HLW Disposal

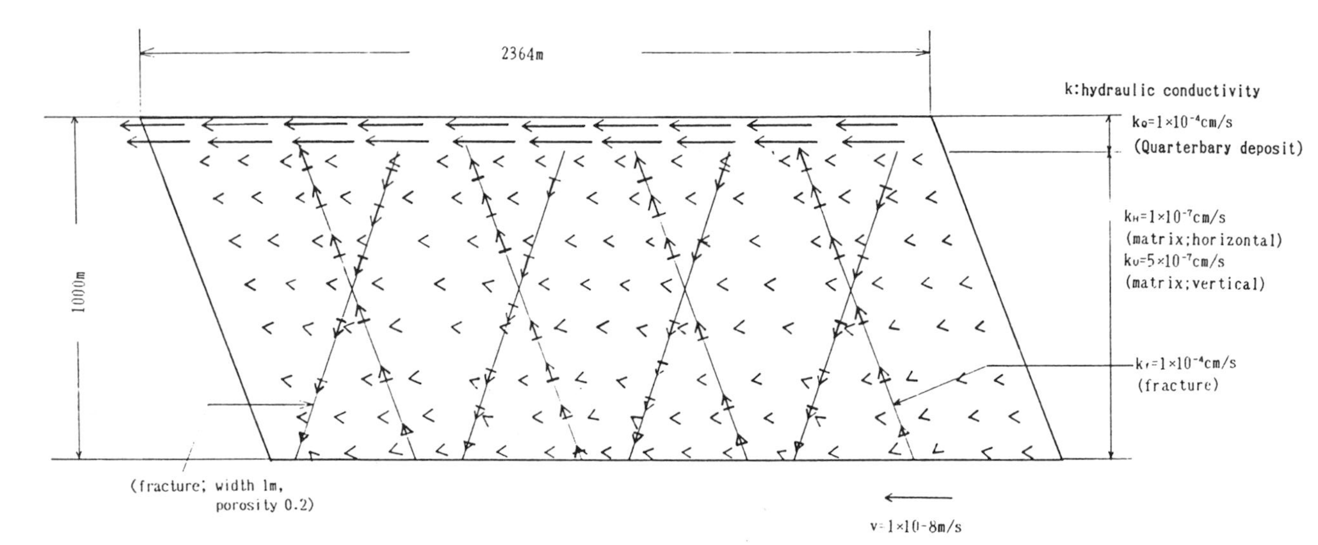

Fig. 5 Groundwater Flow Vectors Simulated for Preliminary Safety Assessment of HLW Disposal

THE DISPERSAL OF COLLOIDS IN FRACTURED ROCK

Peter Grindrod
INTERA-ECL, Henley-on-Thames, United Kingdom

ABSTRACT

How far do colloids migrate in fractured rock?

We present an analysis for colloid migration based on the derivation of macroscopic flow parameters for colloids in single fractures. Dispersion terms, flow rates, and dynamic sorption rates are derived and calculated, depending explicitly upon physical and chemical properties of the colloids, the groundwater, and the fracture. We use such data directly to make predictions of colloid penetration and dispersal. Being based on microscopic theoretical considerations, this approach does not rely on empirically defined K_D's for colloid-rock adsorption, or quantitative estimates of dispersivities.

We discuss a possible experimental approach to the validation of the present model, and hence to refining the current predictions, which are made on the basis of theory alone.

The parameter values obtained will be used as input data for future radionuclide-colloid fracture transport models. The radionuclides are transported both in solution and sorbed to colloids. Thus, reliable estimates for flow rates and dispersion rates are required.

This work has been funded by SKI as part of Project 90.

DISPERSION DES COLLOIDES DANS UNE ROCHE FISSUREE

Peter Grindrod
INTERA-ECL, Henley-on-Thames, United Kingdom

RESUME

Jusqu'où les colloïdes migrent-ils dans une roche fissurée ?

L'auteur présente une analyse de la migration des colloïdes fondée sur la dérivation appliquée à des colloïdes dans des fissures individuelles, des paramètres régissant l'écoulement macroscopique. La dérivation et le calcul des paramètres de dispersion, des débits et des taux de sorption dynamique dépendent explicitement des propriétés physiques et chimiques des colloïdes, de l'eau souterraine et de la fissure. Les résultats obtenus sont utilisés directement pour faire des prévisions concernant la pénétration et la dispersion des colloïdes. Comme elle s'appuie sur des considérations théoriques valables à l'échelle microscopique, cette méthode ne dépend pas de valeurs empiriquement définies de K_D applicables à l'adsorption colloïde-roche ni d'estimations quantitatives de la dispersivité.

L'auteur examine une méthode expérimentale éventuelle permettant de valider le présent modèle et donc d'améliorer la précision des prévisions actuelles qui sont uniquement fondées sur la théorie.

Les valeurs des paramètres obtenues seront utilisées comme données d'entrée pour les prochains modèles de transport des radiocolloïdes dans les fissures. Les radionucléides sont transportés à la fois en solution et associés par sorption aux colloïdes. C'est pourquoi des estimations fiables des débits et des taux de dispersion sont nécessaires.

Ces travaux ont été financés par le SKI dans le cadre du Projet 90.

1 Introduction

The precise role of colloids in the transport and dispersal of radionuclides through fractured and porous rock is, to date, poorly understood. Although laboratory and field data regarding colloid and groundwater chemistry is available, there have been few attempts to incorporate such information into a dynamic colloid migration model.

In this paper, we shall concentrate on colloid dispersal in a single, idealised fracture. This must present the colloids with their best opportunity to migrate over *long* distances, at average flow rates 20-30% in excess of that of any solutes present [1]. Since real flow paths are irregular and more tortuous, this would both perturb the assumed flow rate, and increase the rock surface area available for colloid adsorption.

Radionuclides are transported both in solution and adsorbed to colloids. In order to analyse nuclide-colloid fracture transport models (including matrix diffusion and retardation of the solute in the surrounding rock), robust quantitative values for colloid transport parameters, as compared to the solute, are required as **input data**.

In [1] there is a brief review of nuclide-colloid migration modelling along with the conceptual and mathematical specification of the microscopic models used as the starting point below.

The aim of the current work is to make the calculation of quantitative **macroscopic flow parameters** for colloids explicit, emphasising their dependence upon the chemical and physical data available from laboratory and field studies. We obtain a **colloid migration model**, using such parameters, which may be solved to estimate colloid migration and dispersal within the fracture.

In deriving the dynamic flow properties for colloids (as functions of colloid size, surface charge, composition, groundwater chemistry, fracture width, fluid flow, etc.), we employ a novel mathematical approach [2] which generalises Taylor dispersion theory. Our calculated colloid dispersivities, average flow rates, and dynamic sorption rates reflect the surface forces acting (repelling colloids from, or attracting them towards, the rock faces), as well as the adsorption of colloids in contact with the rock surfaces. Classical dispersion theory for fractures merely accounts for the diffusion of particles across the flow profile and employs inappropriate boundary conditions at the rock surface [2].

We present some predictive calculations for the distribution of distances penetrated by individual colloids within the fracture. The analysis is available in closed form when surface forces are assumed negligible. Accordingly, we have coded (and verified) a numerical algorithm capable of deriving the macroscopic flow parameters from the chemical and physical specifications of the colloids, the groundwater and the fracture [3].

The sorption of radionuclides to colloids is reversible so that even though individual colloids are not transported over large distances, the nuclides may exchange (repeatedly) from the solute to colloid-bound states, taking advantage of the increased colloid flow rates.

2 Physical and chemical data requirements

2.1 The colloids

We consider colloids having hydrodynamic radii, a, in the range 10-1000 nm. We shall assume that the hydrodynamic radius is equal to the average physical radius of the colloids, though the generalisation is straightforward.

Colloids diffuse in suspension with diffusivity D given by the Stokes-Einstein relation

$$D = \frac{k_B T}{6\pi\mu a}$$

as a function of radius a. The remaining constants, k_B (the Boltzman constant), T (the absolute temperature), and μ (groundwater viscosity) are given in table 1 below.

2.2 The fracture

We shall assume that the fracture is planar, of constant half width, L, with uniform Poiseuille flow in a single direction (the x-direction), given by

$$u(z) = \frac{3}{2}u_0\Big(1 - \frac{z^2}{L^2}\Big),$$

where z denotes the cross-fracture distance from the centre plane, see figure 1, and u_0 the cross-fracture average fluid velocity.

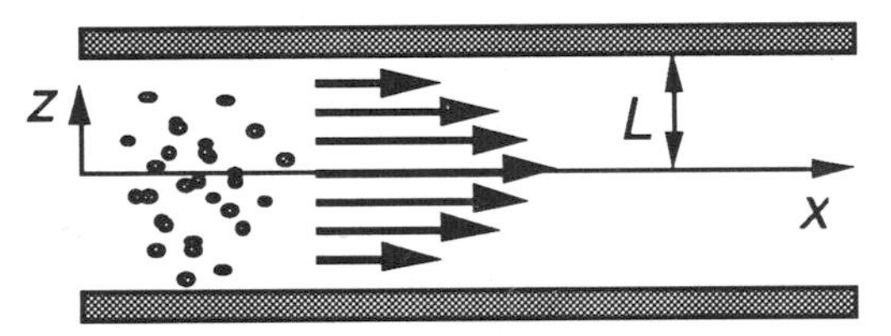

Figure 1: The Fracture

Typical values are given in table 1.

L	10^{-4} m
a	$10^{-8} - 10^{-6}$ m
k_B	1.38×10^{-23} J/K
T	270-280 K
μ	3.8×10^{-11} Nyr/m^2
u_0	$10 - 50$ m/yr
χ	10^8 m^{-1}
η	10^{-20}? J

Table 1: Parameter values.

2.3 Surface forces and adsorption

We shall assume that a colloid of radius a with centre situated at $z = \pm(L - a)$ (that is, in contact with the rock surface) becomes adsorbed and penetrates no further along the fracture.

Colloids are subject to surface forces acting across the fracture. Free colloids are also subject to drag forces opposing their motion relative to the groundwater. As in [1], we shall write

$$v(z) = \frac{F(z,a)}{6\pi\mu a},$$

where $v(z)$ denotes the cross-fracture component of the velocity of colloids, of radius a, as a function of their cross-fracture location, z. $F(z, a)$ is the surface force, acting on colloids in the z-direction, due to the presence of the rock surfaces.

Following [1], $F(z, a)$ is the sum of separate terms. The force on a colloid due to the presence of electric double layers (surface charges) is of the form

$$-\frac{8y_c y_r amk_B T\pi}{\chi}\exp(-\chi(L-a))\sinh\chi z,$$

where y_c and y_r are dimensionless potentials of the colloidal and rock surface (assumed equal for both rock faces); χ is the Debye-Huckel parameter (see table 1); m is the number of ions per unit volume in the groundwater (m^{-3}).

If $y_c y_r > 0$, as for example when the colloids and the rock are of similar composition, then this force is repulsive, causing the colloids to stay more centre stream.

The van der Waals force, due to the interaction of molecules on the colloid and rock surfaces, is attractive but acts over short distances. It is given here by

$$\frac{2}{3}\frac{a(L-a)z\eta}{((L-a)^2 - z^2)},$$

where η is the Hamaker constant, depending upon the composition of the colloid and rock surfaces (see table 1).

Other long or short range forces may be considered within our present approach. We merely require their functional form together with associated parameter values.

3 Macroscopic colloid migration

The models discussed in this section are linear with respect to the colloid density functions, so the distributions of colloids of different sizes may be superimposed. Hence for the present purposes there is no loss of generality in assuming a population of colloids having uniform radius a and allowing this to vary.

In [1] and [3], we use the expressions given in the previous section to formulate a full colloid migration model including axial (through-fracture) fluid advection, cross-fracture advection due to surface forces, diffusion, and sorption at the rock surfaces. In [2], we show how such a **microscopic** model may be approximated by a **macroscopic** model for migration and dispersal in the axial direction alone. In the current case, we obtain

$$\frac{\partial c}{\partial t} = D^* \frac{\partial^2 c}{\partial x^2} - u^* \frac{\partial c}{\partial x} - s^* c. \tag{1}$$

Here $c(x,t)$ is the density of free colloids per unit fracture length; D^* is the dispersivity (including both diffusion and hydrodynamic dispersion); u^* is the average colloid axial flow rate; s^* is the dynamic sorption rate.

D^*, u^*, and s^* are dependent upon D, $u(z)$, $v(z)$, a, and L, defined in the previous section. Their derivation rests on an asymptotic comparison between the spectral properties associated with the full fracture model and (1) [2]. The precise details are not relevant here, save for the fact that the method generalises classical dispersion theory, valid for solutions dominated by small wave numbers in the x-direction (for example, for smooth distributions or large time). Note also that the sorption process (modelled in the boundary condition imposed on the full model) appears as a dynamic loss rate in (1).

A special case arises when surface forces are assumed negligible. We may calculate analytically:

$$\begin{aligned} D^* &= D + \frac{(L-a)^6}{L^4} \frac{u_0^2}{D} \frac{(0.02630399)}{\pi^2}, \\ u^* &= \frac{3}{2} u_0 \left(1 - 2\left(\frac{1}{6} - \frac{1}{\pi^2}\right) \frac{(L-a)^2}{L^2} \right), \\ s^* &= \frac{D\pi^2}{4(L-a)^2}, \end{aligned}$$

where $D = k_B T / 6\pi\mu a$, as before.

In figures 2-5, we depict the calculated values for D^*, u^*, s^* and $x^* \equiv u^*/s^*$ (the mean distance travelled by colloids released into the fracture), as functions of radius a. We used

$$v(z) = -v_0 \exp(-(L-a)\chi) \sinh \chi z,$$

where v_0 =0,1000 or 10000 m/yr. In order to exaggerate the repulsive effect, when $v_0 > 0$, we took $\chi = 1$ m^{-1}. If $\chi = 10^8$ m^{-1}, the effect is restricted to within O(10nm) of the rock surfaces, and diffusive dispersal dominates over the remainder of the fracture width.

In reality colloids may be supplied at all points along the fracture due to a variety of sources [1]. Accordingly, radionuclides migrate against a background of free colloids being supplied to and removed from the fracture. If colloids are supplied in the fracture at a rate $g(x,t)$ per unit length per year, we obtain

$$\frac{\partial c}{\partial t} = D^* \frac{\partial^2 c}{\partial x^2} - u^* \frac{\partial c}{\partial x} - s^* c + g(x,t). \tag{2}$$

4 How far do colloids migrate?

In answering this question, one must take care so as not to diminish prematurely the role played by colloids in the enhancement of radionuclide migration. Here we calculate the penetration of individual colloids within the idealised fracture. Colloids may migrate, become sorbed to the rock, get released by hydrodynamic scouring, migrate, and so on. The effect would be a constant supply of colloids along the fracture (c.f (2)), with each colloid migrating according to the behaviour illustrated in figures 2-5. They provide an alternative medium of transport for sorbing/desorbing nuclides, which may shift from colloid to solute and vice versa. Whilst colloid-bound, the nuclide macroscopic transport properties are altered to those of the host colloid: a radical change from the solute. We shall make calculations for nuclide dispersal based on the present work elsewhere.

The solution of (1) starting from a population of mass M supplied at $x = 0$ and $t = 0$ is

$$c = Me^{-s^*t}\frac{\exp(-(x-u^*t)^2/4D^*t)}{\sqrt{4\pi D^*t}}.$$

As $t \to \infty$, all colloids become sorbed to the rock. A measure of colloid penetration is given by calculating the fraction of colloids which become sorbed to the rock at distances greater than a fixed position, $x = X$ say, down the fracture. Let $H(X)$ denote this quantity (a fraction $H(X)$ of colloids migrate at least as far as $x = X$). We have

$$H(X) = \frac{s^*}{2\sqrt{\mu}(\sqrt{\mu}-\frac{u^*}{\sqrt{4D^*}})}\exp\left(-\frac{X}{\sqrt{D^*}}(\sqrt{\mu}-\frac{u^*}{\sqrt{4D^*}})\right)$$

where $\mu = s^* + u^{*2}/4D^*$.

We show $H(X)$ in figures 6a and 6b. Further analysis and discussion is given in [3].

5 Suggested experiments

Although we have considered a planar fracture, similar calculations can be made for a fracture having a uniform circular cross-section. Clearly there would be much utility in performing laboratory experiments employing capilliary tubes; well documented, uniform (perhaps latex) colloids; slow pump rates. The colloid radii should be varied over a number of experiments, as could the surface charges, the water chemistry, the flow rates, and fracture apertures. Some such data may already be available in the form of hydrodynamic chromatography employing porous capilliary membranes.

Such experiments and data could be used to validate the theoretical model presented here, and refine the predictions made on both the laboratory and field scales of space and time.

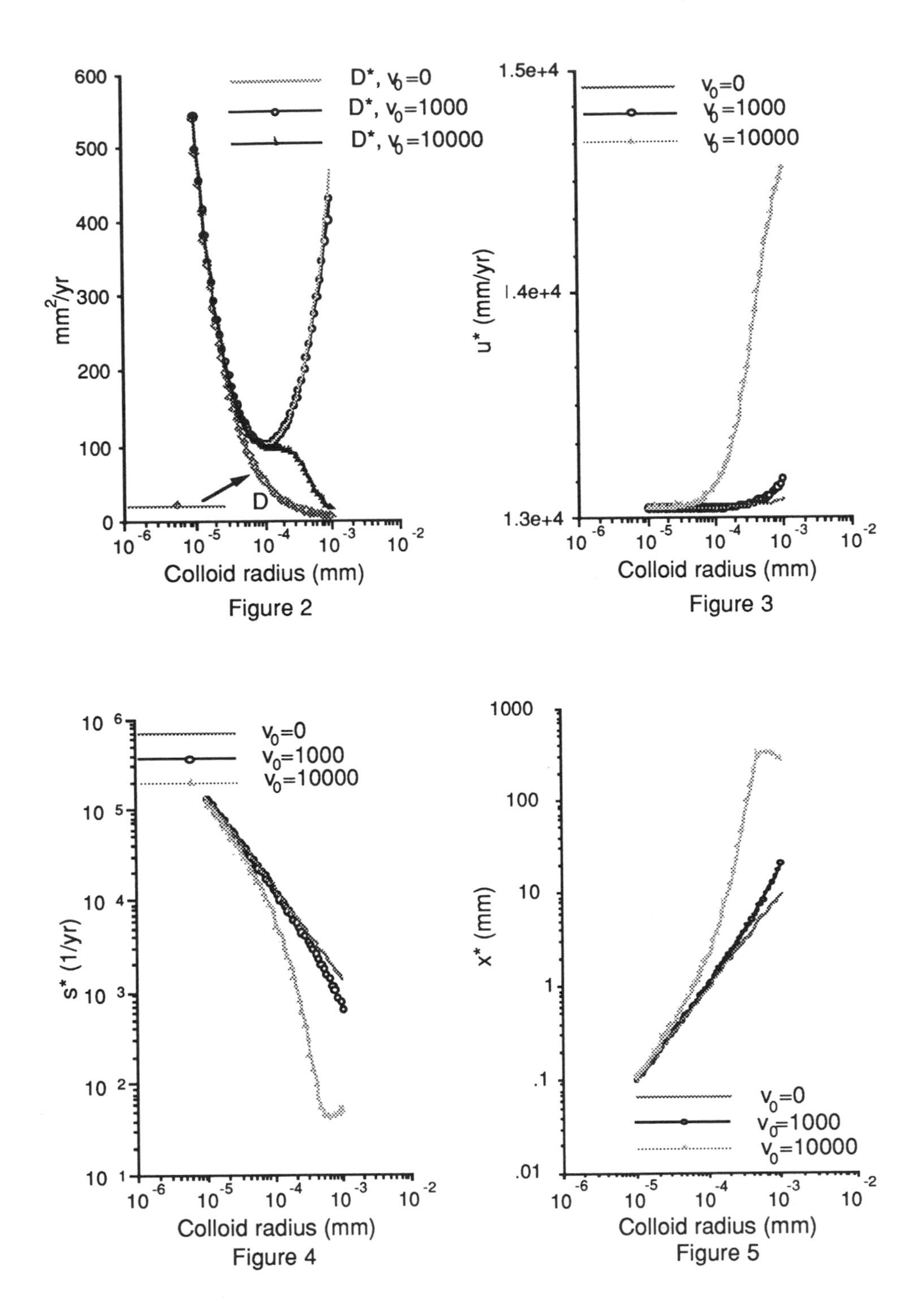

Figure 2

Figure 3

Figure 4

Figure 5

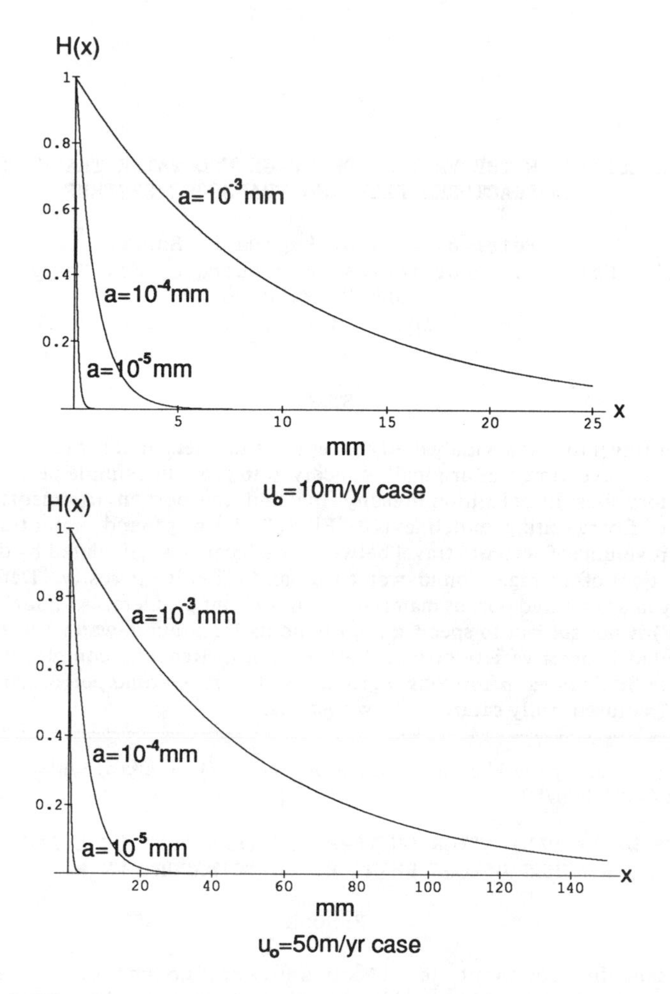

Figure 6: The function H a) v_0=0, u_0=10 m/yr; b) v_0=0, u_0=50 m/yr

References

[1] P. Grindrod, Colloid-nuclide migration in fractured rock: mathematical model specification, INTERA-ECL report I2145-1, Version 2, June 1989.

[2] P. Grindrod, An asymptotic spectral comparative approach to the derivation of one dimensional transport models for solutes and colloids in fractures, INTERA-ECL report I2145-2, Version 2, November 1989.

[3] P. Grindrod and D.J. Worth, The determination of macroscopic flow parameters for colloid migration in fractured rock, INTERA-ECL report I2145-3, Version 1, January 1990.

COMPLEXITY IN THE VALIDATION OF GROUND-WATER TRAVEL TIME IN FRACTURED FLOW AND TRANSPORT SYSTEMS

Peter B. Davies, Regina L. Hunter
Sandia National Laboratories, Albuquerque, New Mexico, USA
John F. Pickens
INTERA Inc., Austin, Texas, USA

ABSTRACT

Ground-water travel time is a widely used concept in site assessment for radioactive waste disposal. While ground-water travel time was originally conceived to provide a simple performance measure for evaluating repository sites, its definition in many flow and transport environments is ambiguous. The U.S. Department of Energy siting guidelines (10 CFR 960) define ground-water travel time as the time required for a unit volume of water to travel between two locations, calculated by dividing travel-path length by the quotient of average ground-water flux and effective porosity. Defining a meaningful effective porosity in a fractured porous material is a significant problem. Although the Waste Isolation Pilot Plant (WIPP) is not subject to specific requirements for ground-water travel time, travel times have been computed under a variety of model assumptions. Recently completed model analyses for WIPP illustrate the difficulties in applying a ground-water travel-time performance measure to flow and transport in fractured, fully saturated flow systems.

This work was supported by the U.S. Department of Energy (U.S. DOE) under contract DE-AC04-76DP00789

COMPLEXITE DE LA VALIDATION DU TEMPS DE TRANSPORT DE L'EAU SOUTERRAINE DANS LES REGIMES D'ECOULEMENT ET DE TRANSPORT EN MILIEU FISSURE

RESUME

Le temps de transport de l'eau souterraine est un paramètre largement utilisé pour évaluer les sites d'évacuation des déchets radioactifs. Bien que le temps de transport de l'eau souterraine ait été retenu à l'origine comme un paramètre de comportement simple pour l'évaluation des sites de dépôt, sa définition est ambiguë dans de nombreuses conditions d'écoulement et de transport. Dans une de ses lignes directrices sur l'implantation des sites (10 CFR 960), le Ministère de l'énergie des Etats-Unis définit le temps de transport de l'eau souterraine comme le temps qu'il faut à un volume unitaire d'eau pour se déplacer d'un point à un autre, le calcul consistant à diviser la longueur du trajet effectué par le quotient du flux moyen d'eau souterraine et de la porosité efficace. Donner une définition satisfaisante de la porosité efficace dans un milieu poreux fissuré n'est pas une tâche aisée. Bien que le Projet pilote de confinement des déchets (WIPP) ne soit pas tenu de respecter des normes arrêtées en matière de temps de transport de l'eau souterraine, des temps de transport ont été calculés compte tenu de diverses hypothèses de modélisation. Des analyses de modèles récemment achevées concernant le WIPP montrent à quel point il est difficile d'appliquer une unité de mesure concernant le temps de transport de l'eau souterraine quand l'écoulement et le transport se produisent dans des milieux fissurés totalement saturés.

INTRODUCTION

The objective of this paper is to provide a broad overview of the regulatory context of the ground-water travel-time concept and to illustrate the difficulties in applying this concept as a performance measure in fractured, fully saturated flow systems. The illustration of complexity and ambiguity in the ground-water travel-time concept comes from recent model analyses of potential contaminant transport along a major segment of the offsite travel path at the Waste Isolation Pilot Plant (WIPP) in the southwestern United States. The WIPP is a U.S. Department of Energy facility designed to provide a repository for emplacement of approximately 180,000 cubic meters of transuranic waste from defense-related activities [1].

REGULATORY CONTEXT OF THE GROUND-WATER TRAVEL-TIME CONCEPT

The ground-water travel-time concept has been formally and informally applied to a variety of potential radioactive waste disposal sites in the United States. In the United States, the generally applicable environmental standard governing geologic repositories for radioactive waste is specified by the Environmental Protection Agency (EPA) in 40 CFR 191 [2]. The EPA standard does not give a rule for ground-water travel time, but rather sets a limit for cumulative release of radioactive waste to the accessible environment (roughly speaking, at the land surface and crossing a boundary located 5 kilometers from the repository) over a 10,000 year period. The EPA standard, 40 CFR 191, has been remanded to the EPA for further consideration, and therefore, this paper discusses this standard as originally promulgated in 1985.

The Nuclear Regulatory Commission (NRC) is responsible for determining that repositories licensed for the disposal of high-level waste will meet the EPA standard. NRC has chosen to carry out this responsibility by writing a second rule, 10 CFR 60 [3], in such a way that compliance with this rule should assure compliance with 40 CFR 191. The NRC's implementation of the standard, 10 CFR 60, sets explicit rules about ground-water travel time. These rules state that a repository must be sited in such a way that the pre-waste-emplacement travel time of ground water along the fastest path of likely radionuclide travel from the projected location of the future repository's disturbed zone to the accessible environment is at least 1,000 years. However, the NRC rule does not specify how ground-water travel time is to be computed.

The Department of Energy is responsible for the siting and construction of radioactive waste repositories and has issued general guidelines for the siting process in 10 CFR 960 [4]. These guidelines define ground-water travel time as being calculated by dividing travel-path length by the quotient of average ground-water flux and effective porosity. The guidelines state that a site shall be disqualified if the pre-waste-emplacement ground-water travel time from the disturbed zone to the accessible environment is expected to be less than 1,000 years along any pathway of likely and significant radionuclide travel.

Most of the waste planned for disposal under the EPA standard, 40 CFR 191, is high-level waste comprising spent fuel from commercial power reactors; repositories for this waste must be licensed by the NRC. Repositories for the disposal of defense-generated transuranic wastes will not be licensed by NRC and such repositories must meet only the requirements of the EPA standard. Therefore, the WIPP is not subject to specific requirements for ground-water travel time. However, WIPP would clearly meet this aspect of the NRC regulation because the total travel path includes vertical flow through approximately 400 meters of undisturbed salt as well as lateral flow in a transmissive dolomite unit that overlies the salt.

FLOW AND TRANSPORT AT THE WIPP SITE UNDER BREACH CONDITIONS

Although WIPP is not subject to specific ground-water travel-time requirements, travel times have been computed under a variety of model assumptions [5,6,7,8,9] for characterizing flow and transport in the Culebra dolomite, which is a shallow, relatively transmissive unit considered the most likely pathway for offsite transport in the event of a repository breach. Recent model analyses for the WIPP illustrate significant difficulties in applying ground-water travel time as a performance measure in fractured, fully saturated flow systems. Difficulties in applying ground-water travel-time as a performance measure in fractured, unsaturated flow systems have been discussed by Kaplan et al., 1989 [10].

The calculations discussed in the remainder of this paper were carried out for the recently completed Supplement Environmental Impact Statement for the WIPP [1,8,9] and for sensitivity analyses of transport under breach conditions [6]. These calculations examine offsite contaminant transport in the Culebra dolomite as a result of a hypothetical penetration of the WIPP repository and an underlying pressurized brine source by a future hydrocarbon exploration well. The following discussion focuses on flow and transport in the Culebra dolomite segment of the total transport pathway.

Conceptual Model

The bedded salt formation containing the WIPP repository is underlain by isolated pockets of pressurized brine in fractured anhydrite units that have been mildly deformed in response to salt flow [8,11]. Recent geophysical measurements at WIPP indicate that pressurized brine may underlie a portion of the waste panels [8]. The human intrusion scenario considered for the Supplement Environmental Impact Statement [1,8,9] consisted of a hypothetical hydrocarbon exploration borehole penetrating a waste panel in the repository and an underlying pressurized brine pocket, thereby allowing release of contaminated brine into the Culebra dolomite (Figure 1).

The Culebra dolomite is an 8-meter-thick, fractured dolomite and is the most transmissive hydrogeologic unit overlying the WIPP repository. The Culebra has been the focus of extensive hydraulic testing at a variety of scales and of several conservative tracer tests [8]. Core analyses indicate that the average matrix porosity of the Culebra is 0.16 and tracer tests along the offsite transport pathway indicate an average fracture porosity of 0.0015 [6,8].

Numerical Implementation of Flow and Transport Models

Numerical simulations of the breach system depicted in Figure 1 have been implemented using the SWIFT II [12] flow and transport code. Three coupled model segments simulate fluid flow from a hypothetical pressurized brine pocket to the Culebra, ground-water flow in the Culebra, and contaminant transport in the Culebra to a boundary located approximately 5 kilometers from the repository. The first model segment is used to generate a fluid-loading and contaminant source term in the Culebra dolomite. In this segment, the pressurized brine pocket is dynamically linked to the Culebra dolomite through a breach borehole. Brine pocket properties are based on hydraulic testing of the brine pocket encountered in borehole WIPP-12, which is located approximately 2 kilometers north of the WIPP waste panels [8]. Brine flowing up the borehole is assumed to dissolve waste up to a specified solubility limit or until all mass for a given waste constituent is dissolved. Simulations were performed for a conservative solute and for four radioactive-decay chains (^{240}Pu, ^{239}Pu, ^{238}Pu, and ^{241}Am).

In the second model segment, ground-water flow in the Culebra is simulated using the model of LaVenue et al. [7], which has been calibrated using transmissivity information from 58 well locations, head information from 61 observations wells, and transient stresses from three large scale pumping tests and from shaft construction. A portion of the undisturbed flow field from the LaVenue et al. model is presented in Figure 2. Two-dimensional transport simulations utilized the undisturbed flow field because the injection of brine from the breach borehole had a negligible effect on the average ground-water flux along the offsite travel path.

In the third model segment, contaminant transport along the fastest offsite travel path in the Culebra (Figure 2) is simulated for two-dimensional, single- or dual-porosity conditions. Dual-porosity transport parameters have been derived from convergent-flow tracer tests at two hydropad locations along the principal offsite transport pathway [6,8]. The contaminant source in the Culebra has been specified as a variable-width strip-source, which is computed assuming an idealized plume from point-source injection rates calculated using the first model segment.

Travel Time Implications of Model Results

This discussion focuses only on travel time in the Culebra dolomite from the breach borehole to a boundary located approximately 5 kilometers from a vertical projection of the waste panels. The location, length, and average flux for the fastest travel path in the Culebra have been determined using a particle tracking code. For the LaVenue et al. flow model [7], the fastest travel path originates above the southwest waste panel, follows flow east-southeastward toward a higher transmissivity zone, and then continues southward along that zone (Figure 2). The length of this travel path is approximately 6000 meters and the average ground-water flux is approximately 1.9×10^{-9} m/s. The next step is to divide the quotient of path length and average flux by effective porosity. However, in a fractured porous medium, the concept of effective porosity is ambiguous. Assuming that all flow occurs through the fractures, with a porosity of 0.0015, yields a ground-water travel time of approximately 150 years. Assuming that no preferential flow occurs in the fractures and that water moves uniformly through a combined medium having a fracture-plus-matrix porosity of 0.1615 yields a ground-water travel time of approximately 16,000 years. Which, if either, of these two travel times should be taken as the performance measure? The fracture-only travel time may be unrealistically short because it assumes that the matrix plays no role in contaminant migration. The fracture-plus-matrix travel time may be unrealistically long because it assumes that the matrix participates fully during transport and contaminant migration is not at all enhanced by the presence of fractures. The two-orders-of-magnitude difference between these travel times illustrates that a flow-based travel time yields significant ambiguity when used as a performance measure.

From both technical and regulatory standpoints, the critical issue is contaminant transport, not ground-water flow. In a fractured porous medium, contaminant-transport time is strongly influenced by the physical properties of the fractures and rock matrix, and by the free-water diffusivity of a given contaminant (in addition to retardation due to chemical interactions and decay of radioactive contaminants). These physical properties control the degree of interaction between fractures and matrix, and therefore, strongly influence offsite contaminant-transport times and rates.

Simulations of the transport of a conservative, nonretarded contaminant in the Culebra dolomite provide an example of the additional information that transport simulation provides. These simulations assume a constant contaminant concentration at the breach well of 2.4×10^{-7} kg/kg and a source-plume width that decreases with time as the driving pressure depletes in the underlying brine pocket. Breakthrough curves at the 5-kilometer boundary are presented for single-porosity, fracture-only transport; single-porosity, fracture-plus-matrix transport; and dual-porosity transport, with transport interaction (matrix diffusion) between fractures and matrix (Figure 3). Single-porosity, fracture-only transport yields a peak concentration of 6×10^{-9} kg/kg at approximately 200 years, which is of the same order as the fracture-only ground-water travel time. Single-porosity, fracture-plus-matrix transport yields a peak concentration of 4×10^{-9} kg/kg at approximately 22,000 years, which is of the same order as the fracture-plus-matrix ground-water travel time.

Actual contaminant breakthrough at the 5-kilometer boundary would be expected to fall some time between these two extremes, and when it will occur depends on the degree of transport interaction between fracture and matrix. Only a dual-porosity simulation provides the critical information needed to determine when contaminant breakthrough will occur. The dual-porosity simulation results (Figure 3) suggest that for the estimated Culebra transport properties, there is a significant degree of interaction between fractures and matrix. For the dual-porosity simulation, the peak concentration of 3×10^{-9} kg/kg occurs at approximately 24,000 years, which is similar to the fracture-plus-matrix peak concentration and breakthrough time. However, the influence of the fractures can be seen in the earlier breakthrough times at lower concentration levels. For example, breakthrough at the 10^{-16} kg/kg concentration level occurs at approximately 5000 years in the fracture-plus-matrix simulation, but at only approximately 2000 years under dual-porosity conditions (Figure 3). Another way to view the difference is to note that at approximately 5500 years, the single-porosity, fracture-plus-matrix and dual-porosity simulations differ in contaminant concentration at the 5-kilometer boundary by approximately four orders of magnitude (Figure 4).

CONCLUSIONS

Ground-water travel time is intended to provide a simple, effective performance measure for characterizing hydrogeologic containment at a potential radioactive waste disposal site. The calculations presented in this paper, which are based on actual site-characterization data from a repository that is currently under development, illustrate that ground-water travel time is an highly ambiguous performance measure and does not provide a meaningful surrogate for characterizing contaminant transport behavior in fractured, fully saturated flow systems. As an alternative, a performance measure based on *transport rather than flow* would provide a more meaningful measure for assessing site suitability for radioactive waste disposal. The discussion presented in this paper focuses only on issues surrounding the physical definition of a performance measure and does not consider issues associated with the characterization of uncertainty, which is also an important component of the total performance assessment.

REFERENCES

[1] U.S. Department of Energy: Final Supplement, Environmental Impact Statement, Waste Isolation Pilot Plant, DOE/EIS-0026-FS, 13 volumes, 1990, Washington, D.C.

[2] U.S. Environmental Protection Agency: "Environmental Standards for the Management and Disposal of Spent Nuclear Fuel, High-Level and Transuranic Radioactive Waste; Final Rule," 40 CFR 191, Federal Register, 1985, vol. 50, pp. 38066-38089.

[3] U.S. Nuclear Regulatory Commission: "Disposal of High-Level Radioactive Waste in Geologic Repositories," 10 CFR 60, Federal Register, 1983, vol. 48, pp. 28194-28228.

[4] U.S. Department of Energy: "Nuclear Waste Policy Act of 1982; General Guidelines for the Recommendation of Sites for the Nuclear Waste Repositories; Final Siting Guidelines," 10 CFR 960, Federal Register, 1984, vol. 49, pp. 47714-47770.

[5] Reeves, M., Kelley, V.A., and Pickens, J.F.: Regional Double-Porosity Solute Transport in the Culebra Dolomite: An Analysis of Parameter Sensitivity and Importance at the Waste Isolation Pilot Plant (WIPP) Site, SAND87-7105, Sandia National Laboratories, Albuquerque, New Mexico, 1986.

[6] Pickens, J.F., Freeze, G.A., Kelley, V.A., Reeves, M., and Upton, D.T.: Regional Double-Porosity Solute Transport in the Culebra Dolomite Under Brine-Reservoir-Release Conditions: An Analysis of Parameter Sensitivity and Importance at the Waste Isolation Pilot Plant (WIPP) Site, SAND89-7069, Sandia National Laboratories, Albuquerque, New Mexico [in preparation].

[7] LaVenue, A.M., Cauffman, T.L., and Pickens, J.F.: Ground-Water Flow Modeling of the Culebra Dolomite: Volume I - Model Calibration, SAND89-7068/1, Sandia National Laboratories, Albuquerque, New Mexico [in preparation].

[8] Lappin, A.R., Hunter, R.L., Garber, D.P., and Davies, P.B., eds.: Systems Analysis, Long-Term Radionuclide Transport, and Dose Assessments, Waste Isolation Pilot Plant (WIPP), Southeastern New Mexico; March, 1989, SAND89-0462, Sandia National Laboratories, Albuquerque, New Mexico, 1989.

[9] Lappin, A.R., Hunter, R.L., Davies, P.B., Reeves, M., Pickens, J.F., and Iuzzolino, H.J.:Systems Analysis, Long-Term Radionuclide Transport, and Dose Assessments, Waste Isolation Pilot Plant (WIPP), Southeastern New Mexico; September, 1989, SAND89-1996, Sandia National Laboratories, Albuquerque, New Mexico [in preparation].

[10] Kaplan, P., Klavetter, E., and Peters, R.: "Approaches to Groundwater Travel Time," Waste Management '89, 1989, vol. 1, pp. 493-495.

[11] Borns, D.J., Barrows, L.J., Powers, D.P., and Snyder, R.P.: Deformation of Evaporites Near the Waste Isolation Pilot Plant (WIPP) Site, SAND82-1069, Sandia National Laboratories, Albuquerque, New Mexico, 1983.

[12] Reeves, M., Ward, D.S., Johns, N.D., and Cranwell, R.M.: Theory and Implementation for SWIFT II, The Sandia Waste-Isolation Flow and Transport Model for Fractured Media, Release 4.84, SAND83-1159, Sandia National Laboratories, Albuquerque, New Mexico, 1986.

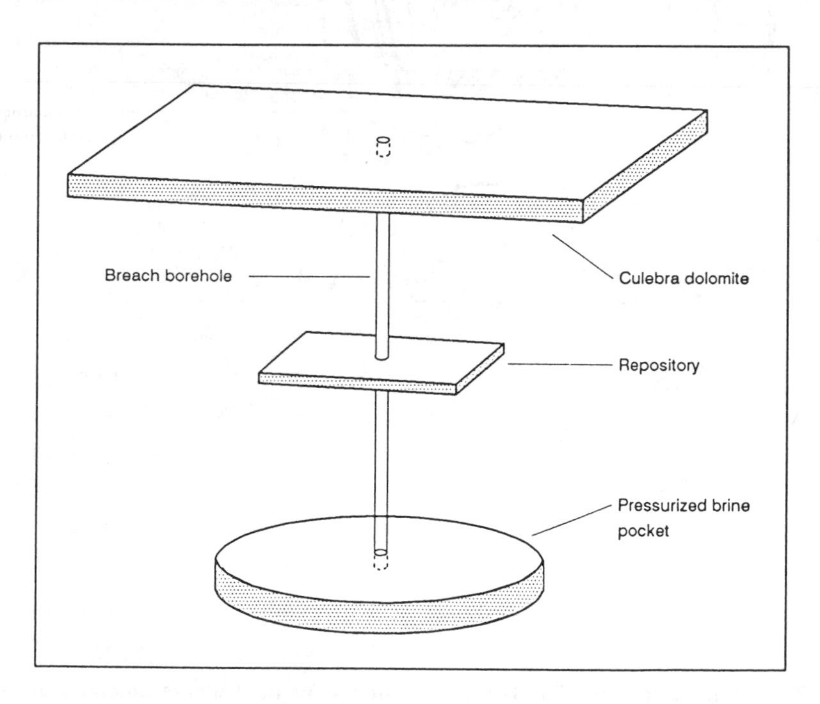

Figure 1. Schematic diagram of flow system considered for breach simulation. This system consists of a borehole penetrating both the repository and an underlying pressurized brine pocket, with contaminant release and offsite transport in the Culebra dolomite.

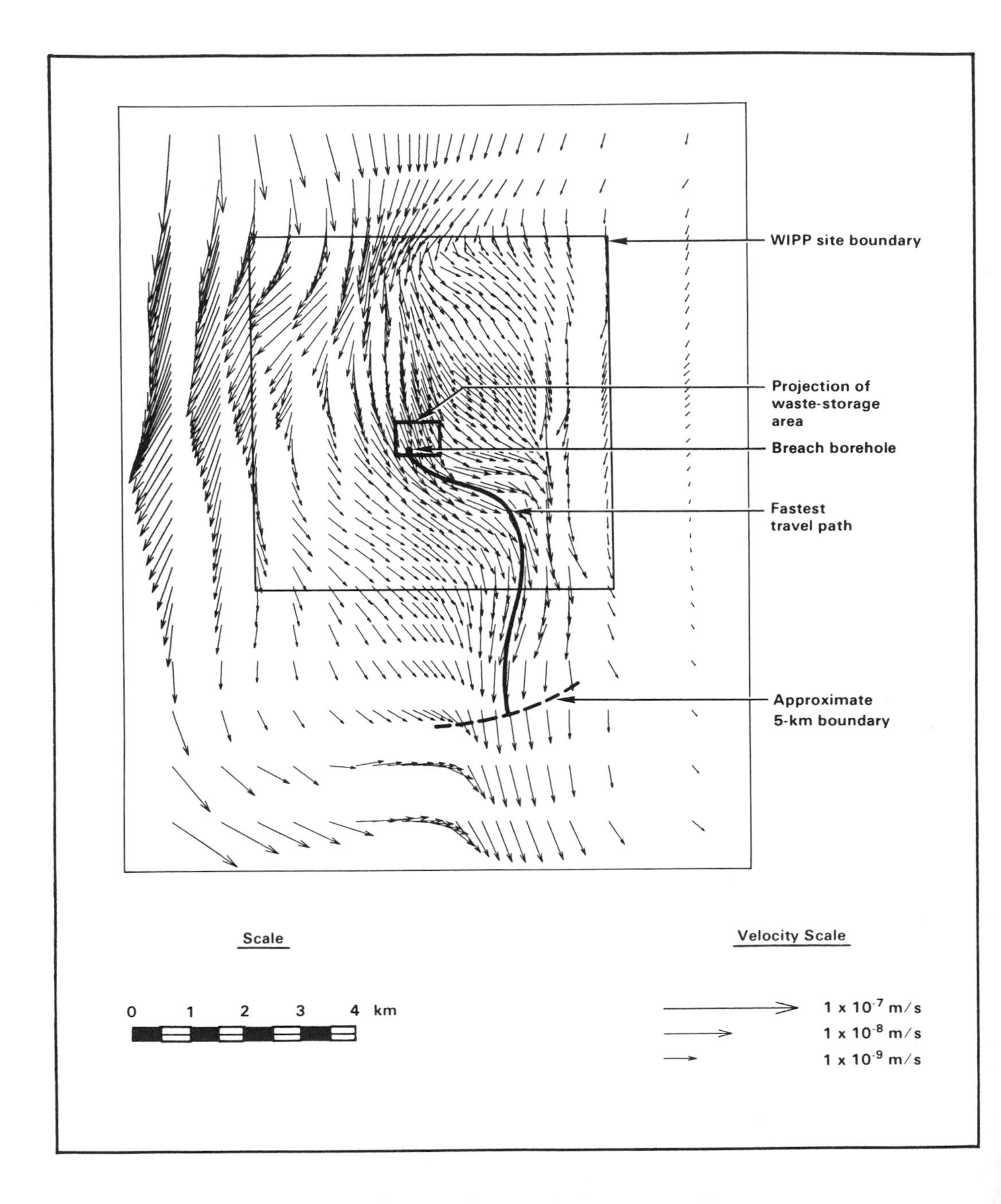

Figure 2. Simulated flow field (Darcy velocities) for the Culebra dolomite under undisturbed conditions [7]. Fastest transport pathway from a breach borehole to the approximate 5-kilometer boundary is also shown.

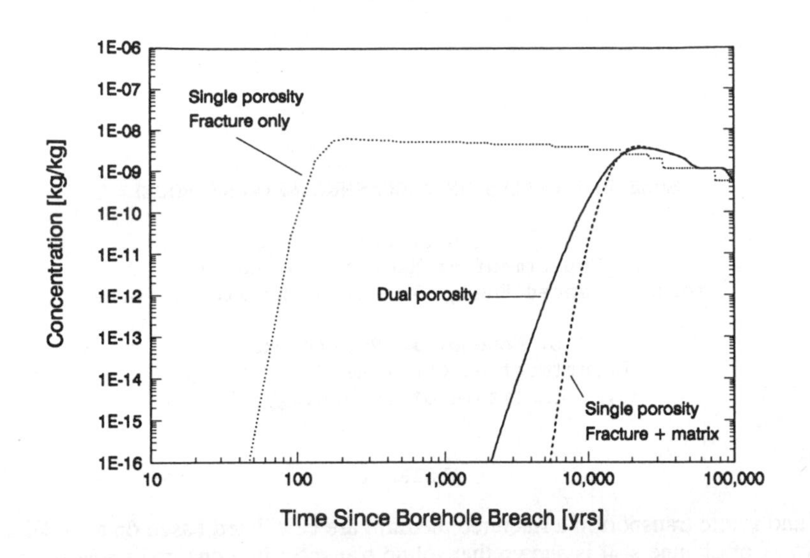

Figure 3. Contaminant concentration versus time at the 5-kilometer boundary for a conservative, nonretarded solute. Curves depict single-porosity, fracture-only transport; single-porosity, fracture-plus-matrix transport; and dual-porosity transport.

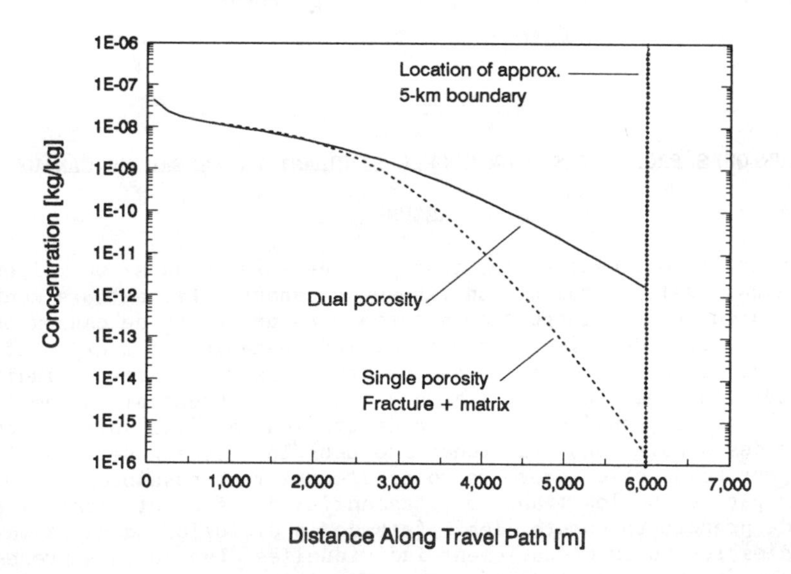

Figure 4. Contaminant concentration profiles along the transport pathway at approximately 5500 years. Curves depict single-porosity, fracture-plus-matrix transport and dual-porosity transport.

SOME PROPERTIES OF A CHANNEL NETWORK MODEL

H.S. Lee
Department of Radwaste Disposal
Korea Advanced Energy Research Institute, Korea

L. Moreno, I. Neretnieks
Department of Chemical Engineering
Royal Institute of Technology, Sweden

ABSTRACT

Flow and solute transport in a fractured medium are calculated based on a model using a network of channels. It is shown that solute transport in a channel network may be calculated considering a set of independent pathways for the solute transport. The division of a channel network into individual pathways which do not depend on other pathways is discussed. The calculation of the effluent concentration can be obtained as the weighted sum of the concentration in the pathways. Simulations show that for large networks the number of possible paths may be reduced by considering only those paths which transport most of the contaminant. The effects of matrix diffusion are readily included once the individual pathways are established. An analytical solution for matrix diffusion from a pathway with variable properties is presented.

QUELQUES PROPRIETES D'UN MODELE SIMULANT UN RESEAU DE CANAUX

RESUME

Pour calculer l'écoulement et le transport de solutés dans un milieu fissuré on utilise un modèle simulant un réseau de canaux. Les auteurs montrent que l'on peut calculer le transport de solutés dans un réseau de canaux en prenant en compte une série de voies d'écoulement indépendantes. Ils exposent le principe d'une division du réseau de canaux en voies d'écoulement individuelles indépendantes les unes des autres. On calcule la concentration de l'effluent en faisant la somme pondérée de la concentration dans les voies d'écoulement. Il ressort des simulations effectuées que pour les réseaux de dimension importante on peut réduire le nombre de voies d'écoulement possibles en ne retenant que celles par lesquelles transite l'essentiel des éléments contaminants. Il est aisé de prendre en compte les effets de la diffusion dans la matrice une fois définies les voies d'écoulement individuelles. Les auteurs présentent une solution analytique pour la diffusion dans la matrice à partir d'une voie d'écoulement ayant des propriétés variables.

INTRODUCTION AND BACKGROUND

There is considerable experimental evidence that water flow in fractured rocks is very unevenly distributed [1]. These observations show that only a small part of the fracture conducts water, and that the water flowrates may vary greatly between different pathways. Moreover, it was found that a few paths conduct most of the water flow in a fractured medium.

Flow and solute transport in fractured media have been modeled using different approaches. Neretnieks [2] and Tsang and Tsang [3] explored the properties of a model where all the flowpaths were independent. Dershowitz et al. [4] fracture network model can accommodate flowpaths in the fracture planes and so form a channel network model. Cacas et al. [5z,6] also used a model where the channels in a channel network are located in the planes of individual fractures.

Flowrates in the individual channels in such channel network models are readily calculated by simultaneously solving the equations for the hydraulic heads in every channel intersection (node). The only information needed is the hydraulic resistance of each channel in addition to the boundary conditions.

To analyze the nuclide migration in a channel network different approaches have been used: the mass lumping, particle following, and the particle tracking algorithms [7]. The advantage of the first technique is that the concentration is known everywhere, and that it may be extended to include sorption and rock matrix diffusion. It requires the extra mass to describe the uptake of mass into the rock matrix. The computing time tends to become large. The particle following or particle tracking algorithms are easy to use. The calculations are faster compared to mass lumping.

Surface sorption is easily included in the models by modifying the water residence time by a retardation factor. Even though matrix diffusion can be included it is somewhat complex to simulate in large networks.

This paper deals primarily with the analysis of matrix diffusion effects in channel networks. It consists of two parts: In the first part the division of the channel network into individual pathways which do not depend on the other pathways is discussed. The individual pathway is a possible pathway traversed by a species, from the injection to the collection point. The individual pathway traverses a number of the channels of the channel network. The mass injected in the network is divided into so called partial concentration to give the concentration in the independent pathway.

The second part of the paper describes the calculation of the concentration in the water in the pathways when surface sorption and matrix diffusion effects are accounted for.

GENERATION OF THE INDIVIDUAL PATHWAYS IN A CHANNEL NETWORK

We assume that there is a network of channels in a fractured medium with known properties. An example of a fracture network composed of channels that have known properties is shown in Figure 1. The nodal points are located at the intersections of the channels. This network may be three dimensional.

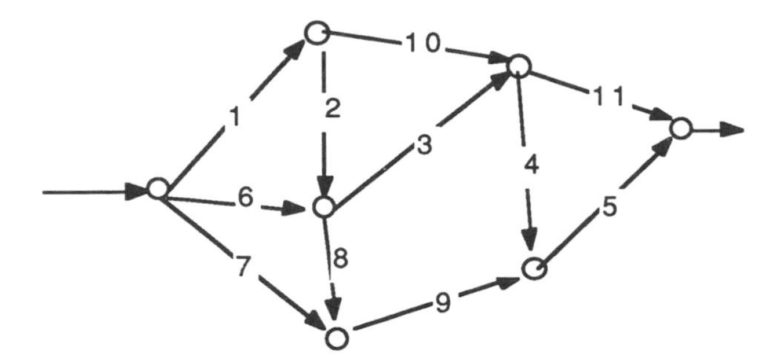

Figure 1. Example of a channel network

Assuming steady state flow, the mass balance at each nodal point can be solved to give the hydraulic head at each node, for given boundary conditions. The water flowrate q_i and the water residence time t_i can be calculated if the volume of each channel is known. Based on these values, the probability of occurrence of each pathway may be obtained. It may also be interpreted as the fraction of total flowrate that moves in any individual pathway. For example, the probability that a tracer may pass through the pathway which has the channels 6, 3, and 11 in Figure 1 is

$$P = \left(\frac{q_6}{q_1 + q_6 + q_7}\right)\left(\frac{q_3}{q_3 + q_8}\right)\left(\frac{q_{11}}{q_{11} + q_4}\right) \tag{1}$$

The total residence time in this pathway is $\tau = \tau_6 + \tau_3 + \tau_{11}$

All possible pathways through which the solute may flow are generated. An algorithm was developed to determine all possible pathways. First, one pathway is chosen from the injection to the collection point, then we return to the starting point of the last channel and take a new pathway and so on. This process is continued until all possibilities are exhausted.

The solute breakthrough curves are formed with the cumulative probability for the different pathways and residence times. The residence time distribution of the water that moves from inlet to outlet can thus be "exactly" established. For every pathway the flowrate and residence time in every channel are established. Even more important, for the study of the fate of solutes which may diffuse into the rock matrix, the "wetted surface" and matrix properties along each pathway are determined.

Assuming no dispersion and a tracer which only can reside in the water, the concentration at the inlet of channel "j" is determined by all the N streams "i" converging to this point. For the initial condition of zero concentration everywhere the concentration at the inlet to channel "j" is

$$c_j = \frac{\sum_{i=1}^{N} c_i(t)\, q_i H(t-t_i)}{q_j} = \sum_{i=1}^{N} c_i(t)\, H(t-t_i) \frac{q_i}{\sum_{i=1}^{N} q_i} \tag{2}$$

where H is the Heaviside function which has the value of one when t is greater than t_i, otherwise zero. In Equation (2), c_i refers to the concentration at the inlet to channel "i", and q_i refers to the flowrate of channel "i" which ends in the starting point of the channel "j". The term $q_i/\Sigma q_i$ in Equation (2) is the fraction of the flow in channel "j" passing through channel "i".

The Heaviside function H is used because the flow passing through channel "i" appears at channel "j" after time t_i which is the actual travel time for the fluid from the origin to the end of channel "i". This equation therefore considers the concentration dependence on time. For instance, the solute which passes through channel "i" reaches channel "j" at time t_i. The concentration in channel "j" before time t_i is computed by putting the function H as zero in Equation (2). This means that the concentration in channel "j" is diluted by the pure water passing through the channel "i" until time t_i. Equation (2) is used recursively to determine the concentration in a pathway.

LINEARITY OF THE CONVECTION-MATRIX DIFFUSION EQUATIONS

In this section a more general form of Equation (2) will be derived. The linearity of the convection-matrix diffusion equations is also discussed. It will be shown that because of the linearity of these equations, the total concentration in any channel may be calculated as a weighted sum of the concentrations in all the pathways which flow in this channel. This property will later be used to include the effects of matrix diffusion.

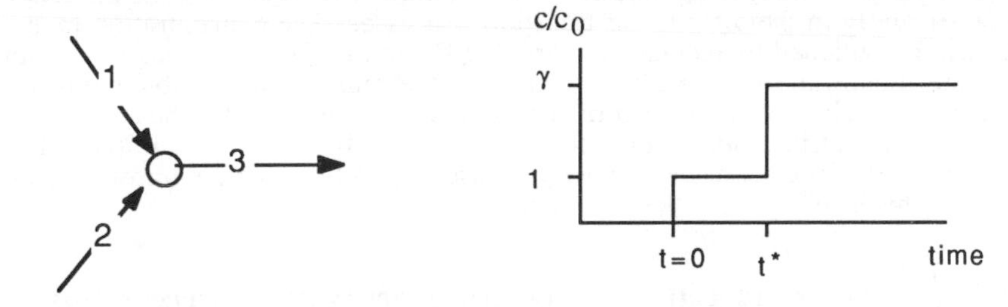

Figure 2. Channels 1 and 2 flow into 3

Figure 3. Concentration at exit of channel 3

Referring to Figure 2 consider a case where the concentration at the outlet of pathways 1-3 and 2-3 is zero. At time zero the concentration at the exit of pathway 1-3 becomes c_0, while it takes a time t^* for the pathway 2-3 to contribute so that the effluent concentration from channel 3 becomes γc_0. Figure 3 shows the effluent concentration from channel 3 as a function of time. This can be generalized. The exit concentration from a channel can be expressed by applying Duhamel´s theorem as follows

$$\frac{c}{c_o} = \int_0^t g(\lambda) f(t-\lambda)\, d\lambda \qquad (3)$$

where f is the solution of a linear solute transport equation for a delta function as input and g is the function of the inlet concentration to the channel. The function g can be divided according to the contribution from each pathway arriving at the inlet to the channel as follows.

$$g(\lambda) = g_1(\lambda) + g_2(\lambda) + g_3(\lambda) + ... = \sum_{i}^{N} g_i(\lambda) \tag{4}$$

where g_i is the contribution to the concentration from pathway "i" entering the channel of interest. Combining Equations (3) and (4) this results in

$$\frac{c}{c_o} = \int_0^t \sum_{i=0}^{N} g_i(\lambda) f(t-\lambda)\, d\lambda = \sum_{i=0}^{N} \int_0^t g_i(\lambda) f(t-\lambda)\, d\lambda \tag{5}$$

N is the number of all possible pathways going through the channel. This means that the overall effect of the inlet concentration can be obtained by summation.

In summary, the composite flow network can be divided into individual flow paths. The overall effluent curve can be obtained by superposition of the breakthrough curves of a solute through the individual flow paths.

It is thus in principle possible to determine all possible pathways and to calculate the solute transport independently in them for linear transport processes. The concentration in any channel can then be obtained by recursively applying Equation (5). In practice this is not possible for other than quite small networks, because the number of possible pathways grows very rapidly with increasing number of channels in the network. Also the complexity of the recursive calculations using Equation (5) quickly becomes prohibitive. The calculations can be simplified considerably for the case of purely advective transport with matrix diffusion. This is discussed in the next section.

Analytical solution for advective solute transport with matrix diffusion through a pathway made up of different channels

Each path traverses channels which have different residence times, flowrates, wetted surfaces, and diffusive and sorptive properties. The mass balance equations in each pathway are, however, similar to the conventional governing equations. This means for example, that the water velocity has no longer a constant value along the flow path but varies with location. The hydrodynamic dispersion is assumed to be negligible. The governing equations in a pathway are [8]

$$R_d(z)\frac{\partial c_f}{\partial t} + u(z)\frac{\partial c_f}{\partial z} = \frac{2D_e(z)}{\delta(z)}\frac{\partial c_p}{\partial x}\Big|_{x=0} \tag{6}$$

$$\frac{\partial c_p}{\partial t} = D_a(z)\frac{\partial^2 c_p}{\partial x^2} \tag{7}$$

where c_f is the concentration in the water in the pathway, c_p is the concentration in the rock matrix adjacent to this pathway, R_a is the retardation factor due to surface sorption, u is the water velocity, D_e is the effective diffusivity, δ is the aperture size of the channel and D_a is the apparent diffusivity. The initial conditions are zero concentration at initial time. The concentration at x = 0 is changed to c_o at t = 0, and equilibrium is assumed between c_f and c_p at the channel surface. Equations (6) and (7) can be solved by the Laplace transform method to give the following results [9].

$$c_f = c_o \text{erfc}\left(\frac{\Psi(z)}{2\sqrt{t-\phi(z)}}\right) \tag{8}$$

where

$$\phi(z) = \int_0^z \left(\frac{R_a(z')}{u(z')}\right) dz' \tag{9}$$

$$\Psi(z) = \int_0^z \left(\frac{2D_e(z')}{\delta(z')}\frac{1}{\sqrt{D_a(z')}}\frac{1}{u(z')}\right) dz' \tag{10}$$

The integration proceeds from the inlet, z = 0, through all channels of the pathway ending at a distance z which is a distance along the pathway.

Referring to Figure 1, along each pathway and channel the values of D_e, D_a, u, δ, R_a are needed to calculate Equations (9) and (10). Because the system is linear, the different pathways may be handled separately. Equation (8) can be used directly for each pathway and added at the outlet, providing there is perfect mixing at the nodes [7].

Having established all possible (or important pathways) the application of Equation (8) is straightforward and is very much simpler than the recursive use of Equation (5).

SIMULATION RESULTS

The models developed above have been used to simulate solute transport in a channel network. Simulations were first performed with only advective transport in a network. A network consisting of a 10 x 10 grid having 100 equally long channels was used. The standard deviation of natural logarithm of the apertures of the channels was taken to be 0.41 and the log mean to be $1.43 \cdot 10^{-4}$ m. The cubic law was assumed to apply. The data were chosen to give breakthrough curves similar to those used by Moreno et al. [10] in their analysis of a tracer experiment in a fracture. The simulated results are shown in Figure 4. In the same Figure also results from the Advection Dispersion model and the Channeling model [2] are shown. The parameters of those models were chosen so that the resulting curves were similar. The data for the different models are shown in Table 1. It is seen that the curves are quite similar but the mean apertures μ_a are quite different. The standard deviation for the channeling model, P-Ch, is less than that for the network.

Calculations were also performed including the effects of matrix diffusion. The same data were used in all three models as previously. The results are shown in Figure 5. The network gives a much larger tail and lower concentration at long times. This is because the mean aperture is much smaller and thus the wetted surface per water volume larger in the dominant pathways.

Table 1. Data for three models; mean aperture size and standard deviation of aperture distribution.

NetW model	$\mu_a = 1.43\ 10^{-4}$ m	$\sigma_a = 0.41$
AD model	$\mu_a = 6.08\ 10^{-4}$ m	Pe = 3.27
P-Ch model	$\mu_a = 4.79\ 10^{-4}$ m	$\sigma_a = 0.35$

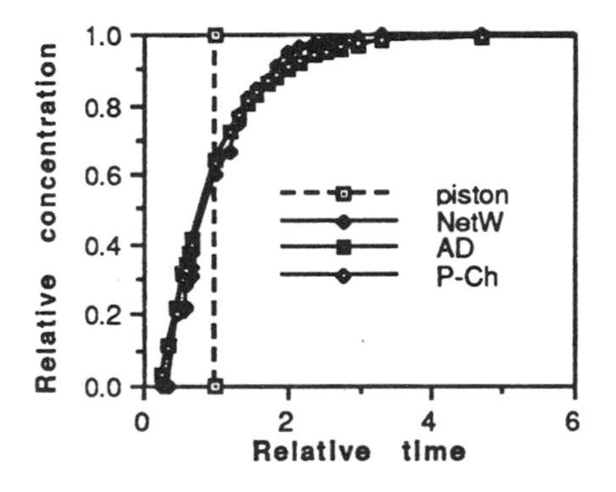

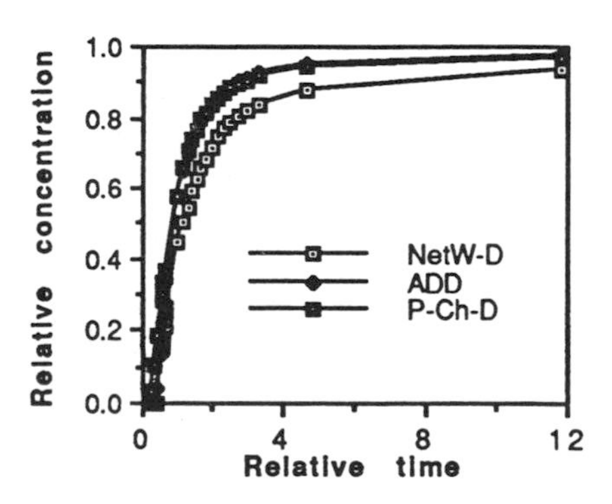

Figures 4 and 5. Effluent curves without (4) and with matrix diffusion (5)

DISCUSSION AND CONCLUSION

The proposed method to account for all possible pathways is possible to use in networks with a small number of channels. For networks with a large number of channels the number of possible pathways becomes too large. We propose that one then uses particle tracking techniques to find the most important pathways. Having found these it is straightforward to integrate the residence time function, Equation (9), and the matrix interaction function, Equation (10), for use in Equation (8). The linearity of the system allows the concentration in any channel to be found by adding the concentrations in all pathways in that channel.

REFERENCES

1. Neretnieks, I., H. Abelin, L. Birgersson, Some recent observations of channeling in fractured rocks. Its potential impact on radionuclide migration., Proceedings DOE/AECL conference, Sep 15-17, 1987, San Francisco, p 387-410, 1987.

2. Neretnieks, I., A Note on Fracture Flow Dispersion Mechanisms in the Ground, Water Resour. Res., 19, p 364, 1983.

3. Tsang, Y.W., C.F. Tsang, Channels model of flow through fractured media, Water Resour. Res., 23, p 467-479, 1987.

4. Dershowitz, W.S., B.M. Gordon, J.C. Kafritsas, A new three dimensional model for flow in fractured rock, IAH Conference, Jan 7-12, 1985, Tucson, Arizona, V. XVII proceedings published by USA IAH, p 449-462, 1985.

5. Cacas, M.C., G. de Marsily, B. Tillie, A. Barbreau, E. Durand, B. Feuga, P. Peaudecerf, Modelling fracture flow with a stochastic discrete fracture network: calibration and validation. The flow model, in print, Water Resour. Res., 1990.

6. Cacas, M.C., E. Ledoux, G. de Marsily, A. Barbreau, E. Durand, P. Calmels, B. Gaillard, and R. Margritta, Modelling fracture flow with a stochastic discrete fracture network: calibration and validation. The transport model, in print, Water Resour. Res., 1990.

7. Robinson, P. C., Connectivity, Flow and Transport in Network Models of Fractured Media, Ph. D. Thesis, Oxford Univ., May 1984.

8. Neretnieks, I., A. Rasmuson, An approach to modelling radionuclide migration in a medium with strongly varying velocity and block sizes along the flow path. Water Resour. Res., 20, p 1823-1836, 1984.

9. Carslaw, H. S., J. C. Jaeger, Conduction of Heat in Solids, 2nd ed., Oxford Univ. Press, 1959.

10. Moreno, L., I. Neretnieks, T. Eriksen, Analysis of Some Laboratory Tracer Runs in Natural Fissures, Water Resour. Res., 21, p 951, 1985.

VALIDATION OF A CONCEPTUAL MODEL FOR FLOW AND TRANSPORT IN FRACTURES

A. Hautojärvi, V. Taivassalo, S. Vuori
Technical Research Centre of Finland
Nuclear Engineering Laboratory, Helsinki, Finland

ABSTRACT

A novel concept of transport in channelled fracture flow has been developed and applied to interpret the results of the radially converging experiment performed at Finnsjön as the test case 5 of the INTRAVAL project. The dispersion of tracers is calculated taking into account the coupling of an assumed velocity profile over the channel width and molecular diffusion. An obtained consistent picture of groundwater flow and tracer break-through curves is encouraging for the validation process. The effects of matrix diffusion and diffusion into stagnant pools in the fracture are also evaluated and compared with the developed theory. In the case of high flow velocities typical for field experiments dispersion is dominated by the velocity profile across the channel width. Diffusion into stagnant areas may also have an important contribution on dispersion. The effects of matrix diffusion are negligible in conditions typical for field experiments. However, with slower flow velocities the tracer break-through behavior is dominated by matrix diffusion.

VALIDATION D'UN MODELE THEORIQUE SIMULANT L'ECOULEMENT ET LE TRANSPORT DANS LES FISSURES

RESUME

Une nouvelle théorie du transport et de l'écoulement par des canaux en milieu fissuré a été élaborée et mise en oeuvre pour interpréter les résultats de l'expérience de convergence radiale réalisée à Finnsjön (cas témoin n° 5 du projet INTRAVAL). On calcule la dispersion de traceurs en prenant en considération le couplage d'un profil de vitesse postulé dans le plan transversal des canaux et la diffusion moléculaire. La cohérence des résultats obtenus en ce qui concerne l'écoulement de l'eau souterraine et les courbes de passage des traceurs est encourageante pour le processus de validation. Les auteurs évaluent et comparent également avec la théorie qui a été élaborée les effets de la diffusion dans la matrice et dans les poches d'eau stagnante qui se trouvent dans la fissure. Dans le cas des vitesses d'écoulement élevées représentatives des expériences menées sur le terrain, la dispersion est dominée par le profil de vitesse au niveau de la section transversale des canaux. La diffusion dans les zones d'eau stagnante peut également avoir un effet important sur la dispersion. Les effets dus à la diffusion dans la matrice sont négligeables dans les conditions représentatives des expériences menées sur le terrain. Toutefois, lorsque les vitesses d'écoulement sont plus faibles le comportement des traceurs est principalement influencé par les phénomènes de diffusion dans la matrice.

1. INTRODUCTION

In crystalline rock groundwater flows in fractures and fissures having varying apertures. Due to the strong dependence of the flow resistance on the fracture aperture the water flows preferentially in a certain part of the fracture plane. These pathways are called channels. A migrating tracer is transported by convection along these channels. During transport the tracer may diffuse into areas of stagnant water and into microfissures of the rock matrix. In flow channels tracer molecules diffuse across streamlines and are thus transported with different velocities in different parts of channels. All these phenomena cause dispersion and affect the break-through behavior of a tracer pulse.

The problem of diffusive convection in a two-dimensional velocity field can be handled analytically using some approximations. We have developed a theory for a linearly varying velocity across the channel width [1]. The theory has been applied to interpret the experimental results of the INTRAVAL test case 5. In the interpretation a concept of channels having a much smaller aperture than width has been used. Although the velocity profile across the channel aperture is likely to be parabolic in nature, the fast diffusion smooths statistically out any velocity differences between the migrating molecules in that direction. Velocity gradients extending to much larger dimensions might, however, exist across the fracture plane. If the characteristic extension of the velocity gradient is comparable or larger than the spread due to diffusion during transport time, then the tracer is dispersed due to the velocity gradient.

As matrix diffusion can also be handled analytically [2], it is relatively easy to compare these two phenomena. Diffusion into stagnant water pools in the fracture can also be calculated with the matrix diffusion theory simply by taking the channel width as the aperture, the porosity to be several tens of per cent and the effective diffusion coefficient equal to the diffusion coefficient in water.

The importance of these three phenomena is studied by assuming that water flows in a channel having dimensions typical for the channels employed in the conceptual modelling of the Finnsjön fracture zone 2. Each phenomenon is studied independently. Different velocities being typical for the Finnsjön experiments and also velocities lower by a factor of 1000 representing typical values in a repository performance analysis have been used in the study.

2. INTERPRETATION OF INTRAVAL TEST CASE 5 RESULTS

In the radially converging experiment at Finnsjön tracers were injected in three boreholes and in three subzones in each hole and sampled in the pumping borehole. Preliminary test results are given in reference [3]. Some calculations for the test case 5 were performed prior to knowing the results conserning the flow through the injection sections in the boreholes and tracer break-through curves. These apriori calculations are discussed in more detail in reference [1]. The predictions of transport times were based on the description of the test area [4] and measured transmissivities in four boreholes. The groundwater flow times were calculated applying an approach of channelled flow. The predicted transport times are presented in Table I together with measured values showing a good accordance with few exceptions. It seems that some high values of transmissivities in single boreholes do not necessarily indicate good connections in a subzone. This is consistent with the recently published interference data [5] which were not available at the time of the apriori calculations.

Table I. Comparison of apriori predicted and measured transport times in the radially converging exepriment at Finnsjön.

Injection section		Predicted time (h)	Measured time (h)
BFi01	U	99	75
BFi01	M	463	700
BFi01	L	140	1250
KFi06	U	110	106
KFi06	M	(525)[1]	1300
KFi06	L	159	194
KFi11	U	53	24
KFi11	M	248	850
KFi11	L	75	70[2]

[1] Prediction was made for a different section as actually was used
[2] The tracer break-through curve was reinterpreted. The recovery is very low.

The assumption that flow through borehole sections would be determined just by the geometry in the radial potential field, i.e. the ratio of the borehole diameter to the arc length of the full circle through the borehole and having its center at the pumping hole, appeared to be wrong typically by a factor of 20. The flow being that much higher could be explained, however, by a skin factor and channeling.

The flow rate affects the concentration in the injection section and after obtaining the measured values an interpretation of the results was performed. The aim was to give a consistent explanation of the measured borehole concentrations and break-through curves with the chosen channel concept and flow rate through the injection section. This was important because the flow rate affects the injection concentration in a time dependent way and has thus a contribution to the break-through curve.

The transport times were fitted to compare in detail calculated and measured dispersions. A good overall agreement was obtained by taking into account just one channel from each injection section to the pumping borehole. A compilation of the rising parts of the break-through curves with high mass recoveries is presented in Figure 1 to show that the dispersion behavior of the chosen concept is consistent with most of the measured curves. Some difficulty arises from the fact that the flow through the injection sections seems to have changed during the experiment. Results are presented fully in reference [6].

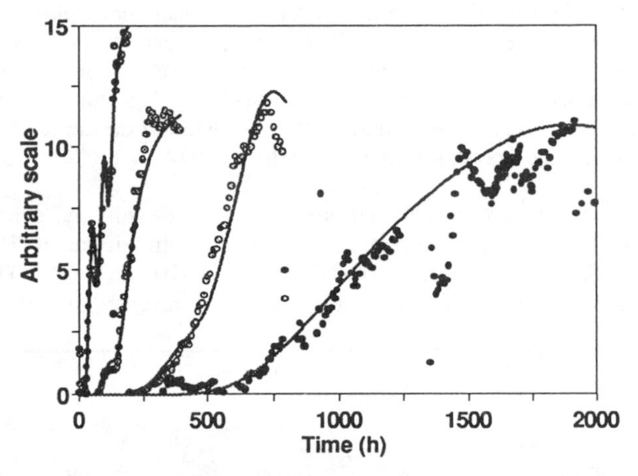

Figure 1. Experimental and theoretical dispersions in the radially converging experiment at Finnsjön. The rising parts of the break-through curves from injection sections KFi11 upper, BFi01 upper, KFi06 lower and BFi01 middle are presented on an arbitrary vertical scale to show all the curves in one figure. The theoretical dispersion is due to molecular diffusion in a linear velocity field extending from zero to the maximum velocity at a distance of 5 cm in each case.

3. EFFECTS OF FLOW VELOCITY ON THE PHENOMENA CAUSING DISPERSION

In order to assess and compare various phenomena causing dispersion in channelled flow, calculations were performed for non-sorbing tracers in two fictitious cases representing field experiments and flow situation in repository performance analyses. An idealized channel with reflecting boundaries was used as a concept. Three phenomena were considered:

- dispersion due to diffusive convection in a linearly varying velocity field in the channel
- matrix diffusion
- diffusion into large stagnant pools outside of the flow channel.

The channel length was taken to be 150 m and the aperture 0.5 mm. The tracer concentration was taken to be constant at the inlet of the channel for 50 h or 50000 h, respectively, and the relative massflux per hour at the outlet was calculated.

The convective diffusion was calculated similar to the test case 5 interepretations, where only the mean velocity and channel width have to be known. The width of the velocity profile was taken to be 5 cm as before.

In the case of matrix diffusion the walls forming the channel were considered to be permeable for diffusion with an effective diffusion coefficient of $5 \cdot 10^{-14}$ m^2/s and porosity of 0.2 %.

In addition to the diffusion into the matrix the tracer molecules may diffuse into areas where the groundwater is nearly stagnant. This situation can be handled analytically just by applying the same matrix diffusion equation as for the case of conventional matrix diffusion. The width of the area where water is flowing was taken as "aperture" (5 cm) and the effective diffusion coefficient was taken to be equal to the diffusion coefficient of the tracer molecules in water ($5 \cdot 10^{-10}$ m^2/s). The "porosity" has taken to be 50 %.

The results of dispersion calculations due to the velocity gradient and molecular diffusion with four different flow velocities are presented in Figure 2. The applied velocities were $0.55 \cdot 10^{-3}$ (A), $0.16 \cdot 10^{-3}$ (B), $0.08 \cdot 10^{-3}$ (C) and $0.05 \cdot 10^{-3}$ m/s (D). The results of matrix diffusion calculations with the same flow velocities are presented in Figure 3.

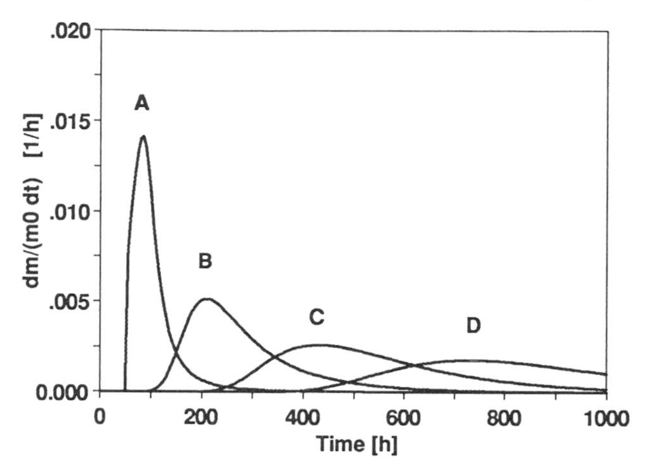

Figure 2. Relative break-through curves with different flow velocities.
A: $0.55 \cdot 10^{-3}$ m/s, B: $0.16 \cdot 10^{-3}$ m/s, C: $0.08 \cdot 10^{-3}$ m/s and D: $0.05 \cdot 10^{-3}$ m/s.
Dispersion is due to diffusive convection in a linearly varying velocity field.

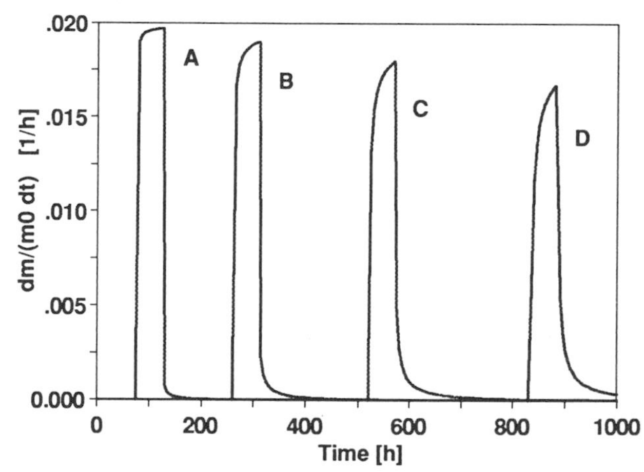

Figure 3. Relative break-through curves with different flow velocities.
A: $0.55 \cdot 10^{-3}$ m/s, B: $0.16 \cdot 10^{-3}$ m/s, C: $0.08 \cdot 10^{-3}$ m/s and D: $0.05 \cdot 10^{-3}$ m/s.
Dispersion is due to matrix diffusion with $D_e = 5 \cdot 10^{-14}$ m^2/s.

The break-through curves in the case where dispersion is assumed to be caused solely by diffusion into stagnant pools in the fracture plane are presented in Figure 4.

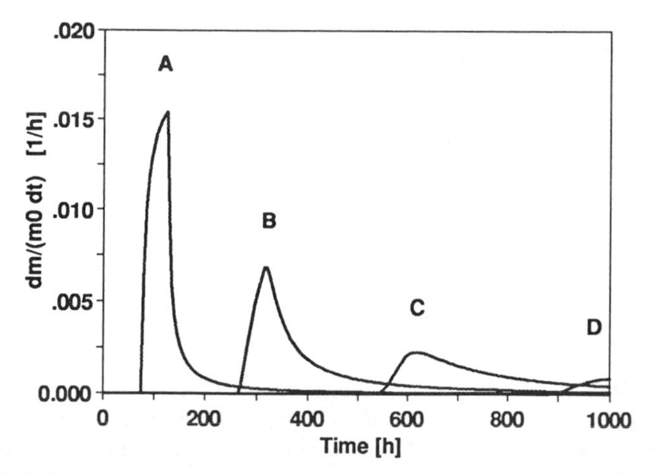

Figure 4. Relative break-through curves with different flow velocities.
A: $0.55 \cdot 10^{-3}$ m/s, B: $0.16 \cdot 10^{-3}$ m/s, C: $0.08 \cdot 10^{-3}$ m/s and D: $0.05 \cdot 10^{-3}$ m/s.
Dispersion is due to diffusion into stagnant pools in the fracture plane.

The flow velocities applied above are typical for the Finnsjön experiments. In a performance analysis the flow velocities due to natural conditions are likely to be much slower. Therefore, the calculations were repeated using flow velocities which were slower by a factor of 1000 (A'-D'). The results for the diffusive convection in a varying velocity field and in the matrix diffusion case are presented in Figures 5 and 6, respectively.

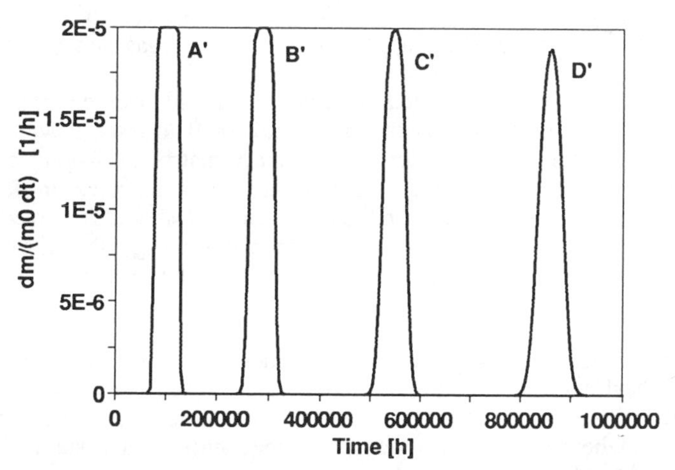

Figure 5. Relative break-through curves with different flow velocities.
A': $0.55 \cdot 10^{-6}$ m/s, B': $0.16 \cdot 10^{-6}$ m/s, C': $0.08 \cdot 10^{-6}$ m/s and D': $0.05 \cdot 10^{-6}$ m/s.
Dispersion is due to diffusive convection in a linearly varying velocity field.

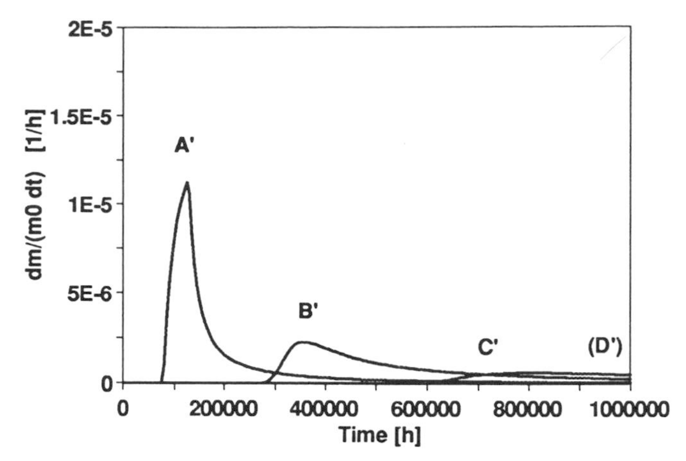

Figure 6. Relative break-through curves with different flow velocities.
A': $0.55 \cdot 10^{-6}$ m/s, B': $0.16 \cdot 10^{-6}$ m/s, C': $0.08 \cdot 10^{-6}$ m/s and D': $0.05 \cdot 10^{-6}$ m/s. Dispersion is due to matrix diffusion with $D_e = 5 \cdot 10^{-14}$ m^2/s. The curve D' cannot be resolved in this scale.

It is not relevant to calculate the diffusion into stagnant areas with the same assumptions as for fast flows, because the semi-infinite boundary conditions in the fracture plane are not likely to be valid for very long times.

4. CONCLUSIONS

The applied concept of flow channels and the developed mathematical model of diffusive convection has proven to be consistent with the experimental results of the test case 5, radially converging experiment, of the INTRAVAL project. It is thus encouraging to continue the validation process of the approach and test its consistency more fully.

It can be seen from Figure 2 that a velocity gradient over the channel width may be the main cause of dispersion in flow situations typical for field tests. It would be very difficult to extract the effects of matrix diffusion and evaluate matrix diffusion parameters from the experimental results. Diffusion into stagnant areas seems to have in fast flows a more significant effect on the migration than matrix diffusion. It should be noted, however, that it has been assumed that diffusion mixes any concentration gradients over the dimension considered (channel width in this case). This may not be true for short times or large dimensions.

Boundary conditions for diffusion are much more important in slow flows and determine the behavior of the migration. With reflective boundary conditions the migration of a tracer will approach a steady state where all the molecules experience statistically the same velocities and hence have equal transport times through a system. This would be the situation e.g. for the diffusion into dead end pores or the so called limited matrix diffusion. The ordinary matrix diffusion does not reach the steady state in a sparse fracture system.

Dispersion causing phenomena which are measured in experiments tend to have only a minor effect in flow conditions typical for repository performance analysis. Furthermore, it is almost impossible to extract the matrix diffusion parameters from experiments where too fast flow velocities have been used.

To learn more about the flow and migration in fractures and to validate concepts it would be necessary to evaluate the effects of all possible phenomena affecting the experimental results and perform additional experiments at most favorable conditions for each of the phenomena to determine the parameters reliably. This would probably mean rather long experiments for matrix diffusion. In any case the effects of other phenomena causing dispersion should be well under control before matrix diffusion can be studied.

The effects of different phenomena can be distinguished best when as short as possible injection duration is used taking of course into account the limits of detection and accuracy of measurements. It would be also necessary to study the behavior of dispersion with different velocities in order to really validate transport models instead of just fitting parameters.

5. REFERENCES

[1] Hautojärvi, A. and Taivassalo, V. : Intraval Project, Test case 5: Tracer tests at Finnsjön, Predictive modelling of the radially converging experiment. Nuclear Waste Commission of Finnish Power Companies, Report YJT-88-13, November 1988.

[2] Neretnieks, I. : Diffusion in the Rock Matrix : An Important Factor in Radionuclide Retardation ?. J. Geophys. Res., **85**, Vol. B8, (1980), 4379-4397.

[3] Gustafsson, E., Andersson, P., Eriksson, C.-O. and Nordqvist, R. : Radially converging tracer experiment in a low-angle fracture zone at the Finnsjön site, Central Sweden, The fracture zone project - Phase 3. Swedish Geological Co., Uppsala, Sweden, Draft report IRAP 88313, March 1989.

[4] Ahlbom, K., Smellie, J.A.T., (Editors) : Characterization of fracture zone 2, Finnsjön study site. SKB, Technical Report 89-19, August 1989.

[5] Andersson, J.-E., Ekman, L., Gustafsson, E., Nordqvist, R. and Tirén, S. : Hydraulic interference tests and tracer tests within the Brändan area, Finnsjön study site, The fracture zone project - Phase 3. SKB, Technical Report 89-12, June 1988.

[6] Hautojärvi, A., Taivassalo, V. and Vuori, S. : Interpretation of test case 5, radially converging experiment. Paper presented at Intraval workshop, Las Vegas, February 1990.

SITE CHARACTERIZATION FOR THE SWEDISH HARD ROCK LABORATORY

G. Bäckblom, P. Wikberg
Swedish Nuclear Fuel and Waste management Co., Stockholm

G. Gustafson
Chalmers University of Technology, Department of Geology, Gotenburg

R. Stanfors
IDEON, Lund

ABSTRACT

Siting and site characterization of a repository for spent fuel is a key issue within the SKB's R&D programme. In order to prepare for the siting and licensing of a spent fuel repository, SKB has decided to construct a new underground research laboratory. Site characterization was initiated in 1986. Predictions have been made prior to construction and they will be compared to reality during the construction of the laboratory. A tunnel will be blasted to a depth of 500 m starting autumn 1990, where a suite of investigations and experiments will be executed.

The paper gives a broad overwiew of the project, focussed on the adopted process of site characterization, conceptual modelling and numerical modelling.

CARACTERISATION D'UN SITE POUR LE LABORATOIRE SUEDOIS DANS DES ROCHES DURES

RESUME

Le choix et la caractérisation d'un site de dépôt de combustible irradié sont un élément essentiel du programme de R&D du SKB. Pour faciliter le choix d'un site de dépôt de combustible irradié et son autorisation, le SKB a décidé de construire un nouveau laboratoire souterrain de recherche. Les travaux de caractérisation du site ont commencé en 1986. Des prévisions ont été effectuées avant la construction et on les comparera à la réalité lors de la construction du laboratoire. A l'automne de 1990 commenceront des travaux d'excavation à l'explosif d'une galerie à 500 m de profondeur où une série d'études et d'expériences sera réalisée.

Les auteurs exposent les grandes lignes de ce projet en mettant l'accent sur le processus retenu pour la caractérisation du site, la modélisation théorique et la modélisation numérique.

CHALLENGES IN SITE CHARACTERIZATION FOR A HIGH-LEVEL WASTE REPOSITORY

The final repository for the Swedish spent nuclear fuel will utilize both natural and man-made, engineered barriers. The design, construction and analysis of the long-term performance of the repository represents new challenges to site characterization.

The rôle of the rock in the overall technical system is more complex for a repository than for ie a water power plant or an underground gas storage. For the outlet tunnel at a water power plant, the rôle of the rock in the system is often only limited to the issue of mechanical stability. In an underground un-lined gas storage mechanical stability is needed, as well as limiting boundary conditions at the cavern roof and wall, so that the cavern will not leak. Site characterization and design must ensure that these boundary conditions can be maintained during the operation of the storage. Site characterization for a high-level waste repository must give design parameters for the engineered barriers so that an appropriate service life for the canisters is assumed. The characterization must also provide necessary data for calculation of nuclide transport in the host rock. These demands entail a thorough description of ground water flow and chemistry not only close to the rock/cavern boundary but a description that encompasses rock volumes.

Site characterization for a repository must give appropriate data for the geometrical lay-out of the repository as well. The design process should emphazise flexibility so that the repository succesively is adapted to the rock conditions that influence the safety of the repository. This approach follows a long tradition within engineering geology. The challenge is to push this philosophy for the design and construction of a repository. The active design philosophy ("design-as-you-go") during construction is a potential for selection of the best near fields available at a chosen site. The licensing process must recognize the benefit of this flexibility in the design.

The most unique aspect of site characterization for a repository is the assessment of changes in the host rock with time. The time elapsed from start of pre-investigation to post-closure may be up to 70 years which is rather unique for a civil engineering (but not a mine!) facility. Small changes in the host rock can be anticipated during this period with respect to ie fracture mineral precipitation. These changes must be measured and evaluated. In the post-closure phase there should be data that describe what effects e g temperature, glaciation, deglaciation will have on the host rock. These data are necessary either to describe how things are (performance assessment) or how things cannot be (the ultimate safety analyses).

THE RATIONALE AND GOALS FOR THE HARD ROCK LABORATORY

The scientific investigations within SKB's research programme are a part of the work of designing a final repository and identifying and investigating a suitable site. This requires extensive field studies regarding the interaction between different engineered barriers and host rock. A balanced appraisal of the facts, requirements and evaluations led to the proposal to construct an underground research laboratory [1]. This proposal was very positively received by the reviewing bodies. In the autumn of 1986, SKB initiated field work for the siting of an underground laboratory, the Hard Rock Laboratory, in the Simpevarp area in the municipality of Oskarshamn. At the end of 1988, SKB decided to locate the facility on southern Äspö about 2 km north of the Oskarshamn nuclear power plant. Construction is planned to start during 1990. Approvals from the authorities is expected during spring 1990. The HRL is not the site for the final repository in Sweden. The site will be needed for research and development for several decades.

The main goals of the R&D work at the Hard Rock Laboratory are [2]:

To test the quality and appropriateness of different methods for characterizing the bedrock with respect to conditions of importance for a final repository,

To refine and demonstrate methods for how to adapt a final repository to the local properties of the rock in connection with planning and construction,

To collect material and data of importance for the safety of the final repository and for confidence in the quality of the safety assessments.

Prior to the siting of the final repository for spent fuel in the mid-1990s, the activities at the Hard Rock Laboratory shall serve to:

Verify pre-investigation methods

Finalize detailed investigation methodology

As a basis for a good optimization of the final repository system and for a safety assessment as a basis for the siting application, which is planned to be submitted a couple of years after 2000, it is necessary to:

Test models for groundwater flow and transport of solutes

In preparation for the construction of the final repository, which is planned to begin in 2010, the following shall be done at planned repository depth and under representative conditions:

Demonstrate construction and handling methods

Test important parts of the repository system

These tests shall be able to be carried out on a sufficient scope as regards time and scale to provide the necessary support material for Government approval of the start of construction. Certain tests may therefore have to be started in the mid-90s.

AN OUTLINE OF THE APPROACH TO SITE CHARACTERIZATION DURING THE PRE-INVESTIGATION STAGE

Site characterization is a multi- and inter-disciplinary task that necessitates integration during data acquisition, evaluation and presentation of results. In order to facilitate such integration, three basic decisions were made for the site characterization.

Investigation stages

The first decision is obvious, to structure the investigation in distinct stages. Site characterization for a final repository can be divided into three parts, characterization from surface and boreholes prior to any construction activities, characterization during access to a potential repository level and characterization during construction of the repository. The same division is made for the HRL. It is planned to do a comprehensive evaluation before construction, after access to the 500 m level and after certain investigations in the Operational phase. The pre-investigations have as well been divided in steps and for every step an imterim evaluation was made on the data available so far [3, 4].

Investigation scales

The second basic decision was to conceptualize the site characterization results to different scales, Figure 1, as it was thought that conceptualizations will be needed in different scales for a real repository. These conceptual models are the bases for appropriate numerical models and for the layout of a repository. The regional scale >> 1000 m forms a basis for later detailed investigations. Assessments in regional scale can be used to select suitable outcrops for the repository, to define areas of recharge and discharge and to give the overall tectonic pattern. The site scale model 100 - 1000m can be used to define major fracture zones and/or major flow paths. These investigations will provide guidance on what depth the repository should be placed at as well as a potential repository volume. Characterization to this scale defines the farfield ground water flux and nuclide transport through the repository and to the biosphere. Block scale assessment 10 - 100 m will be used to position deposition tunnels and later to position canisters. Essential assessments are on the transport of solutes from possibly failed canisters to major flow paths. The detailed scale 1 - 10 m defines the geohydrological, chemical and mechanical nearfield to the waste packages. By proper positioning of the packages it will be possible to influence the overall safety of the repository.

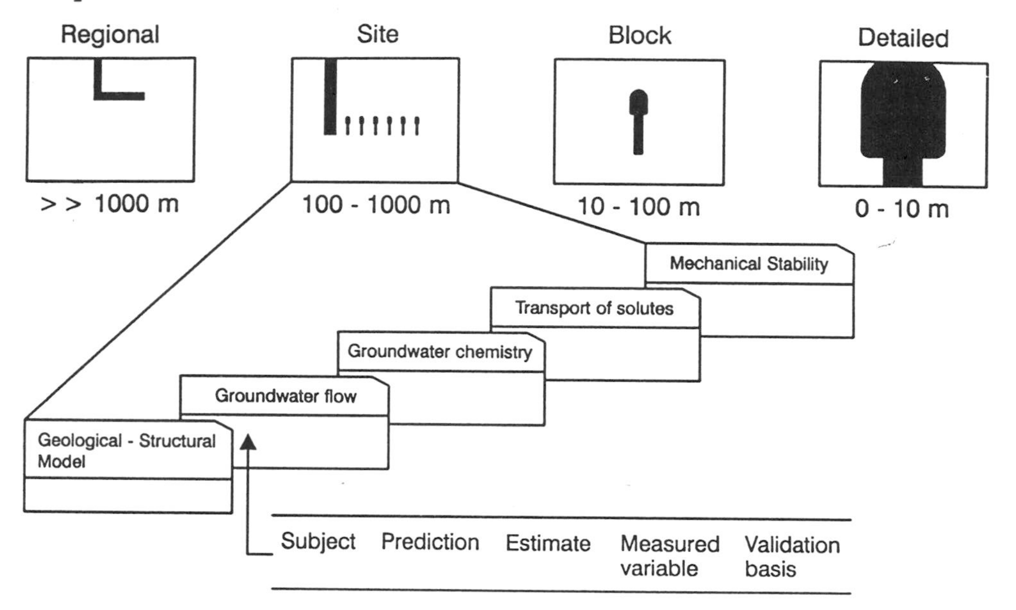

Figure 1 Overview of scales and issues.

Investigation issues

The third decision was to designate five key issues of relevance for design, and/or performance assessment and/or safety analysis.

The **geological-structural model** incorporates the simplified description of lithology and the fracturing. The simplification of the real geological medium to a geological model is one of the most crucial issues in the site characterization as this simplification gives the basis for design and strongly affects the other four key issues. It must however be emphazised that the geological description will evolve along with the characterization programme. The first pre-investigation can be used to set up working hypotheses that can be checked in the next phase of the characterization. If they prove to be wrong they can be modified and enhanced during the more detailed characterization in the coming stages. The ultimate description of the host rock will of course appear before the repository is sealed. These descriptions can be improved if nomenclature on the concepts of zone and reliability of investigations is developed. Such attempts are under way [5]. **Ground water flow** is a key issue as the flow influences the service life of canisters and dissolution of the waste. Description of ground water flow also gives necessary but not sufficient data for calculation of nuclide transport. **Ground water chemistry** is a key issue as it influences corrosion of canisters, dissolution of the waste and provides necessary but not sufficient data for calculation of transport of nuclides from the repository. **Transport of solutes** is a key issue as it gives a necessary basis for the assessment of the safety function of the geological barrier in the multi-barrier system. **Mechanical stability** is important both in a short and a long term assessment. Mechanical stability is needed during construction. The long-term issue is to identify potential zones of movements, so that the canisters are not cut by future movements. The long-term assessment is as well needed as a prime function of the rock is to provide stable geohydrological chemical conditions.

Integration during the data acquisition

The drilling and investigation programme in the boreholes and the cross-hole programmes have been a mutual responsibility for the principal investigators of the project. Four drilling campaigns have been carried through. The first campaign comprised percussion drilling at three candidate sites for the laboratory. The second campaign was used to get a general understanding of the target area Äspö and the reference area Laxemar. The third campaign was devoted to investigation of fracture zones at south part of Äspö. The fourth campaign was not initially planned in 1986, but was started in the autumn of 1989 as a consequence of a re-design of the HRL.

In order to facilitate the site characterization a new methodology was developed for borehole investigations allowing an improved integration of geological, geohydrological and geohydrochemical data.

The borehole coring technique was modified, whereby a drawdown can be created in the hole by pumping. This prevents or at least reduces the penetration of drill cuttings and drilling water into the fractures in the rock. In order to make this possible, the uppermost about 100 m of the borehole is drilled with a larger diameter. Air-lift pumping can then be carried out from this part of the borehole. The hole has a large enough diameter to accomondate a submersible pump, so that pumping test can be carried through. The design also made it possible to use flow-meter logging to detect transmissivities larger than 10^{-7} m^2/s positioned to an accuracy of one meter in the hole. Water sampling and analysis of the chemical content of the water is done at the surface during the air-lift pumping, during breaks in the drilling every 100 meter for short pumping, during long-time pumping tests and at last with a sampling probe in sealed-off sections in the hole.

Integration of data, conceptualizations and numerical models

The overall intention is to conceptualize the rock for different scales and to different issues. Numerical models are used to make the quantitative evaluations of the conceptualizations. All these predictions will be checked during construction of the laboratory in order to refine the description of the site.

Table I provide an overview of current data and conceptual/numerical models. It is foreseen in the programme for the HRL that the site characterization will be more directed towards transport of solutes in the future, that can be implemented at a well characterized site.

CONCLUDING REMARKS

Site characterization needs objectives so that sufficient data are gathered in every stage of the investigation. It is thus recommended that a consensus is developed what should be achieved during pre-investigations, during access to the repository and during construction of the repository. "State-of-the-art" safety analyses can provide support in setting those objectives. It however seems inevitable that characterization must be carried through with a broader perspective, as new findings can change the outcome of a safety analyses. A flexible design of the repository will enhance the safety of the overall system. The licensing process must recognize the benefit of this flexibility in the design and facilitate for this approach. The licensing process must as well recognize and allow for that site characterization is a step-wise procedure and that the safety assessments can be improved during pre-investigations, during access to repository level and during repository construction when the bulk of data for the host rock will be collected.

REFERENCES

[1] SKB: "R&D-Programme 86. Handling and final disposal of nuclear waste. Programme for research, development and other measures." SKB, Stockholm. 138pp.

[2] SKB: "R&D-Programme 89. Handling and final disposal of nuclear waste. Programme for research, development and other measures." SKB, Stockholm. 196pp.

[3] Gustafson G, Stanfors R, Wikberg P: "Swedish Hard Rock Laboratory: First evaluation of pre-investigations 1986 - 1987 and target area characterization." SKB Technical Report TR 88-16, Stockholm. 99pp.

[4] Gustafson G, Stanfors R, Wikberg P: "Swedish Hard Rock Laboratory: Evaluation of 1988 year pre-investigations and description of the target area, the island of Äspö." SKB Technical Report TR 89-16, Stockholm. 152pp.

[5] Bäckblom G: "Guide-lines for use of nomenclature on fractures, fracture zones and other topics." SKB TPM 25-89-007, Stockholm. 6pp.

Table I. Overview of data, conceptual- and numerical models used in the site characterization for the HRL

SCALE	DATA	BASIS FOR CONCEPTUAL MODELS	NUMERICAL MODEL
REGIONAL	Air-borne geophysical survey (magnetic, EM, radiometric, VLF)	Ref [3]	2D and 3D regional flow model (NAMMU)
	Gravity measurements (one station per km^2)		
	Petrophysical measurements (density susceptibility, IP, etc)		
	Interpretation of lineaments (LANDSAT, digital terrain models)		
	Mapping of solid rock		
	Characterization of main tectonic zones (mapping, ground geophysics)		
	General hydrology of the area		
	Regional welldata analysis and water chemistry		
	Compilation of geohydrological data from construction works in the area		
SITE ÄSPÖ	Detailed mapping and petrographic studies along cleaned trenches (1 600 m)	Ref [3]	Generic model of draw-down (analytical element in 2D, 3D)
	Detailed geophysical studies (VLF, resistivity, magnetic, radiometric, seismic refraction, -reflection)	Ref [3]	Generic model of fresh/sa-line water interface and stochastic continuum (2D,
	Detailed study of ductile and brittle structures		
	Borehole investigations (20 percussion drilled holes, 2 200 m, 14 cored holes, 6 600 m)	Ref [4]	Predictive model for long-time pumping test (NAMMU 3D and Phoenics 3D)
	- Lithology (detailed mapping, thin sections, modal- and chemical analyses) - Fractures (frequency, RQD, minerals, surface, relative and absolute orientation) - Geophysics (up to 13 geophysical logs) - Rock stresses (hydraulic fracturing, overcoring, mechanical properties) - Head monitoring	In prep	Predictive model for draw-down (-pressure, inflow, changes in position of interface, scoping calculation for tracers (Phoenics, 3D)
	- Hydraulic tests (air-lift tests every 100 m, 72 h pumping test, flow-meter logging, packer tests in 3 m and 30 m sections) - Interference tests - Long-time pumping test (one month pumping, two months recovery) - Dilution tests to measure ambient flow in cored holes - Tracer tests (non-sorbing) - Sampling of groundwater chemistry during air-lift tests, pumping tests and sealed-off sections (full characterization in 10 points, 30 parameters)	In prep	Predictive model for tracer tests (Phoenics, 3D)
BLOCK	Detailed mapping of fractures at surface (lengths, distributions)	In prep	Fracture flow model (Fracman, 3D)
	Data on transmissivities and orientations in boreholes		

VALIDATION OF TRANSPORT MODELS FOR USE IN REPOSITORY PERFORMANCE ASSESSMENTS: A VIEW

C.P. Jackson, D.A. Lever, P.J. Sumner
AEA Technology, Harwell Laboratory, Oxfordshire, United Kingdom

ABSTRACT

We present our views on validation, stressing the importance of presenting for review the case for (or against) a model. We outline a formal framework for validation, which includes calibration, testing predictions, comparison with alternative models, analysis of discrepancies, presentation, consideration of implications and suggesting improved experiments. We illustrate the approach by application to an INTRAVAL test case based on laboratory experiments. Three models were considered: one with and one without rock-matrix diffusion, and a model with kinetic sorption. We show that the model with rock-matrix diffusion is the only one to provide a good description of the data. We stress the implications of extrapolating to larger length and time scales for repository performance assessments.

REFLEXIONS SUR LA VALIDATION DES MODELES DE TRANSPORT DESTINES AUX EVALUATIONS DU COMPORTEMENT DES DEPOTS

RESUME

Les auteurs exposent leurs réflexions sur la validation, en mettant l'accent sur l'importance de la présentation d'arguments pour (ou contre) un modèle. Ils tracent un cadre précis pour la validation qui prend en compte les éléments suivants : étalonnage ; vérification des prévisions ; comparaison avec d'autres modèles possibles ; analyse des anomalies ; présentation ; enseignements à tirer et suggestions d'expériences mieux adaptées. Pour illustrer leur méthode les auteurs l'appliquent à un cas témoin INTRAVAL fondé sur des expériences de laboratoire. Trois modèles ont été examinés : un avec et un sans diffusion roche-matrice et le troisième avec sorption cinétique. Il ressort que le modèle simulant la diffusion roche-matrice est le seul qui rende compte correctement des données. L'auteur insiste sur les conséquences de l'extrapolation à des échelles de taille et de temps plus grandes pour les évaluations du comportement des dépôts.

1. INTRODUCTION

Radioactive waste arises as a by-product of nuclear generation of electricity and the use of radioisotopes in medicine and industry. Many countries are currently evaluating ways to dispose of radioactive waste safely. It is widely agreed that the preferred option is to process the waste into stable solid forms and place these in an underground repository in a suitable geological formation. This approach aims to keep the radioactive materials away from Man's immediate environment (the biosphere) until decay has reduced their radioactivity to an acceptable level.

The natural pathway for radionuclides from waste in an underground repository to return to the biosphere is dissolution and transport by flowing groundwater. Analysis of this groundwater pathway plays an important role in selection of a suitable site for a repository, and in assessment of the performance of the repository. Models must be used because very long times need to be considered. It is clearly desirable that the models should be checked. Numerical models should be verified, that is shown accurately to represent the intended mathematical models; and these should be validated, that is shown adequately to represent physical phenomena.

In this paper we present our views on validation. We outline a formal framework for validation illustrated by application to one of the test cases from the international INTRAVAL project, which seeks to address the issues of validation of groundwater transport models for use in performance assessments of underground repositories for radioactive waste.

2. VALIDATION

The IAEA definition of validation is[1]:

> "Validation is a process carried out by comparison of model predictions with independent field observations and experimental measurements. A model cannot be considered validated until sufficient testing has been performed to ensure an acceptable level of predictive accuracy. (Note that the acceptable level of accuracy is judgmental and will vary depending on the specific problem or question to be addressed by the model.)"

We consider that validation is not just about comparing predictions of the model with physical observations. It is about establishing whether or not the model is an acceptable representation of physical phenomena. As such it also involves examining the model for consistency with principles that are generally accepted in the scientific community, and presenting the case for (or against) the model for review. Presentation of the case is a very important part of validation, particularly for models that might be used in a performance assessment for a radioactive waste repository, since many people other than technical experts may be involved in decisions relating to such a repository.

Normally a model of a specific system involves many components or submodels. Many of these components are often representations in the context of the system under consideration of general physical laws, such as conservation of mass. Validation can address the general physical laws, or a model of a specific system. The first is really part of the general progress of science. A general physical law can be taken to be validated when it is widely accepted within the scientific community that it provides a good representation of physical phenomena. The general physical laws are validated by demonstrating for many specific systems that models based upon these laws provide good descriptions of the systems. In validation of models of specific systems it is not necessary to validate general physical laws that are widely accepted. Rather validation can concentrate upon those submodels that may not be generally accepted, and of course upon demonstrating that the general physical laws have been sensibly applied in the context of the particular system. It should be noted that a model may fail adequately to represent a given physical system for two reasons: either (i) the general physical laws underlying the model are inappropriate, or (ii) the model adopts an inappropriate representation of the parameters of the model. For example, spatial variation of the parameters may not be taken into account properly.

There are three complementary types of investigation that may be used in model validation: laboratory experiments, field experiments, and natural analogues. All three types of study have their advantages and disadvantages, and none is a perfect source of data. Laboratory experiments can be well controlled and well characterised, but are severely restricted in the length and time scales that can be considered; field experiments address larger scales but cannot be quite as well controlled or characterised as laboratory experiments; and natural analogues provide information on the length and time scales of interest in repository performance assessments, but cannot be controlled and are often difficult to characterise. Used together the different types of investigation provide data that can be used to build confidence in the models used in safety assessments.

3. A FRAMEWORK

A formal framework for validation is useful. It helps to ensure that all the relevant issues have been addressed, provides the basis for presenting the case for the model, and helps to make explicit the subjective aspects of validation. We consider that the following framework is appropriate:

(a) review models
(b) review data
(c) calibrate a specific model
(d) define acceptability of the model (with due regard to its purpose)
(e) predict and test
(f) compare with alternative models
(g) analyse discrepancies
(h) assess parameters
(i) present study for review
(j) consider implications
(k) suggest improved experiments
(l) review consistency

We make some comments on these topics and others below.

Goodness-of-fit measures

The use of goodness-of-fit measures to assess the agreement between values calculated using the model and physical observations helps to reduce and make explicit the subjective element in validation. Different goodness-of-fit measures may be appropriate for calibration and testing predictions. A root mean square weighted error with differences between calculated and measured values weighted according to the experimental error is probably the most appropriate measure.

Acceptability of the model

The purpose of the model should be taken into account when considering what constitutes an acceptable model. In particular, performance assessments for deep underground radioactive waste repositories inherently involve extrapolation of models from length and time scales on which experiments realistically can be carried out to very large length and time scales. Note that the level of detail required in such extrapolations may be less than required in applications that consider shorter scales. In order to have as much confidence as possible in the extrapolations attempts should be made to validate the models using data from as wide a range of length and time scales as possible.

Comparison with alternative models

We think that it is very important to compare a model with alternative models. In particular, this helps to put the extent of validation into context.

Consistency

Models to be used in performance assessments should be consistent with one another, with all the observed data, and with general scientific principles. We think that little confidence can be placed in a model that gives grossly inaccurate predictions for one aspect of the system, even though it predicts another aspect well; although it is sometimes argued that such models can be used to predict the second aspect of the system.

4. GROUNDWATER TRANSPORT MODELS

We think that it is important to make explicit the statement, which we believe to be generally accepted in the scientific community, that the processes of transport by groundwater are understood in a broad sense at the scale of the rock pores (or fractures in a fractured rock). A substance may be transported in solution by the processes of advection and diffusion: that is it may move at the local velocity of the groundwater or it may diffuse relative to this. The substance may also interact (chemically or physically) with the walls of the pore space, and it may interact with other substances being transported by the groundwater. This is not to say that the details of the interactions are known. However, these details are not really of direct interest. What is of interest is the effective behaviour on larger scales that is the consequence of the behaviour on the small scale. The class of possible models of large-scale transport by groundwater is severely restricted by the broad understanding of behaviour on the pore scale, and by general physical principles such as conservation of mass.

There are basically three approaches to modelling groundwater transport: the continuum porous medium approach, the fracture-network approach, and the channelling approach. In the continuum approach equations are derived in terms of quantities that can be regarded as macroscopic averages over many pores. (They can also be regarded as statistical averages over ensembles of many realisations of the rock). In the fracture-network and channelling approaches an attempt is made to represent the system on the scale of the individual pores. The representation is necessarily idealised. However, these approaches allow study of (i) the consequences on the macroscopic scale of the pore scale behaviour, (ii) the way possible limiting macroscopic behaviour is obtained, and (iii) variations in the pore-scale behaviour over the system.

The simplest continuum model that describes the effects of advection, diffusion, dispersion, radioactive decay, sorption and rock-matrix diffusion is governed by the equations

$$\frac{\partial}{\partial t}(\phi RN) + \mathbf{q}.\nabla N + \lambda\phi RN = \nabla.(\phi \mathbf{D}.\nabla N) + F \quad (1)$$

where

- N is the concentration of tracer in the groundwater,
- ϕ is the porosity,
- R is the retardation due to linear equilibrium sorption,
- $\mathbf{q}$ is the Darcy velocity or specific discharge,
- λ is the radioactive decay constant,
- $\mathbf{D}$ is the dispersion tensor,
- F is the flux of tracer from the rock matrix to the groundwater.

Usually $\mathbf{D}$ is modelled as

$$D_{ij} = D_p\, \delta_{ij} + a_T\, v\, \delta_{ij} + (a_L - a_T)\, \frac{v_i v_j}{v}, \quad (2)$$

where

- D_p is the pore-water diffusion coefficient,
- a_L and a_T are the longitudinal and transverse dispersion lengths,
- v is the transport velocity $\mathbf{q}/\phi$.

The flux of tracer from the rock matrix to the flowing groundwater is modelled as

$$F = \left(\frac{\phi^m}{w_{max}}\right) D^m \frac{\partial N^m}{\partial w}(w = w_{max}), \quad (3)$$

where quantities with a superscript m refer to the rock matrix, and w is a coordinate in the rock matrix, directed towards the flowing groundwater, with upper limit w_{max}, and D^m is the molecular diffusion coefficient in the matrix. The concentration of tracer in the rock matrix N^m satisfies a 1D equation similar to (1) without the velocity-dependent terms, that accounts for diffusion and sorption in the rock matrix.

This model can be simplified by neglect of the contribution from rock-matrix diffusion. The effects of kinetic sorption can also be modelled.

5. AN EXAMPLE

We have applied the methodology described above within the INTRAVAL project. The project is structured around a number of test cases. We concentrated our efforts upon test case 1b, which is based upon laboratory tracer experiments on core samples of fractured hard rock. The experiments consist of driving water through small cylindrical samples of rock (with dimensions of order centimeters). A brief pulse of tracer (^{233}U) is added to the water entering the sample and the concentration of tracer in the emerging water measured as a function of time. Data is available for three granite samples and one gneiss sample.

We sought to use the data provided in the test case, (i) to develop the methodology described above, and (ii) to see whether or not it provided support for three different models of transport. The simplest model included the effects of advection, diffusion, dispersion, radioactive decay and equilibrium linear sorption. Models that included the effects of rock-matrix diffusion and kinetic sorption were also considered. We discuss below some aspects of the study.

Review data

Various biases in the experimental data were identified. In particular the experiment was only carried out with a single tracer and for a single flow velocity. This is not really a criticism of the original experiments since these were not carried out with the aim of validation.

Calibrate specific models

Each model was calibrated for each sample using the first few data points. A non-linear least-squares approach was used for the calibration. The outlet concentrations for the models were calculated by numerical inversion of Laplace Transforms using the Talbot algorithm.

Predict and test

For each model and sample the predicted outlet concentrations were compared with the data not used for the calibration. The comparison for one of the samples is shown in Figures 1. The comparison for the other samples is very similar.

Comparison with alternative models

As part of the study several models were compared. Clearly the model with rock-matrix diffusion best represents the data.

One point is worth noting. For each sample and model the best fits to all the data points exhibit very similar behaviour to the best fits to the first few points. The comparison of the models shows that validation is not just about fitting curves to data. The kinetic-sorption model involves the same number of model parameters as the model with rock-matrix diffusion, and yet the kinetic-sorption model cannot represent the data nearly as well as the model with rock-matrix diffusion, which is clearly more realistic.

Assess parameters

The parameters of the model with rock-matrix diffusion were assessed for physical reasonableness. Unfortunately the mathematical model involves fewer parameters than there are physical parameters. Thus even if the experiments defined the model parameters, they would not define the physical parameters uniquely and additional experiments would be needed to define all the physical parameters. In fact the experiments do not actually define all the model parameters for all the samples. For two of the samples the best-fit to the data is completely insensitive to the values of one of the parameters (corresponding to the dispersion) over a wide range. For the other samples the dispersion length was determined. Values of the order of centimeters were found, which are reasonable. Further the retardation was calculated from the best fit and the value of the flow velocity. Values in the range 20 to 200 were determined, which are in accord with other measurements.

Examination of discrepancies

The differences between the best fit model with rock-matrix diffusion and the data were examined. In general these are small, but there is sufficient structure in these differences to suggest that something very minor is being neglected.

Consider implications

The implications of the study for repository performance assessments were considered. Clearly the mathematical model with rock-matrix diffusion discussed has been shown to be a good representation of the physical phenomena in the experiments. Thus some confidence has been gained in applying the model in repository performance assessments. However, it is necessary to consider how the model would be used in assessments. In particular it would be used to model transport over much longer time and length scales than considered in the experiment. Thus it is necessary to consider how the model would be extrapolated to the relevant scales.

It should be noted that models with different physical interpretations may be represented by the same mathematical equations. Clearly the same agreement between model predictions and observations can be obtained with the different models in such a case, and it is not possible to distinguish between the different models on the basis of this agreement. (However, it may be possible to differentiate between the models on the basis of the reasonableness of the parameters). The equations (1), (2) and (3) provide an example of this. The same equations describe a continuum model or a model of a single fracture or channel. However, the different models have to be extrapolated to larger scales in different ways. If the model is regarded as a continuum model then the model should be simply extrapolated taking account of the empirical correlation for dispersion lengths. On the other hand, if the model is regarded as a model of a single fracture or channel, then it is not sensible simply to extrapolate the model. Rather the consequences of flow and transport in a network of fractures or channels should be considered, since it is not reasonable to expect a single fracture or channel to extend over the distances considered. However the extrapolation is made, it is desirable to check it.

Suggest further experiments

We made several suggestions for further experiments, in particular, that the experiments be repeated, that several different tracers be used, and that different flow rates be used. These experiments would provide data to check the model.

6. CONCLUSIONS

We have successfully applied the methodology described above to the INTRAVAL test case. The formal framework was found to be useful.

We consider that loosely speaking, we have validated the rock-matrix diffusion model for the given experiments. More precisely, we consider that we have shown that the model provides a very good description of the physical phenomena that occur in the experiment, and that the parameters of the model are reasonable. In a sense we have also invalidated the models without rock-matrix diffusion, in that given the different levels of agreement between model predictions and observations shown in Figure 1, it would be very difficult to justify a priori using in a performance assessment any model other than the model with rock-matrix diffusion. However, the model with rock-matrix diffusion reduces to the model without rock-matrix diffusion in many circumstances. Thus after suitable analysis it may be possible to justify using the model without rock-matrix diffusion.

We have considered the implications of our analysis for repository performance assessments. In particular we stress the implications of extrapolating to relevant length and time scales.

We have made recommendations for further experiments, both improved laboratory experiments, and experiments on larger scales to test the extrapolation of the model to larger length scales.

ACKNOWLEDGMENTS

This work was funded by UK Nirex Ltd.

REFERENCES

[1] Radioactive Waste Managment Glossary, 2nd Edition, International Atomic Energy Agency Technical Document, IAEA-TECDOC-447,1988.

FIGURE

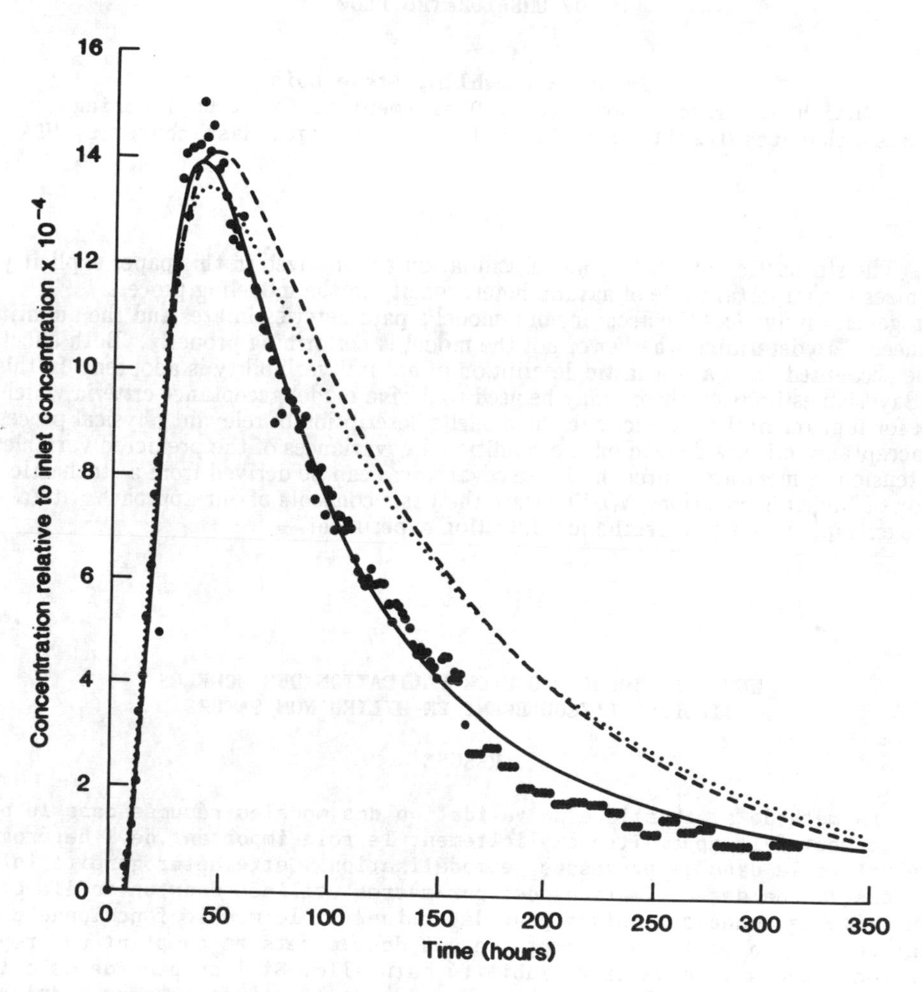

Figure 1. Comparison of experimental data and model predictions for one sample. The graphs show the normalised tracer concentration (relative to the inlet concentration x 10^{-4}) as a function of time. The points are the experimental data, the solid line is the best fit for the model with rock–matrix diffusion, the dotted line is the best fit for the model without rock–matrix diffusion, and the dashed line is the best fit for the model without rock–matrix diffusion but with kinetic sorption. (The first 41 points were used to calibrate the model)

A STOCHASTIC APPROACH FOR VALIDATING MODELS OF UNSATURATED FLOW

Dennis McLaughlin, Steve Luis
Ralph M. Parsons Laboratory, Department of Civil Engineering
Massachusetts Institute of Technology, Cambridge, Massachusetts, USA

ABSTRACT

The stochastic approach to model validation summarized in this paper explicitly recognizes the important role of natural heterogeneity in the modeling process. Heterogeneity influences the accuracy of a model's parameter estimates and the quantity of data needed to determine whether or not the model is performing properly. Both effects can be accounted for if a stochastic description of natural variability is adopted. In this case Bayesian estimation theory may be used to derive model acceptance criteria which check for fundamental deficiencies in the model's description of relevant physical processes. The acceptance criteria depend on the conditional covariances of the predicted variables (e.g. tension or moisture content). These covariances can be derived from a stochastic version of Richards equation. We illustrate the basic concepts of our approach with a simple example based on a synthetic infiltration experiment.

METHODE STOCHASTIQUE DE VALIDATION DES MODELES SIMULANT L'ECOULEMENT EN MILIEU NON SATURE

RESUME

La méthode stochastique de validation des modèles résumée dans le présent document fait apparaître explicitement le rôle important de l'hétérogénéité naturelle dans le processus de modélisation. Cette hétérogénéité influe sur l'exactitude des estimations des paramètres utilisés dans un modèle et sur la quantité de données requises pour déterminer si le modèle fonctionne correctement ou non. On peut tenir compte de ces deux effets en adoptant une représentation stochastique de la variabilité naturelle. Si l'on procède de cette façon, on peut utiliser la théorie bayésienne de l'estimation pour déduire des critères d'acceptation d'un modèle qui permettent de déceler les insuffisances fondamentales dans la description des processus physiques pertinents qui est faite dans ce modèle. Ces critères d'acceptation dépendent des covariances conditionnelles des variables sur lesquelles portent les prévisions (par exemple tension et taux d'humidité). Il est possible de déduire ces covariances d'une version stochastique de l'équation de Richards. Les auteurs illustrent les principes fondamentaux de leur méthode par un exemple simple fondé sur une expérience synthétique d'infiltration.

INTRODUCTION

Comparisons between theory and experimental results in hydrogeology are greatly complicated by the heterogeneity of the natural environment. Heterogeneity, coupled with the difficulty of making subsurface observations, imposes severe limitations on our ability to adequately describe the processes we are trying to predict. When heterogeneity is significant and data are limited there may be no objective way to determine whether or not a model's predictions are satisfactory. This problem is widely recognized by modelers but is rarely confronted directly.

We can obtain a clearer view of the role of heterogeneity in validation if we consider how hydrogeologic models are developed. Most models concerned with the movement of water through the subsurface are based on fundamental physical principles such as conservation of mass, momentum, and energy. These principles are not, in themselves, enough to provide useful predictions of environmental variables. They must be supplemented by constitutive assumptions which provide the closure needed to obtain a solution. Constitutive assumptions introduce additional parameters which frequently vary over space and/or time. Examples relevant to flow through a porous continuum include the various rate coefficients, such as hydraulic and thermal conductivity, which are used to relate mass and energy fluxes to potential gradients. Examples in fractured media include fracture properties and the coefficients which determine interactions between fracture and matrix flow.

It is certainly possible to envision a theory (or model) of subsurface flow which includes all relevant physical principles but cannot provide reliable predictions in a specific application because the theory's constitutive coefficients are unknown. This is a common occurrence in hydrogeology. Researchers have, for example, developed a number of elegant physically–based models of multiphase nonisothermal flow through porous media which cannot be adequately tested until their constitutive parameters are defined. Unfortunately, these constitutive parameters are usually difficult to measure and must be inferred indirectly from measurements of related variables. The resulting parameter estimates are often highly uncertain.

It is important to realize that a hydrogeologic model may be difficult to validate even when its constitutive parameters are well known. If the model attempts to predict detailed variations in heterogeneous variables such as groundwater velocity or solute concentration, it may be impossible to collect enough data to verify whether or not the predictions are correct. Of course, we may not care about predicting detailed distributions of hydrogeologic variables. We may be satisfied with reliable descriptions of larger scale trends (e.g. velocity or concentration averaged over a specified volume and/or time interval). Unfortunately, it may be difficult to estimate such trends from the limited numbers of point measurements which are typically collected in field experiments.

We can summarize the above ideas with a list of the primary reasons why model predictions differ from field observations [1]:

1. The model (and its supporting theory) do not properly describe the physical processes which control subsurface flow at the site of interest. In practice this is most often because important physical processes are either overlooked or neglected when simplifying assumptions are introduced.

2. The model's constitutive parameters are incorrectly specified. This may be because the data used to obtain the parameter estimates are unrepresentative or it may be because the parameters are inherently unobservable.

3. The field data used for validation are unrepresentative of conditions at the site.

We believe that the model validation process should be designed to determine which, if any, of the factors listed above is responsible for differences between predictions and observations. This can be done if the effects of natural heterogeneity are described in a realistic quantitative way. The validation concepts summarized in the following sections indicate how probabilistic methods can be used to make validation more rigorous and more informative. But perhaps more importantly, they suggest how model validation can be expanded to include questions of experimental design as well as more traditional modeling issues.

A STOCHASTIC APPROACH TO MODEL VALIDATION

The basic concept behind our approach to model validation is simple. We attempt to evaluate the impact of parameter estimation and sampling errors on observed differences between model predictions and field measurements. Differences which cannot be attributed to these two error sources are assumed to be caused by fundamental deficiencies in the model. This approach is similar to the one used in classical regression theory to test the validity of linear algebraic models. In hydrogeologic applications, regression methods must be extended to account explicitly for spatial heterogeneity. In order to make our discussion more specific, we limit our attention to flow through unsaturated porous media. The concepts can, however, be readily adapted to other problems.

In most unsaturated flow applications, we are concerned with comparing time and space–dependent predictions of moisture content and/or tension with a relatively small number of field measurements. The predictions are generally based on some version of Richards equation and depend on site–specific soil properties such as the unsaturated hydraulic conductivity and specific capacity. It is unrealistic to expect such a model to predict tension fluctuations in a field soil at scales of a few centimeters. In field applications unsaturated flow models are typically used to predict tension variations over distances of meters to tens of meters. In such cases the model's soil properties are often referred to as "effective parameters" to indicate that they describe bulk flow over a extended region rather than flow at a point.

In field settings where heterogeneity is significant we should expect to observe local differences between bulk flow predictions and point measurements. We can put some probabilistic bounds on these differences if we adopt a stochastic description of point–to–point variability . In particular, we assume that small–scale variations in moisture content and tension are due to small scale variations in the hydraulic (and possibly chemical) properties of the soil. Soil property variations encountered at a particular site are treated as if they were realizations of random fields with known statistical properties. This stochastic description of soil heterogeneity provides a concise way to account for spatial persistence, layering, anisotropy and many other effects observed in field soils (see [2,3,4,5,6,7] for more detailed discussions of the stochastic approach).

The effect of random soil properties on moisture content and tension can be inferred from the three–dimensional Richards equation, which we assume to be valid at small scales (e.g. tens of centimeters). Suppose that the random soil properties of interest are the unsaturated hydraulic conductivity and specific capacity, expressed as functions of tension and location. We assume that these functions can be expressed in terms of a finite number of random parameters which depend on location but not on tension. Examples are the various parameters included in the van Genuchten [8] model of unsaturated soil properties. If the random soil parameters are assembled in a vector $\alpha(x)$, the solution to the stochastic version of Richard's equation may be written formally as:

$$\psi t(x) = \Psi[\alpha(x),x,t] \qquad (1)$$

where x is a three–dimensional spatial coordinate, t is time, and $\Psi[\cdot]$ is an operator which relates the functions $\alpha(x)$ and $\psi(x,t)$. The tension $\psi(x,t)$ is a random field because it depends on the random field $\alpha(x)$.

Now consider a model which attempts to predict large–scale variations (trends) in tension during an infiltration and drainage event. We wish to test the validity of this model by comparing its predictions to a set of tension measurements collected after the model's parameters have been specified. We suppose that the model is based on effective hydraulic conductivity and specific capacity functions which are estimated from a limited number of soil samples. The effective soil property functions depend on a vector $\hat{\alpha}$ of spatially invariant parameters. In this case the tension predicted by the model may be written formally as:

$$\hat{\psi}(x,t) = \hat{\Psi}[\hat{\alpha},x,t] = \hat{\Psi}[\hat{\alpha}(\alpha_1^*,\alpha_2^*,...,\alpha_N^*),x,t] \quad (2)$$

The operator $\hat{\Psi}[\cdot]$ is an operator which relates the vector $\hat{\alpha}$ and the function $\hat{\psi}(x,t)$. The function $\hat{\alpha}(\alpha_1^*,\alpha_2^*,...,\alpha_N^*)$ describes how the model's parameters are actually calculated from field measurements of the soil parameters α_i^* at sampling locations x_i, i=1,...N. If we consider the soil measurements to be given, the model prediction is a deterministic function of space and time.

The difference between the model's prediction and the true tension at any given time t and location x is simply:

$$\epsilon(x,t) = \psi(x,t) - \hat{\psi}(x,t) \quad (3)$$

If the model's objective is to predict the actual value of tension at every point, $\epsilon(x,t)$ is the prediction error. If, however, the model is intended to describe large–scale trends $\epsilon(x,t)$ is not an error but is simply a local fluctuation due to small–scale heterogeneity. The likely scatter of the true tension values about the model's prediction can be measured in terms of the variance of $\epsilon(x,t)$. A well known result of Bayesian estimation theory [9] states that the mean of $\epsilon(x,t)$ is zero and its variance σ_ϵ^2 is minimized when the prediction $\hat{\psi}(x,t)$ is the expected value of $\psi(x,t)$ conditioned on all available measurements (in this case, the soil measurements $\alpha_1^*,...,\alpha_N^*$). It is reasonable to require that a model which is designed to predict bulk trends in the true tension should have the properties of an unbiased minimum variance estimate. Such a model has the greatest chance of providing a reasonable fit to the true spatial distribution of tension. Note that we are not saying that a reasonable model of large–scale tension needs to be based on stochastic concepts. The model may rely on traditional deterministic descriptions of unsaturated flow. We only require that predictions from the model, whatever its origin, approximate the conditional mean.

If we adopt the definition that a structurally correct model of large–scale trends is one which yields predictions equal to the conditional mean, then we can pose model validation as a hypothesis testing problem with the following null hypothesis [9]:

$$H_0\text{: The model is structurally correct, i.e. } \hat{\psi} = E[\psi \mid \alpha_1^*,...\alpha_N^*]$$

In this case, the validation problem is to determine, from tension measurements taken during the predicted infiltration event, whether or not the null hypothesis can be rejected. Although there are many ways to develop statistical tests of such hypotheses, we will focus here on the use of approximate confidence intervals. In this case, the hypothesis will be rejected if a sufficiently large number of measurements fall outside of confidence intervals constructed around the model's prediction, which is hypothesized to be equal to the conditional mean of $\psi(x,t)$.

In order to construct confidence intervals about the conditional mean, we need to know the conditional probability density of $\psi(x,t)$. Since this is difficult to estimate in practice, we take a pragmatic approach and assume normality or, perhaps, log normality. In this case, the required confidence intervals can be derived from the conditional variance of $\psi(x,t)$ — i.e. the variance of ψ given the observations $\alpha_i^*,...\alpha_N^*$. We write this variance as $\sigma_\psi^2(x,t\,|\,\alpha_1^*,...\alpha_N^*)$. A plot of the model prediction (which is the conditional mean when H_0 is true) with the $\pm 2\sigma_\psi(x,t\,|\,\alpha_1^*,...\alpha_N^*)$ bounds superimposed defines an acceptance region which should contain most of the measured tension values if H_0 is true (see Figure 1). If a "significant" number of measurements fall outside of this region H_0 is rejected and we conclude that the model does not properly describe the infiltration event.

The validation approach described above explicitly recognizes the role of the soil samples used to estimate the model's effective parameters and the role of the tension measurements used for validation. Soil sampling enters through the conditioning process while tension sampling enters through the definition of "significance" used to accept or reject H_0. If the soil sampling program is comprehensive, the conditional variance will be small and the acceptance criterion will be relatively stringent. Similarly, if many tension measurements are available, it is less likely that the model will be falsely accepted or rejected. Clearly, the usefulness of a model validation exercise depends strongly on the quality and quantity of the data it relies upon.

Our validation approach requires a plausible estimate of the conditional tension variance $\sigma_\psi^2(x,t\,|\,\alpha_1^*,...\alpha_N^*)$. The concepts of Bayesian probability theory provide a way to obtain such an estimate. If the normality assumption mentioned above is adopted this theory provides expressions which relate $\sigma_\psi^2(x,t\,|\,\alpha_1^*,...\alpha_N^*)$ to the unconditional covariance $P_{\psi\psi}(x,x',t)$ between tensions at any two locations x and x′ [9]. The unconditional covariance can be estimated either from Monte Carlo simulations or from small perturbation analyses such as those proposed by Yeh et al. [3,4] and Mantoglou and Gelhar [5,6,7]. The first approach is computationally intensive and subject to numerical difficulties and the second depends on simplifying assumptions which may not hold in every application. Nevertheless, we believe that available methods for deriving tension covariances can provide useful approximate estimates of the confidence intervals required for model validation. We are currently testing these methods both on synthetic computer experiments and on a field experiment at Las Cruces, New Mexico, USA [10].

AN EXAMPLE

As an example, of our ongoing work in model validation, we show in Figure 1 a comparison of tension predictions from two models with measurements from a three–dimensional synthetic infiltration experiment performed by Ababou [11]. The figure plots tension predictions and measurements (at 5 days) along a vertical profile located below an irrigation strip. The predictions in Figure 1a are from a one–dimensional analytical solution while those in Figure 1b are from a two–dimensional numerical model.

FIGURE 1

Comparison of Tension Predictions and Measurements for Two Unsaturated Flow Models (confidence intervals indicated with dotted lines)

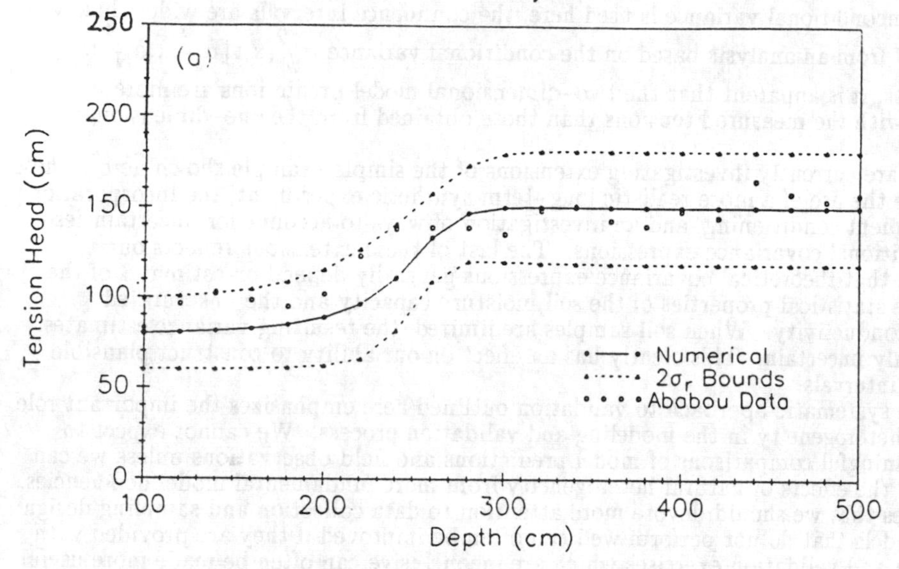

a) Two–Dimensional Model

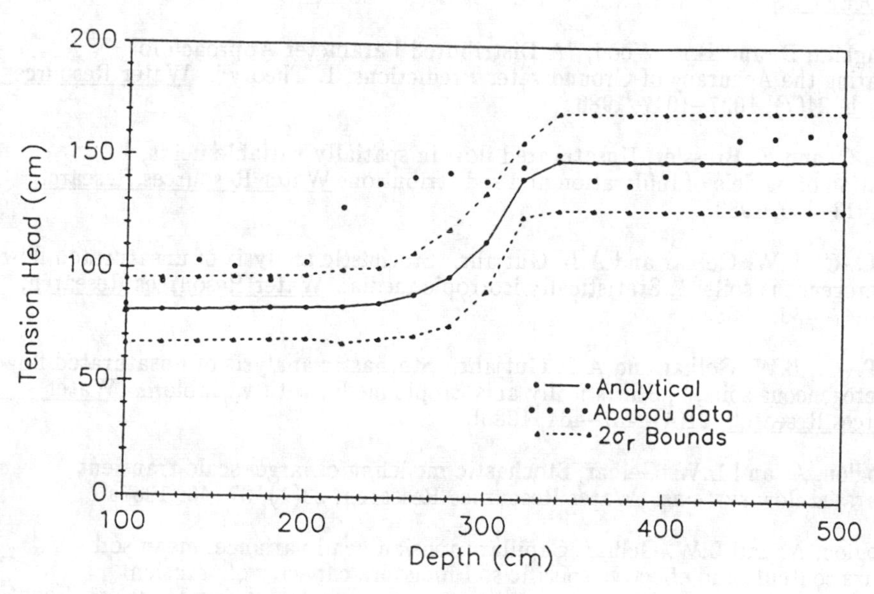

b) One–Dimensional Model

In both cases, effective parameters were estimated from a small number of soil measurements sampled from the heterogeneous fields generated in the synthetic experiment (see [12] for a more extensive discussion). The confidence intervals in the figure are based on the unconditional tension variance $\sigma_{\psi}^{2}(x,t)$ provided by Mantoglou and Gelhar [6]. Since the unconditional variance is used here, the confidence intervals are wider than would be obtained from an analysis based on the conditional variance $\sigma_{\psi}^{2}(x,t \mid \alpha_1^{*},...\alpha_N^{*})$. Nevertheless, it is apparent that the two–dimensional model predictions are more consistent with the measured tensions than those obtained from the one–dimensional model.

We are currently investigating extensions of the simple example shown here. These will include the use of a more realistic long–term synthetic experiment, the incorporation of measurement conditioning, and an investigation of ways to account for uncertainties in the unconditional covariance expressions. The last of these extensions reflects our recognition that theoretical covariance expressions generally depend on estimates of the small–scale statistical properties of the soil moisture capacity and the unsaturated hydraulic conductivity. When soil samples are limited, the resulting variance estimates can be highly uncertain. This clearly has an effect on our ability to construct plausible confidence intervals.

The systematic approach to validation outlined here emphasizes the important role of natural heterogeneity in the modeling and validation process. We cannot expect to obtain meaningful comparisons of model predictions and field observations unless we can distinguish the effects of natural heterogeneity from more fundamental model deficiencies. This implies that we should devote more attention to data collection and sampling design issues. Models that do not perform well can often be improved if they are provided with better data and validation exercises which are inconclusive can often be made more useful if they are based on a more comprehensive monitoring program. These aspects of model validation need to be recognized both in theoretical analyses and in practical applications.

REFERENCES

[1] McLaughlin, D. and E.F. Wood, "A Distributed Parameter Approach for Evaluating the Accuracy of Groundwater Predictions, 1. Theory". Water Resources Research, 24(7), 1037–1047, 1988.

[2] Dagan, G. and E. Bressler, Unsaturated flow in spatially variable fields, 1. Derivation of models of infiltration and redistribution, Water Resources Research, 19(2), 413–320, 1983.

[3] Yeh, T.–C., L.W. Gelhar and A.L. Gutjahr. Stochastic analysis of unsaturated flow in heterogeneous soils, 1, Statistically isotropic media. Water Resources Research, 21(4), 447–456, 1985a

[4] Yeh, T.–C., L.W. Gelhar and A.L. Gutjahr. Stochastic analysis of unsaturated flow in heterogeneous soils, 2, Statistically anisotropic media with variable α. Water Resources Research, 21(4), 457–464, 1985b.

[5] Mantoglou, A. and L.W. Gelhar, Stochastic modeling of large–scale transient unsaturated flow systems. Water Resources Research, 23(1), 37–46, 1987a.

[6] Mantoglou, A. and L.W. Gelhar, Capillary tension head variance, mean soil moisture content, and effective specific soil moisture capacity of transient unsaturated flow in stratified soils. Water Resources Research, 23(1), 47–56, 1987b.

[7] Mantoglou, A. and L.W. Gelhar, Effective hydraulic conductivities of transient unsaturated flow in stratified soils. Water Resources Research, 23(1), 57–68, 1987c.

[8] van Genuchten, M.Th. and W. Jury. Progress in unsaturated flow and transport modeling. Reviews of Geophysics, 25(2), 135–140, 1987.

[9] Schweppe, F.C. Uncertain Dynamic Systems. Prentice Hall, Inc. 1973.

[10] Nicholson, T.J., P.J. Weirenga, G. Gee, E.A. Jacobson, D.J. Polmann, D. McLaughlin, and L.W. Gelhar, "Validation of Stochastic Flow and Transport Models for Unsaturated Soils: Field Study and Preliminary Results", DOE/AECL '87 Conference on Geostatistical Sensitivity, and Uncertainty Methods for Groundwater Flow and Radionuclide Transport Modeling, San Francisco, CA., Sept., 1987.

[11] Ababou, R. Three–dimensional flow in random porous media, Ph.D. Thesis, Dept. of Civil Engineering, Massachusetts Institute of Technology, Cambridge, MA, 1988.

[12] Shea, D., "Spectral Analysis and Performance Evaluation for Unsaturated Flow Modeling", SM Thesis, Department of Civil Engineering, Massachusetts Institute of Technology, Cambridge, MA, June 1989.

HYDROLOGIC MODELING AND FIELD TESTING AT YUCCA MOUNTAIN, NEVADA

D.T. Hoxie
U.S. Geological Survey
Denver, Colorado, USA

ABSTRACT

Yucca Mountain, Nevada, is being evaluated as a possible site for a mined geologic repository for the disposal of high-level nuclear waste. The repository is proposed to be constructed in fractured, densely welded tuff within the thick (500 to 750 meters) unsaturated zone at the site. Characterization of the site unsaturated-zone hydrogeologic system requires quantitative specification of the existing state of the system and the development of numerical hydrologic models to predict probable evolution of the hydrogeologic system over the lifetime of the repository. To support development of hydrologic models for the system, a testing program has been designed to characterize the existing state of the system, to measure hydrologic properties for the system and to identify and quantify those processes that control system dynamics.

MODELISATION HYDROLOGIQUE ET ESSAI SUR LE TERRAIN A YUCCA MOUNTAIN (NEVADA)

RESUME

Le site de Yucca Mountain (Nevada) fait l'objet d'une évaluation en vue de l'implantation éventuelle d'un dépôt excavé en milieu géologique pour l'évacuation de déchets de haute activité. On se propose de construire le dépôt dans un tuf fissuré, fortement cimenté, à l'intérieur de l'épaisse zone non saturée (500 à 750 mètres) du site. Pour connaître les caractéristiques du régime hydrogéologique de la zone non saturée de ce site, il faut donner une description quantitative de son état actuel et élaborer des modèles hydrologiques numériques pour en prévoir l'évolution probable pendant la durée de vie du dépôt. Afin d'étayer la mise au point de modèles hydrologiques adaptés, un programme d'essais a été conçu pour déterminer les caractéristiques actuelles du régime, pour en mesurer les propriétés hydrologiques et pour définir et quantifier les processus qui régissent sa dynamiques.

Introduction

Yucca Mountain, located in the arid southwestern United States approximately 150 km northwest of Las Vegas, Nevada, is being evaluated by the U. S. Department of Energy (DOE) as a possible site for a mined geologic repository for the permanent disposal of high-level nuclear waste. The repository is proposed to be constructed within the unsaturated zone underlying the site at a mean depth of about 200 m below land surface and at a distance ranging from 180 to 400 m above the local water table. The mined repository would encompass a total area of about 5.6 km^2 with a design capacity for storing 70,000 metric tons of heavy-metal waste consisting principally of spent UO_2 fuel assemblies from commercial nuclear-power reactors.

The principal mechanism for possible release of radionuclides to the biosphere from a geologic repository is considered to be radionuclide dissolution and transport by moving ground water [1]. Arguing that the flux of ground water is likely to be small within unsaturated zones in arid regions, Winograd [2] proposed that thick unsaturated zones in such regions may constitute favorable host environments for nuclear-waste disposal. At the Yucca Mountain site, for example, the thickness of the unsaturated zone ranges from 500 to 750 m, and mean annual precipitation at the site probably is only about 150 mm [3]. Because potential evapotranspiration at the site is expected to be about 1600 mm/yr [4], most of the moisture received as precipitation at land surface at Yucca Mountain probably is returned to the atmosphere by evapotranspiration with only a small residuum entering the unsaturated zone as net infiltration. Based on these approximate data and preliminary determinations of unsaturated-zone hydrologic properties, the mean volumetric flux of ground water percolating downward through the unsaturated zone at the Yucca Mountain site under existing conditions is estimated not to exceed 0.5 mm/yr [1].

Although these expected hydrologic conditions may be favorable for siting a repository in the unsaturated zone at Yucca Mountain, present data and understanding of the site geosphere system are not sufficient to complete an adequate assessment of either overall site suitability or long-term (up to 10^5 yrs) waste-isolation capability of a repository system constructed at the site. A credible assessment requires gaining sufficient knowledge of the existing state of the site geosphere system to enable quantitative prediction, within acceptable limits of uncertainty, of probable system evolution over the lifetime of the repository. This assessment is to be accomplished through a site-characterization program encompassing a broad range of geologic, hydrologic, geochemical, and meteorologic studies [1]. Because the repository is proposed to be constructed within the unsaturated zone and because the unsaturated zone is expected to be the primary barrier for radionuclide migration [1], this paper focuses on a specific set of activities necessary for conceptual and quantitative characterization of the site unsaturated-zone hydrogeologic system.

The Yucca Mountain Site

Yucca Mountain is an uplifted, eastward-tilted structural block consisting physiographically of a system of generally north-trending fault-delineated ridges. Topographic relief within the ridge system is about 650 m, and the system attains a maximum altitude of almost 1900 m to the north where the Yucca Mountain block merges with volcanic uplands composed of the eroded remnants of a Miocene caldera complex. The block is bordered on the west, south, and east by alluviated intermontane basins with basin-floor altitudes ranging from 800 to 1100 m.

Stratigraphically, Yucca Mountain is underlain by a sequence of interlayered welded and nonwelded, primarily ash flow, rhyolitic tuffs of Miocene age. As reported by Montazer and Wilson [5], the welded tuffs typically are characterized by low matrix porosity (~0.1), low saturated hydraulic conductivity (~10^{-11} m/s), and high fracture density (~30 fractures/m^3); whereas the nonwelded tuffs, which include bedded reworked tuff and minor ash-fall deposits, typically are more porous (~0.3), more conductive (~10^{-7} m/s), and less fractured (~3 fractures/m^3). Since their deposition during middle to late Miocene time, the tuffs have been extensively devitrified and locally altered to zeolites, especially within nonwelded units [6].

Hydrogeologic Site-Characterization Program

With respect to the site unsaturated-zone hydrogeologic system, the primary objectives of the site-characterization program at Yucca Mountain are to describe quantitatively the existing unsaturated-zone hydrogeologic system and to develop a hydrologic-modeling capability to predict quantitatively expected future states of the system. Description of the existing state of the system entails specifying the spatial distribution of matric potential, saturation, and material properties within the domain enclosed by the system boundaries. The development of a predictive capability entails the construction of mathematically-based models designed to simulate the state and dynamics of the hydrogeologic system. The construction of system state and evolution models requires not only specification of an initial state of the system but also specification of appropriate hydrologic boundary conditions for the system as well as knowledge of those processes that, acting on and within the system, control system response and evolution. The lower hydrologic boundary for the unsaturated-zone hydrogeologic system is defined by the water-table configuration, which is controlled by the regional (saturated-zone) ground-water flow system. The upper hydrologic boundary is land surface at which, in principle, moisture can enter or discharge from the unsaturated zone. However, neither the rates at which nor the mechanisms by which moisture may be moving into or out of the unsaturated system are known. Furthermore, the dominant controlling processes within the system, such as the fundamentals of moisture storage and flow in variably saturated rocks, remain incompletely understood and quantified. Consequently, a major task of the site characterization program is to identify and quantify these boundary mechanisms and internal-system processes and to formulate them mathematically for subsequent implementation in hydrologic models. The site-characterization program also must provide those material-property and other parameter data needed to invoke the models. Field investigations that are intended to address three of these issues include (1) definition of the upper hydrologic boundary condition, (2) characterization of fluid-flow processes occurring within the fractured tuffs of the unsaturated zone, and (3) specification of the existing state and properties of the hydrogeologic system. These field investigations entail conducting tests and experiments to validate specific concepts and hypotheses, collecting hydrogeologic data, and in-situ monitoring of system state and change.

Definition of the Land-Surface Boundary Condition

The rate and areal distribution of net infiltration at land surface is expected to be the primary control on the distribution of moisture flux, matric potential, and saturation within the unsaturated-zone hydrogeologic system [5]. However, neither the mechanisms by which net infiltration occurs nor the areal or temporal distribution of net infiltration is known for the Yucca Mountain site. Consequently, a field-testing program is being implemented to address the issues of how, where, and when net infiltration occurs at the site under present-day climatic conditions. No perennial streams or other surface-water bodies are present at or near the site; therefore net infiltration presumably occurs only as a result of the direct infiltration of water received from precipitation at land surface. Because of the overall aridity of the site, the current conceptual model considers net infiltration most likely to occur where infrequent, episodic runoff is concentrated into stream channels, impounded within alluviated basins, or diverted into open fractures or fault zones [7]. To test this hypothesis and to collect data on the areal distribution of net infiltration, about 100 neutron-access holes are being installed at the site in a variety of topographic, physiographic, and geologic settings. Water-content profiles within the boreholes, as determined from neutron logging, will be obtained periodically and during and following major precipitation events. These water-content data will provide direct, although highly localized, information on the mechanisms controlling moisture storage and movement within that interval of the unsaturated zone extending from land surface to a horizon below the plant-root zone.

Detailed precipitation distributions for the site will be generated by kriging data obtained from a local weather-station network. These distributions will be integrated with the results of surficial-material mapping and the neutron-access hole point data to synthesize geostatistically a model for the areal and short-term temporal distributions of net infiltration at the site. Ponding and artificial-rainfall experiments will be conducted at selected locations to estimate probable maximal rates of net infiltration in order to estimate upper limits on the rate of net infiltration that could occur under wetter climatic conditions. These data will be augmented by evapotranspiration measurements using eddy-correlation and Bowen-ratio techniques to estimate the land-surface moisture-balance distribution for the site. The complete borehole and meteorological data sets are intended to permit estimation of present-day net-infiltration rates at the site. These estimates will be supplemented, for example, by climatic modeling, in order to generate a set of long-term projections of net-infiltration rates needed to specify the upper boundary condition for the system.

Characterization of Water Flow in Variably Saturated Fractured Rocks

The storage and movement of water within the unsaturated zone at Yucca Mountain is presumed to be controlled by combined capillary and gravitational forces [5], and, in particular, the rock matrix is hypothesized to be a porous-medium continuum in which liquid-water flow is Darcian [7]. However, many of the tuff units at the Yucca Mountain site are highly fractured, and the presence of fractures introduces not only a fundamental heterogeneity within the unsaturated-zone flow domain but also the complication that fractures constitute potential pathways for moisture flow. The processes of liquid-water storage and flow in fractures transecting variably saturated porous media remains incompletely understood and quantified. Presumably, if capillary processes dominate within the fractures, matric-potential equilibrium will obtain between the rock matrix and fractures in which case water will be held in fractures at sites where the fracture aperture is less than or equal to the characteristic linear dimension of the largest water-filled pores within the rock matrix. Under these equilibrium conditions, water is expected to move across fractures at virtually all rock-matrix saturation values, but appreciable water flow longitudinally within fractures would not be expected to occur unless the surrounding rock matrix is almost completely saturated [5], [8], [9]. However, nonequilibrium flow of water could occur in fractures at high liquid-water flow rates, such as could be produced for example, by injection of a pulse of water directly into an open vertically oriented fracture. Nonequilibrium flow of water in fractures transecting unsaturated porous media generally is regarded to be a transient phenomenon because moving water will be dispersed through anastomosing fracture channels and eventually drawn into the surrounding (unsaturated) rock matrix by capillary forces. However, preliminary numerical experiments based on an idealized fracture model indicate that pulses of water could be transported over distances exceeding 100 m within the unsaturated zone at the Yucca Mountain site by transient nonequilibrium gravity flow in fractures [10].

Wang and Narasimhan [9] have developed an approximate model based on a statistical description of the internal fracture geometry that accounts for many of the expected aspects of liquid-water flow in single, variably saturated fractures. These single-fracture concepts are not directly extendable, however, to developing computational methodologies to account quantitatively for water flow in variably saturated, hydrologically interconnected fracture systems. Montazer and Wilson [5] proposed and Klavetter and Peters [8] implemented an overlapping continuum a model to describe water flow in variably saturated fractured media in which the fracture system is idealized as a coarse-grained porous medium occupying the same space as the rock matrix. Although this "composite-continuum" model is plausible, the model remains to be validated empirically, and, if validated, field methods remain to be developed to measure or otherwise estimate the equivalent continuum hydrologic properties appropriate to the fracture system.

Conceptual and quantitative characterization of the hydrologic interaction between fractures and the rock matrix is needed because, as indicated above, fractures are potential conduits for possible rapid liquid-phase transport of radionuclides under transient nonequilibrium conditions. In addition, possible gas-phase advective radionuclide transport could be induced within fracture systems as a result of macroscopic convective flow occurring within the air-filled fracture space. Such convective flow also would advect water vapor and, therefore, could affect the overall moisture distribution within the more highly fractured hydrogeologic units. For example, based on observations of air flow in boreholes open to the unsaturated zone at the site, Weeks [11] suggests that topographically induced air circulation may be occurring under natural conditions within the near-surface fractured tuff units. Natural forced convection of pore-gas within the unsaturated zone presumably also could be driven by barometric effects or by the ambient geothermal gradient. In addition, the heat generated by radioactive decay is expected to induce "heat-pipe" convective cells near the repository [12].

An extensive set of both laboratory and field experiments are planned not only to test concepts and enhance understanding of the processes controlling water flow in variably saturated, fractured porous media but also to develop methodologies to estimate fracture hydrologic properties. Specifically, infiltration and desaturation experiments are being performed on naturally fractured cores and blocks in order to characterize the hydrology of discrete fractures under variably saturated conditions. Cross-hole pneumatic testing within packed-off intervals in boreholes is to be conducted to infer the macroscopic hydrologic properties of natural fracture systems within the site unsaturated zone. In-situ testing within a proposed Exploratory Shaft Facility is planned to characterize the hydrologic properties of both discrete fractures and fracture systems within the densely welded tuff currently designated to be the host rock for the repository. The laboratory and field experiments are intended to generate fundamental understanding and data regarding fluid-flow processes in fractured rocks and to provide data by which to test and validate models that simulate these flow processes and effects. The overall objective is to develop and test methodologies for hydrologically characterizing variably saturated fractured tuff and for incorporating fracture-system hydrology in hydrologic models intended to simulate the hydrogeologic system.

Surface-Based Borehole Studies

Specification of the existing state of the site unsaturated-zone hydrogeologic system is intended to provide the set of initial conditions needed to predict system evolution. About 30 surface-based vertical boreholes, in addition to the neutron-access holes described above, are planned to penetrate the unsaturated zone at the site. The locations of the boreholes are being selected to provide sufficient areal coverage of the site for geostatistical analysis and modeling of the borehole data as well as to investigate specific features such as fault zones and potentially favorable infiltration zones. The boreholes will be drilled using air as drilling fluid in order to minimize the impact of drilling on ambient hydrologic conditions. In-situ matric potential, saturation, and rock-matrix hydrologic properties will be measured on cores and cuttings obtained during drilling. The existing state of the hydrogeologic system will be characterized by constructing geostatistical models for the spatial distribution of matric potential and saturation. Classical statistical methods coupled with geostatics will be applied to the hydrologic-property data to define hydrogeologic units and to assign mean properties to these units in order to characterize large-scale spatial heterogeneity within the system.

Standard geophysical logs will be run in the boreholes to provide lithologic, fracture, porosity, and water-content distributions along the well bore. In addition, vertical seismic-profiles will be obtained in boreholes transecting the central part of the site to provide three-dimensional information on lithologic variability and the lateral and vertical extent of fracturing within each hydrogeologic unit. Downhole sensors, including pressure transducers, Peltier psychrometers, and thermocouples will be installed at various horizons within a selected set of boreholes to permit long-term monitoring of pore-gas pressure, matric potential, and temperature within the subsurface. Monitoring ambient conditions within the unsaturated zone is intended not only to refine specification of the existing state of the hydrogeologic system but also to characterize the spatial distribution and amplitudes of secular changes or natural fluctuations occurring within the system. These measured system changes or fluctuations with time provide data against which model predictions of short-term system evolution can be compared to permit partial assessment of model adequacy and accuracy.

Summary and Conclusions

The research and data-collection program at Yucca Mountain is intended to characterize the site unsaturated-zone hydrogeologic system. The hydrologic characterization entails quantitative specification of the existing state of the system and development of a capability to predict probable evolution of the system. Consequently, the program includes collecting hydrogeologic data at the site; identifying and quantifying those processes that, acting within and on the system, control system state and evolution; and, finally, developing mathematically based system-evolution models by which to predict probable as well as possible future system states. Specific tasks described in this paper include the characterization of the rate and surface distribution of net infiltration into the site unsaturated zone, the equilibrium and nonequilibrium flow of water within variably saturated fractures, and the surface-based borehole studies intended to define system state and properties. The site-characterization information to be obtained for the overall Yucca Mountain geosphere system is needed not only to evaluate the suitability of the site for locating a high-level nuclear-waste repository but also to support engineering design of the repository and performance assessments of the repository system.

Acknowledgment

This paper was prepared in cooperation with the Office of Civilian Radioactive Waste Management of the U. S. Department of Energy through Interagency Agreement DE-AI08-78ET44802.

References Cited

[1] DOE (U. S. Department of Energy): "Site Characterization Plan, Yucca Mountain Site, Nevada Research and Development Area, Nevada," DOE/RW-0199 (1988).

[2] Winograd, I. J.: "Radioactive Waste Disposal in Thick Unsaturated Zones," Science, 212, 4502 (1981), 1457-1464.

[3] Quiring, R. F.: "Precipitation Climatology of the Nevada Test Site," National Weather Service Report WSNO 351-88, National Oceanographic and Atmospheric Administration (1983).

[4] Kohler, M. A.; Nordenson, T. J., and Baker, R. D.: Evaporation Maps for the United States, U. S. Weather Service/U. S. Department of Commerce, Washington, D. C. (1959).

[5] Montazer, P., and Wilson, W. E.: "Conceptual Hydrologic Model of Flow in the Unsaturated Zone, Yucca Mountain, Nevada," Water Resources Investigations Report 84-4345, U. S. Geological Survey (1984).

[6] Sheppard, R. A., Gude, 3d, A. J., and Fitzpatrick, J. J.: "Distribution, Characterization, and Genesis of Mordenite in Miocene Silicic Tuffs at Yucca Mountain, Nye County, Nevada," Bulletin 1777, U. S. Geological Survey, (1988).

[7] Hoxie, D. T.: "A Conceptual Model for the Unsaturated-Zone Hydrogeologic System, Yucca Mountain, Nevada," Radioactive Waste Management and the Nuclear Fuel Cycle, 13(1-4) (1989), 63-75.

[8] Klavetter, E. A., and Peters, R. R.: "Fluid Flow in a Fractured Rock Mass," Report SAND85-0855, Sandia National Laboratories (1986).

[9] Wang, J. S. Y., and Narasimhan, T. N.: "Hydrologic Mechanisms Governing Fluid Flow in a Partially Saturated, Fractured, Porous Medium", Water Resources Research, 21, 12 (1985), 1861-1874.

[10] Bodvarsson, G. D.; Niemi, A.; Spencer, A., and Attanayake, M. P.: "Preliminary Calculations of the Effects of Air and Liquid Water-Drilling on Moisture Conditions in Unsaturated Rocks," Report LBL-25073, Lawrence Berkeley Laboratory (1988).

[11] Weeks, E. P.: "Effect of Topography on Gas Flow in Unsaturated Fractured Rock: Concepts and Observations," Geophysical Monograph 42, American Geophysical Union (1987), 165-170.

[12] Tsang, Y. W., and Pruess, K.: "A Study of Thermally Induced Convection Near a High-Level Nuclear Waste Repository in Partially Saturated Fractured Tuff," Water Resources Research, 23, 10, (1987), 1958-1966.

LABORATORY RESEARCH PROGRAM TO AID IN DEVELOPING AND TESTING THE VALIDITY OF CONCEPTUAL MODELS FOR FLOW AND TRANSPORT THROUGH UNSATURATED POROUS MEDIA*

R.J. Glass
Geoscience Analysis Division, Sandia National Laboratories
Albuquerque, NM, USA

ABSTRACT

As part of the Yucca Mountain Project, a laboratory research program is being developed at Sandia National Laboratories that will integrate fundamental physical experimentation with conceptual model formulation and mathematical modeling and aid in subsequent model validation for unsaturated zone water and contaminant transport. Experimental systems are being developed to explore flow and transport processes and assumptions of fundamental importance to various conceptual models. Experimentation will run concurrently in two types of systems: fractured and nonfractured tuffaceous systems; and analogue systems having specific characteristics of the tuff systems but designed to maximize experimental control and resolution of data measurement. Areas in which experimentation currently is directed include infiltration flow instability, water and solute movement in unsaturated fractures, fracture-matrix interaction, and scaling laws to define effective large-scale properties for heterogeneous, fractured media.

*This work was performed under the auspices of the U.S. Department of Energy, Office of Civilian Radioactive Waste Management, Yucca Mountain Project, under contract DE-AC04-76DP00789. This document was prepared under Quality Assurance Level III and WBS 1.2.1.4.6.

PROGRAMME DE RECHERCHE EN LABORATOIRE EN VUE DE FACILITER LA MISE AU POINT ET LA VERIFICATION DE LA VALIDITE DE MODELES THEORIQUES SIMULANT L'ECOULEMENT ET LE TRANSPORT A TRAVERS DES MILIEUX POREUX NON SATURES*

R.J. Glass
Geoscience Analysis Division, Sandia National Laboratories
Albuquerque, NM, USA

RESUME

Un programme de recherches en laboratoire est en cours d'élaboration aux Laboratoires nationaux de Sandia dans le cadre du Projet de Yucca Mountain. Ce programme conjuguera des travaux d'expérimentation physique fondamentale avec la formulation de modèles théoriques et la modélisation mathématique et contribuera à la validation ultérieure de modèles simulant l'écoulement de l'eau et le transport de contaminants dans une zone non saturée. Des dispositifs expérimentaux sont mis au point en vue d'étudier les mécanismes d'écoulement et de transport et les hypothèses à la base de divers modèles théoriques. Les expériences seront menées parallèlement dans des tufs fissurés et non fissurés dans des dispositifs analogues ayant les mêmes caractéristiques que les tufs, mais conçus de façon à maîtriser au mieux les conditions d'expérience et à exploiter pleinement les mesures réalisées. Les domaines dans lesquels des expériences sont actuellement en cours sont les suivants : instabilité de l'écoulement d'infiltration, circulation de l'eau et des solutés dans des fissures non saturées, interaction fissure-matrice et effets d'échelle en vue de définir des propriétés effectives à grande échelle applicables à des milieux fissurés hétérogènes.

* Ces travaux ont été réalisés sous les auspices du Ministère de l'énergie des Etats-Unis, Bureau de la gestion des déchets radioactifs civils, Projet Yucca Mountain, au titre du contrat DE-AC04-76DP00789. Conditions d'établissement du présent document : Quality Assurance Level III et WBS 1.2.1.4.6.

INTRODUCTION

The U.S. Department of Energy is investigating a prospective site for a high-level nuclear waste repository located in unsaturated volcanic ash (tuff) deposits hundreds of meters thick at Yucca Mountain, Nevada. As part of this investigation, the ability of the natural system to restrict water movement and to retard the migration of radionuclides if released must be assessed. While this assessment will be made with respect to specific regulations formulated by the U.S. Nuclear Regulatory Commission and the Environmental Protection Agency, the analysis must be made with a sound scientific understanding of the flow and transport processes that occur at the site. Because there always will be uncertainty with respect to the flow and transport processes, our modeling procedures, and the temporal and spatial variation of model parameters, the final assessment will have to be cast in terms of probability. To develop this general understanding, we must systematically

1. identify all processes by which radionuclides could migrate through the rock formation to the accessible environment;
2. develop basic scientific understanding of these processes through fundamental conceptual and mathematical modeling, controlled experimentation, and model validation (invalidation) exercises at both the laboratory and field scales;
3. bound various processes in terms of system parameters such as initial conditions, boundary conditions, and distribution of properties in both time and space; and
4. provide informational needs for site characterization so that the probability of occurrence for each process can be assessed and appropriate model parameters measured.

Fundamental questions concerning flow and transport in unsaturated porous media can be addressed in part through controlled laboratory experimentation. A laboratory research program is being developed at Sandia National Laboratories for the Yucca Mountain Project that will integrate fundamental physical experimentation with conceptual model formulation and mathematical modeling and aid in testing the validity of models at the meter scale. The laboratory program is part of a broader effort currently being planned that will address the validity of conceptual models used in calculation of groundwater travel time and radionuclide transport through the unsaturated zone at Yucca Mountain. Questions raised in modeling exercises and field studies will be used to direct laboratory experimentation. In general, research in the laboratory is prioritized with respect to understanding flow processes which could cause the site to fail to meet regulatory requirements (i.e., decrease water or radionuclide travel time), testing key assumptions in models for processes considered to be important, and developing new conceptual models as found to be necessary. The research program stresses fundamental research and in this sense will have broad applicability within the general field of flow and transport through unsaturated porous media.

In this paper, the laboratory research program is outlined. The general approach for laboratory experimentation will be presented followed by an outline of several areas of research where studies are currently underway. The program is designed to be flexible. Future studies will be defined as the broader validation effort develops and key informational needs are determined that require a further understanding of flow and transport mechanisms.

APPROACH AND METHODS FOR LABORATORY RESEARCH PROGRAM

Laboratory studies test our understanding of basic processes. Simple qualitative experimentation demonstrating a process is necessary as a first step toward understanding. The approach for the laboratory research program, however, emphasizes systematic quantitative experimentation, conceptual modeling, and model validation (invalidation) exercises directed to achieve fundamental understanding.

Two types of experimental systems will be used. The first is in tuffaceous systems and thus contains all the natural complexity of the rock. Rock types will vary from bedded nonwelded to nonbedded nonwelded, partially welded and welded tuff with and without fractures. Experimental samples will be taken from either Yucca Mountain or natural analogue sites. The second type of experimental system is analogous to the tuff system but simpler, having only certain predetermined attributes of the tuff. These analogue systems are designed to maximize experimental control (i.e., ability to systematically vary system parameters) and resolution of data measurement. To allow systematic variation of hydraulic properties, the analogue systems will be composed of unconsolidated sand, glass beads, porous glass beads, or "rocks" fabricated to specification (e.g., ceramics, sintered glass, or sintered metal). Rough-walled fractures will be simulated with roughened glass plates or fabricated rocks held together at different spacings. Experimentation in both types of systems will run concurrently, with experimentation in analogues driven by what we know about or discover from work in tuff systems and vice versa.

The experimental systems will strive to acquire high-resolution temporal and spatial data to allow the possibility of identifying additional flow and transport mechanisms. In experiments where full three-dimensional data acquisition is required, tomographic techniques using either x-ray or gamma-ray transmission, nuclear magnetic resonance, positron emission, or other methods will be developed and applied. Most of these methods, however, are limited in the size of system to which they can be applied. For many of the questions we are currently investigating, experiments on the scale of a meter are required. To obtain high-quality data at the meter-scale, two-dimensional experiments are conducted in extensive (1x1m) but thin (0.01m) slabs of material. Data measurement techniques for thin slabs include optical, x-ray, and gamma-ray transmission techniques. For many experiments we will concentrate on the first two as they are rapid and can be used as "field" measurement techniques while the gamma-ray densitometer is much slower and is primarily a point measurement technique.

Analogue systems will be designed to take full advantage of optical techniques. Optical techniques for visualizing moisture content make use of the fact that transmission of light through translucent media, such as silica sand or glass beads, increases with an increase in moisture content. By illuminating the back of a thin slab of media, the moisture content integrated over the thickness of the slab is visualized as light intensity that varies from point to point at the front of the slab. Intensity fields can be recorded up to 30 times a second and digitized into an array of 512 x 512 or more points using video imaging technology. Currently the optical technique is being used qualitatively to determine "relative" moisture content (one location is wetter or drier relative to another). To further develop the technique, we are developing calibration methods and comparing the optical technique with a standard gamma-ray densitometer. An adaption of the technique will also be used to visualize transient dye concentration in steady-state flow fields. Calibration methods to allow quantitative measurement of concentration are currently in development. Preliminary results indicate the optical technique is also useful in visualizing packing-induced heterogeneity and thus it may be possible to use it to characterize heterogeneity as well.

For opaque tuffaceous systems, x-rays replace light and their attenuation is used to measure moisture content in extensive thin slabs. X-ray fluorescing film placed on the back side of the slab transforms the x-ray intensity field into a visible light intensity field which again is visualized, recorded, and digitized using video imaging technology. For situations that do not require high spatial and temporal resolution, a standard gamma-ray densitometer is used.

In experimental design, concepts of dimensional analysis, scaling, and similitude are developed and applied to increase understanding and generalize results [1, 2, 3, 4]. For systems where these concepts are applicable, once a physical experiment is conducted or a solution of the dimensionless form of the governing equation has been calculated for one porous medium, the results apply immediately to all similar porous media and flow systems through scaling relations. Systematic exploration of dimensionless parameter space allows the efficient characterization of system response for all possibilities of the dimensional parameters. The concept of similar porous media can also be exploited to allow experimentation in porous media similar to the one of primary interest. This can minimize the experimental difficulties of working with some porous materials where the time scale of the flow process is either too short or too long to make measurement practical.

CURRENT STUDIES

A number of studies, in various stages of planning or completion, are underway to aid in the development of the laboratory research program. These studies are being used not only to develop techniques but to increase our understanding of several flow and transport processes and to challenge several key assumptions embodied in many currently accepted conceptual models. The studies can be grouped into four main areas of research: infiltration flow instability; water and solute movement in unsaturated fractures; fracture/matrix interaction; and scaling laws to define effective large-scale properties for heterogeneous, fractured media.

1. Infiltration flow instability

Most conceptual models assume that infiltration flows are essentially stable with any irregularity in the flow field caused by spatial variability in hydraulic properties, initial conditions, or boundary conditions. Yet, gravity-driven instability of an infiltration flow or "wetting front instability" can cause the flat wetting front moving downward through an unsaturated porous medium to break into fingers which move vertically, bypassing a large portion of the vadose zone. Wetting front instability within porous media has been demonstrated in both laboratory and field settings and has been shown to have a dramatic effect on water and solute transport [5, 6, 7, 8, 9]. The development of a two-zone moisture content field consisting of high moisture content finger cores surrounded by lower moisture content fringe regions and the persistence of this structure from infiltration cycle to infiltration cycle has been demonstrated and explained with a simple theory based on hysteresis in the moisture characteristic relations [10]. The dependence of finger properties on system parameters for initially dry, coarse, nearly uniform sand has been determined through dimensional analysis and experimentation [11, 12]. Stability criteria and relations for finger width or diameter have been formulated through linear stability analysis and compared to experimental data showing remarkably good agreement for homogeneous media where the analysis applies [12, 13].

Generalization of the results obtained from these previous theoretical and experimental studies suggests many situations, such as an increase in conductivity with depth, unsaturated infiltration from a boundary held at less than saturation, and redistribution following an infiltration event, can cause a wetting front to become unstable and form persistent fingers. Because all of these situations potentially occur at Yucca Mountain, the process of wetting front instability must be understood and bounded. To accomplish this, several complicating factors that may stabilize most situations must be explored. The most important of these factors are pore size distribution, contact angle (wettability), heterogeneity, and initial moisture content. A series of experiments to investigate the effects of these factors is currently underway in a meter-scale slab chamber using optical techniques to follow the evolution of the moisture content fields in silica sands. The grain size distribution of the sand and thus the pore size distribution of the media are being varied systematically. Similitude theory applied to finger properties is used to design the grain size distributions. Several preliminary experiments in horizontally microlayered sand systems suggest that fingers widen and perhaps are suppressed as the amplitude and spatial frequency of the property oscillation between layers increase. A series of experiments where microlayering is systematically varied is planned as are experiments where the effects of contact angle and initial moisture content will be systematically explored.

Invasion percolation theory modified to include contact angle, buoyancy, multiple neck pore filling facilitation, and initial moisture content is being used to build a conceptual model which incorporates the essentials of the pore scale mechanism for finger formation. Combination of experimentation and modeling should allow the bounding of gravity-driven fingering in porous media and thus the ability to assess its occurrence at Yucca Mountain. Our current work investigating flow through rough-walled fractures (2 below) has demonstrated that gravity-driven fingers may occur in vertical fractures as well.

2. Water and solute movement in unsaturated fractures

Many of the units of tuff composing both the saturated and unsaturated zones at Yucca Mountain are considered to be highly fractured [14]. Therefore, an understanding of the effects of fractures on water and solute transport within these zones is crucial. In the extreme of very low permeability matrix, such as in highly welded, vitric tuff, or for matrix which is near saturation, the effect of the matrix on flow and transport through a conducting fracture will be of second order. As a first step toward understanding the more difficult problem where the influence of the matrix on flow through the fracture cannot be neglected, we will study unsaturated flow within a fracture in impermeable media.

Little is known concerning the distribution of water and air in an unsaturated fracture and its influence on flow and transport through the fracture. In both analogue and welded tuff systems the effects of fracture surface roughness and orientation in the gravity field on the unsaturated fracture flow field structure and solute transport will be studied systematically. In particular, gravity-driven instability causing the formation of downward-moving fingers within the fracture should occur in nonhorizontal unsaturated fractures. Preliminary analogue systems consist of two roughened glass plates held together to form a simulated fracture. Six different roughnesses are being used, and the angle of the fracture plane with respect to gravity is varied. The structure of transient and steady-state flow fields in the glass fractures is recorded photographically or on video and analyzed using digital image analysis. The use of dye pulses in the water supplied to the fracture may allow the characterization of solute transport through these systems as well and will be explored in future studies. Subsequent studies will be made in molds of natural fractures, natural fractures in welded tuff, and fabricated, fractally generated fractures machined into surfaces.

3. Fracture-matrix interaction studies

In fractured, permeable rock formations, the movement of water and solutes between fractures and porous matrix (fracture-matrix interaction) can have a profound influence on the rate at which water and solutes will migrate through the formation. Models of flow through fractured rocks are based on assumptions concerning the fluid and solute transfer between fractures and adjacent porous matrix. Basic research will be performed to understand fracture-matrix interaction (for both water and solute) and challenge our current assumptions concerning the process.

The influence of flow field structure within unsaturated fractures on fracture-matrix interaction will be studied. The influence of gravity-driven instability to cause fingers within a fracture that may persist and greatly influence fracture-matrix interaction will also be considered. We will begin with a simulated fracture, one side of which will be impermeable and clear (glass) and the other side will be porous (ceramic, sintered glass beads or tuff). Such a system will allow us to document carefully the structure of water contained in the fracture while tensiometers installed in the porous side will monitor the transient pressure field.

The physics of fluid and dissolved contaminant transfer between a fracture and the surrounding porous matrix in the presence of a fracture coating or alteration zone will also be explored. Porous media composing the matrix will be fabricated by sintering packs of glass beads. After homogeneous matrix blocks have been made, material will be added to a side of the block to constitute a coating. The thickness of the coating material and its properties will be varied. To understand altered surface chemical properties of a fracture coating requires the systematic variation of the surface chemical properties. Technology exists to alter the surface properties of glass through a number of processes developed for chromatographic analysis of chemical solutions. These techniques may be applied to the fabricated coating layers to allow a systematic exploration of geochemical processes.

4. Scaling laws to define effective large-scale hydraulic properties for heterogeneous, fractured media

Experimentation and subsequent modeling of water movement in a small unsaturated core of tuff have shown the matrix properties of tuff to be highly variable on the centimeter scale [15, 16]. In addition, fractures and microfractures are present in many tuff formations. The definition of equivalent or effective properties on the scale of a meter to tens of meters which embody these smaller scale heterogeneities is essential for repository-scale calculations of water and radionuclide transport. To aid in the formulation and testing of scaling laws for equivalent media property models from the centimeter through the meter scale, we will conduct experimental studies in both analogue and tuff systems. These studies must subsequently be augmented with in situ field experiments of varying scale to extend the relationships to the field.

In analogue systems, porous media with different heterogeneity structure and with and without high permeability fractures will be generated in sand and fabricated rock material. Transient infiltration experiments will be conducted in extensive slab systems composed of these materials and the moisture content within the flow field will be recorded in time using either optical or x-ray techniques and video/digital imaging. Boundary conditions around the edge of the slab will be controlled using porous pressure plates to supply either known pressure or flux. Steady-state moisture flow with transient solute transport experiments will also be conducted. Data will allow the evaluation of equivalent porous media concepts in well parameterized systems.

In tuff systems, thin slabs of tuff up to 1 m square will be cut and ground smooth. Impermeable material will be contact cemented to the sides of the slabs and porous pressure plates will be installed around the edges of the slabs to impose known boundary conditions. Because most slabs will contain naturally occurring fractures, their influence on the developing flow field can be evaluated. Transient infiltration experiments will be conducted and moisture contents within the flow field will be recorded in time using x-ray techniques and video/digital imaging. By cutting slabs along the principal axes of visual bedding and supplying water to a small hole in the center of the slab, anisotropy on the scale of the experiment can be evaluated. Transient solute transport experiments will also be conducted using x-ray absorbing solute or radioactive tracers.

CONCLUSION

Conceptual models applied to predict long-term transport of water and radionuclides at Yucca Mountain or elsewhere must be evaluated critically. Conceptual model formulation begins by making simplifying assumptions. For a model to accurately predict physical system response, the physics of the major processes that occur for the range of parameter space, physical scale and boundary conditions within which the model will be applied must be represented adequately. The goal of the laboratory research program being developed at Sandia National Laboratories is to acquire the fundamental scientific understanding of flow and transport processes that may occur in the unsaturated zone at Yucca Mountain and thereby assist in developing and testing the validity of our conceptual models for performance assessment.

REFERENCES

1. Miller, E.E., and R.D. Miller : "Physical theory for capillary flow phenomena", J. Appl. Phys. (1956) 27:324-332.

2. Kline, S.J. : Similitude and Approximation Theory. McGraw-Hill Inc., New York, 1965.

3. Tillotson P.M. and D.M. Nielsen : "Scale factors in soil science", Soil Sci. Soc. Am. J. (1984) 48:953-959.

4. Sposito, G., and W.A. Jury : "Inspectional analysis in the theory of water flow through unsaturated soil", Soil Sci. Soc. Am. J. (1985) 49:791-798.

5. Hill, D.E., and J.-Y. Parlange : "Wetting front instability in layered soils", Soil Sci. Soc. Am. Proc. (1972) 36:697-702.

6. Starr, J.L., H.C. DeRoo, C.R. Frink, and J.-Y. Parlange : "Leaching characteristics of a layered field soil", Soil Sci. Soc. Am. J. (1978) 42:376-391.

7. Glass, R.J., T.S. Steenhuis, G.H. Oosting, and J-Y Parlange : "Uncertainty in model calibration and validation for the convection-dispersion process in the layered vadose zone", Proc. Int. Conf. and Workshop on Validation of Flow and Transport Models for the Unsaturated Zone, Ruidoso, NM, 1988, 119-130.

8. Glass, R.J., T.S. Steenhuis, and J.-Y. Parlange : "Wetting front instability as a rapid and far-reaching hydrologic process in the vadose zone", J. of Contam. Hydrol. (1988) 3:207-226.

9. Glass, R.J., G.H. Oosting, and T.S. Steenhuis : "Preferential solute transport in layered homogeneous sands as a consequence of wetting front instability", J. of Hydrol. (1989) 110:87-105.

10. Glass, R.J., T.S. Steenhuis, and J.-Y. Parlange : "Mechanism for finger persistence in homogeneous unsaturated porous media: Theory and verification", Soil Sci. (1989) 148:60-70.

11. Glass, R.J., J.-Y. Parlange, and T.S. Steenhuis : "Wetting front instability I: Theoretical discussion and dimensional analysis", Water Resour. Res. (1989) 25:1187-1194.

12. Glass, R.J., T.S. Steenhuis, and J.-Y. Parlange : "Wetting front instability II: Experimental determination of relationships between system parameters and two-dimensional unstable flow field behavior in initially dry porous media", Water Resour. Res. (1989) 25:1195-1207.

13. Parlange, J.-Y., and D.E. Hill : "Theoretical analysis of wetting front instability in soils", Soil Sci. (1976) 122:236-239.

14. Montazer, P., and W.E. Wilson : "Conceptual hydrologic model of flow in the unsaturated zone, Yucca Mountain, Nevada", USGS-WRI-84-4345 (1984) U.S. Geological Survey, Lakewood CO.

15. Reda, D.C. : "Influence of transverse microfractures on the imbibition of water into initially dry tuffaceous rock", Flow and Transport Through Unsaturated Fractured Rock, Geophysical Monograph 42, eds. D.D. Evans and T.J. Nicholson, American Geophysical Union, Washington DC, 1987, 83-90.

16. Eaton, R.R., and N.E. Bixler : "Analysis of a multiphase, porous-flow imbibition experiment in fractured volcanic tuff", Flow and Transport Through Unsaturated Fractured Rock, Geophysical Monograph 42, eds. D.D. Evans and T.J. Nicholson, American Geophysical Union, Washington DC, 1987, 91-98.

VALIDATION EFFORTS IN MODELING FLOW AND TRANSPORT THROUGH PARTIALLY SATURATED FRACTURED ROCK - THE APACHE LEAP TUFF STUDIES

T.C. Rasmussen, D.D. Evans
Department of Hydrology and Water Resources
University of Arizona, Tucson, USA
T.J. Nicholson
U.S. Nuclear Regulatory Commission
Washington, USA

ABSTRACT

A study designed to assess field and laboratory characterization techniques and to generate data sets for analytic and numeric models of ground-water flow and solute transport in unsaturated fractured rock is described. The generated data sets result from both simple and more complex experiments performed in support of the INTRAVAL program for validation of flow and transport models. Simple experiments focus on interstitial, hydraulic, pneumatic, and thermal properties of 105 oriented rock core segments extracted at selected depths. Additional experiments are being performed on fractured rock blocks as well as on field scales. Data sets incorporating nonisothermal processes are being obtained from core and field experiments. The modeling issues being addressed are scaling effects, spatial heterogeneities, and coupled (e.g., liquid, vapor, thermal, and solute) effects for flow and transport.

TRAVAUX VISANT LA MODELISATION DE L'ECOULEMENT ET DU TRANSPORT A TRAVERS UNE ROCHE FISSUREE PARTIELLEMENT SATUREE - ETUDES MENEES SUR LE TUF D'APACHE LEAP

RESUME

Les auteurs présentent une étude visant à évaluer les techniques de caractérisation sur le terrain et en laboratoire et à obtenir des séries de données pour des modèles analytiques et numériques simulant l'écoulement d'eau souterraine et le transport de solutés dans une roche fissurée non saturée. Les séries de données obtenues proviennent d'expériences simples ou relativement complexes réalisées pour étayer le programme INTRAVAL de validation des modèles d'écoulement et de transport. Les expériences simples portent sur les propriétés interstitielles, hydrauliques, pneumatiques et thermiques de 105 carottes de roche orientées, prélevées à diverses profondeurs. Des expériences supplémentaires sont en cours de réalisation sur des blocs de roches fissurées ainsi que sur les effets d'échelle. Des données prenant en compte les processus non isothermiques sont obtenues à partir d'expériences réalisées sur des carottes et in situ. Les problèmes de modélisation étudiés sont les effets d'échelle, les hétérogénéités spatiales et les effets couplés (par exemple phase liquide, phase vapeur, effets thermiques et solutés) intervenant dans l'écoulement et le transport.

INTRODUCTION

Field hydraulic, pneumatic, thermal, and solute experiments are being conducted at the Apache Leap Tuff Site (ALTS) in tuffaceous rock (dated approximately 19 m.y. B.P.) near Superior, Arizona, located approximately 160 km north of Tucson, and 100 km east of Phoenix. Like the unsaturated zone in many other tuff deposits in the American Southwest [1], the unsaturated zone at ALTS extends to great depth due to topography. Dewatering at a nearby underground mine also contributes to the depth of the unsaturated zone. Vertical and horizontal fracture traces are evident on the land surface and can extend for several hundred meters. Laboratory experiments are conducted using oriented cores obtained during inclined borehole drilling and using shaped blocks removed from the field site.

ALTS is being used to gather data and evaluate methods for the purpose of meeting regulatory research objectives related to high-level nuclear waste disposal. ALTS is located near the extreme western edge of the Pinal Mountains, which rise to over 2100 m in elevation. Lying immediately east of Superior, Arizona is the Apache Leap which forms a 600 m west-facing escarpment that exposes a sequence of volcanic ash-flow tuffs, welded to partially welded, overlaying a mineralized carbonate sequence. ALTS is approximately one km to the east of the escarpment at an approximate elevation of 1200 m on the uppermost unit of a sequence of ash-flow tuff sheets. The tuff is a consolidated deposit of volcanic ash, with particle diameters less than 0.4 mm, resulting a high temperature mixture of gas and pyroclastic materials about 19 million years ago. The ash-flow deposits at one time covered an area of approximately 1000 km^2 with a maximum thickness of 600 m but have been eroded in some places to 150 m in thickness [2].

Atmospheric precipitation has been recorded near the site, with the long-term average estimated to be approximately 538 mm/year. Most of the precipitation occurs during two distinct periods, from mid-July to late-September, and from mid-November to late-March. Summer storms are characterized by high-intensity, short-duration thunderstorms during periods of both high temperature and evapotranspiration demand. Winter storms are of longer duration and lower intensity and coincide with cooler periods which result in much lower evapotranspiration demands.

Regional ground-water levels below the site have been substantially modified by dewatering activities at the nearby Magma mine which extends to a depth of 1500 m below the site. Perched water has been observed at several locations near the site, notably in shallow alluvial aquifers along major washes and from seeps near washes. A hydrologic response in the Magma mine is observed within days following streamflow in Queen Creek, an intermittent stream located approximately 100 m above the mine's haulage tunnel. Increased inflows up to several weeks following streamflow events have been observed at deeper levels.

Five sets of boreholes have been installed at the site. Each set consists of three boreholes drilled in a plane. Four of the five sets are inclined at an angle of 45°, while the fifth set is vertical and extends to a depth of 30 m. The lengths of the inclined boreholes in each set are 15, 30 and 45 m. The bottom of each borehole in a series lies along the same vertical line. The planes of two inclined borehole sets are parallel to each other and the other two inclined sets are oriented 90° and 180° to the planes of the first two sets. A plastic cover has been placed over the rock surface to cover an area of approximately 30 x 50 m. The cover is designed to prevent evaporation from the rock to the atmosphere as well as to prevent precipitation from infiltrating into the rock.

OBJECTIVE

The ultimate objective of the work described in this paper is to examine and possibly predict water flow and solute transport through unsaturated fractured rock under field scale conditions in support of high-level nuclear waste isolation regulatory research activities. In the spirit of the INTRAVAL study, interaction between theoreticians, field investigators and modelers is encouraged in the ALTS work. Specific sub-objectives in support of the ultimate objective are:

- To assess appropriate technology for characterizing and monitoring water flow and transport through unsaturated fractured rock, including interaction between the rock matrix and the fracture system in the tuff.

- To examine relevant uncoupled and coupled hydraulic, pneumatic, thermal and solute transport processes at laboratory scales (1 to 60 cm) and at field scales (1 to 30 m).

- To generate data sets for uncoupled and coupled flow and transport systems which will be used to test the validity of unsaturated flow and transport models related to the INTRAVAL program.

- To assess, cooperatively with other INTRAVAL participants, various experimental and modeling approaches and their limitations in predicting flow and transport through unsaturated fractured rock.

These sub-objectives are related to a USNRC-supported assessment of characterization methods based primarily on soil science techniques for determining the hydraulic, pneumatic, thermal, and solute transport properties of rock matrix cores, and the comparison with properties determined from *in-situ* experiments. These sub-objectives also include evaluation of the determination and use of moisture characteristic and unsaturated transmissivity and conductance curves for fractures. Both the matrix and fracture characterization methods are being examined for various flow and transport processes at selected scales.

The processes include hydraulic, pneumatic, thermal, and solute transport components [3]. Possible coupling of the processes are being examined in experiments conducted over several length scales, including: (1) unfractured tuff cores measuring 6 and 10 cm in length, (2) rock blocks with a single discrete fracture measuring approximately 50 cm in length, and (3) *in-situ* measurements taken over distances of from 1 to 30 m. Time scales in the experiments range from minutes for small core samples, through months for discrete fracture block experiments, to years for large-scale field experiments.

Coupled processes are being examined for various temporal and spatial scales under conditions of partial liquid saturation using non-isothermal experiments on rock cores as well as field heater experiments. The determination of the significance of material heterogeneities and multiple discontinuities is also being made at the field site. Due to the presence of multiple porosities (including microporosities, fractures, and vugs), hydraulic properties can vary seven orders of magnitude as a function of position.

The sub-objectives also incorporate the evaluation of the predictive capability of various alternative modeling strategies (including the equivalent porous continua representation of fractured flow, as opposed to discrete fracture network flow representation within a porous matrix) for their ability to represent accurately fluid flow and solute transport processes in unsaturated fractured rock. NUREG/CR-5097 by Yeh et al. [4] discusses the various conceptual models being considered along with a review of selected computer implementations of numerical methods for providing simulation results in support of experimental design activities.

VALIDATION ASPECTS

Predictions of solute transport from a high-level waste repository to the accessible environment are uncertain due to the possible exclusion of relevant processes and the fact that site characterization activities cannot identify and estimate material properties that may be highly variable. The INTRAVAL study described here is designed to evaluate the relative significance associated with excluding various processes, and to evaluate scale-dependent procedures used to estimate material properties. One aspect of validation activities is the attempt to extrapolate data from laboratory tests conducted over short distances and time scales interpreted using simple models, to large field tests conducted over much larger distances and longer time scales interpreted using stochastic models.

The approximation of multiple porosities and multiple dimension flow using an effective porosity and dimensionality must be validated prior to application of effective parameters for predictive purposes [5]. Repository performance assessment issues include the influence of high permeability zones of reduced dimensionality (e.g., channels within fractures) on the ability to contain high-level waste, and the importance of non-isothermal conditions on near-field hydraulic conditions. From the proposed experiments, useful performance-assessment-related information can be extracted directly. Modeling of the experimental results is one component of validation activities and is a principal reason for conducting the tests. Ideally, the tests are designed using calibrated models, with calibration data sets having been obtained using laboratory and field tests. Once calibrated, the model will be validated by proposing a perturbation of the system not related to calibration experiments for the purpose of evaluating the assumptions inherent in the model.

Specific validation exercises are being conducted with respect to the conceptual models described in the experimental plan below. The validation exercises include:

- Use of an equivalent porous medium model for characterizing the rock matrix properties with respect to hydraulic, pneumatic, thermal and solute transport;
- Evaluation of various parametric relationships between material properties and water content, relative saturation and matric potential;
- Assessment of the ability to parameterize and conceptualize flow through a discrete fracture bounded by a porous matrix;
- Identification of relevant processes related to non-isothermal flow through an unfractured porous tuff rock; and
- Ability of existing mathematical methods to predict flow and transport through nonisothermal, unsaturated fractured rock.

The validation data sets are not limited to these exercises alone, and can be expected to yield additional value as other validation exercises are identified.

DESCRIPTION OF THE EXPERIMENT

A multistage experimental plan is being used to estimate parameters related to hydraulic, pneumatic, thermal, and solute transport. The effect of variable saturation on the parameters is being evaluated for a wide range of matric suctions, from oven-dry to saturated. Also, the effect of fractures on bulk properties at various scales is being estimated. Table 1 presents the stages, scales and processes to be investigated.

Table 1: Experimental stages designed to predict fluid flow and solute transport through unsaturated fractured rock. Each stage introduces additional complexity.

Stage	Scale	Medium	Processes
I	cm	Matrix (core)	- Liquid, Gas, Heat Flow - Gas Diffusion
II	cm	Matrix/Fracture (block)	- Matrix, Fracture Flow - Solute Transport
III	m	Matrix/Fracture (*in situ*)	- Matrix, Fracture Flow - Solute Transport
IV	cm	Matrix (core)	- Nonisothermal matrix flow - Coupled liquid, vapor, air, solute, and heat flow
V	m	Matrix/Fracture (*in situ*)	- Nonisothermal matrix/fracture flow - Coupled liquid, vapor, air, solute, and heat flow

Specific parameters from rock core segments (Stage I experiments) include:
- Rock physical properties, including porosity, bulk and grain density, total and incremental pore surface area, and pore size distributions.
- Moisture characteristic curves.
- Rock matrix hydraulic, pneumatic, and thermal conductivity relationships as a function of matric potential.

Parameters from fractured blocks (Stage II experiments) include:
- Fracture surface roughness, aperture, and characteristic curves.
- Fracture transmissivity as a function of matric suction.
- Solute transport properties of individual fractures.

Parameters estimated from *in-situ* borehole measurements (Stage III experiments) include:
- Hydraulic, pneumatic and thermal conductivities at the same lithologic interval used in the Stage I experiments to determine matrix properties.
- Ambient borehole water contents obtained at three-meter intervals.

Laboratory core heater measurements (Stage IV experiments) are being used to estimate parameters including coupled liquid, vapor, air, thermal energy, and solute transport coefficients for the rock matrix. Some of these coefficients are being estimated independently before or after the experiments on the sample or on rock immediately surrounding the core samples. Others are being estimated during the course of the experiment.

Field heater measurements (Stage V experiments) will be used to determine the effects of fractures on coupled liquid, vapor, air, thermal and solute transport coupling coefficients. The effect of material property heterogeneity on field scales will be examined with respect to estimates based on laboratory samples.

Spatial and Temporal Scales

The core and borehole sampling strategy is to measure material properties repetitively at the same lithologic interval. Comparisons of matrix and bulk properties can then be compared to infer the influence of heterogeneities and scale effects. Various temporal scales are being employed at the different physical scales, ranging from minutes for core experiments to years for field experiments.

Experimental Setup

A multistage experimental plan is being conducted to increase incrementally the complexity from simple systems to complicated ones. The focus of the study is two related field experiments, one with and the other without imposed temperature gradients and in similar geologic settings (slightly welded unsaturated fractured tuff). The experimental site with an undisturbed thermal regime has been under study for three years. The fifteen boreholes were installed and used for *in situ* site characterization, while the obtained cores are being used to log the fractures and to characterize the rock matrix. Water and gas injection-recovery tests were made at the site to determine temporal and spatial responses. A long-term water injection and recovery experiment is planned.

Complementary Experiments

Additional experiments in an abandoned road tunnel and mine haulage tunnel have also provided important data for welded tuff. Comparison of parameters between sites should allow models developed at one site to be verified at another. Regional geochemical experiments are being performed to provide data sets. Natural analogue experiments would extend the time scales of current analyses, but are beyond the scope of the present study. Nonisothermal studies in the G-tunnel at the Nevada Test Site [6] also provide data sets for comparison with results from ALTS.

Modeling Activities

Interpretations of field observations are being performed using of various types of models, including equivalent porous media analogues and discrete-fracture models. Predictions of flow and transport through three-dimensional networks of discrete fractures have been made for flow through saturated discrete fracture networks [7,8], variably saturated discrete fractures [9], and single discrete fractures embedded in a porous matrix [10,11]. Laboratory experiments on heat flow through a rock core have been modeled by the USNRC's Mr. Timothy McCartin who used the TOUGH computer program [12]. McCartin also has modeled field scale injection and redistribution processes the computer program VAM2D [13]. The USNRC's Dr. Richard Codell presented a particle tracking model to evaluate solute transport at the Third INTRAVAL Workshop at Helsinki [14]. Flow through a fractured rock block has been modeled by Mr. Paul Davis and Ms. Natalie Olague at Sandia National Laboratories using VAM2D [13]. Table 2 lists personnel and institutions involved in modeling phenomena at ALTS.

Table 2: List of modelers and institutions involved with model validation using Apache Leap Tuff experiments.

Stage		Simulation Modeling Groups
I	Core	- T. Rasmussen (UAz)
II	Fractured Block	- T. Rasmussen (UAz) - P. Davis & N. Olague (SNL) - T. McCartin (NRC) - R. Ababou (CNWRA) - J. Smoot (PNL) - J. Wang et al. (LBL)
III	Field Imb./Redist.	- T. Rasmussen (UAz) - T. McCartin & R. Codell (NRC) - J. Wang et al. (LBL)
IV	Nonisothermal Core	- J. Yeh and A. Guzman (UAz) - T. McCartin & R. Codell (NRC)
V	Field Heater	- J. Yeh and M. Shaikh (UAz) - T. McCartin (NRC) - K. Pruess et al. (LBL)

NRC: U.S. Nuclear Regulatory Commission, Washington, D.C. 20555
SNL: Division 6416, Sandia National Laboratories, Albuquerque, NM 87185
UAz: Dept. of Hydrology & Water Resources, Univ. of Arizona, Tucson, 85721
LBL: Earth Sciences Div., Lawrence Berkeley Lab., Univ. of California, Berkeley, CA 94720
PNL: Pacific Northwest Labs, P.O. Box 999, Richland, WA 99352
CNWRA: Center for Nuclear Waste Regulatory Analyses, 6220 Culebra Rd., P.O. Drawer 28510, San Antonio, TX 78284

CURRENT STATUS AND EXPERIMENTAL SCHEDULE

Data sets related to Stage I activities (i.e., matrix core properties) have been completed and are currently available in draft form. Data sets for Stage II activities (i.e., block experiment) are partially completed and additional data sets will be completed by August 1990. Stage III activities (i.e., field injection tests) are also partially completed with additional experiments in progress. Prototype experiments for Stage IV activities (i.e., non-isothermal core experiments) are currently under way, with a final experiment contemplated for completion by August 1990. Stage V activities (i.e., field heater experiment) are in the planning stages. The plan should be completed by October 1990 with borehole construction and instrument emplacement scheduled for completion by October 1991.

REFERENCES

[1] Winograd, I.J., 1971, "Hydrogeology of Ash Flow Tuff: A Preliminary Statement", Water Resour. Res., 7(4):994-1006.

[2] Peterson, D.W., 1961, "Dacitic Ash-Flow Sheet Near Superior and Globe, Arizona", Ph.D. Dissertation, Stanford University, Palo Alto, CA.

[3] Pruess, K. and J.S.Y. Wang, 1987, "Numerical Modeling of Isothermal and Nonisothermal Flow in Unsaturated Fractured Rock - A Review", in Flow and Transport Through Unsaturated Fractured Rock, Geophysical Monograph 42, p. 11-22.

[4] Yeh, T.C.J., T.C. Rasmussen and D.D. Evans, 1988, "Simulation of Liquid and Vapor Movement in Unsaturated Fractured Rock at the Apache Leap Tuff Site: Models and Strategies", NUREG/CR-5097, 73 pp.

[5] Barker, J.A., "A Generalized Radial Flow Model for Hydraulic Tests in Fractured Rock", Water Resour. Res., 24(10):1796:1804.

[6] Daily, W. and A. Ramirez, 1989, "Evaluation of Electromagnetic Tomography to Map in Situ Water in Heated Welded Tuff", Water Resour. Res., 25(6):1083-1096.

[7] Huang, C. and D.D. Evans, 1985, "A 3-Dimensional Computer Model to Simulate Fluid Flow and Contaminant Transport Through a Rock Fracture System", NUREG/CR-4042, 109 pp.

[8] Rasmussen, T.C., C. Huang, and D.D. Evans, 1985, "Numerical Experiments on Artificially Generated, Three-Dimensional Fracture Networks: An Examination of Scale and Aggregation Effects", Memoirs of Congress on Hydrology of Rocks of Low Permeability, International Association of Hydrogeologists, Tucson, Az., 17(2):676-682.

[9] Rasmussen, T.C., 1987, "Computer Simulation of Steady Fluid Flow and Solute Transport Through Three-Dimensional Networks of Variably Saturated, Discrete Fractures", in Flow and Transport Through Unsaturated Fractured Rock, AGU Geophys. Monograph 42, p. 107-114.

[10] Rasmussen, T.C, T.-C.J. Yeh, and D.D. Evans, 1989, "Effect of Variable Fracture Permeability/Matrix Permeability Ratios on Three-Dimensional Fractured Rock Hydraulic Conductivity", in Geostatistical, Sensitivity, and Uncertainty Methods for Ground-Water Flow and Radionuclide Transport Modeling, by B.E. Buxton (ed.), Battelle Press, p. 337-358.

[11] Rasmussen, T.C. and D.D. Evans, 1989, "Fluid Flow and Solute Transport Modeling Through Three-Dimensional Networks of Variably Saturated Discrete Fractures", NUREG/CR-5239.

[12] Pruess, K., 1987, "TOUGH User's Guide", NUREG/CR-4645, 78 pp.

[13] Huyakorn, P.S., J.B. Kool, J.B. Robertson, 1989, "VAM2D - Variably Saturated Analysis Model in Two Dimensions", NUREG/CR-5352.

[14] INTRAVAL, 1989, "Third INTRAVAL Workshop", Helsinki, Finland, June 12-16.

VALIDATION EFFORTS IN MODELING PARTIALLY SATURATED FLOW & TRANSPORT IN HETEROGENEOUS POROUS MEDIA - THE LAS CRUCES TRENCH STUDIES

Thomas J. Nicholson
Office of Nuclear Regulatory Research
U.S. Nuclear Regulatory Commission, Washington, USA

Richard G. Hills
Department of Mechanical Engineering
New Mexico State University, Las Cruces, NM, USA

Peter J. Wierenga
Department of Soil & Water Science
University of Arizona, Tucson, AZ, USA

ABSTRACT

In order to examine conceptual models of partially saturated flow and transport through heterogeneous porous media, a series of experiments have been performed at the Las Cruces Trench Site. The site has been characterized extensively using in-situ measurements, and core and disturbed soil samples. Observations of water content, tension, and solute movement from a series of large-scale field infiltration and transport experiments have been made and model validation efforts are currently underway. A new large scale field experiment is planned for 1990. Water flow and transport for this experiment will be modeled prior to the release of experimental data. Simple statistical comparisons between model predictions and field observations will be made when the experimental data are released.

TRAVAUX DE VALIDATION VISANT LA MODELISATION DE L'ECOULEMENT ET DU TRANSPORT DANS DES MILIEUX POREUX HETEROGENES PARTIELLEMENT SATURES - ETUDES MENEES AU SITE DE LAS CRUCES TRENCH

RESUME

Une série d'expériences a été exécutée au site de Las Cruces Trench pour étudier des modèles théoriques de l'écoulement et du transport dans des milieux poreux hétérogènes partiellement saturés. Des mesures in situ ainsi que des échantillons de carottes et de sol remanié ont permis de définir dans le détail les caractéristiques du site. Des observations de la teneur en eau, de la tension et de la circulation des solutés ont été excécutées à partir d'une série d'expériences à grande échelle d'infiltration et de transport in situ et des travaux sont actuellement en cours pour valider le modèle. En 1990, il est prévu de mener une nouvelle expérience à grande échelle in situ l'écoulement et le transport de l'eau seront modélisés avant la diffusion des données expérimentales. Une fois les données expérimentales diffusées, des calculs statistiques simples seront effectués pour comparer les prévisions du modèle avec les observations faites sur le terrain.

OBJECTIVE AND BACKGROUND

The purpose of the Las Cruces Trench Site experiments is to test deterministic and stochastic theories of flow and transport in partially saturated heterogeneous porous media by comparing model predictions with observed measured flow and transport parameters. To do this, a set of characterization experiments and two infiltration/redistribution experiments are being designed and executed. Water and tracers were applied in a carefully controlled fashion to the surface of the experimental plots. The motion of water and the transport of various tracers through the vadose zone were monitored. The emphasis of the testing is on natural soil systems that are heterogeneous but are well characterized hydrologically. The scales of the experiments range from several cm (core disturbed soil samples) to several tens of meters (field experiments). Temporal scales range from several hours to over a year.

The experimental site is located at the New Mexico State University College Ranch, 40 km northeast of Las Cruces, NM, USA. The field site is situated on a basin slope of Mount Summerford at the north end of the Dona Ana Mountains. The geologic features, geomorphic surfaces, soil series and vegetation types found in the area around the field test area are typical of many areas of southern New Mexico (Gile et al., 1970), and are similar to other arid and semiarid areas of the southwestern United States. The climate in the region is characterized by an abundance of sunshine, low relative humidity, and an average Class A pan evaporation of 239 cm/yr. Average annual precipitation is 23 cm with 52% of the rainfall occurring between July 1 and September 30. The average monthly maximum air temperature is highest in June at 36 C and lowest in January at 13 C.

A trench 16.5 m long, 4.8 m wide and 6.0 m deep was excavated in undisturbed soil. Originally, the trench was unlined which allowed for visual monitoring of the wetting front advances. A series of photographs were taken to record the vertical and horizontal movements of the wetting front over a period of months. More recently, the trench walls were structurally reinforced which prevents visual monitoring.

During the construction of the site, characterization experiments were performed. Approximately 600 core and 600 disturbed soil samples were collected and taken to the laboratory for determining soil-water retention and saturated hydraulic conductivity values. In addition, the saturated hydraulic conductivity values were determined in situ using borehole permeameter tests at a similar number of locations.

The physical properties of the soil were determined in sufficient detail to allow estimation of many of the statistical parameters needed in the stochastic and deterministic models. Spatial scales for the soil heterogeneities range from less than a meter to several meters. Based on approximately 450 in situ and laboratory based measurements, the saturated hydraulic conductivity ranges over three orders of magnitude. The van Genuchten water retention parameters α and n range over three and one order of magnitude, respectively. There appear to be no complex discontinuities such as faults or fracture networks.

PHASE I CALIBRATION EXPERIMENTS

Plot 1 Experiment

The purpose of the first experiment (denoted as Plot 1 experiment) was to test the tensiometers and solute samplers during infiltration for very dry soils and to develop and test the data acquisition methods used in later experiments. Water containing tritium was applied at a rate of 1.76 cm/day to an area adjacent to the south side of the trench. Tritium movement and tensions were monitored during infiltration, and water content was monitored during infiltration and redistribution. The movement of water below the soil surface was monitored using neutron probes and tensiometers. The neutron probe monitoring and tensiometer sampling periods are on the order of hours during initial infiltration, on the order of days as time progresses and the water infiltrates deeper into the soil profile, and finally on the order of months during the later stages of redistribution.

Soil solution samples were taken to determine the movement of tritium and bromide below the soil surface during the Plot 1 calibration experiment. The trench and irrigated area were covered to prevent infiltration from local runoff and to minimize evaporation. The subsequent database has been used for model calibration for high infiltration and percolation rates.

The Plot 1 experimental results indicated that the use of solution samplers was a viable technique for monitoring solute transport in unsaturated soils, particularly when the tensions were less than 150 cm. At higher suctions the water transmission rate is too low to obtain sufficient quantities of sample to monitor. Tensiometers, placed adjacent to suction samplers operated reasonably well after the wetting front had passed. These units do not work in the dry soil. Neutron probes are used to monitor water contents in three planes located parallel to the trench face and in one plane located perpendicular to the trench face.

Plot 2a Experiment

The experience gained in performing the Plot 1 experiment was used to design and run the Plot 2a experiment. For the Plot 2a experiment, an irrigation area measuring 1.2 m by 12 m located adjacent and perpendicular to the trench was constructed. The drip irrigation system consists of 40 irrigation lines placed perpendicular from the trench face with each irrigation line containing 4 drippers. Laboratory and field uniformity tests of the drip irrigation system prior to the Plot 2a experiment indicated extremely small non-uniformities and a random distribution to the non-uniformities. The applied water was metered onto the plot to allow for water balance calculations. During this experiment, water was applied at 0.43 cm/day for 75.5 days. Tritium at a final concentration of 0.1m Ci/l and bromide at a final concentration of 0.8 g/l were applied with the water during the first 11.5 days of irrigation.

The movement of water was monitored through approximately 50 tensiometers inserted through the north face of the trench wall and through neutron probe readings taken at approximately 1000 locations in four different vertical planes. The movement of tritium and bromide was monitored during both irrigation and redistribution through approximately 50 solute samplers inserted through the north face of the trench wall. Tritium and bromide concentrations were sampled on a regular basis.

The data analysis for the Plot 2a experiment is complete and the data are presently in the Las Cruces database and available for the Phase I effort of INTRAVAL.

Available Experimental Data

The following parameters and key measurements are available from the characterization and Phase I experiments:

- Saturated hydraulic conductivity (both in situ and laboratory measurements) and $\theta(h)$ data have been determined for approximately 600 locations in a vertical plane immediately adjacent to the test plots.

- Measurements of θ_r and θ_s and estimates of the van Genuchten water retention parameters α and n have been determined for each of the 600 locations. Specific water capacity, C, can be estimated from the resulting van Genuchten curves.

- Moisture contents or moisture profiles as a function of time are monitored using downwell neutron-probe logging.

- Tension data obtained from tensiometers placed on a two dimensional grid on the north trench wall are monitored as a function of time.

- Concentrations of the tracers are sampled through the trench wall on a two dimensional grid as a function of time.

(1) Raw Data

Water retention data, density profiles, particle-size analysis data, and saturated hydraulic conductivities measured through both laboratory and in situ techniques during the characterization experiments, and water content data, tensiometer data, tritium and bromide concentrations, and water applications rates measured during the Plot 1 and Plot 2a experiments are contained in the Las Cruces Database. The Plot 2b data will also be incorporated into the database using the same format. In addition, morphological mapping of the trench face, drip irrigation distribution tests, and outflow and inflow measurements from large and small column-support tests are on file at New Mexico State University and at the University of Arizona.

(2) Processed Data

Processed data have been reported in two annual reports (NMSU to NRC 1985, 1986), in Wierenga et al. (1986), in Nicholson et al. (1987), and in Wierenga et al. (1989). These data consist of moisture profiles determined from neutron probe calibration curves; tension profiles from tensiometer data mapped as a function of time; tritium and bromide profiles and breakthrough curves (obtained from solute sample analysis); and wetting front observations from the first trench experiment from visual mapping of the wetting front advance with time. In addition, the van Genuchten water retention parameters α and n for each of the 600 core sample locations and the parameters for uniform and layered soil models are included in the Las Cruces Database and discussed in Wierenga et al. (1989). Information on small and large soil columns to examine dispersivity data are also available in Wierenga and van Genuchten (1989).

(3) Data Storage

Data are stored in Las Cruces, New Mexico in laboratory notebooks, on IBM (5 1/4" floppy) disks, and on 10 MByte Bernoulli disks. Redundant copies of the digital data are stored in Tucson on 20 MByte Bernoulli disks and in the Las Cruces Database. The Las Cruces Database is installed on a Digital VAX and is backed up regularly by computer center personnel. The database consists of 27 files in ASCII format, is in the public domain, and is accessible at no cost through the international computer networks using FTP. The database is updated on a regular basis.

PHASE II VALIDATION EXPERIMENT

The objective of the Phase II validation experiment is to further test deterministic and stochastic water flow and transport theories by comparing model results with experimental observations from a field-scale natural heterogeneous soil system. During Phase II, a new experiment will be performed using the same experimental set-up as was used during the Phase I, Plot 2a experiment. However, the new experiment (denoted the Plot 2b experiment) will use different tracers and irrigation rates. Water flow and solute transport for the Plot 2b experiment will be modeled prior to the release of the Plot 2b experimental data. The data obtained from the Characterization, Plot 1, and Plot 2b experiments are presently available for model calibration. Model predictions will be compared to the Plot 2b experimental data when the Plot 2b data are released.

Configuration of the Plot 2b Experiment

The irrigation area adjacent to the trench used in the Plot 2a will be used again. The applied water will be metered onto the plot to allow for water balance calculations as was done for Plot 1 and 2a. Water will be applied at a different rate during the Plot 2b model validation experiment to represent a new stress in addition to the new initial and boundary conditions. The trench and adjacent irrigated area are covered to prevent surface runoff and recharge due to precipitation and to minimize evaporation.

Sampling Strategy

The neutron probe data, the tensionmeter data, and the solute samples taken through the trench wall will be taken for the same locations as for Plot 2a experiment. However, additional solution samplers and tensiometers will be installed as the water and solute front move through the soil profile. The experiments will be at different flow rates using different tracers. Thus, the Plot 2b data will be independent of the Plot 2a data in the sense that different experiments are run but dependent in the sense that the experiments are run at the same location using the same monitoring system.

Cores for solute sampling taken during and after the Plot 2b experiment will be taken from locations not previously sampled. Thus this data will be independent of the solute concentration data obtained through the trench face during the Plot 2a experiment.

MODEL VALIDATION STRATEGY

Model validation will be addressed on two levels. On the first level, model predictions will simply be compared to field observations graphically and through simple statistical measures of correlation. While this level of model validation is not rigorous, it does allow the results of many models to be compared with relatively little effort. The following approaches are being considered:

1. Using simple deterministic models, various scenarios for the Plot 2b experiment will be simulated by NMSU to help optimize the water/tracer application rates and the choice of tracers.

2. The Plot 2b experiment will then be run and the actual water/tracer application rates will be provided to the modelers.

3. Using the actual water/tracer applications, the Plot 2b experiment will be modeled.

4. After modeling is complete, the modelers will provide NMSU with ASCII files of the predicted water contents, tensions, and solute concentrations at the measurement locations in a predetermined format. In addition, the stochastic modelers will provide NMSU with the predicted statistics of their models. NMSU will tabulate simple statistical measures of correlation between model prediction and observation and will include the predictive statistics provided by the stochastic modelers.

5. Following receipt of the model predictions, data on the observed water contents, tensions, and solute concentrations, as well as all of the model predictions, will be distributed in ASCII format to the participants (through the Las Cruces Database). This will allow the individual participants to address the issues of model validation in more detail.

At present the following models are anticipated to be used:

- Sandia National Laboratories (SNL) will utilize the VAM2D finite element code developed by Huyakorn et al. (1989). VAM2D can model flow and transport in variably saturated anisotropic soil as well as single-species and chained decay transport.

- Massachusetts Institute of Technology (MIT) will use stochastic flow models (Gelhar (1984), Mantoglou et al., (1987a, b, c), Polmann et al. (1988), Yeh et al. (1985a,b,c)) to simulate the long term movement in the Plot 1 experiment. These models generate effective media approximations to the important hydraulic parameters allowing non-Monte Carlo simulation of flow through heterogeneous and anisotropic soils.

- Pacific Northwest Laboratory (PNL) will model transport of both inert and chemically active tracers (if applied) for the Plot 2b experiment. Initially, the transport analysis will be performed using one or more deterministic codes (e.g., TRACR3D, PORFLO3). Multiple realizations will be simulated using Monte Carlo methods to estimate the variation in the solute front due to spatially variable soil properties. Later, if chemically active tracers are used in the transport experiments, a coupled geochemical code (e.g., FASTCHEM) will be used to predict tracer plume movement in the trench soil profile.

- Center for Nuclear Waste Regulatory Analysis (CNWRA) will utilize the three-dimensional flow model of Ababou (1988). This code models the nonlinear effects of partially saturated flow in a randomly heterogeneous and anisotropic soil. The input parameters will be conditioned using the Las Cruces database. The CNWRA algorithm has been used to generate three-dimensional high resolution flow simulations, as described in Ababou and Gelhar (1988).

- NMSU will use the water content based algorithm of Hills et al. (1989a,1990) to model flow and transport using isotropic uniform and layered soil models.

On the second level of model validation, statistically rigorous quantitative validation methodologies are under development. Their use in the Plot 2b and 1 experiments is under consideration.

PREVIOUS MODELING

The modeling that has been documented to date using the field data sets are described in MIT reports [e.g., Ababou (1988), Polmann et al., (1988)], Nicholson et al. (1987), and Hills et al. (1989a,b). Additional comparisons between experimental and predicted data for the Las Cruces Trench site have recently been presented at the 4th INTRAVAL Workshop by Ababou (CNWRA), Goodrich (SNL), Kool (HydroGeologic Inc.), Bensabat (MIT), Rasmusson (KEMAKTA), Smyth (PNL), and Hills (NMSU). These comparisons are presently being documented in the INTRAVAL Phase I report.

EXPECTED RESULTS

Experimental View

The criteria for validation and the requirements for measurement precision will be established in discussions with scientists who are running similar tests. It is expected that much of the validation testing may well be aided by improved techniques that become available and are used by countries outside the USA, but not currently being tested (because of budget constraints, etc.) in the USA. These ideas, discussed and shared, will be valuable for enhancement of the study. New methods of tracer sampling techniques as well as development and testing of transport codes for unsaturated soil systems will aid the experimentalist in designing appropriate tests to evaluate the transport parameters needed in the modeling component of the study.

Modeler's View

The modeler's view is quite similar to that of the experimentalist in that the exchange of ideas and modeling techniques with INTRAVAL participants from the USA and other countries will be useful. For example, as stochastic codes become available from other countries, these codes could be used/tested against data available from this project. The geostatistical approaches can be enhanced by discussion with experts from INTRAVAL projects that routinely use spectral analysis and other techniques to obtain the statistical parameters needed in the stochastic model(s). The information exchange will be valuable during all stages of this project.

REFERENCES

Ababou, R., "Three-Dimensional Flow in Random Porous Media," Ph.D. Thesis, Department of Civil Engineering, Massachusetts Institute of Technology, Cambridge, Massachusetts, January 1988, 2 volumes, (1988).

Ababou, R. and L.W. Gelhar, "A High-Resolution Finite Difference Simulator for 3D Unsaturated Flow in Heterogeneous Media," Computational Methods in Water Resources, Elsevier and Computational Mechanics Publications, Vol. 1, pp. 173-178, (1988).

Gelhar, L.W., "Stochastic Analysis of Flow in Heterogeneous Porous Media." In Fundamentals of Transport Phenomena in Porous Media, J. Bear and M. Corapcioglu, eds. Martinus Nijhoff Publ. Dordrecht, Netherlands, pp. 675-717, (1984).

Gile, L.U., J.H. Hawley, and R.B. Grossman, "Distribution and Genesis of Soils and Gemorphic Surfaces in a Desert Region of Southwestern New Mexico, Guidebook," pp. 156. Soil Sciences Society of America and New Mexico State University, Las Cruces, (1970).

Hills, R.G., I. Porro, D.B. Hudson and P.J. Wierenga, "Modelling One-Dimensional Infiltration into Very Dry Soils - Part 1: Model Development and Evaluation," Water Resources Research, Vol. 25, No. 6. pp. 1259-1269, (1989a).

Hills, R.G., D.B. Hudson, I. Porro, and P.J. Wierenga, "Modelling One-Dimensional Infiltration Into Very Dry Soils - Part 2: Estimation of the Soil-Water Parameters and Model Predictions," Water Resources Research, Vol. 25, No. 6, pp. 1271-1282, (1989b).

Hills, R.G., P.J. Wierenga, D.B Hudson, M.R. Kirkland, "Site Description, Experimental Results, and Two-Dimensional Flow Predictions for the Las Cruces Trench Experiment 2," in preparation for submittal to Water Resources Research, (1990).

Huyakorn, P.S., J.B. Kool, and J.B. Robertson, "VAM2D - Variable Saturated Analysis Model in Two Dimensions," NUREG/CR-5352, U.S. Nuclear Regulatory Commission, Washington, DC, (1989).

Mantoglou, Aristotelis, and Lynn W. Gelhar. "Stochastic Modeling of Large Scale Transient Unsaturated Flow Systems." Water Resources Research, Vol. 23, No. 1. pp. 37-46, (1987a).

Mantoglou, Aristotelis, and Lynn W. Gelhar. "Capillary Tension Head Variance, Mean Soil Moisture Content, and Effective Specific Soil Moisture Capacity of Transient Unsaturated Flow in Stratified Soils," Water Resources Research, Vol. 23, No. 1, pp. 47-56, (1987b).

Mantoglou, Aristotelis, and Lynn W. Gelhar. "Effective Hydraulic Conductivities of Transient Unsaturated Flow in Stratified Soils," Water Resources Research, Vol. 23, No. 1, pp. 56-68, (1987c).

Nicholson, T.J., and others, "Validation of Stochastic Flow and Transport Models for Unsaturated Soils: A Comprehensive Field Study, "Proceedings of USDOE-AECL Symposium on Geostatistical Sensitivity and Uncertainty Methods for Ground water Flow and Radionuclide Transport Modeling, San Francisco, CA, (1987).

Polmann, D.J., E.G. Vomvoris, D. McLaughlin, E.M. Hammick, and L.W. Gelhar, "Application of Stochastic Methods to the Simulation of Large-Scale Unsaturated Flow and Transport," NUREG/CR-5094, U.S. Nuclear Regulatory Commission, Washington, DC, (1988).

Wierenga, P.J. and others, "Validation of Stochastic Flow and Transport Models for Unsaturated Soils: A Comprehensive Field Study," NUREG/CR-4622, U.S. Nuclear Regulatory Commission, Washington, DC, (1986).

Wierenga, P.J. and M.Th. van Genuchten, "Solute Transport through Small and Large Unsaturated Soil Columns," Groundwater, Vol. 27, No. 1, pp. 35-42, (1989).

Wierenga, P.J., A. Toorman, D. Hudson, J. Vinson, M. Nash, and R.G. Hills, "Soil Physical Properties at the Las Cruces Trench Site," NUREG/CR-5441, U.S. Nuclear Regulatory Commission, Washington, DC, (1989).

Yeh, T.-C. Jim, Gelhar, L.W., and Gutjahr, A.L. "Stochastic Analysis of Unsaturated Flow in Heterogeneous Soils 1. Statistically Isotropic Media." Water Resour. Res., Vol. 21, No. 4, pp. 447-456, (1985a).

Yeh, T.-C. Jim, Gelhar, L.W., and Gutjahr, A.L. "Stochastic Analysis of Unsaturated Flow in Heterogeneous Soils 2. Statistically Anisotropic Media." Water Resour. Res., Vol. 21, No. 2, pp. 457-464 (1985b).

Yeh, T.-C. Jim, Gelhar. L.W., and Gutjahr, A.L., "Stochastic Analysis of Unsaturated Flow in Heterogeneous Soils 3. Observations and Applications." Water Resour. Res., Vol. 21, No. 4, pp. 465-471, (1985b).

THE WATERTIGHTNESS TO UNSATURATED SEEPAGE OF UNDERGROUND CAVITIES AND TUNNELS

J.R. Philip
CSIRO Centre for Environmental Mechanics
Canberra, Australia

ABSTRACT

The theory of water exclusion from (or entry into) subterranean holes from downward unsaturated seepage is outlined, together with results for various hole configurations. Holes serve as obstacles to the flow, increasing water pressure at the hole surface. When the seepage is fast enough and/or the hole is large enough, water pressure exceeds soil air pressure and water enters the hole. Cavity shape is important also. Applications include the design of tunnels and underground repositories for nuclear wastes against the entry of seepage water.

IMPERMEABILITE A L'INFILTRATION DES CAVITES ET GALERIES EN MILIEU NON SATURE

RESUME

L'auteur esquisse une théorie concernant l'écoulement, en milieu non saturé, des eaux d'infiltration en présence d'une cavité souterraine : contournement ou entrée dans la cavité. Il présente des résultats obtenus avec des cavités de formes variées. Les cavités font obstacle à l'écoulement car il s'y produit une augmentation de la pression de l'eau à leur surface. Quand le taux d'infiltration dépasse un certain seuil ou la cavité une certaine taille, la pression de l'eau devient plus forte que celle de l'air dans le sol et l'eau pénètre dans la cavité. La forme de la cavité joue également un rôle important. Les principes de cette théorie pourraient être appliquées à la conception de galeries et de dépôts souterrains "anti-infiltration" pour l'évacuation des déchets nucléaires.

1. INTRODUCTION

Consider a downward unsaturated seepage flow. What happens when it encounters a subterranean hole? Does water seep into the hole or does it simply flow around it, leaving the interior of the hole dry? This is the question underlying the work reported here. Conventional wisdom, based on capillary statics, says that water does not enter the hole. But our hydrodynamic studies show that, depending on seepage rate, soil (or rock) properties, and hole size and shape, water may or may not enter.

The applications are many. For GEOVAL the primary context is the watertightness (or otherwise) against downward unsaturated seepage of tunnels and subterranean repositories for nuclear wastes deep underground in arid regions.

This class of problem is at present under active study. Here we outline the theory and current results. Full treatment of many aspects mentioned only briefly or omitted in this account are given in References [1]-[7].

2. PROBLEM FORMULATION

The problems we study may be set up in terms of the full nonlinear steady unsaturated flow equation in a homogeneous isotropic porous medium, together with the appropriate boundary conditions.

2.1 *Quasilinear flow equation*

We have done some work with the full nonlinear equation, but we have worked mostly with a special form of it, the quasilinear flow equation. Exponential representation of the dependence of the hydraulic conductivity on the moisture potential gives this very convenient form [8]-[10]. Our work with the full nonlinear equation [3] indicates that the quasilinear analysis gives ample accuracy for most practical purposes.

This is most helpful, since the quasilinear analysis is mathematically much simpler and leads readily to a very useful dimensionless formulation. Results for a particular cavity shape are found in a dimensionless form embracing in one hit the various combinations of downward seepage rate, cavity size, and soil hydraulic and capillary properties.

2.2 *Boundary conditions*

For the *seepage exclusion problem* the boundary conditions are: (i) the downward seepage velocity *far from the cavity* has a given constant value; (ii) the normal flow velocity *at the cavity surface* is zero. There is a *supplementary condition* at the cavity surface that the water pressure there nowhere exceeds the cavity air pressure. If it did, a seepage surface would form on the wall, and water would leak into the cavity. We then no longer have a seepage exclusion problem, but a *seepage entry problem*. This, more complicated, problem is discussed in 5 below.

2.3 *The dimensionless relation* $\vartheta_{max}(s)$

The dimensionless parameter s enters the analysis. It is essentially the ratio of a length characteristic of the *cavity geometry* to a length characteristic of the *capillary flow properties* of the soil. When s is small, capillarity dominates the flow process. When s is large, gravity dominates.

A second important dimensionless quantity is ϑ_{max}. This is the dimensionless potential corresponding to the maximum water pressure at the cavity surface.

Then the critical element of the solution of the seepage exclusion problem, for a given cavity shape, is the function $\vartheta_{max}(s)$. When

$$\vartheta_{max}(s) \leq K_1/K_0 ,$$

where K_1 is the saturated hydraulic conductivity and K_0 is the seepage velocity far from the cavity, water is *excluded* from the cavity. When

$$\vartheta_{max}(s) > K_1/K_0 ,$$

water *enters*.

3. SOLUTIONS OF THE EXCLUSION PROBLEM

3.1 *Exact solutions*

Exact solutions of the exclusion problem have been established for circular cylinders [1], spheres [2], parabolic cylinders [3], and paraboloids [3]. For the parabolic configurations, exact nonlinear solutions as well as exact quasilinear ones are available.

Apart from the parabolic configurations, exact solutions are found as series which are rapidly convergent and easily summable at small s, but summation becomes more difficult and less reliable as s increases. A further limitation is that, although exact solutions are available in principle for geometries such as elliptic cylinders, spheroids, and ellipsoids, these lead to exotic functions, the properties of which have been only partly explored, and which remain for the most part neither evaluated nor tabulated.

As indicated in 2.3, the most practically important relation found from the analysis is the function $\vartheta_{max}(s)$. It might be thought that numerical analysis would readily give $\vartheta_{max}(s)$ in cases where exact solutions are unavailable. It happens, however, that ϑ tends to vary most rapidly near the cavity apex, with numerical methods least efficient in precisely that region.

Both numerical solutions and the summation of exact solutions give decreasing accuracy as s grows large. Fortunately roof boundary-layer solutions are available which are usefully accurate at relatively small s and grow increasingly exact as s increases.

3.2 *Roof boundary-layer solution*

For s large enough, a boundary layer develops on the upper parts of the cavity roof. Across the layer, normal to the roof, ϑ decreases rapidly from a maximum at the roof surface towards 1 outside the layer. On the other hand, the variation of ϑ parallel to the roof is relatively slow. Retaining from the full flow equation only the variation with respect to the appropriate normal coordinate, we secure an ordinary D.E., which is the required boundary-layer equation. The equation specific to each point on the roof, subject to appropriate boundary conditions, is then amenable to solution.

The resulting boundary-layer solution is exact in the singular cases of parabolic-cylindrical and paraboloidal cavities. For circular-cylindrical [1] and spherical [2] cavities the boundary-layer solutions have been compared with the exact solutions. The boundary-layer solutions give excellent results for values of ϑ on the uppermost $\frac{1}{4}$ to $\frac{1}{2}$ of the roof surface. In particular, results for the function $\vartheta_{max}(s)$ agree well with the exact results even for quite small s with the error decreasing rapidly as s increases.

A helpful outcome of the boundary-layer solutions is that, for apically blunt cavities, they yield simple and useful asymptotic expansions for $\vartheta_{max}(s)$ in descending powers of s.

A great advantage of the boundary-layer method is that it readily yields solutions for more complicated cavity configurations. We instance elliptic cylinders [4], spheroids [4], and ellipsoids [5]. Degenerate forms of these solutions apply to cavities with strip-shaped [4], and circular [4] and elliptic [5] disc-shaped, roofs. A central aspect of the boundary-layer solutions for these configurations, which may involve 2 or 3 independent characteristic cavity lengths, is use of the appropriate coordinate system which ensures the retention of all relevant length scales in the bounday-layer formulation.

The boundary-layer analysis has been extended recently to more general shapes, such as various cavities with flat roofs [7]. For these problems we use appropriate orthogonal curvilinear coordinate systems. For two-dimensional and axisymmetric configurations, these are readily generated through conformal mapping.

4. SOME ASPECTS OF EXCLUSION SOLUTIONS

4.1 *Circular cylinders*

The solution for circular cylindrical cavities [1] typifies seepage exclusion from all cavities which are apically blunt. The cavity characteristic length is taken as the cavity radius.

Figure 1 is a schematic diagram illustrating the critical points and regions of the flow field about the cavity. We note the *upstream stagnation point* where the potential ϑ has its maximum value ϑ_{max}, *and the roof boundary layer*. Figure 2 maps the dimensionless potential ϑ and the dimensionless stream function Φ for values of s from 0.25 to 8. Note how the distribution of ϑ becomes increasingly asymmetric with respect to z as s increases. One striking manifestation of this is the emergence of lobes with ϑ large, emanating from upper mid-latitudes of the cavity. The size of the lobes, the values of ϑ therein, and their downward distortion, all grow markedly as s increases. These *roof-drip lobes* are produced by run-off of water unable to penetrate the cavity roof. Note that $\vartheta < 1$ in a region extending beneath the cavity, the *dry shadow*. Throughout the remainder of the flow region the presence of the cavity produces a positive perturbation in ϑ, i.e. $\vartheta > 1$.

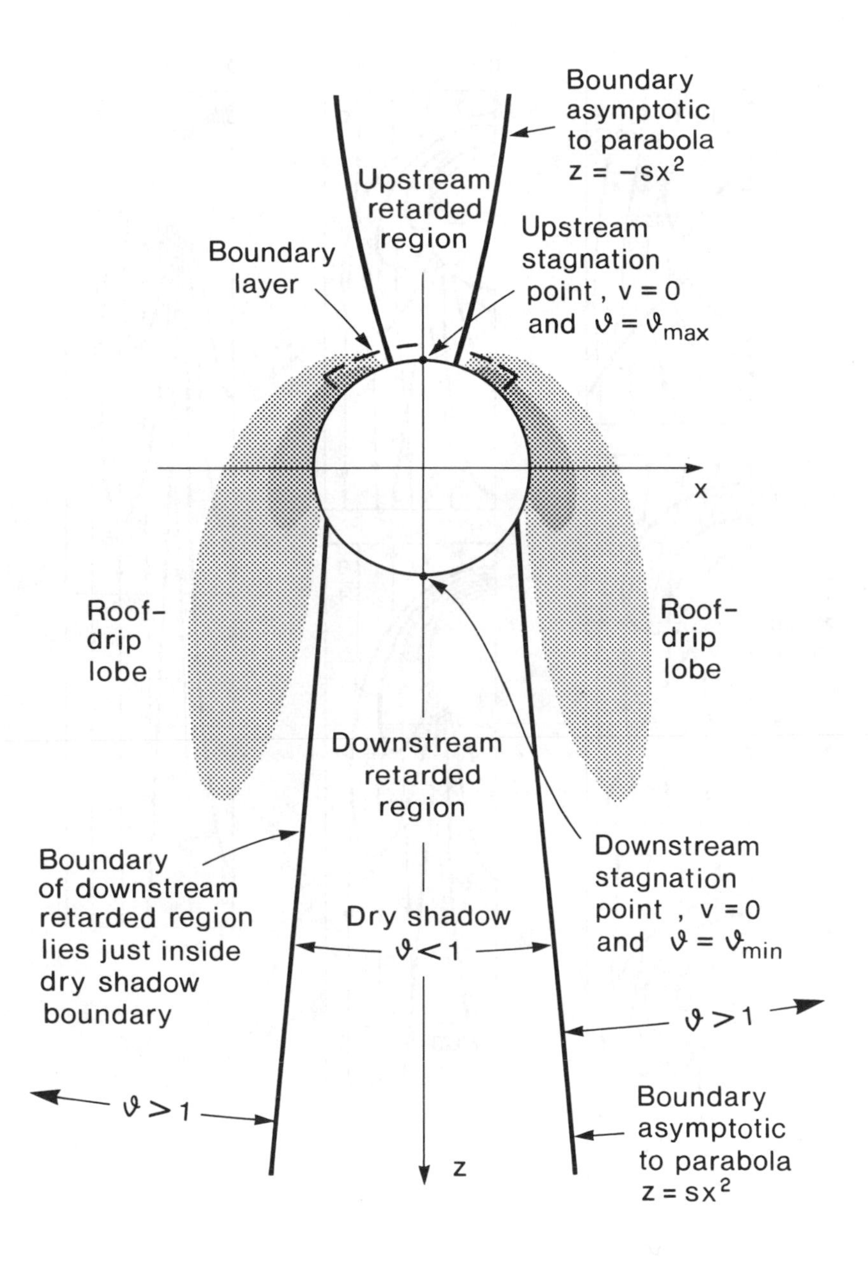

Figure 1. Seepage about cylindrical cavities. Schematic diagram illustrating critical points and regions of the flow field.

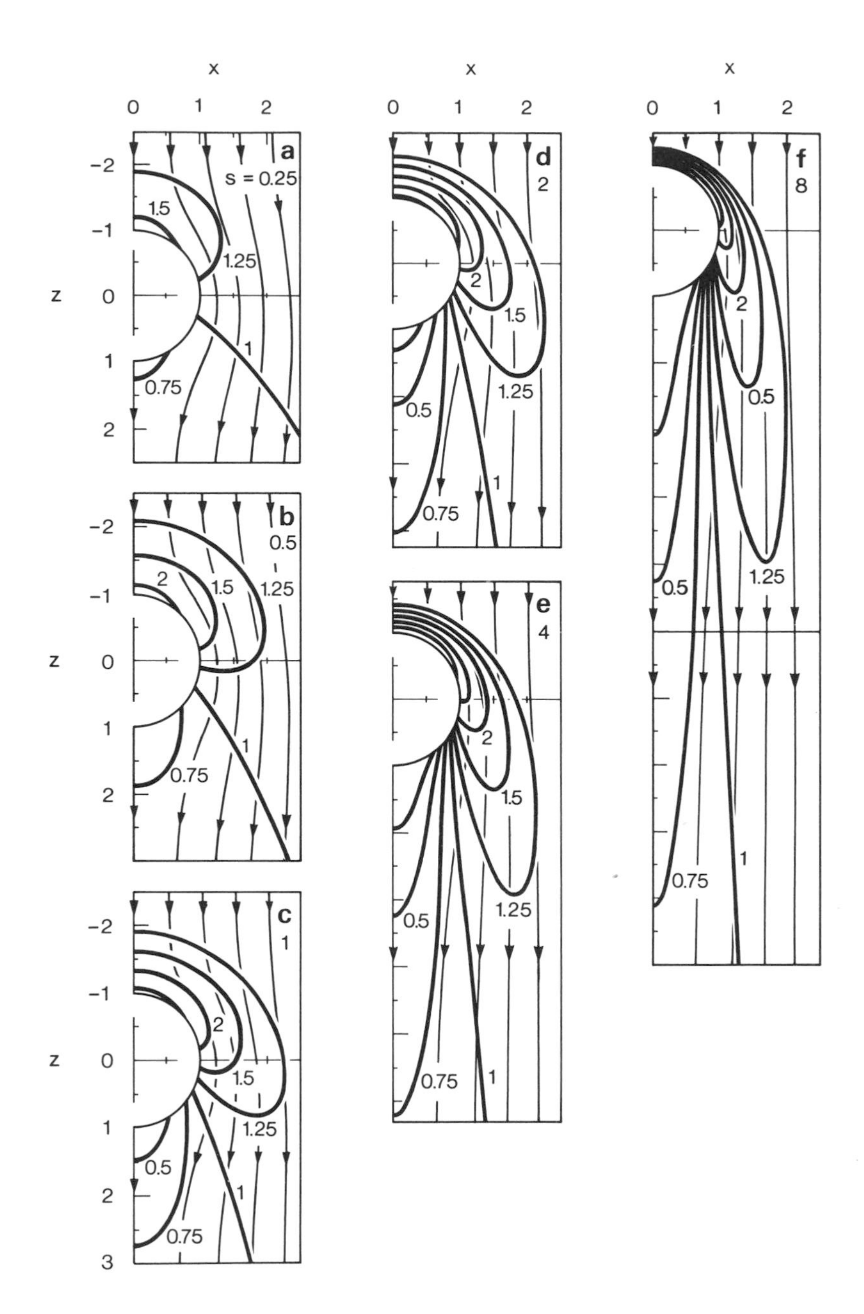

Figure 2. Seepage about cylindrical cavities. Maps of potential ϑ and stream function Φ for indicated values of dimensionless parameter s. Bold curves are equipotentials, with the numerals values of ϑ. The sequence of values is 0.25, 0.5, 0.75, 1, 1.25, 1.5, 2, 3, 5, 9, 17. Fine curves with arrows are streamlines. Φ-values on streamlines increase successively to the right by 0.5.

4.2 *Hierarchy of geometrical parameters*

A very general analysis for ellipsoidal cavities [5] indicates that there is a hierarchy of physical importance of various geometrical parameters, as measured by their influence on the value of ϑ_{max}. The dominant parameter is based on *apical total curvature*; next in importance is a horizontal/vertical *aspect ratio*; and a parameter based on the ratio of cavity lengths in the two horizontal principal directions has a trivial effect. As one expects on physical grounds, the major influence is cavity geometry in the immediate neighbourhood of the stagnation point.

4.3 *Sloping cylinders*

To this point our discussion of cylinders (including those of arbitrary cross-section) has referred to cylinders with axis horizontal. We have shown, however, that solutions for cylinders with axis at an angle to the horizontal are very simply inferred from the corresponding solutions for horizontal cylinders [6].

5. THE SEEPAGE ENTRY PROBLEM

From one viewpoint the question of the watertightness, or otherwise, of underground cavities is settled by solving the exclusion problem. On the other hand, in certain contexts we need to know how the area of the seepage surface and the rate of water entry increase as the seepage velocity K_0 increases beyond its critical value, K_1/ϑ_{max}. That information is needed, for example, in relation to "through-flow" in worst case scenarios for underground hazardous waste repositories.

The seepage entry problem is mathematically more difficult than the exclusion problem. The boundary conditions on the cavity surface are mixed, and the problem is analogous to a free-surface problem in that the seepage surface is unknown a priori and its delineation for a given cavity configuration and parameter combination is part of the problem.

Entry problems are so complicated that a general search for exact solutions holds little promise as a fruitful way to go. Numerical methods will be inevitably elaborate, and accurate and reliable results will demand very careful programming and large computing resources.

The roof boundary-layer method developed for exclusion problems fortunately carries over to entry problems, and work is in progress. A limitation is that the method promises to be accurate only so long as the seepage surface occupies no more than $\frac{1}{4}$ to $\frac{1}{2}$ of the cavity roof area.

The only configurations yielding complete and exact results for the seepage entry problem are the singular cases of parabolic-cylindrical and paraboloidal cavities. For both the whole cavity wall becomes the seepage surface and water enters at the uniform rate (per unit plan area) $(K_0 - K_1)/\vartheta_{max}$. If the seepage were to drip from its point of entry, there would be spatially uniform "rainfall" in a leaking parabolic or paraboloidal repository.

6. REFERENCES

[1] Philip, J.R., Knight, J.H., and Waechter, R.T.: "Unsaturated Seepage and Subterranean Holes: Conspectus, and Exclusion Problem for Circular Cylindrical Cavities", Water Resour. Res., 25 (1989) 16-28.

[2] Knight, J.H., Philip, J.R., and Waechter, R.T.: "The Seepage Exclusion Problem for Spherical Cavities", Water Resour. Res., 25 (1989) 29-37.

[3] Philip, J.R., Knight, J.H., and Waechter, R.T.: "The Seepage Exclusion Problem for Parabolic and Paraboloidal Cavities", Water Resour. Res., 25 (1989) 605-618.

[4] Philip, J.R.: "Asymptotic Solutions of the Seepage Exclusion Problem for Elliptic-Cylindrical, Spheroidal, and Strip- and Disc-Shaped Cavities", Water Resour. Res., 25 (1989) 1531-1540.

[5] Philip, J.R.: "Some General Results on the Seepage Exclusion Problem", Water Resour. Res., 26 (1990), No. 3, in press.

[6] Philip, J.R.: "The Seepage Exclusion Problem for Sloping Cylindrical Cavities", Water Resour. Res., 25 (1989) 1447--1448.

[7] Philip, J.R.: "Conjectures on Certain Boundary-Layer Equations and Natural Coodinates", Proc. R. Soc. Lond., A (1990) in press.

[8] Philip, J.R.: "Steady Infiltration from Buried Point Sources and Spherical Cavities", Water Resour Res., 4 (1968) 1039-1047.

[9] Philip, J.R.: "The Quasilinear Analysis, the Scattering Analog, and other Aspects of Infiltration and Seepage", in Infiltration Development and Application, (ed. Y.-S. Fok.) 1987, 1-27 (Water Resources Research Centre: Honolulu.)

[10] Philip, J.R.: "The Scattering Analog for Infiltration in Porous Media", Rev. Geophys., 27 (1989) 431-448.

APPROACHES TO THE VALIDATION OF MODELLING OF GROUNDWATER FLOW RELATED TO THE PERFORMANCE ASSESSMENT OF RADIOACTIVE WASTE DISPOSAL SITES IN ROCK SALT

P. Glasbergen
National Institute of Public health and Environmental Protection (RIVM)
Bilthoven, The Netherlands

R. Storck
Gesellschaft für Strahlen- und Umweltforschung (GSF)
Braunschweig, Federal Republic of Germany

ABSTRACT

Strongly varying densities form a specific and essential problem related to repositories in rock salt with respect to site characterization as well as simulation of the performance of a repository in a salt formation. During the execution of the HYDROCOIN and INTRAVAL projects, several test cases related to disposal in rock salt were extensively discussed. Although the results of calculations for these cases seemed to converge, validation in the strict sense was not obtained.

After a review of the status of model validation by the end of phase 1 of the INTRAVAL project, a discussion of the need for validation from the angle of performance assessment is given. Proposals for continuation in INTRAVAL phase 2 are presented, as well as their limitations, and some recommendations are given for further work on the validation of the modelling of radionuclide migration under saline conditions.

DEMARCHES VISANT A LA VALIDATION DES MODELES D'ECOULEMENT D'EAU SOUTERRAINE DANS LE CONTEXTE DE L'EVALUATION DES PERFORMANCES DES SITES D'EVACUATION DES DECHETS RADIOACTIFS DANS LE SEL

RESUME

Des variations importantes de densité constituent un problème particulier et essentiel, propre aux dépôts dans le sel, en matière de caractérisation du site et de simulation des performances d'un dépôt situé dans une formation saline. Au cours de l'exécution des projets HYDROCOIN et INTRAVAL, plusieurs cas d'étude liés à l'évacuation dans le sel ont fait l'objet d'examens approfondis. Bien que les résultats des calculs ainsi effectués semblent converger, une validation au sens strict n'a pas été obtenue.

Après examen de l'état de validation des modèles à l'issue de la phase 1 du projet INTRAVAL, cette communication discute de la nécessité d'une validation dans la perspective de l'évaluation des performances. Des propositions pour la poursuite des travaux au cours de la phase 2 d'INTRAVAL sont présentées de même que leurs limitations, et quelques recommandations sont avancées pour de nouvelles activités sur la validation des modèles de migration des radionucléides dans des conditions de salinité.

1. INTRODUCTION

Mathematical models are employed in various stages of the performance assessment of radioactive waste disposal in rock salt. First, large-scale modelling of the groundwater flow at a disposal site is part of the site characterization programme. This can lead to the indentification of areas where radionuclide migration could occur. Second, because groundwater is the main transport medium enabling radionuclides to return to the biosphere, modelling of flow and transport of contaminated groundwater through the overlying formations characterizes the geosphere in its barrier function within the safety assessment. For the latter type of model the emphasis falls on the calculation of potential migration for a postulated radionuclide release from the salt dome after invasion of the repository area by water and the transport of radionuclides through the salt formation.

A fundamental problem related to the study of disposal options in rock salt is how to simulate the natural groundwater system and predict radionuclide migration under strongly varying salt concentrations.

Groundwater-flow and transport models have been constructed in several countries for site characterization, and as parts of the model chain for the evaluation of scenarios considered in safety assessment of radioactive waste disposal in salt formations [1-5]. In most of these studies, possible effects of density variations on the flow system were not taken into account.

Elements of these studies have been used in the formulation of two test cases for the HYDROCOIN project. These test cases were originally linked to siting studies, such as the test case simulating regional groundwater flow in an area with bedded salt (comparable to the situation in parts of Texas) and the test case for a simplified representation of density-dependent flow above a salt dome comparable to the aquifer system overlying the Gorleben salt dome (FRG) [6]. Although these test cases seemed to be rather simple, they gave rise to a number of difficulties that led initially to unsatisfactory results. Later, some degree of consensus seems to have been reached. However, with respect to the nature of these test cases there was no proof that the finally converging results constituted an accurate representation of a real flow system. In the final part of the HYDROCOIN study, some work was done on the sensitivity of the first two test cases, but the basic problems were not solved.

When the INTRAVAL project started, the situation concerning the validation of flow models with strongly varying densities was still so uncertain, that the time was not considered appropriate for starting work on the validation of nuclide migration models through such systems. Indeed a step backward was even taken by trying to design a well-defined laboratory experiment that would show whether the classical mathematical formulations for flow and transport are applicable to brine transport or require modification.

The present paper starts with an overview of the situation at the end of the HYDROCOIN exercise, continues with the progress made in INTRAVAL phase 1, and ends with plans for future work formulated on the basis of a proposed validation strategy.

2. EVALUATION OF THE SALT-RELATED CASES OF THE HYDROCOIN PROJECT

According to the overall objective of HYDROCOIN, three elements of the study are recognized [7]:

- the impact of different solution algorithms on the groundwater flow calculations;

- the ability of different models to describe field and laboratory experiments; and
- the impact of various physical phenomena on the groundwater flow calculations.

The project addressed problems on three levels: code verification, model validation, and sensitivity and uncertainty analysis. Code verification can be accomplished by comparing numerical results with analytical solutions. For the salt-related test cases, however, no analytical solutions exist. Test case 5 of level 1 was performed as a code intercomparison exercise. A relatively good agreement between the calculated results of the participating groups was be observed after thorough discussion [8]. This only means that the codes have been verified against each other but that the models have not yet been validated.

In test-case 6 of level 1 a regional flow system related to a bedded salt layer in Texas was simulated. For the calculations constant density was assummed. Although most of the results of the participating groups were comparable [9], the validation value of this test case is debatable in view of the simplifying basic assumptions. In level 3, as an extension to this test case, one of the output options was to assess travel times and particle trajectories.

The comparison with temperature-induced flow, as performed in level 2 (test-case 2), could only serve qualitatively, because the original temperature experiments were based on photographs and not on measurements. It was concluded that validation of models of variable-density fluid flow can only be properly performed if well-established experimental data are available [10]. This test-case did not lead to firm conclusions about the ability of the models to describe flow under strongly varying density differences, because there were doubts about the full analogy of heat and salt transport.

The uncertainty and sensitivity analyses in level 3 of HYDROCOIN increased insight into the importance of the criteria for the choice of grid size, and time steps.

The HYDROCOIN results showed that is was necessary that experiments be designed in close interaction between modellers and experimenters if progress in validation is to be attained. It was felt that in view of the extremely long travel times of groundwater particles, the scaling-up to real field situations and the problem of transient boundary conditions should only be addressed after the basic problems of flow and transport with strongly varying densities had been solved.

3. PROGRESS MADE IN THE INTRAVAL PROJECT ON THE SALT-RELATED TEST CASES

The purpose of the INTRAVAL project is to increase our understanding of how various geophysical, hydrogeological, and geochemical processes of importance to radionuclide transport can be described by specially formulated numerical models. In view of the difficulties encountered in HYDROCOIN, it was considered to be the best approach to start with a laboratory experiment dealing with flow at various densities and later to make use of data from a pumping test in the Gorleben area. These test-cases do not take radionuclide migration into account in contrast with other INTRAVAL test-cases, which consider migration of radionuclides under natural conditions or in laboratory experiments.

A laboratory experiment simulating front migration of a salt solution through a vertical plate model, filled with glass beads, yielded a number of breakthrough curves for low and high salt concentrations [11]. Although an attempt was made to make the set-up of the experiment as well defined as possible, the porosity was not equal over the whole model and not constant for all steps of the experimental investigations. The porosity was calculated from low-concentration experiments but might be influenced by other factors such as inter-plate distance variations. Dispersivity could also be calculated from the low-concentration experiments. To obtain an acceptable fit, the teams involved in this test case had to perform several calculations with decreasing grid size and smaller time steps.

The use of porosities and dispersivities established by modelling of low-concentration experiments intended to simulate the high concentration experiments did not yield to satisfactory results [12 - 14]. All codes applied to this test case indicate that a better simulation of the slope of the breakthrough curves requires a lower dispersivity than that used in the low concentration experiments. Note also that, with respect to the breakthrough time, all calculations give curves displaced from the observed one. This would imply, according to ref. [13], that the classical formulation of dispersion based on a Fickian relation is not valid in a situation with high concentration gradients. An extension of Fick's law with a density-dependent term as proposed in ref. [13] could provide an explanation for the observed discrepancies. However, a repetition of the experiment with an improved set-up is needed, according to ref. [12] before such a formulation can be accepted. The applicability of Darcy's law in situations with high concentration gradients remains uncertain.

4. NEEDS FOR VALIDATION RELATED TO PERFORMANCE ASSESSMENTS OF REPOSITORIES IN ROCK SALT

A generic picture of the situation around a rock salt repository is depicted in Figure 1, illustrating processes that might induce and affect the radionuclide transport. In many safety assessment studies of disposal in rock-salt, these processes were considered as part of the scenarios leading to release into the biosphere [15, 16].

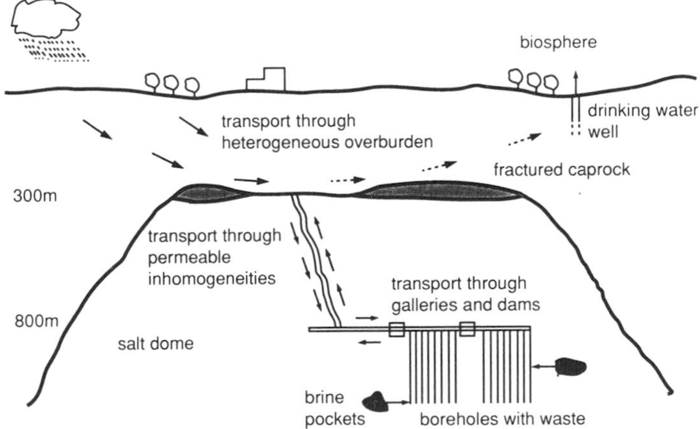

Figure 1. Potential fluid movements relevant to radionuclide transport inside and outside a salt structure that contains a radioactive waste repository.

On the basis of laboratory, field, and in-situ investigations related to disposal in rock salt, a large number of aspects, affecting radionuclide migration through the geosphere, can be identified, such as:

- hydraulic and transport parameters
- dispersion mechanisms

- scaling properties (space and time)
- spatial variability
- non-steady hydrological boundary conditions
- density-driven flow and density effects on transport
- solution of evaporites at the interface of rock salt and overlying porous strata and its effect on density-driven flow
- fracture flow along preferential pathways through a fractured caprock
- retention of radionuclides as a function of the geochemical environment
- variable geochemical factors such as: Eh, pH, Kd
- complexation of radionuclides with dissolved salts
- colloid formation with organic material in groundwater in overlying aquifers
- matrix diffusion, of special relevance for bedded salt covered by chalk, limestone, or clay layers

However, due to limitations imposed by the available codes, lack of data, and computer capacity, in general, the safety assessment studies only cover a limited number of processes considered to be of importance. In that case it is assumed that neglect of several processes would only lead to higher dose estimates. Although this might be the case for the density flow, which is often neglected in the published studies, transport-influencing processes, mainly those of a hydrochemical nature, could enhance transport.

Sofar, only the front migration of a fluid with high-density replacing fresh water has been studied on a laboratory scale [11]. Because these sharp fronts seem not to occur in field situations, see e.g. the Gorleben cross-section [14], the question arises whether the observed discrepancies between model simulations and measured front migration also occur in the field. Several items relevant for this question will be covered by the two large salt test-cases proposed for INTRAVAL phase 2. The Gorleben test case will deal with density effects and solution phenomena in two steps. The scale of the problem is greater and the flow rate is much lower relative to those in the laboratory experiment. The first step toward scaling this problem up is the simulation of a pumping test where density variations have been measured in a number of observation wells. The second step is the simulation of the solution of rock salt at an aquifer/salt boundary: Here, the natural time scale of the problem is of primary importance. It is not known whether the process reached a steady state after the last glaciation of NW Europe. The Gorleben case might be extended to take into account ^{14}C migration patterns and the past variations in hydrological boundary conditions.

The WIPP-site test-case will cover deformation and pressure-driven flow effects within a bedded salt formation itself. In rock salt repositories, as schematically illustrated in Fig. 1, a near field-domain might be defined as a zone consisting of the canister and the salt directly surrounding the waste where radiation, corrosion, high temperatures, and convergence will dominantly influence the processes potentially leading to release. The intermediate field consists of the remainder of the rock-salt formation including its inhomogeneities. The WIPP-site test case will cover this field for a bedded salt site. Here strong hydraulic gradients are introduced by the excavation of galleries. Important problems include the description of fluid migration under these conditions. The main questions are whether the classic flow equations are still applicable and, whether the deformation of the medium and the solution and precipitation of salt must be taken into account.

For geological reasons, there is a difference between salt domes and salt layers. Deformation of a salt dome during its diapiric stage leads to a more strongly compacted rock. The degree of impurity and of inhomogeneous inclusions are also important factors. The fluids present in salt domes can in many cases be regarded as being concentrated in micro-inclusions between salt cristals. Larger brine inclusions have rarely been encountered [17]. The inhomogeneous beds within salt layers could conduct water if there were a sufficient gradient.

Some work has been done outside the INTRAVAL framework on the modelling of large-scale systems and the effects of high densities. A regional 2-D cross-section with a length of 46.5 km was studied in Germany [14]. This section was assumed to be underlain by a bedded salt layer at 2000 m below surface. The simulations done with the SUTRA code indicate a much lower velocity in the saline environment at greater depth compared to the fresh water simulations, which might point to a diffusion-dominated system, whereas convective transport is observed in the upper fresh-water system.

A review of the items considered in the two proposed test cases, makes it clear that the validation of radionuclide migration in the geosphere, which is of course essential for the safety assessment modelling, is still lacking. At several places, laboratory studies dealing with the geochemistry of radionuclides in saline environments are being performed. The relation between sorption of radionuclides and salt concentration, as well as colloid formation, and complexation are important items. These experiments all have in common that their time and space scales differ greatly from the expected natural scales. Besides ^{14}C dating, no studies have been carried out to investigate the migration of natural radionuclides in groundwater systems with a high salt concentration. The available information about the behaviour of uranium in closed basins and terminal lake situations has neither been quantified, nor the observed processes modelled [18]. For this reason, a partial validation of transport models must for the time being be achieved with laboratory experiments. Because transport models must be extended to take into account transport of colloids and organic complexes, well-defined laboratory experiments might be a good starting point for the development of conceptual models.

With respect to the difficulties encountered in assessing the flow path for a particle leaving a salt formation and entering an aquifer system with strongly varying densities, a detailed study of natural tracers might be advisable. For example, sulphur isotopes are indicators for the presence of sulphates dissolved from caprock, which in many cases covers salt domes. Isotopes such as ^{18}O offer are a more qualitative indication of the origin of the groundwater. Differentiation between marine intrusions and meteoric water containing dissolved rock salt is possible. Indications of the paleohydrological boundary conditions, which are essential for any reconstruction of the development of a system with variable density, can also be obtained by combining the results of various isotope studies [19]. Chloride isotopes must be evaluated with respect to their usefulness in aquifer systems with high salinities.

5. CONCLUSIONS

By the end of the INTRAVAL phase 1 project consensus had not been reached about the explanation of the discrepancies between laboratory observations on brine-front migration and the model simulations. The proposal to add a non-linear term to Fick's law has, however, provided satisfactory agreement between calculated and experimental results of the test-case experiment. Investigation of this problem will require additional laboratory experiments .

It is generally felt that the scale of the problems to be studied should be brought to real flow conditions which would permit investigation of the relevance of the problems arising from the laboratory experiment. Pumping tests are considered as the first step for investigation of the ability of existing models to simulate the movement of transition zones of saline and fresh water. On a regional scale the spatial density of the available data as well as the uncertainty as to whether the data represent a steady hydrological system could lead to the conclusion that models based on various concepts of the basic equations do not show a significantly different better fit with the observations.

After analysis of the pumping tests on the basis of the available models, the question of validation strategy for density-affected groundwater-flow models must be re-examined. Especially the indication for a modified theoretical basis originating from the laboratory experiments and the large scaling effect between a laboratory experiment on a 1 m scale and a pumping test on a 1000 m scale might make additional experiments on an intermediate scale necessary.

In terms of the models needed for site characterization and safety assessment, an attempt to develop and validate models for nuclide migration in a medium of high ionic strength taking into account complexation and colloid migration as well as adsorption should be encouraged.
There is increasing interest those working on projects dealing with disposal in deep geological formations, other than rock salt, in the results of studies on density effects on flow and transport, because a strong increase of salinity with depth has been observed in many places and outside evaporite basins as well.

Current attempts at regional modelling point to large differences between models taking into account the assumed density increase with depth and those neglecting density differences. Substantial difficulties are encountered in calibration of this type of model due to lack of both data and knowledge about the dynamics of a large-scale groundwater system. The validation of models describing flow on a regional scale in deep formations of low permeability is needed in order to make it possible to set the hydrological boundary conditions for local radionuclide transport modelling. Major climatic fluctuations affecting hydrology have to be evaluated too. Methods to assess deep circulation of groundwater, e.g. arising from the developments in isotope hydrology, should be considered as valuable information for model validation.

ACKNOWLEDGMENT

The ideas expressed and conclusions reached in this paper are based mainly on extensive discussions in the INTRAVAL Working Group on Salt-Related Test Cases.

REFERENCES

1. Storck R., and S.Hossain, 1986. Coupling of near-field processes in salt dome repositories for radioactive wastes. In: System Performance Assessment for Radioactive Waste Disposal. NEA, OECD, Paris, p. 115-123.
2. Bütow E., L.Heredia, S.Lütkemeier-Hosseinipour, S.Struck, 1986. Approaches to model groundwater flow and radionuclide transport at a German salt dome. In: Proceedings of the Symposium on groundwater flow and transport modeling for performance assessment of deep geologic disposal of radioactive waste: A critical evaluation of the state of the art. NUREG/CP-0079, Battelle, Richland, p. 3-14.
3. Atwood H., L.Picking, 1986. A preliminary simulation model to determine groundwater flow and ages within the Palo Duro Basin hydrogeologic province. In: Proceedings of the symposium on groundwater flow and transport modeling for performance assessment of deep geologic disposal of radioactive waste: A critical evaluation of the state of art. NUREG/CP-0079, Battelle, Richland, p. 257-275 (3 enclosures).

4. Slot A.F.M., P.Glasbergen, I.Nijhoff-Pan, 1990. Geosphere migration of radionuclides released from a rock-salt repository. In: Safety Assessment of Radioactive Waste Repositories. OECD, Paris, p. 920-927.
5. Glasbergen P., G.Englund-Borowiec, 1989. Groundwater models as part of the safety assessment of underground disposal of hazardous waste in The Netherlands. In: G.Jousma, J.Bear, Y.Y.Haimes and F.Walter, eds. Groundwater contamination: use of models in decision-making. Kluwer Ac. Publ., Dordrecht, p. 577-586.
6. Bütow E., E.Holzbecher, 1989. On the modelling of groundwater flow under the influence of salinity. In: Groundwater contamination: use of models in decision-making. Kluwer Ac.Publ., Dordrecht. p. 263-271.
7. Grundfelt B., B.Lindbom, K.Andersson, 1989. Hydrocoin-findings from level 2; model validation. In: Safety Assessment of Radioactive Waste Repositories. OECD, Paris. p.717-724.
8. The Coordinating Group of the HYDROCOIN project, Swedish Nuclear Power Inspectorate, 1988. The International HYDROCOIN project, Level 1: Code Verification. NEA, SKI, OECD, Paris.
9. Kimmeier F., C.Wacker, P.Perrochet, P.Hufschmied, L.Kirary, 1986. Code verification of the groundwater flow model FEM301, HYDROCOIN level 1-case 2 and case 6. NAGRA, technical report 86-22, Baden.
10. NEA and SKI, 1990. The International HYDROCOIN project, level 2: Model Validation. OECD, Paris.
11. Hassanizadeh S.M., 1990. Experimental set-up for brine transport in porous media. National Institute of Public Health and Environmental Protection, Bilthoven.
12. Arens G., E.Fein, 1990. Flow and transport at high salinity. In: Geoval, 1990.
13. Hassanizadeh S.M. 1990. Verification and validation of coupled flow and transport models. In: Geoval 1990.
14. Schelkes K, P.Vogel, H.Klinge, R.M.Knoop, 1990. Modelling of variable-density groundwater flow with respect to planned radioactive waste disposal sites in West-Germany - Validation activities and first results. In: Geoval 1990.
15. Storck R. 1989. Performance assessment for nuclear waste repositories in salt domes; results and experiences. In: Safety Assessment of Radioactive Waste Repositories. OECD, Paris, p. 237-246.
16. Glasbergen P., S.M.Hassanizadeh, 1988. Some considerations on the disposal of hazardous waste in salt formations. In: Land disposal of hazardous waste. Ellis Harwood, Cambridge. p. 269-276.
17. Nies A. 1989. State of the art on modelling repository performance in salt domes. In: Safety Assessment of Radioactive Waste Repositories. OECD, Paris, p. 569-580.
18. VanLuik A.E., 1987. Uranium in selected endorheic basins as partial analogue for spent fuel behavior in salt. In: B.Côme, N.A.Chapman (eds.) Natural analogues in radioactive waste disposal. Graham and Trotman, London.
19. Soreau S., P.Glasbergen, 1989. Preliminary indications from natural isotopes on the dissolution of rock-salt in deep circulating groundwater of The Netherlands. In: CEC Natural Analogue Working Group, Third Meeting. EUR 11725, Luxemburg p. 210-220.

VERIFICATION AND VALIDATION OF COUPLED FLOW AND TRANSPORT MODELS

S.M. Hassanizadeh
National Institute of Public Health and Environmental Protection (RIVM)
Bilthoven, The Netherlands

ABSTRACT

In addition to difficulties experienced in the verification and validation of geohydrological models, there exist complications specific to coupled flow and transport models. For example, a classical verification exercise is often not possible because there are no analytic solutions to the equations embedded in coupled models. Intercomparison of codes is also often complicated by the effect of temporal and spatial discretizations. Nevertheless, partial verification of codes may be obtained subject to certain restrictions. There are also validation difficulties specific to coupled models. They may have a *theoretical* as well as an *experimental* nature. Examples are the uncertainty in the choice of the conceptual model which may best describe the coupled processes and the problem of resolving effects of various processes in a given experiment. These issues are discussed in detail and illustrated by means of previous and ongoing experiences related to a mathematical model called METROPOL–3.

VERIFICATION ET VALIDATION DES MODELES COUPLES DE L'ECOULEMENT ET DU TRANSPORT

RESUME

Aux difficultés habituelles que l'on rencontre dans la vérification et la validation des modèles hydrogéologiques viennent s'ajouter des complications particulières, dans le cas des modèles couplés de l'écoulement et du transport. Ainsi, il est souvent impossible de procéder à une vérification classique car il n'existe pas de solutions analytiques aux équations introduites dans les modèles couplés. Il arrive aussi souvent que l'emploi de valeurs discrètes pour les paramètres spatiaux et temporels complique la comparaison des programmes de calcul entre eux. Néanmoins, sous réserve de certaines restrictions, une vérification partielle des programmes de calcul est réalisée. La validation des modèles couplés présente également des difficultés particulières qui peuvent avoir une double nature à la fois théorique et expérimentale. A titre d'exemples, on peut citer l'incertitude dans le choix du modèle théorique le mieux à même de décrire les processus couplés et les problèmes que l'on rencontre pour comprendre les effets des divers mécanismes qui interviennent dans une expérience donnée. L'auteur examine toutes ces questions en détail en s'appuyant sur diverses expériences passées et en cours liées à un modèle mathématique appelé METROPOL-3.

1. INTRODUCTION

In most geohydrological problems one is interested to obtain the fluid pressure and the distribution of solutes. In mathematical terms, one has to solve two equations: the flow equation (for obtaining the pressure) and the solute transport equation (for obtaining the solute concentration distribution). Often, one is able to solve the flow equation first and calculate the fluid velocity field, and then solve the transport equation subsequently. In such cases, there exists only a one–way coupling; that of dependence of transport on the flow velocity. However, there exist many practical situations where variations of solutes concentration directly influence the fluid density and thereby affect the pressure and flow fields. In such cases, the processes of fluid flow and solute transport are mutually coupled and the above–mentioned equations have to be solved simultaneously.

Problems and issues related to the modelling of coupled flow and solute transport are discussed in detail by Hassanizadeh and Leijnse [1]. They point out that three main issues require special attention when constructing coupled flow and transport models. These are related to: i) the basic (governing) equations, ii) the boundary conditions, and iii) the numerical methods. Hassanizadeh and Leijnse [1] discuss a few alternative conceptual models and certain boundary conditions generally applicable to coupled flow and transport problems. They also compare three different numerical methods regarding computational efficiency and the ability to account for nonlinearities. Their conclusions, however, remain somewhat speculative and they point out that physical experiments are required to establish the validity of proposed conceptual models and boundary conditions. A suitable framework to carry out such investigations, and draw more definite conclusions about conceptual models and numerical methods employed in coupled models, is the verification and validation framework. Although verification may be viewed simply as part of a validation exercise, it is treated here separately to facilitate the discussion. In this study, we use the word "verification" in a more general sense, to stand not only for comparison of models with analytic solutions but also to include intercomparison of various numerical codes. The validation exercise reported here has had a limited scope. It is more intended to validate a nonlinear dispersion theory (conceptual model) applicable to high–concentration–gradient situations, and cannot be considered as validation of a given mathematical model.

2. GOVERNING EQUATIONS AND CONCEPTUAL MODELS

The basic equations describing fluid flow and solute transport in porous media are the equations of conservation of mass. These are [2]:

$$\frac{\partial n\rho}{\partial t} + \nabla.(\rho \boldsymbol{q}) = 0 \qquad (1)$$

$$n\rho\,\frac{\partial \omega}{\partial t} + \rho \boldsymbol{q}.\nabla\omega + \nabla.\boldsymbol{J} = 0 \qquad (2)$$

where n is the medium porosity, ρ is the fluid mass density, $\boldsymbol{q}$ is the fluid velocity vector, ∇ is the divergence vector, ω is the solute mass fraction and $\boldsymbol{J}$ is the solute dispersive mass flux. These relations must be supplemented by equations of conservation of momentum. For creeping (or slow) isothermal flow in a porous medium, simplified forms of equation of motion for a fluid and a dissolved solute are [2]:

$$\nabla p - \rho \boldsymbol{g} = \hat{\boldsymbol{\tau}}^{f}(\rho,\ \omega,\ \boldsymbol{q},\ \boldsymbol{J},\ T) \qquad (3)$$

$$\overset{*}{\nabla\mu} = \overset{*}{\boldsymbol{\tau}}{}^{s}(\rho,\ \omega,\ \boldsymbol{q},\ \boldsymbol{J},\ T) \qquad (4)$$

where p is the fluid pressure, $\boldsymbol{g}$ is the gravity vector, T is temperature, $\hat{\boldsymbol{\tau}}^{f}$ is the resistance (force) of the system to the fluid flow, $\overset{*}{\boldsymbol{\tau}}{}^{s}$ is the resistance (force) of the system to the solute dispersion, and $\overset{*}{\mu}$ is the solute chemical potential, which accounts for the increase in the free energy of the solution as a result of increase in the solute mass fraction [2]. Equations (1)–(2) are physical laws and are considered to be valid for the full range of thermodynamic processes and for all kinds of solutes and fluids. Equations (3)–(4), on the other hand, contain certain conceptual simplifications and constitutive assumptions and are not necessarily valid for the full range of thermodynamic processes. Actually, one has to employ even simpler functional relationships for $\hat{\boldsymbol{\tau}}^f$ and $\overset{*}{\boldsymbol{\tau}}^s$, in terms of $\boldsymbol{q}$ and $\boldsymbol{J}$, in order to render (3)–(4) amenable to mathematical analysis and suitable for practical usage. Commonly, a linear dependence of $\hat{\boldsymbol{\tau}}^f$ on $\boldsymbol{q}$ and of $\overset{*}{\boldsymbol{\tau}}^s$ on $\boldsymbol{J}$ are employed such that:

$$\hat{\boldsymbol{\tau}}^f = - \boldsymbol{R}^f . \boldsymbol{q} \tag{5}$$

$$\overset{*}{\boldsymbol{\tau}}^s = - \boldsymbol{R}^s(\boldsymbol{q}) . \boldsymbol{J} \tag{6}$$

Substituting these relationships into (3) and (4) and rearranging, one obtains Darcy's law and Fick's law, respectively. Obviously, constitutive equations embedded in (3)–(6) are not valid for the full range of pressure and concentration variations. The range of applicability of any geohydrological model based on (1)–(6) has to be determined with the aid of experiments. An example based on a laboratory experiment is given in section 4.

The three–dimensional finite element model METROPOL–3 [Ref. 3] developed at RIVM for simulating coupled flow and transport is based on equations (1)–(6). METROPOL–3 has been employed in HYDROCOIN and INTRAVAL studies and results of some of those studies are reported in this work.

3. VERIFICATION OF COUPLED MODELS

It is necessary that we acquire a very good knowledge of the uncertainties associated with the numerical implementation of our conceptual model. As appropriately said by Niederer [4], "before a model can reasonably be validated it must have reached a certain stage of maturity". This is best achieved through verification exercises. Verification of a model must be regarded as a prerequisite of any validation exercise. Problems and issues associated with verification of uncoupled and coupled models are discussed by Cole et al. [5]. Here we refer to some of those issues as we discuss various examples.

We have carried out our verification exercise in steps of increasing difficulty. First we have compared results of METROPOL–3 with those of an analytical solution for the case of low–concentration solute transport. One of the difficulties in verification of coupled models is that it is often impossible to obtain analytical solutions for the full set of governing equations. If, however, in equation (2) ρ and $\boldsymbol{q}$ are assumed to be constant, and $\boldsymbol{J}$ is given by a Fickian relationship, one can obtain analytic solutions for a large class of boundary conditions. These assumptions correspond to a low–concentration–gradient situation. The solutions mostly contain error functions and/or infinite series. A computer package called TRANS [6] is developed for evaluating these solutions. Breakthrough curves calculated with METROPOL–3 agree very well with those obtained with TRANS (Figure 1). It appeared that in order to obtain a perfect fit, one must use a grid size of 2 mm and a maximum time step of 1.0 seconds. Larger grid sizes and/or time steps resulted in a poor agreement between METROPOL–3 and TRANS.

Next, METROPOL–3 was compared with a one–dimensional model, with moving grid, called SPRINT [7]. The SPRINT model solves a coupled set of nonlinear partial differential equations and/or algebraic equations. Therefore, it is possible to solve the full set of (1)–(6) subject to time–dependant boundary conditions. The model is robust and dispersion and convergence problems can be overcome by proper choice of the numerical integration scheme. Comparisons were made between METROPOL–3 and SPRINT for both low– and high–concentration–gradient situations and good agreement was obtained between the two models However, there is still some numerical dispersion in METROPOL–3 at high concentrations (see e.g. Figure 2).

Next we have compared our model with other models which solve the full set of governing equations in two or three dimensions. In such exercises, one has to be very careful in drawing conclusions from the comparison results. The problem is that often in 2– and 3–D simulations, the model is sensitive to temporal and spatial discretizations [8 and 5]. It is not always possible to cut back on the time step and grid size values to eliminate numerical dispersion because of computer time and simulation costs. Nevertheless, the intercomparison of various codes provides valuable insight into the working of models and helps with identifying their potential shortcomings of models [5].

An interesting and very useful code comparison exercise for coupled models was performed within the HYDROCOIN Level 1 framework. In test case 1.5, simulations were performed for two–dimensional steady state flow and brine transport around a hypothetical salt dome [9]. In the beginning all models had difficulty with numerical artifacts due to temporal and spatial discretization effects, and agreement between various models was poor. Modifications to the codes and refinement of the grids helped to remove some of the discrepancies so that at the end, various models compared, in a global sense, reasonably well with each other. Some examples are shown in Figure 3. However, closer inspection of results show that large local differences exist in the calculated salt mass fraction distribution, especially in the vicinity of the salt dome boundary. Two explanations can be offered: i) The steady state calculations are obtained as solutions of the transient problem at long times and it is possible that the simulation time must still be larger. Because the velocities are very small near the salt dome boundary, variations in mass fraction between consecutive time steps will be small too. Therefore, the observation that "very little happens in a time step" is not a sufficient proof that steady state is reached. ii) Because of the sharp variations in salt mass fraction near the salt dome boundary, a very fine mesh is needed there in order to avoid any numerical dispersion.

In order to illustrate the latter point, RIVM calculations with METROPOL–3 for case 1.5 were performed with a number of different grids. In Figure 4a, the results obtained with 675 elements are shown. Obviously, these are not anywhere near results obtained by other models in Figure 3. Refining the grid to 2700 elements produces the graph in Figure 4b which shows improvement over the coarse grid. Now if still 2700 elements but an irregular grid, with refinement near the salt dome, is employed, then graph in Figure 4c is obtained which is in reasonable agreement with those in Figure 3.

Nevertheless, there is yet no proof that any of these models runs with a grid fine enough for the particular problem under study. This point deserves further investigations subject to the availability of larger and faster computers.

4. VALIDATION OF COUPLED MODELS

Experiences learned from HYDROCOIN and INTRAVAL studies tell us that validation of models is a complicated, multi–sided and at times impossible task. In recent years, we have learnt much about the intricacies of the job and have become more and more aware of the limitations of our present resources and abilities in carrying out an "ideal" and classical validation exercise. This has led to much scepticism about feasibility of our validation goals, and some consider such goals completely out of reach. This has resulted in a general revision and/or partition of validation goals. Nowadays, it is becoming more and more common to talk about "partial validation" and even "invalidation" of models.

Above considerations are specially applicable to coupled flow and transport models. Two major sets of difficulties can be identified in the road to validation of such models.

1) **Theoretical Complications:** Coupled phenomena are often accompanied with large variations in variables such as temperature, concentration, and/or deformation. These large variations render invalid some of the basic assumptions behind such conceptual models as Darcy's law, Fick's law and Fourier's law. As a result, the processes can no longer be

modelled as isothermal, constant density, constant viscosity, constant dispersivity and so on. One has to come up with appropriate modifications to the conceptual models. This is not a straightforward task. Theoretical considerations may provide a number of alternatives but often a decision cannot be made from behind the desk about the dominant mechanism. This has to be determined by experiments. In interpreting experimental data, we need to employ a computer model. But then the questions are: is that model validated, and does it give us reliable results? This gives rise to a mixing of "physical research" with "validation process".

2) **Experimental Complications:** As discussed above, in coupled problems, there are often two or more major processes which influence each other. For example in the case of brine transport, fluid flow and solute dispersion mutually affect each other. Ideally, one would like to design experiments so that these effects can be resolved. Then one should be able to attribute the discrepancy between model calculations and experimental data to only one of the major processes described by the model. This would make it easier to choose among the alternatives for modifications to the basic equations. Unfortunately, designing such an experiment for two– and three–dimensional situations is very difficult if not impossible. Another point of interest related to experimental complications is that of using the physical analogy between different processes. One would be tempted, for example, to make use of the analogy between heat transport and solute transport phenomena and try to use experimental data obtained for one process to validate a model describing the other process. Such an approach was incorporated in the HYDROCOIN study as test case 2.2. But, it appeared that if variations in concentration are large, the analogy breaks down (for results of case 2.2, see Refs. [10] and [11]).

Recently, an experimental study of brine transport in a porous medium was carried out at RIVM which aimed at identifying the physical processes which come into play when large concentration variations occur (see Ref. [12] for details of the experiment). Two sets of displacement experiments have been carried out: i) displacing a low–concentration solution with a slightly higher concentration such that density remains practically constant, and ii) displacing a low–concentration solution with a high–concentration solutions. Coupling effects are not present in the low–concentration experiments. These are used for evaluating parameters such as porosity and dispersivity. These values are fixed and additional low–concentration experiments are simulated. In Figure 5a, the calibration fit based on experiment L1D01, and in Figure 5b the "validation" fit based on experiment L1D02 are presented. The slight time shift in the arrival time for L1D02 is due to the change in porosity of the medium during the four–month period between the two experiments. One may conclude that METROPOL–3 is "validated" for the conditions under which low–concentration experiments are carried out. Next, still using the same parameter values obtained above, the high–concentration experiments are simulated. Results are shown in Figure 5c and it is apparent that no satisfactory fit is obtained here. Similar results were obtained with the model SPRINT. The conclusion is that at high concentrations, the conceptual model based on equations (1)–(6) is not valid. Thus, it is necessary to modify our basic equations and in particular relations (5) and (6). That is, modifications to both Fick's law and Darcy's law may be necessary and thus we are confronted with the difficulty described above. Fortunately, the boundary conditions and the set–up of the experiments was such that these two effects could be resolved. The present data set allows us to switch off the influence of Darcy's law and to study possible modifications to Fick's law. Because METROPOL–3 is the model to be validated and because it is quite a large and elaborate model, it was decided to use the simpler model SPRINT to investigate various modification to our conceptual theories. These studies led to the following modification of equation (6).

$$\overset{*}{\boldsymbol{\tau}}^S = - \boldsymbol{R}_1^S . \boldsymbol{J} - \boldsymbol{R}_2^S . (|\boldsymbol{J}|) \boldsymbol{J} \qquad (7)$$

which when combined with (4) provides the following nonlinear extension of the classical Fickian equation for hydrodynamic dispersion:

$$(1+\beta|J|)J = -\rho D.\nabla\omega \qquad (8)$$

This equation was incorporated into the SPRINT model and the data set of one of the high–concentration experiments (H1D02) was used to evaluate the new coefficient β (Figure 6a). Afterwards, β was kept fixed and another high–concentration experiment (H1D01) was simulated and again very satisfactory results were obtained as illustrated in Figure 6b. It must be emphasized that this exercise does not amount to the validity of certain mathematical model, but only confirms the validity of the conceptual model (8) for high– as well as low–concentration situations. To study the "validity" of METROPOL–3 one needs to incorporate (8) in the model and simulate additional (2 or 3–D) experiments.

5. CONCLUSIONS

Performing a classical (and "ideal") verification and validation exercise for coupled flow and transport models is an impossible task mainly because of the complexity of the basic equations and the numerical code. Partial verification of METROPOL–3, a 3–D finite element model for density–dependent flow and transport, is obtained by comparing it with an analytic solution for constant density transport, with a one–dimensional model of coupled flow and transport, and with other two– and three–dimensional codes. The comparison in the latter case is complicated by the effect of temporal and spatial discretizations. Validation of METROPOL–3 based on experimental data is possible only for low–concentration–gradient situations. A new conceptual model given by equation (8) appears to describe nonlinear dispersion in high–concentration–gradient situations satisfactorily.

REFERENCES

(1) Hassanizadeh, S.M. and Leijnse T.: "On the modelling of Brine Transport in Porous Media", *Water Resources Research*, **24**, 1988, 321–330.

(2) Hassanizadeh, S.M.: "Derivation of Basic Equations of Mass Transport in Porous Media, 2. Generalized Darcy's and Fick's Laws", *Advances in Water Resources*, **9**, 1986, 207–222.

(3) Sauter F. *et al.*: "METROPOL, a computer code for the simulation of transport of contaminants with ground water", Report no. 728514007, RIVM, Bilthoven , The Netherlands, 1990.

(4) Niederer, U.: "Perception of Safety in Waste Disposal: The Review of the Swiss Project Gewahr 1985", GEOVAL 1987, SKI, Stockholm, 1987, 11–26.

(5) Cole, C.R. et al.: "Lessons from the HYDROCOIN Experience", GEOVAL 1987, SKI, Stockholm, 1987, 269–285.

(6) Veling, E.J.M.: "TRANS: A Program for Analyzing One–Dimensional Convection–Diffusion Equations with Adsorption and Decay", Report no. 958611001, RIVM, Bilthoven, The Netherlands, 1986.

(7) Berzins, M., Dew, P.M. and Furzeland, R.M.: "Developing Software for Time–Dependent Problems Using the Method of Lines and Differential–Algebraic Integrates", *Applied Numerical Mathematics*, **5**, 1989, 375–397.

(8) Leijnse, A. and Hassanizadeh, S.M.: "HYDROCOIN Project Level 1, Case 5 and Level 3, Case 4", Report no. 728528004, RIVM, Bilthoven, The Netherlands, 1989.

(9) HYDROCOIN Secretariat: "HYDROCOIN Report Level 1: Code Verification", OECD, Paris, 1988, 110–132.

(10) Leijnse, A. and Hassanizadeh, S.M.: "HYDROCOIN Project Level 2, Case 2", Report no. 728528005, RIVM, Bilthoven, The Netherlands, 1989.

(11) HYDROCOIN Secretariat: "HYDROCOIN Report Level 2: Model Validation", OECD, Paris, 1990.

(12) Hassanizadeh, S.M. *et. al.*: "Experimental Study of Brine Transport in Porous Media", Report no. 728514005, RIVM, Bilthoven, The Netherlands, 1988.

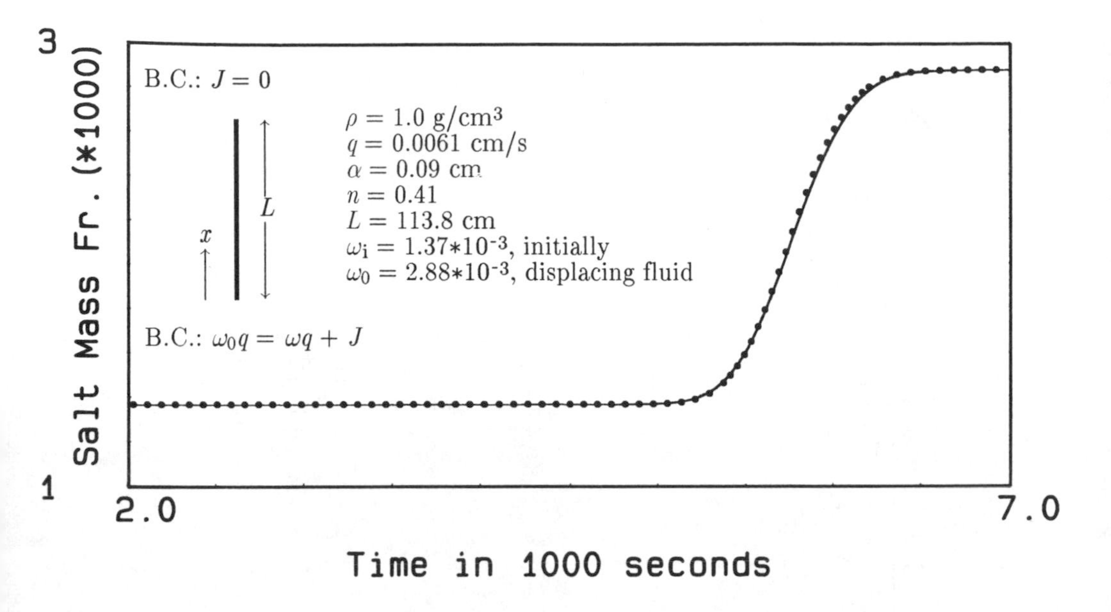

Figure 1. Comparison of METROPOL–3 with TRANS; breakthrough curve at x = 86 cm.

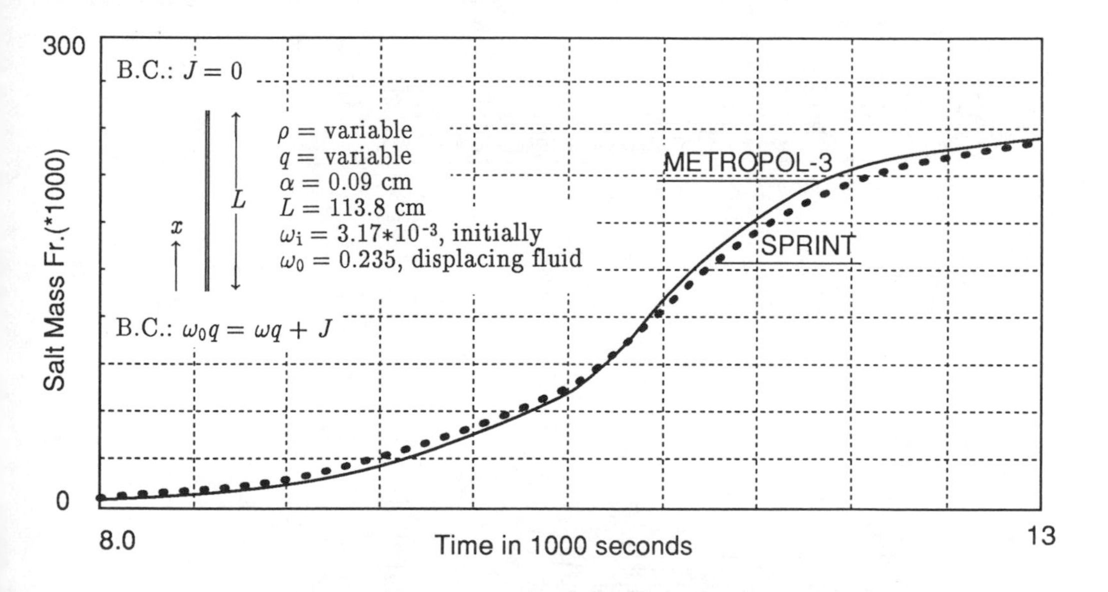

Figure 2. Comparison of METROPOL–3 with SPRINT; breakthrough curve at x = 86 cm.

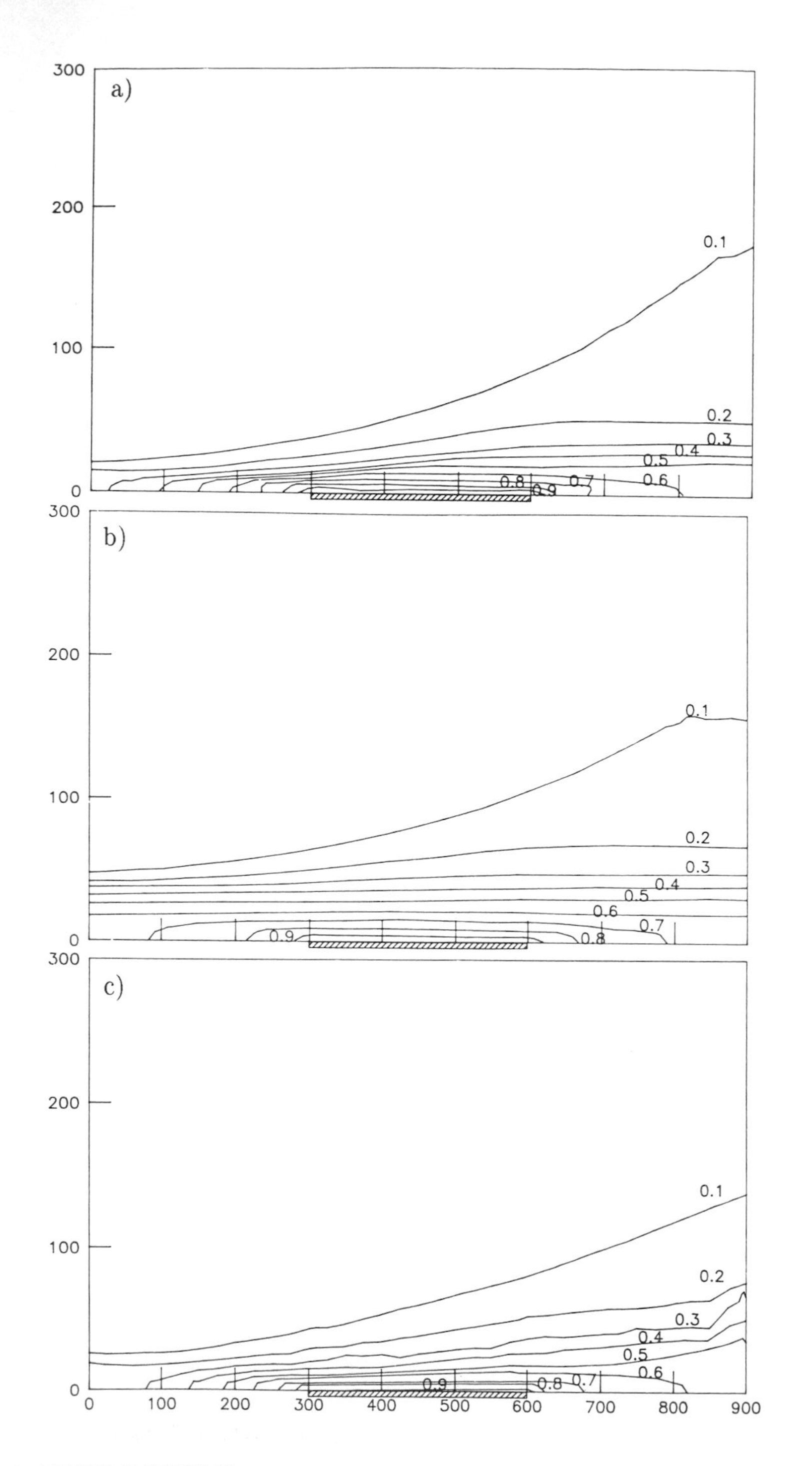

Figure 3. HYDROCOIN Case 1.5, Mass fraction from: a) SUTRA, b) NAMMU, c) CFEST

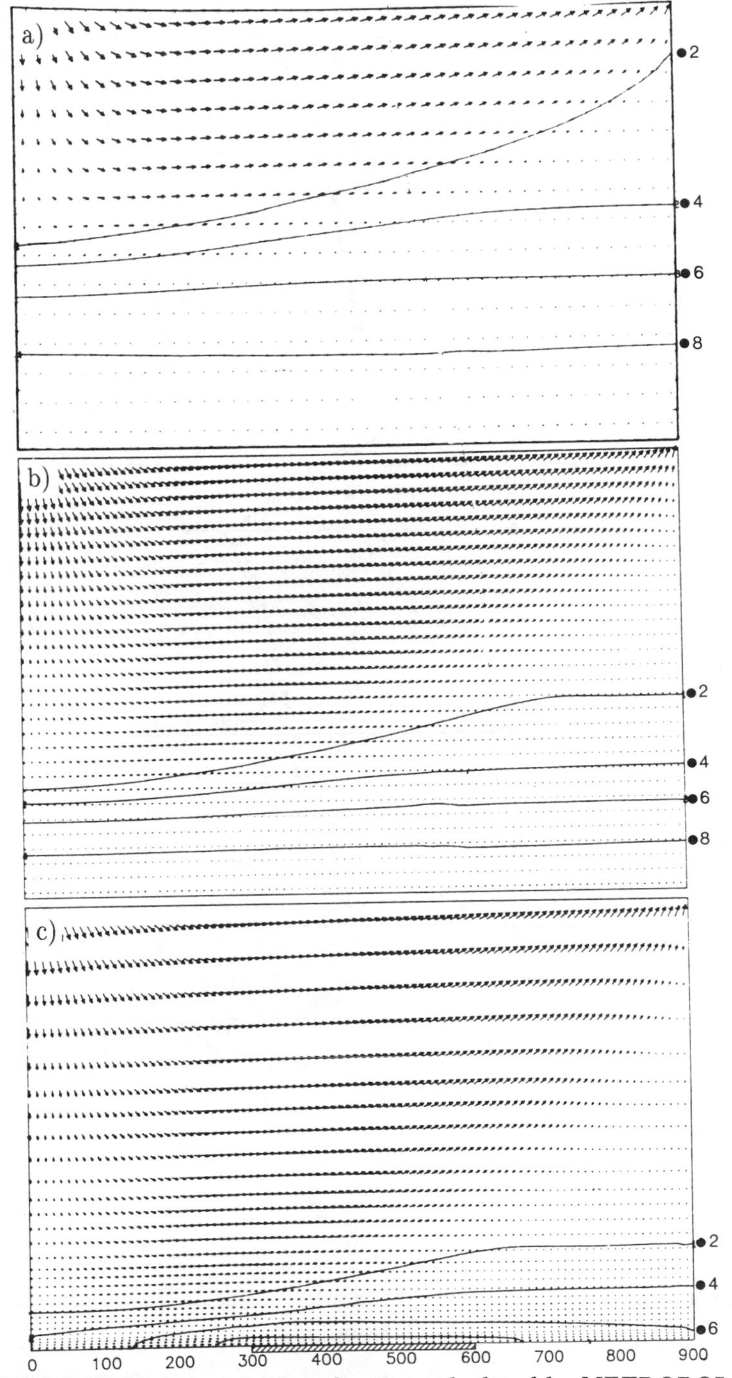

Figure 4. HYDROCOIN Case 1.5, Mass fraction calculated by METROPOL–3: a) regular mesh, 675 elements b) regular mesh, 2700 elements, c) irregular mesh, 2700 elements.

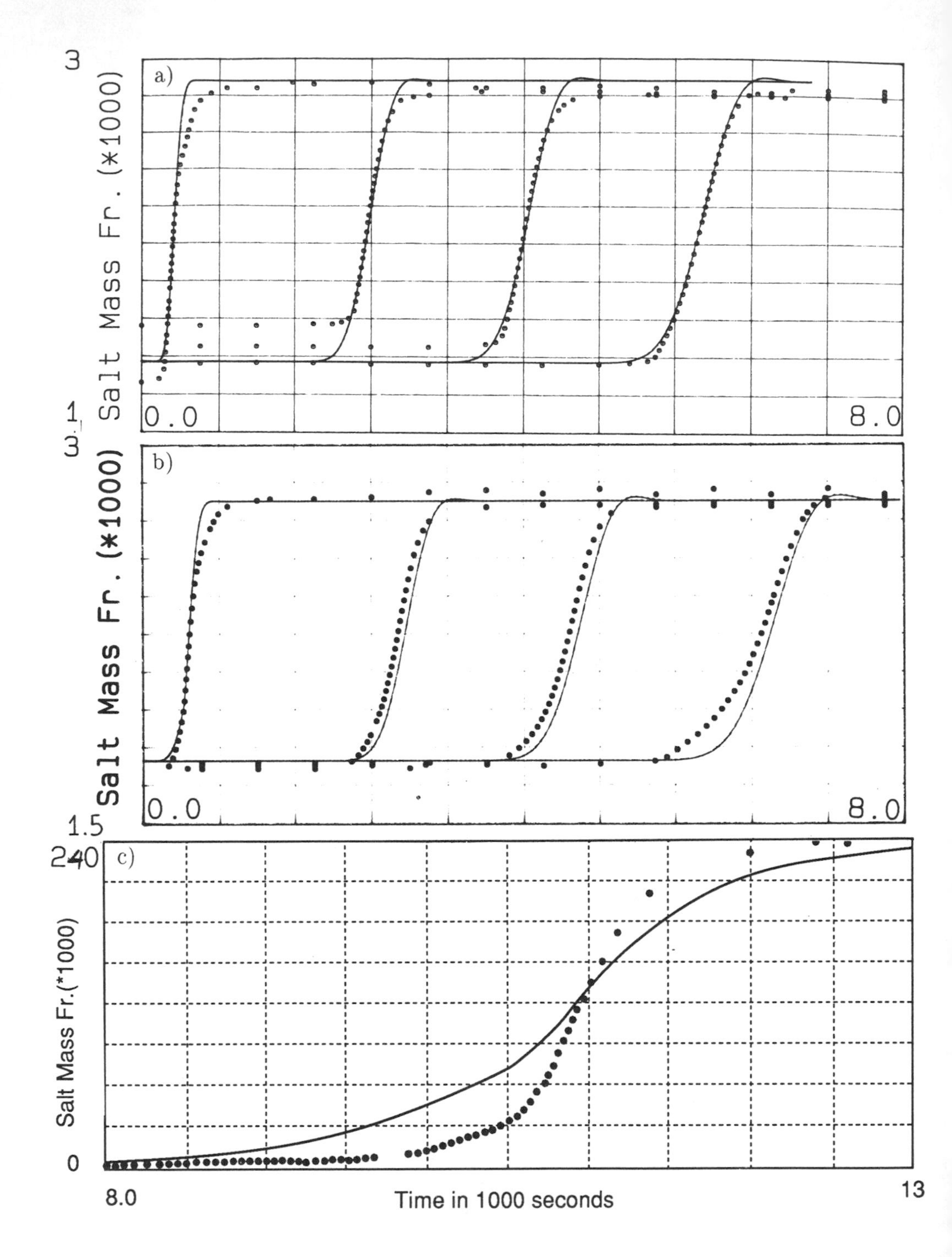

Figure 5. Simulation of brine transport experiment with METROPOL–3: a) and b) low–concentration experiments L1D01 and L1D02, c) high–concentration experiment H1D02

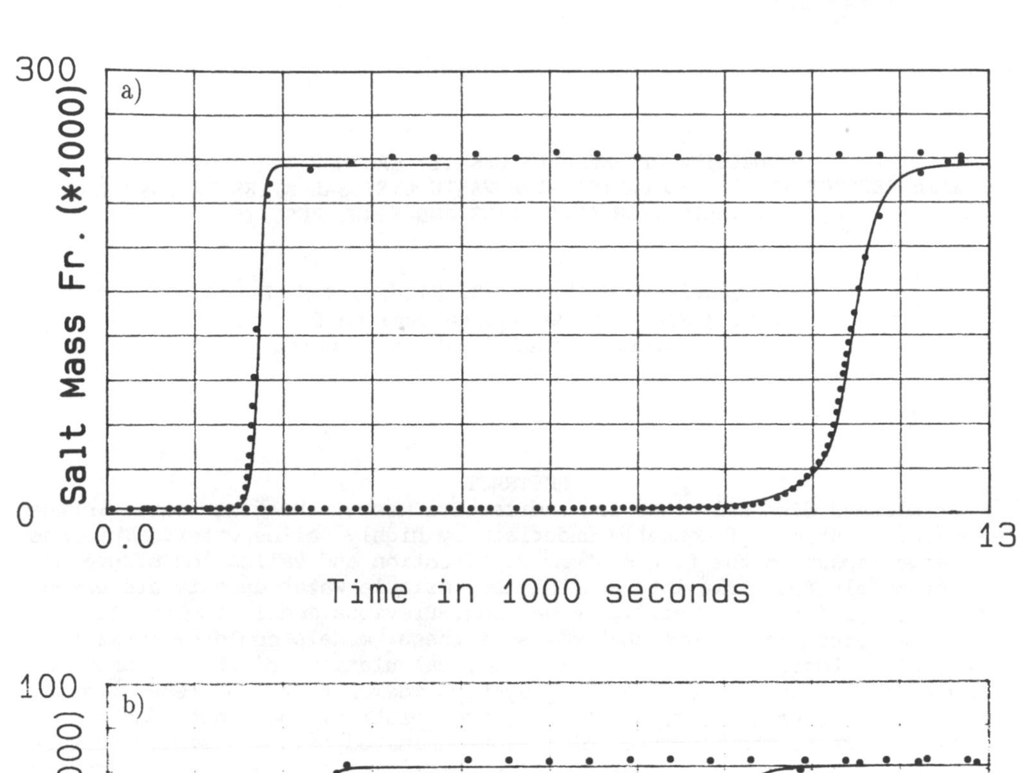

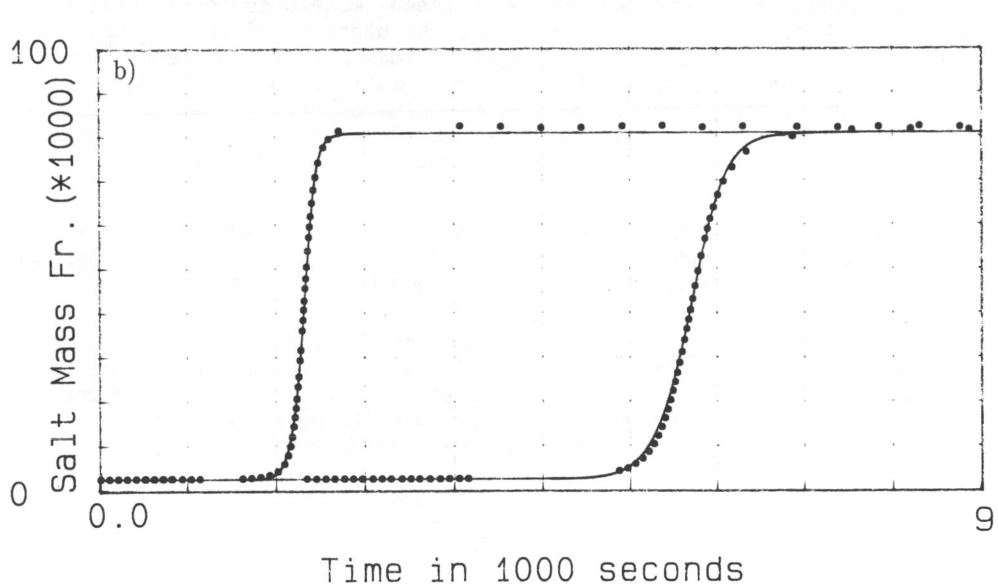

Figure 6. Simulation of high–concentartion–gradient experiments with SPRINT incorporating the nonlinear dispersion theory: a)H1D02, b)H1D01.

MODELLING OF VARIABLE-DENSITY GROUNDWATER FLOW WITH RESPECT TO PLANNED RADIOACTIVE WASTE DISPOSAL SITES IN WEST GERMANY - VALIDATION ACTIVITIES AND FIRST RESULTS -

K. Schelkes, P. Vogel, H. Klinge, R-M. Knoop
Bundesanstalt für Geowissenschaften & Rohstoffe
Hannover, Federal Republic of Germany

ABSTRACT

At planned disposal sites for radioactive waste, the fresh water in the aquifer system is frequently underlain by highly saline water. This has a large impact on the flow system. Verification and validation of groundwater models that take into account the variable water density are essential steps for site specific modelling. Previous model results from an INTRAVAL test case prove that parts of these models could be invalid in special situations. On the other hand, calculations of the groundwater movement in deep, layered aquifer systems demonstrate the general suitability of these models, as shown by the relatively good accordance with measured data. They suggest diffusion-dominated salt transport in such systems and indicate changes in flow velocities and pattern, which are important for a long-term safety assessment.

MODELISATION DE L'ECOULEMENT D'EAUX SOUTERRAINES DE DIVERSES DENSITES POUR L'ETUDE DES SITES D'EVACUATION DE DECHETS RADIOACTIFS PREVUS EN RFA - ACTIVITES DE VALIDATION ET PREMIERS RESULTATS -

RESUME

Aux sites prévus pour le stockage des déchets redioactifs, on trouve souvent, dans les systèmes aquifères, des eaux douces au-dessus des eaux fortement salées. Ceci a une grande influence sur le système d'écoulement. La vérification et validation des modèles mathématiques de l'eau souterraine, tenant compte des densités variables de l'eau, sont des étapes essentielles pour une modélisation appropriée aux sites specifiques. Les résultats des modèles antérieures provenant d'un test INTRA-VAL prouvent que ces modèles peuvent être partiellement faux dans des situations particulières. Cependant, les calculs de l'écoulement de l'eau souterraine dans des systèmes stratifiés profonds montrent, à cause de la concordance relativement bonne des valeurs calculées avec celles mesurées, l'utilité générale de ces modèles. Ils suggèrent un transport, surtout par diffusion, du sel dans ces systèmes et ils indiquent des changements dans les vitesses et la distribution d'écoulement qui sont importants pour une estimation de la sécurité à long terme.

INTRODUCTION

The behaviour of the groundwater flow system in the strata surrounding a radioactive waste disposal site in a deep geological formation is one of the most important factors for the long-term safety assessment of the site. At all planned disposal sites in the Federal Republic of Germany, the fresh water in the uppermost parts of the aquifer system is underlain by highly saline groundwater. The salt concentration usually increases with depth and reaches in some parts that of saturated brine, especially near the top or flanks of salt domes and near bedded salt.

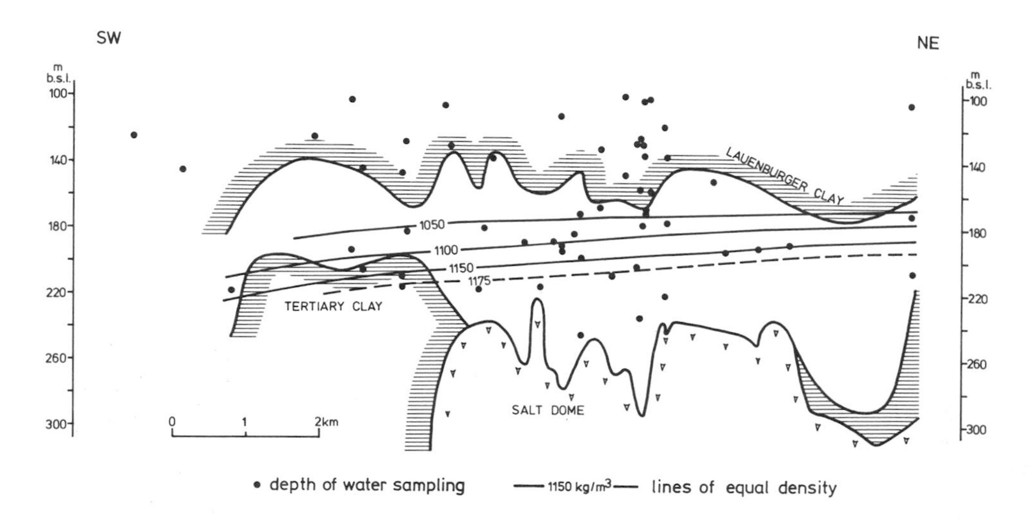

Fig.1: Lines of equal groundwater density in a cross section through the deeper aquifer in a subglacial channel crossing the Gorleben salt dome

As an example, Figure 1 shows a cross section through a mainly sandy aquifer at a depth of 100 – 300 m below m.s.l. in a subglacial channel across the Gorleben salt dome. Lines of equal density obtained from measurements on water samples are shown. The densities increase towards the northeast, the direction of freshwater flow in the overlying aquifer. Groundwater movement in such an aquifer system depends to a large degree on its salinity. Therefore, special attention is given to the verification and validation of models that take into account the variable density of groundwater.

In addition to tests on the applicability of some of these models, the effects of the variable density on groundwater movement are being studied. Some of the work for this project is being carried out within the scope of the INTRAVAL study. The validation strategy, especially with respect to the "Gorleben" site, is to progress from laboratory scale to small test areas in the field and ultimately to large test areas. At the same time particular attention is being paid to problems of variable-density groundwater movement in very deep, layered-aquifer systems.

The following discussion can only give an overview of our present applications of variable density models and summarize some results of the model calculations.

ACTIVITIES IN THE INTRAVAL STUDY

The validation activities started with calculations for the laboratory scale experiments within the INTRAVAL project. These 1-D and 2-D experiments are described in [1]. In these experiments, carried out at RIVM in the Netherlands, salt water flowing into the bottom of a column filled with glass beads displaces salt water with a lower salinity. The mass fraction of the salt in the displacing fluid is several g/kg for the low concentration experiments and of up to 236 g/kg for the high concentration experiments.

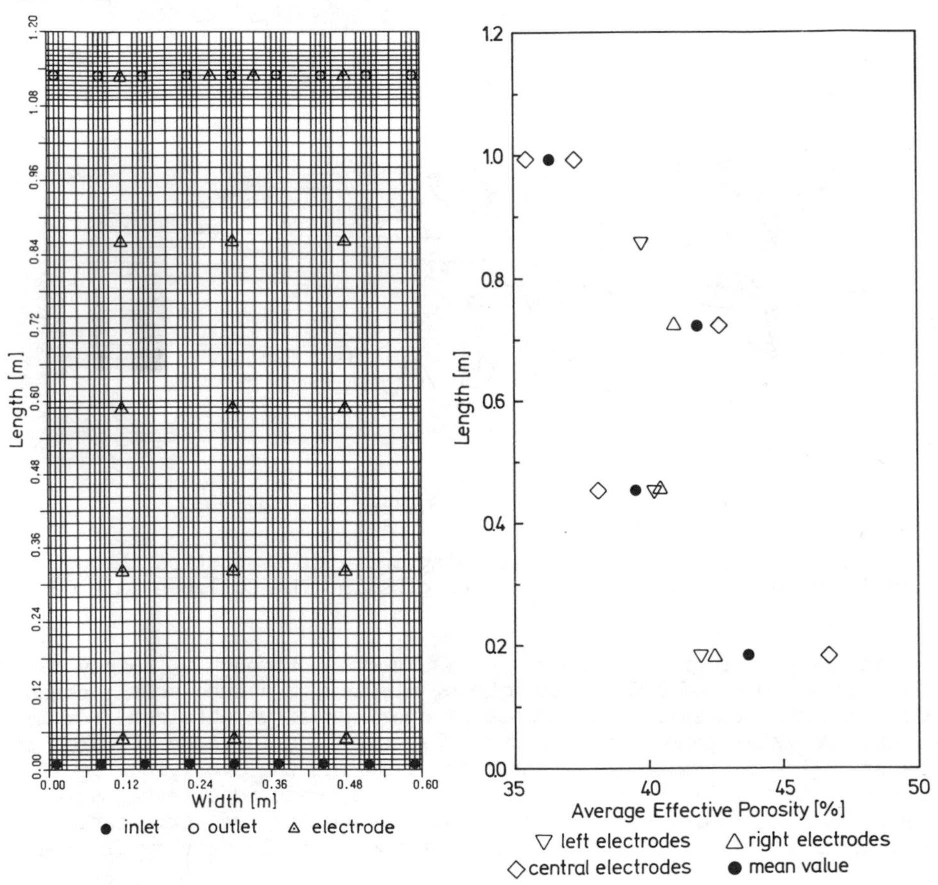

Fig.2: A two-dimensional grid for model calculations on the laboratory experiments showing the positions of inlets, outlets, and electrodes

Fig.3: Effective porosities between electrodes derived from empirical breakthrough curves from a low-concentration experiment

The largest of the grids modeled and the positions of the inlets, outlets, and electrodes of the column are shown in Figure 2. In the model calculation each inlet is represented by one node, temperature is neglected, and the permeability is given a constant value of 1.7×10^{-10} m².

The effective porosities (Fig. 3) for the model calculations had to be derived from measured breakthrough curves from the low concentration experiment. These calculated porosities vary laterally and show a decreasing trend towards the top of the column. This is physically not plausible, but the calculated porosities are only approximate values, that take unknown effects into account, for instance, the exact spacing between the two plexiglass plates of the column.

Two numerical models have been used, the 2-D finite-element model SUTRA [2] and the 3-D finite-difference model HST3D [3]. The two-dimensional model calculations were carried out for different grid sizes using a minimum longitudional dispersivity of 1 mm and time steps of 5 – 20 s. The fit obtained was not considered satisfactory. The slopes of the calculated and measured breakthrough curves more or less agree in the case of the low-concentration experiments, whereas for the high-concentration case, the slopes of the measured curves are much steeper.

Various parameter studies were made to obtain a better fit to the experimental data. For instance, modifications of the SUTRA code making viscosity and density nonlinear functions of mass fraction hardly affect the calculated breakthrough curves, owing to the one-dimensionality of the experiments and the fact that the flow rate is dictated by the boundary conditions.

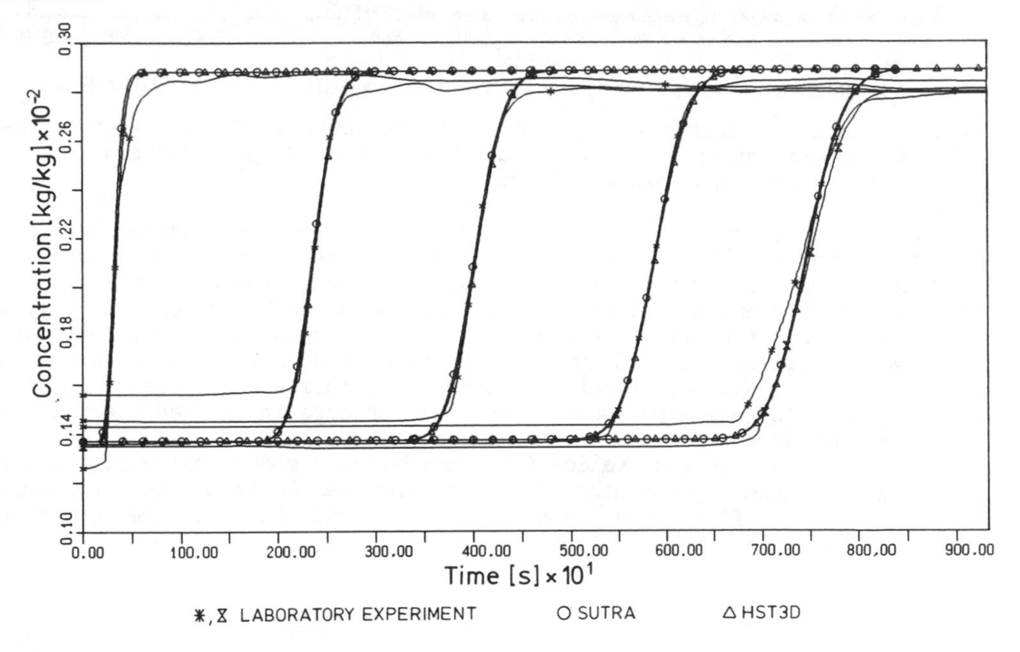

Fig.4: Empirical and theoretical breakthrough curves for a low-concentration experiment (grid length 2 mm, long. disp. 1 mm, time step 0.5 s)

The most effective variation in the parameters was to decrease the time step, which produced a very strong effect on the slope of the breakthrough curves. New calculations with a one-dimensional grid for the low concentration experiment with a grid length of 2 mm, a time step of 0.5 s and a dispersivity of 1 mm gave very good agreement with the empirical data (Fig.4).

The calculations with HST3D showed the same breakthrough curves, but shifted by several seconds owing to differences in the handling of the inflow and observation points (Fig. 4).

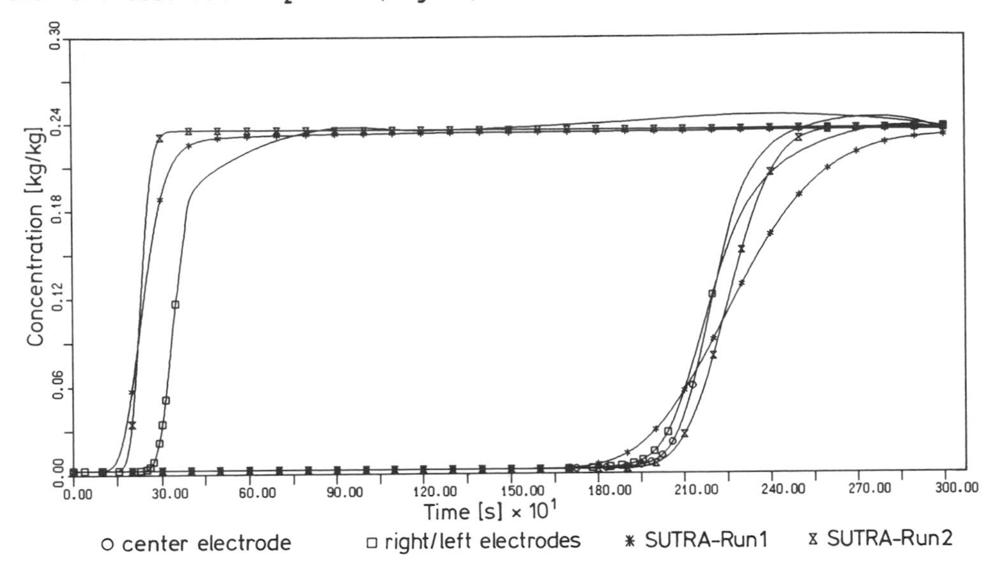

Fig.5: Empirical and theoretical breakthrough curves of a high-concentration experiment (run 1: grid length 2 mm, long. disp. 1 mm; run 2: grid length 0.5 mm, long. disp. 0.25 mm)

The calculations for the high-concentration experiment using best-fit parameter values obtained from the low-concentration experiment were unfortunately unsatisfactory (Fig. 5). Changing of the time step to a smaller value had hardly any effect on the slope of the breakthrough curves. Only a decrease in the dispersivity value gave better results. The result obtained with a dispersivity of 0.25 mm (Fig. 5) gives a good fit to the measured breakthrough curves, but offset. Consequently, the same dispersivity value is not valid for all experiments. Possible changes in the basic equations are being discussed.

The next steps in the validation procedure are model calculations for small-scale field experiments and for saline/freshwater aquifer systems. The database and first model runs for a test case based on the Gorleben situation (see also Fig. 1) is being prepared.

GROUNDWATER MOVEMENT IN DEEP, LAYERED AQUIFER SYSTEMS

Another study area is in the northeastern part of the Federal Republic of Germany. Triassic and Permian salt deposits are overlain by a 2000-m-thick sequence of sediments ranging from Upper Triassic to Upper Cretaceous. The sediments consist of low permeability marls and claystones with intercalations of permeable sandstone and limestone layers. Oolithic iron ores are present in the Upper Jurrassic. The Jurrassic sediments are covered by low permeability Lower Cretaceous claystones up to 500 m thick. Triassic sediments are exposed in a hillsite area in the southern part of the study area.

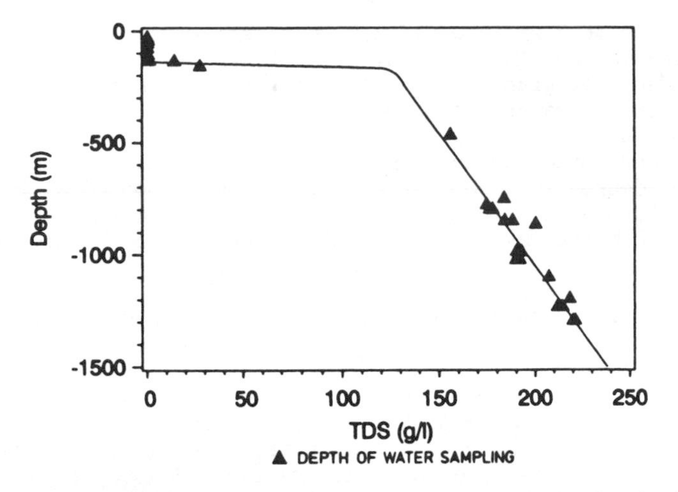

Fig.6: Water salinity (TDS) versus depth derived from chemical analysis of pore water samples

The upper regime of freshwater, 150 – 200 m thick, is controlled by convection and recharge of meteoric water. The deep aquifers contain highly saline connate waters. Studies in an abandoned iron mine indicate a linear increase of porewater salinity with depth from 160 g/l at 500 m to 220 g/l at a depth of 1300 m (Fig. 6).

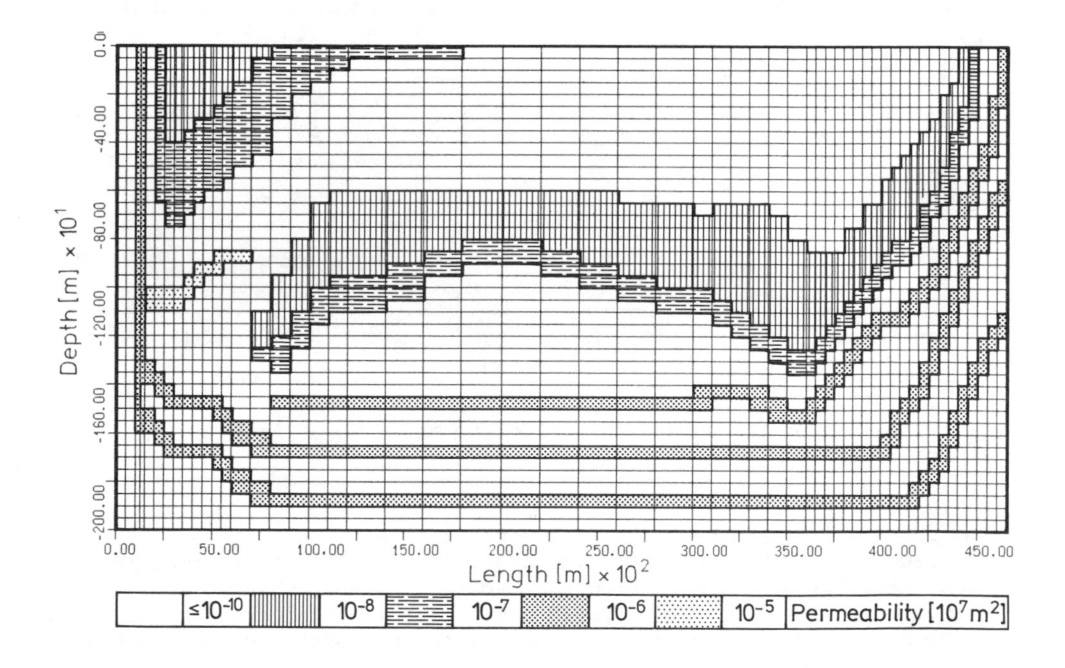

Fig.7: Model grid with permeability distribution for a deep, layered aquifer system

A cross section 46.5 km long and 2000 m deep across the study area was taken for the simulation. The permeability, thickness, and spatial orientation of the hydrostratigraphic units are presented in Figure 7. The hydrologic model comprises several high permeability aquifers surrounded by low permeability aquitards.

The lateral and bottom boundaries are no-flow boundaries. A specified hydraulic head, decreasing from left to right, was imposed at the top of the model. The lateral no-flow boundaries also are no-solute-flux boundaries. A saturated brine was specified along the bottom, representing the dissolving of bedded salt. A solute mass fraction of zero was specified along the top, representing freshwater aquifers near the land surface. Fluid densities were allowed to vary linearly from 1000 to 1200 kg/m³ for solute mass fractions of 0.0 – 0.285. These values correspond to freshwater and saturated brine, respectively.

The SUTRA code was used to study the dynamics of density-driven groundwater flow. Dynamic equilibrium of the salt-water system implies that pressure and concentration do not change with time. Because no steady-state option is available for this code to handle variable density flow, a transient solution method had to be selected. Dynamic equilibrium was approximated by starting from physically reasonable initial conditions and running the model until only small changes in salinity and pressure occurred.

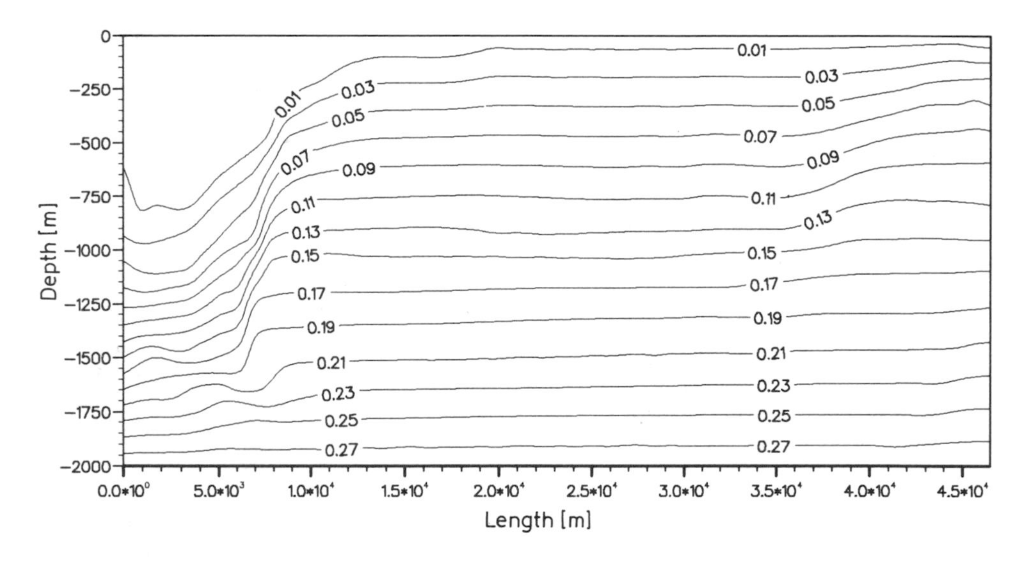

Fig.8: Isolines of the solute mass fraction after about four million years (model time)

The results indicate that the salt-water system is in a state of dynamic equilibrium. The solute mass fraction distribution after approximately four million years is depicted in Figure 8. Apart from a freshwater lens in the recharge area, the salt-water system exhibits a linear increase in density with depth. A comparison of these results with the corresponding freshwater system reveals a major difference in the deep groundwater hydraulics of the two systems. The pure freshwater system is dominated by convective groundwater transport. The salt-water system, however, is characterized by quite different groundwater flow patterns and a reduction of the interstitial velocity by one to two orders of magnitude relative to the freshwater system. These results, as well as the empirical data shown in Figure 6, indicate that in the saltwater system, diffusion is the dominant mechanism of solute transport from the bedded salt to the land surface.

CONCLUSIONS

The calculations of the INTRAVAL study on the laboratory experiment have shown that the empirical data can be reproduced with the existing model codes. They demonstrated the importance of small grid lengths and very small time steps (and therefore extremely large computer memory and very long CPU times) for the calculation of such saline/freshwater systems. The fact that dispersion length for the one system is not applicable to the other system has led to discussion of possible changes in the basic transport equation. But it seems to be important only for very high concentration gradients. Therefore, the calculations carried out for more natural conditions will be hardly affected by these findings.

The first results of the two-dimensional model studies on a deep, layered aquifer system (one example is described here) are in relatively good agreement with the empirical data. The mainly linear increase in density with depth in both the calculated and empirical data suggests diffusion-dominated transport of salt from the bottom of the system. Differences between the two data sets in the salinity gradient to the top could be caused by the neglecting of salt domes on the flanks of the model system. The reduction in the flow velocities and the presence of salt-water convection cells in the saline/freshwater system show their importance for determination of groundwater flow paths and traveltimes in a safety assessment of radioactive waste disposal sites.

REFERENCES

[1] Hassanizadeh, S.M., Leijnse, T., de Vries, W.J., Gray, W.G. : "Experimental Study of Brine Transport in Porous Media", RIVM Internal Report No. 728717006, Bilthoven, The Netherlands, June 1988

[2] Voss, C.I. : "SUTRA: A Finite-Element Simulation Model for Saturated-Unsaturated, Fluid-Density-Dependent Ground-Water Flow with Energy Transport or Chemically-Reactive Single Species Solute Transport", U.S. Geol. Survey, Water Resour. Invest. Rep. 84-4369, 1984

[3] Kipp, K.L. : "HST3D: A Computer Code for Simulation of Heat and Solute Transport in Three-Dimensional Ground-Water Flow Systems", U.S. Geol. Survey, Water Resour. Invest. Rep. 86-4095, 1987

FLOW AND TRANSPORT AT HIGH SALINITY

Georg Arens, Eckhard Fein
Gesellschaft für Strahlen- und Umweltforschung mbH München
Institut für Tieflagerung, Braunschweig, Federal Republic of Germany

ABSTRACT

Salt has been selected as potential host medium for radioactive waste disposal in the F.R. of Germany. It is therefore important to study groundwater flow and radionuclide transport at high salt concentrations. In this case the flow and transport equations of brine are strongly coupled by fluid density and viscosity. As a first attempt to validate our models, we try to simulate a simple laboratory experiment. It has been carried out in the Geotechnique Laboratory of Delft University of Technology, The Netherlands, and is part of the international project INTRAVAL, (test case 13). Due to the experimental set-up our main subject of interest is the dispersive flux of brine. Since the simulations are very time consuming we restricted ourselves to one-dimensional cases. Results of different calculations with the finite-difference codes SWIFT and CHET will be represented. It will be shown that the Fickian ansatz is inadequate to describe dispersion at high concentrations or high concentration gradients, respectively.

ECOULEMENT ET TRANSPORT EN CAS DE SALINITE ELEVEE

RESUME

La République fédérale d'Allemagne a choisi le sel comme milieu d'accueil possible pour l'évacuation des déchets radioactifs. Il est donc important d'étudier l'écoulement de l'eau souterraine et le transport des radionucléides en cas de concentration élevée de sel. Dans cette hypothèse, les équations décrivant l'écoulement et le transport de saumure sont étroitement couplées par la densité et la viscosité des fluides. Pour valider ses modèles, l'auteur essaie dans un premier temps de simuler une expérience simple de laboratoire. Cet essai a été réalisé au Laboratoire de géotechnique de l'Université de technologie de Delft (Pays-Bas) et s'inscrit dans le cadre du projet international INTRAVAL (cas témoin n° 13). Compte tenu du montage expérimental, le principal sujet d'intérêt est la dispersion du flux de saumure. Comme les simulations prennent beaucoup de temps, elles ne portent que sur des cas unidimensionnels. L'auteur présente les résultats de divers calculs obtenus au moyen des programmes SWIFT et CHET basés sur la méthode des différences finies. Il ressort de ces calculs que la théorie de Fick ne rend pas compte correctement de la dispersion à concentrations élevées, ni à forts gradients de concentration.

Introduction

In the F.R. of Germany salt has been selected as potential host medium for radioactive desposal. Since the salinity of groundwater varies from zero, i.e. sweet water at the surface, to saturated brine above the salt dome, it is important to study flow of groundwater under the influence of high salt concentrations. In that case flow and transport equations are strongly coupled by the density and the viscosity of the fluid. This coupling requires an immense numerical expense. In order to have an example as simple as possible without the influence of geological structures and their uncertainties we tried to predict the result of a laboratory experiment.

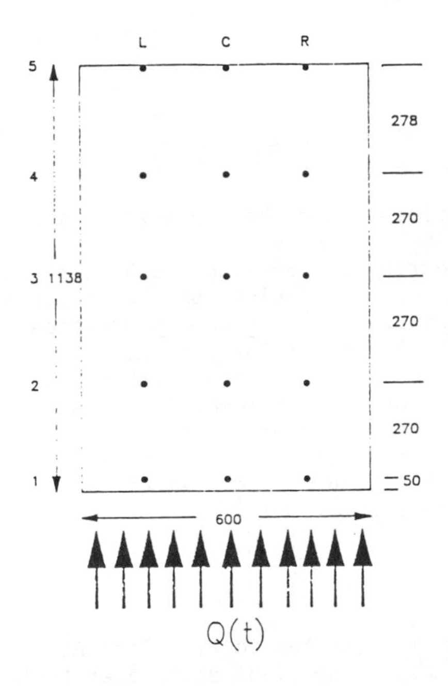

Fig. 1: Schematical set-up of the RIVM experiment.
All dimensions in mm. L, C, R stand for left, central and right electrode, respectively. Q(t) is the time dependent inflow rate.

The experiment was designed and performed in collaboration of the National Institute of Public Health and Environmental Protection (RIVM), The Netherlands, and the Geotechnique Laboratory of Delft University of Technology, The Netherlands /1/. The experimental set-up consists of a column which is filled with glass beads and water of very low salt concentration. Water of higher concentration flows through the column in vertical direction from the bottom to the top. At different locations the time dependent concentration in the column is measured by electrodes. In addition the time dependent inflow rate Q(t) of brine is measured. The setup is schematically shown in fig. 1.
At the bottom of the column there are nine inlet holes which are uni-

formly distributed. If all inlet holes are open we get a good approximation of one-dimensional flow. Otherwise if, for example, only one inlet hole is open we have a two-dimensional flow. Altogether seven experiments with different salt concentrations have been performed. Runs with a small change in concentration are labelled L (low) in contrast to label H (high) which indicates a large change in concentration. As well in low as in high concentration experiments both one- and two-dimensional flows are induced.

experiment label	mass fraction [10^{-3} g/g] initial	final	dimension
L1D01	1.37	2.88	1
L1D02	1.75	2.78	1
L1D03	2.97	3.70	1
L2D01	3.84	4.50	2
H1D01	2.78	81.03	1
H1D02	3.17	235.6	1
H2D01	4.65	176.66	2

Tab. 1: Names of the different experiments and their concentrations

Due to limitations in computer resources we decided, as a first step, to examine the one-dimensional cases only. Calculations for these cases are considerably simpler since Darcy velocity is then proportional to the measured inflow rate. Thus, calculation of Darcy velocity by solution of the flow equation is not necessary, and the effort in solving the coupled flow and transport equations is avoided. To solve the remaining transport equation we still have to determine the values of two parameters: porosity n and the dispersion length α. Assuming that neither porosity nor dispersion length depend on the salt concentration, we determine these parameters by experiment L1D01.

Determination of Parameters

- Porosity n

Consider an electrode with distance z from the inlet holes. At breakthrough time T for that electrode, the volume V of salt water that has already flown into the column is given as

$$V = \int_0^T dt \, Q(t) \tag{1}$$

On the assumption that porosity is only a function of distance from the column's bottom, a second way of calculating V is

$$V = a \int_0^z dz' \, n(z') \tag{2}$$

with the cross section of the column $a = 65.4 \text{ cm}^2$.
Assuming further that n is constant in each of the sections between two rows of electrodes, the porosities of these sections can be obtained from equation (2) where V is known from equation (1).

- Dispersion length α

In experiment L1D01 the inflow rate is nearly constant. For this case there exists an analytical solution of the transport equation:

$$C(z,t) \simeq \frac{C0}{2} \, \text{erfc} \left(\frac{z - ut}{2\sqrt{\alpha\, ut}} \right)$$

with
- mass fraction — $C(z,t)$
- initial mass fraction — $C0$
- fluid velocity — $u(t) = \frac{Q(t)/a}{n}$
- compl. error function — $\text{erfc}(x)$

By these procedures the parameters are determined without solving the transport equation numerically. The porosities n and the dispersion length α as obtained from L1D01 are displayed in tab. 2.

region z [mm]	porosity	dispersion length α [mm]
0 - 50	.416	
50 - 320	.466	
320 - 590	.380	.8
590 - 860	.423	
860 - 1138	.353	

Tab. 2: Porosities and dispersion length determined from L1D01

Computer Codes

Considering the enormous CP times which are needed by the well known computer codes like SWIFT, CFEST and SUTRA in solving the coupled equations, we decided to use the transport module of the code CHET (**CHE**mistry and Transport) which is under development in our institute. CHET is an one-dimensional FD code which uses backward-difference and forward-difference approximation in space and time, respectively. In our version neither adsorption nor radioactive decay are taken into account. In order to verify this version of CHET we compared results of CHET and SUTRA and obtained excellent agreement.

Results

All calculations are performed with the following grid block length and time steps:

$\Delta z = .25$ mm $\qquad Pe = \Delta z / \alpha \simeq .3$

$\Delta t \leq .1$ s $\qquad Co = |u| \, \Delta t / \Delta z \leq .1$

In fig. 2 our results for case L1D01 are displayed. The agreement with the experimental data is not surprising since our parameters have been determined using this set of data. The experimental results of L1D02 are compared to our predictions in fig.3. Except for the first electrode, simulation and experiment disagree as is in L1D03 which is not shown in this paper. Disagreement, however, is restricted to the breakthrough times, while the steepness of the curves seems to fit. So we conclude that the dispersion length is of the right order of magnitude at least for the low concentration experiments.

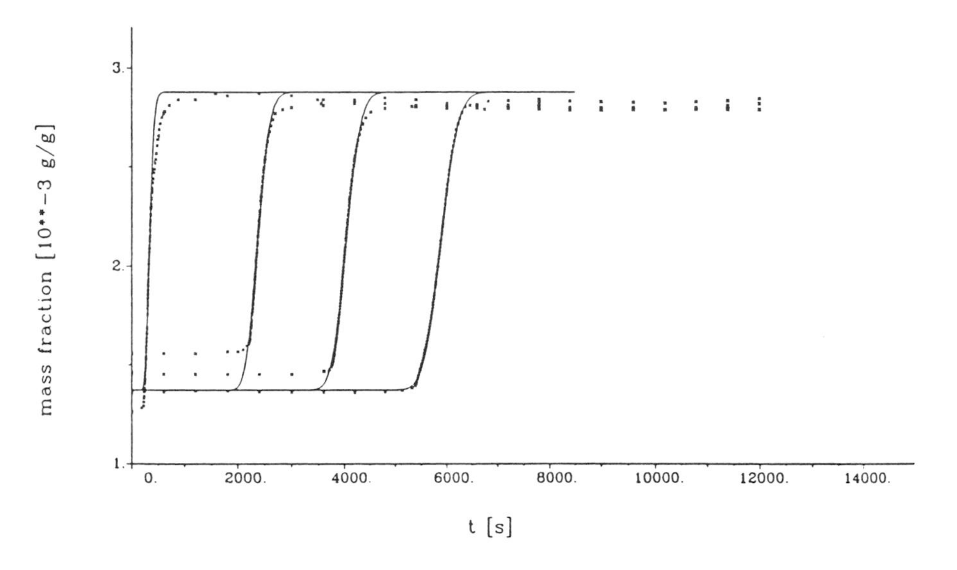

Fig. 2: Experiment L1D01, CHET calculations (solid line), exp. data (x)

Fig. 4 and fig. 5 show the results of the cases with high concentration H1D01 and H1D02, respectively. Now neither the breakthrough times nor the dispersion length are describing the experimental data. We repeated the simulations of L1D01 and H1D02 with the computer code SWIFT. The latter results are shown in fig. 6 and agree completely with our former results. To be sure that we got rid of numerical dispersion we repeated the calculation of H1D02 with SWIFT by using a blocklength of Δz = .125 mm. The differences are negligible. Since there are no numerical errors in our simulation we conclude that at high concentrations we are not able to predict the experimental data by using the Fickian ansatz for dispersion.

Conclusion

Since we only wanted to validate the transport equation at least in the one-dimensional cases, we do not worry at the moment about the wrong breakthrough times. Maybe one has to fit the porosities for each experiment separately. But one should keep in mind that one has to explain this dependence. Since the Fickian ansatz for dispersion is not able to describe the experiments under the condition of high salt concentra-

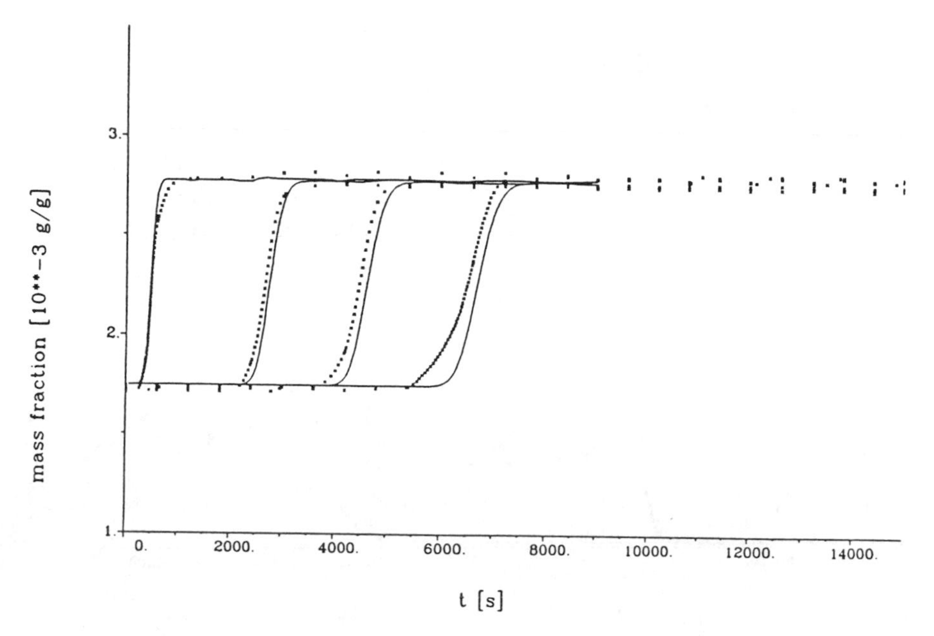

Fig. 3: Experiment L1DO2 (low concentration), CHET simulation

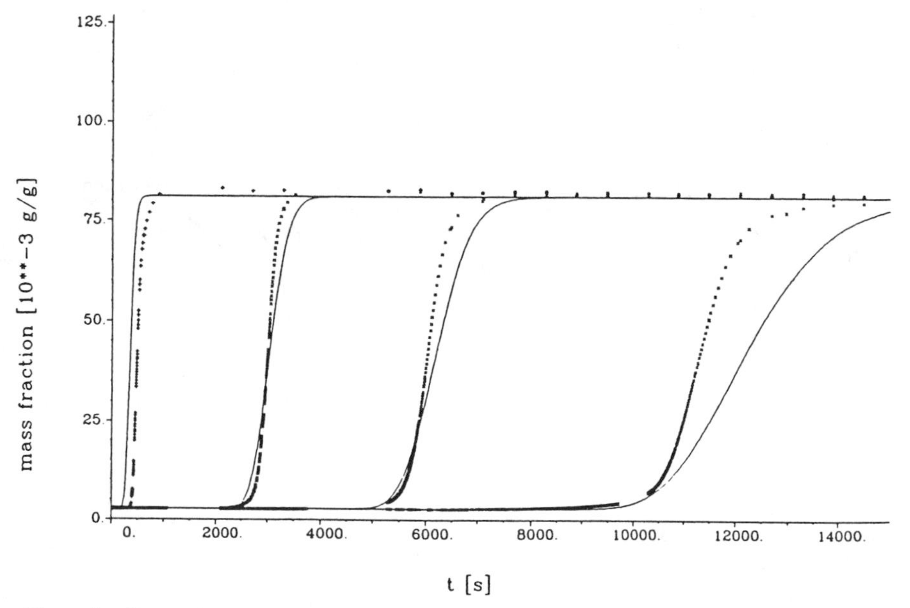

Fig. 4: Experiment H1D01 (high concentration), CHET simulation

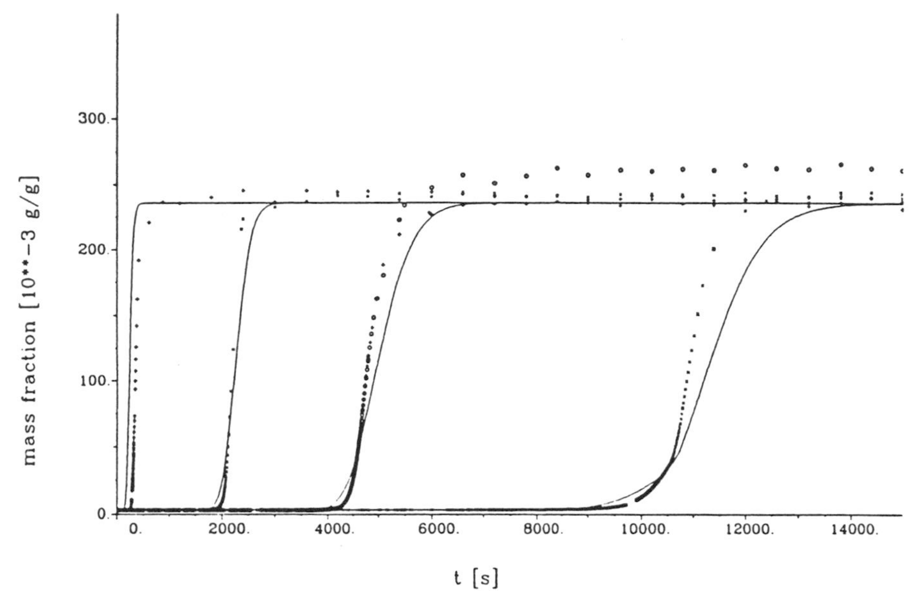

Fig. 5: Experiment H1D02 (high concentration), CHET simulation

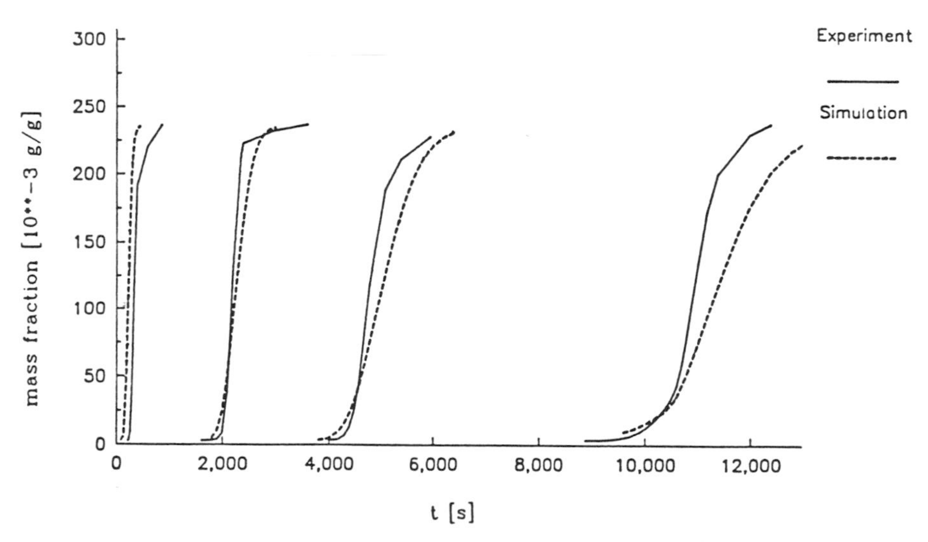

Fig. 6: Experiment H1D02 (high concentration), SWIFT simulation

tions we have to improve Fick's law. Additional terms in the transport (and flow) equation, which have been proposed by M. Hassanizadeh /2/ do not seem to be helpful as is shown by himself at the Bilthoven meeting of the Salt Working Group of INTRAVAL /3/. So far dispersion length is a property of the porous medium only. Similiar to the case of viscosity for example, a possible dependence on fluid properties may solve the problem. Thus dispersion length will become a function of concentration or of concentration gradient, respectively. Since there is no evidence for a the functional relation $\alpha = \alpha(C, \nabla C)$, we need more and better data. We therefore propose to repeat the experiments with an improved experimental set-up.

References

/1/ Hassanizadeh S.M., Leijnse T., de Vries W. J., Gray W. G.
Experimental Study of Brine Transport in Porous Media
RIVM Internal Report 728717006, June 1988, first draft

/2/ Hassanizadeh S.M. Derivation of basic equations of mass transport in porous media, Part 1. Macroscopic balance laws
Adv. Water Resources, 1986, Volume 9, December, 196 - 206
Part 2. Generalized Darcy's and Fick's law
Adv. Water Resources, 1986, Volume 9, December, 207 - 222

/3/ Minutes of the Third Meeting of the Salt Working Group of INTRAVAL
Bilthoven, 1-2 November 1989

MIGRATION EXPERIMENTS IN THE UNDERGROUND FACILITY AT MOL TO VALIDATE SAFETY ASSESSMENT MODEL

M. Monsecour, M. Put, A. Fonteyne
SCK/CEN, Mol, Belgium

H. Yoshida
PNC, Tokyo, Japan

ABSTRACT

To validate predictive calculations for the migration of water (HTO) in a deep clay formation a large scale in-situ migration experiment has been set up in the underground facility installed in the Boom clay at the Mol site. A piezometernest containing several filters at 1 m intervals is emplaced in a single borehole drilled from the gallery of the underground laboratory. Tritiated water is injected through one of the filters and its migration is followed by sampling the porewater in the neighbouring filters. The results obtained since the start of the experiment two years ago are compared with the predicted values.

EXPERIENCES DE MIGRATION REALISEES DANS LE LABORATOIRE SOUTERRAIN DE MOL POUR VALIDER LE MODELE D'EVALUATION DE LA SURETE

RESUME

Pour valider les valeurs prédites par calcul pour la migration de l'eau (THO) dans une formation argileuse profonde, une expérience "in-situ" a été réalisée dans le laboratoire souterrain installé dans l'argile de Boom au site de Mol. Un nid piézométrique contenant plusieurs filtres, espacés de 1 m, est placé dans un seul trou de forage creusé à partir de la galerie du laboratoire souterrain. De l'eau tritiée est injectée à travers un des filtres et sa migration est suivie par échantillonnage de l'eau interstitielle dans les filtres adjacents. Les rsultats obtenus depuis le début de l'expérience, il y a deux ans, sont comparés avec les valeurs prédites.

1. INTRODUCTION

Since 1974 a Belgian research and development programme is carried out by SCK/CEN on the geological disposal of radioactive waste. Its main objectives are to develop the concept and the techniques, to demonstrate the feasibility and to assess the safety and performance of the geological disposal of conditioned radioactive waste in a deep argillaceous formation.

The parameters required in safety assessment studies are determined in the laboratory on clay samples taken from the underground. Different types of migration experiments are performed on the samples. Reshaped and reconsolidated clay plugs are used for flow through type diffusion experiments [1,2]. Clay cores drilled parallel with and perpendicular to the stratification of the formation are used for percolation experiments which revealed the anisotropic properties of the clay formation. The values of the migration parameters obtained from the small scale laboratory experiments are used for long term safety assessment calculations.

Despite the precautions taken during sample collection and diffusion experiments, parameters determined in the laboratory are subject to uncertainty because removal of samples from the ground might produce irreversible changes.

To improve the confidence in the laboratory data and to validate the safety assessment model a large scale in-situ experiment has been set up.

In 1982 SCK/CEN has build an underground research laboratory (URL) in the Boom clay formation underlying the Belgian nuclear research facilities at Mol at a depth of 220 meter. Figure 1 shows the underground facility as build up to date. In 1985 a piezometernest has been installed in the clay formation through access hole No. 1 in the concrete plug (CP1) at the right end of the URL.

The aim of this piezometernest was to get a good estimation of the in-situ hydraulic parameters of the clay formation and to start a large scale in-situ migration test. The purpose of the large scale migration test is to validate the safety assessment model and the migration parameters used for the long time predictive calculations. The experiment is a joint effort between SCK/CEN and the Power Reactor and Nuclear Fuel Development Corporation (PNC) of Japan.

2. DESCRIPTION OF THE EXPERIMENT

Figure 2 gives a schematic representation of the piezometernest installed in the URL. The all stainless steel system contains 9 piezometers interspaced by 0.9 m tubes. Each piezometer consists of two concentric tubes, the outer one being made of sintered stainless steel (L = 85 mm, ϕ = 46 mm). The distance between the filters from center to center is 1 meter. A standpipe of 2 mm inner diameter is connected to the space separating the concentric tubes. The whole system is assembled in the gallery before it is mounted in a borehole.

A horizontal borehole with a diameter of 50 mm and a depth of 10 m is drilled in the clay formation by rotary drilling, the cuttings being removed by compressed air. Immediately after drilling, the completely assembled piezometernest is pushed in the borehole and inert gas is flushed through the filters to prevent oxidation of the clay. After ∿ 2 days the small gap separating the tubing and the borehole wall is completely sealed by convergence creep of the clay ; the gas flow stops automatically as soon as the porewater pressure exceeds the gas pressure. The clay formation acts as a packer and isolates each of the 9 piezometers. The pressure build-up due to the water rising in the piezometer is measured by Bourdon manometers coupled with snap-tite connections to the 2 mm standpipes. The presence of the underground laboratory at atmospheric pressure creates a hydraulic pressure gradient driving the porewater towards the URL. Figure 3 shows the porewater pressure distribution as a function of distance from the URL lining.

From 20.01.1988 to 10.03.1988, a stainless steel vessel containing 925 Mega Becquerel of HTO has been connected between filters 2 and 5 of CP1. A pressure difference of 0.35 MPa drives the tritiated water in the direction of filter 5 at a flowrate of 5 ml per day. The migration of HTO in this large scale experiment is 3-dimensional. Monitoring of the tracer is accomplished by sampling the porewater in the adjacent filters 4 and 6, situated at one meter intervals from the point of tracer injection.

3. PARAMETER VALUES USED FOR THE PREDICTIVE CALCULATIONS

The migration of radionuclides in the Boom clay formation is described by a transport equation including the following main parameters for the species under consideration [1] :

Parameter	Symbol	Value	Unit
- hydraulic conductivity	K	$= 3.2 \times 10^{-12}$	m/s
- hydraulic gradient	dh/dx	$= -18.9$	m/m
- diffusion accessible porosity	η	$= 0.35$	
- dispersion length	a	$= 0.002$	m
- retardation factor	R	$= 1$	
- half live HTO	T	$= 12.28$	years
- Darcy velocity	V_d	$= 6.0 \times 10^{-11}$	m/s
- convection velocity	$V_x = V_d/\eta R$	$= 1.7 \times 10^{-10}$	m/s
- dispersion constant	D_x	$= 4.1 \times 10^{-10}$	m^2/s
- dispersion constant	D_y	$= 4.1 \times 10^{-10}$	m^2/s
- dispersion constant	D_z	$= 2.0 \times 10^{-10}$	m^2/s

The diffusion parameters have been determined in flow through type diffusion experiments on reconsolidated clay plugs [2]. The apparent dispersion constant and the dispersion length have been measured by percolation experiments on clay cores drilled parallel and perpendicular to the stratification of the Boom clay formation [3]. The dispersion constant perpendicular to the stratification is about half the value of the one in the direction of the stratification.

The hydraulic conductivity has been determined in-situ in the piezometernest used for migration experiments. The Darcy velocity is calculated from the pressure profile (Figure 3) and the knowledge of the hydraulic conductivity.

For the Boom clay the following relation exists between the diffusion accessible porosity η and diffusion constant D [2]

$$\eta RD = 6.84\times10^{-10} \times \eta^{3/2} \qquad (m^2/s)$$

The value used for the diffusion accessible porosity is 0.35 at an estimated in-situ consolidation pressure of 2.4 MPa.

4. COMPARISON OF THE CALCULATED AND THE EXPERIMENTAL CONCENTRATION PROFILE

For the design of the experiment 3 dimensional simulations have been done with the MICOF program [4]. The first aim of the simulation was to verify if a migration experiment with a non-retarded tracer (HTO) on the meter scale was possible within a reasonable time scale.

Figure 4 gives the results of the simulation of the concentration in the interstitial liquid, at a distance of 1, 2 and 3 meter from the injection point, as a function of time. The figure shows that after a time of about 1.2 year the concentration at the first filter at a distance of 1 meter from the injection point increases above the detection limit. For the second filter at a distance of 2 meter it takes about 4.9 years, and for the third filter we have to wait 12 years.

To appreciate the sensitivity of the results to the diffusion accessible porosity, calculations have been done for values of 0.3, 0.35 and 0.40. The results are given on the figure.

The figure shows a very good agreement between the predicted values and the experimental points measured in filter number 4. It gives confidence in the conceptual model used for the calculations and in the value of the parameters determined in laboratory experiments.

The activity profile measured in filter No. 6 is identical to the one observed in filter No. 4, with the exception of the results obtained in the beginning of the experiment. Tritium concentrations of 150 - 33 - 3.7 and 0 $Bq.l^{-1}$ have been measured after respectively 82 - 138 - 209 and 306 days since the start of the experiment. These unexpected results are supposed to be due to a slight contamination occurring during the loading of the tracer.

Figure 5 gives the sensitivity of the calculated concentrations in the liquid to the value of the hydraulic conductivity. The figure shows clearly that the in-situ determined value of 3.2×10^{-12} m/s matches very good with the experimental results. The results of the simulations show a high sensitivity to the value of the hydraulic conductivity.

5. FINAL REMARKS

For the assessment of the safety of deep geological disposal of radioactive waste reliable models and data are needed. The reported comparison of the results of a large scale in-situ migration experiment and the 3-dimensional simulations with the MICOF programme shows that good predictive calculations are possible. Laboratory data obtained from short term experiments (a few days) on small-scale samples (30 cm3) prove to be reliable for the prediction of a field experiment lasting for several years and involving several tens of m^3 of clay.

The experimental results at a distance of 1 meter have been reported, while those at a distance of 2 meter are expected within a few years.

Five other piezometernest have been installed but have not yet been loaded with tracer. Two of them are intended for the in-situ determination of the anisotropy of permeability and diffusion parameters (boreholes drilled perpendicular and parallel to the bedding plane). Two others were installed in a combined heat and radiation field. Preliminary tests are actually in progress to measure the hydraulic properties and to evaluate if any disturbance of the migration pattern occurs due to injection and sampling procedures.

6. ACKNOWLEDGEMENTS

J. Mermans is gratefully acknowledged for the tritium measurements. The financial support for this work was provided by NIRAS/ONDRAF and CEC.

7. REFERENCES

[1] Put M. and Henrion P. : "Chemistry and Migration Behaviour of Actinides and Fission Products in the Geosphere", Special Issue of Radiochimica Acta, 1988.

[2] Henrion, P., Put, M., Van Gompel, M. : "The influence of compaction on the diffusion of non-sorbed species in Boom clay", presented at "Migration 89", USA, 1989.

[3] R&D programme on radioactive waste disposal. SCK/CEN, 2nd semester 1989.

[4] Put, M. : "Radioactive Waste Management and the nuclear fuel cycle", Volume 6, No. 3-4 (1985), pp. 361-390.

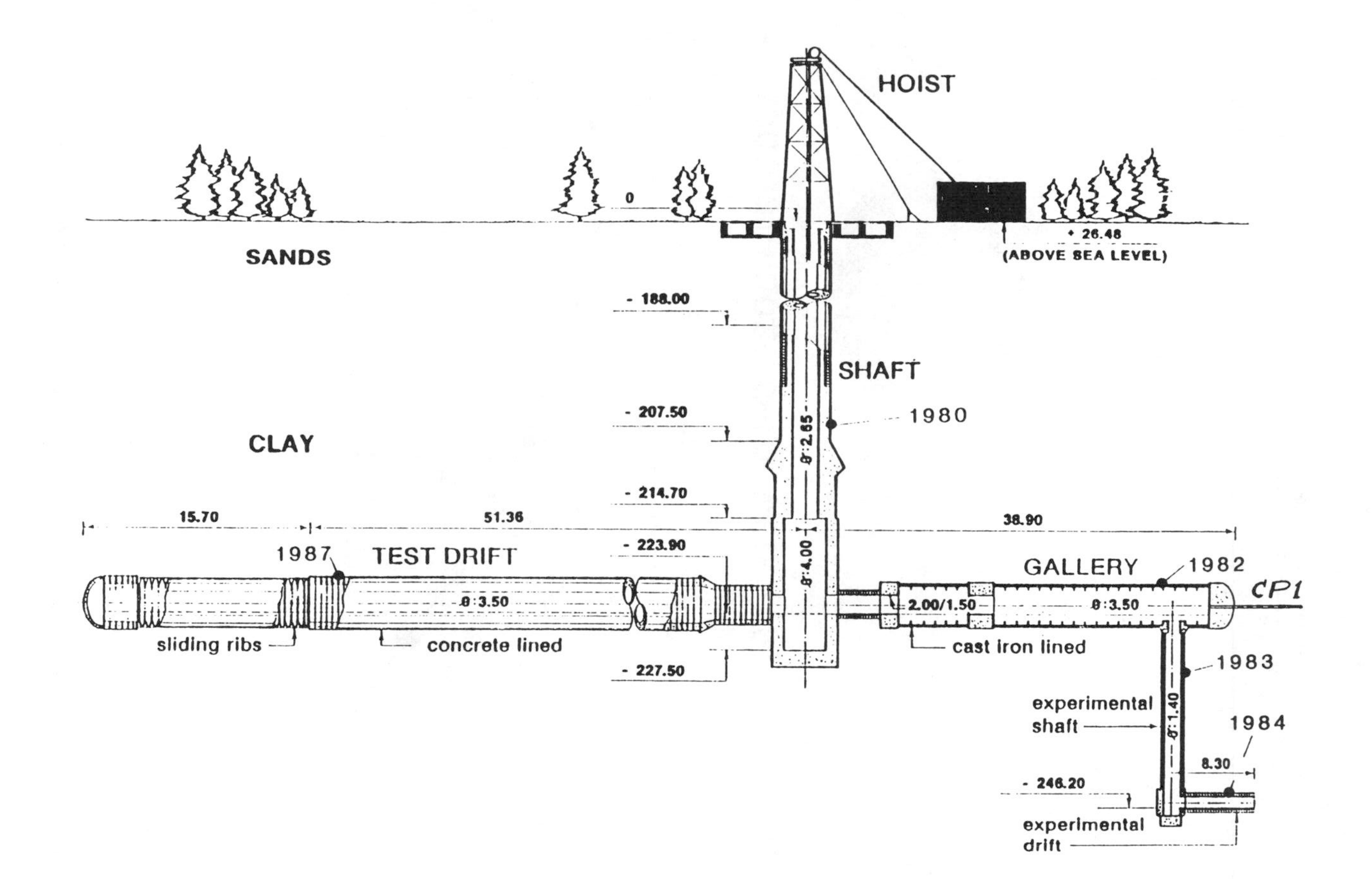

Figure 1: Scheme of the as built underground facility at SCK/CEN MOL Belgium.

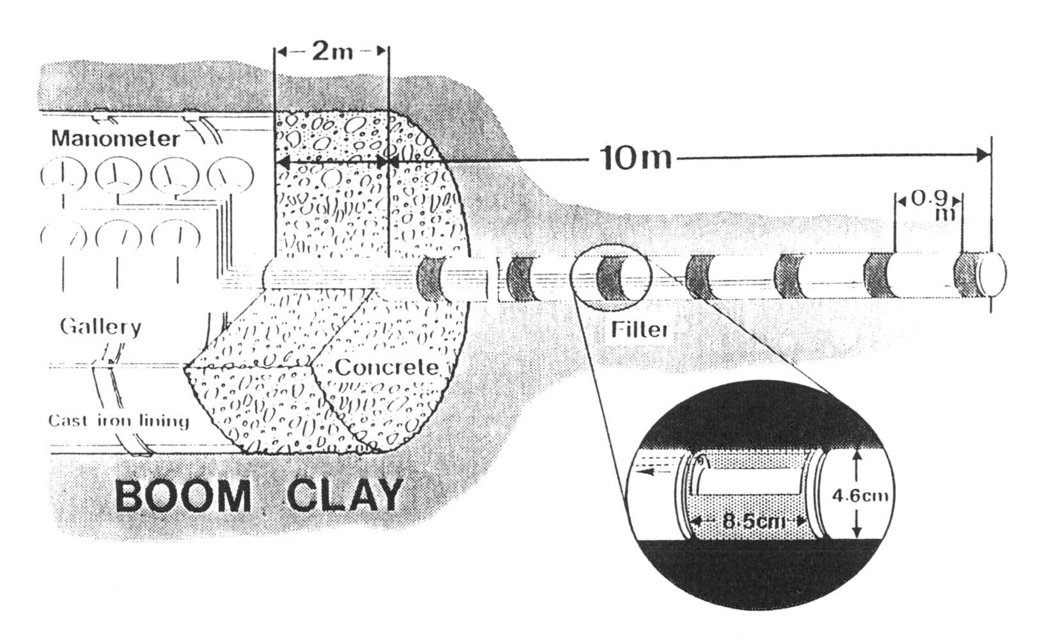

Figure 2: Conceptual view of piezometernest.

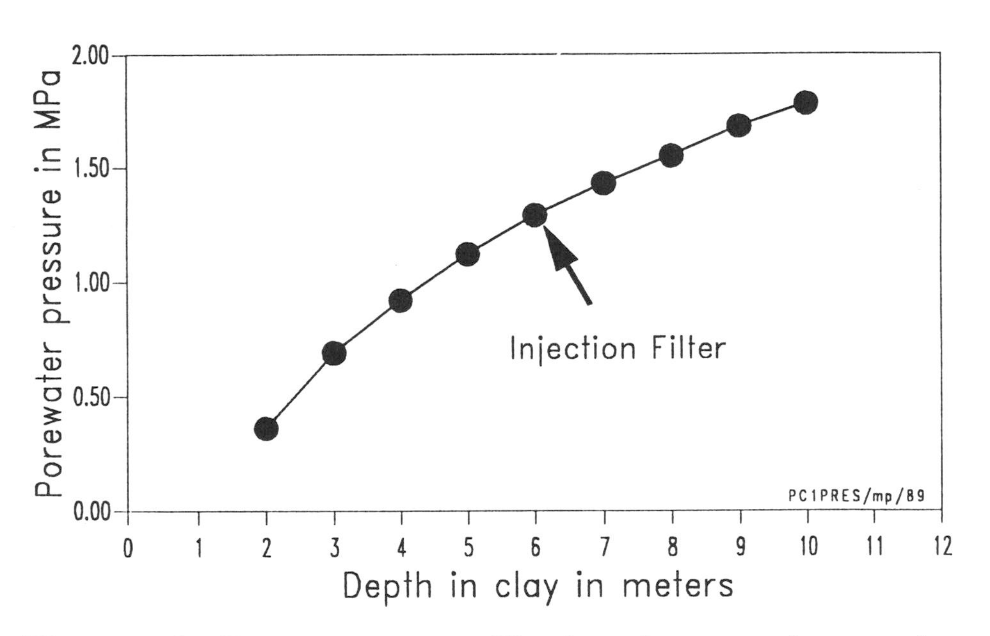

Figure 3: Pressure profile in piezometernest.

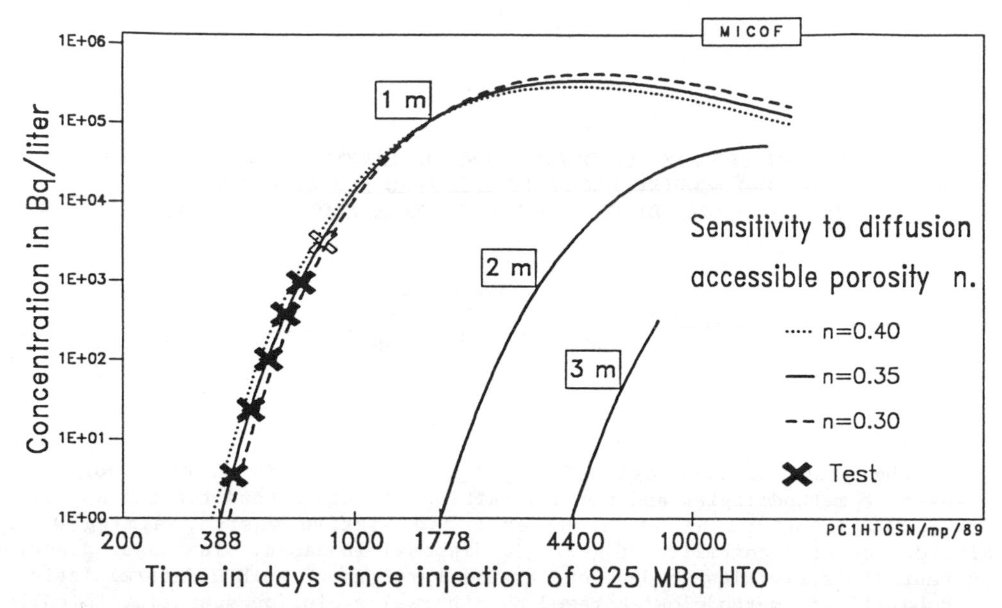

Figure 4: Concentration in the liquid at distances of 1, 2 and 3 meter from the injection point.

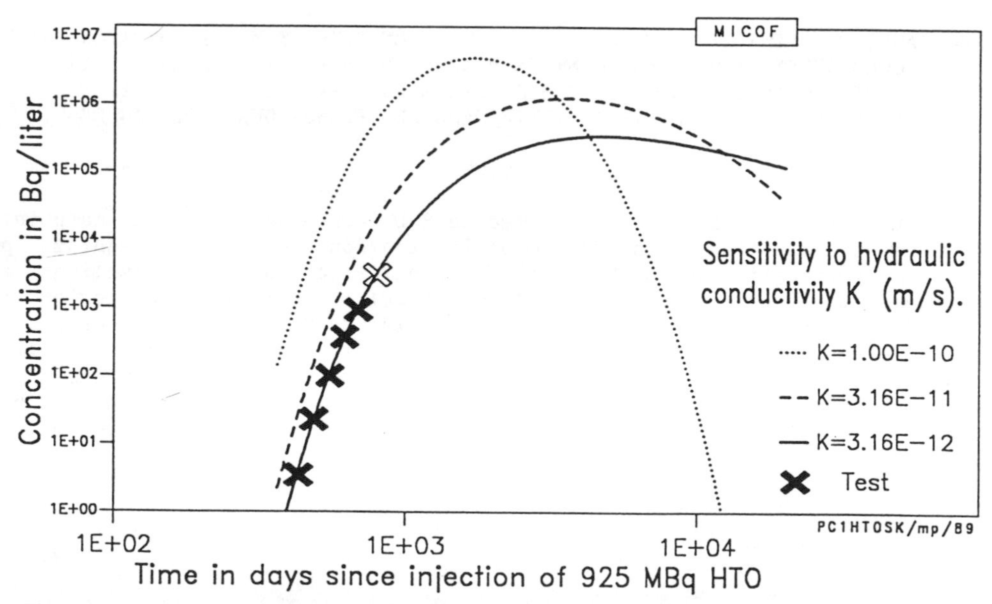

Figure 5: Concentration in the liquid at a distance of 1 meter from the injection point.

THE JAERI PROGRAM FOR DEVELOPMENT OF SAFETY ASSESSMENT MODELS AND ACQUISITION OF DATA NEEDED FOR ASSESSMENT OF GEOLOGICAL DISPOSAL OF HIGH-LEVEL RADIOACTIVE WASTES

Hideo Matsuzuru
Japan Atomic Energy Research Institute
Tokai, Ibaraki, Japan

ABSTRACT

The JAERI is conducting R&D program for the development of safety assessment methodologies and the acquisition of data needed for the assessment of geologic disposal of high-level radioactive wastes, aiming at the elucidation of feasibility of geologic disposal in Japan. The paper describes current R&D activities to develop interim versions of both a deterministic and a probabilistic methodologies based on a normal evolution scenario, to collect data concerning engineered barriers and geologic media through field and laboratory experiments, and to validate the models used in the methodologies.

PROGRAMME DE L'INSTITUT DE RECHERCHES SUR L'ENERGIE ATOMIQUE DU JAPON (JAERI) CONCERNANT LA MISE AU POINT DE MODELES D'EVALUATION DE LA SURETE ET L'OBTENTION DES DONNEES REQUISES POUR EVALUER L'EVACUATION DES DECHETS DE HAUTE ACTIVITE DANS DES FORMATIONS GEOLOGIQUES

RESUME

Le JAERI réalise un programme de R-D concernant la mise au point de méthodes d'évaluation de la sûreté et l'obtention des données requises pour évaluer l'évacuation des déchets de haute activité dans des formations géologiques, l'objectif étant de déterminer si cette forme d'évacuation est faisable au Japon. L'auteur expose les activités de R-D en cours pour élaborer des versions provisoires de méthodologies déterministes et probabilistes fondées sur un scénario d'évolution normale, pour réunir des données concernant les barrières artificielles et les milieux géologiques par le biais d'expériences in situ et en laboratoire et pour valider les modèles utilisés dans les méthodologies.

1. INTRODUCTION

High-level radioactive wastes generated at a nuclear fuel reprocessing plant will be disposed of in a deep geologic disposal system consisting of engineered barriers and geologic barriers, in line of the policy established by the Atomic Energy Commission of Japan. The JAERI is, therefore, conducting R&D program for the development of safety assessment methodologies and the acquisition of data required for safety assessment of the disposal system.

The JAERI is developing interim versions of both a deterministic and a probabilistic assessment methodologies, taking into account a normal evolution scenario. The objectives of the study are:

- to extract the important pathways and parameters in safety assessment,
- to identify the major uncertainties involved in the results of assessment,
- to indicate the effects of various barriers to be involved in the disposal system,
- to assign a priority of future research,
- to provide information which may be used to derive criteria and guidance for safety regulation of the disposal system,
- and to provide methodologies for licensing procedures.

The deterministic methodology involves such models as a source term model to evaluate amounts of radionuclides released from a disposal facility, water flow and solute transport models for transportation of radionuclides in geologic media, a biosphere model and a health effect model. The probabilistic methodology, which can incorporate a distribution density function of the values of model parameters, consists of the models mentioned above but in simplified forms applicable to probabilistic calculations, and various sub-modules for data sampling, sensitivity analysis, uncertainty analysis and so on.

The JAERI is carrying out extensive experimental works to prepare data base needed for safety assessment, to support the development of the models and to provide the data required for model validation studies. Items involved are:

- leaching of radionuclides from a vitrified product,
- long-term stability of a vitrified product,
- corrosion of metals,
- diffusion and adsorption of radionuclides in/on buffer materials,
- fundamental geochemical study on the interaction between geologic media and radionuclides including mineralization of radionuclides,
- fundamental radiochemical study on the behavior of radionuclides in an aquatic solution,
- geological and hydrological field-experiments,
- and natural analogue study.

The JAERI is participating in international co-operative program to enhance R&D activities to develop the safety assessment methodologies and to obtain data for safety assessment works and model validation studies:

- INTRAVAL for the geosphere models,
- PSAC for the probabilistic system assessment methodologies,
- BIOMOVS for the biosphere models,
- and ALAP for natural analogue studies.

2. CURRENT DEVELOPMENT OF SAFETY ASSESSMENT METHODOLOGIES

Deterministic Methodology

A deterministic safety assessment methodology for geologic disposal of high-level radioactive wastes has being developed aiming mainly at the contribution to the preparation of regulatory criteria, standards and guidances, and to the establishment of the methodologies for licensing procedures. A preliminary version of the methodology, which based on a normal evolution scenario, will be completed by the end of 1991. The safety of a disposal system depends on the ability of engineered barriers consisting of a container, a vitrified product and a buffer material, and natural barriers in the geosphere and the biosphere. The maulti-barrier concept of a disposal system assumed here is:

- the waste container plays as a barrier to radionuclide release during the initial period after the closure of a facility,
- the vitrified matrix acts as a barrier to confine radionuclides therein, however, once the container has failed significantly, the matrix may be subjected to a dissolution process which initiates the release of radionuclides to the subsequent barriers,
- a low permeable buffer zone in which radionuclides are transported through diffusion and retention mechanisms plays as a barrier to retard the migration of radionuclides to a host rock,
- the host rock and the surrounding geosphere where most portions of the radionuclides released are effectively retained and delayed act as a predominant natural barrier to retard the transportation of radionuclides to the biosphere,
- and finally the radionuclides which enter the biosphere may be diluted with a large volume of surface water bodies such as ocean, rivers, lakes, and so on before coming in contact with the man.

This concept may realize the isolation of most of the radionuclides contained in the waste during a long time, allow the reduction of radionuclide concentrations at the biosphere, and finally maintain radiological consequences to individuals or populations at acceptable levels, even in the case where the ability of some of barriers would be reduced as is expected in some altered evolutions. The concept should be incorporated in the models used in the methodology, thus the methodology consists mainly of:

- a source term evaluation model for a disposal facility,
- radionuclide transport models for the geosphere,
- a radionuclide transport model for the biosphere,
- and a dose evaluation model.

The disposal facility (engineered barriers) is assumed to be composed of a vitrified product encapsulated in a metallic container including a canister and an over-pack, and a buffer material such as bentonite. The source term evaluation model, therefore, includes the following processes:

- the dissolution of a vitrified matrix controlled by a film-diffusion and a specific solubility, in order to determine a concentration of a specific radionuclide at the interface between the vitrified product and the buffer zone,
- the corrosion of the metallic container to determine the pint of time when the dissolution of the matrix initiates,
- the diffusion-controlled transport of radionuclides through the buffer zone, taking into account retardation processes, to determine the source term to the geosphere.

Geologic media assumed here are homogeneous porous media connected with several dominant fractures which further connect directly with underground water in a sedimentary soil layer or with ocean. The migration of radionuclides in the homogeneous porous media surrounding the disposal facility is simulated by an analytical solution of of a mass transport equation involving one-dimensional advection, three-dimensional dispersion, retention and decay chains. The transport of radionuclides through the fractures modeled as a pipe is analyzed by the analytical solution of the mass transport equation containing one-dimensional advection, one-dimensional dispersion, retention and decay chains. Although the retention involves numerous physico-chemical mechanisms such as reversible and irreversible chemical adsorption, physical adsorption, molecular diffusion into immobile water, filtration, precipitation, aggregation and so on, the identification of all of the mechanisms that take place, however, is so far still worked out. Even if each potential mechanism is successfully identified and modeled precisely, the computation with the model thus developed requires a much greater data base than that currently available (e.g., radiochemical speciation data, kinetic parameters corresponding to each interaction mechanism, geochemical equilibrium data, a complete mineralogic composition of the geologic media, and also the local variation of these data). Therefore, the basic retention mechanisms were not specifically distinguished in the model used here, but their total effects that retard the migration of radionuclides were assumed to be empirically represented by a distribution coefficient, Kd, defined as the ratio of the amount of radionuclide retained on the media to the amount in the solution. The use of Kd in the model is based on the following assumptions that Kd is independent of the concentration of the radionuclide, that each radionuclide migrates independently of the others, and that each retention occurs instantaneously and reversibly.

Although the geosphere model used in the safety assessment methodology is rather simple, the following models have been developed to support the analysis with the simplified model:

- two-dimensional water flow model in the porous media (2DSEEP, SPMIX),
- three-dimensional water flow model in the porous media (3DSEEP),
- two-dimensional solute transport model in the porous media (MIG2DF),
- two-dimensional solute transport model in the fractured media (DBP),
- and three-dimensional solute transport model in the porous media (MIG3D).

Current efforts are being focused on the development of models for water flow and solute transport in fracture network media, and methodologies for a sensitivity analysis an inverse analysis.

The biosphere model combined with the dose evaluation model, which based on a time-dependent compartment model, simulates environmental transport of radionuclides through the environment. The model evaluates the internal doses in a term of the committed effective dose equivalent resulting from both inhalation of contaminated suspended materials and ingestion of contaminated food and drinking water, and the external doses in a term of the effective dose equivalent due to the immersion in gamma radiation fields. As shown in Fig.1, the exposure pathways assumed to be involved are:

- internal exposure from inhalation of suspended materials,
- internal exposure from ingestion of food (farm products, aquatic products) and drinking water,

- external exposure from radionuclides suspended from the surface,
- and external exposure from radionuclides deposited on the surface.

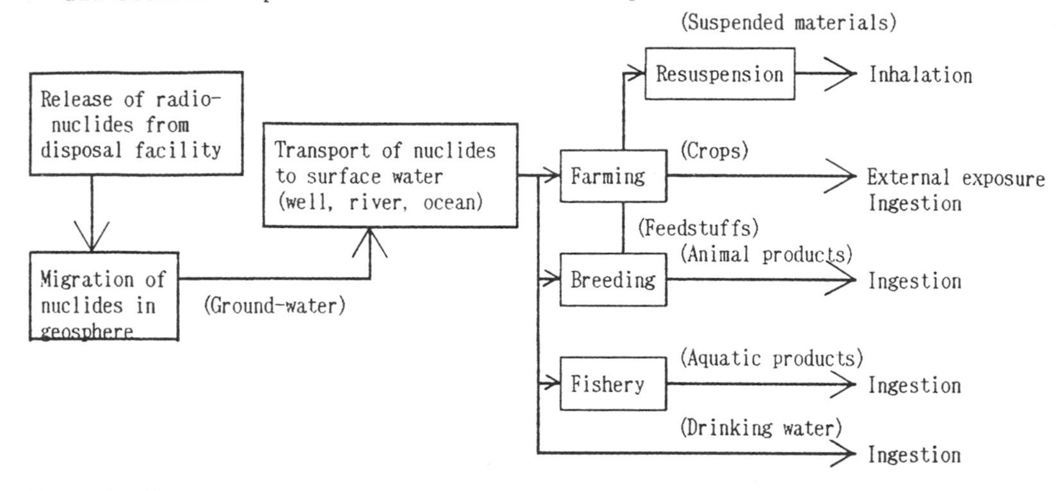

Figure 1 Transport and exposure pathways involved in normal evolution scenario

Probabilistic Methodology

Long term prediction of the safety of the disposal system contains inevitably a significant variability and uncertainty due to altered evolution. For the purpose of safety regulation, therefore, safety analyses are often carried out based on a conservative case or a reasonably conservative case, taking into account the probability of its occurrence. However, the definition of a conservative case or a reasonably conservative case and also its probability of occurrence is highly subjective and is subjected to so-called "an expert judgment". This situation may require a methodology that models explicitly the variability and uncertainty involved in the disposal system so as to reflect them in the results of safety assessment directly. The analysis with the methodology provides a quantitative information for the definition of a conservative case or a reasonably conservative case.

In addition to the deterministic methodology, therefore, we are developing a probabilistic (stochastic) safety assessment methodology that accounts for the variability and uncertainty associated with a long term prediction by incorporating the probabilistic distribution of values of model parameters, aiming at sensitivity and uncertainty analyses. An interim version of the methodology will complete by the end of 1991. The computer code system involves assessment models analogous to those used in the deterministic methodology but in simplified forms applicable to stochastic calculations, and various sub-modules for data sampling, sensitivity analyses, uncertainty analyses, and so on. The code system gives a distribution of estimates of radiological consequence that corresponds distributions of the parameter values.

The probabilistic code system may require many thousands of times of execution to obtain a satisfactory convergence in the consequence distribution, that leads to the use of highly simplified models in the system with sacrifice of physical reality to some extent. This might result in that radiological consequence obtained with the methodology racks the significance in radiation protection aspects. Therefore, the methodology may be mainly used for the following purposes:

- to support the deterministic safety assessment methodology that will be used in licensing procedures,
- to extract the important parameters that have the possibility to affect significantly the result of the assessment,
- to show the major variability and uncertainty involved in the result,
- and to indicate the relative importance of each elementary barrier to be involved in the disposal system.

Figure 2 shows the relationship between the deterministic and the probabilistic approaches.

Characterization of disposal system

Development of scenarios

Probabilistic approach

Deterministic approach

Estimation of probability of occurrence

Selection of a scenario

Specific scenario

Sampling of parameter values

Data base

Selection of parameter values

Assessment models (simplified)

Disposal facility (engineered barriers)

Geosphere

Biosphere

Dose estimation

Assessment models

Disposal facility (engineered barriers)

Geosphere

Biosphere

Dose estimation

Sensitivity and Uncertainty analysis

Risk estimation

Establishment of criteria, satndards, guidances

Safety Assessment

Comparison with a dose limit

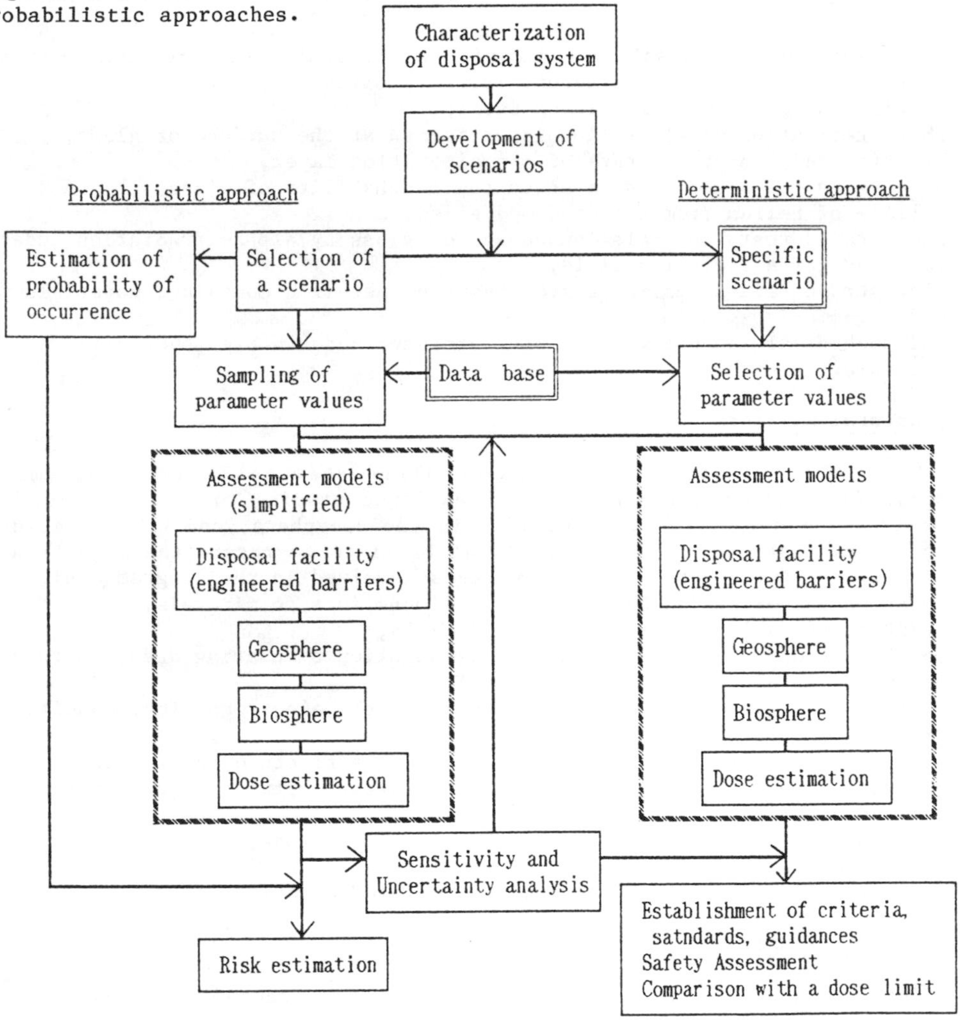

Figure 2 Relationship between the deterministic and probabilistic approaches

3. CURRENT DATA ACQUISITION PROGRAM[1],[2]

The JAERI is conducting extensive experimental works on both engineered barriers and natural barriers to elucidate mechanism that governs the transport of radionuclides in and through man-made materials and the geosphere, and to obtain relevant data. The mechanism thus elucidated supports the development of conceptual and mathematical models to be incorporated in the safety assessment methodologies and to be used in a supplemental analysis. Experimental data obtained contribute to the development of data base needed for the assessment work and to the validation of the models.

Engineered Barriers

Current efforts on safety evaluation studies of engineered barriers are concentrated on a vitrified product. Items involved are:
- static and continuous-flow leach tests,
- characterization of alteration layer formed at the surface of glass,
- kinetic study on growth rate of the alteration layer,
- effects on the alteration layer on the leachability of glass,
- release of helium from the Cm-doped glass,
- accelerated tests on radiation damage of glass by alpha-irradiation under beta and gamma radiation field,
- slow strain rate stress-corrosion tests on metallic container materials under gamma-irradiation,
- and mechanical integrity tests on a canister under a high hydraulic pressure.

Natural Barriers

The JAERI is conducting fundamental geochemical and radiochemical studies, field experiments and natural analogue studies to obtain mechanisms governing transport of radionuclides in the geosphere, and to provide data needed to evaluate the ability of natural barriers to retard the migration of radionuclides in the geosphere. Main items involved in the program are:
- sorption and diffusion behaviors of radionuclides in minerals,
- fixation of radionuclides in geologic media,
- solubility of radionuclides in aqueous solution simulating deep underground water,
- migration of radionuclides in granite under the electrochemical conditions simulating those in deep geologic media,
- migration of radionuclides in simulated single fracture in granite,
- alteration mechanism of minerals and interaction between altered minerals and radionuclides,
- behavior of iron during weathering of a granitic rock,
- geological and hydrological field experiments,
- and natural analogue study at Koongara uranium deposit.

4. Model Validation Study

Current model validation studies are being focused on water flow and solute transport models in the geosphere. The validation studies have being conducted mainly in a realm of international co-operative projects such as HYDROCOIN and INTRAVAL. A generic procedure and a standard for the model validation, especially for complicated geosphere models, is still far from established, and most of the approaches used to date intend to a curve fitting exercise. This is in contrast to model validation studies on the biosphere models which are rather simple than the geosphere models, in which the validation of models and model parameter values is effectively undertaken through a blank test, as is achieved in BIOMOVS.

The reasons of the disagreement, if obtained, between experimental data to be compared and the results predicted by the model to be validate may be ascribes to:

- an inadequate model used,
- insufficient or wrong data to define the initial and boundary conditions for describing the system to be analyzed,
- an unsatisfactory quality of model parameter values,
- and an insufficient quality of experimental data to be compared.

This makes the definition of the concept of model validation for the geosphere models difficult and complex. To make clear the meaning of model validation processes, therefore, any data to be used in the procedure should be subjected to a rigorous quality assurance program.

5. Future Works

The JAERI is conducting R&D program on geologic disposal of high-level radioactive wastes to develop safety assessment methodologies and to develop data base need for the assessment. In addition to these studies, future works are concentrated on:

- development of a scenario analysis methodology,
- and model validation studies, especially on the geosphere models.

REFERENCES

(1) Nakamura, H and Tashiro, S. (Ed.):"Progress Report on Safety Research of High-Level Waste Management for the Period April 1987 to March 1988", JAERI-M 88-201, 1988.

(2) Nakamura, H and Muraoka, S. (Ed.):"Progress Report on Safety Research of High-Level Waste Management for the Period April 1988 to March 1989", JAERI-M 89-192, 1989.

SENSITIVITY ANALYSIS OF GROUNDWATER FLOW MODEL USING THE DIFFERENTIAL ALGEBRA METHOD

H. Kimura
Department of Environmental Safety Research
JAERI, Ibaraki-ken, Japan
A. Isono
Information and Mathematical Science Lab., Inc.
Tokyo, Japan

ABSTRACT

The techiques of sensitivity and uncertainty analyses are necessary for the performance assessment of the waste repositories and validations of mathematical and numerical models to determine key parameters in the models. An automated procedure is developed for performing large-scale sensitivity studies based on the use of computer tools. The procedure is composed of a FORTRAN precompiler called SANA and a preprocessor called PRESANA. The computer code SANA translates a FORTRAN code into its sensitivity analysis code using the Differential Algebra method, and the PRESANA is the computer tool to help the translation of the FORTRAN code. In this presentation, the automated procedure of sensitivity analysis and its application to the 3D-SEEP code which simulates 3-dimensional groundwater flow in the porous media are described. The sensitivity calculations for a sample problem are performed, and the applicability of SANA and PRESANA codes is discussed.

ANALYSE DE SENSIBILITE DES MODELES D'ECOULEMENT DE L'EAU SOUTERRAINE PAR LA METHODE "DIFFERENTIAL ALGEBRA"

RESUME

Les techniques d'analyse de la sensibilité et de l'incertitude sont nécessaires pour évaluer le comportement des dépôts de déchet et valider les modèles mathématiques et numériques en vue d'en déterminer les paramètres essentiels. Les auteurs ont établi une procédure automatique pour réaliser des études de sensibilité à grande échelle fondées sur l'utilisation d'outils informatiques. Cette procédure met en jeu un pré-compilateur FORTRAN appelé SANA et un pré-processeur appelé PRESANA. Le programme de calcul SANA transcode un programme FORTRAN en son programme d'analyse de sensibilité au moyen de la méthode "Differential Algebra", le PRESANA étant l'outil informatique utilisé pour faciliter le transcodage du programme FORTRAN. Les auteurs décrivent la procédure automatique d'analyse de sensibilité et son application au programme de calcul 3D-SEEP qui simule un écoulement tridimensionnel d'eau souterraine dans les milieux poreux. Les auteurs présentent des calculs de sensibilité appliqués à un problème représentatif et examinent les applications éventuelles des programmes SANA et PRESANA.

1. INTRODUCTION

Mathematical and numerical models have been developed to describe physical and chemical processes that are expected to occur in geologic repositories for high-level radioactive wastes. However, the parameters of these models are generally not well known. Sometimes there are not even sufficient data to determine which of these models is applicable. The techniques of sensitivity and uncertainty analyses are necessary for the performance assessment of the waste repositories and validations of the models to determine key parameters in the models.

In the past, analytical methods based on standard statistical method and direct numerical method based on perturbation method have been used for sensitivity analysis. For large-scale problem, however, these methods have computational difficulties. Oblow et al. [1,2,3] developed an automated procedure for sensitivity analysis using computer calculus. The procedure is embodied in a FORTRAN precompiler called GRESS, which automatically processes computer models adding derivative-taking to the normal calulated results. We also developed the automated procedure for performing large-scale sensitivity studies based on the use of computer tools. The procedure is composed of a FORTRAN precompiler called SANA and a preprocessor called PRESANA. The major differences of both procedures can be seen the following points. One is the difference of computational methods to get the derivatives of parameters. GRESS computes the derivatives analytically, while SANA computes the derivatives using the Differential Algebra (DA) method which was developed by M. Berz [4]. Differential algebras are related to the theories of nonstandard analysis, formal power series and automated differentiation. Another difference is an addition of preprocessor computer tool PRESANA to help the translation of FORTRAN codes. In order to precompile FORTRAN codes, preprocessing operations are necessary and require not a few man powers. The reduction of man power is accomplished by the preprocessor PRESANA.

In this presentation, we describe the automated procedure of sensitivity analysis and its application to the groudwater flow model, and discuss the applicability of SANA and PRESANA codes. The 3D-SEEP [5] code developed to simulate saturated-unsaturated groundwater flow in the 3-dimensional porous media was chosen as the target of this procedure.

2. DIFFERENTIAL ALGEBRA METHOD

The DA-method was first applied for the description of beam dynamics. Any order derivatives of parameters can be obtained by this method according to the power and the memory of computer system. Parameters related to sensitivity study in a given FORTRAN program are treated as vector variables (DA variables) in the DA-method. Computation of high order (>2) derivatives requires large memory and "Supercomputing" technique.

A brief description of DA-method is given here with the help of a simple example (0-th and first order derivatives). Consider the vector space R^2 of ordered pairs (a_0, a_1), $a_0, a_1 \in R$ in which an addition and a scalar multiplication are defined in the usual way:

$$(a_0, a_1) + (b_0, b_1) = (a_0 + b_0, a_1 + b_1) \qquad (1)$$

$$t \cdot (a_0, a_1) = (ta_0, ta_1) \qquad (2)$$

for $a_0, a_1, b_0, b_1 \in R$. Components a_0 and b_0 denote 0-th order derivatives (real part), and components a_1 and b_1 denote first order derivatives (differential part). Besides the above addition and scalar multiplication a multiplication and a quotient between vectors are introduced in the following way:

$$(a_0, a_1) \cdot (b_0, b_1) = (a_0 \cdot b_0, a_0 \cdot b_1 + a_1 \cdot b_0) \qquad (3)$$

$$\frac{(a_0, a_1)}{(b_0, b_1)} = (\frac{a_0}{b_0}, \frac{a_1}{b_0} - \frac{a_0 \cdot b_1}{b_0^2}) \quad (\text{if } b_0 \neq 0) \tag{4}$$

Note that the multiplication has a certain similarity to complex multiplication; in fact, the second component of product is exactly the same as in the case of complex multiplication. In the same way as in the case of the complex numbers, one can identify $(a_0, 0)$ as real number a_0. Where in the complex numbers, $(0, 1)$ was a root of -1, here it has another interesting property :

$$(0, 1) \cdot (0, 1) = (0, 0) \tag{5}$$

which follows directly from equation (3). So $(0, 1)$ is a root of 0. Such a property suggests to think about $d = (0, 1)$ as something infinitely small and small enough that its square vanishes. Because of this we call $d = (0, 1)$ the differential unit. After having introduced the basic arithmetic, we will show a remarkable property of this Differential Algebra. As an example, consider the following function:

$$f(x) = \frac{1}{x + \frac{1}{x}} \tag{6}$$

The derivative of the function is

$$f'(x) = \frac{\frac{1}{x^2} - 1}{(x + \frac{1}{x})^2} \tag{7}$$

Suppose we are interested in the value of the function and its derivative at $x = 2$. We obtain

$$f(2) = \frac{2}{5}, f'(2) = -\frac{3}{25} \tag{8}$$

Now take the definition of the function f in the equation (6) and evaluate it at $(2, 1)$. Because the function f is composed of the basic function x, and its derivative at $x = 2$ is 1. We obtain:

$$\begin{aligned} f[(2,1)] &= \frac{1}{(2,1) + \frac{(1,0)}{(2,1)}} \\ &= \frac{1}{(2,1) + (\frac{1}{2}, -\frac{1}{4})} \\ &= \frac{1}{(\frac{5}{2}, \frac{3}{4})} \\ &= (\frac{2}{5}, -\frac{3}{4}/\frac{25}{4}) \\ &= (\frac{2}{5}, -\frac{3}{25}) \end{aligned} \tag{9}$$

As we can see, after the evaluation of the function the first component is just the value of the function at $x = 2$, whereas the second component is the derivative of the function at $x = 2$. For a function including higher order derivatives, the derivatives of the function is obtained in a similar way.

3. SANA PRECOMPILER AND PRESANA PREPROCESSOR

We assume that a FORTRAN program is computing a vector quantity **F**. This vector quantity will depend on a set of input parameters denoted by **p**. The basic problem in any sensitivity study is to find the rate of change (d**F**/d**p**) in the result **F** arising from changes in input model parameters **p**. The output of the program will be calculated by the rules of arithmetics, FORTRAN intrinsic functions and/or user subroutines. Therefore, if the FORTAN program and intrinsic functions are expressed in the rules of arithmetics defined by the Differential Algebra, we can compute the derivatives of **F** about **p**. The purpose of the SANA precompiler is to help us writing such FORTRAN program. The SANA code reads the FORTRAN source code text, replaces any differentiable operation by an operation involving the computation of the original quantity and the derivatives of that quantity with respect to the chosen independent parameters **p**. In practice this is achieved by replacing the original FORTRAN statement by a series of CALL statements to DA-library. The DA- library is composed of the subroutines and the FORTRAN intrinsic functions which perform various DA operations. It is necessary to precompile the FORTRAN code by the SANA code that user assigns the names and the array sizes of DA variables and DA statements need DA operations in the original FORTRAN source. This work could be done by the preprocessor PRESANA. The following preprocessing works are sequentially performed by the PRESANA code.

- search of the main program
- preservation of informations about the variables appeared in the subroutines
- preservation of COMMON informations
- pursuit how the DA variables propagate in the original FORTRAN program
- pick up non-DA variables need not differentiation (RI variables ; scalar variables)
- various checking work about type, variables and statement
- preparation of main program for the DA-program (sensitivity analysis program)
- generate DA-commands specifying the names and the array size of DA and RI variables, and remove type statements and array declarators of DA variables in the original program
- generate DA-commands in the assignment statements including the DA variables
- convert the DA variables appeared in the WRITE and IF statements to the RI variables
- specification of input parameters for sensitivity analysis
- specification of output variables for sensitivity analysis

The target DA program (sensitivity analysis code) will be obtained by precompiling the source program preprocessed with the PRESANA using the SANA code. All these procedures are illustrated in Fig. 1.

4. NUMERICAL RESULTS AND DISCUSSIONS

We apply the automated procedure of sensitivity analysis for the groundwater flow code 3D–SEEP. The 3D–SEEP code simulates saturated-unsaturated flow of groundwater in the permeable geologic media for the safety evaluation of nuclear wastes disposal. The 3D–SEEP code is based on the 3-dimensional Galerkin finite element method. The groundwater flow is modelled by single phase flow governed by Darcy's law, and the simplified double porosity model is introduced to consider fractured media. This code can handle non-uniform flow regions having irregular boundaries and arbitrary degree of local anisotropy.

A sample problem of HYDROCOIN (an international project for studying groundwater hydrology modelling strategies) [6] Level 2 study is used for the sensitivity study of the 3D–SEEP code.

The present test problem is based on a small groundwater flow system in a fractured monzonitic gneiss block at the Chalk River Research Area in Ontario, Canada. The test area named the CRNL groundwater flow study site is constituted by the upper half of the rectangular area marked in the attached topographical map (Fig. 2). The performance and evalution of the field experiments are described in Raven's report [7]. The study site is a 200 m by 150 m area of predominantly quartz monzonite which is overlain and underlain by paragneiss and inclusions of other rock types. The site deemed to be of relatively uniform fracturing and presumably also the subsurface fluid flow properties are relatively uniform. Five major discontinuities including four fracture zones and one diabase dyke have been identified on the site. Two of the fracture zones are subhorizontal, one is inclined and one is vertical (see Fig.3 and 4).

The modelled area in this study is shown as a hatched rectangular portion in Fig. 2. Boundary conditions are displayed in Fig. 5. South vertical boundary were taken to be no-flow boundaries, supposing that diabase dyke is impermeable. Western and eastern boundaries were also taken to be no-flow. The following equation was adopted to prescribe hydraulic heads on northern boundary.

$$h = \frac{h_{top} - h_{bottom}}{z_{top} - z_{bottom}} \cdot (z - z_{bottom}) + h_{bottom} \qquad (10)$$

where h_{top} is given by the elevation of upper base lake and h_{bottom} is given by the measured head of fracture zone-1 (FZ-1). Hydraulic heads on the ground surface were also prescribed by the elevations. Bottom boundary is formed by FZ-1 and is taken to be no-flow. The steady state solutions were obtained by using a mesh of 1716 nodes and 1340 elements. This model consists of 10 layers and the lowest layer is assigned to FZ-1. Hydraulic conductivities of each layer and fracture zone used for caluclations are shown in Table 1.

In the present case, the application of SANA and PRESANA to the 3D–SEEP code allowed simulations which the original results and first order derivatives of almost output parameters with respect to 7 × 13 of input parameters (hydraulic conductivity and porosity) were calculated. The present version of PRESANA code can not automatically preprocess codes involving file input-output of the DA variables. A set of equations of this Finite Element code is solved by the blocked Skyline method, therefore the FORTRAN statements involving such I/O processes were manually rewritten. The CPU time of this simulation was 170 times (1.8 times per one parameter) as long as that of a single run in the original 3D–SEEP code, and the required memory size was about 6 times as large as that of original code. The increment of the required memory size was mainly caused by the supplement of the DA-library. The memory size of DA-library is increasing in proportion to the number of DA variables. In this sensitivity analysis code, almost output parameters were treated as DA-variables. If number of DA-variables is reduced, more economical run is possible. In addition, the DA-library contains many CALL and IF statements for error check, the improvement of DA-library is necessary to save the CPU time. The differential values calculated from this method were compared with those from the perturbation method, and very good agreements were obtained.

In this sensitivity calculation, the hydraulic heads of boreholes were chosen as the performance measure. The sensitivity of the calculated heads to an input parameter can be expressed by a sensitivity coefficient defined by the rate of change in head (**dh/dk**) for input parameter multiplied the value of input parameter (**k**). The results of sensitivity analysis at borehole FS-10 to important 5 parameters concerning to the hydraulic conductivities of fracture zones are shown in Fig. 6. The results of sensitivity analysis at borehole FS-10 to important 5 parameters concerning to the hydraulic conductivities of rock mass are also shown in Fig. 7. The hydraulic conductivities of fracture zone FZ-1, 3 and 4 are sensitive to the head distribution of the borehole FS-10, while that of FZ-2 is not sensitive. These are reasonable results considering the connections of fracture zones (see Fig. 3 and 4). The vertical components of hydraulic conductivities of rock mass are only sensitive to the head distribution of FS-10. An example using the results of sensitivity analysis is shown in Fig. 8. In this figure, the hydraulic conductivity of FZ-3 is one order of magnitude smaller than original run. As expected from the results of figure 6, the hydraulic heads of new run are increased in the region of depth 100 ~ 130 m. Important informations about the parameters of model can be obtained from the sensitivity analysis using the DA method.

5. CONCLUSION

The SANA and PRESANA automated sensitivity analysis procedure based on DA method has been presented. The differential values of output parameters about selected input parameters are automatically obtained by applying this procedure to FORTRAN computer code except file I/O of DA-variables. The SANA and PRESANA codes have been applied to the 3D–SEEP code, and sensitivity results have been verified using the perturbation method. The results applied for the 3D–SEEP code show that this procedure is accurate and useful tool to perform sensitivity analysis of all parameters. The test problem of the FEM code was a large-scale steady state simulation of groundwater flow. The CPU time and memory size for this problem reveal a 1.8 : 1 (per one parameter) run-time overhead cost and a 6 : 1 memory size overhead for sensitivity code compared to an original code run. This increment of CPU time is resulted from the number of DA-operations, in addition the power of DA-library. Future works will include improvements of the SANA and PRESANA code for the economical run using supercomputing technique and application to the inverse problem.

REFERENCES

1. Oblow, E. M., " An automated procedure for sensitivity analysis using computer calculus," ORNL/TM-8776, ORNL (1983)
2. Oblow, E. M., " GRESS, Gradient-Enhanced Software System, Version D User's Guide," ORNL/TM-9658, ORNL (1985)
3. Oblow, E. M. et al., " Sensitivity Analysis Using Computer Calculus : A Nuclear Waste Isolation Application," Nucl. Sci. Eng. : 94, 46-65 (1986)
4. Berz, M., " Differential Algebraic Description of Beam Dynamics to Very High Orders," SSC-152, LBL (1988)
5. Kimura, H. and S. Muraoka, " The 3D–SEEP Computer Code User's Manual," JAERI-M 86-091, JAERI (1986)
6. HYDROCOIN progress report No 1 ~ 6, SKI, Sweden (1984 ~ 1987)
7. Raven, K. G., " Hydraulic Charcterization of a Small Groundwater Flow System in Fractured Monzonitic Gneiss," Atomic Energy of Canada Limited, Pinawa, Manitoba, Canada, (1985)

Table 1 Hydraulic conductivities used for the simulation

	k_{xx}	k_{yy}	k_{zz}	k_{xy}	k_{yz}	k_{zx}
FZ-1	1.0×10^{-6}	1.0×10^{-6}	1.0×10^{-6}	0.0	0.0	0.0
FZ-2	1.0×10^{-7}	1.0×10^{-7}	1.0×10^{-7}	0.0	0.0	0.0
FZ-3	1.0×10^{-7}	1.0×10^{-7}	1.0×10^{-7}	0.0	0.0	0.0
FZ-4	4.0×10^{-7}	4.0×10^{-7}	4.0×10^{-7}	0.0	0.0	0.0
layer 2	1.0×10^{-11}	6.0×10^{-12}	1.0×10^{-11}	0.0	0.0	0.0
layer 3	1.0×10^{-10}	6.0×10^{-11}	1.0×10^{-10}	0.0	0.0	0.0
layer 4	1.0×10^{-10}	6.0×10^{-11}	1.0×10^{-10}	0.0	0.0	0.0
layer 5	1.0×10^{-10}	6.0×10^{-11}	1.0×10^{-10}	0.0	0.0	0.0
layer 6	1.0×10^{-10}	6.0×10^{-11}	1.0×10^{-10}	0.0	0.0	0.0
layer 7	1.0×10^{-10}	6.0×10^{-11}	1.0×10^{-10}	0.0	0.0	0.0
layer 8	1.0×10^{-10}	6.0×10^{-11}	1.0×10^{-10}	0.0	0.0	0.0
layer 9	1.0×10^{-10}	6.0×10^{-11}	1.0×10^{-10}	0.0	0.0	0.0
layer 10	1.0×10^{-9}	6.0×10^{-10}	1.0×10^{-9}	0.0	0.0	0.0

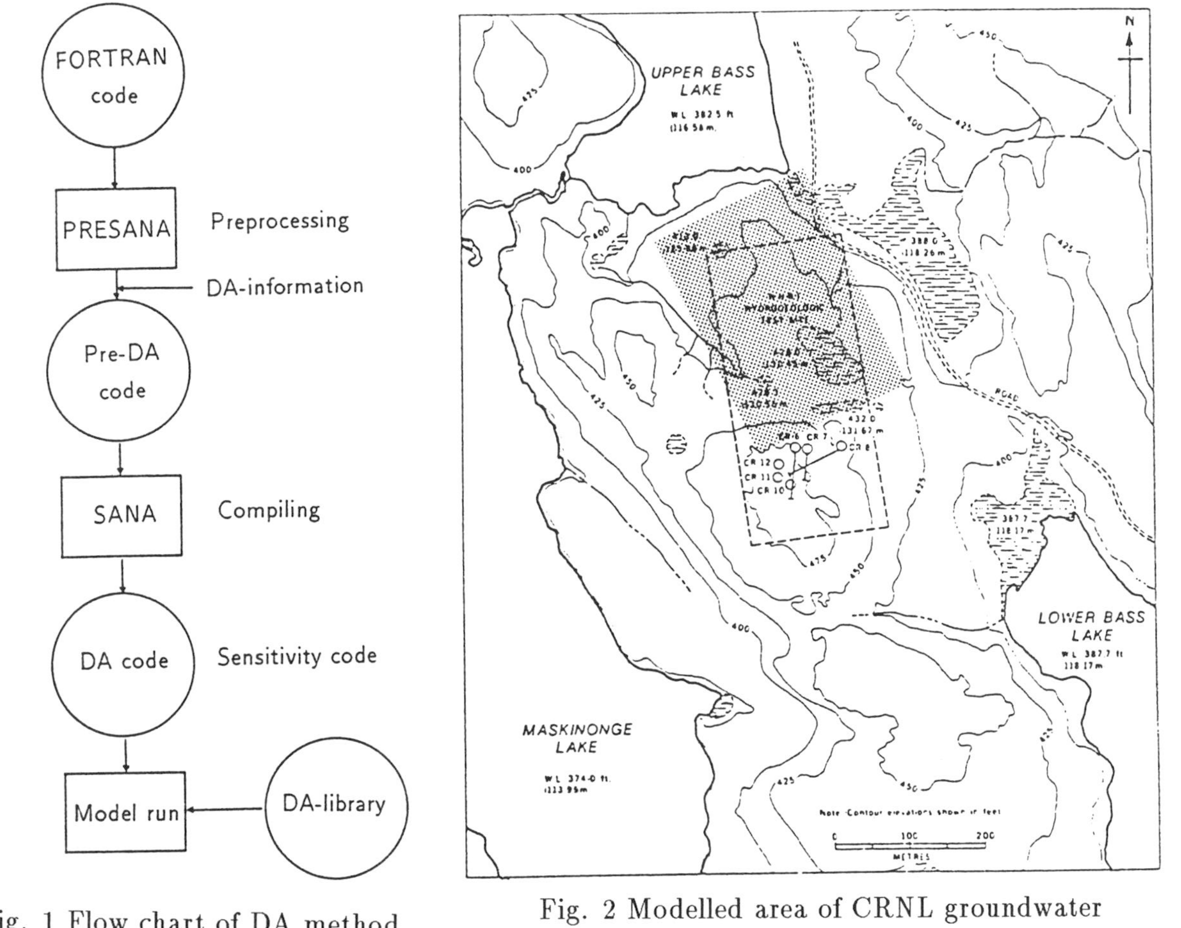

Fig. 1 Flow chart of DA method

Fig. 2 Modelled area of CRNL groundwater flow study site

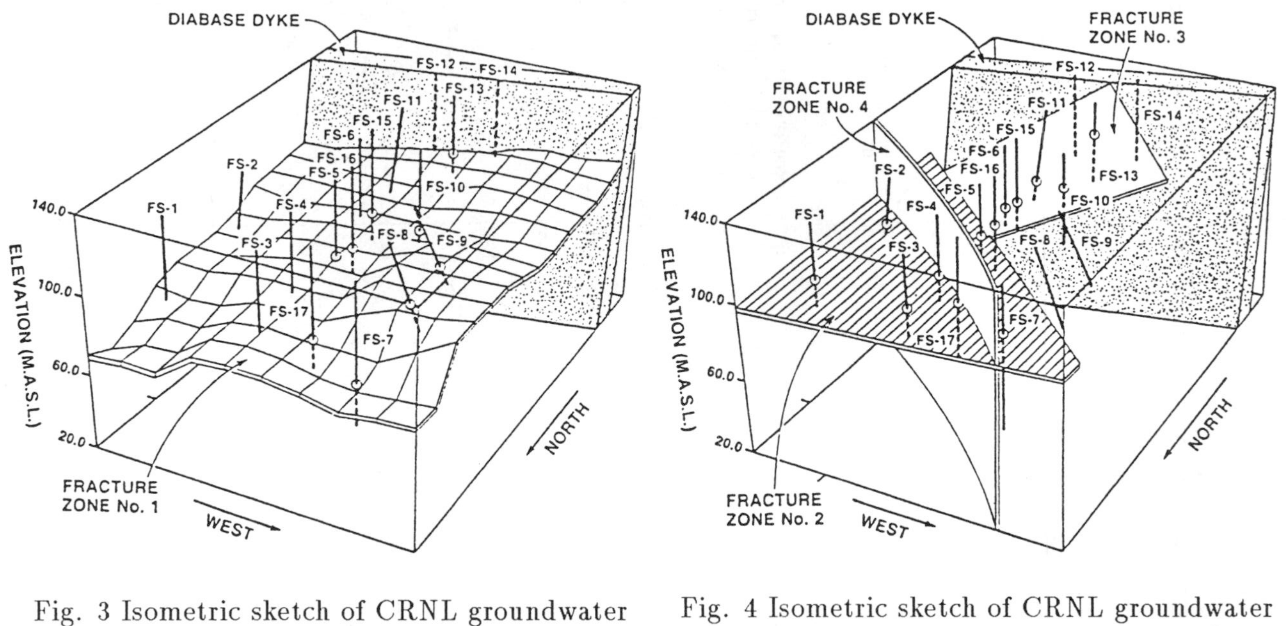

Fig. 3 Isometric sketch of CRNL groundwater flow study site showing the location of test boreholes, fracture zone 1, and diabase dyke

Fig. 4 Isometric sketch of CRNL groundwater flow study site showing the location of test boreholes, fracture zone 2,3 and 4 and diabase dyke

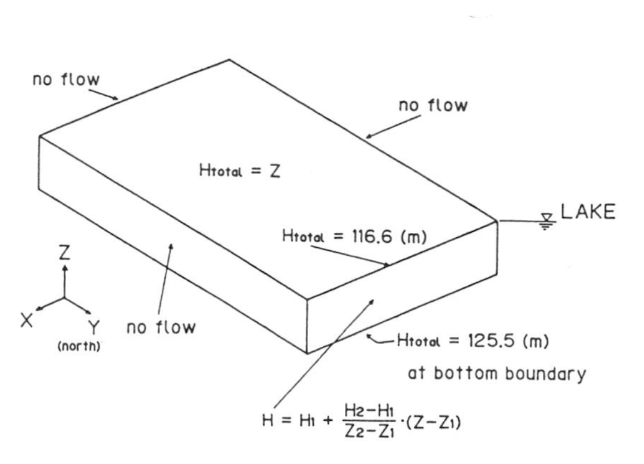

Fig. 5 Boundary conditions

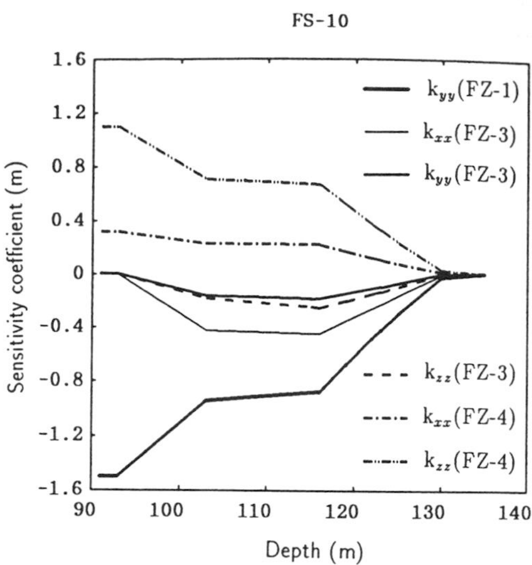

Fig. 6 Sensitivity results at borehole FS-10 to the hydraulic conductivities of fracture zone FZ-1, 3 and 4

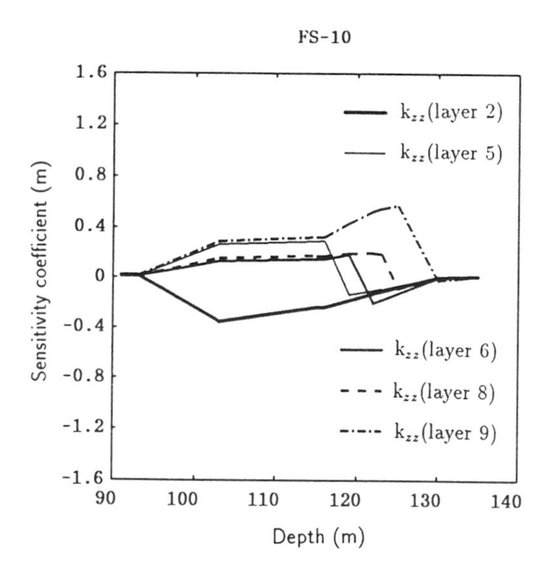

Fig. 7 Sensitivity results at borehole FS-10 to the hydraulic conductivities of rock mass

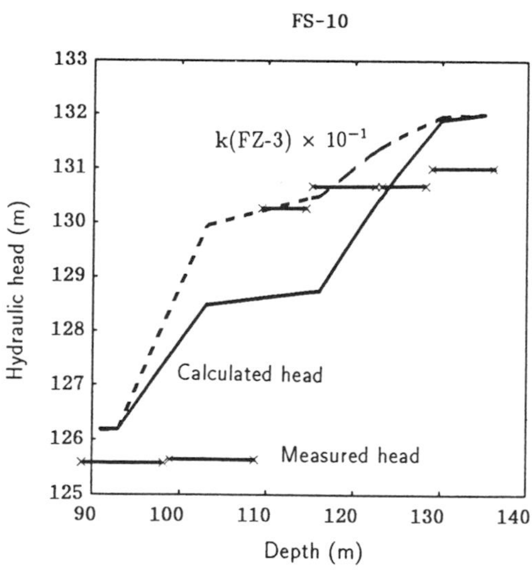

Fig. 8 Measured and calculated hydraulic heads at borehole FS-10

USE OF LARGE-SCALE TRANSIENT STRESSES AND A COUPLED ADJOINT-SENSITIVITY/ KRIGING APPROACH TO CALIBRATE A GROUNDWATER-FLOW MODEL AT THE WIPP SITE

Richard L. Beauheim
Sandia National Laboratories, Albuquerque, NM, USA
A. Marsh LaVenue
INTERA, Inc., Albuquerque, NM, USA

ABSTRACT

A coupled adjoint-sensitivity/kriging approach was used to calibrate a groundwater-flow model to 10 years of human-induced transient hydraulic stresses at the WIPP site in New Mexico, USA. Transmissivity data obtained from local-scale hydraulic tests were first kriged to define an initial transmissivity distribution. Steady-state model calibration was then performed employing adjoint-sensitivity techniques to identify regions where transmissivity changes would improve the model fit to the observed steady-state heads. Subsequent transient calibration to large-scale hydraulic stresses created by shaft construction and long-term pumping tests aided in identifying smaller scale features not detected during steady-state calibration. This transient calibration resulted in a much more reliable and defendable model for use in performance-assessment calculations.

UTILISATION DE CONTRAINTES TRANSITOIRES A GRANDE ECHELLE ET D'UNE METHODE COUPLEE ASSOCIANT SENSIBILITE ET KRIGEAGE POUR ETALONNER UN MODELE D'ECOULEMENT D'EAU SOUTERRAINE AU SITE DE WIPP

RESUME

Une méthode couplée associant sensibilité et krigeage a été utilisée pour étalonner un modèle d'écoulement d'eau souterraine simulant une période de dix ans de contraintes hydrauliques transitoires provoquées au site de WIPP (Nouveau Mexique, Etats-Unis). On a commencé par kriger les données relatives à la transmissivité obtenues à partir d'essais hydrauliques à l'échelle locale pour définir une distribution initiale de la transmissivité. On a procédé ensuite à un étalonnage du modèle en régime permanent en employant des techniques de sensibilité pour délimiter des zones où des modifications de la transmissivité amélioreraient la concordance du modèle avec les charges hydrauliques observées en régime permanent. L'étalonnage transitoire qui a été opéré ensuite pour les contraintes hydrauliques à grande échelle provoquées par la construction de galeries et les essais de pompage de longue durée ont contribué à déterminer des caractéristiques à plus petites échelles qui n'avaient pas été decelées pendant l'étalonnage en régime permanent. Cet étalonnage en régime transitoire permet d'obtenir un modèle beaucoup plus fiable et justifiable pour les calculs d'évaluation du comportement des dépôts.

The Waste Isolation Pilot Plant (WIPP), located near Carlsbad, New Mexico (Fig. 1), is intended to be the United States' permanent repository for low-level and transuranic wastes generated by the nation's defense programs. The repository is under construction 655 m below ground surface within bedded evaporites of the Permian Salado Formation. Performance-assessment calculations for the WIPP require a reliable groundwater-flow model for the Culebra dolomite, the most transmissive water-bearing unit overlying the repository horizon. Since 1976, 58 wells have been completed to the Culebra to provide data on hydraulic head, transmissivity, and storativity. Testing at these wells has indicated that the Culebra is a locally fractured medium exhibiting double-porosity hydraulic behavior. Variations in the degree of fracturing of the Culebra in the vicinity of the WIPP site have resulted in a range of transmissivities spanning six orders of magnitude. This high degree of heterogeneity has complicated flow-model calibration by increasing the uncertainty in transmissivity estimates between measurement points.

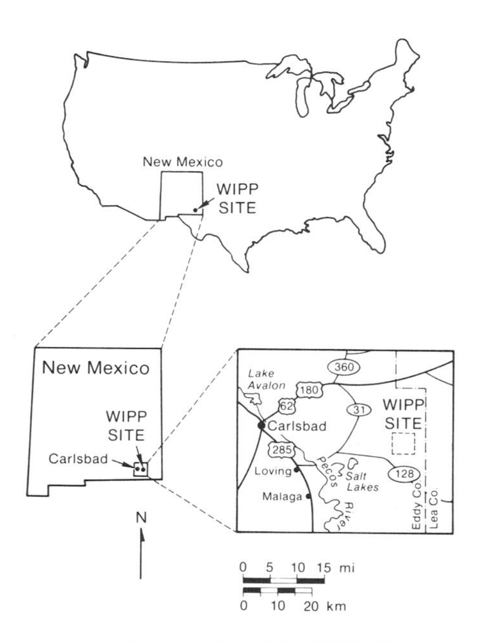

Figure 1. Location of the WIPP Site.

A recent modeling study conducted to estimate the transmissivity distribution of the Culebra used a 28 x 48 x 1 finite-difference grid to represent an area of 21.3 x 30.6 km, roughly centered on the 6.4 x 6.4-km WIPP site [1]. SWIFT II, a groundwater-flow and transport code [2], was employed during this study to solve for the steady-state and transient pressures over the model region. Transmissivity data provided by hydraulic tests performed at individual boreholes were kriged to define an initial grid-block-averaged transmissivity distribution over the model domain. The initial transmissivity distribution was then modified to bring the simulated heads into agreement with estimates of steady-state and transient heads at the WIPP-site boreholes.

The procedure used to modify the transmissivities employed GRASP II, a post-processor for steady-state or transient SWIFT II simulations which couples adjoint-sensitivity and kriging techniques [3]. GRASP II was used to determine the locations at which changes to the transmissivity field would provide the greatest reduction to a cumulative weighted least-squares performance measure, J, defined by:

$$J = \sum_{t} \sum_{i} W_{i,t} \left(H_{i,t} - Hob_{i,t}\right)^2 \qquad (1)$$

where

J = the performance measure
w = relative weight selected by modeler
H = calculated head
Hob = observed head
t = time window for summation
i = borehole designation .

Transmissivities are then manually adjusted at the locations identified as most sensitive by GRASP II in order to improve the fit between the model-calculated and the observed heads. These synthetic transmissivity data points are referred to as "pilot points" after Marsily et al. [4]. The general equation coupling the adjoint-sensitivity and kriging techniques is:

$$\frac{dJ}{dT_p} = \sum_m \frac{dJ}{dT_m} \frac{dT_m}{dT_p} \qquad (2)$$

where

T_p = $\log_{10}$ pilot-point transmissivity
T_m = kriged estimate of $\log_{10}$ grid-block-averaged transmissivity
m = number of grid blocks in finite-difference model .

As Eq. 2 illustrates, the sensitivity of a performance measure J to the addition of a pilot point with transmissivity T_p into the modeled system has two components. The first component represents the dependence of the performance measure J on the kriged estimates of the grid-block-averaged transmissivities. The second component represents the sensitivity of the kriged estimate of the grid-block-averaged transmissivity to the addition of a pilot point. Eq. 2 is evaluated at locations coincident with a selected grid of potential pilot-point locations. The calculated sensitivities are normalized and contoured to identify the highest sensitivity region(s) with respect to modifying the transmissivity field (Fig. 2). The kriged estimate of transmissivity at that location is then manually adjusted by the modeler.

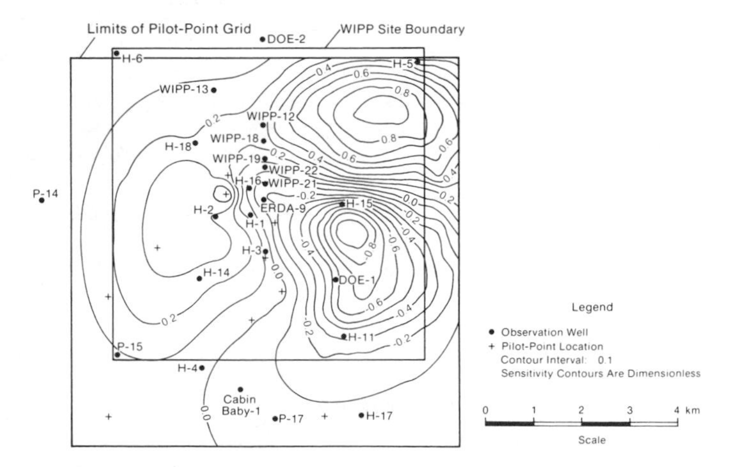

Figure 2. Normalized Sensitivities of H-15 and DOE-1 Transient Pressure Performance Measure to Changes in Transmissivities at Potential Pilot-Point Locations.

Manual adjustment of the pilot-point transmissivities is preferred to an automatic inversion algorithm because it allows the modeler to use his physical judgement or experience, based on site hydrogeologic information, in the adjustment. The pilot-point transmissivities are included in the transmissivity database for kriging to generate a revised transmissivity field. The process of identifying pilot-point locations, assigning synthetic transmissivity values at those locations, and rekriging the transmissivity field is repeated until the simulated heads fall within the uncertainties of the observed heads. The same procedure may be applied to the definition of specified-head boundary conditions. The approach allows for the specification of both steady-state and transient performance measures to permit simultaneous calibration to steady-state and transient head data. Separate weights may be assigned to the steady-state and transient head differences to ensure improvement of the fit to observed transient responses while maintaining the fit to the steady-state heads.

After calibrating the model to the observed steady-state heads by introducing pilot points to the observed transmissivity field, the initial transient simulation was performed. The steady-state calibrated model adequately predicted the transient responses at many boreholes, even when those responses were complex (Fig. 3a). However, the predicted responses at those boreholes believed to lie within or near areas of localized transmissivity features, such as narrow fracture zones, did not represent the observed responses well (e.g., see the simulated response at H-15 to a large-scale pumping test performed in 1988 in Fig. 3b). Small-scale transmissivity features do not generally dominate steady-state flow on a regional scale and, therefore, are not typically detected during steady-state calibration. The GRASP II code was utilized to determine the most sensitive locations for small-scale features necessary to reduce the differences between the model-calculated and observed transient responses. Improving the fit to the observed transient responses (Fig. 4) was then accomplished by introducing pilot points in the sensitive areas. The efficiency of the coupled adjoint-sensitivity/kriging approach allowed transient calibration of the model for approximately the same effort as required for steady-state calibration. Typically, calibration to multiple transient events is much more demanding than calibration to steady-state conditions.

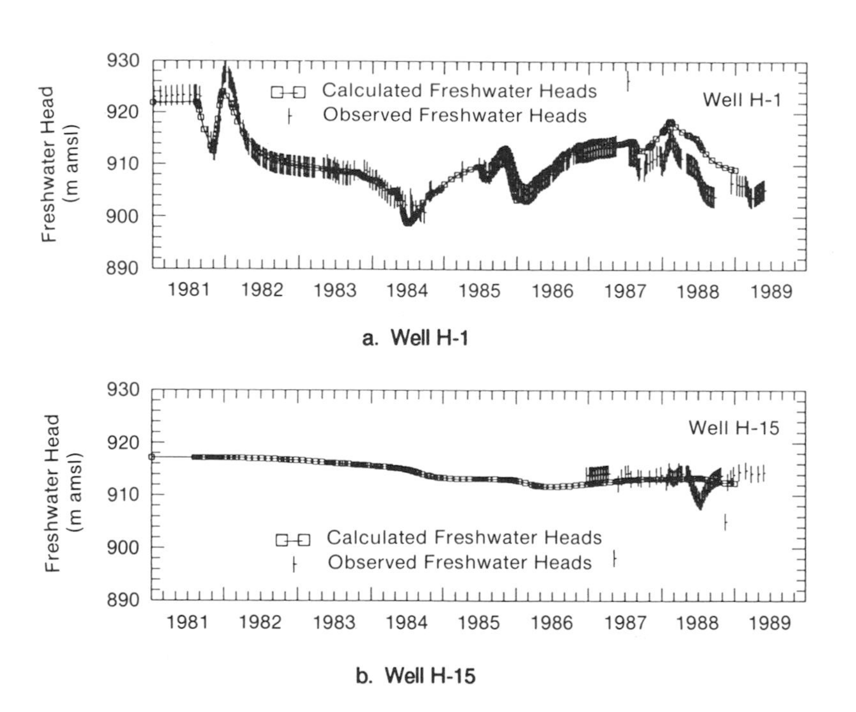

Figure 3. Calculated and Observed Transient Heads at H-1 and H-15 Using the Steady-State Calibrated Transmissivity Field.

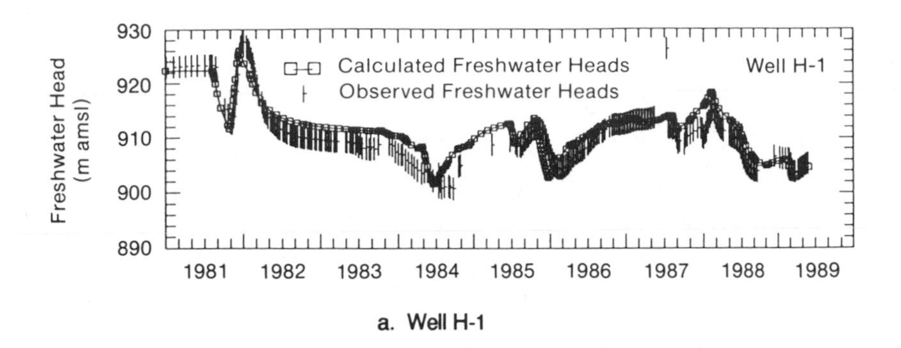

a. Well H-1

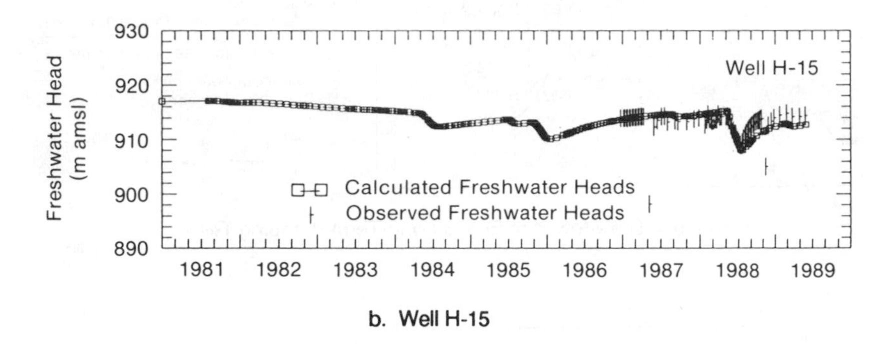

b. Well H-15

Figure 4. Calculated and Observed Transient Heads at H-1 and H-15 Using the Transient Calibrated Transmissivity Field.

A key factor permitting the defendable calibration of the flow model was the existence and documentation of large-scale transient hydraulic stresses on the Culebra. These stresses were caused by construction of four shafts at the WIPP site and by three large-scale pumping tests lasting one to two months each [5,6,7]. The pumping tests were performed at different locations around the WIPP site, and were designed to have spatially overlapping fields of influence. Each test produced observable hydraulic responses over an area of 10 to 20 km^2 (Fig. 5). Calibration to the large-scale transient events required significant changes to the transmissivity field (Fig. 6), providing information on the distribution of fracturing that would never have been obtained from steady-state calibration or from calibration to smaller scale transient events alone. The required changes to the transmissivity field also changed the estimated flow paths used for solute-transport modeling (Fig. 7). The results of the transient calibration show the importance of testing a heterogeneous hydrologic system on the same scale as that at which one wishes to predict (model) the system's behavior. A model can only be considered to be valid on the scale at which it has been calibrated, and a model can only be calibrated on the scale at which the physical system has been tested.

This study has shown: 1) that coupling adjoint-sensitivity techniques with kriging is a highly efficient approach to combined steady-state and transient model calibration; and 2) that calibration of a regional-scale model of a heterogeneous system requires calibration to regional-scale transient stresses. Steady-state calibration or transient calibration to non-overlapping local-scale stresses cannot provide as reliable a regional-scale model for performance-assessment calculations.

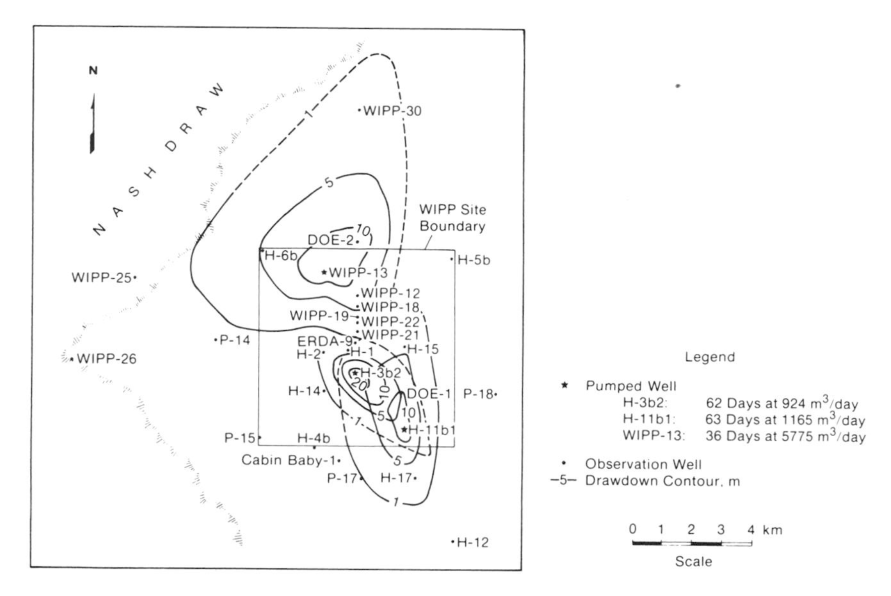

Figure 5. Drawdowns at End of Long-Term Pumping Tests.

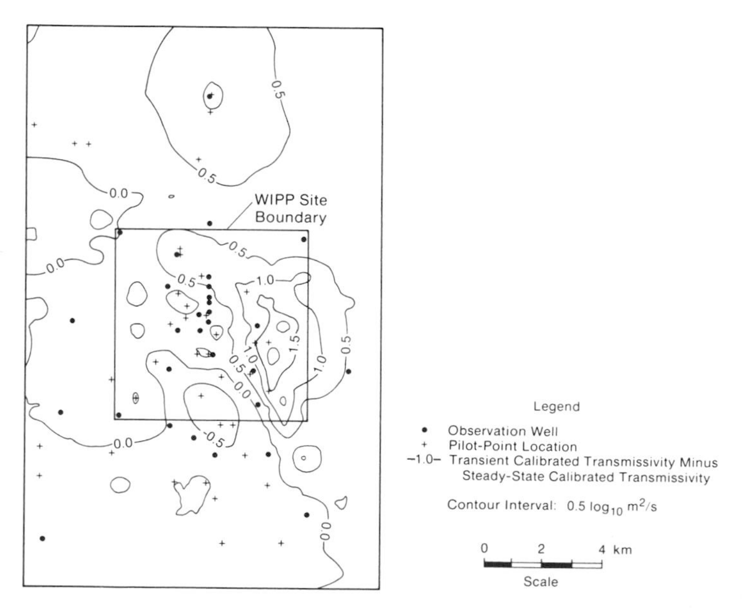

Figure 6. Differences Between the Steady-State Calibrated and Transient Calibrated Transmissivity Fields in the Central Model Region.

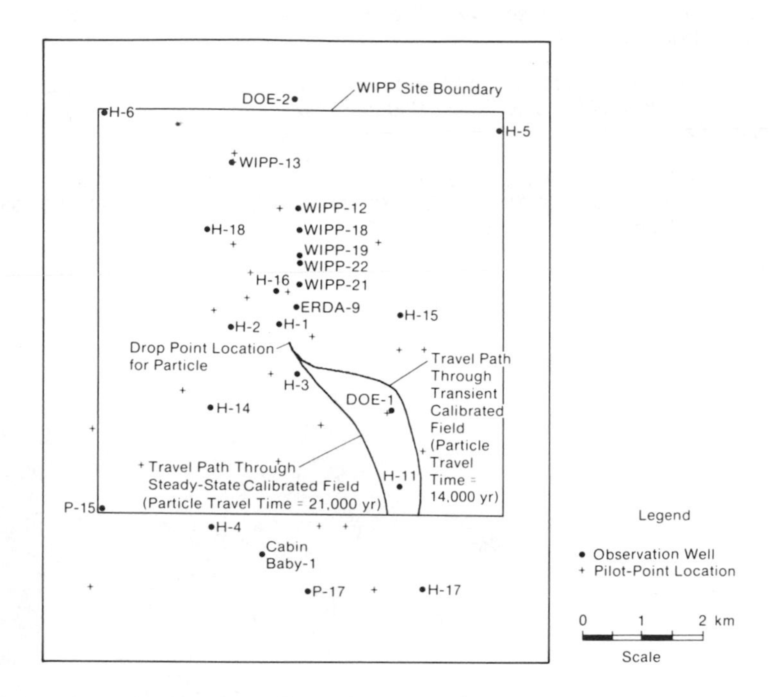

Figure 7. Comparison of Particle Travel Paths through the Steady-State Calibrated and Transient Calibrated Transmissivity Fields.

ACKNOWLEDGEMENTS: This work was supported by the U.S. Department of Energy under contract DE-AC04-76DP00789.

REFERENCES

[1] LaVenue, A.M., T.L. Cauffman, and J.F. Pickens, 1990. *Ground-Water Flow Modeling of the Culebra Dolomite: Volume I - Model Calibration.* Contractor Report SAND89-7068/1. Sandia National Laboratories, Albuquerque, NM, USA.

[2] Reeves, M., D.S. Ward, N.D. Johns, and R.M Cranwell, 1986. *Theory and Implementation for SWIFT II, the Sandia Waste-Isolation Flow and Transport Model for Fractured Media, Release 4.84.* U.S. Nuclear Regulatory Commission Report NUREG/CR-3328, Sandia Report SAND83-1159. Sandia National Laboratories, Albuquerque, NM, USA.

[3] RamaRao, B.S., and M. Reeves, 1990. *Theory and Verification for the GRASP II Code for Adjoint-Sensitivity Analysis of Steady-State and Transient Ground-Water Flow.* Contractor Report SAND89-7143. Sandia National Laboratories, Albuquerque, NM, USA.

[4]Marsily, G. de, G. Lavedan, M. Boucher, and G. Fasanino, 1984. "Interpretation of Interference Tests in a Well Field Using Geostatistical Techniques to Fit the Permeability Distribution in a Reservoir Model," *Proc. Geostatistics for Natural Resources Characterization*, NATO-ASI, GEOSTAT 1983, Tahoe, CA, USA; G. Verly, M. David, A.G. Journel, and A. Marechal, eds., pp. 831-849, D. Reidel, Hingham, MA, USA.

[5] Beauheim, R.L., 1987. *Analysis of Pumping Tests of the Culebra Dolomite Conducted at the H-3 Hydropad at the Waste Isolation Pilot Plant (WIPP) Site.* Sandia Report SAND86-2311. Sandia National Laboratories, Albuquerque, NM, USA.

[6] Beauheim, R.L., 1987. *Interpretation of the WIPP-13 Multipad Pumping Test of the Culebra Dolomite at the Waste Isolation Pilot Plant (WIPP) Site.* Sandia Report SAND87-2456. Sandia National Laboratories, Albuquerque, NM, USA.

[7] Beauheim, R.L., 1989. *Interpretation of H-11b4 Hydraulic Tests and the H-11 Multipad Pumping Test of the Culebra Dolomite at the Waste Isolation Pilot Plant (WIPP) Site.* Sandia Report SAND89-0536. Sandia National Laboratories, Albuquerque, NM, USA.

UNCERTAINTIES IN GROUNDWATER TRANSPORT MODELLING - A COMPONENT OF UNCERTAINTY IN THE PERFORMANCE ASSESSMENT OF LOW-LEVEL RADIOACTIVE WASTE DISPOSAL SITES

Cheng Y. Hung
U.S. Environmental Protection Agency
Washington, USA

ABSTRACT

This paper presents an analytical method of uncertainty analysis and its application to the groundwater transport modelling, particularly for the performance assessment of low-level radioactive waste (LLW) disposal facilities. The model imposes a quasi-steady state simplification to obtain an analytical solution which is used for the uncertainty analysis. This method is developed to reduce the computation time compared with a simulation method, such as the Monte Carlo method. The analysis concludes that the uncertainties of performance assessment due to the groundwater transport model are generally small.

INCERTITUDES DANS LA MODELISATION DU TRANSPORT PAR L'EAU SOUTERRAINE - UN ELEMENT D'INCERTITUDE DANS L'EVALUATION DU COMPORTEMENT DES SITES D'EVACUATION DES DECHETS DE FAIBLE ACTIVITE

RESUME

L'auteur présente une méthode d'analyse des incertitudes et son application à la modélisation du transport par l'eau souterraine, en particulier pour l'évaluation du comportement des installations d'évacuation des déchets de faible activité. Ce modèle impose de faire l'hypothèse simplificatrice d'un régime quasi-permanent afin d'obtenir une solution analytique qui est utilisée pour l'analyse d'incertitude. Cette méthode vise à réduire le temps de calcul par rapport à une méthode fondée sur une simulation, comme la méthode de type Monte Carlo. Il ressort de cette analyse que le modèle de transport par l'eau souterraine ne contribue généralement que faiblement aux incertitudes de l'évaluation du comportement des sites d'évacuation.

1.0 INTRODUCTION

Performance assessments of a variety of low-level radioactive waste (LLW) disposal sites were evaluated as an important part of the U.S. Environmental Protection Agency's (EPA) program to develop generally applicable environmental standards for the land disposal of LLW [1, 2, and 3]. The development of the performance requirements for the LLW standard incorporates the results of performance assessments.

The EPA family of risk assessment models, Prediction of the Radiological Effects from Shallow Trench Operation (PRESTO-EPA), were used for the performance assessment of reference sites to demonstrate that the proposed numerical standards can be met by the waste disposal industry [2 and 4].

This paper intends to discuss the theoretical development of the analytical method used to obtain the results of an uncertainty analysis for several critical mobile radionuclides commonly found in low-level radioactive wastes.

For the purpose of this uncertainty analysis, the overall uncertainties of the performance assessment were divided into five distinct components [4]: the waste source term, groundwater transport, food chain transport, organ dosimetry, and health effects conversion factors. This study will be limited to the discussion of the uncertainty due to the groundwater transport only.

The EPA's performance assessment of LLW disposal sites was conducted for the total health effects to the downstream population and the maximum individual dose to the critical population group. The latter analysis was emphasized in EPA's analysis because it is the primary criterion specified in EPA's LLW standards. The following discussion covers these two analyses.

A commonly used method of uncertainty analysis is a simulation method, such as the Monte Carlo method. This method uses a straight theoretical base but generally requires considerably longer computer process time. Therefore, an analytical method was developed by EPA to avoid the use of such a time consuming simulation model and to reduce the cost of analysis.

2.0 METHOD OF ANALYSIS

The PRESTO-EPA model is a complex dynamic model which requires considerable time to complete a single computer run. The application of a simulation type of uncertainty model, such as Monte Carlo, for the PRESTO-EPA code could require thousands of runs resulting in enormous computer process time. Two analytical models were therefore developed to avoid a time consuming Monte Carlo analysis.

2.1 DEVELOPMENT OF ANALYTICAL SOLUTION

The geosphere transport portion of the PRESTO-EPA model assumes that the trench cap failure will develop starting from year y_1, linearly increase to its maximum level at year y_2, and maintain this level for the rest of the time of analysis. Therefore, the flow condition represented by the rate of infiltration through the trench cap will be in an unsteady state and no general analytical solution is obtainable. However, an analytical solution can be obtained if a quasi-steady state simplification is imposed on the flow condition. The quasi-steady state model assumes that the trench cap failure reaches its maximum level at the very beginning of the analysis and the same level is maintained through out the entire period of analysis. A preliminary analysis indicated that such an approximation will not introduce severe errors in the risk assessment results if the ultimate trench cap failure develops within 50 or 100 years as assumed in EPA's analyses [6].

Although there are three major radionuclide release pathways considered in the PRESTO-EPA model, the EPA's extensive risk analysis indicated that the contribution from the groundwater release pathway is predominant over the other two release pathways. Therefore, for the purpose of the uncertainty analysis, only the groundwater release pathway is considered in the analytical model.

2.1.1 Analytical Solution for Total Health Effects

Radionuclide Release Equation

Taking the entire trench as a control volume, the basic one-dimensional sorption-desorption mass balance equation was derived by Hung [5 and 6] as:

$$\partial C/\partial t = -\lambda_d C - (E/\epsilon V R_t)C \qquad (1)$$

In which, ϵ is the porosity of the trench material; C is the concentration of radionuclide in the leachate; V is the total volume of the trench material; λ_d is the decay constant for the radionuclide; E is the rate of exfiltration; t is the time; R is the retardation factor; and subscript t denotes trench materials.

When the initial condition: $C = C_0$, is applied to the basic equation, the following analytical solution is obtained:

$$C = C_0 Exp(-\lambda_d - E/\epsilon V R_t)t \qquad (2)$$

Equation 2 can also be rewritten in terms of the initial radionuclide inventory, I_0, which yields:

$$C = (I_0/\epsilon V R_t) Exp\{-(\lambda_d + E/\epsilon V R_t)\}t, \qquad (3)$$

in which I_0 is the total radionuclide inventory of a specific radionuclide at the initial stage.

When Equation 3 is multiplied by the rate of exfiltration, the rate of radionuclide transport can be rewritten as:

$$Q = (E I_0/\epsilon V R_t) Exp\{-(\lambda_d + E/\epsilon V R_t)\}t \qquad (4)$$

where Q denotes the rate of radionuclide leaching out of the trench bottom.

Groundwater Transport Equation

Equation 4 represents the rate of radionuclide transport out of the trench bottom. The released radionuclides are subsequently transported vertically through the saturated or unsaturated soil toward the aquifer. When the radionuclides reach the aquifer, they will be mingled into the groundwater flow and subsequently be transported horizontally to the environmental receptors. A generalized analytical solution for an one-dimensional mass balance equation expressing the transport of radionuclides at the downstream end of the aquifer is also derived by Hung [7] as:

$$Q(t) = \eta Q_0(t - RL/v)\mathrm{Exp}(-\lambda_d RL/v)\bullet U(t - RL/v), \qquad (5)$$

where

$$\eta = \mathrm{Exp}[(P/2) - (P/2)\surd 1 + (4\lambda_d RL/(Pv)]/\mathrm{Exp}(-\lambda_d RL/v) \qquad (6)$$

in which, η denotes the Hung correction factor compensating for the effect of longitudinal dispersion; L is the length of aquifer; U is the unit step function and P is the Peclet number expressed by vL/D; D is the coefficient of dispersion; v is the interstitial velocity of the groundwater.

Substituting Equation 4 into Equation 5 and using the subscript v to represent the vertical reach yields:

$$Q(t) = \eta_v(I_0E/\epsilon VR_t)\mathrm{Exp}\{-\lambda_d R_v L_v/v_v$$
$$-(\lambda_d + E/\epsilon VR_t)(t - R_v L_v/v_v)\}\bullet U(t - R_v L_v/v_v) \qquad (7)$$

Equation 7 represents the rate of radionuclide transport at the lower end section entering the aquifer. For the analysis of the total health effects, the PRESTO-EPA model assumes that the disposal site is a point source. Therefore, the rate of radionuclide transport through the horizontal aquifer can also be obtained in the same manner as discussed for the vertical reach. Using Equation 7 as the boundary condition and applying Equations 5 and 6 to the horizontal reach, one obtains:

$$Q(t) = \eta_v\eta_h(I_0E/\epsilon VR_t)\mathrm{Exp}\{-\lambda_d(R_v L_v/v_v + R_h L_h/v_h)\}$$
$$\mathrm{Exp}\{-\lambda_d - E/\epsilon VR_t)(t - R_v L_v/v_v - R_h L_h/v_h)\}$$
$$U(t - R_v L_v/v_v - R_h L_h/v_h), \qquad (8)$$

in which subscripts v and h denote parameters for the vertical host soil reach and horizontal aquifer reach, respectively.

To simplify the equation, one may set:

$$T_l = R_v L_v/v_v + R_h L_h/v_h; \text{ and } \eta_l = \eta_v\eta_h \qquad (9)$$

Equation 8 can then be rewritten as:

$$Q(t) = \eta_l(I_0E/\epsilon VR_t)\mathrm{Exp}\{-\lambda_d T_l - (\lambda_d + E/\epsilon VR_t)(t - T_l)\}U(t - T_l), \qquad (10)$$

In the above equation, T_l denotes the total transit time required for the radionuclide to migrate through the host soil and the aquifer.

The total health effects indicator will be the cumulative radionuclide release at a discharge point over a period of 10,000 years. This can be obtained by integrating Equation 10 over the period of release, i.e.,

$$Q_T = \int_{T_l}^{10000} \eta_1 (I_0 E/\epsilon V R_t) \mathrm{Exp}\{-\lambda_d T_l - (\lambda_d + E/\epsilon V R_t)(t - T_l)\} dt \qquad (11)$$

The integration yields:

$$Q_T = \eta_1 (I_0 E/\epsilon V R_t) [1/\{\lambda_d + E/\epsilon V R_t\}] [\mathrm{Exp}(-\lambda_d T_l) - \mathrm{Exp}\{-(\lambda_d + E/\epsilon V R_t) \cdot 10000 + (E/\epsilon V R_t) T_l\}] \qquad (12)$$

where Q_T is the cumulative radionuclide inventory released at the discharge point.

Since our main focus is on the long half-life and mobile radionuclides which contribute the major health effects to the humans, the following relationship exists:

$$T_l \ll 10000, \text{ and } \lambda_d \ll E/\epsilon V R_t \qquad (13)$$

Consequently, Equation 12 can be approximated by:

$$Q_T = \eta_1 (I_0 E/\epsilon V R_t) [1/\{\lambda_d + E/\epsilon V R_t\}] \mathrm{Exp}(-\lambda_d T_l) \qquad (14)$$

Equation 14 is the basic dynamic equation specifically developed for the purpose of uncertainty analysis.

2.2.2 Analytical Solution for the Maximum Individual Dose

Radionuclide Release and Vertical Transport Equations

In analyzing the maximum individual dose indicator, it is assumed that the access well from which the postulated individual obtains his water is located near the fence line of a disposal facility. Furthermore, the disposal site is considered as an area source to maintain the accuracy of the simulation. Therefore, the radionuclide release and the vertical groundwater transport equations should be identical to those developed for the total health effects indicator analysis, Equations 4 and 7.

Groundwater Transport Equation

The rate of radionuclide transport per unit width of aquifer flow can be calculated by integrating the radionuclide transport per unit area over the entire distance of influence, that is:

$$q_w = \int_0^X q_1\{t - R_h(x + L_w)/v_h\} \mathrm{Exp}\{-\lambda_d R_h(x + L_w)/v_h\} dx \qquad (15)$$

in which x is the distance from the edge of disposal trenches; and X is the distance of influence. The distance of influence is defined here as the distance from the edge of trenches to the furthest point under the disposal site from where the transport of radionuclide reach the well at the time of interest.

It was found that the critical radionuclides contributing the health effects are predominated by those radionuclides having extremely long half-life and are mobile. If the analysis is focused on these radionuclides, the maximum radionuclide concentration will occur when the distance of the influence reaches the other edge of the disposal site.

When this condition is applied, the maximum radionuclide transport rate at the access well can be obtained by integrating the mass transport over the entire length of the site. Mathematically it may be expressed in:

$$q_{max} = \int_0^{L_h} q_1\{t - R_h(x + L_w)/v_h\} \, Exp\{-\lambda_d R_h(x + L_w)/v_h\} dx \qquad (16)$$

in which, L_h is the distance from one edge of trenches to the other edge.

Substituting Equation 14 into Equation 16 yields:

$$q_{max} = (\eta_v I_0 v_h/AR_h) Exp\{-\lambda_d(T_v + T_h + T_w)\}$$

$$[1 - Exp\{-(E/\epsilon VR_w)(R_h L_h/v_h)\}] \qquad (17)$$

where: $T_v = R_v L_v/v_v$; $T_h = R_h L_h/v_h$; $T_w = R_h L_w/v_h$; and T is the transit time and subscripts v, h, and w are the vertical reach, horizontal reach and well reach, respectively.

The maximum individual dose indicator, represented by the maximum concentration, can be calculated by dividing Equation 17 by the unit width flow rate of groundwater. When this is done the result is:

$$C_{max} = (\eta_v I_0/AH_h \epsilon_h R_h) Exp\{-\lambda_d(T_v + T_h + T_w)\}$$

$$[1 - Exp\{-(E/\epsilon VR_t)(R_h L/v_h)\}] \qquad (18)$$

where H_h is the thickness of the aquifer and ϵ_h is the porosity of the aquifer. Equation 18 is the dynamic equation for the calculation of the maximum individual dose indicator.

2.2 CALCULATION OF JOINT PROBABILITY DISTRIBUTION

The probability distribution function of the health effect indicators can be calculated from the consecutive arithmetic operations of a pair of random input parameters as defined in the dynamic equation of the indicator. These arithmetic operations include addition, subtraction, multiplication, and division. The basic equations for these operations are discussed below:

2.2.1 Addition and Subtraction

When the probability density function for the two independent random variables X and Y is given, the cumulative probability of the sum of the two random variables can be written based on the joint-probability theorem as [8]:

$$F_z(Z) = \int_0^{Z} f_x(X) \int_0^{Z-X} f_y(Y)\,dy\,dx \qquad (19)$$

where Z is the random variable representing the sum of random variables X and Y; F_z is the cumulative density function for Z; and f_x and f_y are the probability density function for the random variables X and Y.

The same equation can also be used to evaluate the probability density function for subtracting random variable Y from variable X by transforming the subtrahend Y into its image function.

2.2.2 Multiplication and Division

Using the same approach as that used in the addition and subtraction, the cumulative probability of the product of two random variables can be written based on the joint-probability theorem as [8]:

$$F_u(U) = \int_0^{u} f_x(X) \int_0^{u/x} f_y(Y)\,dy\,dx \qquad (20)$$

where U is the random variable representing the product of random variables X and Y, and F_u is the cumulative probability density function for the random variables U. This equation can also be used for the division of the random variable X divided by random variable Y after transforming the divisor into its inverse function.

3.0 CASE ANALYSES

3.1 SCENARIO ANALYZED

In order to grasp the general idea of the uncertainty of risk assessments due to the groundwater transport portion of the analysis, a scenario which imposes the maximum health impacts to the downstream population and critical population group was selected for demonstration. Among the sites that EPA analyzed, a site located in the southeastern part of the United States imposes the highest impacts to the down stream population and was selected for this purpose. The site is exposed to humid meteorological conditions and the waste, with little pretreatment, is disposed of in a moderately porous soil formation.

3.2 RANDOM INPUT PARAMETERS

Among those input parameters required to calculate the health impacts indicators, eight parameters which were judged to be the most sensitive and have the largest uncertainty in the evaluation were selected as random inputs to the model. They are: 1) distribution coefficient for the waste, 2) distribution coefficient for the host soil, 3) distribution coefficient for the aquifer, 4) rate of trench cap failure, 5) distance from trench bottom to aquifer, 6) distance from disposal site to the discharge point or environmental access point, 7) groundwater velocity, and 8) percolation velocity.

Since extensive field and laboratory measurement data for above parameters were not available, the probability density distribution of the random input parameters were assumed arbitrarily. The standard deviation for input parameters are maintained between 12% and 25% of their mean values.

3.3 RESULTS OF ANALYSES AND DISCUSSION

The uncertainty analyses were conducted for the total health effects and maximum individual dose indicators. They were analyzed for three radionuclides: H-3, C-14, and I-129. The results of probability density distribution analyses are normalized with their mean values and are presented in Table 1

Table 1. Results of Standard Deviation Calculations

Radio-nuclide	Inputs	Outputs	
		Total H. E.	Max. Indiv. Dose
	(%)	(%)	(%)
C-14	12 - 25	0.	8.
I-129	12 - 25	0.	26.
H-3	12 - 25	65.	47.

In spite of a large standard deviation assumed for the input parameters, the output parameters for the total health effects indicator analysis converge into a very small standard deviation except for the short half-life radionuclide, H-3. The small uncertainties obtained are due to the fact that the radionuclides are mobile and have a long half-life. Therefore the change in the transit time due to the uncertainty in the input parameters will not significantly affect the cumulative mass of radionuclide released to the downstream basin.

Although the standard deviations for the maximum individual dose indicators are in general greater than that for the total health effects indicators, some of the standard deviations for the long half-life radionuclide still converge from the standard deviations for the input parameters. This tendency is due to the fact that the change in transit time is more sensitive to the maximum individual dose indicator than the total health effects indicator.

Figures 1 and 2 show typical inputs and outputs probability density distribution for C-14 respectively for the total health effects indicator and the maximum individual dose indicator analyses.

4. UNCERTAINTIES OF RISK ASSESSMENT

The results of risk assessments conducted by the EPA for the national low-level radioactive waste disposal management systems indicate that the health impacts for a short term analysis are predominated by those long half-life and mobile radionuclides, such as C-14 and I-129 [3 and 4]. The short term analysis as defined here denotes 10,000 years for the total health effects analysis and 1000 years for the maximum individual dose analysis. As indicated in the previous analysis, the uncertainties of the health impacts analyses for these radionuclides are generally small. Therefore, one may conclude that the uncertainties of the risk assessments due to groundwater transport of radionuclides are not excessive.

One should also note that the shorter half-life radionuclides and/or immobile radionuclides will impose much larger uncertainties to the health effects indicators. However, these radionuclides will not, in general, impose significant health effects during a short term risk assessment.

5. CONCLUSIONS

For the purpose of uncertainty analysis, the proposed approach, using an approximated analytical solution in place of the sophisticated PRESTO-EPA risk assessment model, will provide a reasonable approximation. The same simplification can also be applied for a rough risk assessment of a LLW disposal site.

The proposed approach can also considerably reduce the time of computer simulation from that of a Monte Carlo approach. Furthermore, the proposed approach takes any irregular form of probability density distribution for the input parameters. Therefore, the distribution of the input parameters do not have to be converted into a selected mathematical formula. This avoids some unnecessary errors.

The results of these analyses indicate that the uncertainties of the total health effects analyses are negligibly small. As to the uncertainties of the maximum individual dose indicator, they are somewhat larger than that for the total health effects analyses but are not excessive.

In general, the health impacts from a LLW disposal site are predominated by long half-life and mobile radionuclides in a short term analysis, 10,000 years for the total health effects analysis and 1000 years for the maximum individual dose analysis. Furthermore, since the uncertainties of the health impact analyses for these radionuclides are not excessive it also implies that the uncertainties of the performance assessment due to groundwater transport will not be excessive.

REFERENCES

1. Gruhlke, J.M., et. al., "US EPA's Proposed Environmental Standards for the Management and Land Disposal of LLW and NARM Waste," Proceedings of the Waste Management Symposium, February 26 - March 2, 1989, Tucson, Arizona.

2. Hung, C.Y., et. al., "Use of PRESTO-EPA Model in Assessing Health Effects from Land Disposal of LLW to support EPA's Environmental Standards," Proceedings of the 5th Annual Participants' Information Meeting on DOE LLW Management Program, Denver, Colorado, August 30, 1983.

3. Hung, C.Y., et. al., "Results of EPA's Risk Assessments of Alternative Methods of LLW Disposal," Proceedings of the 8th Annual DOE LLW Management Forum, Denver, Colorado, September 22 - 26, 1986.

4. U.S. Environmental Protection Agency, "Draft Environmental Impact Statement for Proposed Rules, Volume 1: Background Information Document," EPA Report, EPA 520/1-87-012-1, June 1988.

5. Hung, C.Y., "Prediction of Long Term Leachability of a Solidified Radioactive Waste from a Short-term Leachability Test by A Similitude Law for Leaching Systems," Proceedings of the Nuclear and Chemical Waste Management, Vol. 3, pp 253-243, 1982.

6. Hung, C.Y., "A Simplex Method of Estimating the Uncertainty of Simulating the Release and Transport of Radionuclides from a LLW Disposal Site for Health Risk Assessment," EPA Interim report, Waste Management Standards Branch, September, 1985.

7. Hung, C.Y., "An Optimum Groundwater Transport Model for Application to the Assessment of Health Effects Due to Land Disposal of Radioactive Waste," Proceeding of Nuclear and Chemical Waste Management, Vol. 6, PP 41-50, 1986.

8. Larsen, R.J., et. al., "An Introduction to Mathematical Statistics and its Applications," Prentice-Hall, Inc., Englewood Cliffs, N.J., 1981.

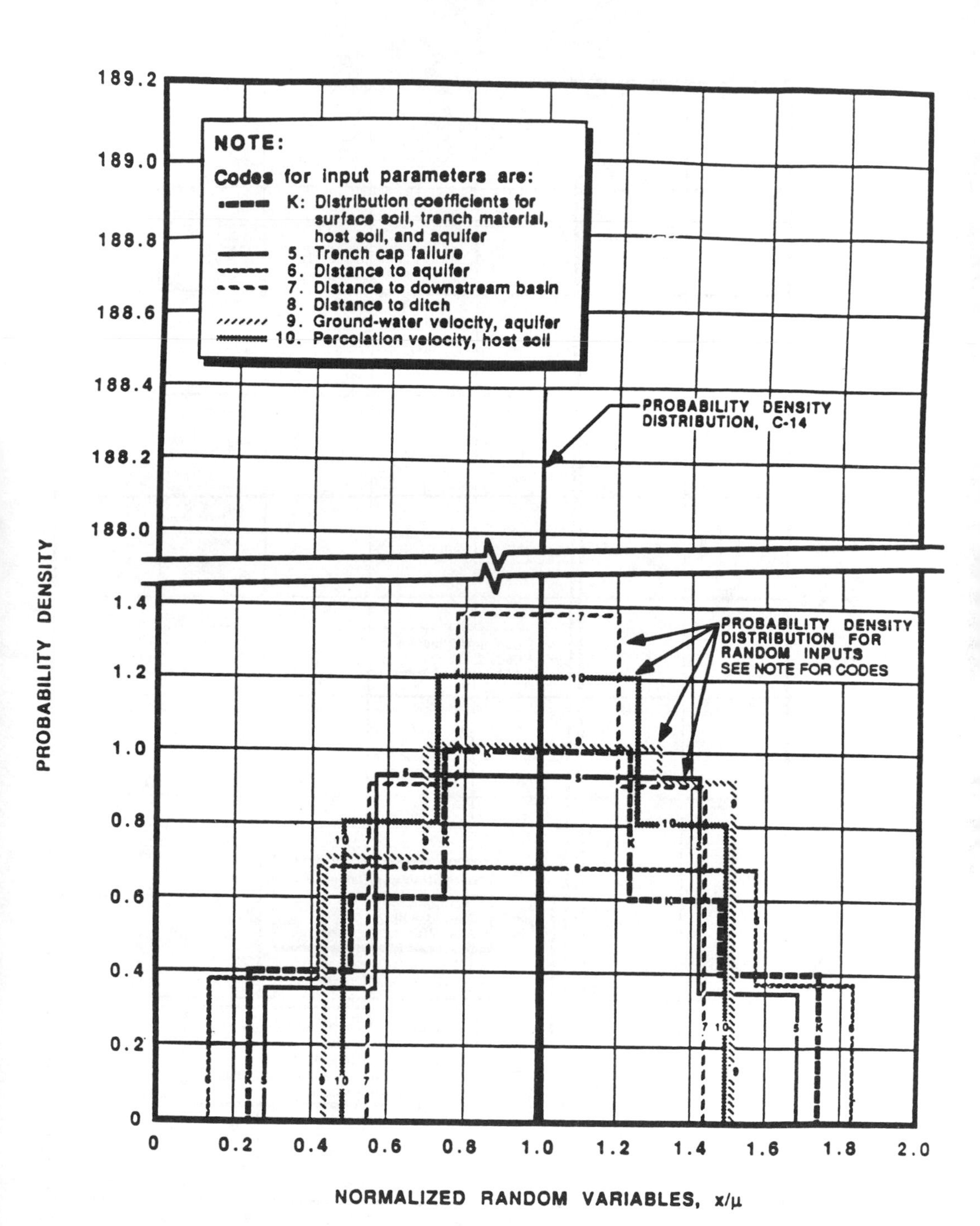

Figure 1. Results of Uncertainty Analysis for the Total Health Effects indicator, C-14

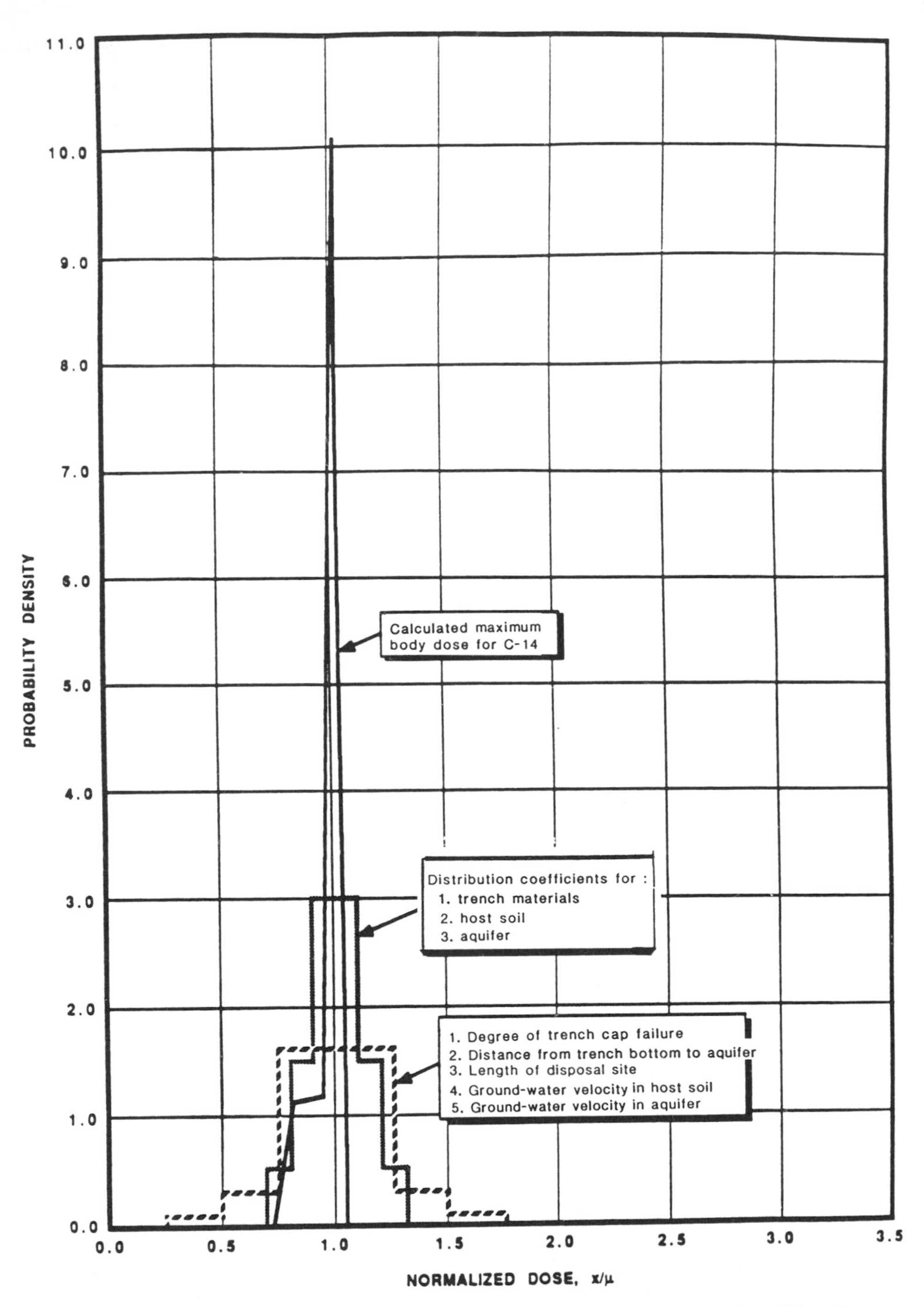

Figure 2. Results of Uncertainty Analysis for the Maximum Individual Dose indicator, C-14

A COMPARISON OF RESULTS FROM GROUNDWATER FLOW MODELLING FOR TWO CONCEPTUAL HYDROGEOLOGICAL MODELS FOR THE KONRAD SITE

Georg Arens, Eckhard Fein, Richard Storck
Gesellschaft für Strahlen- und Umweltforschung MbH München
Federal Republic of Germany

ABSTRACT

Radioactive wastes with negligible heat production are planned to be disposed of into a deep iron ore formation at the Konrad site. This repository will be bedded in a low permeable formation called Oxfordian in a depth of 800 - 1300 m below the surface. The host formation is largely covered with clay of a few hundred meters thickness. The hydrogeological model area has an extension of 14 km in the west-east and 47 km in the north-south direction.

The geological formations within the model area are disturbed by several fractured zones with a vertical extension of several hundred meters intersecting different horizontal layers. Due to this fact two hydrogeological models have been developed: The first one handles the fractured zones by globally increased permeabilities of the geological formations. The second handles the fractured zones by locally increased permeabilities, leaving the permeabilities of undisturbed areas unchanged. For both models, groundwater flow calculations have been carried out including parameter variations of permeability values. The results of the calculations are presented as flow paths which are compared for both models.

COMPARAISON DES RESULTATS DE LA MODELISATION DE L'ECOULEMENT DE L'EAU SOUTERRAINE CONCERNANT DEUX MODELES HYDROGEOLOGIQUES THEORIQUES APPLICABLES AU SITE KONRAD

Georg Arens, Eckhard Fein, Richard Storck
Gesellschaft für Strahlen- und Umweltforschung MbH München
Federal Republic of Germany

RESUME

On prévoit d'évacuer des déchets radioactifs très faiblement calogènes dans une formation profonde riche en fer au site Konrad. Ce dépôt sera aménagé dans une formation faiblement perméable appelée Oxfordien à une profondeur comprise entre 800 et 1 300 mètres. La formation d'accueil est en grande partie surmontée par une couche d'argile de quelques centaines de mètres d'épaisseur. La zone hydrogéologique modélisée a 14 km dans le sens est-ouest et 47 km dans le sens nord-sud.

Les formations géologiques à l'intérieur de la zone modélisée sont affectées par plusieurs zones fissurées d'une extension verticale de plusieurs centaines de mètres qui recoupent diverses couches horizontales. Cela explique pourquoi on a élaboré deux modèles hydrogéologiques. Dans le premier, on a pris en compte les zones fissurées en accroissant globalement la perméabilité des formations géologiques. Dans le second, on a pris en compte les zones fissurées en accroissant localement la perméabilité sans modifier celle des zones non affectées. Dans les deux cas, on a exécuté les calculs relatifs à l'écoulement de l'eau souterraine en faisant varier les paramètres des valeurs de perméabilité. Les résultats des calculs sont présentés sous forme de voies d'écoulement qui sont comparées entre elles dans les deux modèles.

Introduction

One of the most important subjects of a performance assessment of geological isolation systems for radioactive wastes is the calculation of groundwater flow in the environment of a repository. The lengths and the travel times of flow paths from the repository to the surface are decisive for a potential contamination of the biosphere. Due to relatively large uncertainties of hydrogeological parameters it is often impossible to develop a realistic hydrogeological model. Instead of this an idealized or simplified model is elaborated using conservative assumptions.

Under contract to the PTB (Physikalisch-Technische Bundesanstalt) as part of the performance assessment for the Konrad site, two hydrogeological models were developed: the layer model /1/ which is a simplified and idealized model, and the fracture model which is a more realistic one. The validity of the simplifying assumptions of the layer model has been proved with the more realistic fracture model /1/. For example, the shortest travel time calculated with the idealized model should be less than the shortest travel time expected from a realistic model.

It should be pointed out that the fracture model is almost beyond the scope of the SWIFT code. The calculations yield travel times which are somewhat smaller but still of the same order of magnitude. They are performed only for comparison and not as part of the performance assessment for the Konrad site. For the latter purpose the simulations are repeated with a finite-element code /1/. The results of the finite-element simulation agree with the SWIFT results for the layer model.

The hydrogeological models

The Konrad site is located in Lower Saxony between the cities Braunschweig and Salzgitter. The repository will be bedded in a low permeable formation called Oxfordian in a depth of 800 - 1300 m below the surface. The host formation belongs to the Upper Jurassic and is largely covered with clay of the Lower Cretaceous of a few hundred meters thickness. The repository area has an extension of 1.8 km in the west-east and 3 km in the north-south direction. The hydrogeological model area is orientated along natural boundaries. It has an extension of 14 km in the west-east, 47 km in the north-south direction and a depth of 2.3 km. The model area is bounded in the west and east by salt domes, in the south by a salt dome and a watersheed and in the depth by the Zechstein. The boundary condition in the north direction is not very well known. A closed boundary is assumed, because of a very week watershed and the outcrop of the Oxfordian in the southern part of the model area. The geological structure is divided into more than 10 geological formations and eight fractured zones. The location of the Konrad site and the extension of the model area is shown in Fig. 1 .

The layer model is an idealized and simplified model. Due to the lack of data of propagation and permeabilities of the fractured zones, the zones are handled by globally increased permeabilities of the Oxfordian, Cornbrash sandstone, Rhaetian and the Upper shell limestone. It is assumed, that these geological formations are likely ways for propagation of radioactive wastes from the repository to the surface. Two variations of the layer model were performed. To take into account a disturbance of the covering clay, in the first variation the clay has a much higher permeability than in the second variation. A comparison between permeabilities of the layer models and the fracture model is shown in Table 1.

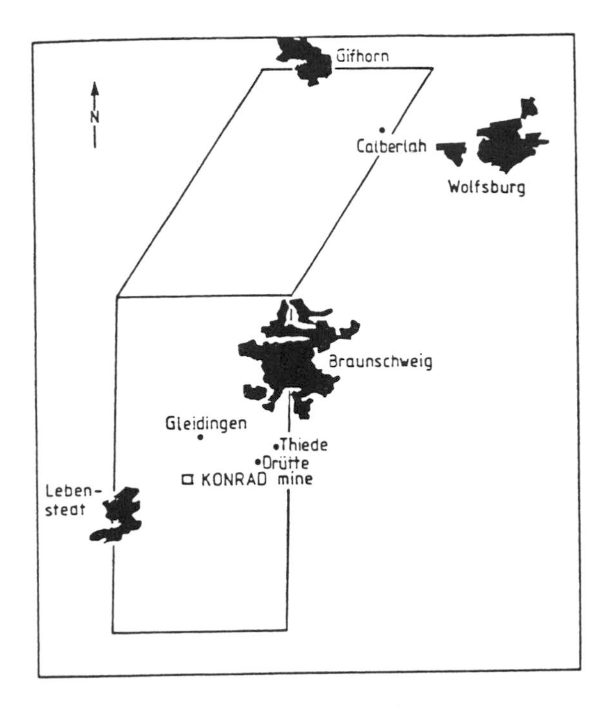

Fig. 1: Location of the KONRAD site

The geological structure of the fracture model is the same as in the layer model. Only the Albian is distinguished from the clay of the Lower Cretaceous. The fractured zones are now handled by locally increased permeabilities, leaving the permeabilities of the undisturbed areas unchanged. A factor for increasing or decreasing the permeabilities is given for each zone. The range of the factor is adjusted to the size of the fault and the amount of montmorillonites. Four variations were examined, which differ from each other by the value of these factors. The variation with the so-called best estimate values for the factors is called reference case. The range and the values of the factors for the variations are shown in Table 2.

The numerical method

The groundwater flow was simulated with the finite-difference code SWIFT (Simulator for Waste Injection, Flow and Transport). Due to limi-

geological formation	permeabilities (m/s)		
	layer model		fracture model
	Var 1	Var 2	
Quaternary, Tertiary	10^{-5}	10^{-5}	10^{-5}
Emscherian marl	10^{-8}	10^{-8}	10^{-8}
Upper Cretaceous, Pläner limestone	10^{-7}	10^{-7}	$5. \times 10^{-7}$
Albian			10^{-12}
Lower Cretaceous	10^{-10}	10^{-12}	10^{-11}
Hils sandstone	10^{-5}	10^{-5}	10^{-5}
Kimmeridge, Portland, Wealden	10^{-8}	10^{-8}	$5. \times 10^{-9}$
Oxfordian	10^{-7}	10^{-7}	10^{-8}
Cornbrash sandstone	10^{-6}	10^{-6}	$5. \times 10^{-7}$
Rhaetian	10^{-6}	10^{-6}	10^{-7}
Upper shell limestone	10^{-6}	10^{-6}	10^{-7}
Clay and marlstones of the Jurassic and Keuper	10^{-10}	10^{-10}	10^{-11}

Tab. 1: Permeabilities of the layer and the fracture model

fractured zones	factors for changing the permeabilities				range of the factor
	reference case	Var 1	Var 2	Var 3	
A	50	50	50	50	0.1 – 75
B	25	25	25	25	0.1 – 50
C	10	10	20	0.5	0.5 – 20
D	10	10	10	10	0.5 – 15
E	50	50	50	50	0.1 – 75
G	10	10	10	10	1.0 – 50
H	25	25	25	25	0.5 – 25
V	60	500	500	0.5	0.5 – 500

Tab. 2: Factors for changing the permeabilities

tations of computer resources, the complex geological structure of the Konrad site was only coarsely modelled. A rectangular finite difference grid was generated with 30 blocks in the south-north, 15 blocks in the west-east and 23 blocks in the vertical direction. The block sizes in the horizontal directions are 0.75 - 2.5 km and in the vertical direction 100 m. The boundaries of the model area are closed except at the surface. An annual averaged contour map of the water table was used to determine the pressure boundary condition at the surface. Streamlines were calculated in a semianalytical way using a special interpolation scheme of the flow velocities inside a grid cell /2/ and /3/. Since the geological structure is represented very rough the calculations of streamlines, especially for the fracture model poses a lot of problems. The travel times of streamlines which cross an aquitard are in many cases too low.

Results

To determine the flow paths from the repository to the surface, streamlines from every grid cell which is part of the repository area were started. To analyze the large number of streamlines, they were classified in four representative flow paths. A flow path with a relatively short traveltime or a flow path which is used very frequently was called representative. The results of all variations of the two models were summarized to define a representative flow path.

The first and most important representative flow path of the layer and the fracture model reaches the surface in the north-east of the model area. The streamlines pass through the Oxfordian and the Cornbrash sandstone and require more than 200 000 years to reach the surface near by the town Calberlah.

The second and most important representative flow path of the layer model variation 1 reaches the surface above the repository. Due to the decreased permeability of the clay of the Lower Cretaceous the streamlines pass through the clay and require more than 400 000 years to reach the surface near by the town Gr. Gleidingen.

flow path	shortest traveltime in millions of years					
	layer model		fracture model			
	Var. 1	Var. 2	ref. C.	Var. 1	Var. 2	Var. 3
1		0.3	0.2	0.2	0.3	0.4
2	0.4		3.2	0.9	0.5	
3			1.2	1.4		3.9
4			5.		0.6	

Tab. 3: Traveltimes of representative flow paths

The third representative flow path crosses the underlaying clay to enter the Rhaetian. Passing along this layer it finally reaches the surface near by the town Thiede.

Near by the town Drütte the fourth flow path reaches the surface crossing the overlaying clay and passing through the Hils sandstone and the Pläner limestone.

The third and the fourth representative flow paths are results of the fracture model. They are caused by a fault which crosses the repository area. The increased permeabilities of the fractured zone enables the streamlines to cross the overlaying and underlaying clays. The traveltimes of representative flow paths three and four are very dependent on the permeability of the fractured zone, but they are not shorter than 600 000 years. The representative flow paths are shown in Table 3.

Conclusions

The results of the simplified and idealized hydrogeolgical model show two important flow paths. The first passes through the host formation the Oxfordian and reaches the surfrace after more than 200 000 years. The second crosses the clay of the Lower Cretaceous and reaches the surface after more than 400 000 years. These flow paths are confirmed by results of the more realistic hydrogeological model. The increased permeabilities of the fractured zones do not cause flow paths with significantly shorter traveltimes than the flow paths of the layer model.

A main conclusion of the performance assessment of the Konrad site is: the traveltimes of flow paths from the repository to the surface are much longer than 100 000 years. This statement was confirmed with two conceptual different hydrogeological models. The locally increased permeabilities of the fractured zones of the fracture model do not influence this main statement. Therefore it is not necassary to start a very expensive research program to determine the permeabilities of the fractured zones.

References

/1/ Illi H., Langzeitsicherheit des Endlagers Konrad, 13. Tagung Radioaktiver Abfall - Langzeitsicherheitsnachweis bei der Endlagerung radioaktiver Abfälle, Haus der Technik, Essen, 31.10.1989

/2/ Pollock D.W., 1988, Semianalytical Computation of Path Lines for Finite-Difference Models, Groundwater, Vol 26, No. 6

/3/ GSF: SAPT - SWIFT Analytical Particle Tracking, Version I, Braunschweig, August 1988

ARE PARTICLE-TRACKING ALGORITHMS THE ADEQUATE METHOD TO DETERMINE MIGRATION PATHS FOR PERFORMANCE ASSESSMENT?

P. Bogorinski
Gesellschaft für Reaktorsicherheit (GRS) mbH, Köln, FRG

A. Boghammar
Kemakta Konsult AB, Stockholm, Sweden

ABSTRACT

Nuclear waste repositories have to be constructed in such a way that no hazardous effects on the environment do occur even after closure of the facility. To show compliance with this requirement longterm safety assessments simulate potential releases of radionuclides from the repository, their migration through the geosphere and their potential hazards in the biosphere. An integral part of these simulations is the calculation of goundwater flow and nuclide transport in the rocks surrounding the repository. Since transport calculations require great computational efforts these should be restricted to the regions of interest. These are usually derived from the groundwater flow models by means of particle-tracking algorithms. However, these do not necessarily represent the fastest travel path to the environment. Therefore one has to be aware of the limitations of pathlines for use in longterm safety assessments.

COMMENT DETERMINER LES VOIES DE MIGRATION DANS L'EVALUATION DU COMPORTEMENT DES SITES D'EVACUATION : LES ALGORITHMES DE TRACAGE DE PARTICULES SONT-ILS LA METHODE APPROPRIEE ?

RESUME

Il faut que les dépôts de déchets nucléaires soient construits de telle façon qu'aucun effet dangereux pour l'environnement ne se produise même après la fermeture de l'installation. Pour montrer que cet impératif est respecté, les évaluations de la sûreté à long terme simulent les rejets potentiels de radionucléides hors des dépôts, leur migration à travers la géosphère et les dangers éventuels qu'ils présentent dans la biosphère. Le calcul de l'écoulement de l'eau souterraine et du transport des radionucléides dans les roches encaissant le dépôt font partie intégrante de ces simulations. Etant donné que les calculs relatifs au transport demandent beaucoup de temps d'ordinateur, on doit se cantonner aux zones revêtant de l'intérêt. Celles-ci sont généralement déterminées à partir des modèles d'écoulement de l'eau souterraine par le biais d'algorithme de traçage des particules. Toutefois, ces derniers ne représentent pas forcément le trajet le plus rapide jusqu'à la biosphère. Il importe donc de garder à l'esprit la valeur relative de ces trajets quand on les utilise pour évaluer la sûreté à long terme.

1. INTRODUCTION

Longterm safety assessments of nuclear waste repositories have to show, that even after closure of the repository no hazardous effect on the environment will occur due to a potential release of radionuclides. One of the most likely postulated scenarios considers the transport of radionuclides from the repository to the surface by flowing groundwater. Safety analyses based on this scenario involve the simulation of the groundwater flow and nuclide transport. In most cases simulation of groundwater flow requires the definition of an extensive model region in order to account for natural boundary conditions or to avoid errors due to the influence of uncertain boundary conditions. Because of limitations in computational resources (time and memory), big models only allow relatively coarse dicretisation. This may be sufficient for steady state groundwater flow calculations. However, radionuclide transport models require a high degree of discretisation to avoid large numerical dispersion. Therefore the model for transport calculations should contain only those regions where the nuclides are expected to migrate to minimise the computational effort. Under certain conditions it may even be sufficient to carry out the nuclide transport simulations using 1D-models.

Particle tracking algorithms, which calculate the trajectories of water particles starting from the repository within the potential field, provide the link between groundwater flow and nuclide transport models. These trajectories or pathlines are assumed to be the main migration paths for the radionuclides. Therefore nuclide transport simulations have to consider only those regions of the model area that are reached by pathlines. If 1D-calculations are intended it is very important to be sure that trajectories provide the realistic migration path. However, this may not always be the case. Our paper will show that one has to be careful to select the appropriate 1D-modell for transport calculations.

First we will describe the model that has been set up to simulate the HYDROCOIN level 1 case 6 with the ground water flow and transport code NAMMU. For further investigations this model has been modified to allow more realistic calculations and to show more clearly the effects mentioned above. Finally we will describe the result of nuclide transport calculations and the comparison to the pathline calculations.

2. COMPUTER CODES USED

2.1 Simulation Codes

For our study of ground water flow and nuclide transport we used the NAMMU code (Numerical Assessment Method for Migration Underground) [1] that had been developed by AEA Technology at Harwell. The code is based on the finite-element method.

NAMMU models groundwater flow as single phase flow in a porous medium governed by Darcy's equation. As a result of the calculation one obtains the potential field depending on the properties of the rocks and the fluid and on the boundary conditions. Furthermore, transport phenomena for heat and radionuclides are included in the code. They consider advective and dispersive/diffusive transport. Heat transport may feed back to the potential field because of temperature dependent fluid density. The results of transport simulations are the time dependent temperature and concentration fields.

The numerical solution method is provided by the Galerkin finite-element subroutine library TGSL. NAMMU uses the TGIN package which allows free-format input language. It is logically structured in a tree-like manner.

2.2 Pre- and Postprocessors

To generate the model for the original test case the HYPAC program and subroutine library [2] was used developed by Kemakta under contract from the Swedish Nuclear Fuel and Waste Management Company (SKB). It was designed to facilitate the pre- and post-processing procedures of finite-element modelling, mainly in conjunction with groundwater simulations. HYPAC designs the finite-element mesh, checks it, assigns material properties and boundary conditions to the elements and optimises the mesh. In order to be able to use a pre-processed grid generated with HYPAC for NAMMU simulations, an interface subroutine was supplied.

3. HYDROCOIN-TEST-CASE

3.1 Description of the Model

In the first step of our study described here NAMMU was applied to HYDROCOIN test case 6 of level 1. The case was designed to test the ability of codes to model steady-state 3-dimensional groundwater flow in a regional system which is represented by a sequence of rock layers with highly contrasting permeabilities. Details of the model can be found in the HYDROCOIN report [3].

Figure 1 shows a cross-section through the model which was set up for the NAMMU simulations. The rocks on the left side are formed by a highly permeable alluvial system. The sequence of layers on the right side are from top to bottom a highly permeable fluvial system, a low permeable fluvial-lacustrine system, a nearly impermeable salt-shale system, a moderate permeable limestone-shale system and a relatively high permeable limestone-shale-granite system. The limestone systems and the alluvial system are separated by a transition zone with moderately low permeability.

There are four types of boundary conditions represented in the model. At the top surface a recharge rate is applied, except along surface rivers where a fixed head boundary condition corresponding to their elevation is used. The rest of the model boundaries are regarded as no-flow boundaries, except for the lower part of the left side boundary at the depth of the limestone systems, where a fixed head is applied.

Figure 1 also shows the finite element mesh. It was made up of 20-nodal cuboid elements. The finer grid in the upper part of the model was required by the higher pressure gradients caused by the recharge boundary condition. In order to save on computer memory and computing time the discretisation was reduced in the salt-shale system, since it virtually forms an aquiclude.

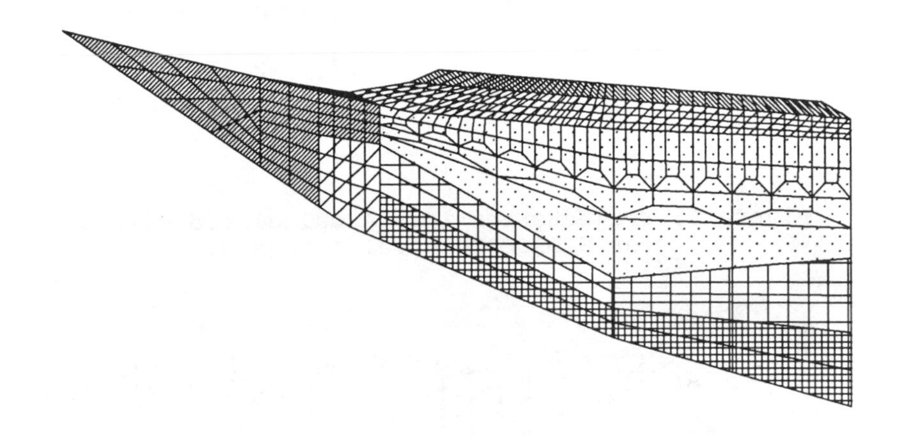

Figure 1: Cross-section through the NAMMU finite element grid for simulation of HYDROCOIN Level 1 Case 6.

3.2 Modelling Results

The results required for the HYDROCOIN-study were calculated heads along different lines, fluxes in a horizontal plane crossing the salt-shale system and trajectories starting from given points. In this paper, only the last group of results are reported. However, it should be mentioned that the other results agree very well with those obtained with other codes during the HYDROCOIN study.

The trajectories required were computed with two different procedures, namely with the HYPAC post-processor and with the NAMMU post-processor. Figure 2 shows the HYPAC results on a zoom of the region of interest, while figure 3 shows the NAMMU results.

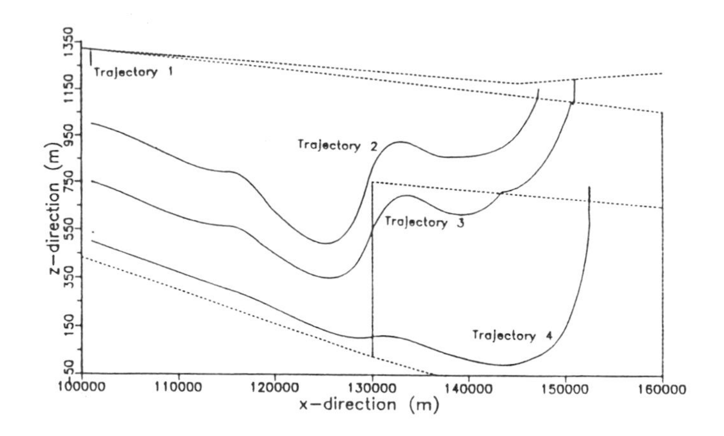

Figure 2: Trajectories 1 to 4 calculated with HYPAC for HYDROCOIN Level 1 Case 6.

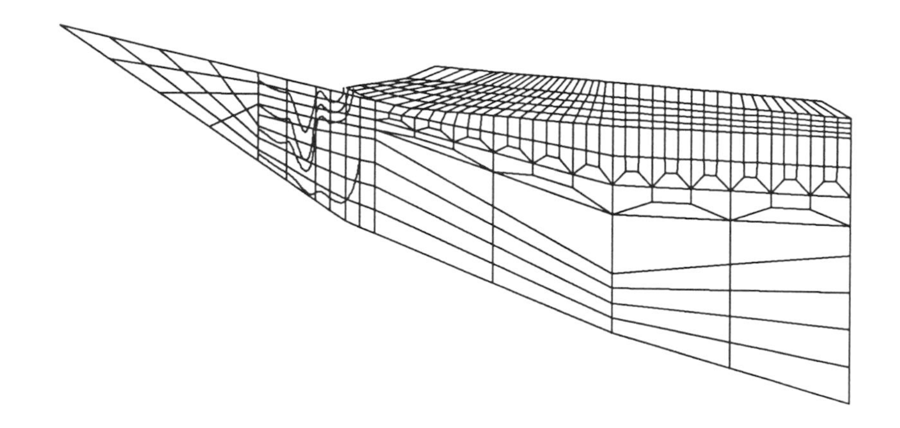

Figure 3: Trajectories 1 to 4 calculated with NAMMU for HYDROCOIN Level 1 Case 6.

Comparison of both results shows comparatively good agreement. The most obvious difference can be observed for trajectory 1. The HYPAC pathline goes straight up to the surface whereas the NAMMU pathline follows the general flow field in a similar way as the other three trajectories. This may be due to a different interpretation of the potential field influenced by the recharge boundary condition.

The results from both post-processors agree very well with one group of results from HYDROCOIN. However, it should be noted that trajectory 4 gets stuck at the interface between the transition zone and the alluvial system. This can be explained by the forced inflow boundary condition at the surface which distorts the potential field in that region.

4. MODIFIED HYDROGEOLOGICAL MODEL

4.1 Description of Modification

To obtain more realistic results the boundary condition at the top surface was modified. Instead of the recharge boundary condition which resulted in some unphysical calculated heads at the surface, as reported by HYDROCOIN [3], a pressure boundary condition was applied corresponding to the elevation of the top surface, which was thus assumed to be the groundwater table.

A second modification was due to the intention to carry out nuclide transport simulations. For such calculations the grid was thought to be too coarse resulting in a large amount of numerical dispersion. Therefore the grid had to be refined in the region of the salt-shale system as well as in lateral direction. The modified model was generated with the NAMMU pre-processor. The type of elements used for these simulations are 27-nodal cuboid elements.

A cross-sectional view is shown in figure 4. The difference in discretisation to the original model is obvious.

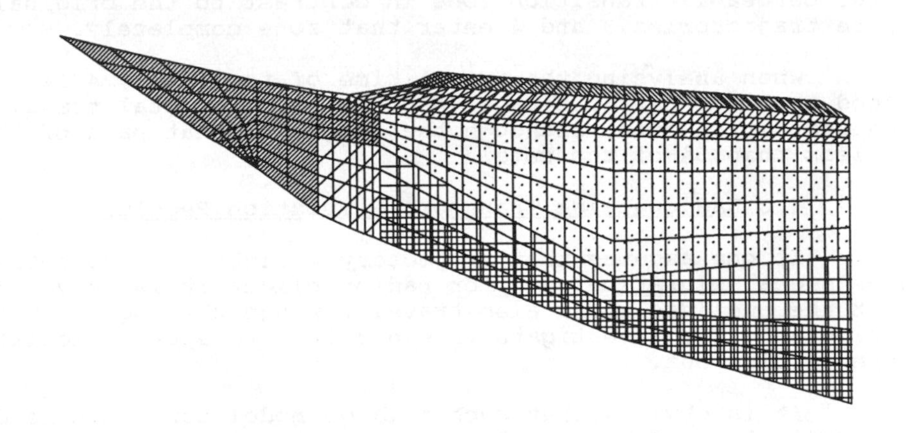

Figure 4: Cross-section through the NAMMU finite element grid of the modified model.

4.2 Groundwater Modelling Results

Since the potential field differs remarkably from the original model due to the modified boundary condition the trajectories are expected to differ as well. This is true as figure 5 shows.

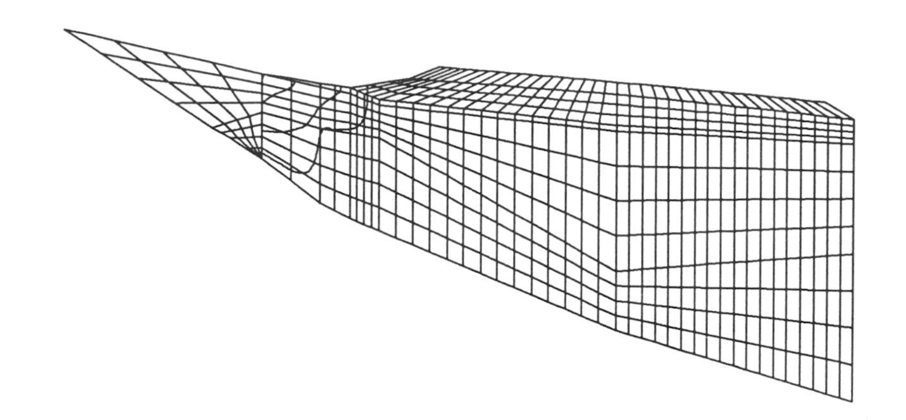

Figure 5: Trajectories 1 to 4 calculated with NAMMU for the modified model.

All trajectories follow immediately an upward path to the surface and none gets stuck. Only pathline 4 cuts across an edge of the lower permeable transition zone in contrast to the original model where trajectories 3 and 4 enter that zone completely.

When analysing the travel time of trajectory 4 it could be found that the particle uses up to 50% of its total travel time in the transition zone, although the length of that part of the path is less than 5% of the total.

4.3 Radionuclide Transport Simulation Results

If one assumes that trajectory 4 might be a potential 1-dimensional migration path for radionuclides it is very unlikely that the nuclides would also travel through the edge of the transition zone. To investigate this nuclide transport calculations were carried out.

It is obvious that such a large model could not be discretised finely enough to apply a reasonable dispersion length. Numerical criteria require that the dispersion length is not smaller than half the distance between two nodes to avoid a large amount of numerical dispersion. However, such a large dispersion spreads the nuclides over most of the model region but the maximum of the concentration should follow the main migration path.

The nuclide transport calculations that were carried out on the basis of the model used for the modified HYDROCOIN case showed the expected results. The nuclides migrate more or less straight up to the surface avoiding the lower permeable transition zone. Travel times were faster than calculated with the particle tracking procedure. Figure 6 shows the concentration contours after approximately one half of the time period calculated as travel time in the pathline calculation.

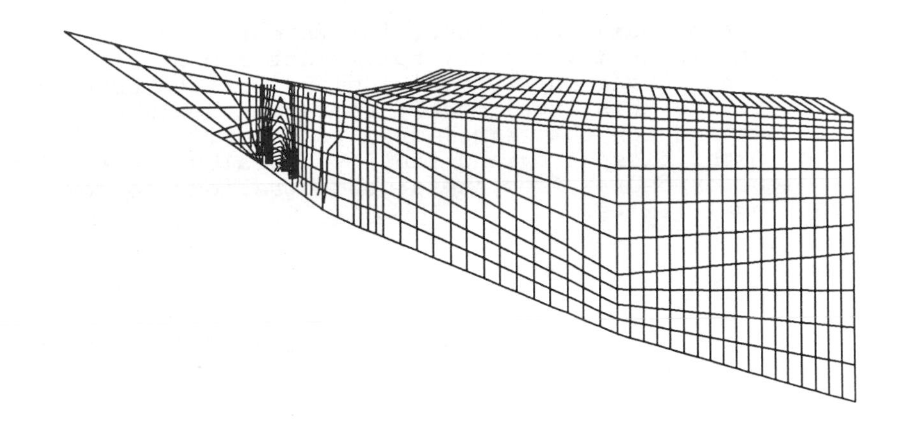

Figure 6: Nuclide concentration contours calculated with NAMMU for the modified model.

However, as anticipated, numerical dispersion causes a wide spread of nuclides. Hence, the results regarding the nuclide concentrations have to be considered to be unrealistic. Therefore the next step in our study will be to cut off most of the model leaving only that part extending from 100 km to 200 km. Then the grid will be refined, the potential field will be interpolated on the refined nodes and pressure boundary conditions will be derived from the original potential field, and finally appropriate boundary conditions will be assigned for the transport equation. For that model the transport calculation will be repeated expecting more realistic results.

5. CONCLUSIONS

Longterm safety assessments for nuclear waste repositories are carried out to show that even after long time periods exposures resulting from a release of radionuclides from the waste and penetration the engineered barriers are well below acceptable limits. Generally, migration of the radionuclides with the groundwater is assumed to be one of the most likely exposure pathways. Since these assessments can only be made by means of analytical tools, uncertainty of the results, due to the used numerical models, has to be reduced as much as possible. Therefore the potential travel paths for radionuclides have to be selected carefully. Under certain conditions post-processing of groundwater calculations by means of particle-tracking algorithms may not be sufficient. A careful analysis of calculated travel paths is necessary, especially in heterogeneous media. Our study has shown that otherwise the assessments will result in a false picture of long travel times if pathlines enter low permeable regions. There the nuclides may be retarded extensively if these paths are taken as migration paths. In those cases it may be worthwhile to carry out 3-dimensional nuclide transport simulations on a coarse grid accepting a larger amount of numerical dispersion to determine the regions where radionuclides may travel; to extract a model for these regions from the larger one by deriving appropriate boundary conditions; to refine this local model; to repeat 3-dimensional transport calculations on that finer grid to determine nuclide concentrations in the biosphere; and finally to calculate the potential radiological exposures from these concentrations to compare them with the legal limits.

REFERENCES

[1] Atkinson, R.; Herbert, A.W.; Jackson, C.P.; Robinson, P.C.: NAMMU User Guide DOE/RW/85.065, May 1985

[2] Grundfelt, B.: GWHRT - a Finite Element Solution to the Coupled Ground Water Flow and Heat Transport Problems in Three Dimensions SKB Technical Report KBS TR 83-51, 1983.

[3] The International HYDROCOIN Project. Groundwater Hydrology Modelling Strategies for Performance Assessments of Nuclear Waste Disposal Level 1: Code Verification OECD, Paris 1988.

SKB - SWEDISH HARD ROCK LABORATORY

PREDICTIVE GROUNDWATER FLOW MODELLING OF A LONG TIME PUMPING TEST AT ÄSPÖ

B. Lindbom, B. Grundfelt, A. Boghammar
Kemakta Consultants Co., Stockholm, Sweden

M. Liedholm, I. Rhén
VIAK AB, Gothenburg, Sweden

ABSTRACT

The Swedish Nuclear Fuel and Waste Management Co. (SKB) plans to establish an underground research facility located in the archipelago north of Oscarshamn in southern Sweden. As part of the research programme prior to the construction phase, efforts have been put into investigations aiming at describing the natural groundwater conditions, as well as developing numerical tools for the prediction of the impact from the laboratory in terms of groundwater flow. One step in this process was devoted the development of methods for the forecasting of drawdown responses caused by pumping activities in the surroundings of the potential laboratory.

The present project was concerned with the predictive groundwater flow modelling of a long time pumping test at Äspö. The predictive modelling was mainly oriented towards forecasting of expected drawdowns monitored in a number of boreholes as a response to the pumping in one of the core drill holes in the southern part of Äspö. Prior to the prediction, a number of calibration exercises were run through. These were concerned with the prescribed groundwater formation rate applied at the top boundary of Äspö, as well as calibration of hydraulic conductivities of the rock matrix. The hydraulic conductivities were calibrated against drawdown responses from three short term pumping tests.

The results to be predicted were the drawdowns in a number of boreholes responding to a long time pumping test in a core drill hole in the southern part of Äspö. The predicted results were subject to comparisons with the ones measured in field. The validity of the model set-up was judged by means of a goodness-of-fit function. This was applied to the calculated drawdowns, which corresponded to the final time-step in the transient calculation of the drawdown responses.

The groundwater flow code NAMMU, which is based on the finite-element method, was used for the simulations. The finite-element mesh that was generated to describe the modelled domain consisted of 20-noded three dimensional brick elements. The version of NAMMU used for this project, was installed on a Convex computer, model C-210. The project has been carried out under contract from SKB, and was reported jointly by representatives from Kemakta Consultants Co. and VIAK. Kemakta was responsible for the numerical modelling, while VIAK was responsible for the evaluation of field data as an input to the modelling.

SKB - LABORATOIRE SUEDOIS DANS DES ROCHES DURES

MODELISATION PREVISIONNELLE DE L'ECOULEMENT D'EAU SOUTERRAINE PROVOQUE PAR UN ESSAI DE POMPAGE DE LONGUE DUREE A ÄSPÖ

B. Lindbom, B. Grundfelt, A. Boghammar
Kemakta Consultants Co., Stockholm, Suède

M. Liedholm, I. Rhén
VIAK AB, Gothenburg, Suède

RESUME

La Société suédoise de gestion du combustible et des déchets nucléaires (SKB) prévoit de construire une installation de recherche souterraine dans l'archipel situé au nord d'Oscarshamn dans le sud de la Suède. Dans le cadre du programme de recherche préalable à la phase de construction, des études ont été réalisées en vue de définir les conditions naturelles de l'eau souterraine et d'élaborer des outils mathématiques pour prévoir les conséquences de l'installation du laboratoire sur le régime hydrogéologique. Une des étapes de ce processus a consisté à mettre au point des méthodes permettant de prévoir les rabattements provoqués par des pompages aux alentours du laboratoire envisagé.

Dans le projet décrit ci-après, il s'agissait de modéliser les effets sur l'écoulement de l'eau souterraine d'un essai de pompage de longue durée à Äspö. La modélisation portait essentiellement sur la prévision des rabattements escomptés dans un certain nombre de piézomètres d'observation à la suite d'un pompage dans l'un des trous de forage réalisés dans la partie sud d'Äspö. Avant de faire des prévisions, on a procédé à un certain nombre d'essais d'étalonnage qui ont porté sur le débit prescrit de la formation aquifère appliqué à la limite supérieure d'Äspö ainsi que sur l'étalonnage des conductivités hydrauliques de la matrice rocheuse. Les conductivités hydrauliques ont été étalonnées à partir des rabattements provoqués par trois essais de pompage de courte durée.

Il s'agissait de prévoir les rabattements dans un certain nombre de piézomètres à la suite d'un essai de pompage de longue durée dans un trou de forage réalisé dans la partie sud d'Äspö. On a ensuite comparé les résultats obtenus avec les mesures faites sur le terrain. On a jugé de la validité du modèle au moyen d'une fonction permettant d'apprécier le degré de concordance. Cette méthode a été appliquée aux rabattements calculés qui correspondaient au pas de temps final dans le calcul transitoire des rabattements induits.

Pour les simulations on a utilisé le programme de calcul NAMMU, qui simule l'écoulement des eaux souterraines par la méthode des éléments finis. La maille d'éléments finis qui a été créée pour décrire le domaine modélisé se compose de 20 volumes unitaires tridimentionnels nodélisés. La version du programme NAMMU utilisée pour ce projet a été installée sur un ordinateur Convex, modèle C-210. Ce projet a été sous-traité par le SKB et présenté conjointement par des représentants de Kemakta Consultants Co. et de la firme VIAK. Kemakta était responsable de la modélisation numérique alors que VIAK était chargée de l'évaluation des données in situ destinées à alimenter le modèle.

1. Introduction

The Swedish Nuclear Fuel and Waste Management CO (SKB) plans to establish an underground research laboratory – the Swedish Hard Rock Laboratory – to be sited close to the CLAB–facility in southern Sweden, at the isle Äspö located close to the nuclear power plant Simpevarp, see Figure 1 for the approximate location.

The primary objectives of the laboratory are to:

- Test and verify rock investigation methods of importance for the safety assessment of a final repository of high level waste.
- Develop and test methods for adjustments of the design and construction of the final repository to the rock conditions.
- Produce basic facts and data of importance for the safety and its analysis for the final repository.

The current phase – Preinvestigation – is aimed at siting the laboratory, describing the natural conditions in the bedrock and predicting the changes that will occur during construction of the laboratory. In a project preceding the present one [1], feasible boundary conditions and influence area for a smaller scale model were investigated in a rough manner. The present project therefore aimed at:

- establishing a 3D numerical model of Äspö,
- calibration of recharge rates (or infiltration capacities) and hydraulic conductivities,
- prediction of the pressure responses from a long time pumping test, and
- validation of the model set-up.

The numerical code used within this project is NAMMU [2], with the use of the program package HYPAC [3] for pre- and post processing purposes.

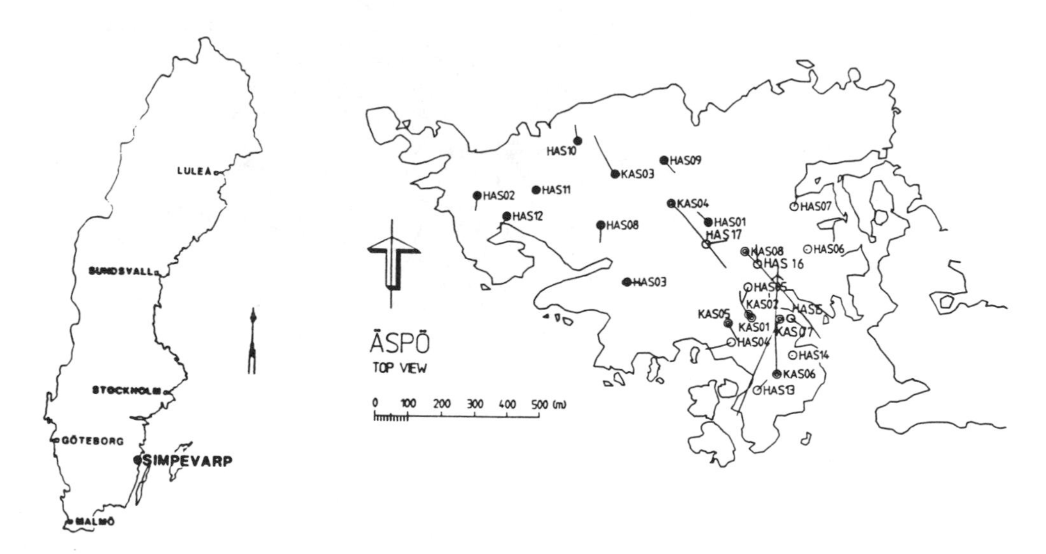

Figure 1 *Map of Sweden indicating the approximate location of Äspö (left) and borehole map of Äspö (right).*

2. Conceptual Model – Numerical Model – Input Data

Based on results reported in [4], the rock mass in the domain to be modelled was divided into five General Property Areas (GPA1-GPA5) horizontally, divided into 10 layers. These blocks thus defined represent different hydrological units, with individually assigned hydraulic conductivities. In addition, the domain contains some conductive structures which have been identified during the field work. The finite element mesh describing Äspö and its vicinity is shown in a plan view in Figure 2, where the conductive structures are denoted FW1-FW4, and the major discontinuity dividing the isle in two parts is called Äspö Shear Zone. The hydraulic conductivities of the General Property Areas are indicated layer-wise in Table I. These values correspond to the initial values and may change during the forthcoming calibration procedure.

The lateral boundaries were of no-flow type, as well as the bottom boundary. The mesh was extended far enough not to have any undesired boundary effects introduced in the model. The depth of the model is 2000 m. The boundary condition along the top surface of the model was a constant recharge rate prescribed over the isle Äspö (except for spot-wise appearing marshes where atmospheric pressure was prescribed). The part of the domain located outside Äspö was regarded as "zero-contours" coinciding with the sea-level, i.e. atmospheric pressure was prescribed to this area as well. The nodal points at the top surface of the mesh were vertically adjusted to coincide with the natural groundwater table.

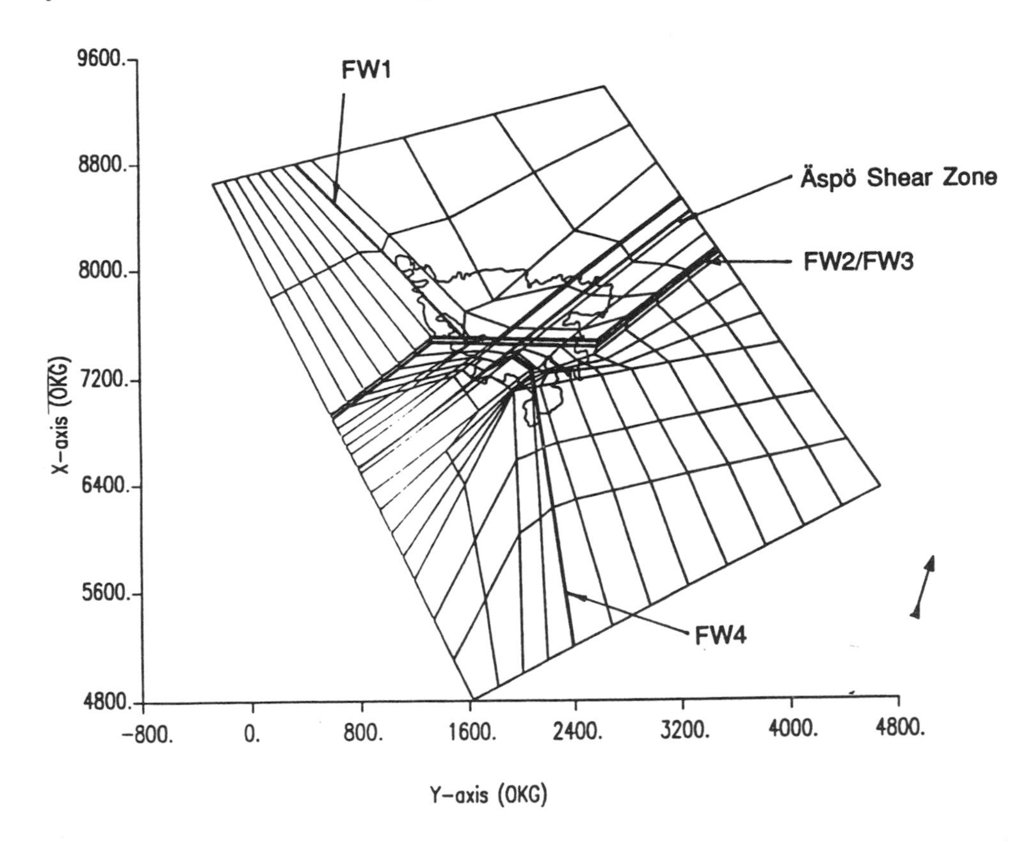

Figure 2 *A plan view of the mesh. The solid line corresponds to the shore-line of Äspö.*

Table I Initial hydraulic conductivities (K) of the General Property Blocks (log K m/s).

Layer	Level(m)	General Property Areas GPA 1	GPA 2	GPA 3	GPA 4	GPA 5
A	0-50	-7.40	-8.02	-8.69	-10.20	-9.41
B	50-100	-7.53	-8.12	-8.76	-10.18	-9.44
C	100-300	-8.50	-7.98	-8.00	-10.14	-8.72
D	300-500	-8.00	-8.57	-9.17	-10.50	-10.50
E	500-600	-10.50	-10.37	-10.25	-10.00	-10.00
F	600-800	-9.00	-9.00	-9.00	-9.00	-9.00
G	800-850	-10.00	-10.00	-10.00	-10.00	-10.00
H	850-1200	-9.20	-9.37	-9.55	-9.91	-9.73
I	1200-2000	-10.03	-9.96	-9.90	-9.77	-9.83

3. Calibration of Infiltration Capacity

Since the calculations would involve drawdowns, i.e. unsaturated conditions in the uppermost part of the model, a held potential boundary condition was not feasible along the top boundary of Äspö. Rather, a reasonable recharge rate (10 mm/year) was prescribed to the model, and the hydraulic conductivities of the two topmost layers were adjusted to make the calculated groundwater table coincide with the measured one. The natural groundwater table was based on piezometric measurements in the uppermost sections of the boreholes within the site. The calculated piezometric levels were compared to the natural by step-wise increasing the conductivities in the uppermost layers. The fit with observed data was judged by means of a goodness-of-fit function calculated according to the formula:

$$D_h = \sqrt{\frac{\sum_{i=1}^{n} (h_i^m - h_i^c - \Delta h)^2}{(n-1)}}, \quad \Delta h = \frac{\sum_{i=1}^{n} (h_i^m - h_i^c)}{n}, \qquad (1)$$

where D_h is the goodness-of-fit value, n is the number of observed data, h^m is the measured piezometric level, and h^c is the calculated piezometric level at a specific point.

The two layers of concern for the calibration of the recharge rate, were layers A and B. The simulations involved changes over the entire modelled domain, i.e. all general property areas (GPA1-GPA5) were affected. However, a strong emphasis was put on GPA4 and GPA5, since these were the areas where the largest deviations from the natural groundwater table were observed using the initial data from Table I.

The goodness-of-fit values obtained with the initial and the hydraulic conductivities (of the two topmost layers) calibrated with respect to their infiltration capacities and corresponding best-fit values are shown in Table II. Note that only the initial and final values are shown in this table; a number of intermediate simulations were also performed in order to reach the final values. Figure 3 displays the natural groundwater table and the one obtained with the calibrated hydraulic conductivities.

Table II Initial and calibrated conductivities (log K) and corresponding goodness-of-fit-value.

Layer	Level below sea level	GPA1	GPA2	GPA3	GPA4	GPA5	D_h
A	50 m	-7.40	-8.02	-8.69	-10.20	-9.41	141.3
B	50–100m	-7.53	-8.12	-8.76	-10.18	-9.44	
A	50 m	-7.40	-7.82	-7.19	-7.80	-7.60	0.65
B	50–100m	-7.53	-7.92	-7.26	-7.80	-7.60	

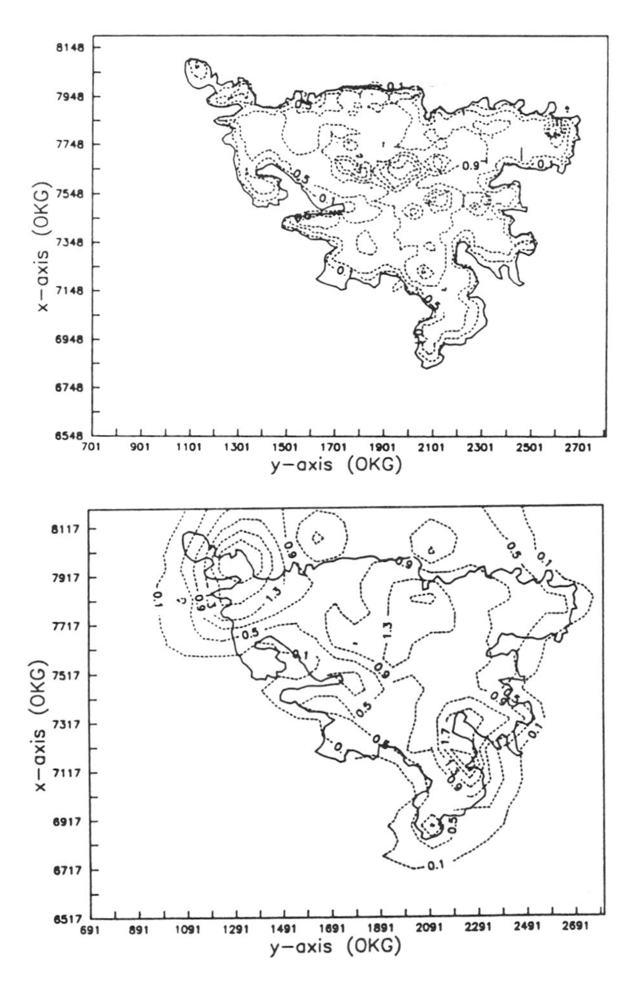

Figure 3 *Natural groundwater table (upper) and best-fitted calculated groundwater table, (lower). The artificial build-up of pressure contours in the north-western corner was accounted for by prescribing a constant head boundary condition (p=0kPa) to this node.*

4. Calibration of Hydraulic Conductivities

This phase involved calibration of the hydraulic conductivities under steady state conditions against the drawdowns in boreholes responding on short time pumping tests conducted in three core drill holes. Based on the assumption that measuring errors and evaluation errors are log-normally distributed, a Monte Carlo sampling routine was applied to the model in order to sample the conductivities in the domain. A standard deviation from the initial values of $\sigma_{\log k}=0.4$ was used. The sampled values were restricted to be within a range of 2σ on both sides of the mean value, which corresponds to a confidence interval for the sampled value of 95 %. The fits of the model results (i.e. the calculated drawdowns) to the monitored values were judged by means of a goodness-of-fit function formulated as follows:

$$D_s = \sqrt{\frac{\sum_{i=1}^{n} (s_i^m - s_i^c - \Delta s)^2}{(n-1)}}, \qquad \Delta s = \frac{\sum_{i=1}^{n} (s_i^m - s_i^c)}{n}, \qquad (2)$$

where D_s is the goodness-of-fit value, n is the number of observed data, s^m is the measured drawdown, and s^c is the calculated drawdown in a specific borehole section.

The pump tests were simulated by prescribing sink terms to nodal points, the positions of which corresponded to borehole sections in which the pumping equipment was installed. The recharge rate of 10 mm/year that was calibrated in the previous step was kept constant, as well as the hydraulic conductivities that were assigned to the two topmost layers.

The unsaturated conditions caused by the drawdowns were handled with a "sharp-interface" approach, which means that the derivative of the saturation with respect to the pressure in the capacity term will be effectively nil and that the relative permeability will be identical to one over the whole saturation range. This assumption is possible since the groundwater flow is assumed to take place in fissures which are expected to be so wide that the corresponding capillary forces can be neglected. Also, the feed of water from the partially saturated parts of the rock to the fissures is, because of its low permeability and strong capillarity, expected to be so slow compared to the duration of the pumping tests to be simulated that it can be neglected.

Eighteen simulations were performed with simulations of the three short time pumping tests one at a time. The goodness-of-fit function was applied to the calculational results. The combination that yielded the best fit value globally was considered as the best one, and was consequently used for the prediction of the long time pumping test. The last line in Table III corresponds to the global best-fit value. The calibrated conductivities are shown in Table IV.

Table III *Goodness-of-fit value (acc. to Eq 2) for the pump tests. Best-fits are underlined.*

Simul. #	Test 1	Test 2	Test 3
1-3	0.2969	0.2860	2.881
4-6	<u>0.2896</u>	<u>0.2564</u>	2.939
7-9	0.2974	0.2705	2.911
10-12	0.3030	0.3410	2.948
13-15	0.3065	0.2842	<u>2.780</u>
16-18	0.2921	0.3442	2.793

Table IV *The hydraulic conductivities (log K m/s) used to obtain the best-fit simulation according to Table III; cf Table I, which shows the initial values prior to the calibrations.*

Layer	Level below sea level	GPA1	GPA2	GPA3	GPA4	GPA5
A	0-50 m	-7.40	-7.82	-7.19	-7.80	-7.60
B	50-100 m	-7.53	-7.92	-7.26	-7.80	-7.60
C	100-300 m	-8.54	-8.22	-7.88	-10.18	-8.20
D	300-500 m	-8.76	-8.67	-9.28	-10.72	-10.74
E	500-600 m	-10.24	-10.45	-9.52	-9.60	-9.40
F	600-800 m	-9.58	-8.95	-8.64	-9.01	-8.45
G	800-850 m	-10.49	-9.38	-10.49	-10.49	-9.49
H	850-1200 m	-9.20	-9.37	-9.55	-9.91	-9.73
I	1200-2000 m	-10.03	-9.96	-9.90	-9.77	-9.83

5. Prediction of the Pumping Test – Validation of the Model Set–up

The model was now prepared for the transient prediction of the long time pumping test performed in a core drill hole in the southern part of Äspö. The conductivities shown in Table IV were assigned to the model. The duration of the pumping test was 55 days with an assumed drawdown of about 50-60 m in the pumping hole. The simulation was carried out with 54 time-steps with roughly 15 hours consumption of CPU-time on the computer used.

The main entities to be predicted were drawdown responses in the surrounding boreholes, and pathlines. Since the system was under heavy stress due to the sharp gradients introduced by the pumping, it proved to be extremely difficult to generate particle tracks that were reliable. It would probably have required a substantial increase of the grid resolution. This presentation will be focussed on the prediction of the responding drawdowns.

The drawdowns were to be predicted in roughly 45 open or packed-off borehole sections. The calculations indicated that the flow regime was divided into a northern part not affected by the pumping, and a southern part strongly influenced by the pumping. (The filled boreholes in Figure 1 correspond to boreholes not affected by the pumping.) This regional view of the flow system was also confirmed by the measurements. The goodness-of-fit function in Eq 2 was applied to the computed final drawdowns, i.e. the final time-step of the transient simulation.

The arithmetic value of the deviations between calculated and measured values was evaluated to 1.64 m taking all borehole sections into account, while the corresponding goodness-of-fit value was evaluated to D_s = 2.85 m. However, the evaluation of the goodness-of-fit function does not take the prediction of the transient phase into account. The calculated drawdowns were all predicted to appear too late compared to the measured values, whereas the shape of the calculated drawdown transients were similar to the measured when plotted log-linearly. This lead to the conclusion that the assumed specific storage used was too high, but that the calibrated conductivities were "correct" in a rough sense. Figure 4 illustrates this, where the solid line corresponds to the calculated drawdown, the dotted line is the monitored drawdown, and the thick dashed line corresponds to calculated drawdown with a correction factor of 100 in the time space. The thick dashed line thus illustrates how the model would have predicted the drawdown in this particular borehole, had a 100 times lower specific storage been assigned.

A change of the specific storage would affect the value of the goodness-of-fit function favourably, since an earlier response would have allowed the groundwater table to drop more than it did under the present circumstances (the calculation of the transient phase was interrupted at the time of the end of the duration of the pumping test). To cope with these tasks properly, a calibration of the storage capacities would have been necessary. This would have required transient calibrations, which was beyond the scope of the presented project.

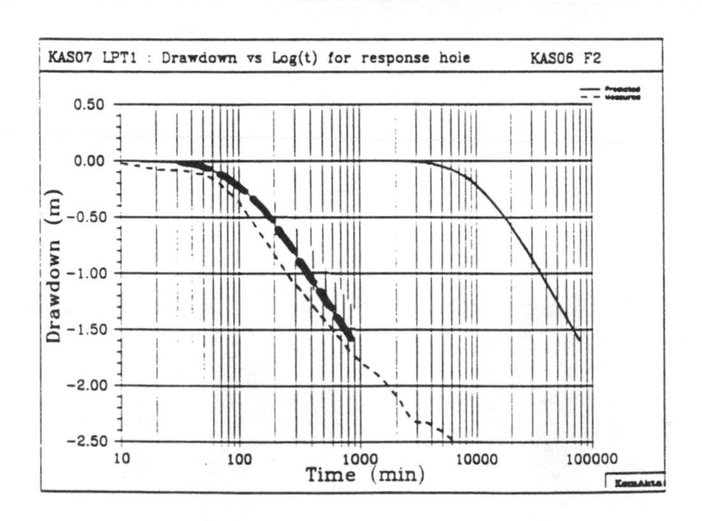

Figure 4 *Drawdown curve for one of the (packed-off) borehole sections. Solid line=predicted curve; thin dashed line=measured curve; thick dashed line=predicted curve corrected with a factor of 100 in time, i.e. the specific storage has been reduced with a factor 100.*

This paper is based on results reported by:
Grundfelt, B. M. Liedholm, B. Lindbom, I. Rhén, 1990,
"Predictive Groundwater Flow Modelling of a Long Time Pumping Test (LPT1) at Äspö", Progress Report xx-xx, SKB, Stockholm, Sweden. (Report in print)

References:

[1] Gustafson G., M. Liedholm, B. Lindbom, K. Lundblad, 1989, "Groundwater Flow Calculations on a Regional Scale at the Swedish Hard Rock Laboratory", Progress Report 25-88-17, SKB, Stockholm, Sweden.

[2] Atkinson R., A.W. Herbert, C.P. Jackson, P.C. Robinson, 1985, "NAMMU User Guide", Report AERE R–11364, U.K. Atomic Energy Research Establishment, Harwell Laboratory, United Kingdom.

[3] Grundfelt, B., A. Boghammar, H. Lindberg, 1989, "HYPAC User's Guide", SKB Working Report 89–22, SKB, Stockholm, Sweden.

[4] Liedholm, M., 1989, "Combined evaluation of geological, hydrogeological and geophysical information I", Progress Report 25–9–03, SKB, Stockholm Sweden.

Session 3

PROGRESS IN VALIDATION OF GEOCHEMICAL MODELS

Séance 3

ETAT D'AVANCEMENT DES TAVAUX DE VALIDATION DES MODELES GEOCHIMIQUES

Chairman - Président

T. NICHOLSON
(United States)

THE CHEMVAL PROJECT - AN INTERNATIONAL STUDY AIMED AT THE VERIFICATION AND VALIDATION OF EQUILIBRIUM SPECIATION AND CHEMICAL TRANSPORT MODELS

D Read[1], T W Broyd[1] and B Come[2]

[1] WS Atkins Engineering Sciences, Epsom, UK
[2] Commission of the European Communities, Brussels, Belgium

ABSTRACT

This paper describes activities within CHEMVAL, a three year project concerned with the verification and validation of geochemical models. Each of the four main project stages is described both in terms of the modelling work undertaken and the accompanying effort to provide a reviewed thermodynamic database for use in radiological assessment. Seventeen organisations from eight countries are participating in CHEMVAL, which is being undertaken within the framework of the Commission of European Communities MIRAGE2 programme of research.

1. INTRODUCTION

Increasing confidence in the capabilities of geochemical models has led to them being used to support radiological safety assessments in a number of countries around the world. Such usage naturally requires a clear demonstration that both computer programs and associated databases have been subjected to rigorous testing procedures. Verification and validation of chemical equilibrium models form the basis of CHEMVAL [1] which aims to assess their performance when applied to actual field and laboratory data sets.

The CHEMVAL project consists of four main stages:

STAGE 1 : verification of chemical equilibrium models

STAGE 2 : attempted validation of speciation-solubility models by comparison with experimental field and laboratory data.

STAGE 3 : verification of coupled chemical transport models

STAGE 4 : attempted validation of coupled models

The current phase of CHEMVAL is now complete. This paper summarises the main findings of CHEMVAL Stages 1 and 2 and highlights selected results from the coupled model verification/validation exercises. The outcome of the parallel database review is also discussed in view of its potential role in radiological risk assessment studies.

2. CHEMVAL STAGES 1 AND 2 : Verification/validation of speciation models

Owing to the large number of codes and thermodynamic databases available to CHEMVAL participants [1] an extensive verification programme was felt to be essential prior to attempting any validation studies. Twenty eight test cases based on five separate model systems were devised and modelled by participants in CHEMVAL Stage 1 [2]. An agreed methodology was followed throughout in order to distinguish variation in results caused by differences in thermodynamic data from that caused by other factors such as coding errors or procedure.

Sufficient results were obtained to demonstrate the ability of current tools for solving the diverse set of geochemical problems considered [1]. No serious "bugs", as such, were identified in any of the codes used, though each had its shortcomings with respect to particular modelling functions. For further information on the verification exercise the reader is referred to the CHEMVAL Stage 1 report [2].

Whereas the verification study considered realistic but largely hypothetical test cases, CHEMVAL Stage 2 was concerned with the performance of equilibrium codes and data when simulating real field and laboratory data sets. Four candidate test sites were selected for model validation studies, based on data availability and relevance to radiological risk assessment; namely Mol, Gorleben, Maxey Flats and Oman [3]. Nineteen test cases of varying complexity were devised to allow comparison between modelling results and experimental measurements. However, since none of the four sites had been fully characterised, field analyses from Mol and Maxey Flats were supplemented by data from controlled experiments to ensure adequate representation of relevant processes.

For the purposes of CHEMVAL Stage 2, validation was defined as *qualitative or quantitative evidence that the models accurately reflect laboratory or field observations*. In these terms, positive evidence of equilibrium model validation was obtained from three of the four sites studied, specifically:

Mol - actinide solubility in synthetic clay water (Figure 1)

Gorleben - americium solubility and (indirectly) actinide complexation by EDTA and citrate [3]

Maxey Flats - iron solubility and pH changes during controlled oxidation of anoxic trench leachates [3].

Although these results are encouraging, they all relate to laboratory studies and cannot be extended to the field situation at present. In the case of the Boom Clay, this clearly reflects the dominant role played by high molecular weight (HMW) organics in heavy metal immobilisation [4]. Such processes could not be simulated rigorously using existing codes and data.

3. CHEMVAL STAGES 3 AND 4 : VERIFICATION/VALIDATION OF COUPLED MODELS

In comparison to equilibrium speciation packages, coupled chemical transport codes are new and relatively untested. Since CHEMVAL constitutes the first large-scale verification exercise involving coupled codes, test cases were kept fairly simple and verification approached in a stepwise manner [5].

Six computer programs were employed during this stage of CHEMVAL [5]. Three are "directly-coupled", whereby mass action terms are incorporated within transport equations [6], and three "iteratively-coupled", relying on iteration between discrete speciation-solubility and transport packages [7]. Owing to the preliminary nature of this work, it was decided to investigate the results obtained for the separate aspects of equilibrium chemistry and hydrodynamic transport before proceeding with fully coupled problems. In all, five "speciation" and six "migration" test cases were posed, followed by a study of three coupled cases [5].

Results for calculations involving full coupling of chemical equilibria to hydrodynamic transport are available for all three test problems, dealing with, respectively cement dissolution, bentonite clay alteration and sodium hydroxide injection into a siliceous aquifer. The outcome of the clay alteration study is discussed below.

Case 2 of the verification exercise was derived from published information [8], revised where necessary to make it amenable to all coupled codes used within CHEMVAL [5]. It involves the alteration of sodium bentonite by inward diffusion of calcium-rich groundwater. The conversion proceeds by ion-exchange of Ca^{2+} for Na^{+} and participants were asked to predict the progress of the alteration front as a function of distance and time. Details of the problem formulation are given in the relevant CHEMVAL report [5].

Figure 2 plots concentration versus distance profiles for aqueous sodium and calcium at 200 years. The results given by the two participants are in excellent agreement and clearly show the depletion of sodium caused by calcium substitution near the boundary. The slight variation between the CHEMTARD [6] and CHMTRNS [9] profiles reflect the different time-space grids employed by the two modelling groups.

Similarly encouraging verification results were obtained for test cases 1 and 3 [5].

Given the data requirements of coupled chemical transport models and the scarcity of suitably complete experimental information, it proved difficult to identify appropriate "validation" test cases for use within CHEMVAL Stage 4. Efforts were concentrated, therefore, on two well-defined column experiments [5].

- neptunium migration through glauconitic Mol sands

- heating and acidification of Fontainebleau sands.

Few results were obtained for the two cases reflecting perhaps, the premature nature of this item of work. Nevertheless, the three organisations who did attempt coupled modelling succeeded in reproducing the main features of the experimental systems [5].

The CHEMTARD code [6] was used to simulate neptunium migration through the glauconite sand column [10]. Local equilibrium was assumed and the Triple Layer Model [11] employed to account for adsorption of Np(V) species (principally NpO_2^{+}, $NpO_2CO_3^{-}$) onto the sand. Results were compared to best-fit analytical solutions of the transport equation derived by Bidoglio et al [10]. Experimental and calculated profiles are compared in Figure 3. Although the results in no way constitute model validation they provide a useful test of the accuracy of the CHEMTARD numerical scheme. Full details are given in the CHEMVAL Stages 3/4 report [5].

4. THERMODYNAMIC DATA REVIEW

The need for a standard thermodynamic database to support CHEMVAL modelling studies was recognised at the inception of the project. It provides a reference point for CHEMVAL participants and greatly facilitates the interpretation of modelling results. Review of the CHEMVAL database has been continuing since 1987. Three versions of this compilation were issued to participants at appropriate stages to coincide with modelling tasks (Table I). The fourth and final listing is currently being checked prior to distribution.

The database now contains constants for more than 1000 aqueous and solid species, with emphasis on major groundwater components and elements of radiological importance [12]. All entries have been reviewed independently and traced back to source. It is readily apparent, however, that the CHEMVAL database is far from complete. Specific deficiencies have been identified in reports on CHEMVAL Stages 1 and 2 [2,3] and include several actinide phases of potential relevance to risk assessment. Nevertheless, the database is internally consistent, well documented and widely used by research groups, both in EC countries and elsewhere. For these reasons, the decision to promote database activities to run in conjunction with the verification/validation studies has clearly been justified.

5. CONCLUDING DISCUSSION

International verification-validation studies have become an established method for generating confidence in the tools used to formulate a radiological safety case. CHEMVAL has made a significant contribution in this respect over the past three years. Specific achievements include the following:

i) production of a reviewed and standardised thermodynamic database.

ii) identification of weaknesses in thermodynamic compilations and priority areas for database enhancement.

iii) verification of both equilibrium speciation and coupled chemical transport codes.

iv) assessment of the capabilities and limitations of extant models.

In addition, greater awareness of related work has removed much of the uncertainty surrounding alternative approaches to geochemical modelling. Not suprisingly, a large number of uncertainties remain, particularly regarding the adequacy of the models when applied to field data as opposed to laboratory experiments. Without doubt, much remains to be done in terms of quantifying geochemical interactions and key areas, such as actinide complexation by humics, have already been highlighted as priorities for future work.

Owing to timescale and resource limitations, modelling exercises attempted within CHEMVAL were based on published experimental data. It was not possible to carry out a major analytical programme in parallel and involve experimentalists in joint definition of objectives at the start of the project. As a direct consequence, the products of the analytical work were not entirely compatible with the requirements of numerical modelling. This, in turn, limited the opportunities for quantitative comparison of results.

It is hoped that closer collaboration between "modellers" and "measurers" will substantially increase the scope for model validation studies in the future.

ACKNOWLEDGEMENTS

This paper is based on work funded jointly by the Commission of the European Communities and the United Kingdom Department of the Environment. We are grateful to Mr Nick Harrison for his support and throughout the project. All participants are warmly thanked for their hard work in making the CHEMVAL project both enjoyable and successful.

REFERENCES

1. Read, D. and Broyd, T.W. : "Verification and validation of predictive computer programs describing the near and far-field chemistry of radioactive waste disposal systems". Radiochim Acta 44/45, 407-415, 1988.
2. Read, D. and Broyd, T.W. : "CHEMVAL Project - Report on Stage 1 : Verification of speciation models". CEC Report No. EUR12237EN, 1989.
3. Read, D. et al : "CHEMVAL Project - Report on Stage 2 : "Application of speciation models to laboratory and field data sets" (Draft) 1990.
4. Carlsen, L. : "The role of organics on the migration of radionuclides in the geosphere". CEC Report No. EUR12024EN, 1989.
5. Read, D. et al : Report on CHEMVAL Stages 3 and 4 (manuscript in preparation 1990).
6. Liew, S.K. and Read, D. : "Development of the CHEMTARD coupled process simulator for use in risk assessment". DOE Report No. DOE/RW/88.051, 1988.
7. Coudrain-Ribstein, A. and Jamet, Ph. : "Le modele geochimique CHIMERE; Principes et notice d'emploi". EMP Report No. LHM/RD/88/35, 1988.
8. Jacobsen, J.S. and Carnahan, C.L. : "Numerical simulation of alteration of sodium bentonite by diffusion of ionic groundwater components". LBL Report No. LBL-24494, 1987.
9. Noorishad, J. et al : "Development of the non-equilibrium reactive chemical transport code CHMTRNS". LBL Report No. LBL-22361, 1987.
10. Bidoglio, G. et al : "Neptunium migration in oxidizing clayey sand". App. Geochem. 2, 275-284.
11. Kent, D.B. et al : "Surface complexation modelling of radionuclide adsorption in sub-surface environments". Stanford University Technical Report No. 294, 1986.
12. Chandratillake, M. et al : Final Report on University of Manchester Contribution to CHEMVAL (manuscript in preparation), 1990.

Table I Status of CHEMVAL Thermodynamic Database

STAGE	RELEASE DATE	AQUEOUS SPECIES	SOLID SPECIES	FORMAT
1	JUNE 1987	376	259	PHREEQE MINEQL
2	MAY 1988	521	327	PHREEQE MINEQL
3	JUNE 1989	525	319	PHREEQE MINEQL EQ3/6
4 (Draft)	APRIL 1990	742	407	PHREEQE MINEQL DBASEIII+

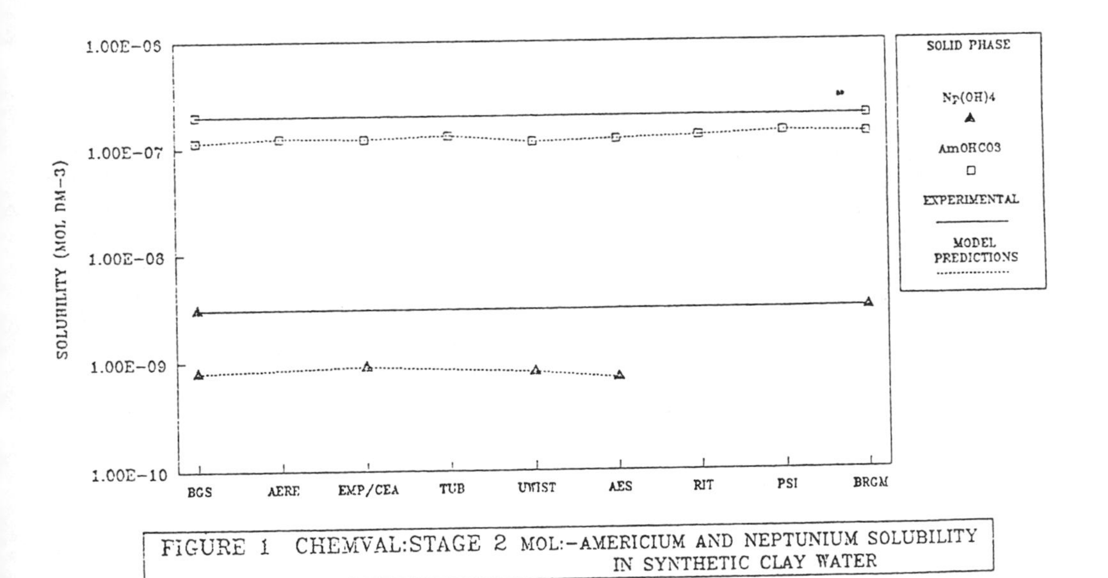

FIGURE 1 CHEMVAL:STAGE 2 MOL:-AMERICIUM AND NEPTUNIUM SOLUBILITY IN SYNTHETIC CLAY WATER

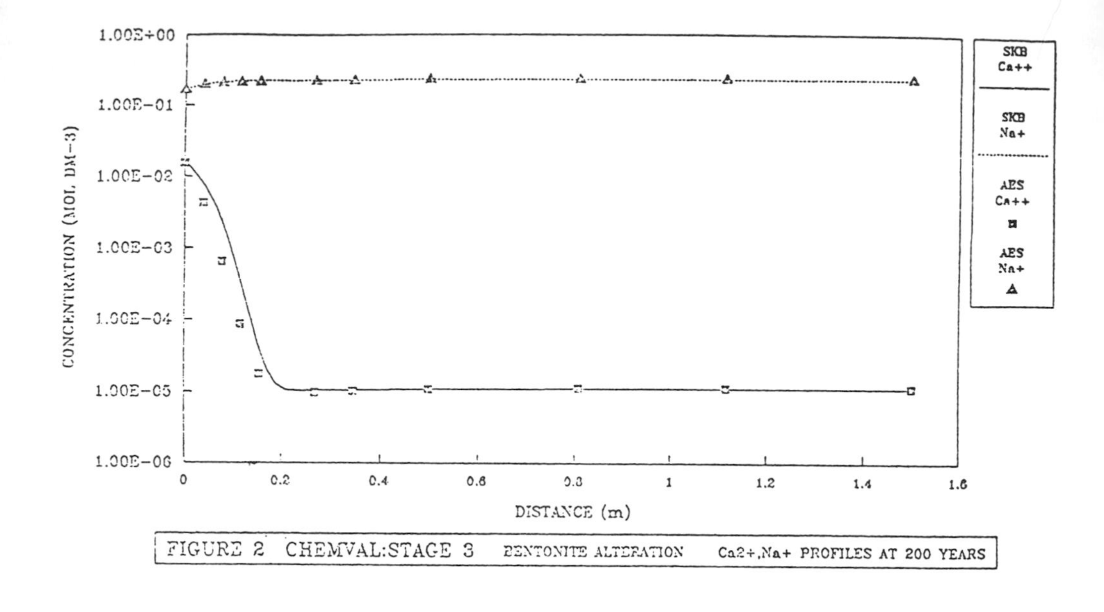

FIGURE 2 CHEMVAL:STAGE 3 BENTONITE ALTERATION Ca2+,Na+ PROFILES AT 200 YEARS

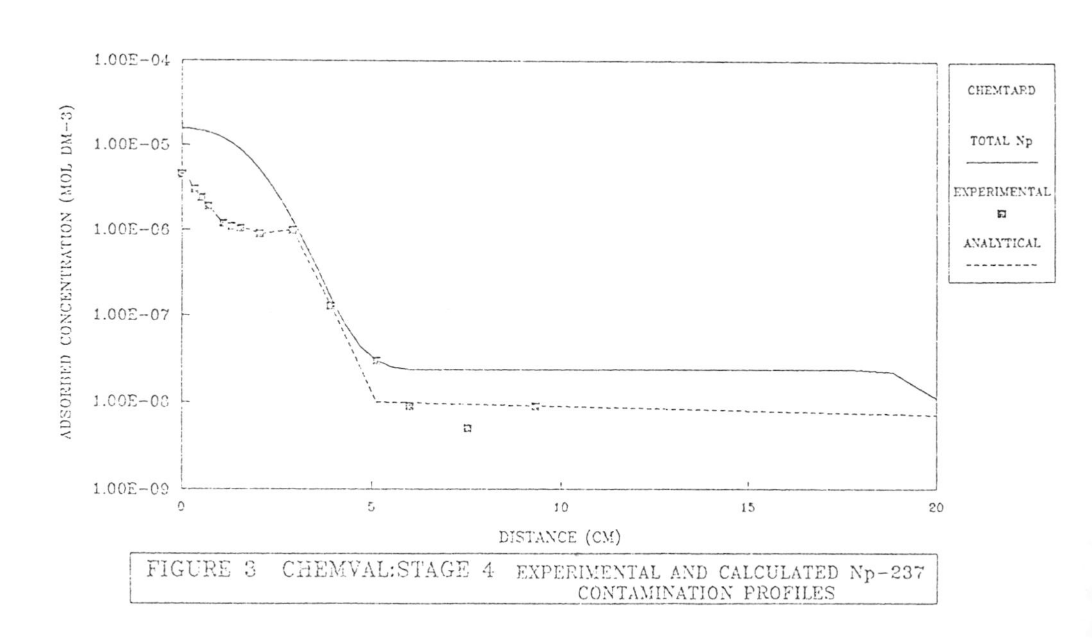

FIGURE 3 CHEMVAL:STAGE 4 EXPERIMENTAL AND CALCULATED Np-237 CONTAMINATION PROFILES

HYDROGEOCHEMICAL MODELLING OF THE NEEDLE'S EYE NATURAL ANALOGUE (SCOTLAND)

Ph. Jamet*, P. Lachassagne*, E. Ledoux*, P.J. Hooker**, P. Escalier des Orres***

*Ecole des Mines de Paris, 77305 Fontainebleau, FRANCE

**British Geological Survey, Keyworth, Nottingham NG12 5GG, UNITED KINGDOM

***Commissariat à l'Energie Atomique. BP 6, 92260 Fontenay-aux-Roses, FRANCE

ABSTRACT

First a hydrodynamic model of the natural analogue site of Needle's Eye is built with the help of the *METIS* code. Then the transport of the uranium is simulated from its natural deposits through sediments rich in humic facies. The use of a distribution coefficient (K_d) in the treatment of the behaviour of the solute does not make it possible to reproduce the observed concentrations. With the coupled model *STELE* the precipitation of UO_2 in the reducing environment of the sediments can be simulated. The results agree with observations in the aqueous and the solid phase.

RESUME

On construit d'abord à l'aide du programme *METIS* un modèle hydrodynamique du site d'analogue naturel de Needle's Eye. Le transport de l'uranium provenant de minéralisations naturelles à travers des sédiments riches en faciès humiques est ensuite simulé. L'approche du comportement du soluté par un coefficient de distribution (K_d) ne permet pas de rendre compte des concentrations observées. L'emploi du modèle couplé *STELE* permet de simuler un phénomène de précipitation de UO_2 dans l'environnement réducteur des sédiments. Les résultats sont concordants avec les observations en phases aqueuse et solide

INTRODUCTION.

The study of natural geochemical systems ("natural analogues") is useful in furthering the understanding of the processes that set in motion, transport and accumulate radionuclides in the area surrounding an underground waste disposal facility and it provides a means of testing the validity of coupled transport models and of the thermodynamic databases on which these models are founded.

The work described here concerns the modelling of uranium migration at the Needle's Eye site. This study is the result of a cooperation between the BGS (British Geological Survey), the SURRC (Scottish Universities Research and Reactor Centre) for the field aspects and the Centre d'Informatique Géologique (Ecole des Mines de Paris) for the modelling aspects. It was commissioned by the U.K. Department of the Environment, the French Commissariat à l'Energie Atomique (Institut de Protection et de Sûreté Nucléaire) and the Commission of European Communities (DG XII).

GEOLOGY AND URANIFEROUS DEPOSITS AT THE NEEDLE'S EYE SITE.

The study area is situated in the vicinity of the town of Dalbeattie (Kirkcudbrightshire) on the north bank of the Solway Firth. The geology of this region was formed during the Palaeozoic as follows :

To the north : Ordovician and Silurian terrains folded during the Caledonian orogenesis and marked by a great many Devonian batholith intrusions (granodiorites) which have given rise to contact metamorphism (Silurian hornfelses).

To the south : Carboniferous limestones in abnormal contact with the former through a major E-W Hercynian fault. The fault is marked by an approximately 40 m-high paleocliff at the foot of which there are deposits of Quaternary (from -10 000 to the present) silty sediments interbedded with peat. This 100-meter long slope toward the sea is, strictly speaking, the Needle's Eye site. The geology of the region is shown in figure 1.

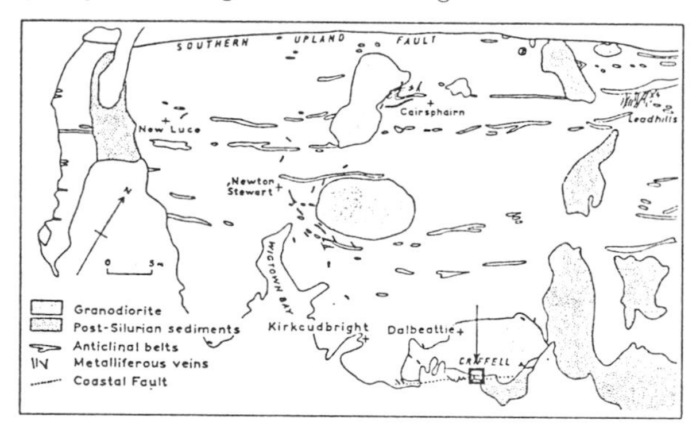

Figure 1. Geology of SW-Scotland (from Miller and Taylor, 1966).

Systematic uranium prospecting south of the Criffell batholith instigated by the Atomic Energy Division (1957- 1961) brought to light signs of uranium in the vicinity of the coastal zone. In the Needle's Eye sector two large veins are thus known to cut across the hornfelses in the cliff.

These veins, 10 to 50 cm thick, are particularly rich in uranium in the form of pitchblende (or uraninite) in the middle sector of the cliff [1] [2]. The mineral deposits are thought to be between 165 and 205 million years old.

STUDY AREA.

Radiometric investigations of the sediments have made it possible to distinguish areas that seem rich in radionuclides although the precise reasons for their presence have not been determined (primary deposits in the substratum, secondary in the sediments, etc.) [3].

The most thoroughly studied zone consists of a band of sediments, 85 m long and around 50 m wide, from the cliff to a tidal channel (the Southwick Water) in a NW-SE direction.

In this area the team from the BGS has carried out several measurement expeditions mainly concerned with the chemical compositon of the water and the permeability of recent deposits. 87 water samples were taken, 46 of them on the same N-S line ("line 18"). Eleven permeabilities were measured in the sediments.

HYDROGEOLOGICAL MODELLING.

The first part of the study was devoted to the building of a flow model of the Merse sector since this is a necessary prerequisite to the use of the model coupling geochemistry and transport.

The modelling was done in two stages : first with a regional model into which the studied zone was integrated in its regional hydrogeological context and then with a local model which uses the boundary conditions calculated in the preceding one as a basis for calculating the flows in the recent sediments only. The computer model used for both these simulations is the *METIS* code [4].

Regional hydrogeological model.

It is evident from the topographical map (figure 2) that the regional hydrogeological unit in question corresponds to the basin situated between the summit of Clifton Craig (alt. 183 m) and the Southwick Water. Its lateral boundaries are marked by two streams running parallel to the main extension of this structure.

The flow is conditioned mainly by the topography. A two-dimensional representation along a NW-SE axis therefore constitutes a correct approximation.

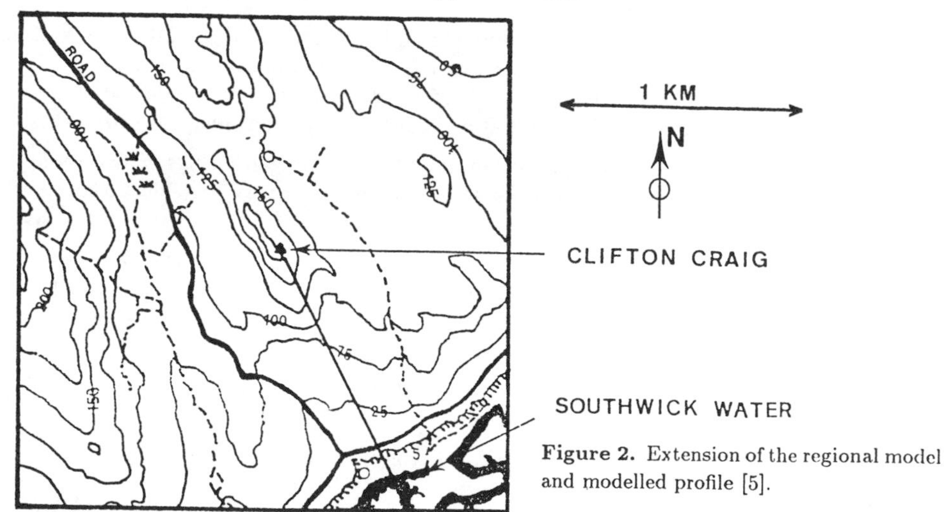

Figure 2. Extension of the regional model and modelled profile [5].

The geological units included in the regional domain are therefore, from NW to SE, Criffell granodiorite, Silurian hornfelses, Carboniferous limestones, recent sedimentary deposits (silts and peatbog) and the EW fault zone sealed by silts (figure 3).

With the exception of the Quaternary sediments, these media have a large-scale permeability, i.e. the flow occurs mainly in the fractures which are assumed to cease to be conductive below a depth of 30 m. However, the scale of the studied domain makes it possible to represent these units by their equivalent continuous media [6].

The parameters used in the modelling are shown in table 1 [5].

parameter	value
K_{ah}	$7\,10^{-6}$ m/s
K_h	10^{-6} m/s
K_l	$4,5\,10^{-6}$ m/s
K_g	$5\,10^{-8}$ m/s
K_s^x	$3\,10^{-6}$ m/s
K_s^y	$3\,10^{-7}$ m/s
K_{hu}	$3\,10^{-7}$ m/s
H_M	2,4 m
R	320 mm/year

Table 1. Parameters of the regional model (see Table of Notations).

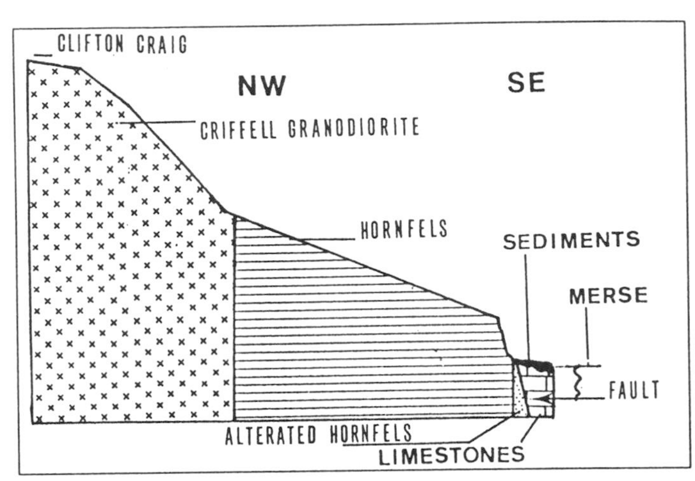

Figure 3. Geological units of the regional model.

Local hydrogeological model.

This is a detail of the preceding model (figure 4) and it includes all the Quaternary sediments found on line 18. The points on this line are localized by the coordinates on their horizontal axis which vary from north to south between +30 m and -55 m. Eight piezometers were installed making it possible to carry out stratified sampling and, at certain points, permeability measurements.

The outline of the bottom of the series was obtained from the auger borings made by the BGS. Because of the shape of this erosion surface the recent series has a very variable thickness which ranges from 1 to 4 meters.

In this series there are two main lithological facies, a humic facies which is abundant to the north (peatbog) and to which we have given an isotropic permeability of $3\,10^{-7}$ m/s and silty facies including sandy deposits which take up the rest of the domain and for which we have maintained the permeabilities of the regional model.

From observations in the field, it seems that hydraulically, the sediments are divided into two sectors:

To the north : a zone where the water seeps out (peatbog).

To the south : a relatively well-drained zone.

The boundary between the two zones falls approximately above the fault. Furthermore, an exploration ditch has revealed water flowing into the bottom of the recent series which has been explained by the presence of a lithologic stratum with a higher conductivity ("beach level").

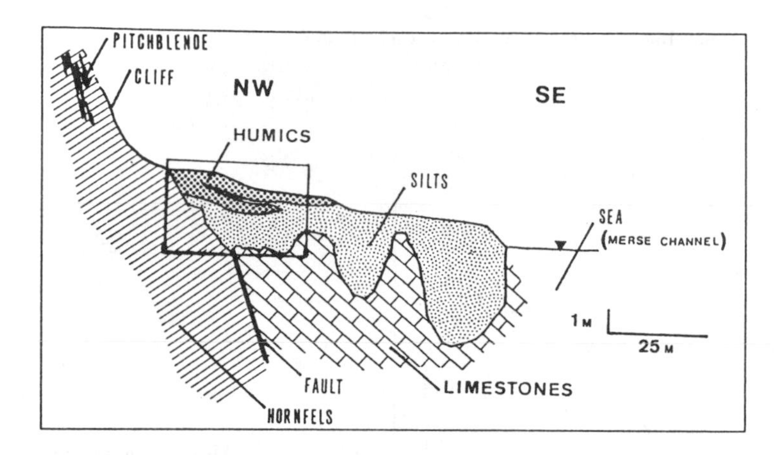

Figure 4. Sediments of the Merse (local domain) and the extension of the coupled model (framed).

Results

Figure 5 shows the shape of the calculated piezometry in the sediments. The flow in the peatbog proves to be vertical. Beyond the fault the underlying limestones drain the Quaternary sediments. The major flow from the hornfelses toward the sediments is reproduced by the model without recourse to individual lithological treatments of the silts.

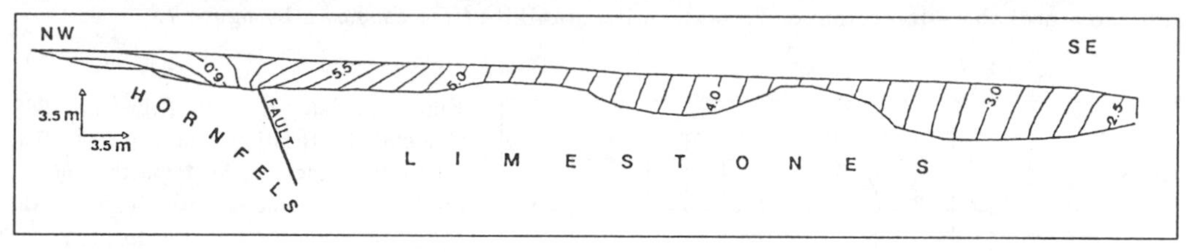

Figure 5. Calculated piezometry in the local domain.

GEOCHEMICAL MODELLING.

The object of this work was to interpret the chemical data gathered by the BGS during the four measurement expeditions. This was done with the speciation code *CHIMERE* which was developed at the CIG and is capable of calculating, among other things, the mineral saturation indices which may, for example, control the concentration of dissolved uranium.

Results

Uranium source term (figure 6). The map of the mean concentrations in aqueous U in the domain shows that there is a major transfer from the north. These concentrations decrease generally from the bottom upwards in the series, which proves that water with very high concentrations enters through the substratum of the hornfelses. The runoff from the northern part of the site is also heavily charged.

Consequently, the scheme of the uranium transport through the sediments agrees entirely with the local hydrogeological model. The meteoric water passing through the hornfelses leaches the deposits in the cliff containing pitchblende and then, they either trickle down the cliff and into the peatbog or they pass into the lower part of the sediments north of the fault.

Further south, the water passing through only silts retains a low uranium content except in the band of terrain that is in contact with seawater (3 ppb of dissolved uranium) and in the upper middle section where the runoff water from the north is in part infiltrated.

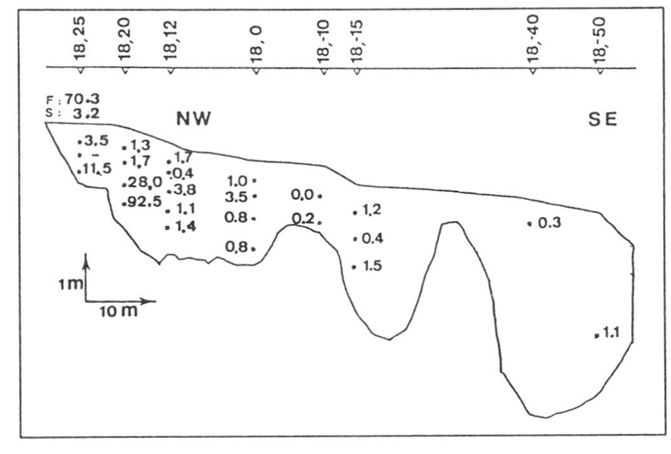

Figure 6. Mean uranium aqueous concentrations (ppb) in the local domain (F : in the cliff, S : in the surface flow).

Geochemical behaviour of the uranium. The *CHIMERE* model made it possible to test the saturation indices of the uranium minerals included in the *CHEMVAL* database. The calculations indicate a probable equilibrium of the water with uraninite UO_2 as shown by figure 7.

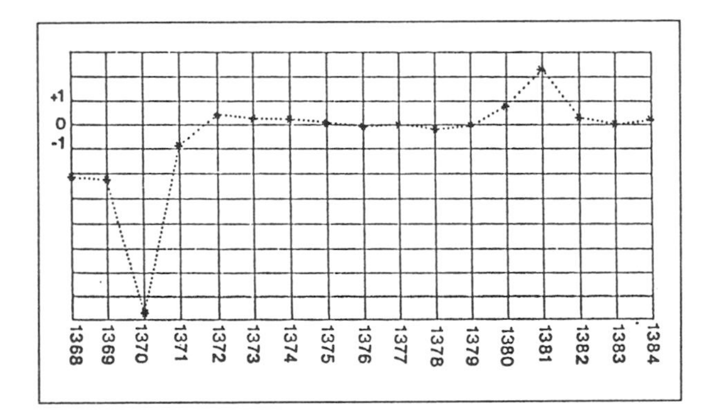

Figure 7. Samples from February 1988. Calculated saturation indices of UO_2. The saturation index is the logarithm of the product of the ionic activity divided by the equilibrium constant of the formation of the solid. A value of 0 corresponds to equilibrium. On the horizontal axis, the numbers of the samples. The low concentrations to the left of the diagramme are values from water dripping off the cliff. The heavily concentrated solutes on the right are from the northern part of the site in contact with the hornfelses.

The contrast in redox potential between the bottom of the recent series (hornfels water) (+350/+400 mV) and its interior which corresponds to the humic facies (+150/+250 mV) strengthens this assumption. The *CHIMERE* model brings out the influence of the Eh on the concentrations of dissolved uranium which vary between 10^{-7} mol/l at +400 mV and 10^{-12} mol/l at +200 mV.

HYDROGEOCHEMICAL MODELLING.

The modelling of the uranium transport was first done with a simple model based on the coefficient (K_d) of the distribution of uranium between the aqueous and the solid phase, the value of which, measured on humic samples collected on the site, is close to 15 000 ml/g (under oxic conditions). The *METIS* model has shown that this hypothesis is unable to explain the massive disappearance of the uranium, north of the site, in the upper part of the piezometers 18,+25 and 18,+20 (table 2). Moreover, this result is not sensitive to the K_d value.

Sampling point	Depth	Measured C	Calculated C (in ppb)		
BGS grid	(in cms)	(in ppb)	1000 years	2000 years	5000 years
18,25	100	11,5	11,26	11,42	11,51
"	65	?	7,85	8,28	8,55
"	40	3,5	2,01	3,55	4,82
18,20	156	92,5	95,00	95,00	95,00
"	120	28,00	80,83	82,03	82,90
"	70	1,7	45,96	53,72	59,75
"	40	1,3	12,09	25,68	40,98

Table 2.K_d-model - comparison of the measured and the calculated uranium concentrations in the piezometers 18,+20 and 18,+25.

Data in the coupled calculation.

The modelled domain covers the northern and middle zones ($y < -5$), where the precipitation occurs.

Bearing in mind the Eh domain on the site, the major species of dissolved uranium are uranyle carbonates and hydroxycarbonates. A uniform source term with a value of 100 ppb is fixed at the outcrop of the hornfelses. In the middle section the concentration of the reinfiltrated flow is fixed at 2 ppb. The redox potential of the water from the hornfelses is estimated at +400 mV and of the water from the sediments at +250 mV. The pH is uniform over the whole domain (≈ 7.8).

Results.

Great care was taken to reproduce through the calculations the observed uranium concentrations in the piezometers 18,+20 and 18,+25 which create problems under the Kd assumption.

The adjusted UO_2 precipitation kinetics are estimated at a rate constant of 1 day^{-1}. It has not been possible to prove that this is a realistic value. The concentration pattern becomes stationary after approximately 1 month (figure 8). Table 3 compares the measured and the calculated values in the two piezometers.

The agreement is excellent in piezometer 18,+20 and seems poor in piezometer 18,+25. This we believe to be due to the heterogeneity of the source term which is consistent with the fractured nature of the substratum. This aspect has been taken into account in the K_d approach but not here for simplicity. However, if a value of 30 ppb is attributed to the water introduced in this area, one see that the observed concentrations would be fairly well reproduced.

The uranium accumulated in the sediments (figure 9) is on the order of 500 ppm when the system such as it is now has worked for 5 000 years. This is entirely plausible in view of the supposed age of the series and of the available reserve of oxidizable matter. This content is close to that measured locally in the sediments and frequently found in exploited uranium deposits [7].

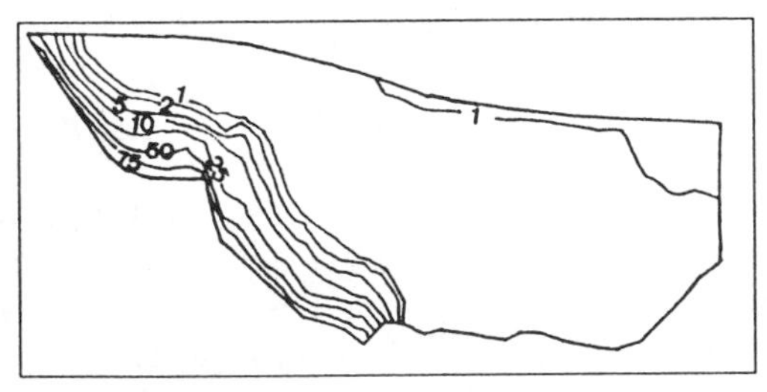

Figure 8. Distribution of uranium concentrations in the sediment water calculated in a stationary state.

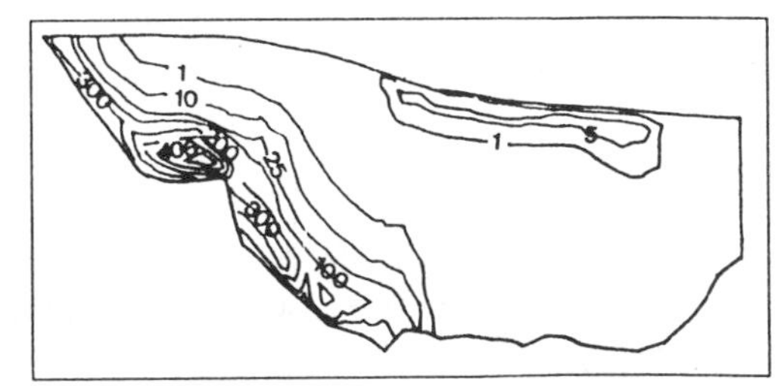

Figure 9. Fixed uranium after one month (concentrations in ppb expressed as in the fluid phase).

Localisation	Depth	Measured C (ppb)	Calculated C (ppb)
18,25	100	11,5	34,20
"	65	?	12,87
"	40	3,5	0,86
18,20	156	92,5	100
"	120	28,00	28,70
"	70	1,7	1,98
"	40	1,3	0,18

Table 3. Model of uranium precipitation. Comparison between measured and calculated values in the piezometers 18,+20 and 18,+25.

CONCLUSION.

The results of the modelling of the natural analogue site of Needle's Eye demonstrate that a model of thermodynamic equilibrium goes farther than a K_d model toward an understanding of the transfer processes. However, this study does not constitute a proof of the validity of the *STELE* model since we do not know if the geochemical process that has been modelled (precipitation of uraninite) is, in reality, the phenomenon that ought to be taken into account [8].

Therefore, the next stage of this study will mainly be concerned with the geochemistry in order to determine which phenomena to model (other mineral phase of uranium or fixation on humic matter). Moreover, the acquisition of new hydrological data would make it possible to improve the calibration of the hydrogeological model.

TABLE OF NOTATIONS.

K_g permeability of the Criffell granodiorite
K_{ah} permeability of the alterated hornfelses
K_s^x longitudinal permeability of the silts
K_{hu} permeability of the humic layers
R mean infiltration on the site
K_h permeability of sound hornfelses
K_l permeability of the limestones
K_s^y transverse permeability of the silts
H_M average level of the Southwick Water.

REFERENCES

[1] Miller, J.M., Taylor, K. : "Uranium Mineralization near Dalbeattie, Kirkcudbrightshire", Bull. Geological Survey of Great Britain, 25, 1966, 1-18.

[2] Hooker, P.J. et al. : "Natural Analogue of Radionuclide Migration : Reconnaissance Study of Sites", 1986, NERC Report FLPU 86-6, 26p.

[3] Soubeyran, R., Ledoux, E., Marsily (de) G. : "Modélisation du Transfert d'Analogues Naturels", 1988, Rapport CIG LHM/RD/88/11, 42p.

[4] Goblet, P. : "Programme *METIS*, Notice de Conception, Rapport ENSMP-CIG LHM/RD/89/23, 1989, 97p.

[5] Jamet, Ph, Lachassagne, P., Doublet, R., Ledoux, E. : "Modelling of the Needle's Eye Natural Analogue", 1989, Rapport CIG LHM/RD/89/81, 49p.

[6] Marsily (de), G. : "Quantitative Hydrogeology, Grounwater Hydrology for Engineers", Academic Press, 1986, 440p.

[7] Basham, I.R., Milodowski, A.E., Hyslop, E.K., Pearce, J.M. : "The Location of Uranium in Source Rocks and Sites of Secondary Deposition at the Needle's Eye Natural Analogue Site, Dumfries and Galloway", 1989, BGS Technical Report WE/89/56, 54p.

[8] Hooker, P.J. : "An Overview and Assessment of the British Geological Survey's Research Work on Natural Analogue Studies in Great Britain", Proc. Migration'89 Meeting, Monterey, 1989, 22p.

THE RELAY SUBSTANCE METHOD: A NEW CONCEPT FOR PREDICTING POLLUTANT TRANSPORT IN SOILS

M. Jauzein, C. André, R. Margrita
DAMRI-SAR-SAT, CENG, BP 85X, F-38041 Grenoble, France

M. Sardin, D. Schweich
Laboratoire des Sciences du Génie Chimique, CNRS-ENSIC, 1 rue Grandville, BP 451, F-54001 Nancy, France

ABSTRACT

Relay substances are reacting solutes which allow one to in-situ determine site-specific parameters governing pollutant transport. The method is illustrated by the prediction of cesium transport at field scale. Preliminary laboratory experiments showed that cesium transport was affected by the exchange capacity of the soil. Because of the poor representativity of laboratory soil samples, the effective exchange capacity under field conditions was obtained using lithium as a relay substance. Prediction of cesium transport is then made possible, although it is found to depend on dilution by field water. The concept of "dilution earliness" is used to predict limiting situations of dilution and the corresponding breakthrough curves.

RESUME

Les substances relais sont des solutés réactifs permettant la mesure in-situ de paramètres physico-chimiques spécifiques d'un site et gouvernant le transport d'un polluant. La méthode est illustrée par la prédiction du transport du cesium à l'échelle du site. Des expériences préliminaires de laboratoire ont montré que le transport du cesium dépendait de la capacité d'échange du sol. Du fait du peu de représentativité des échantillons de sol, la capacité d'échange effective à l'échelle du site a été déterminée en utilisant le lithium comme substance relais. La prédiction du transport du césium est alors possible bien qu'elle soit trouvée fonction de la dilution par l'eau du site. Le concept de "précocité de dilution" est utilisé pour prévoir des situations limites de dilution et les courbes de restitution associées.

INTRODUCTION

Prediction of solute transport in soils requires a model accounting for homogeneous and solid-fluid (i.e., heterogeneous) reactions and flow pattern. We will speak of an interaction mechanism for the set of homogeneous and heterogeneous reactions. As soon as the structure of the interaction mechanism is known, the parameters of each reaction must be determined. These parameters pertain to two categories:

-Those which are interaction-specific and can accurately be determined at laboratory scale. Rate constants of some reactions or equilibrium constants of any reaction are among these parameters which can be safely extrapolated from laboratory to field conditions.

-Those which are site-specific, owing to spatial variability or water saturation. In addition to flow parameters, available (or effective) adsorption or exchange capacity, mass transfer parameters depending on flow characteristics are among the site-specific data which must be measured at field scale.

PRINCIPLE OF THE RELAY SUBSTANCE METHOD

Let us assume that the interaction mechanism between a pollutant and a soil sample has been determined by preliminary laboratory studies, and that the retention of the pollutant by the soil constituents has been found to depend on the exchange capacity. Owing to spatial variability, or water saturation of soil, the exchange capacity measured under laboratory condition is generally different from that observed under field condition. The pollutant being forbidden in field experiments, it must be replaced by a suitable species which may be used to measure the effective exchange capacity at field scale. Similarly to water tracers which are flow tracers, relay substances are "tracers of chemical interactions".

Selecting a relay substance requires to know the interaction mechanism governing the transport of the pollutant and the relay substance. Moreover, the interaction mechanism of the relay substance must involve the site-specific parameter(s). Mechanism elucidation and relay substance selection typically pertain to laboratory experiments. When field experiments with the relay substance are available, they are interpreted to recover the site-specific parameter. Finally, a computer model can be fed with the site- and interaction-specific parameters to predict pollutant transport. This progressive method is illustrated below by the prediction of cesium transport.

EXAMPLE OF TRANSPORT OF DISSOLVED CESIUM

Cesium transport in a sample of sediments of Isère river (Grenoble, France) were studied by column experiments [1]. Table I summarizes the soil and field water properties. Because of clays, calcium-cesium cation exchange was found to be the main heterogeneous interaction responsible for cesium retention. Cation exchange was also coupled with calcite dissolution and ionization of water and carbonic acid. Column experiments at various composition of the feed solution showed that the exchange process must be accounted for by a two-site model [2, 3], the exchange capacities and the Gaines-Thomas selectivity coefficients of each site being:

$CEC_1 = 10\ meq.kg^{-1}$, $CEC_2 = 8\ meq.kg^{-1}$

$K_{1Ca/Cs} = 2.2\ mol.l^{-1}$, $K_{2Ca/Cs} = 1.3\ 10^{-4}\ mol.l^{-1}$

The selectivity coefficients are defined by:

$$K_{Ca/Cs} = \frac{N_{Ca}}{C_{Ca}} \left(\frac{C_{Cs}}{N_{Cs}} \right)^2 \qquad (1)$$

where C is the concentration ($mol.l^{-1}$) in the liquid phase and N the ionic fraction with respect to the site considered. Later on, we will use the total exchange capacity per unit volume of liquid phase defined by:

$$N_E = (CEC_1 + CEC_2) \frac{M}{V_0} \qquad (2)$$

where M is the mass of soil and V_0 the pore volume. Typically, N_E is of the order of 60 $meq.l^{-1}$.

Figure 1 illustrates the agreement between experimental and theoretical breakthrough curves (BTC) obtained with IMPACT computer code [4, 5]. As told above, the exchange capacities cannot be extrapolated to field condition, and a relay substance must be sought after.

Table I: Properties of soil and field water

Soil			
Sand and silt (>0.002mm): ≈ 98% w/w		Clays (<0.002mm): ≈ 1 to 2% w/w	
Calcium carbonate: ≈ 17% w/w		Exchange capacity: ≈20±10 $meq.kg^{-1}$	
Field water			
Cations	$meq.l^{-1}$	Anions	$meq.l^{-1}$
Ca^{2+}	7.3	HCO_3^-	7.7
Mg^{2+}	1.9	Cl^-	0.9
Na^+	0.49	SO_4^{2-}	1.2
K^+	0.044	NO_3^-	0.027

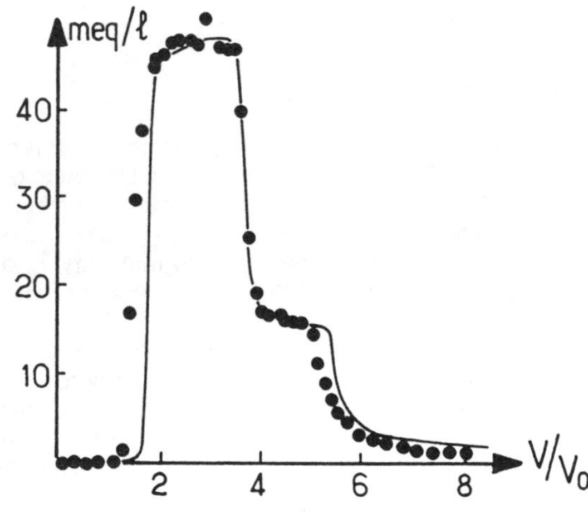

Figure 1: Calcium and cesium BTCs resulting from a 3 pore volume injection of CsCl 50 $meq.l^{-1}$, driven by $CaCl_2$ 15 $meq.l^{-1}$.

The relay substance

Since cesium is an alkali metal, relay substances must be found among alkali metals, such as lithium, sodium or potassium. Potassium has been excluded because the exchange process is known to be complex and sometimes irreversible [2, 3]. Column experiments were thus performed with lithium and sodium. A common value of the exchange capacity, equal to $CEC_1 + CEC_2 = 18$

meq.kg^{-1} was found. Figure 2 illustrates the agreement between experimental and theoretical curves obtained by IMPACT computer code.

Sodium and lithium were therefore potential relay substances which were sensitive to the overall exchange capacity $CEC=CEC_1+CEC_2$. Their respective Gaines-Thomas selectivity coefficients were:

$$K_{Ca/Li} = 32 \text{ mol.l}^{-1}, \quad K_{Ca/Na} = 8 \text{ mol.l}^{-1} \tag{3}$$

These values qualitatively agree with those proposed by Bruggenwert and Kamphorst [3] for various clays.

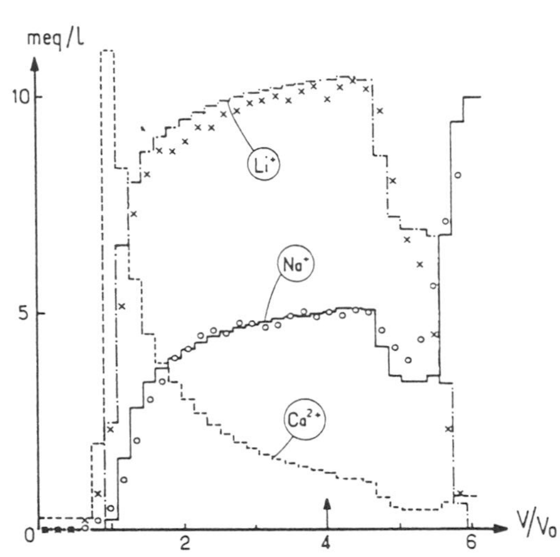

Figure 2: Calcium, lithium and sodium BTCs resulting from a 4 pore volume injection of NaCl 4.5 meq.l^{-1} and LiCl 10 meq.l^{-1}, driven by $CaCl_2$ 5 meq.l^{-1} and NaCl 4.5 meq.l^{-1}. Crosses and dots are experimental results. Curves are theoretical predictions.

Field experiments with the relay substance

Lithium was chosen as the relay substance since it has a smaller affinity for soil constituents than sodium, and since it is absent of field water. The injection consisted of 180 liters of a solution of iodide (water tracer) and lithium chloride (0.39 mol.l^{-1}). Figure 3 illustrates the responses recorded at a piezometer located six meters downstream the injection point. Only 1% of the injected tracers were recovered. Iodide BTC suggests that water flew through two main layers of different permeabilities which were already known from geological study.

Iodide and lithium BTCs were found to be superimposed upon normalzation of the time axis by the mean residence time, and the concentration axis by the amount of tracer recovered. This behavior is typical of linearly interacting solutes. Although cation exchange is a non-linear process, the linear behavior can be observed at trace level of the reacting solute. Using equation (1) for an exchange process between a divalent cation C and lithium, the non-dimensional distribution coefficient of lithium is:

$$K_d = \frac{C_{Lis}}{C_{Lil}} = N_E \left(\frac{N_C}{K_{C/Li}\, C_C}\right)^{1/2} \tag{4}$$

where C_{Lis} and C_{Lil} are the equilibrium concentrations (mol per liter of aqueous phase) of lithium on the solid and in the liquid phases respectively. Provided that N_C and C_C remain constant during the transport process, K_d is constant. To calculate the distribution coefficient from lithium BTC, two supplementary assumptions were made:

-Since calcium and magnesium have close exchange properties [3], these two cations are lumped together as cation C. Its concentration C_C is the sum of the calcium and magnesium concentrations. The selectivity factor $K_{C/Li}$ was taken equal to $K_{Ca/Li}$.

-C_C and N_C were given by the sodium-calcium exchange equilibrium at the composition of field water, i.e., $C_C = 4.6$ mmol.l^{-1} and $C_{Na} = 0.49$ mmol.l^{-1} (See Table I). With $K_{C/Na} = K_{Ca/Na}$ (See equation (1)), one obtains N_C very close to 1. As a result, $K_d = 2.6\ N_E$, and the retardation factor for lithium is:

$$R = 1 + K_d = 1 + 2.6\ N_E \qquad (5)$$

Interpreting Figure 3 gives R = 1.14, and $N_E = 54$ meq.l^{-1}. Iodide and lithium BTCs obtained at other detection piezometers gave N_E ranging from 42 to 62 meq.l^{-1}, showing thus the spatial variability of this parameter.

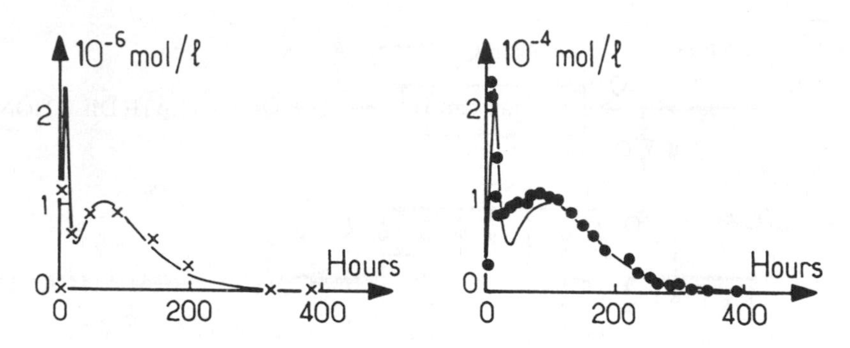

Figure 3: Iodide (left) and lithium (right) BTCs. Dots and crosses are experimental results whereas continuous curves are theoretical predictions.

The apparent linearity of lithium adsorption was checked using IMPACT computer code for simulating the true non-linear transport process. The following simplifying assumptions were made:

-Chloride sulfate and nitrate were lumped into a single inert monovalent anion of concentration $C_A = 2.1$ mmol.l^{-1} (See Table I).

-Lithium was driven by a solution of sodium at 0.5 mmol.l^{-1} and lumped divalent cation having the exchange properties of calcium and a concentration of 4.6 mmol.l^{-1}.

-The observed concentrations of dissolved calcium and bicarbonate were accounted for by an equilibrating partial pressure of carbon dioxide 500 times higher than in atmosphere. This can be explained by the microbiological and root activity in the upper layer of the aquifer.

-Water flow was modeled by two parrallel convective-dispersive streams. Their properties are summarized in Table II.

Table II: Properties of the parallel convective-dispersive flows.

Stream	Fraction of feed flow-rate	Fraction of pore volume	Péclet number
Fast	0.053	0.0043	23
Slow	0.947	0.9957	5

The continuous curve of Figure 3 (right) was obtained with the previous assumptions and the exchange capacity determined by equation (5). The good agreement prooves that lithium actually behaves as a trace species and that the interpretation in terms of linear adsorption is valid.

Prediction of cesium transport

Knowing the site- and interaction-specific parameters, various scenarios can be simulated with IMPACT computer code on the basis of the flow model described above and the properties of the calcium-cesium and calcium-sodium exchange processes.

In the presented prediction, we assumed that the molar amount of cesium injected was the same as lithium (180 liters at 0.39 $mol.l^{-1}$). Since cesium is much strongly adsorbed than lithium, cesium transport is expected to be non-linear because of the exchange processes, and the shape of the theoretical BTC is expected to depend on the dilution of cesium by field water. Unfortunately, nothing was known about the local dilution process between the injection and detection piezometers. Therefore, the dilution process was modeled by the "dilution earliness" concept [1, 6] illustrated by Figure 4.

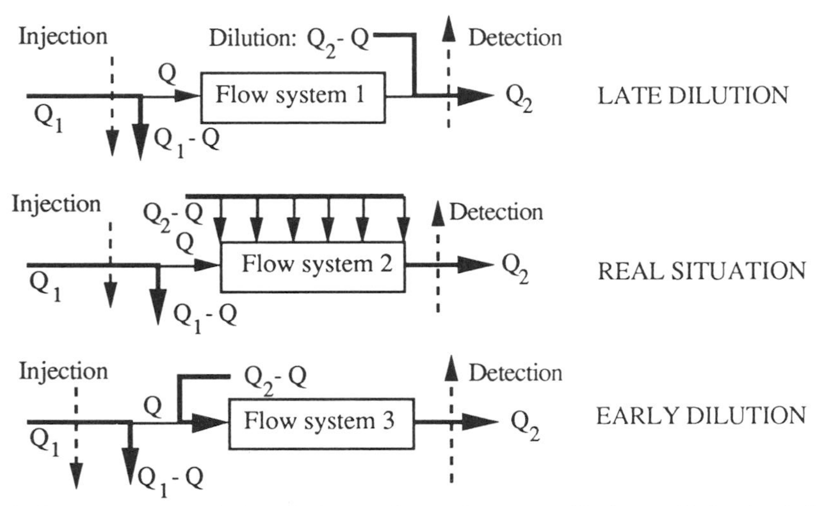

Figure 4: Conceptual models for dilution earliness. Top: Late dilution. Middle: Real situation. Bottom: Early dilution. The three fow systems have the same residence time distribution.

Q_1 and Q_2 are the volumetric flow-rates through the injection and detection piezometers respectively. They can be determined measuring the internal age distribution in each piezometer. Q is the volumetric flow-rate between the two piezometers, and it is determined from iodide BTC. At the injection point, Q_1-Q is responsible for the fraction of lost tracer, whereas it is not responsible for dilution. Letting n_I and n_D be the amounts of injected and detected tracer, and $C_D(t)$ the tracer concentration at the detection piezometer, we have:

$$n_D = Q_2 \int_0^\infty C_D(t)dt\,, \qquad \frac{n_D}{n_I} = \frac{Q}{Q_1}, \qquad Q = \frac{Q_1 Q_2}{n_I}\int_0^\infty C_D(t)dt \tag{6}$$

Q_2-Q is the dilution flow-rate. Although dilution can occur all along the transport process (Figure 4, middle), two limiting situations are of special interest: Either "early dilution" close to the injection piezometer (Figure 4, bottom), or "late dilution" close to the detection piezometer (Figure

4, top). In case of late dilution, tracer enters the porous soil at the injection concentration C_I, whereas in case of early dilution, it enters at concentration $C_I Q/Q_2$. The experimental flow-rates were:

$$Q_1 = 3.7 \text{ l.h}^{-1}, \quad Q_2 = 15 \text{ l.h}^{-1}, \quad Q = 1.41\ 10^{-2} \text{ l.h}^{-1}$$

Consequently, the inlet concentrations were:

$C = C_I = 0.39 \text{ mol.l}^{-1}$ for late dilution.
$C = C_I Q/Q_2 = 3.7\ 10^{-4} \text{ mol.l}^{-1}$ for early dilution.

Dilution earliness has no consequence on the transport of linearly interacting solutes because interaction does not depend on the concentration level. This means that dilution earliness cannot be observed from residence time distribution measurement. Iodide and lithium BTCs are thus useless for estimating dilution earliness. Consequently, only the two limiting dilution situations can be considered for prediction purpose. Figure 5 illustrates the predicted cesium BTC for early and late dilutions, assuming the exchanging sites are uniformly distributed throughout the soil system.

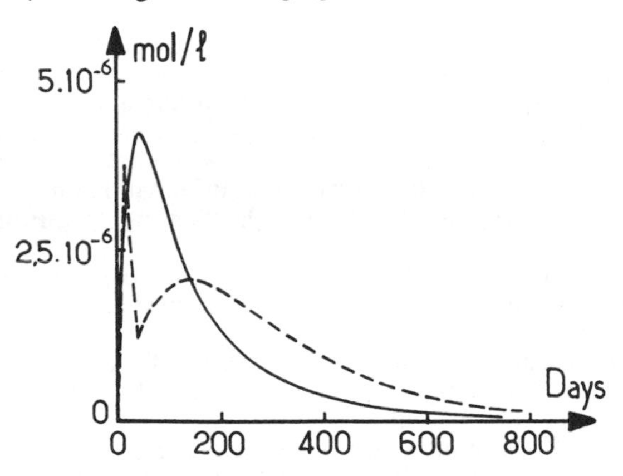

Figure 5: Cesium BTC. Continuous line: Late dilution. Broken line: Early dilution.

According to dilution earliness, BTCs are different. This behavior is typical of a non-linear transport process, although the concentration level is low.

These theoretical predictions were not checked against experiment because BTC spreads over more than one year and it was impossible to have a steady field water flow over such a long period. Nevertheles, new field experiments are planned to check the reliability of the prediction and to investigate the dilution process.

CONCLUSIONS

The experiments described above show that the relay substance method is a powerful tool for determining site-specific parameters. Although it has been applied to the effective exchange capacity, it could be useful for in-situ determination of other parameters such as adsorption surface area, Eh and pH, biological activity, mass transfer constants, etc., with suitably chosen species which are sensitive to the parameter considered. However, let us emphasize that the choice of a relay substance rely upon a reliable interaction mechanism in order to recover meaningful parameters. When several relay substances are available, one must find a compromise between the species having the lowest detection level and the weakest interaction strength to save time during the field experiment.

The relay substance method can only yield parameters averaged over the soil system located between the injection and detection points. Let us emphasize that the resulting mean parameter is determined by a complex competition between flow and chemical processes. Therefore, in addition to spatial distribution, the mean parameter can depend on the relay substance. This is illustrated by cesium and lithium which are unequally affected by the dilution process. Assuming that cesium is a relay substance, a field BTC could be interpreted according to various dilution assumptions. Consequently, various effective exchange capacities would result, although it is not known whether they would be significantly different. To overcome this problem, it is advisable to find experimental conditions under which the relay substance behaves as a linearly interacting solute.

Dilution earliness is a new concept originating from studies of mixing in chemical reactors [1, 6]. The dilution process is important in case of non-linear interactions, and theoretical predictions can be much different according to the assumptions made (See Figure 5). This suggests that non-linearly interacting solutes could be used as relay substances to get information on dilution along the transport process.

REFERENCES

[1] Jauzein M.: Méthodologie d'étude du transport transitoire de solutés dans les milieux poreux". PhD Thesis, INPL, Nancy, France, 1988.

[2] Harmsen K.: "Theories of cation adsorption by soil constituents: discrete-site models". In Soil chemistry, Part B, Physico-chemical models, Bolt G.H., Ed., Elsevier, Amsterdam, 1982, 77-140.

[3] Bruggenwert M.G.M., Kamphorst A.: "Survey of experimental information on cation exchange in soil systems". In Soil chemistry, Part B, Physico-chemical models, Bolt G.H., Ed., Elsevier, Amsterdam, 1982, 141-204.

[4] Jauzein M. , André C., Margrita R., Sardin M. and Schweich D.: "A flexible computer code for modelling transport in porous media: IMPACT", Geoderma special issue, 44, 2-3 (1989), 93-113.

[5] Jauzein M;,André C., Margrita R., Sardin M., Schweich D.: "IMPACT computer code: a help in design and interpretation of laboratory experiments. Proc. GEOVAL 90, SKI-OECD/NEA, Stockholm, Sweden, May 14-17, 1990.

[6] Schweich D., Jauzein M., Sardin M.: "Consequence of physico-chemistry on transient concentration wave propagation in steady flow". Proc. International Conference and Workshop on the Validation of Flow and Transport Models for the Unsaturated Zone, Wierenga P.J. and Bachelet D. Ed., La Cruces, NM State University Pub., 1988, 370-380.

LESSONS LEARNED FROM MODEL VALIDATION - A REGULATORY PERSPECTIVE

P. Flavelle, S. Nguyen, W. Napier,
Atomic Energy Control Board

and

D. Lafleur
Intera Technologies Ltd.
Canada

ABSTRACT

The Atomic Energy Control Board, the regulatory authority of the nuclear industry in Canada, has undertaken a project to validate the computer codes Femwater/ Femwaste. The strategy adopted to perform the validation was to calibrate the model based on historic monitoring data, predict subsequent behavior and then confirm the 1989 position of contaminant plumes from a uranium tailings impoundment.

The existence of a large body of published data from the site (one of the most thoroughly studied uranium tailings sites in Canada) did not provide an adequate calibration of the site model. An initial field trip was necessary to collect additional calibration data. The sources of discrepancy between the measured and predicted positions of the contaminant plumes in 1989 are identified and quantified. The reliability of the predictions is demonstrated to be equal to the accuracy (bias plus uncertainty) of the model calibration. Recommendations for improvements in the validation strategy are made, including the need to begin the validation with an unambiguous problem definition (objective, variables to be assessed and validation criteria).

RESUME

La Commission de contrôle de l'énergie atomique (CCEA) est responsable de la réglementation de l'industrie nucléaire au Canada. La CCEA a entrepris un projet de validation des logiciels Femwater/Femwaste. La stratégie adoptée consiste à choisir un site d'enfouissement de résidus de l'uranium, calibrer le modèle du site avec les informations existantes, utiliser le modèle calibré pour prédire la position du front de contamination en 1989, et enfin recueillir des informations in-situ en 1989 pour verifier les prédictions.

Même si le site en question a été étudié par de nombreux chercheurs , les informations qui en resultent se sont révélé insuffisantes pour permetter une calibration adéquate du modèle du site. Des informations in-situ additionnelles ont dû être receuillies afin d' améliorer la calibration. Les déviations entre les positions mesurée et prédite du front de contamination sont expliquées.Il est montré que la qualité de la prediction est équivalente à celle de la calibration. Des améliorations du processus de validation sont aussi proposées. On insiste sur l'importance d'une definition claire du problème (objectif, variables à comparer, critères de validation) dans tout exercice de validation.

Introduction

The Atomic Energy Control Board (AECB) has the responsibility to ensure that the use of nuclear energy in Canada does not pose undue risk to health, safety, security and the environment. This is accomplished through a comprehensive licensing program, including the licensing of radioactive waste management facilities.

Licensing submissions concerning waste management facilities often include the results of computer predictions of contaminant migration from the facilities. As part of the licensing process, it is necessary to review those calculations critically and independently. The staff of the AECB have undertaken to develop some limited modelling capabilities. The Femwater/Femwaste package for two-dimensional flow is one of the modelling tools considered for use by the AECB staff. Femwater is a finite element code that solves the flow equation in saturated or unsaturated porous media. Nodal velocities are calculated in Femwater for use in Femwaste, a finite element code that solves the contaminant transport equation. The transport mechanisms included are advection, dispersion, diffusion, adsorption (via a linear, Freundlich or Langmuir isotherm) and radioactive decay (of a single species only).

Verification of Femwater/Femwaste has been reported in the literature [1,2,3]. To our knowledge, however, little work has been published on the validation of the codes on large-scale field problems, and in particular on contaminant migration from tailings sites. This project was undertaken to evaluate the reliability of Femwater/Femwaste as a performance assessment tool to predict conservatively the position as a function of time of acidic, contaminated seepage from a uranium tailings impoundment. Not only will this project determine the capabilities and limitations of the codes to make such calculations, it will provide AECB staff with experience in computer code validation. For conservative calculations as a basis for regulatory decisions, the position and rate of movement of the contaminant plumes should never be under-predicted.

Contaminant plumes identified as seepage area A at the Nordic tailings impoundment, located near the town of Elliot Lake, Ontario, Canada, were modelled in this project . There is a large volume of published data on the physical, hydrogeological and geochemical properties of those tailings and the underlying geological formations [4-8]. Interpretations of the movement of the contaminant plumes for the period 1979-1984 have also been published, minimizing the effort and cost of characterizing the site for calibrating the model.

Only the effort to predict the movement of the pH plume is reported and discussed in this paper. The validation strategy consisted of the following steps:

- All pertinent existing data were compiled in order to develop a model of the site. Missing or uncertain input information were identified.
- A field program was performed in June, 1989 to resolve as many of the uncertainties in input data identified in the previous step as possible . No data on plume positions were gathered during that field trip. The field program results were combined with the 1983-1984 plume data to develop a final model of the site.
- The final model was used to make predictions of the plume positions in 1989
- A second field program was performed in September, 1989 to collect plume position data. The measurements were then compared critically with the predictions.

Existing information on the site

The Nordic tailings mineralogy consists mainly of quartz, feldspar, gypsum and about 7 weight% pyrite. The tailings were deposited in a glaciated valley between bedrock outcrops, on a spruce bog underlain by a glacio-fluvial sand, a relatively impermeable glacial till and bedrock. Containment is provided by permeable dams built with waste rock. A discontinuous peat layer 0.5 to 1m thick constitutes the top horizon of the aquifer underlying the tailings. It is thought that the discontinuities in the peat layer act as "windows" through which tailings porewater drains into the sand aquifer. One such "window" is thought to be beneath the tailings dam (Figure 1).

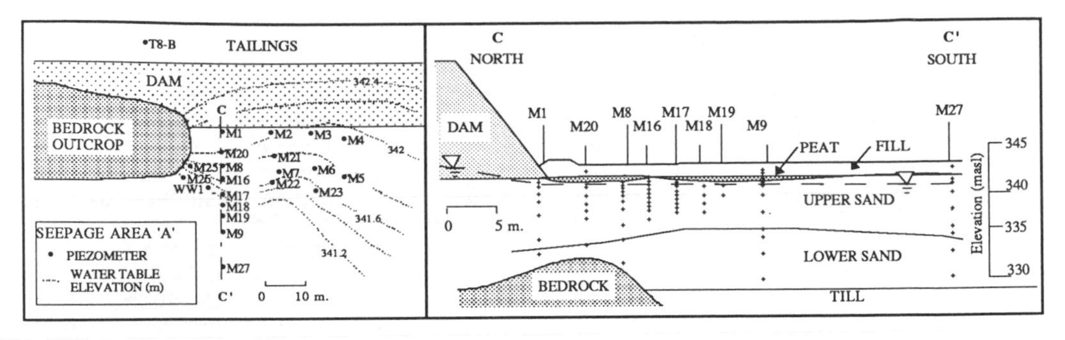

Figure 1 : Plan and vertical section view of Seepage Area A, Nordic Tailings, Elliot Lake, Canada

The hydraulic conductivity of the tailings, determined by both field tests (rising head and pump tests) and laboratory tests (constant head permeameter), is reported as 10^{-8} to 10^{-6} m/s [7,8]. The water table in the tailings at the dam at seepage area A is near the base of the tailings.

The aquifer material varies from a fine sand to a coarse sand with pebbles, divided into an upper medium sand and a lower very fine sand [4,5,6]. Some authors have speculated on the existence of a thin silt layer separating the sand strata [4,7]. Most of the groundwater flow and contaminant migration occur in the sand aquifer. An extensive array of piezometers were installed to delineate the contaminant plume at seepage area A, and section C-C' was interpreted to be the plume centreline (Figure 1). It is this section that is used for this validation of Femwater/ Femwaste, assuming that there is no flow normal to the section. The phreatic surface in the sand aquifer coincides with the bottom of the peat layer. Hydraulic conductivity of the sand aquifer ranges from 10^{-5} to 10^{-4} m/s, for the lower and upper sands, respectively. No conductivity tests were performed in the till or the peat. Nevertheless, based on measured gradients and flows and tests performed at neighboring sites, it was estimated that the peat has a hydraulic conductivity of 4×10^{-9} m/s and the till has a hydraulic conductivity of 10^{-8} m/s [7]. It has been suggested that the hydraulic conductivity of the sand aquifer has an anisotropy ratio of 10H:1V [4].

The mechanisms of contaminant leaching and transport at this site have been studied extensively [4-6, 9]. Oxidation of the pyrite in unsaturated tailings near the surface and adjacent to the permeable dam releases H^+ ions which are flushed into the saturated zone by infiltrating precipitation. This lowers the pH of the saturated tailings and leaches contaminants, including Pb, Th, Ra, and U, from the tailings solids. The acidified tailings porewater seeps through the discontinuous peat layer and into the underlying sand aquifer, where it is subject to a number of neutralization reactions. The resulting contaminant plume has discrete zones identifiable by their pH and mineralogical content. These are separated into the inner core of low pH, the outer zone of near-neutral pH, and the neutralization zone separating the two.

Theoretical calculations have suggested that the neutralization zone can be further divided into three sub-regions according to reactions taking place there. In the outer sub-region (at the front of the plume) calcite in the sand dissolves, neutralizing the acidity to about pH 5.73 and precipitating siderite. In the next sub-region the calcite has been consumed, and the siderite re-dissolves, neutralizing the incoming acidic porewater to about pH 5.27 and precipitating $Al(OH)_3$. The sub-region upstream of this one has the $AL(OH)_3$ being consumed to neutralize the porewater to pH 5.02. The zones and sub-regions of the plume are not evident as well defined plateaus in the observed pH, with distinct mineralization. Rather, there is continuous variation in pH and mineralization. The inner core has been identified with pH < 4.8, and the outer zone with pH $>$ 5.5. The outer zone is background to the plume, and has highly variable pH due to the influence of acid rain infiltration and organic acids from the peat. The better defined inner core, where the contamination occurs, is the part of the seepage that is of regulatory concern, and so is the focus of the modelling. The velocity of the inner core of the pH plume is 1-2 m/a, and the velocity of the outer edge of the neutralization zone is about 10 m/a.

Model development and calibration

No comprehensive set of simultaneous water level readings, suitable to calibrate Femwater, is reported in the literature. The initial field work in 1989 collected such a set of water levels and performed some in situ hydraulic conductivity tests. Analysis of the water level data confirmed that section C-C' is on a flow line. The hydraulic conductivities were found to be in the same range as reported in the literature. The June 1989 piezometric data were used in Femwater to model a steady-state two-dimensional flow field in the section C-C'.

The finite element model of section C-C' (Figure 2) had an impermeable lower boundary at the base of the lower sand and constant head boundaries on both ends of the modelled domain. The upper surface was assumed to correspond to the phreatic surface (also to the discontinuous peat layer) with a constant flux infiltration from above. The tailings and unsaturated zone were thus decoupled from the saturated part of the aquifer, and were not included in the analysis. The contaminant source term was simulated by a vertical flux to part of the model's upper boundary.

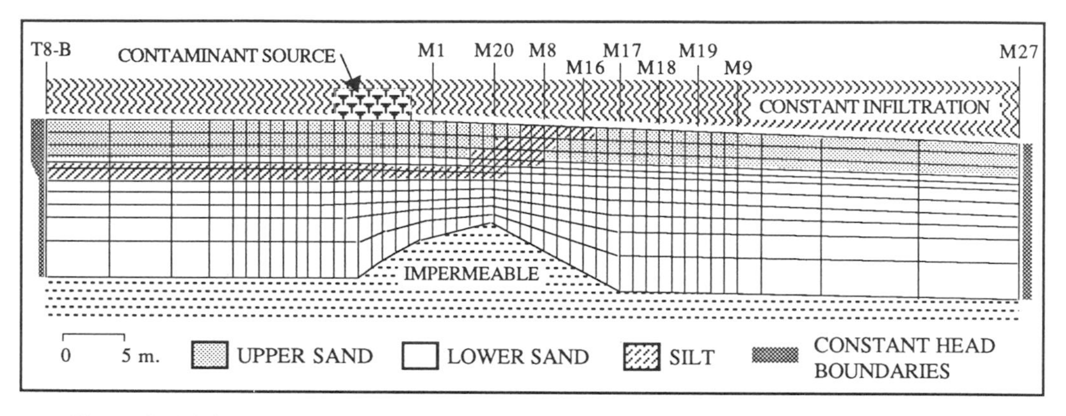

Figure 2 : Finite element mesh and boundary conditions for flow model of Section C-C'

The flow model was calibrated by adjusting the hydraulic conductivities and anisotropy and the hydraulic head distribution at the left-hand (upgradient) boundary of the model to reproduce the observed head distribution in the region of the inner core for 1989. The match between the observed and calculated heads was evaluated by inspection. The primary area of interest, between M1 and M18, had significant downward gradients. This required the introduction of a low-permeability silt layer separating the upper and lower sands beneath the dam (similar to the stratigraphic interpretation of Blair [7]) and rising to the surface between M8 and M16 . The calibrated flow model had uniform surface recharge, horizontal hydraulic conductivities of 1×10^{-4} m/s, 5×10^{-5} m/s and 1×10^{-6} m/s in the upper sand, lower sand and silt layer, respectively, conductivity anisotropy of 10H:1V, and constant head on the upstream boundary which is 1 m less in the lower sand than in the upper sand.

The transport model was calibrated by adjusting the location and width of the source (simulated by a contaminant flux to the upper surface), the source concentration with time, the dispersivity of the aquifer and the retarded interstitial solute velocity. The solute velocity was adjusted to match that of the inner core by varying the K_d applied to the steady-state groundwater velocity field (transferred from the calibrated Femwater model). The calibration was done to match the 1983 and the 1984 plume positions for a simulation time of 17.5 a., and was evaluated by inspection. The calibrated model has a constant source strength of pH 3, located from 2 to 6 m upgradient of piezometer M1. The dispersivity of the sand aquifer is 1 m and the K_d is 0.0175 m^3/kg to correspond to the observed velocity of the inner core.

Model prediction

Prior to the final field trip to measure the 1989 position of the contaminant plumes, the calibrated models were used to predict the distribution of contaminant concentrations (including pH) along Section C-C'. Since a steady-state flow field and a single retardation was assumed, the prediction entailed only translating the contaminant plumes from the position calculated by the model calibration for a further five year period using the steady-state solute velocities. The predicted and measured positions of the pH plume are compared in Figure 3. It can be seen that, as desired, there is a reasonable agreement between the position of the contours of predicted and measured pH in the area between M1 and M16, where pH values correspond to the inner core.

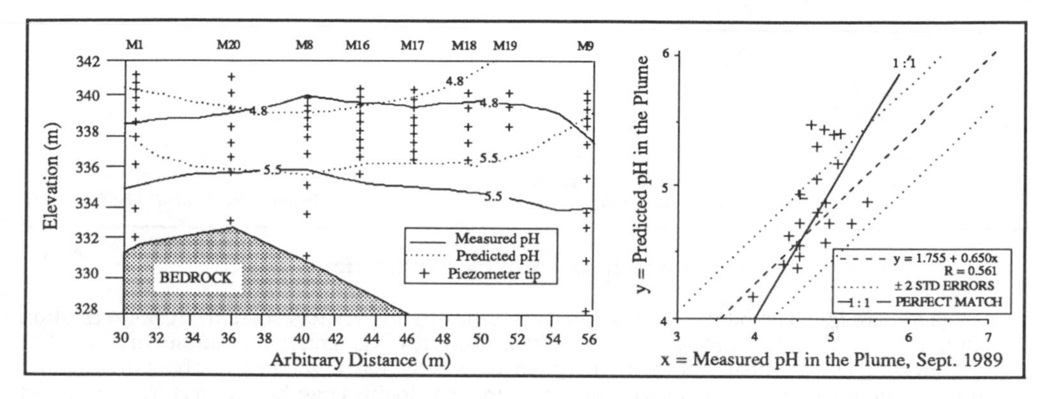

Figure 3 : Comparison of predicted and measured pH (September, 1989)

Evaluation of the validation

A more detailed examination of the calculations performed in this exercise reveals the source and magnitude of biases and uncertainties which are responsible for the deviations between the predicted and measured pH. The assumptions of a linear sorption isotherm and steady-state two-dimensional flow, the quality of the calibrations of hydraulic head, groundwater velocity and pH, and the numerical accuracy of the computer codes are examined.

A single K_d value corresponding to the observed velocity of the inner core was used, in keeping with the regulatory interest in the highly contaminated region. It was found that the movement of the inner core was reasonably well predicted, while the velocity of the outer edge of the neutralization zone was under-predicted. Although predicting the background pH is not of concern here, the poor match at higher pH highlights the problem of trying to simulate effects of a range of geochemical reactions using a single value of contaminant retardation.

The assumption of steady-state groundwater flow was evaluated by comparing two sets of piezometric measurements. Point-by-point comparison of the piezometric data collected in the two 1989 field trips (x=June and y=September) gives a regression line $y = 0.86x + 46.6$; $R=0.985$; standard error of the slope = 0.026 m. The 95% confidence interval on the slope, then, is 0.86 ± 0.05. This $(14 \pm 5)\%$ deviation from a perfect fit (slope of 1.00, meaning no change in the hydraulic gradient distribution with time) is the measured uncertainty in the steady state assumption for the three month period.

Figure 4 shows the match between the piezometric heads calculated using the calibrated Femwater model and the measured data used in the calibration. The fit is quite good for the higher heads and higher gradients (in the region of interest, the inner core), but not as good downstream. Regression analysis of the calculated vs measured data has a slope of 0.776 (R=0.95, standard error = 0.043). The deviation from a perfect fit (slope of a perfect fit = 1.000) is the bias in the hydraulic head calibration. Heads less than 341.4 m will tend to be over-predicted and heads greater than 341.4 m will tend to be under-predicted. This results in calculated hydraulic gradients lower than those measured, as shown by the contour spacing in Figure 4. The scatter of the data about the regression line (measured by the standard error, assuming the deviations from the regression line are normally distributed) reflects the uncertainty in the hydraulic head calibration.

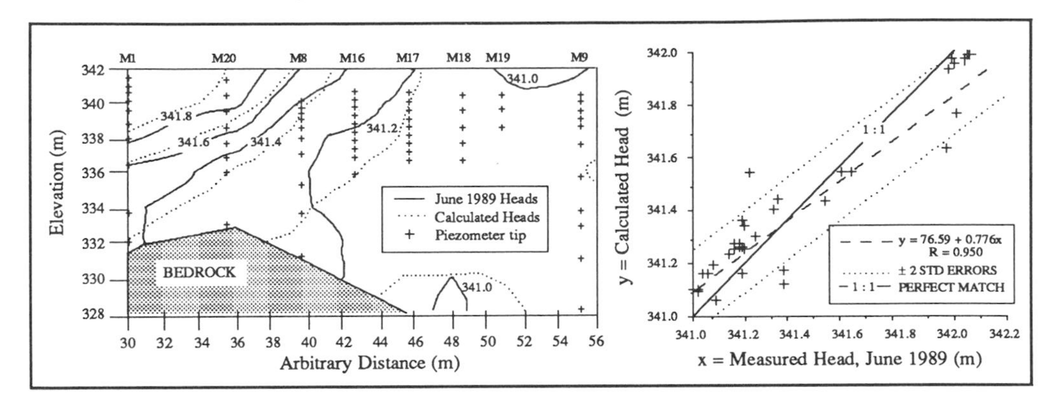

Figure 4 : Calibration of Piezometric Head

There are few data on the groundwater velocities in this vertical section. A point dilution measurement in well WW1 gave a linear velocity of 440 m/a [4]. Tracer measurements from M1 to M19 during the September 1989 field trip gave an average linear velocity of about 40 m/a, but the some of the tracer moved at about 400 m/a. Linear velocities can be inferred from reported hydraulic gradients and porosities [4] and hydraulic conductivities to be 450 m/a near the dam, 180 m/a 30 m downstream from the dam (around M9) and 45 m/a 100 m downstream from the dam. Linear velocities from the calibrated Femwater model are on the order of 100 m/a in both the upper and lower sand, and less than 20 m/a in the assumed silt layer (Figure 2). Although there is insufficient information to make critical comparisons, any error in calculated linear groundwater velocity will be compensated by the calibration of the K_d to get the retarded solute velocity.

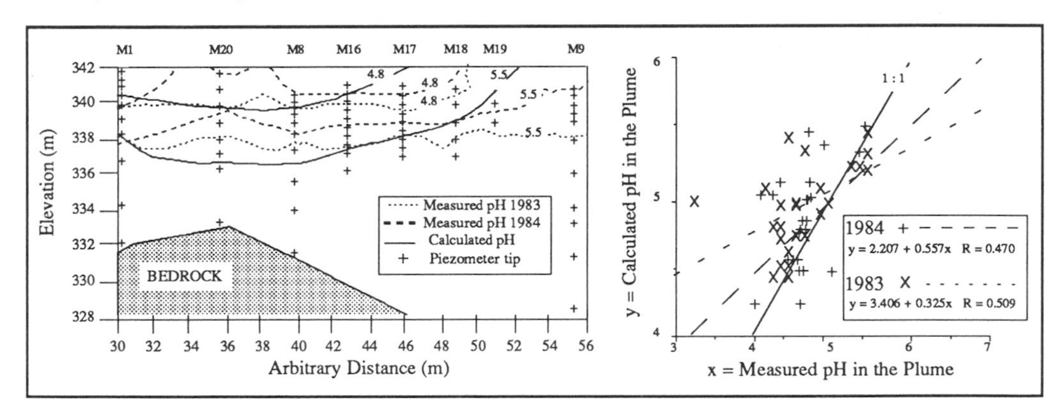

Figure 5 : Calibration of pH

The pH calibration was performed using the hydrogen ion activities from the 1983 and 1984 pH measurements in the region of the plume (pH > 5.5). As illustrated in Figure 5, the depth of the plume is matched well, but the horizontal extent of the plume is under-predicted. This comes from assuming a single retardation coefficient (the solute velocity of the front of the plume has been observed to be greater than of the inner core) coupled with the poorer fit of hydraulic heads downgradient from M17. The variation in the observed position of the plume from 1983 to 1984 is the result of the data being collected by different researchers, at different times of the year and using different procedures. The regression analysis of the plume pH shows the bias in the calibration : for both 1983 and 1984 data the calibrated model tends to under-predict above pH 5 and tends to over-predict below pH 5. This bias applies only to calculated pH less than 5.5, since calculated pH > 5.5 is beyond the range of the model's numerical stability (see below). Since the pH in the inner core of the plume tends to be over-predicted, these model results are not conservative (in spite of the apparently conservative match of the contour depths).

The data scatter about the regression lines for the calibration is measured by the standard error (assuming normal distribution of deviations from the regression line) : 0.28 for 1983 data and 0.34 for 1984 data. The data scatter about the regression line of predictions (Figure 3) is similar, having a standard error of 0.32. The slopes of the regression lines are also similar, indicating that the reliability of the predictions is determined by the accuracy of the calibration.

In order to assess numerical accuracy, Femwaste was run for several one dimensional problems, including a point source problem with the Sauty analytical solution [3]. The relative error (in concentration) was more than ±50% at relative concentrations less than 10^{-3}. This establishes an accuracy limit due to numerical instability of the model to be ±0.3 pH units at pH 6, with increasing error for pH >6. Accordingly, calculated pH greater than 5.5 (corresponding to the highly variable background) were considered unreliable.

Discussion and conclusions

The lack of an appropriate conservative contaminant plume precluded validating the flow field predicted by Femwater. Also, we were unable to resolve (by field work) all of the uncertainties regarding the physical characteristics of the site, notably, the position and strength of the contaminant source as a function of time, the spatial distributions of porosity, hydraulic conductivity and hydrodynamic dispersivity of the aquifer, and the temporal variation of piezometric head and hydraulic gradient. Hence these unknowns contributed to the uncertainty in the flow and transport model calibrations. Additional uncertainty arises from using a two-dimensional section to model three-dimensional flow, precluding accounting for the divergence of the flow downstream of the bedrock outcrop or for flow and dispersion normal to the section.

It should be recognized that the evaluated uncertainties in calibrating the groundwater flow (<±50% numerical error, ±(14±5)% error from the assumption of steady state, (-23±9)% error in the hydraulic gradient calibration) are small compared to the uncertainties in the H+ calibration, which are expressed as uncertainty in the log of the hydrogen ion activity (pH). The ±0.7 pH units data scatter about the calibration regression lines corresponds to an uncertainty band of -80% /+500%. Causes of the large data scatter are discussed in the preceeding section. The statistics of the match of the predicted to measured pH are very similar to those of the pH calibration, implying that the accuracy of the predictions cannot exceed the accuracy of the calibration, but also, encouragingly, that there was no degradation of accuracy from the calibration to the predictions.

A fundamental assumption adopted for these simulations is that the migration of H+ can be simulated as a diffusing species with a sorption isotherm. The varied and complex neutralization and precipitation/dissolution reactions occurring in the aquifer, the variable infiltration of acid rain and the variable distribution of peat (giving rise to organic acids) in the aquifer all contribute to changes in pH at this site. However, for the purposes of a performance assessment calculation, where a limiting value of pH is wanted, it may be possible to simulate the pH migration with a phenomenological sorption isotherm. This validation project has demonstrated that, for the part of the seepage that had a reasonable fit by the calibrations, a reasonable fit in the predictions can be obtained. However, the reasonable fit of the calibration did not include the full horizontal extent of the plume, so the predictions of the horizontal migration of the plume were not conservative. Since the plume will eventually migrate into the region of background pH levels, the contaminant transport calculations in this region should be made with reasonable accuracy to predict such migration. Further work has begun to examine more of the capabilities of Femwater/Femwaste to improve the calibration.

Based on this work it is difficult to make a definitive statement regarding the validation of Femwater/Femwaste for this site. This is because the results are satisfactory for one part of the site and unsatisfactory for another. One lesson learned from this work is that parameter specification is critical in the project design stage. If further work produces promising results, the project will begin another iteration, this time with a better understanding of the specifications needed for a validation. These specifications will include :

- A clear definition of the problem, including what is the purpose of the model to be validated, what parameters will be used to make comparisons, what constitutes acceptable agreement between predictions and reality , what accuracy of measurements is required, and the range of conditions for which validation is needed.
- How the data uncertainties will be incorporated in the comparisons, and what confidence is needed at each step of the validation.

We have found that published site data may not be sufficient for a model validation unless they have been collected specifically for the validation study. Although this is one of the most thoroughly studied uranium tailings sites in Canada, the published data (collected for a variety of purposes) were not complete enough for the validation. It appears that applying this modelling approach to other, less well studied sites would be problematic. However, this may not be the case, since from our experience at this site we caution anyone embarking on a model validation study to be prepared to perform a complete site characterization, based on, but probably independent of, published site data.

References

[1] Yeh, G.T., D.T. Ward : "FEMWATER: A Finite Element Model of WATER flow through saturated-unsaturated porous media", ORNL-5567, Oak Ridges National Laboratory, Oak Ridge, Tennessee (1980).

[2] Water and Earth Science Associates Ltd. : "A survey of computer codes for flow and contaminant transport", internal report prepared for the AECB, Ottawa, Canada (1987).

[3] Intera Technologies Ltd. : "Calibration and validation of FEMWATER/FEMWASTE", to be published as AECB Information Series report INFO- , Ottawa, Canada (1990).

[4] Morin, K.A., J.A. Cherry, N.K. Dave, T.P. Lim, and A.J. Vivyurka : "Migration of acidic groundwater seepage from uranium-tailings impoundments, 1. Field study and conceptual hydrogeochemical model", J. Contam. Hydrol., 2 (1988), 271-303.

[5] Morin, K.A., J.A. Cherry, N.K. Dave, T.P. Lim, and A.J. Vivyurka : "Migration of acidic groundwater seepage from uranium-tailings impoundments, 2. Geochemical behavior of radionuclides in groundwater", J. Contam. Hydrol., 2 (1988), 305-332.

[6] Morin, K.A., J.A. Cherry : "Migration of acidic groundwater seepage from uranium-tailings impoundments, 3. Simulation of the conceptual model with application to Seepage Area A", J. Contam. Hydrol., 2 (1988), 323-342.

[7] Blair, R.D., J.A. Cherry, T.P. Lim and A.J. Vivyurka : "Groundwater monitoring and contaminant occurrence at an abandoned tailings area, Elliot Lake, Ontario", Proceedings of the First International Conference on Uranium Mine Waste Disposal, Vancouver, Society of Mining Engineers of America, Institute of Mining Engineers, New York (1980), 411-444.

[8] Blowes, D.W. and R.W. Gillham : "The generation and quality of streamflow on inactive uranium tailings near Elliot Lake", J. Contam. Hydrol., 1 (1988),

[9] Dubrovsky, N.M., J.A. Cherry, E. J. Reardon and A.J. Vivyurka : "Geochemical evolution of inactive pyritic tailings in Elliot Lake Uranium District", Can. Geotech. J., 22 (1985), 110-127.

RELATIONS BETWEEN GROUNDWATER COMPOSITION AND FRACTURE FILLING MINERALOGY IN SWEDISH DEEP ROCKS

Karin Andersson
Lindgren o Andersson HB, Surte, Sweden

ABSTRACT

Groundwater flowing through the rock will undergo a continous change in composition along the flowpath due to chemical reactions with the surrounding minerals. The surfaces in contact with the water do not have the same composition as the rock matrix, but fracture filling minerals of different origin are present. Here recent data from the SKB investigations of deep groundwater and fracture filling mineralogy have been reviewed and equilibrium calculations using the PHREEQE code performed.

Most waters seem to be close to equilibrium with respect to calcite. For other minerals the saturation state is more varying. If this is due to the real conditions or to inaccurate modeling is not quite clear. Although investigations have been performed at a number of sites during the years, relevant chemical data are few, and it is difficult to couple water chemistry data and fracture filling mineralogy from the same fracture. This is due to the difficulty in sampling deep waters without disturbing the composition. However, ranges of probable composition of major ions and redox potentials in deep granitic groundwaters may be defined.

LIENS ENTRE LA COMPOSITION DE L'EAU SOUTERRAINE ET LA NATURE DES MINERAUX DE REMPLISSAGE DES FISSURES DANS LES ROCHES PROFONDES EN SUEDE

RESUME

La composition de l'eau souterraine qui traverse les roches se modifie constamment le long de son trajet car elle réagit chimiquement avec les minéraux qui l'entourent. Les surfaces en contact avec l'eau n'ont pas la même composition que la matrice de la roche, car les minéraux qui remplissent les fissures ont des origines variées. L'auteur examine des données récentes tirées des études du SKB sur l'eau souterraine profonde et la minéralogie des matériaux de remplissage des fissures, puis présente le résultat de calculs d'équilibre réalisés au moyen du programme PHREEQE.

La plupart des eaux semblent être proches de l'équilibre en ce qui concerne la calcite. Pour les autres minéraux, l'état de saturation varie davantage. On ne sait pas très bien si ces variations sont dues aux conditions naturelles ou à un manque de précision dans la modélisation. Bien que des études aient été menées sur un certain nombre de sites au cours des années, on ne possède que peu de données chimiques utilisables et il est difficile, pour une fissure déterminée, d'établir un lien entre les données hydrochimiques et la nature des minéraux de remplissage de la fissure. Cela tient à la difficulté de prélever des eaux profondes sans modifier leur composition chimique. Toutefois, il est possible de définir des fourchettes probables en ce qui concerne la teneur en ions principaux et les potentiels rédox dans les eaux souterraines des formations granitiques profondes.

INTRODUCTION

The composition of deep groundwaters is the result of interactions between water and minerals in soil and bedrock during a very long time. When a groundwater system is disturbed, eg by the construction of an underground facility like a waste repository, the water composition may undergo local changes.

In order to be able to predict the effects of a repository as well as to predict the chemical environment for radionuclides released to the groundwater after very long times, it is necessary to have a knowledge of the behaviour of the present natural groundwater system. In this report some investigations of groundwater composition and rock mineralogy are evaluated using geochemical equilibrium modeling. The impact of groundwater composition and rock mineralogy on the transport properties of radionuclides in groundwater is also discussed.

AVAILABLE DATA FROM SKB TEST SITES

Investigations of geology, hydrology, geochemistry in deep rock formations have been performed by SKB for a number of years at different Swedish sites. A large material on fracture distribution and mineralogy has been collected. However, the amount of data on groundwater chemistry is not as large as that of geological and hydrological data.

The sampling has mainly been performed in core drilled holes, where sections have been sealed off with packers and water has been pumped out until equilibrium conditions have been reached. The chemical analyses have been performed at a conventional laboratory as well as in a mobile field laboratory (since 1984). Some measurements (pH, Eh) have also been performed in situ, downhole. In the field laboratory the water from the bore hole is led directly into the various analysis equipment without contact with the atmosphere. pH, Eh, conductivity and dissolved oxygen are measured on line. Also major ions and redox sensitive trace elements are analyzed in the mobile laboratory. The equipment is further described in a technical report [1]. Methods and detection limits for the various analyses are given in another report [2].

EQUILIBRIUM CALCULATIONS

Some recently investigated areas have been selected for calculations on groundwater saturation state. These are:

- Klipperås, where the most recent investigations that have been published in technical reports were performed. Here the groundwater data have been thoroughly discussed [3]. Data on fracture filling mineralogy are available to some extent.

- Finnsjön. A series of holes were investigated at an early stage, and new holes have been drilled and investigated more recently [4], [5]. The fracture filling mineralogy here is not described in too much detail.

The data have been used for calculations of speciation and saturation state using the PHREEQE code [6]. Two different data bases have been used for this purpose; one mainly based on the original PHREEQE data base with some mineral phases added within Project 90 [7] and the other is the HATCHES data base, developed at Harwell [8]. The results are compared to published data calculated with the EQ3/6 code [3].

The calculation results are given as saturation index = log [IAP/K_s], where IAP = ion activity product and K_s = solubility product. At equilibrium, IAP = K_s. SI>0 for supersaturated solution and SI<0 for non saturated solution.

Klipperås

Rock type: Granite [9]. The area is intensly fractured, especially at shallow depth. Deeper down the conductivity is low. [10].

Fissure fillings: Calcite, chlorite, epidote, hematite, muscovite/illite, quartz, andularia, pyrite, Fe(III)–oxyhydroxide [10]

Water: Cf Table I. Most data are within KBS–3 range. Neglible tendency to saline water. Strongly reducing.

The data used here are from three different boreholes, and the sampling that is considered representative in SKB's evaluation of the results [3] has been used. Other data from the holes are available as well.

The water is very close to saturation with respect to calcite, Figure 1 and Table II, and also to epidote (not included in the Figure). Amorphous Fe(III)–hydroxide is undersaturated, while pyrite is supersaturated. In this site the "FeS precipitate" is close to saturation. Also hematite and magnetite (not included in the Figure) are quite close to saturation. The saturation state for the different minerals is quite similar for the three holes.

Finnsjön

There are several types of rock within the Finnsjön site. The hole Fi9 is drilled in a granodiorite, which is rock type quite similar to granite but where the feldspars are constituted mostly of plagioclase instead of K–feldspar [11].

Rock type: Granodiorite (32% plagioclase, 30% quartz, 18% microcline, 11% hornblende, 9% biotite/chlorite) [4]

Fissure fillings: Hematite, laumontite, epidote, prehnite, calcite, quartz, chlorite [12]

Water: Saline groundwater, c f Table I. Reducing.

The salinity increases with depth, while pH is not varying very much. Eh data are only available for the two upper levels, where the potential is quite low. The dominating cations are sodium and calcium, which are present in approximately equal concentrations. Since no Eh data are available for the bottom level, –250 mV has been assumed in the calculation.

The Finnsjön water is very close to saturation with respect to calcite and also to the FeS precipitate. For the other minerals the saturation state changes with depth, c f Figure 2. The amorphous Fe(III)–hydroxide is predicted to be very undersaturated at all levels, while pyrite

TABLE I. DEEP GROUNDWATER COMPOSITIONS, Klipperås [3], Finnsjön [5], and ranges used in KBS–3 [13]

mg/l	Kl 1, Kl 2, Kl 9	Fi 9,	Ranges KBS-3
HCO_3^-	80-140	32-285	90 - 275
SO_4^{-2}	0.1-4.3	175-326	0.5 - 15
HPO_4^{-2}	0.001-0.003	0.001-0.004	0.01 - 0.2
NO_3^-	0.005-0.02	0.01-0.14	0.01 - 0.5
F^-	2.9-3.8	1.1-7.2	0.5 - 5
Cl^-	5.9-45	680-5150	4 - 15
HS^-	0.01-0.10	-	0 - 0.5
Ca^{+2}	14-31	115-1691	10 - 40
Mg^{+2}	1.0-2.3	16-91	1 - 10
Na^+	15-47	415-1510	10 - 100
K^+	1.0-1.3	5.8-15	1 - 5
Fe^{+2}	0.008-0.13	0.05-0.19	0.02 - 5
Fe_{tot}	0.010-0.10	0.35-0.94	1 - 5
Mn^{+2}	0.03-0.04	0.19-0.82	0.01 - 0.5
Al^{+3}	0.03-0.07	0.02-0.18	0 - 0.02
NH_4^+	0.01-0.08	-	0.05 - 0.2
SiO_2 tot	9.4-21.2	9.9-16.2	3 - 14
TOC	1.2-3.7	-	1 - 8
pH	7.6-8.3	7.3-7.7	7 - 9
Eh, V	(-0.30)-(-0.27)	0.01-(-0.29)	(-0.45)-0

is supersaturated. Laumontite (not included in the Figure) is close to saturation at the upper level, but supersaturated at lower levels. The higher saturation indices of all Fe minerals at the middle level is due to a measured total iron concentration that is almost twice as high here as at the two other levels. The tendency is the same for Fe(II) and Fe(III) minerals, which possibly is expected since the Eh of the waters is quite equal.

DISCUSSION

Water chemistry data

A "normal fracture" in a granite or gneiss contains a non–saline water mainly within the composition range given in KBS–3. Deviations from the range have been encountered at some points for the Klipperåsen water:

- lower values: HCO_3^-, SO_4^{2-}, HPO_4^-, NO_3^-, Fe^{+2}, Fe_{tot}, NH_4^+
- higher values: Al^{+3}, Cl^-, SiO_2.

A special case is the saline groundwater that is encountered as inclusions at many sites, eg in Finnsjön. It is not clear if this water is in contact with the surrounding water, but the existence of water of such different composition at the same site implies that the contact between the water types occurs to a very small extent if at all.

TABLE II. SATURATION INDICES OF KLIPPERÅS WATERS CALCULATED WITH EQ3/6 AND PHREEQE WITH DATABASES SKI2 AND HATCHES. EQ3/6 DATA FROM [3]

Mineral		Borehole, vertical depth K11, 398 m	KL2, 320 m	K19, 581 m
Calcite				
CaCO3	EQ3/6	-0,19	0,15	-0,27
	PHREEQE SKI2	-0,14	-0,52	0,01
	PHREEQE HATCHES	-0,50	-0,41	0,08
Siderite				
FeCO3	EQ3/6	-1,79	-0,69	-1,09
	PHREEQE SKI2	-1,97	-1,18	-1,49
	PHREEQE HATCHES	-1,85	-1,03	-1,39
Fe(OH)3				
amorph	EQ3/6	0,08	0,21	0,31
	PHREEQE SKI2	-5,82	-6,67	-5,72
	PHREEQE HATCHES	-5,82	-6,67	-5,72
Hematite				
Fe2O3	EQ3/6	10,14	10,43	10,55
	PHREEQE SKI2	1,13	-0,65	1,55
	PHREEQE HATCHES	1,13	-0,64	1,55
Goethite				
FeOOH	EQ3/6	4,61	4,76	4,81
	PHREEQE SKI2	-1,89	-2,78	-1,69
	PHREEQE HATCHES	-1,91	-2,79	-1,70
Pyrite				
FeS2	EQ3/6	4,76	4,81	2,86
	PHREEQE SKI2	7,65	5,81	9,28
	PHREEQE HATCHES	7,65	5,82	9,27

The measurements of pH and Eh downhole give a valuble contribution to the data, since these parameters are very sensitive to changes that occur when the water is exposed to the atmosphere. pH is correlated to the partial pressure of carbon dioxide and this may be difficult to maintain even when pumping the water to the mobile laboratory. Eh is strongly affected by intrusion of air, which also may be difficult to avoid when pumping.

Saturation Index – fracture mineralogy

Water samples are taken from a section of the borehole where the conductivity is high. In the section several different fractures of different conductivity and fracture filling mineralogy may be found. The mineralogy in the fractures can be assessed in the drill core but not further along the flow path of the water. In some cases it may also be difficult to identify the real water–bearing fracture. This means that the fracture filling mineralogy will be based on a statistical treatment of the data from the fracture intersections with the drill core rather than on the real conditions in the water bearing fracture. Fortunately, quite few minerals are dominating the fracture filling mineralogy and the water bearing fractures may in most cases be identified among low conductive and sealed fractures [14].

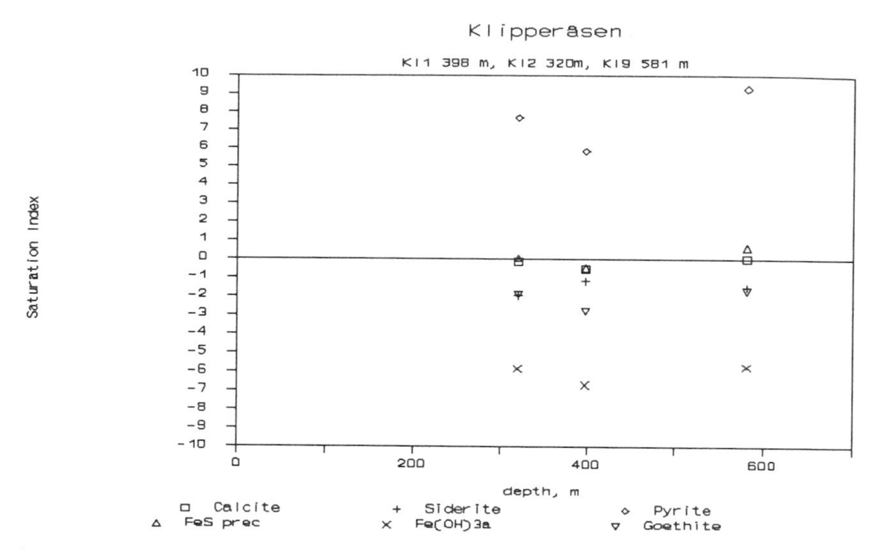

Figure 1. Saturation index vs depth, Klipperåsen

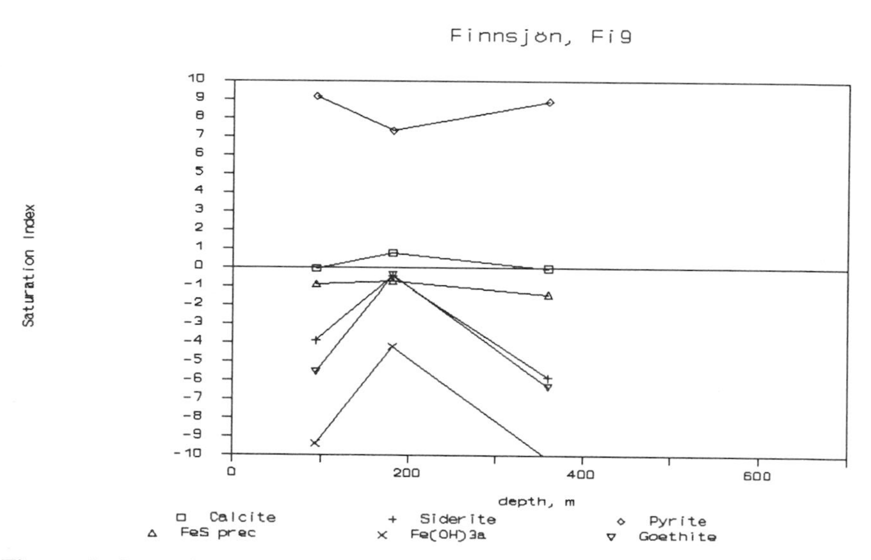

Figure 2. Saturation index vs depth, Finnsjön

Common minerals coating fracture surfaces are calcite, iron minerals like pyrite/hematite/Fe–oxyhydroxide, clay minerals like chlorite/illite/smectite, and also quartz, feldspars and zeolites. In high conductive fractures calcite and iron minerals often dominate. Chlorite and other clay minerals are mainly found in low conductive or sealed fractures [14]. Most waters also seem to be close to equilibrium with respect to calcite.

For the iron minerals the saturation situation is somewhat difficult to interpret. Pyrite, hematite and Fe–oxyhydroxide may be observed in the same fracture. The equilibrium calculations give varying results, often very far from saturation. The very low saturation index for pyrite at Klipperås is due to a prediction of very small amounts of sulfide in the groundwater. This may be caused by a too high Eh value given in the analysis. Since pyrite is present in fractures all along the borehole, this would be expected to influence the redox chemistry as well as the saturation state. However, pyrite reacts very slowly in many systems and it may be observed together with iron hydroxides [14]. A possible solid in equilibrium with the reducing water would be a "FeS precipitate". It has a solubility product ca twelve orders of magnitude higher than that of pyrite, but the equilibrium Eh is quite similar. This solid is close to equilibrium at both Klipperåsen and Finnsjön.

The difficulties with the iron minerals may be due to errors in the water sampling and analysis (Eh, Fe(II)/Fe(III), sulfide) as well as to errors due to the equilibrium calculation itself (thermodynamic data, definition of formation reaction of minerals, non–equilibrium). It is thus difficult to evaluate which reactions that are actually determining Eh and concentrations of iron species and sulfide.

The aluminium and silicate concentrations should be determined by the silicate and aluminum silicate minerals, although it is not obvious that any single mineral is in equilibrium with the water. Quartz is generally somewhat supersaturated, while the amorphous SiO_2 is undersaturated.

Redox potentials

The exact value of the redox potential is difficult to determine, but all investigated deep groundwaters do have a low potential. The measured redox potentials may be too high – it is likely that an error is caused by intrusion of air which gives a too high value. The measured potentials also seems to be lower at each new site, which may be an effect of improved methods but it is clear that the potential in a undisturbed deep groundwater always is very low.

The redox buffering capacity is more difficult to evaluate, especially since equilibrium obviously not is attained for all redox sensitive minerals present. This is an important parameter to restoring of reducing conditions in a closed waste repository.

Geochemical modeling

Apart from the obvious problems with obtaining relevant analysis data for groundwater composition and mineralogy in the fractures, the relevance of geochemical equilibrium calculations may also be discussed.

The groundwater/bedrock is modeled as a system where instantaneous (in geological timescale) equilibrium is attained with some of the mineral phases present. Alternative models, where reaction kinetics is an important process, or where the minerals are treated as solid solutions instead of as well defined, pure, crystalline solids may be more relevant. Eg epidote is treated as a Fe(III) mineral in the data base but a better representation would be a solid solution of Fe(III) and Al(III) silicates. Chlorite is here represented by two end members of a solid solution between the Mg(II)– and the Fe(II)– form. The Mg–chlorite is indicated to be close to saturation while the Fe– form is predicted to be strongly supersaturated.

Models taking both these processes into account exist (kinetics, eg [15] and [16], solid solutions, eg models developed by B Fritz [17]). For the modeling of kinetic processes, there is a large problem in the lack of relevant kinetic data.

The data base for the geochemical equilibrium calcualtions is of large importance. For common, non redox sensitive minerals, the calculations show a quite good agreement between different data bases for PHREEQE and for non redox sensitive elements also with EQ3/6, cf Table II. For redox sensitive systems there are large differences when using EQ3/6. Possibly the EQ3/6 calculations were made at another Eh value than the measured since the differences in log K values are not as large as the difference in calcualted saturation index.

The formation reactions used for a solid or a species may also affect the calculation result, since the method of solving the equilibrium equations in PHREEQE is based on the use of "master species", i e the concentrations of different species of an element are summed in each step. Also electrons are treated as a "master species".

The iron minerals are in many cases predicted to be supersaturated. This may be due to:

- data base (for both minerals and solution species)
- analysis of iron species
- for sulfide minerals also analysis of sulfide
- measured Eh
- method of calculating concentrations

Most data bases agree at least within one order of magnitude, and the predicted supersaturation is often much larger than this. For sulfide there may also be an error in the equilibrium constants in the data base used.

CONCLUSIONS

The composition of deep granitic groundwaters is quite well-known what concerns ranges in concentrations of major ions, pH etc. For the redox system there are difficulties in determining the actual redox potential, although it is clear that the potential is low.

The redox buffering capacity of the rock/groundwater may be difficult to evaluate. Pyrite is not always oxidized completely even in strongly oxidizing environments. This may have an impact on the restoring of reducing conditions after the closure of a repository.

Most deep groundwaters are predicted bo be close to saturation with respect to calcite. This seems reasonable, since calcite is present in most fracture systems. The redox sensitive minerals are often predicted to be far from equilibrium also in water from fractures where these minerals have been observed. The minerals seem to coexist in nature. Possibly there is an equilibrium with "FeS precipitate" in some waters. The silicate concentrations found in the Klipperås and Finnsjö waters give supersaturation with respect to quartz and a number of aluminium silicate minerals. However, the waters are undersaturated with respect to amorphous SiO_2.

The problem of ensuring that the sampled water is representative of that in an identified fracture still seems to remain. The water volumes pumped out for the sampling are very large compared to the low flow rate at the investigated depths, and the water used for a sampling may represent the natural flow during several years. This may affect the equilibrium conditions in the sampled water.

REFERENCES

1. K–E Almén, O Andersson, B Fridh, B–E Johansson, M Sehlstedt, E Gustafsson, K Hansson, O Olsson, G Nilsson, K Axelsen, P Wikberg, "Site investigation equipment for geological, geophysical, hydrological and hydrochemical characterization", SKB TR 86–16, Svensk Kärnbränslehantering AB, Stockholm 1986

2. P Wikberg, K Axelsen, F Fredlund, "Deep groundwater chemistry", SKB TR 87–07, Svensk Kärnbränslehantering AB, Stockholm 1987

3. J Smellie, N–Å Larsson, P Wikberg, I Puigdomènech, E–L Tullborg, "Hydrochemical investigations in crystalline bedrock in relation to existing hydraulic conditions: Klipperås test–site, Småland, Southern Sweden", SKB TR 87–21, Svensk Kärnbränslehantering AB, Stockholm 1987

4. K Ahlbom, P Andersson, L Ekman, E Gustafsson, J Smellie, E–L Tullborg, "Preliminary investigations of fracture zones in the Brändan area, Finnsjön study site", SKB TR 86–05, Svensk Kärnbränslehantering AB, Stockholm 1986

5. K Ahlbom, J A T Smellie (eds), "Characterization of fracture zone 2, Finnsjön study site", SKB TR 89–19, Svensk Kärnbränslehantering AB, Stockholm 1989

6. D L Parkhurst, D C Thorstenson, L N Plummer, "PHREEQE computer code, version current, August 3, 1984", obtained from OECD/NEA data bank, Paris

7. K Andersson, "Documentation of thermodynamic database "SKI". Data for mineral/rock calculations", Technical report in preparation

8. J E Cross, F T Ewart, C J Tweed, "Thermochemical modelling with application to nuclear waste processing and disposal", AERE R 12324, UKAEA, Harwell 1987

9. A Olkiewicz, V Stejskal, "Geological and tectonic description of the Klipperås study site", SKB TR 86–06, Svensk Kärnbränslehantering AB, Stockholm 1986

10. E–L Tullborg, "Fissure fillings from the Klipperås study site", SKB TR 86–10, Svensk Kärnbränslehantering AB, Stockholm 1986

11. C Klein, C S Hurlbut Jr, "Manual of Mineralogy", 20th ed, John Wiley, New York 1985

12. S Sehlstedt, L Stenberg, "Geophysical investigations at the Klipperås study site", SKB TR 86–07, Svensk Kärnbränslehantering AB, Stockholm 1986

13. "Final Storage of Spent Nuclear Fuel - KBS–3", Svensk Kärnbränsleförsörjning AB, Stockholm 1983

14. E–L Tullborg, SGAB, Göteborg, Personal communication, 1989

15. D J Kirkner, A A Jennings, T L Theis, "Multisolute mass transport with chemical interaction kinetics", J Hydrology, 76 (1985), pp 107 - 117

16. J Noorishad, C L Carnahan, L V Benson, "Development of a kinetic–equilibrium chemical transport code", Trans Am Geophys Union, 66 , 18, (1985), p 274

17. B Fritz, "Étude thermodynamique et modelisation des reactions hydrothermales et diagénétiques", Memoire N° 65, Université Louis Pasteur, Strasbourg 1981

ALTERATION OF CHLORITE AND ITS RELEVANCE TO THE URANIUM REDISTRIBUTION IN THE VICINITY OF THE ORE DEPOSIT AT KOONGARRA, NORTHERN TERRITORY, AUSTRALIA*

T. Murakami, H. Isobe
Japan Atomic Energy Research Institute, Ibaraki, Japan

ABSTRACT

Successive stages of the alteration of chlorite, one of the major minerals of the ore host rock in Koongarra, and the uranium distribution among chlorite and its alteration products have been examined. Water has penetrated into the chlorite structure preferentially through its domain boundaries to alter chlorite to vermiculite, and then to kaolinite and possibly smectite. During the alteration, iron minerals are formed. The uranium concentration increases in the minerals as the alteration proceeds. The abundance of chlorite, vermiculite, and kaolinite corresponds well to the uranium concentration on the meter scale and this strongly suggests that the alteration of chlorite affects the uranium migration in Koongarra.

ROLE DE L'ALTERATION DE LA CHLORITE DANS LA REDISTRIBUTION DE L'URANIUM A PROXIMITE DU GISEMENT DE KOONGARRA (TERRITOIRE DU NORD, AUSTRALIE)

RESUME

Les auteurs ont étudié les étapes successives de l'altération de la chlorite, un des principaux minéraux de la formation d'accueil du gisement de Koongarra, ainsi que la distribution de l'uranium associé à la chlorite et à ses produits d'altération. L'eau a pénétré dans la structure de la chlorite en priorité aux limites de son domaine d'extension et l'a altérée en vermiculite puis en kaolinite et éventuellement en smectite. Au cours de cette altération, des minéraux contenant du fer se sont formés. Plus l'altération des minéraux est profonde plus la concentration d'uranium augmente. L'abondance de chlorite, vermiculite et kaolinite correspond bien à la concentration d'uranium sur l'échelle de mesure. Il y a donc de fortes raisons de penser que l'altération de la chlorite joue un rôle dans la migration de l'uranium à Koongarra.

---

* Part of this study was done at ANSTO, PMB 1, Menai, NSW 2234, Australia.

INTRODUCTION

Transuranic elements are a major concern for the disposal of high level radioactive waste because of their long-lasting (tens of millions of years) hazard. If we consider radionuclide migration over this period, the host rocks are not static, but rather dynamic. The rock-forming minerals will be changed structurally and chemically with time by the water-rock interaction, "alteration". The authigenic or secondary minerals formed by the alteration are usually metastable. The attainment of equilibrium may be delayed by as much as 10^7 years (1), longer than the time required for the safety assesment of high level waste disposal. The transuranic elements react with then-coexisted minerals which vary with time by the alteration mechanisms and kinetics. This suggests that the retardation of the transuranic elements varies with time, that is, with co-existed minerals. A knowledge of the alteration mechanisms and kinetics is, therefore, indispensable if we are to understand the radionuclide migration over a long, geologic time.

Koongarra, in the Northern Territory of Australia, provides us with a good testing site in terms of the relevance of the radionuclide migration to alteration (2,3). Quartz-chlorite schist, the ore host rock, has been subjected to alteration (see (4) for a detailed description of the geology and mineralogy of Koongarra). Quartz persists, even on the surface, while chlorite has been altered to clay minerals and iron minerals in the weathered zone, which extends below the surface to a depth of about 20 m (5). Uranium is found to be associated with the alteration products of chlorite in the downstream of the secondary ore deposit which is located in the weathered zone (5). The alteration products of chlorite are vermiculite, kaolinite, smectite, ferrihydrite, goethite and hematite (6).

The present report describes the alteration mechanisms and cation redistribution during the alteration in Koongarra. Our work focuses on the relationships between the alteration and uranium redistribution.

SAMPLES AND EXPERIMENTAL TECHNIQUES

Fourteen samples were used for the present study. All but one were obtained by diamond drilling (referred to as DDH); the remaining sample resulted from percussion drilling. Nine samples were selected from the DDH3 core to examine a mineral profile with depth. One should note that the weathering along depth is slightly different between locations around the ore deposit. The sample names comprise the corehole name and the the distance from the surface along the core in feet, eg. DDH4-100 (23.3m deep).

Part of the samples of the DDH3 suite was powdered and then centrifuged to obtain fine-grained clay fractions less than 2 μm in size for clay minerals identification by X-ray diffraction analysis (XRD). Each sample was treated with ethylene glycol, and heated at 450 and 600°C to distinguish one clay mineral species from another. The textures of the samples were examined by optical microscopy (OM), scanning electron microscopy (SEM), and their compositions, by electron microprobe analysis (EMPA). The details of the XRD, SEM, and EMPA procedures will be described elsewhere (7). After examination by SEM and EMPA, one of the altered chlorite grains of DDH3-143 was prepared for transmission electron microscopy (TEM) observation by ultramicrotomy to further examine changes in microstructures of chlorite at an early stage of the alteration. An alpha-track technique was used to check uranium distribution between the constituent minerals of the samples.

RESULTS AND DISCUSSION

A series of XRD patterns of the clay fractions in the DDH 3 samples indicate the possible existence of chlorite, vermiculite, mica, smectite, and kaolinite for all specimens. On the basis of the criteria used to identify the clay minerals (7), the results of the XRD studies can be summarized as followings; the alteration of chlorite occurs as a function of depth; chlorite is transformed to vermiculite through regularly interstratified chlorite/vermiculite; vermiculite is converted to kaolinite and probably smectite; an iron mineral, possibly ferrihydrite, is formed during the alteration; and goethite and hematite are found in the weathered zone. The successive alteration process as a function of depth may correspond exactly to that as a function of time in Koongarra.

One chlorite grain of a few tenth mm is not a single phase but consists of lath-like domains of 10-100 nm in width, normal to the c* axis, and 100-1000 nm in length. The domains are attached to each other with the c* axis in common or by low angle domain boundaries with layer terminations. At an early stage of the alteration, when most of the original chlorite crystal structure is retained, some regions at the domain boundaries become amorphous, suggesting that they can be preferential paths for water molecules, and possibly ions such as iron.

The first indication of chlorite alteration is its change in color from green to brown, which is due to oxidation of Fe^{2+} (8, 9). With proceeding alteration, the grains of altered chlorite, showing layer microstructures in SEM images, consist of slabs of chlorite and vermiculite 1 μm thick. Between the slabs, an iron phase (possibly ferrihydrite) occurs as a short slab. The layer microstructures suggest that the crystallographic orientation of the chlorite structure is retained during the transformation of chlorite to vermiculite. In a later stage of the transformation, the iron phase is accumulated and precipitates between the vermiculite slabs (Fig. 1). The occurrence of the iron phase reveals that Fe reprecipitation takes place out of the original chlorite and that the iron phase diffuses between the chlorite and vermiculite slabs during the transformation. Because uranium in the weathered zone is highly associated with the iron minerals (2, 3), their distribution in the microstructures at the various alteration stages should be known more precisely to better understand the uranium redistribution.

The chlorite compositions are rather consistent; MgO, 9-12; Al_2O_3, 15-19; SiO_2, 24-27; and FeO, 26-34 (in wt%). The chlorite compositions in the present study are lower in Mg and higher in Fe than that given by Gray and Davey (10). Higher Fe content in chlorite may cause faster vermiculitization because of greater charge imbalance and/or structural distortion owing to the Fe oxidation. This suggestion accounts for the narrow zone (1 m thick) of co-existed chlorite and vermiculite, below the weathered zone. The significant decrease of iron during the transformation from chlorite to vermiculite is remarkable while the decrease of Si, Al, and Mg is much smaller. On the other hand, Ti and K, which are rarely contained in chlorite, are accommodated in vermiculite. If we use average compositions of chlorite and vermiculite to express the reaction from chlorite to vermiculite, we can write:

$$\underset{\text{Chlorite}}{(Al_{1.11}Fe^{2+}_{2.93}Mg_{1.97})(Si_{2.87}Al_{1.13})O_{10}(OH)_8} + 0.24K^{+} + 0.06Ti^{4+} \longrightarrow$$

$$\underset{\text{Vermiculite}}{K_{0.24}(Al_{0.34}Fe^{3+}_{1.01}Mg_{1.55}Ti_{0.06})(Si_{2.61}Al_{1.39})O_{10}(OH)_2}$$

$$+ \underset{\text{Ferrihydrite?}}{1.82FeO(OH).xH_2O} + 0.42Mg^{2+} + 0.26Si^{4+} + 0.29Al^{3+} + (3-1.82x)H_2O.$$

During the transformation, excess Fe forms ferrihydrite (possibly) around vermiculite, which is perfectly consistent with the SEM observation. K and Ti, required to form vermiculite, may be released from mica and anatase of the quartz-chlorite schist, respectively. Excess Mg, Si, and Al may be consumed in the formation of kaolinite and possibly smectite. Additional Fe, Mg, Si, and Al are released during the decomposition of vermiculite, and these may be used again for the formation of ferrihydrite, kaolinite, and smectite. The layered microstructures of altered chlorite observed by SEM and TEM, and the similar crystal structures of chlorite and vermiculite (11) suggest that the formation of vermiculite from chlorite may be explained by cation diffusion with remaining a unit layer as thick as 1.4 nm unchanged. On the other hand, in the SEM images we have not found any evidence of microstructural relationships between kaolinite and chlorite or vermiculite, and furthermore there is a difference in their crystal structure. This suggests that the formation of kaolinite may be by dissolution and reprecipitation.

The alpha-track technique qualitatively shows how uranium is distributed between the constituent minerals. The DDH4-100 thin section has a fracture surface at the upper end and a strongly altered zone next to the fracture (denoted as F and A in Fig. 2a, respectively). Iron phases, possibly ferrihydrite, goethite, and hematite, are abundant in the strongly altered zone. Next to the strongly altered zone, there is an intermediately altered zone, where chlorite is altered to vermiculite and the iron phases are present between grain boundaries (B in Fig. 2a). In a slightly altered zone which is in the lowest part of the figure (C in Fig. 2a), most chlorite grains are not altered. The features of the thin section reveal that fractures are preferential paths for water and that the alteration starts at the fractures and then expands in the rock. The alpha-track map shows that uranium concentrations vary with the extent of the alteration (Fig. 2b); the strongly altered zone has a high uranium concentration, the uranium concentration decreases in the intermediately altered zone, and uranium is much less associated with chlorite in the slightly altered zone. The comparison strongly suggests that although chlorite itself is not a good adsorbent of uranium, it becomes a good adsorbent if it is altered to vermiculite and further altered. This means that the uranium redistribution in Koongarra is closely related to the alteration of chlorite.

The relative abundances of chlorite, vermiculite, and kaolinite were qualitatively estimated as a function of depth on the basis of the XRD intensities (Fig. 3). The amount of chlorite, which is not altered even at 111 feet along the DDH3 core (25.9 m deep from the surface vertically), rapidly decreases between 107 and 103 feet (25 and 24 m deep) and chlorite disappears above 103 feet. Vermiculite appears at 107 feet (25 m deep) and is at a maximum at 103 feet (24 m deep) where kaolinite begins to persist. Vermiculite disappears at 86 feet (20 m deep) where kaolinite is predominant as it is in the shallower zone. The mineral abundances were then compared to the uranium concentrations (12) along the DDH3 corehole (Fig. 3). The chlorite predominant zone corresponds to the zone of lower U concentrations, the vermiculite predominant zone, to the zone of intermediate U concentrations, and the kaolinite predominant zone, to the zone of highest U concentrations. The relationship of the uranium concentrations with the mineral abundances on the meter scale is completely consistent with the different uranium concentrations in the alteration products on the sub-mm scale. The facts confirm that the uranium redistribution has been affected by the chlorite alteration over geologic time.

Acknowledgements

The authors are indebted to A. A. Snelling, and P. Duerden and R. Edghill of the Australian Nuclear Science and Technology Organisation for assistance in the sample collection and the alpha-track measurement. We are also grateful for the discussions throughout the present study with the staff members of the Environmental Radiochemistry Laboratory of Japan Atomic Energy Research Institute.

References

(1) Dibble, Jr., W. E., Tiller, W. A.: Kinetic model of zeolite paragenesis in tuffaceous sediments. Clays and Clay Minerals 29, 323 (1981).
(2) Airey, P. L.: Radionuclide migration around uranium ore bodies in the Alligator Rivers Region of the Northern Territory of Australia-Analogue of radioactive waste repositories. Chem. Geol. 55, 255 (1986).
(3) Airey, P. L., Ivanovich, M.: Geochemical analogues of high-level radioactive waste repositories. Chem. Geol. 55, 203 (1986).
(4) Snelling, A. A.: Koongarra uranium deposits. in: The Geology of the Mineral Deposits of Australia and Papua New Guinea, Monograph 14 (F. E. Hughes, ed.) The Australian Institute of Mining and Metallurgy, Melbourne (in press).
(5) Snelling, A. A.: Uraninite and its alteration products, Koongarra uramium deposit. in: Proceedings of IAEA International Symposium Uranium in the Pine Creek Geosyncline, 487,(J. Ferguson and A. B. Goleby, eds.), IAEA, Vienna 1980.
(6) Airey, P. L., Roman, D., Golian, C., Short, S,. Nightingale, T., Lowson, R. T., Calf, G. E.: Radionuclide migration around uranium ore bodies - Analogues of radioactive waste repositories. United States Nuclear Regulatory Commission Contract NRC-04-81-172, Annual Report 1982-1983, AAEC Report C40 (NUREG/CR-3941, Vol. 1), Australian Nuclear Science and Technology Organisation, Lucas Heights, NSW, Australia (1984).
(7) Isobe, H., Murakami, T.: Alteration of chlorite and its relevance to uranium migration (in preparation).
(8) Ross, G. J.: Experimental alteration of chlorites into vermiculites by chemical oxidation. Nature 255, 133 (1975).
(9) Ross G. J., Kodama, H.: Experimental alteration of a chlorite into a regularly interstratified chlorite-vermiculite by chemical oxidation. Clays and Clay Minerals 24, 183 (1976).
(10) Gray, D. J., Davey, B. G.: Interstratified minerals formed during weathering of chlorite at the Alligator Rivers uranium province Northern Territory. Submitted to Australian Journal of Soil Research.
(11) Baily, S. W.: Structures of layer silicates. in: Crystal Structures of Clay Minerals and Their X-ray Identification (G. W. Brindley and G. Brown eds.) Mineralogical Society, London 1980.
(12) Ohnuki, T., Murakami, T., Sekine, K., Yanase, N., Isobe, H., Kobayashi, Y.: Migration behavior of uranium and thorium series nuclides in altered quartz-chlorite schist (in preparation).

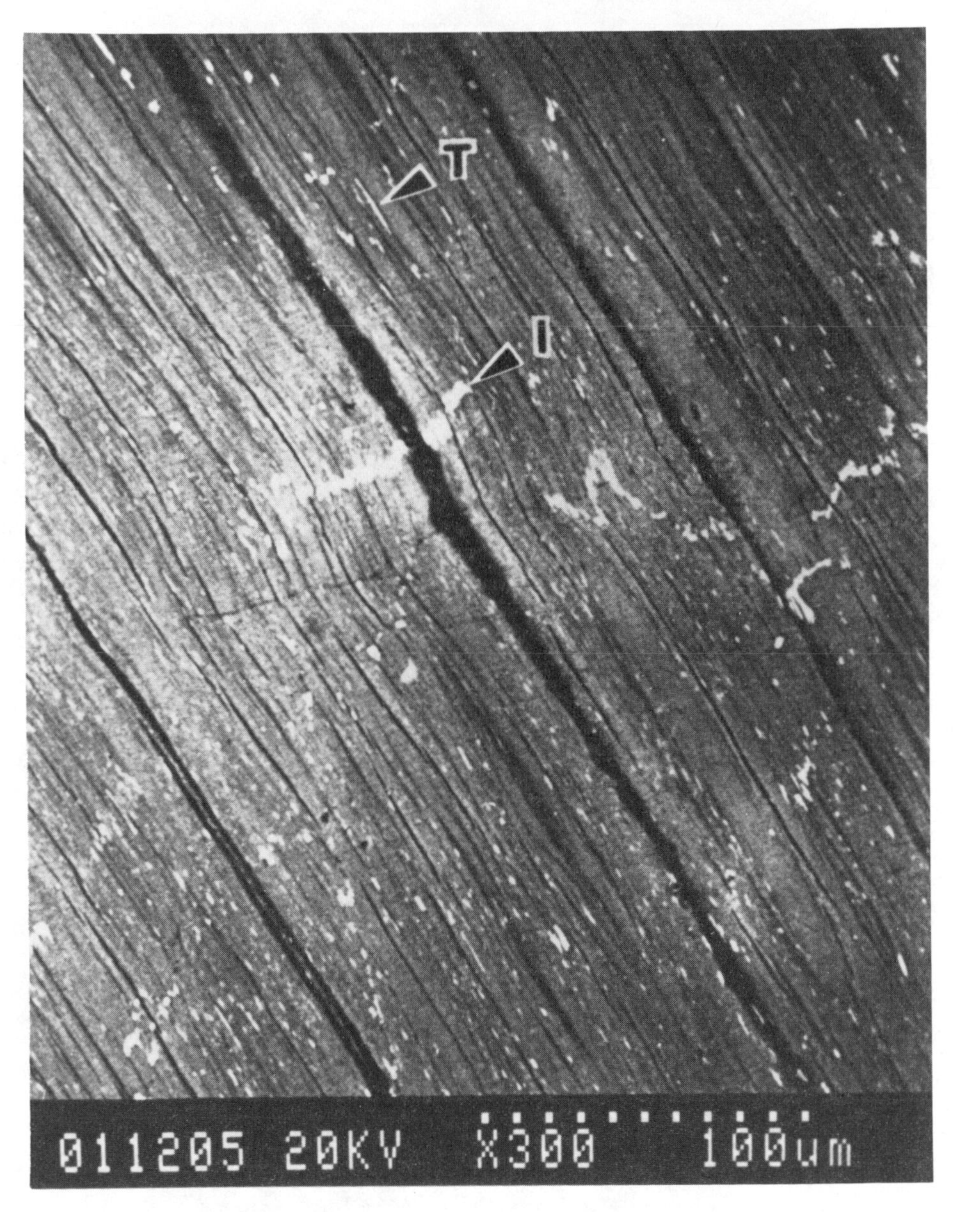

Fig. 1. Backscattered electron image of DDH3-103 showing that an iron phase (white areas, for instance indicated by I), probably ferrihydrite, precipitates between vermiculite slabs (gray in color). Black lines parallel to each other are microfissures. Some of the white areas are those of a titanium phase (indicated, for instance, by T), probably anatase.

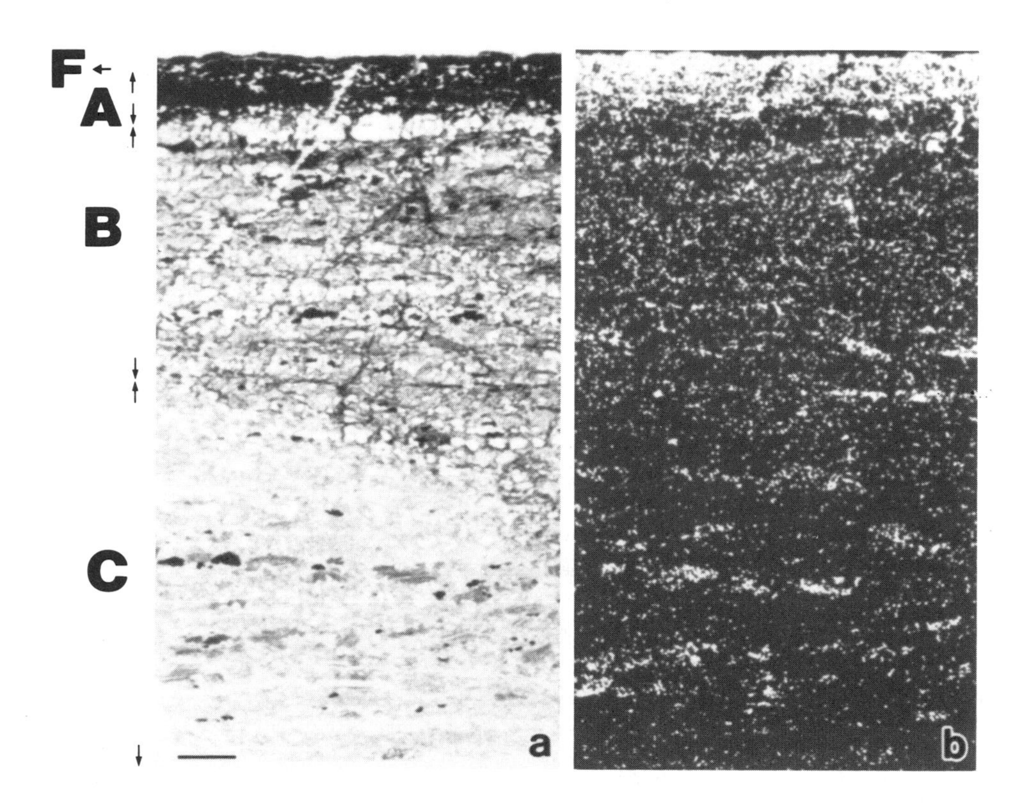

Fig. 2. Comparison of a DDH4-100 thin section (a) and its alpha-track map (b). F denotes a fracture, and A indicates the strongly altered zone (black in color), B, the intermediately altered zone (gray), and C, the slightly altered zone (lighter gray). White areas are those of quartz. The density of alpha tracks differs between the zones. The bar indicates 1 mm.

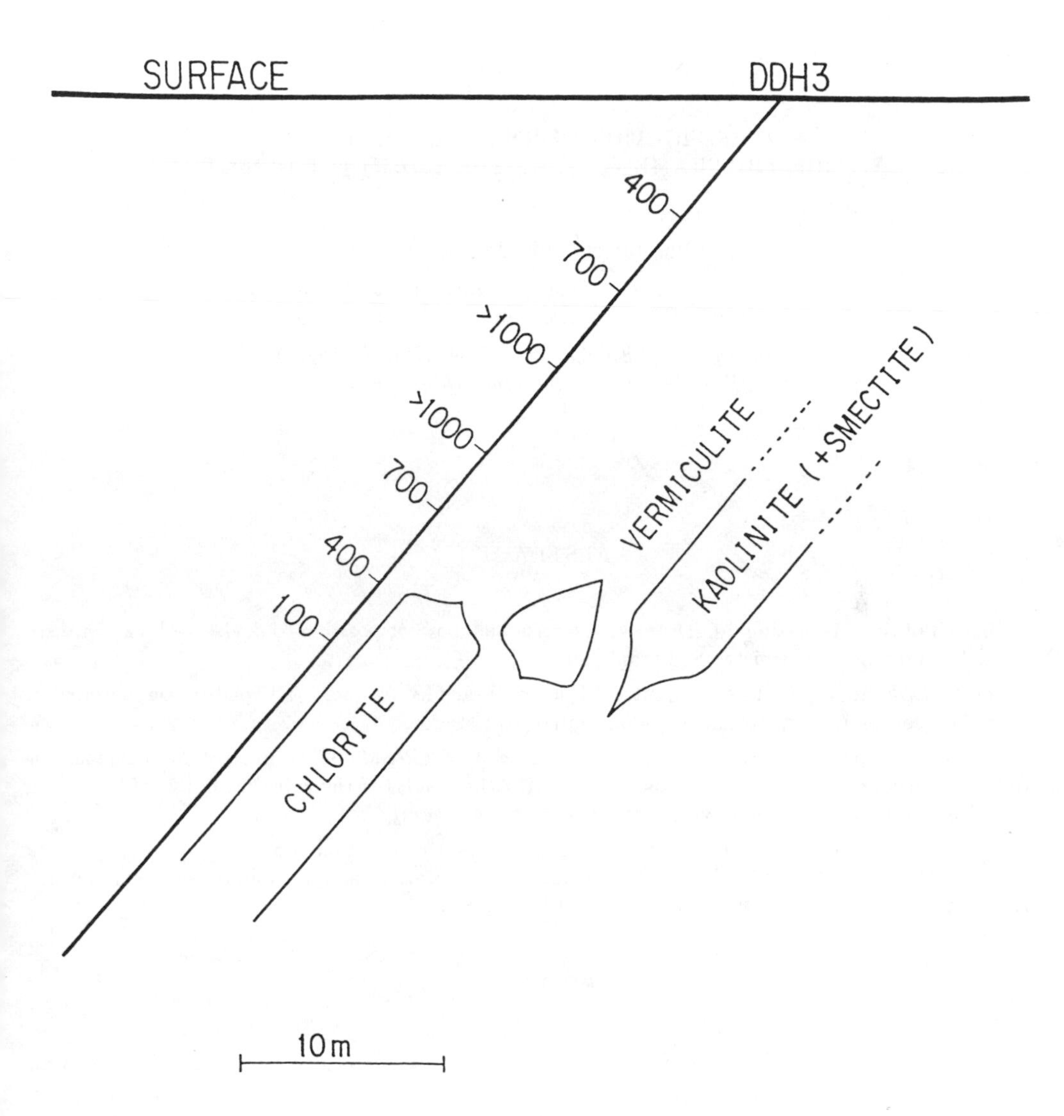

Fig. 3. Comparison of U concentrations and mineral abundances. Uranium concentrations (ppm) along the DDH 3 core are shown along with a schematic representation of the relative abundances of chlorite, vermiculite, and kaolinite (and smectite).

VERIFICATION OF THE *STELE* CODE WITHIN THE FRAMEWORK OF THE *CHEMVAL* PROJECT

Ph. Jamet[*], Ph. Jacquier[**]

[*] *Centre d'Informatique Géologique, Ecole des Mines de Paris*
35, rue St Honoré, 77305 Fontainebleau, France

[**] *Commissariat à l'Energie Atomique, IPSN/DAS/SAED*
BP 6, 92260 Fontenay-aux-Roses, France

ABSTRACT

Phase 3 of the international *CHEMVAL* exercise has made it possible to devise verification tests for models coupling geochemistry and transport.

The *STELE* model, built at the Centre d'Informatique Géologique, is a two-step coupled model, solving the geochemical and transport equations in two stages.

The solution of two of the exercises from phase 3 of *CHEMVAL* is presented. One has an analytical solution which compares satisfactorily with the results of the calculation by *STELE*. The other raises numerical problems typical of a two-step algorithm.

In its present state the *STELE* model is nevertheless a calculation tool well suited to systems with weak geochemical contrasts. Future development of *STELE* should mainly focus on the numerical aspects in order to extend the applicability of the model.

RESUME

La phase 3 de l'exercice international *CHEMVAL* a permis de concevoir des tests de vérification des modèles couplés géochimie-transport.

Le modèle couplé *STELE*, construit au Centre d'Informatique Géologique, est un modèle couplé à deux pas, résolvant en deux étapes les équations de la géochimie et du transport.

La résolution de deux des exercices de la phase 3 de *CHEMVAL* est présentée. L'un possède une solution analytique dont la comparaison avec les résultats du calcul par *STELE* est satisfaisante. Le second soulève des difficultés numériques propres à l'algorithme à deux pas.

En l'état actuel, le modèle *STELE* reste cependant un outil de calcul bien adapté à des configurations présentant de faibles contrastes géochimiques. Le développement futur du modèle *STELE* devra principalement porter sur des aspects numériques pour étendre ses possibilités d'application.

PRESENTATION.

When studying the safety of waste disposal facilities or analysing the movements of pollutants one regularly makes use of numerical transport models in order to simulate the migration of dissolved contaminants in the geosphere.

Classical models such as the finite element model *METIS* [1] solve the dispersion equation [2] in which a great number of mechanisms can be included, e.g. interaction with the matrix (K_d model), chain or radioactive decay, matrix diffusion or various source terms.

The K_d concept is a simple means of taking into account phenomena that delay the migration, e.g. adsorption or desorption phenomena or geochemical transfers (precipitation, dissolution, etc.). Unfortunately, this method is not easy to use because the K_d value does not only depend on the contaminant concerned but to a great extent on characteristics of the terrain which may sometimes be totally unknown.

A few years ago modellers started to design more deterministic tools that would take into account the chemical and geochemical mechanisms as such. These models are called "coupled models" because they solve in tandem the dispersion equation and homogeneous or heterogeneous chemical reactions.

The *STELE* model [3] has been developed at the Centre d'Informatique Géologique for more than three years. Like all tools of this type it is still at verification and development stage.

The aim of the European *CHEMVAL* project has been to verify and to validate speciation and coupled models as well as to assemble a thermodynamic database that might serve as a general reference. Phase 3 of this programme consisted of exercises aimed at verifying coupled models. This study, which concerns the use of the *STELE* code, was commissioned by the Institut de Protection et de Sûreté Nucléaire du Commissariat à l'Energie Atomique.

THE *STELE* COUPLED CODE.

The *STELE* model makes it possible to solve the equations governing the transport through a medium with variable temperature of several chemical species involved in thermodynamic equilibria either inside the fluid phase or with solids or gases.

The general architecture is the same as that of the *METIS* model, the main improvement being the addition of a speciation module, *CHIMERE* [4]. The latter is a model of the same type as *MINEQL* [5] based on the method of chemical components [6].

For each component B_i of the system the dispersion equation for the global balance is solved :

$$\mathrm{div}\left(\overline{\overline{D}}\,\overrightarrow{\mathrm{grad}}[TotB_i] - [TotB_i]\vec{U}\right) + \epsilon\Phi_i = \epsilon\frac{\partial TotB_i}{\partial t} \qquad (1)$$

The term Φ_i is a source term which accounts for the transfers between the aqueous phase and the solid or gaseous phases that may be present in the system.

The method used by *STELE* is said to be a "two-step" solution. In the first step, the speciation is calculated at each node in order to deduce the Φ_i's of the evolution of the defined heterogeneous equilibria. In the second step, the equation (1) is solved for each transported component in the system. The whole process is summarized in figure 1.

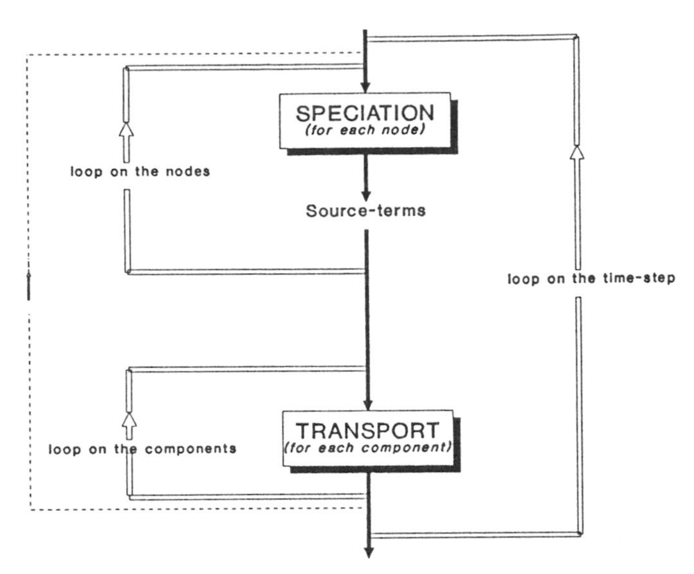

Figure 1. Diagramme of the principle of the calculation algorithm in *STELE*.

"TWO-STEP" AND "ONE-STEP" MODELS.

Another way of solving the coupled geochemical and transport equations is the "one-step" algorithm. Here, the equilibrium equations, expressed as functions of the components in the system, are included in equation 1. All the transport equations must be solved together which makes the process numerically unwieldy. This unwieldiness partly explains why such codes are very often limited to one-dimensional representations. (*CHEMTREN* code, *CHEMTARD* code, etc.)

This algorithm is, however, more rigorous than that of *STELE*. In fact, in the "two-step" algorithm, although the system is at equilibrium before the transport phase, this is not necessarily true afterwards. Therefore, a calibration loop is, in theory, necessary (dotted line in figure 1). The lack of this loop in *STELE* makes it necessary to choose a sufficiently small time step.

It is clear that both methods have advantages and disadvantages. The practical examples that follow, taken from phase 3 of *CHEMVAL*, will illustrate both aspects in the case of the *STELE* model.

VERIFICATION OF THE *STELE* CODE ON AN ANALYTICAL GEOCHEMISTRY-TRANSPORT SOLUTION.

Introduction.

In the course of the verification of a mathematical model the confrontation with a problem that has an analytical solution is a valuable means of testing the coherence of the equations and the numerical processes used to solve them. In real cases the complexity of the geochemical and transport phenomena involved precludes the use of an analytical approach.

The aim of the verification exercises is not to define a real context but to test the main functions of the model on (coupled) systems that may be entirely fictitious. This is the case of the following exercise.

Analytical formulation of the problem.

This analytical case was put together by the "French team" in the *CHEMVAL* group (BRGM, CEA/IPSN, CEA/ANDRA and Ecole des Mines) and concerns a point injection of a mass dM of soda ($NaOH$) into an aquifer where the rock material is pure silica (SiO_2). The flow in the aquifer is uniform along the x axis.

The species figuring in this case are :

aqueous phase : $H_2O, H^+, OH^-, Na^+, H_4SiO_4, H_3SiO_4^-$
solid phase : SiO_2 (as chalcedony)

The chemical equilibria are represented by the following equations :

equilibrium with chalcedony $\quad SiO_2(s) + 2H_2O \rightleftharpoons H_4SiO_4^0 \quad K_s = (H_4SiO_4)$

first acidity of the silicic acid $\quad H_4SiO_4^0 \rightleftharpoons H_3SiO_4^- + H^+ \quad K_1 = \frac{(H^+)(H_3SiO_4^-)}{(H_4SiO_4^0)}$

ionic product of the water $\quad H_2O \rightleftharpoons H^+ + OH^- \quad K_e = (H^+)(OH^-)$

These three equations are combined with the electroneutrality equation of the system :

$$(Na^+) + (H^+) = (H_3SiO_4^-) + (OH^-)$$

Now we make sure that the concentrations of all the chemical species in the system are expressed as functions of the single concentration (C) of Na^+, in particular that of H^+ :

$$(H^+) = \frac{-C + \sqrt{C^2 + 4(K_e + K_1 K_s)}}{2}$$

The concentration C, a function of the time and space coordinates, is that of the perfect tracer since (Na^+) is not involved in any of the chemical equilibria. The analytical solution of the transport of Na^+ is [7] :

one-dimensional case infinite in x $\quad C(x,t) = \frac{dM/\epsilon}{\sqrt{4\pi D_L t}} exp(-\frac{x - U/\epsilon t}{\sqrt{4\pi D_L t}})$

two-dimensional case infinite in x and y $\quad C(x,y,t) = \frac{dM}{4\pi t\sqrt{D_L D_T}} exp[-\frac{(x - U/\epsilon t)^2}{4D_L t} - \frac{y^2}{4D_T t}]$

The pH of the system is therefore known explicitly at each point in space and as a function of time.

Results.

The parameters of the simulation are listed in table 1.

parameter	value	parameter	value
U	0.1 m/day	D_L	0.5 m^2/day
D_T	0.1 m^2/day	ϵ	0.1
dM	10^{-2} mol/m	pK_e	13.998
pK_1	9.77	pK_s	3.554

Table 1. Parameters of chemistry and transport used in the simulations.

Figures 1 and 2 reproduce the results of the calculations with the *STELE* model. Note that the agreement between the simulation and the analytical model is excellent. The calculation was also made in two dimensions. The analytical distribution of the pH (figure 3) and the calculated one (figure 4) have very similar shapes although with slight local divergencies due to the influence of the grid edge.

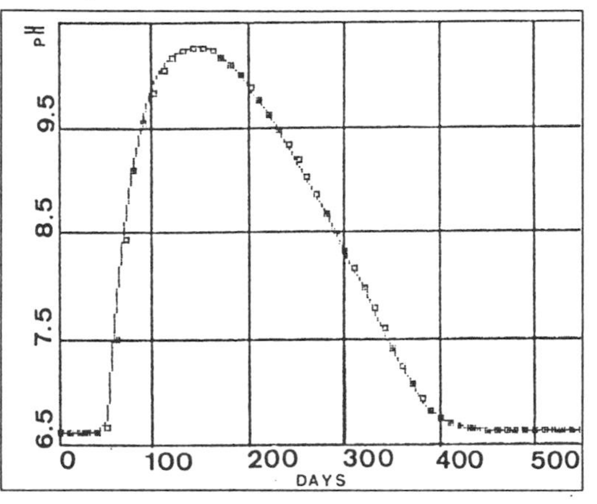

Figure 1. One-dimensional case : pH at 150 m as a function of time (solid line : analytical, squares : *STELE*).

Figure 2. One-dimensional case : pH at 150 days as a function of x (solid line : analytical, squares : *STELE*).

The time needed for the calculations on the CIG mini-computer Bull DPX 5000 is :

- one-dimensional case : 5 minutes (110 meshes, 222 nodes)
- two-dimensional case : 34 minutes (1376 meshes, 2452 nodes).

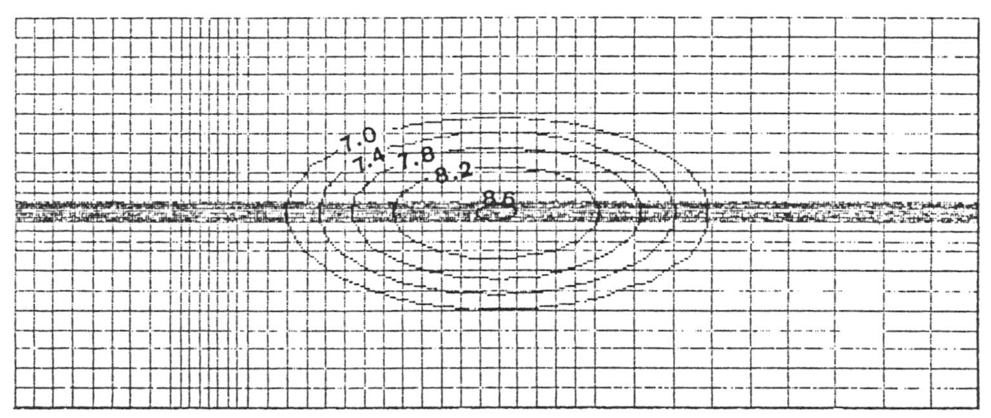

Figure 3. Two-dimensional case : pH at 150 days (analytical) $(-100 < x < 400, -100 < y < 100)$.

Thus the *STELE* model is verified on this analytical solution in which the "two-step" algorithm has worked properly. The equilibrium with SiO_2 was successfully obtained because the transport of this component was not calculated. In certain cases the construction of the *STELE* algorithm as well as the special features of the problem concerned and of the method of components make it necessary to calculate the transport of species at equilibrium. This is where numerical problems may crop up as a consequence of big differences in composition on either side of a dissolution or precipitation front as will be shown in the following case.

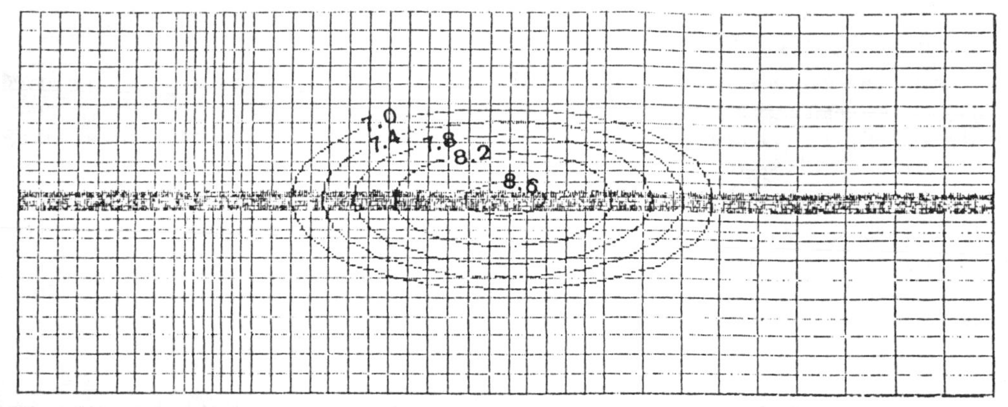

Figure 4. Two-dimensional case : pH at 150 days (*STELE* code) ($-100 < x < 400, -100 < y < 100$).

USE OF THE *STELE* CODE TO SIMULATE A CEMENT DISSOLUTION PROCESS.

The object of interest here is a cement mineralogically represented by portlandite ($Ca(OH)_2$) only. At the beginning of time, groundwater at equilibrium with calcite is made to percolate through it. The geochemical processes taken into account are the dissolution of the portlandite and the complexation in the aqueous phase. The precipitation of the calcite is disregarded. The composition of the waters is detailed in table 2.

	Cement Water	Clay Groundwater
Ca	$2.01\,10^{-2}$	$5.5\,10^{-3}$
Na	$8.01\,10^{-3}$	$8.0\,10^{-3}$
CO_3	$1.01\,10^{-10}$	$4.5\,10^{-3}$
Cl	$4.4\,10^{-2}$	$1.0\,10^{-2}$
pH	12.5	7.5

Table 2. Composition of the cement and groundwater (total concentrations in mol/l).

The system, outlined in figure 5, is one-dimensional.

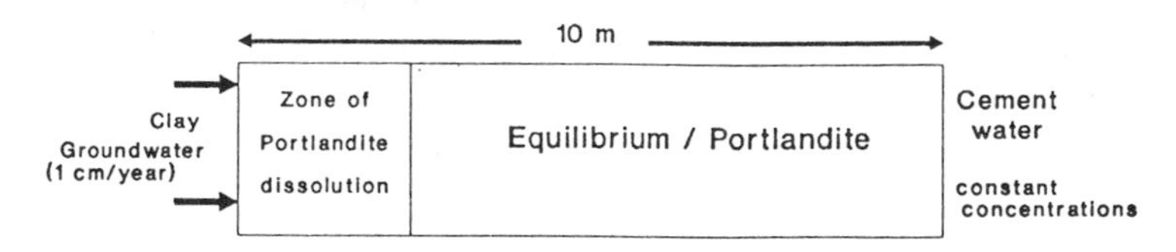

Figure 5. Diagramme of the cement-dissolution problem.

In this problem we have studied the concentrations of the species containing calcium and carbonate which vary a great deal depending on the differences in pH and alkalinity between the two types of water.

The steep dissolution front of the portlandite, a mineral which, in addition, is numerically very sensitive to the pH value, makes it difficult to use to advantage the "two-step" procedure of the *STELE* model. The source-term Φ_i relative to the portlandite is too large to be handled in a numerically satisfactory manner if one chooses time steps of such a length as to make it possible to complete the calculation within a reasonable period of time.

Figures 6 and 7 show an example of numerical anomalies caused by this problem. The observed tendency at the beginning of the Ca^{+2} graph agrees with the one calculated by the direct model *CHEMTARD.* There are, however, oscillations due to the fact that it was necessary to choose a large enough discretisation in space, and thus in time, to make the calculation feasible.

In fact, this zone at the beginning is the one that warrants the use of a coupled model whereas the remainder of the curve is a classical dispersive solution with weak source terms. The same type of incident appears on the graph of $CaHCO_3^+$.

These results demonstrate the shortcomings of the "two-step" algorithm as compared to direct models. It is the very notion of an explicit source term (calculated before the transport equations are solved) that must be questioned because of the numerical instabilities arising from the too large Φ_i's.

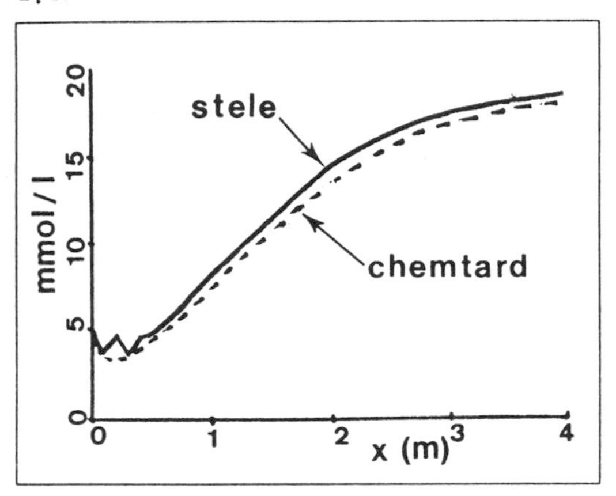

Figure 6. Cement dissolution : Ca^{+2} profile at $t = 100$ years

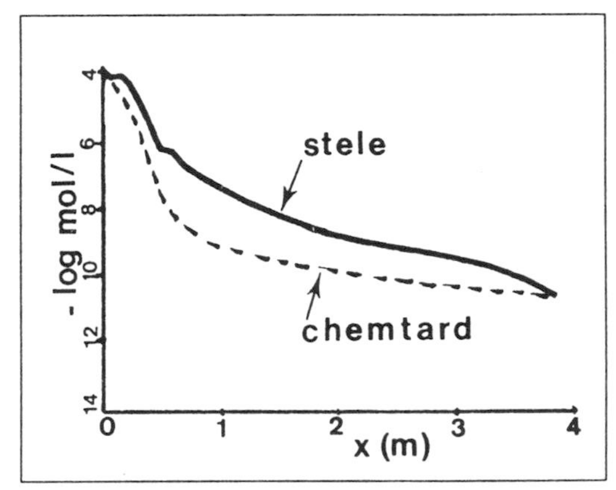

Figure 7. Cement dissolution : $CaHCO_3^+$ profile at $t = 100$ years

CONCLUSION.

At its present stage of development the coupled model *STELE* can be said to be suitable for the solution of coupled problems that are fairly simple or where the differences in chemical composition are fairly small. Moreover, the calculations are relatively fast.

However, cases where such differences are greater or generally, those where one of the features, either the transport or the geochemistry, is not dominant give rise to numerical problems. These problems can only be solved by reviewing the whole question of the numerical solution of coupled problems.

The *STELE* code has other limitations of which the main one is the concept of components forcing the user to attribute the same transport characteristics to a whole family of chemical species which might be totally wrong for some systems.

The "two-step" procedure does, however, offer a great deal of flexibility which makes models such as *STELE* easy to adapt to the solution of problems in two or even three dimensions with complex geochemical effects, e.g. ion exchange [8] and adsorption on metallic oxides. The use of this tool in the natural analogue programme [9] has shown its potential in field studies. It would appear therefore that efforts to improve the model should mainly be concentrated on the numerical aspects.

TABLE OF NOTATIONS.

B_i chemical components
C tracer concentration $[M][L]^{-3}$
dM injected mass of tracer $[M]$
D_L longitudinal dispersivity $[L]^2[T]^{-1}$
D_T transverse dispersivity $[L]^2[T]^{-1}$
$\overline{\overline{D}}$ dispersivity tensor $[L]^2[T]^{-1}$
ϵ kinematic porosity
K_d water-matrix partition coefficient $[L]^3[M]^{-1}$
Tot total concentration of a component
$\vec{U}$ Darcy's velocity $[L][T]^{-1}$

REFERENCES.

[1] Goblet, P. : "Modèle *METIS*, Notice de Conception", Rapport ENSMP-CIG LHM/RD/89/23, 1989, 97p.

[2] Bear, J. : "Dynamics of fluids in porous media", American Elsevier Publishing Company, 1972, 764p.

[3] Coudrain-Ribstein, A. : "Transport d'Eléments et Réactions Géochimiques dans les Aquifères", Th. Doct Etat, 1988, 381p.

[4] Jamet, Ph. : "Programme *CHEMVAL*, Rapport Final CEA/IPSN 1988", Rapport ENSMP-CIG LHM/RD/89/6, 1989, 264p.

[5] Westall, J.C., Zachary, J.L., Morel F.M.M. : "*MINEQL*, a Computer Program for the Calculation of Chemical Equilibrium Composition of Aqueous Systems", M.I.T. Department of Civil Engineering Technical Note n°18, 1976, 91p.

[6] Morel, F.M.M. : "Principles of Aquatic Chemistry", Wiley and Sons, 1983, 446p.

[7] Marsily (de), Gh. : "Quantitative Hydrogeology, Groundwater Hydrology for Engineers", Academic Press, 440p.

[8] Jamet, Ph. : "Programme *CHEMVAL*, Rapport Final CEA/IPSN 1989", Rapport ENSMP-CIG LHM/RD/89/95, 1989, 96p.

[9] Jamet, Ph., Lachassagne, P., Doublet, R., Ledoux, E. : "Modélisation Hydrogéochimique de l'Analogue Naturel de Needle's Eye", Rapport ENSMP-CIG LHM/RD/89/65, 1989, 49p.

IMPACT COMPUTER CODE: A HELP IN DESIGN AND INTERPRETATION OF LABORATORY EXPERIMENTS

M. Jauzein, C. André, R. Margrita
DAMRI-SAR-SAT, CENG, BP 85X, F-38041 Grenoble, France

M. Sardin, D. Schweich
Laboratoire des Sciences du Génie Chimique, CNRS-ENSIC, 1 rue Grandville, BP 451, F-54001 Nancy, France

ABSTRACT

Pollutant transport in soil is generally governed by several competing elementary interactions, either in homogeneous phase or in heterogeneous phase. Interpretation of Breakthough Curves (BTC's) stemming from transport experiments at laboratory scale requires a flexible computer code especially designed when the true nature of the interactions are not accurately known. IMPACT is a new code which can quickly generate BTC's on the basis of assumed interactions at equilibrium. It is based on the concept of "phenomenological mechanism", which allows to account for any interactions governed by mass action law, such as ion exchange, surface complexation, precipitation, gas dissolution, homogeneous reactions. Use of IMPACT is illustrated by three examples, describing the transport of sodium carbonate in presence or not of gaseous CO_2, the transport of competing cations, and the transport of sodium chromate in a argilaceous and calcareous sandy soil.

RESUME

Le transport de polluant en milieu poreux est en général gouverné par plusieurs interactions élémentaires compétitives en phase homogène et en phase hétérogène. L'interprétation des courbes de perçage issues d'expériences de transport menées en colonne de laboratoire requière un code de calcul flexible adapté à la recherche des interactions qui contrôlent le transport des espèces chimiques étudiées. IMPACT est un code flexible fondé sur le concept de "mécanisme phénoménologique" qui permet de prendre en compte n'importe quelles interactions à condition qu'elles puissent être exprimées en terme de loi d'action de masse: échange d'ions, complexation de surface, précipitation, dissolution de gaz ou réactions homogènes. L'utilisation d'IMPACT est illustrée par trois exemples d'étude de transport de solution ionique (carbonate de sodium, cations en compétition et chromate de sodium) dans un sable argilo-calcaire.

INTRODUCTION

Pollutant transport in soil is generally governed by several competing elementary interactions, either homogeneous in the liquid phase or heterogeneous (adsorption, precipitation...). Column experiments at laboratory scale yield Breakthough Curves (BTC's) the shape of which reflects the elementary interactions. Although the physico-chemical parameters of the interactions in the homogeneous liquid phase are well-known, the situation is different for heterogeneous interactions. In many cases, heterogeneous interactions are accounted for by empirical laws to model the transport of solutes (often at trace level) in porous media. The development of computers and numerical methods allows one to use new models such as CHEMTRN[1], PHASEQL/FLOW[2], TRANQL[3] involving more realistic descriptions of the heterogeneous interactions based on thermodynamic principles [4], and assuming local equilibrium. However, interpreting experimental BTC's to recover the elementary interactions requires a flexible computer code to quickly compare the experimental BTC's with theroretical curves obtained with various descriptions of the elementary interactions. IMPACT pertains to this category of codes resting on local equilibrium assumption, and which can quickly generate BTC's on the basis of assumed interactions. It can account for any interactions governed by mass action law, such as ion exchange, surface complexation, precipitation, gas dissolution, homogeneous reactions.

PRESENTATION OF THE CODE IMPACT

Description of physicochemical interactions

The code IMPACT is based on the concept of "phenomenological mechanism" [5]. A "phenomenological mechanism" is composed of a set of stoichiometric relationships, so called elementary interactions, and the associated mass action laws assuming the phases are ideal. It emphasizes that the equilibrium constant is supposed to be a true constant, dependent on the temperature only, and not a conditional constant depending on some concentrations. To be easily included in a transport model, a "phenomenological mechanism" must be composed of the smallest set of elementary interactions, and the latter must account for predominant and truly different phenomena.

The general stoichiometry accounting for an elementary reaction is:

$$\sum_{j=1}^{N} v_{ij} A_j = 0 \qquad \text{with} \qquad K_i = \prod_{j=1}^{N} A_j^{v_{cij}} \qquad (1)$$

A_j is a chemical species in a given phase and v_{ij} the stoichiometric coefficient of A_j in reaction i (v_{ij} is negative for a reactant and positive for a product). In the mass action laws, chemical symbols stand for concentrations. Equation (1) may represent either a homogeneous reaction or a heterogeneous reaction. In many situations the stoichiometric relationship and the associated mass action law are redundant since the exponents and the stoichiometric numbers are related. To avoid redundancy, the code IMPACT is fed with a "reduced stoichiometry" (see below) and the numerical value of the thermodynamic constant only. After elementary reactions are defined, a "phenomenological mechanism" is composed of R independent equations such as (1), involving N species of variable concentrations. The set of stoichiometric coefficients v_{ij} forms a so called stoichiometric matrix, some properties of which have been described by Schweich et al. [5]. To feed the code with the stoichiometric description, two important cases have to be distinguished:

1)Reactions for which the stoichiometric coefficients are equal to the exponents in the mass action law ($v_{ij} = v_{cij}$).

In this case stoichiometric coefficients and mass action law exponents are strictly identical. Giving the v_{ij} to the code is sufficient to recover the mass action law. For instance, homogeneous reactions in liquid phase, such as carbonic acid equilibria, are written as

$$H_2CO_3 = HCO_3^- + H^+ \quad \text{with} \quad K_1 = \frac{HCO_3^-.H^+}{H_2CO_3} = 4.5\ 10^{-7}\ \text{mol/l} \tag{2}$$

$$HCO_3^- = CO_3^{--} + H^+ \quad \text{with} \quad K_2 = \frac{CO_3^{--}.H^+}{HCO_3^-} = 4.7\ 10^{-11}\ \text{mol/l} \tag{3}$$

(The values of thermodynamic constants are given at 25°C, and are used in the simulations presented below).

Heterogeneous reactions which do not create or destroy a phase, such as ion exchange, pertain to this class. The Ca/Na exchange is thus described by

$$Ca^{++} + 2Na_f^+ = Ca_f^{++} + 2\ Na^{++} \quad \text{with} \quad K_{Na/Ca} = \frac{Ca_f^{++}.\ Na^{+2}}{Ca^{++}.\ Na_f^{+2}} \tag{4}$$

Subscript f stands for adsorbed (fixed) species. The sum of the concentrations of fixed cations is the Cation Exchange Capacity (CEC). For convenience the concentrations of species both in solution and on surface are expressed in mole per litre of aqueous phase. In other words, the equilibrium constant incorporates the ratio of the volume of solid to the volume of aqueous solution for heterogeneous equilibria. As a result, the selectivity coefficient $K_{Na/Ca}$ is non dimensional. Let us remark that different A_j in equation (1) can symbolize the same chemical in different phases: for instance Ca^{++} symbolyzes calcium in solution and Ca_f^{++}, the same cation on the surface.

In this first class of reactions the reduced stoichiometry is identical with the true stoichiometry.

2) Elementary interactions with $\nu_{cij} \neq \nu_{ij}$.

This second class is composed of reactions involving a pure phase or reactants in excess. These heterogeneous or homogeneous reactions may be described by a mass action law where the activity of the pure phase component or the species in excess is equal to unity. For instance, water dissociation is commonly represented by the relationships

$$H_2O = H^+ + OH^- \quad \text{with} \quad K_w = H^+.OH^- = 10^{-14} \tag{5a}$$

where the activity of water (solvent in excess) is 1. Consequently, water concentration is not involved in the mass action law. This reaction can be made similar to those of the first class provided that the species in excess is cancelled from the stoichiometric relationship. For instance equation (5a) becomes

$$0 = H^+ + OH^- \quad \text{with} \quad K_w = H^+.OH^- = 10^{-14} \tag{5b}$$

Equation (5b) defines the reduced stoichiometry in which mass action law exponents and stoichiometric coefficients are equal. This reduced stoichiometry is used to feed code IMPACT together with the equilibrium constant.

Similarly, for a pure mineral, the concentrations of dissolved species are governed by a mass action law, a solubility product, independant of the amount of mineral in contact with the solution. For instance, the solubility of calcium carbonate is represented by

$$CaCO_3 = Ca^{++} + CO_3^{--} \quad \text{with} \quad K_s = Ca^{++}.\ CO_3^{--} = 4.5\ 10^{-9} \tag{6a}$$

The activity of the dissolving pure mineral being unity (solid solutions are not accounted for in IMPACT), the stoichiometry (6a) can be reduced to:

$$0 = Ca^{++} + CO_3^{--} \quad \text{with} \quad K_s = Ca^{++}.\ CO_3^{--} = 4.5\ 10^{-9} \tag{6b}$$

Only the two products of the reaction are entered in the code with stoichiometric coefficients equal to 1. In addition to avoid redundancy, the reduced stoichiometry involves the species of variable concentration only. If species in excess can be considered of known concentration, the problem is different for pure mineral phases. Equation (6b) is only valid when the pure phase is present, or in immediately the equilibrium compositions as a function of the mechanism defined in step 2. 4) The initial conditions are defined from the mechanism governing equilibrium inside the column and from the composition of a preconditioning solution which feeds the column. 5) The flow model is defined by the pore volume and the number of mixing cells. 6)The desired results (for instance, BTC's and/or concentration profiles of selected species at given times) are chosen. Then IMPACT automatically generates and solves the mass balance equations relevant to the defined transport problem [6]. In addition, the code feeds a library of reactive species and reactions with thermodynamic contants for subsequent computations. Finally, let us add that IMPACT can be used to calculate the equilibrium composition of a batch system.

EXAMPLES OF THE USE OF IMPACT

Injection of sodium carbonate in a argilaceous and calcareous sandy soil [5]

The aim of this example is to show the flexibility of IMPACT and the effects of the nature of the phenomenological mechanism on the shape of BTC's. The transport of alkali cations in the presence of clay minerals and calcium carbonate in a sample of soil is controlled by cation exchange on clays, calcite dissolution and carbonic equilibria. Since calcium carbonate is in excess, the number of phases is constant. The feed solutions are in contact with atmosphere where the partial pressure of CO_2 is constant. Consequently, the phenomenological mechanism which governs the composition of feed solutions, includes the acid carbonic equilibria (eq.2 and 3), the ionization of water (eq.5b) and the dissolution of CO_2 at constant pressure (eq.8). Inside the column, the essential problem lies in the fact that the phenomenological mechanism are different if the solution is in contact with CO_2 or not.:

(1)No gas phase in the column (saturated medium). The mechanism consists of five reactions : acid carbonic equilibria (eq.2 and 3), water ionization (eq.5b), dissolution of calcite in excess (eq.6b), and cation exchange (eq.4). These five reactions constitute a complex system certain properties of which have been described by Schweich et al.[5,8]. Figure 1A shows the

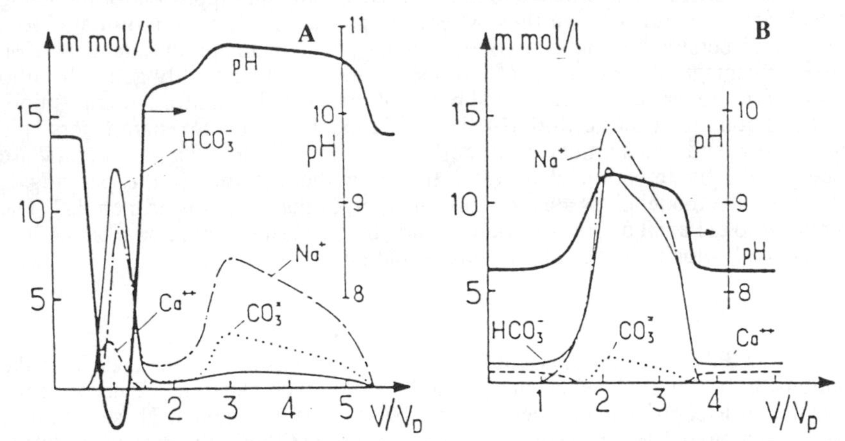

Figure 1: Injection of $NaHCO_3$ into a sandstone containing clays and calcite in excess. A)No dissolution nor degassing of CO_2. B)Constant partial pressure of dissolving/degassing CO_2 inside the porous medium. Experimental conditions: injection of 0.2 pore volume of $NaHCO_3$ 100 meq.litre^{-1}. CEC =32 meq/l of solution. $K_{Na/Ca}$= 220.

theoretical BTC's resulting from the injection of 0.2 pore volume of sodium carbonate 0.01 eq/l in water, driven by water equilibrated with atmospheric CO_2.

other words when the pure phase is in excess. When the pure phase can appear or disappear, the code IMPACT must be fed with the complete stoichiometry (6a), and the user indicates that $CaCO_3$ is a pure mineral.

The concept of reduced stoichiometry allows the user to account for external constraints ; This case is encoutered when a species is maintained at a constant concentration by an artificial mean. Equilibration of a solution with carbon dioxide from the atmosphere is an example.Using Henry's law, the dissolution of CO_2 in water is described by the following equilibrium:

$$CO_2(g) + H_2O = H_2CO_3 \qquad \text{with} \qquad K_H = \frac{H_2CO_3}{P_{CO_2}} \qquad (7)$$

where K_H is Henry'sconstant. The equilibrium composition of a solution in contact with a gas phase where P_{CO2} is constant and equal to the partial pressure of carbon dioxide in atmosphere, can be written as a reaction with one product and no reactant since the latter are assumed to be in excess:

$$0 = H_2CO_3 \qquad \text{with} \qquad K'_H = P_{CO_2}.K_H = H_2CO_3 = 1.2\ 10^{-5} \qquad (8)$$

Simple activitiy correction is possible for ionic species in the aqueous phase only. The activity coefficient γ_i is calculated by the Güntelberg approximation [4],

$$\log_{10}\gamma_j = -0.5\ z_j^2 \frac{\sqrt{I}}{1+\sqrt{I}} \qquad \text{with I(ionic strengh)} = \frac{1}{2} \sum_{j=ions} A_j z_j^2 \qquad (9)$$

To use this relationship, inert ions which are not involved in a reaction, such as Cl^- for the equilibria previously described, must be declared.

The equations (2), (3), (4), (5), (6), and (7) and the 9 species of variable concentrations, which are involved in them, are recorded in a reactions file and a species file, respectively. They are available to define phenomenological mechanisms which is a set of elementary reactions.

Description of flow

IMPACT employs a mixing cells in series model (MC model) to describe the flow in column experiments. This model is an alternative to the classical continuous approach for modeling the one-dimensionnal convective-dispersive flow where the porous medium is assimilated to a continuous medium and dispersion is assumed to obey a Fick's law with an appropriate dispersion coefficient. The main advantage of the discrete MC model is that it can be readily generalized to more complex flow patterns, in using the ideal stirred reactor (mixing cell of uniform composition) as the basic element of a complex flow network [6]. The MC model contains two parameters : the volume of porous media and the number J of cells in series. These two parameters are experimentally determined by injection of a water tracer in the column. If the packing is homogeneous and regular the model gives a good representation of the tracer experiment [7]. The number of cells is related to the axial dispersion coefficient D by the relationship, $Pe = uL/D = 2(J - 1)$, where u is the interstitial velocity, L the column length, and Pe the Péclet number.

Working with IMPACT

A simulation with IMPACT requires several steps: 1)The user has to feed the code with the stoichiometric relationships defining the elementary reactions. For each reaction, the name of the species and the stoichiometric coefficient attached to each species must be entered. These data feed the species file and the reactions file. 2)To simulate a transport experiment, the user must define two phenomenological mechanisms from the elementary reactions: the first mechanism is used to define the composition of the feed solutions, and the second mechanism defines the interactions in the porous medium.3)The feed to the system (limit conditions) is defined by a sequence of solutions of different composition. The user has to enter the number of solutions, the volumes injected, and reference compositions. The reference composition may be considered as the initial non equilibrium composition which leads to the desired equilibrium state. IMPACT gives

(2)Equilibrium with an atmosphere containing CO_2 inside the porous medium (unsatured medium) The phenomenological mechanism governing the composition inside the column is now composed by the five previous reactions to which the equilibration with CO_2 at constant pressure (3.2 10^{-4} atm, same as for feed solution) must be added (eq. 8). This new assumption leads to the simulation of the figure 1B.

These simulations are easily performed with IMPACT and require about 5 minutes to define the mechanisms from the reactions file, and about 45 minutes for each simulation on a mini-computer such as a SUN 3/60 workstation. The simulations also show that equilibration of feed and soil solution with atmospheric carbon dioxide is a key factor which drastically affects the BTC's. The mesure of pH at the outlet of the column appears as a simple means to discriminate the two assumptions. Generally, the first situation is encountered in column experiments. But in an unsaturated medium, the second mechanism controls the transport.

Transport of several competing cations

To carry out the Relay Substance Method [9] we studied the transient transport of several cations in a sandy soil containing calcareous and clay minerals. In this soil the major cations are sodium and calcium, the transport of which is well described by the interactions described above. To predict the transport of radioactive cesium, lithium was used as the relay substance. The main problem which the code may help to solve, is to know whether it is sufficient to add the Li/Ca cation exchange to previous mechanism to describe the transport of lithium.

In a first step sodium/calcium and lithium/calcium ion exchange parameters were determined by column experiments [6]. Figure 2 (A) illustrates an example of experimental BTC's of calcium and lithium. Similar curves were obtained either for Ca/Na or Ca/Li exchange at various concentrations.

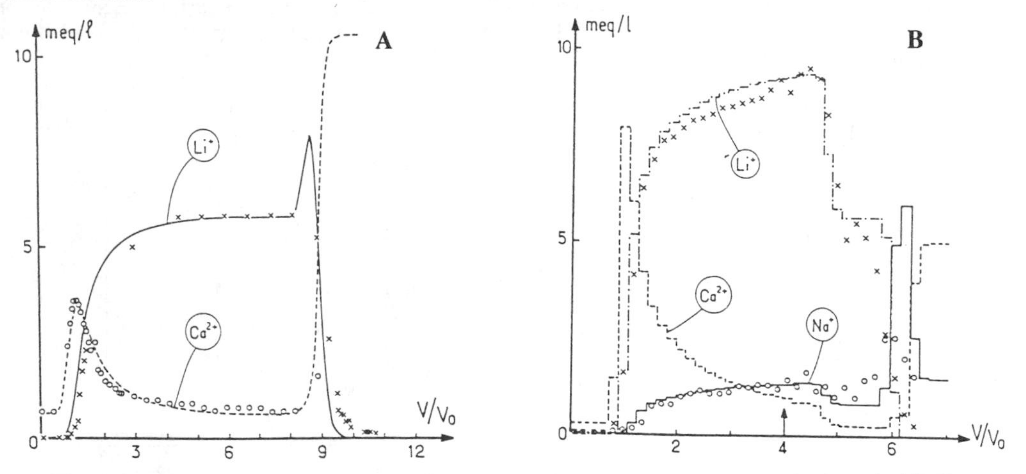

Figure 2: A)Simulation of an experiment involving lithium and calcium cations. Injection of LiCl 54 meq/l, then injection of $CaCl_2$ 10 meq/l. B)Calcium, lithium and sodium BTC's resulting from a 4 pore volume injection of NaCl 4.5 meq.l-1 and LiCl 10 meq.l-1, driven by CaCl2 5 meq.l-1 and NaCl 1.5 meq.l-1. Crosses and dots are experimental results. Curves are theoretical predictions. An other experiment in different conditions is available in [9].

Then, IMPACT was used for simulating the BTC's. All the experiments were well predicted with a common value of the exchange capacity and appropriate values of the selectivity coefficients. In this case, IMPACT was used as a preliminary validation of the interaction mechanism, and to fit the exchange exchange parameters on independent binary exchange experiments. The results were:

$K_{Li/Ca}$= 236 , $K_{Na/Ca}$= 59 , and CEC= 18 meq/kg

In term of Gaines-Thomas selectivity coefficient, which is independant of the mass/volume ratio, M/V and of the exchange capacity ($K_\sigma = 2K.CEC.M/V$) we obtained:

$K_{gLi/Ca} = 32$ mole.l^{-1} , $K_{gNa/Ca} = 8$ mole.l^{-1}

The full validation of the phenomenological mechanism was obtained with an experiment, involving the three cations, which is the case encountered in situ. Figure 2B represents the experimental results and the simulations without any curve fitting. The quality of the prediction confirms the validity of the chosen mechanism and the reliability of the method for describing and determining the elementary interactions step by step. Let us remark that the simulation presented in Figure 2B can be performed prior to the experiments, and can then be used to calibrate the sensors and/or the fraction collector. In this case IMPACT is a useful tool for designing laboratory experiments.

Transport of a potentially precipitating species and application to chromate anion.

This example is developed in the paper of Tevissen et al.[10] where IMPACT is used for speciation calculations in a batch system and to validate a model for chromate adsorption. The speciation calculations show that chromate transport should be essentially affected by anion exchange between chromate, bicarbonate and sulphate, and by precipitation of barium chromate due to dissolved barium in field water.

Anion exchange parameters being known [10] and barium chromate, barium carbonate and barium sulphate solubility products being available from the literature, the code IMPACT could be used to design a complex transport experiment allowing one to display the behaviors induced by the previous interactions, taking place simultaneously. The phenomenological mechanism contains 13 reactions and 22 reacting species. The presented simulation models consecutive pulse injections of $BaCl_2$ followed by Na_2CrO_4 in a sample of the soil from the site. Preconditioning and driving water contains sodium sulfate and is assumed to be equilibrated with the calcium carbonate and carbon dioxide for obtaining a pH equal to 7.65. In figure 3 chromate and barium BTC's (solid lines) exhibit important retentions.

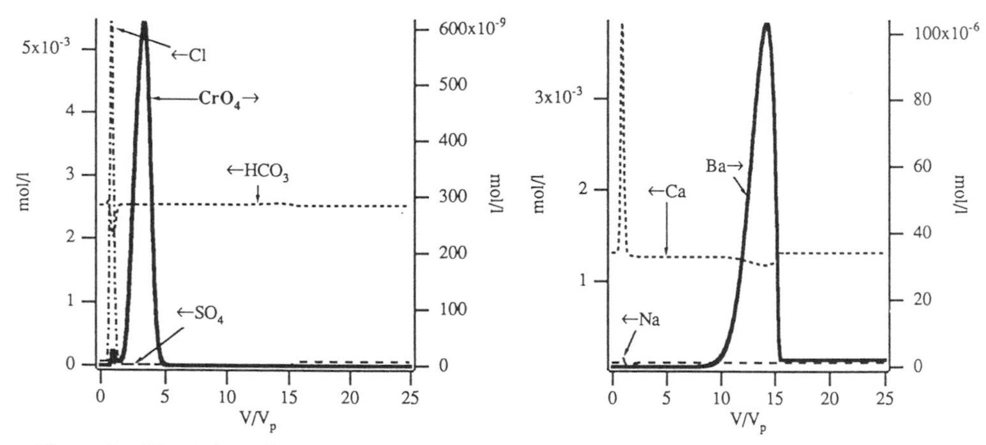

Figure 3 : Simulation of transient transport of chromium (under chromate form) in a argicaleous and calcareous sandy soil at pH 7.65 in presence of $BaCO_3$ and sodium sulfate. Pulse of $BaCl_2$ 0.1 Mol/l followed by a pulse of Na_2CrO_4 10^{-4} mol/l, and driven by Na_2SO_4 0.5 10^{-4} mole/l. The values of thermodynamic constants are available in [10].

These are essentially due to the simultaneous precipitation of barium carbonate and barium chromate at the inlet of the column. Chromate is then dissolved by the driving water and the transport of chromate is governed by the anion exchange process. Simultaneously barium carbonate dissolves, and then barium/calcium cation exchange controls the transport process. The retentions observed by simulation are sufficiently significant to implement an experiment with the same conditions and to check whether the proposed mechanism is reliable. This experiment is presently in progress.

CONCLUSIONS

IMPACT is especially suited for quickly calculating BTC's or batch speciations with rival interaction mechanisms. Changing a mechanism requires to enter the reduced stoichiometries and the associated equilibrium constants. For standard problems involving about 20 reacting species, about 10 minutes are necessary for defining the reactions and a mechanism. Let us emphasize that the maximum numbers of species and reactions are limited by the computer only.

The high sensitivity of BTC's to elementary interactions and the possibility to easily change the interaction mechanism in IMPACT enable the user to quickly test many rival models for explaining experimental results. Alternatively, deleting reactions which are assumed to be secondary enables the user to progressively reduce a complex mechanism to the predominant interactions.

Finally IMPACT can be used to design complex validation experiments of a model. Simulations prior to the validation experiments allow one to determine the best experimental conditions and to calibrate the experiment.

In a future code version, the flow model will be extended to field situation using a cell network for the flow model. Consequently, it will be possible to use the code IMPACT to validate field experiments [6].

REFERENCES :

[1] Stumm, W. and Morgan, J.J., 1970. Aquatic chemistry. Wiley Interscience, New York, 780 pp..

[2] Miller, C.W. and Benson, L.V., "Simulation of solute transport in a chemically reactive heterogeneous system : Model development and application". Water.Res.Res.,1983,19: 381-391.

[3] Walsh, M.P., Bryant, S.L., Schechter, R.S. and Lake, L.W., "Precipitation and dissolution of solids attending flow through porous media". AIChE J., 1984,30: 317-328.

[4] Cerdeberg, G.A., Street, R.L., and Leckie,J.O., "A groundwater mass transport and equilibrium chemistry model for multicomponent systems". Water Resour. Res.,1985, 21, 8: 1095-1104.

[5] Schweich, D., Sardin, M., Jauzein, M., "Consequence of physicochemistry on transient concentration wave propagation in steady flow", Proc. International Conference and Workshop on the Validation of Flow and Transport Models for the Unsaturated zone, Wirenga P.J. and Bachelet D. Ed., La Cruses, NM State University Pub., 1988, 370-380.

[6] Jauzein M., André C., Margrita R., Sardin M. and Schweich D., "A flexible computer code for modelling transport in porous media: IMPACT", Geoderma special issue, 44, 2-3(1989),93-113.

[7] Sardin, M., Krebs, R. and Schweich, D., "Transient mass-transport in the presence of non-linear physico-chemical interaction laws: progressive modelling and appropriate experimental procedures". Geoderma special issue, 38, 1986, 115-130.

[8] Schweich, D. and Sardin, M., "Transient ion exchange and solubilization of limestone in an oil field sandstone: Experimental and theoretical wavefront analysis". AIChE J., 1985, 31: 1882-1890.

[9] Jauzein M., André C., Margrita R., Sardin M. and Schweich D., "The Relay Substance method: a new concept for predicting pollutant transport in soils", Proc.GEOVAL 90, SKI-OECD/NEA, Stockholm, Sweden, May 14-17, 1990.

[10] Tevissen E., Jauzein M., Andre C., Margrita R., Sardin M., Schweich D.,"Migration of chromium in the saturated zone. Simulation of column experiments", Proc.GEOVAL 90, SKI-OECD/NEA, Stockholm, Sweden, May 14-17, 1990.

MIGRATION OF CHROMIUM IN THE SATURATED ZONE. SIMULATION OF COLUMN EXPERIMENTS

E. Tevissen, M. Jauzein, C. André, R. Margrita
DAMRI-SAR-SAT, CENG, BP 85X, F-38041 Grenoble, France

M. Sardin, D. Schweich
Laboratoire des Sciences du Génie Chimique, CNRS-ENSIC, 1 rue Grandville, BP 451, F-54001 Nancy, France

ABSTRACT

The main interactions governing chromate transport in a sediment are presented. Speciation calculations based on field water composition show that calco-carbonic equilibria, cation exchange, anion exchange and barium chromate precipitation are the main interactions. Under field conditions chromium III is negligible. Anion exchange between chromate, sulphate and bicarbonate anions is studied by column experiments. The pH dependent anion exchange capacity and the selectivity factors are determined. Based on literature data and experimentally determined parameters, a chromate breakthrough curve is predicted without any adjustable parameter. Although the prediction is not perfect, the model reproduces the mean location and the maximum concentration of the chromate peak.

RESUME

On présente les interactions prépondérantes gouvernant le transport d'ions chromate dans un sol sédimentaire. Des calculs de spéciation reposant sur la composition de l'eau du site, montrent que les équilibres calco-carboniques, les échanges de cations et d'anions, et la précipitation du chromate de baryum constituent les interactions principales. Dans les conditions du site, le chrome III est en quantité négligeable. L'échange d'anions entre chromate, sulfate et bicarbonate a été étudié en colonne de laboratoire. La capacité d'échange anionique (dépendante du pH) et les coefficients de sélectivité ont été déterminés. A partir de données de la littérature et des paramètres déterminés expérimentalement on prévoit une courbe de perçage de chromate avec un modèle sans paramètre ajustable. Bien que l'ajustement ne soit pas parfait, le modèle reproduit la position moyenne et la concentration maximum du pic de chromate.

INTRODUCTION

This paper is focused on the description of the main homogeneous and heterogeneous interactions governing chromium transport in a saturated layer of sediment. The progressive method described by Jauzein et al. [1] is used to recover the interaction mechanism from breakthrough curves (BTC's) obtained with a laboratory column packed with a sample of sediment from the experimental field of the "Centre d'étude nucléaire de Grenoble" (France). The compositions of the sediment and of the soil water are given in Table 1. Because of the presence of clays and barium, ion exchange and barium chromate precipitation are expected, as already mentionned by Rai and Zachara [2].

Table 1: Composition of soil and field water.

Soil			
Sand & silt (>0.002mm): 96%w/w		Clays (<0.002mm): 4%w/w	
Calcium carbonate: 25%w/w		Cation exchange capacity: $23 meq.kg^{-1}$	
Water			
pH = 7.5 to 7.7, Eh = 430 mV		Total ionic normality: $4.8 meq.l^{-1}$	
Cations	$mmol.l^{-1}$	anions	$mmol.l^{-1}$
Ca^{++}	1.35	HCO_3^-	2.6
Mg^{++}	0.53	Cl^-	1.2
Na^+	1.00	SO_4^{--}	0.42
K^+	0.04		
Ba^{++}	0.0003		

The column experiments consist of injecting pulses containing various reactive species, driven by a water solution as similar as possible to field water. Quantitative analysis of the resulting BTC's with IMPACT computer code allows one to recover and quantify the main interaction responsible for the retention of the reactive solutes.

CHROMIUM SPECIATION UNDER FIELD CONDITION

Field and standard synthetic water solutions

Preliminary speciation calculations with IMPACT code showed that the field water composition is governed by calco-carbonic equilibria, considering that Mg^{++}, K^+, Na^+, Ba^{++}, Cl^-, and SO_4^{--} are inert ions. Using a mean pH of 7.6, and the equilibrium constants given in Table 2 for water and carbonic acid ionizations and calcium carbonate dissolution, the following equilibrium composition is calculated:

$$Ca^{++} = 1.11\ 10^{-3},\ HCO_3^- = 2.14\ 10^{-3},\ H_2CO_3 = 0.12\ 10^{-3},\ CO_3^{--} = 4.0\ 10^{-6}\ mol/l,\ pH = 7.6$$

The concentration of dissolved calcium is found close to the experimental value (Table 1). The equilibrating partial pressure of carbon dioxide is found 10 times higher than in the atmosphere, and it can be explained by the biological activity in the upper layer of the aquifer. This equilibrating partial pressure will be used in further calculations.

Driving or pulse solutions for column experiments were then prepared with a standard synthetic solution simulating field water. This standard synthetic solution was obtained by contacting deionized water with calcium carbonate in the presence of carbon dioxide in a reactor located upstream of the column. Continuous pH measurement allowed one to monitor the CO_2 partial pressure. Experiments showed that the standard solution had a composition close to the field water, except for Mg^{++}, K^+, Ba^{++}, Cl^-, and SO_4^{--}. Adding either barium, sulphate or chromate to the standard solution gave the pulse and driving solutions for the columns experiments described below. Varying the CO_2 partial pressure of the gas phase above the standard solution allowed one to vary the pH of the influent solutions.

Table 2: Reaction stoichiometries and equilibrium constants.

Reaction stoichiometry[1] and equilibrium constant or standard potential[2]			
Homogeneous reactions in the liquid phase			
$H_2O = H^+ + OH^-$	$1.0\ 10^{-14}$	$H_2CO_3 = HCO_3^- + H^+$	$4.5\ 10^{-7}$
$HCO_3^- = CO_3^{--} + H^+$	$4.7\ 10^{-11}$	$Cr(OH)_3^0 + OH^- = Cr(OH)_4^-$	$3.5\ 10^2$
$Cr(OH)_2^+ + OH^- = Cr(OH)_3^0$	$4.5\ 10^7$	$Cr(OH)^{++} + OH^- = Cr(OH)_2^+$	$5.4\ 10^7$
$Cr^{+++} + OH^- = Cr(OH)^{++}$	$2.7\ 10^{10}$	$CrO_4^{--} + H^+ = HCrO_4^-$	$3.2\ 10^6$
$HCrO_4^- + H^+ = H_2CrO_4$	$8.7\ 10^1$		
Precipitation reactions			
$CaCO_{3s} = Ca^{++} + CO_3^{--}$	$4.5\ 10^{-9}$	$BaCrO_{4s} = Ba^{++} + CrO_4^{--}$	$1.1\ 10^{-10}$
$BaSO_{4s} = Ba^{++} + SO_4^{--}$	$1.2\ 10^{-10}$	$BaCO_{3s} = Ba^{++} + CO_3^{--}$	$5.1\ 10^{-9}$
$Cr(OH)_3^0 = Cr(OH)_{3s}$	$6.9\ 10^6$		
Oxido-reduction reactions			
$Cr(OH)_3 + 5OH^- = CrO_4^{--} + 4H_2O + 3e^-$	$E_0 = -0.13$ v	$2H_2O = 4H^+ + O_2 + 4e^-$	$E_0 = 1.229$v[3]
Exchange reactions			
$SO_{4f}^{--} + CrO_4^{--} = SO_4^{--} + CrO_{4f}^{--}$	1.00	$HCO_{3f}^- + CrO_4^{--} = HCO_3^- + CrO_{4f}^{--}$	$3.50\ 10^2$
$Ba^{++} + Ca_f^{++} = Ba_f^{++} + Ca^{++}$	1.00		

[1]Subscript f stands for adsorbed or precipitated species.

[2]Concentrations in mass action laws are in mol.l^{-1} of aqueous solution whatever the phase where the species is present.

[3]O_2 is accounted for by its partial pressure in Nernst law.

Further calculations with IMPACT showed that the experimental Eh (see Table 1) must be accounted for by an oxygen partial pressure about 10^{-21} atmosphere. This very low equilibrating partial pressure shows that dissolution of oxygen only cannot account for the observed Eh. The precise origin of this experimental Eh was not determined and column experiments were performed at the Eh of the standard solution.

Finally, remark that dissolved barium and sulphate in field water (Table 1) are in equilibrium with solid barium sulphate (Table 2). This indicates that solid barium sulphate can be present in the soil although it has not been detected in the studied sample.

Chromium speciation in field water

Speciation of chromium is governed by the elementary reactions given in Table 2, assuming that the equilibrating pressure of carbon dioxide and oxygen are those given in the previous section. Table 3 summarizes the equilibrium composition of a solution containing 10^{-3} mol.l^{-1} of sodium chromate in contact with an excess of calcium carbonate, and in the absence of a gas phase. Two calculations were made, one assuming there is no barium sulphate, and another assuming there is an excess of solid barium sulphate in the solution.

As expected, Eh is sufficiently high to make chromium III negligible. Precipitation of barium chromate occurs whatever the amount of solid barium sulphate. However, when the latter solid is present, a greater amount of barium chromate precipitates.

Consequences of chromium speciation in the liquid phase on chromium transport

The speciation calculations shown in Table 3 and the work of Rai and Zachara [2] suggest that the main heterogeneous interactions of chromium are precipitation with barium and competitive sorption of chromate with other anions (Cl^-, SO_4^{--}, HCO_3^-, and CO_3^{--}). In addition, cation exchange on clays between the cation accompanying chromates and other cations of the soil-water system (Ba^{++}, Ca^{++}, and Na^+) should be expected. Chromium transport in the studied soil-water system is thus affected by anion sorption, precipitation of barium chromate and possibly cation exchange.

Since barium chromate precipitation is only governed by the solubility product, and cation exchange can be characterized following the method described by Jauzein et al. [3], we will now focus on the anion exchange process.

Table 3: Calculated equilibrium composition of a solution containing 10^{-3} mol.l^{-1} of Na_2CrO_4 dissolved in field water.

Species	No $BaSO_{4s}$ mol.l^{-1}	$BaSO_{4s}$ in excess mol.l^{-1}	Species	No $BaSO_{4s}$ mol.l^{-1}	$BaSO_{4s}$ in excess mol.l^{-1}
Ca^{++}	1.21 10^{-3}	1.23 10^{-3}		Chromium VI	
Mg^{++}	5.30 10^{-4}	5.30 10^{-4}	CrO_4^{--}	9.04 10^{-4}	6.76 10^{-4}
Na^+	1.00 10^{-3}	1.00 10^{-3}	$HCrO_4^-$	9.05 10^{-5}	6.86 10^{-5}
K^+	4.00 10^{-5}	4.00 10^{-5}	H_2CrO_4	2.42 10^{-11}	1.87 10^{-11}
Ba^{++}	1.32 10^{-7}	1.78 10^{-7}	$BaCrO_{4s}$	1.68 10^{-7}	2.56 10^{-4}
H^+	3.07 10^{-8}	3.13 10^{-8}			
H_2CO_3	1.66 10^{-4}	1.70 10^{-4}		Chromium III	
HCO_3^-	2.42 10^{-3}	2.43 10^{-3}	$Cr(OH)_4^-$	0.00	0.00
CO_3^{--}	3.69 10^{-6}	3.63 10^{-6}	$Cr(OH)_3^0$	2.56 10^{-19}	1.98 10^{-19}
Cl^-	1.16 10^{-3}	1.16 10^{-3}	$Cr(OH)_2^+$	0.00	0.00
SO_4^{--}	4.2 10^{-4}	6.76 10^{-4}	$Cr(OH)_{3s}$	0.00	0.00
OH^-	3.26 10^{-7}	3.19 10^{-7}			

TRANSIENT ANION EXCHANGE OF CHROMATE WITH BICARBONATE AND SULPHATE

Although ion exchange selectivity coefficients can often be considered constant, exchange capacity on minerals depends on the pH of the aqueous phase. In the range of pH's encoutered in natural systems, cationic exchange capacity can be considered constant, whereas anionic exchange capacity is generally much variable. We will speak of the available anionic exchange capacity for the fraction of potentially exchanging sites which are ionized. We will further assume that the anion exchange process involves chromate, sulphate and bicarbonate anions only (carbonates are excluded), and that the selectivity factors are constant.

Column experiments were used to measure the anion exchange capacity versus pH, and the selectivity factors for CrO_4^{--}/SO_4^{--} and CrO_4^{--}/HCO_3^- exchanges. The glass column, 26 mm i.d.

diameter was packed with 248 g of soil over about 30 cm. The column was fed by a peristaltic pump at a volumetric flow rate of 80 $ml.h^{-1}$. HTO (water tracer) pulse injection gave the pore volume, V_p = 68 ml, and the dispersion was characterized by the column Péclet number, Pe = 160.

Determination of the anion exchange capacity

CrO_4^{--}/HCO_3^- exchange was first studied. The anion exchange capacity was determined using the following method. The column was first fed with a solution of sodium chromate dissolved in a standard solution of known pH until a steady-state composition of the effluent was reached. Then, chromate was eluted by a solution of sodium sulphate in the same standard solution. Sodium chromate and sulphate were at the same concentration, so that no cation exchange occured. During this experiment, the available anion exchange capacity is constant since pH is kept constant. The amount of chromate recovered is thus only dependent on the anion exchange capacity and the HCO_3^-/CrO_4^{--} selectivity factor. Results of these experiments are given in Table 4.

Table 4: Amount of chromate recovered versus pH and concentration of injected chromate.

Run	pH	Feed concentration (mmol.$^{-1}$)		Fixed chromate
		HCO_3^-	CrO_4^{--}	$meq.l^{-1}$
1	8.0	1.2	0.58	0.0
2	7.65	2.17	0.044	0.028
3	7.65	2.17	0.58	0.0935
4	7.65	2.17	1.2	0.0932
5	7.09	4.1	0.58	0.225

At high pH (Run 1) no significant amount of sorbed chromate is detected. This suggests that the available exchange capacity is very low. At a constant pH (Runs 2, 3, and 4, pH = 7.65) the amount of fixed chromate increases and reaches a constant value which can be considered as an estimate of the available exchange capacity (9.3 10^{-5} $eq.l^{-1}$).

Determination of the HCO_3^-/CrO_4^{--} selectivity factor

The HCO_3^-/CrO_4^{--} selectivity factor is then deduced from run 2 and from knowledge of the concentrations of HCO_3^- and CrO_4^{--} in the aqueous phase and of the previously determined exchange capacity. The resulting value is given in Table 2. In terms of Gaines-Thomas constant (independent of the fluid/solid ratio), we obtained K_{gHCO_3/CrO_4} = 0.065 $mol.l^{-1}$. The distribution of chromate between the aqueous and solid phases could then be calculated with this selectivity factor and the exchange capacity. We found that the exchanging sites were not fully saturated by chromate in runs 3 and 4, and thus the estimated exchange capacity and selectivity factor are probably inaccurate. Further experiments would be required to obtain better estimates.

Determination of the SO_4^{--}/CrO_4^{--} selectivity factor

The SO_4^{--}/CrO_4^{--} exchange was studied with experiments consisting of injections of mixtures of sulphate and chromate in a standard solution at pH = 7.65. Using the previously determined exchange capacity and HCO_3^-/CrO_4^{--} selectivity factor, the SO_4^{--}/CrO_4^{--} was found close to unity.

In order to validate the description of the ternary anion exchange mechanism coupled with the calco-carbonic equilibria, we performed a pulse injection of chromate 10^{-4} M. The column was first stabilized under flow of a standard solution at pH = 7.65 containing 4.4 10^{-5} M of sulphate. The same solution was used to drive the chromate pulse. Figure 2 shows the experimental and predicted curves together with the HTO (water tracer) curve. The mean retardation and the concentration maximum of chromate are well predicted, whereas the shapes of the curves are rather different.

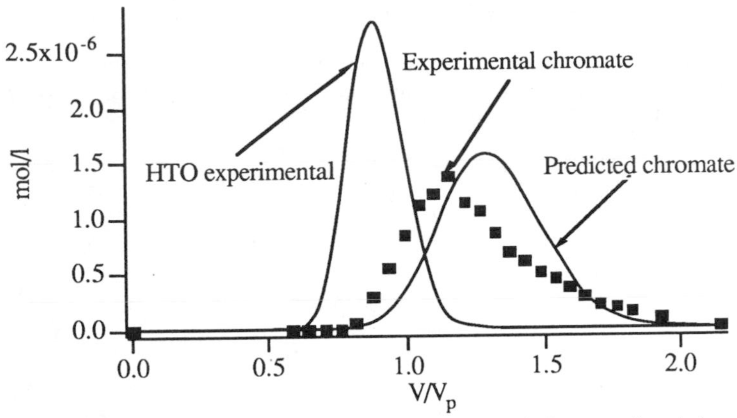

Figure 1: Comparison between experimental and predicted chromate breakthrough curves.

CONCLUSIONS

Chromate transport in the Grenoble sediment is mainly governed by anion exchange coupled with calco-carbonic equilibria, cation exchange and precipitation of barium chromate. Under the field experimental conditions chromium III is found negligible. The experimental method described above allows one to progressively determine the parameters of the main interactions responsible for chromate retention avoiding using some empirical adsorption law.

However, the results are not fully satisfactory as illustrated by Figure 1. The difference between the experimental and theoreticalbreakthrough curves should probably be attributed to the poor accuracy of the anion exchange capacity and selectivity factors of the SO_4^{--}/CrO_4^{--} and HCO^- ,3)/CrO_4^{--} exchange processes. Further experiments are required to obtain better estimates and also to test the ability of the mechanism to account for the dependence of the retention on the pH of the aqueous phase.

Nevertheless the proposed interaction mechanism can still be used to numerically investigate chromium transport in the presence of cation exchange, anion exchange and barium chromate precipitation as illustrated by Jauzein et al. [4].

Acknowledgement: The Comission of European Community is gratefully acknowledged for its financial support.

REFERENCES

[1] Jauzein M., Andre C., Margrita R., Schweich D., Sardin M.: "IMPACT computer code: a help in design and interpretation of laboratory experiments". Proc. GEOVAL 90, SKI-OECD-NEA congress, Stockholm, Sweden, May 14-17, 1990.

[2] Rai D., Zachara J.M.: "Crhromium reactions in geologic materials". Interim report EPRI-EA-5741, Palo Alto, California, USA, 1988.

[3] Jauzein M. , André C., Margrita R., Sardin M. and Schweich D.: "A flexible computer code for modelling transport in porous media: IMPACT", Geoderma special issue, 44, 2-3 (1989), 93-113.

[4] Jauzein M;,André C., Margrita R., Sardin M., Schweich D.: "IMPACT computer code: a help in design and interpretation of laboratory experiments. Proc. GEOVAL 90, SKI-OECD/NEA, Stockholm, Sweden, May 14-17, 1990.

APPLICATION OF THE ALLIGATOR RIVERS ANALOGUE FOR VALIDATION OF SAFETY ASSESSMENT METHODOLOGIES

K. Skagius, K. Pers, F. Brandberg, Kemakta, Stockholm, Sweden
S. Wingefors, Swedish Nuclear Power Inspectorate, Stockholm, Sweden
P. Duerden, Australian Nuclear Science and Technology Organisation, Sydney, Australia

ABSTRACT

Natural analogues play an important role in validation of performance assessment models by providing the possibility to study processes that have been active over scales in time and space relevant for repositories in the long term. The Swedish Nuclear Power Inspectorate (SKI) has initiated a project with the aim of validating performance assessment models based on information and data from the Alligator Rivers Analogue Project. This paper describes the initial phases and planning of the work. Some general viewpoints on validation and application of natural analogues are given. Especially the importance of uncertainties is stressed. Integrated with the validation study a scenario development exercise is performed with the purpose to describe possible external conditions for evolution of the analogue.

1. INTRODUCTION

1.1 Background

In performance and safety assessment of nuclear waste repositories, models and combinations of models are used to predict the long-term behaviour of a complex system of natural and engineered barriers. Validation of performance assessment methods and models are presently of high concern among organisations responsible for nuclear waste disposal programmes and regulatory authorities.

The procedure used in validation exercises is to compare model predictions with experimental observations. Here, information from laboratory experiments, field experiments and natural analogue studies is needed. Laboratory and field experiments yield information on relatively short time scales. Natural geological systems have, however, developed over longer time scales and can therefore provide important information regarding the behaviour of radionuclides over times that is more relevant to radioactive waste repository assessment.

The Koongarra Uranium Deposit in the Alligator Rivers Region in the Northern Territory of Australia is one natural analogue that is investigated with the aim to contribute to the understanding of the scientific basis for the long term prediction of radionuclide migration within geological environments relevant to radioactive waste repositories. The International Alligator Rivers Analogue Project (ARAP) was set up in 1987. The project is sponsored by the OECD Nuclear Energy Agency and the Agreement is signed by the Australian Nuclear Science and Technology Organisation (ANSTO), the Japan Atomic Energy Research Institute (JAERI), the Power Reactor and Nuclear Fuel Development Corporation of Japan (PNC), the Swedish Nuclear Fuel Inspectorate (SKI), the UK Department of Environment (UKDoE) and the US Nuclear Regulatory Commission (USNRC).

The Koongarra uranium mineralisation occurs in two distinct orebodies separated by a barren gap. The main orebody (No 1) has a strike length of 450 m and persists to 100 m depth. The primary ore consists of uraninite veins that crosscut the quartz-chlorite schist. Alteration and oxidation of uraninite within the primary zone have produced secondary uranium minerals, particularly uranyl silicates. From near the surface down to a depth of 25-30 m the schist is weathered, and in this weathered zone another secondary uranium mineralisation, uranyl phosphates, forms a tongue-like fan. Away from the tail of the fan uranium is dispersed in the weathered schist and adsorbed onto clays and iron oxides. The dispersion fan of ore grade material extends down-slope for about 80 m.

A comprehensive experimental and modelling program has been carried out within the Alligator Rivers Analogue Project. A substantial amount of information has been collected on geology, hydrology, geochemistry and radionuclide migration processes. The establishment of a large quality-assured data base has been initiated as well as the provision of sectional contour representations of the data.

In the continuation of ARAP more effort will be given to the evaluation and testing of performance assessment models and methods, using the information available on the primary uranium mineralisation and the formation and evolution of the secondary dispersion fan. In line with this, the Swedish Nuclear Power Inspectorate (SKI) has contracted Kemakta for a project focussing on the application of performance assessment methodology for evaluation of the Koongarra Analogue.

1.2 Objectives

The objective of this project is to use the Koongarra Analogue to test and to evaluate models and model systems used in performance assessment of radioactive waste repository concepts with focus on the Swedish situation. The work is performed in two steps:

1. A review of validation issues in the Swedish Performance Assessment Programme which are covered by ARAP. This step will result in proposals for calculations.

2. Model calculations using the data collected within ARAP.

Integrated with the validation study a scenario identification and development exercise will be performed with the purpose to describe possible scenarios for the evolution of the Koongarra Analogue.

In this document a short description is given of the role of natural analogues in validation of performance assessment models. In addition, some preliminary ideas regarding validation issues in the Swedish Programme that could be covered by ARAP are presented. The results are based on a first brief review of the ARAP documentation. A short description and present status of the scenario development study will also be given.

2. VALIDATION OF PERFORMANCE ASSESSMENT CONCEPTS AND MODELS BY NATURAL ANALOGUES

2.1 General

First of all, we must try to more clearly define what should be meant by validation. For example, it should be recognised that a model (or concept, database etc.) only can be validated in its proper context, i.e. with respect to a given set of scenarios, site and disposal method. We must also be aware of the fact that validation never can be an absolute proof in the philosophical sense. Instead we should strive for an acceptable validation. This also means that we need a measure for the degree of validation and criteria for judgement of its acceptability. These tools are still not available, or rather, consensus about how to handle these questions is not yet developed.

We could start by asking ourselves why we need validation. The obvious answer is that we want assurance that the models (etc.) we employ in a safety analysis provide a sufficient accuracy in prediction of future consequences. Or we can put it the other way and say that we want an acceptable uncertainty in these predictions. Thus, uncertainty might be regarded as a measure of (in)validation. For a given model, its contribution to the total uncertainty in prediction of consequences would then be our measure of its degree of validation.

One problem with this approach is the need for quantification, not only of parameter uncertainties, but also of conceptual uncertainties. In a safety assessment the alternative concepts are - or should be - compared, and the total uncertainty might be evaluated simply by merging the uncertainty ranges for the different concepts. The real problem might be the weighing of such conceptual uncertainties. Thus, quantified uncertainty as a measure of the degree of validation might not be possible at all times. On the other hand, it is a goal that should be strived for and kept in mind for all work on validation of performance assessment models.

Based on the reasoning above the following definition of validation is proposed:

> The assurance that a model provides an acceptable accuracy in the prediction of consequence within the framework of a total safety analysis, considering all input uncertainties and the necessary extrapolations in scale with respect to space and time.

These lines of thought also correspond to the recently increased emphasis on uncertainty analysis as the most important part of the safety analysis.

2.2 Application of Natural Analogues

In the broadest sense the role of natural analogues in validation might be described as follows. The evaluation of natural analogues should provide
- ourselves,
- the scientific community,
- decision makers,
- politicians, and
- the public

with confidence in our ability to describe and predict the future development of a repository, based on our ability to describe the past.

The following issues have to be resolved or dealt with in transferring this confidence from evaluation of natural analogues (NA) to performance assessment (PA):

- the distinction between calibration and validation of models;

- the distinction between (explanatory) research models and (transparent, bounding, "simplistic") PA models;

- the lack of uncertainty analysis in evaluation of NA (and in validation exercises in general!);

- the lack of scenario analysis (as a part of uncertainty analysis);

- a more comprehensive discussion of diversity of phenomena and variability of features;

- inclusion of negative evidence in the discussion, i.e. features that might disprove the case ("negative analogues");

- recognition that not one but many analogues will be needed both for building of general consensus and for gaining acceptable validation in a more narrow field.

In addition, we must strive for methods - or a logic - that allows us to translate the uncertainty in evaluation of a natural analogue to a corresponding uncertainty in performance assessment. Among other things, this means that one should aim at an integrated evaluation, where models etc. are coupled just like in a performance assessment. For the same reason it is important to try the application of PA models and methodologies, e.g. scenario analysis, in the evaluation of an analogue.

The methods applied for evaluation of natural analogues should of course focus on the important features displayed by analogues in comparison to laboratory and field experiments, such as:

- studying the influence of slow microscopic processes, e.g. kinetics, matrix diffusion etc.;

- studying importance of macroscopic features, e.g. flow path characteristics and their variability:

- simulation of large scale (in space and time) processes with simple performance assessment models.

Whenever possible, analogues should be evaluated in order to provide insight into model relevance in an integrated system of models for performance assessment. In this context a scenario analysis of the analogue will provide both a basis for including estimation of uncertainties in the evaluation and a test of our ability to describe natural phenomena over time scales relevant for nuclear waste repositories.

The present work, as a part and contribution to ARAP, is an attempt to include some of the above mentioned aspects in the evaluation of a natural analogue.

2.3 Uncertainties

There are different levels of uncertainties involved in a performance assessment, which will influence the overall uncertainty in the consequence analysis. These corresponding uncertainties should also be considered in the evaluation of a natural analogue.

1. Scenarios; uncertainties associated with the combinations of external conditions and other phenomena that will influence the performance and safety of a repository over long time and large geometrical scales.

2. Conceptual Models; uncertainties associated with the identification and description of processes, geometrical structures and other conditions that will influence the performance and safety of a repository over long time and large space scales.

3. Calculation Models; uncertainties introduced by the simplifications that have to be made
    - in the transformation of conceptual models to mathematical models,
    - due to lack of input data,
    - due to limitations in computer capacity.

4. Input Data; uncertainties associated with the method used to measure or to extract the data, and uncertainties in the representativeness of the data.

One method to estimate and successively reduce the uncertainties in Scenarios is to continuously perform a systematic scenario development. Site investigations and evaluation of natural analogues and well designed field and laboratory experiments are methods used to try to quantify and reduce the uncertainties in Conceptual and Calculation Models as well as in Input Data.

3 VALIDATION ISSUES COVERED BY ARAP

The geological environment considered in Sweden for radioactive waste repositories is totally different from the situation in and around the Koongarra uranium deposit. Also external conditions that indirectly may influence the performance of a repository in Sweden, are different from those that have had impact on the evolution of the Koongarra deposit. Despite these differences an evaluation of Koongarra can contribute to meet some of the needs of validation in performance assessment.

A schematic description of the chain of events releasing and transporting radionuclides from the spent fuel stored in canisters in the repository to the intake of nuclides by man is shown in Figure 1. Based on this flow sheet and a first brief review of the ARAP documentation, some validation issues concerning conceptual models, calculation models and input data have been identified. These are discussed in the following subsections. The application of the scenario development methodology is discussed in the next chapter.

3.1 Conceptual Models

The main validation issue on the conceptual level is to increase the understanding of how processes and geometrical structures under long time effect the mobilisation and migration of radionuclides. The advantage in using a natural analogue in this sense is, of course, the possibility to study processes in a natural system that have been active over long times. This could give evidence for the importance or unimportance of processes identified in laboratory and field experiments, and also provide understanding how to extrapolate from small scale experiments to the large scales in time and space that are relevant in performance assessment.

Figure 1 shows a simplified flow sheet of the main pathway considered in performance assessment of Swedish repository concepts, by which radionuclides in the repository is transferred to man.

Canister failure. The radioactive material is contained in canister which are placed in the repository. Failure of the canisters is caused by corrosion and/or mechanical effects. Phenomena causing the canister failure is a validation issue but could not be covered by ARAP.

Fuel dissolution. Dissolution of the fuel will start when groundwater has penetrated the canister. The spent fuel consists of uranium dioxide and the Swedish groundwater contains silicates and phosphates. In contact with the groundwater the uranium dioxide might be oxidised by oxidants produced by α-radiolysis of the water. In Koongarra the primary uranium mineral uraninite has been oxidised and secondary uranium minerals such as uranyl silicates and uranyl phosphates have been formed. A validation issue in this context could then be to increase the understanding of the uranium chemistry in a system comprised of oxides, silicates and phosphates.

Nuclide transport in man-made barriers. Dissolved radionuclides will migrate out through the ruptured canister and further out through the bentonite backfill surrounding the canister. If the canister failure is caused by corrosion, the corrosion products may constitute a barrier to the migrating nuclides. To increase the understanding of the processes and geometrical structures determining the migration rate in the corrosion products is here identified as a validation issue. With the present proposed canister material, copper, this aspect is not covered by ARAP. If, however, iron is to be considered as canister material, it may be of interest to study the interaction between uranium (and decay products) and ferric hydroxides which are present in the weathered zone in Koongarra, both in crystalline and amorphous form.

Validation issues related to the nuclide migration through the bentonite back fill are similar to those mentioned above for the canister corrosion products, i.e. to increase the understanding of the behaviour of the clay barrier over long times in terms of transport properties and chemistry. Clay materials, vermiculite, kaolinite and smectite formed by the alteration of chlorite, are present in the weathered zone in Koongarra. A study of the interaction between uranium and these clay materials in Koongarra may contribute to the understanding of the mechanisms of interaction between radionuclides and the bentonite backfill in a repository.

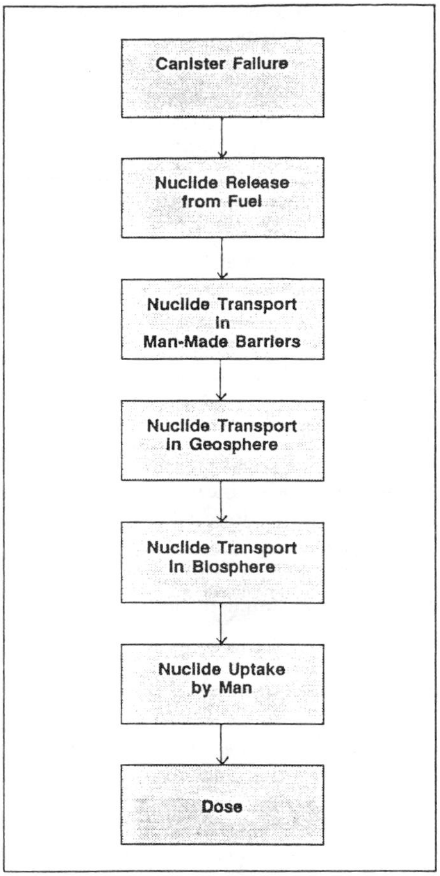

Canister Failure
Nuclide Release from Fuel
Nuclide Transport in Man-Made Barriers
Nuclide Transport in Geosphere
Nuclide Transport in Biosphere
Nuclide Uptake by Man
Dose

Figure 1 Schematic illustration of the groundwater pathway for radionuclide release from a repository.

Nuclide transport in the geosphere. Radionuclides released from the bentonite surrounding the canister will be transported away from the repository by groundwater flowing in fractures in the rock. A validation issue in this context is to increase the understanding of the nature of the flow paths in the fractured rock. This is a basic requirement since the nature of the flow paths has implications on the conceptual understanding of migration processes such as advection, dispersion and matrix diffusion. In addition to this, more has to be learned about the mechanisms of sorption and of transport of radionuclides as colloids/particles and by colloids/particles.

The nature of the groundwater flow in a geological environment is site specific, and the flow situation at Koongarra is most likely quite different from what could be expected in a fractured rock in Sweden. There are, however, indications of fracture flow in Koongarra, both in the weathered and the unweathered zone, so the implications of fracture flow on nuclide transport is a validation issue that may be covered by ARAP. Another validation aspect in this context is to increase the understanding of causes and effects of changes in time and space of the flow situation in a geological environment. In addition to this, the experience gained during the field work aimed at characterise the hydrology in Koongarra may be valuable input for future hydrological investigations of potential repository sites in Sweden.

Despite the differences in groundwater flow situation between a crystalline rock environment in Sweden and Koongarra, similar processes will be important for the migration of radionuclides away from the source. An evaluation of the transport mechanisms active in Koongarra could then contribute to an increased understanding of the transport processes considered to be potentially important for Swedish conditions, i.e. advection-dispersion, sorption, matrix diffusion and colloid transport.

Nuclide transport in the biosphere. Radionuclides released from the geosphere will eventually reach man via transport in the biosphere. This transport is either driven by hydrology or by features specific for the biosphere. For the transport driven by hydrology considerations of conceptual uncertainty are similar to those of the transport in the geosphere, i.e. uncertainties associated with the transport properties of the biosphere and transport processes such as advection-dispersion, diffusion and sorption. For transport by flow in soil an additional conceptual uncertainty is to be found if the soil is unsaturated.

Examples of biosphere features which affect the transport in the biosphere are soil turnover due to water and wind erosion and bioturbation of soil and sediments. Here an increased understanding is needed of how these features actively contributes to the transport of nuclides in the biosphere.

Among the above mentioned validation issues regarding the biosphere transport those connected to the transport driven by hydrology may be covered by ARAP since the transport processes in this case are the same as in the geosphere. Also the validation aspect of transport in unsaturated media could maybe be covered by ARAP since the surface part of the weathered zone is unsaturated during the dry season.

3.2 Calculation Models

A major part of a performance assessment of a nuclear waste repository is concerned with the estimation of the amount and rate of movement of radionuclides in different environments. For this purpose three main types of calculation models are used; hydrology models, geochemical models and transport models. These models are most often used one by one, but models where transport and geochemistry is treated simultaneously are more and more applied.

Calculation models can be divided into research models and assessment models. Research models are detailed complex models by which single phenomena or a combination of few phenomena is studied, with the attempt to obtain a relatively accurate representation of reality. Performance assessment models are used to study the whole repository system considering combinations of processes and geometrical structures of importance for the mobilisation and migration of radionuclides. Assessment models need therefore to be rather simple in order to allow for numerous calculations within a realistic time. Another reason for applying simplified assessment models is the limitation in input data available.

One important validation issue in this context lies in the level of detail that processes and geometrical structures must be considered in performance assessment in order to achieve acceptable results. One example of this concerns the process of sorption. In assessment models sorption is typically modelled using the K_d-approach. The validation aspect in this case then is whether this approach is sufficient enough or if a more detailed mechanistic modelling is needed to assess the effects of sorption. Another example concerns processes where kinetics may be of importance, for example in dissolution/precipitation reactions, which in performance assessment usually is treated with thermodynamic models. The change in geometrical structure and other properties of the barriers with time and space is an additional example of events that normally not are considered in performance assessment models.

By applying performance assessment models in evaluating the migration of uranium and its decay products in the weathered zone in Koongarra it may be possible to assess the effects of using simplified descriptions of processes involved in the transport, for example the K_d-approach, and also to get an indication of the uncertainties introduced by averaging variations in transport properties over time and space.

3.3 Input Data

Data acquisition is a very important task in performance assessment since without high quality data models can neither be developed nor properly applied. Data can be site-specific or system-specific. For example, data describing the geometrical structures in the rock surrounding a repository are specific for that location, while data describing the chemical interaction between radionuclides and rock are defined by the composition of the water and the composition of the rock.

Validation issues in this context concern both methods to derive the necessary data and the representativeness of the acquired data for the desired application. These aspects can be covered by ARAP for certain types of system-specific data. An example is to compare results from laboratory experiments studying the distribution of uranium (and decay products) between Koongarra water and ferric hydroxides and/or clay material representative for the site, with concentrations in porewater and in ferric hydroxides and/or clay materials sampled in the weathered zone in Koongarra. A study of the geochemistry in Koongarra may also be valuable in terms of checking existing thermodynamic data and also to contribute with new data to existing thermodynamic databases.

In order to be able to model the radionuclide transport in Koongarra some site-specific data are needed. These data are not representative for repository locations in Sweden. The methods used in collecting those data are, however, similar and applying those methods at Koongarra may make it possible to evaluate and decrease the uncertainties associated with field and laboratory measurements.

4. SCENARIO DEVELOPMENT

The Swedish Nuclear Power Inspectorate (SKI) and the Swedish Nuclear Waste Management Co. (SKB) have carried out a joint scenario development exercise of a hypothetical repository for spent fuel based on the Swedish KBS-3 disposal concept [1]. A similar approach for a systematic overall evaluation of the Koongarra natural analogue has been suggested by SKI. Such an exercise will not only be valuable for the evaluation of the analogue. It will also give experience for future use of this approach in performance assessment of repository concepts. A short summary of the joint SKI/SKB scenario project and an introduction to the application of the methodology on Koongarra have been given elsewhere [2].

Applying the methodology used in the scenario project in the evaluation of the Koongarra analogue will mean that the following steps have to be carried out.

1. An initial comprehensive identification of FEP's (Features, Events, Processes) that might have had an influence on the evolution and behaviour of the analogue. This step should involve all research groups engaged in ARAP in order to cover all relevant disciplines.

2. Documentation of each FEP in a short memo-text. This text should contain:
   - A short description of the FEP
   - The causes and effects of the FEP
   - Modelling aspects of the FEP

3. Screening of FEP's to obtain three groups of FEP's; screened out FEP's (removal according to well defined criteria), FEP's belonging to the Process System, and the remaining FEP's grouped together to form primary FEP's. In order to be able to do this screening the Process System for an analogue has to be defined. In accordance with the definition of the Process System for a spent fuel repository, the Process System for the Koongarra analogue should comprise the complete set of deterministic chemical and physical processes that might have influenced the formation of the secondary uranium mineralisation and the subsequent dispersion of uranium in the weathered zone.

4. Formulation of scenarios, which is carried out by taking all possible combinations of the remaining FEP's. A scenario is then defined by a set of external conditions which have influenced processes in the Process System.

This scenario identification and development of the Koongarra analogue should be performed in several iterations and should continue to the conclusion of ARAP. A first list of identified FEP's has been prepared and the preparation of a memo-text for each FEP has started. In the continued work with step 2, help from the different groups participating in ARAP will be needed.

5. CONTINUATION OF WORK

The work with the scenario development will continue as presented above. This will be performed integrated with modelling work focussing on the radionuclide transport in the weathered zone.

The first step in the model validation exercise will be to review and evaluate the data available. This has to be done in order to be able to get an overall picture of the phenomena involved in the mobilisation and migration of uranium and its decay products in the weathered zone. The next step will be to formulate a conceptual model for the radionuclide transport in the weathered zone in terms of identifying relevant processes and defining geometrical structures and scales. Based on the conceptual model an appropriate application of a numerical model (code) will be chosen.

To begin with, very simple model calculations studying the radionuclide movement in the weathered zone will be performed. This incudes 1-D transport modelling and geochemical modelling. No modelling of the hydrology is planned since other groups within ARAP are dealing with that. Based on the results of the first series of calculations and the availability of necessary data it may be relevant to continue by increasing the complexity of the models.

5. REFERENCES

[1] Andersson J., Carlsson C., Eng T., Kautsky F., Söderman E., and Wingefors S.: The joint SKI/SKB Scenario Development Project, SKI Technical Report 89:14, 1989.

[2] Wingefors S.: "A Systematic Approach to the Overall Evaluation of a Natural Analogue: Objectives and Planning", Alligator Rivers Analogue Project, Progress Report 1 June 1989 - 31 August 1989, Australian Nuclear Science and Technology Organisation (ANSTO), 1989, 101-118.

Session 4

PROGRESS IN VALIDATION
OF COUPLED THERMO-HYDRO-MECHANICAL EFFECTS

Séance 4

ETAT D'AVANCEMENT DES TRAVAUX DE VALIDATION
DES EFFETS COUPLES THERMO-HYDRO-MECANIQUES

Chairman - Président

B. CÔME
(CEC)

VALIDATION OF TWO ROCK MECHANICS CODES AGAINST COLORADO SCHOOL OF MINES BLOCK TEST DATA

O. Stephansson, T. Savilahti
Division of Rock Mechanics, Luleå University of Technology, Sweden

N. Barton, P. Chryssanthakis
Norwegian Geotechnical Institute, Oslo, Norway

ABSTRACT

The finite element code HNFEMP and the micro universal distinct element code MUDEC were tested against the Colorado School of Mines (CSM) block test data. HNFEMP uses a material consitutive model that accounts for the properties of the intact rock and the discontinuities without including the discontinuities as separate entities. In MUDEC the rock mass is modelled by describing the response of every discontinuity and deformable block separately, with linear or non-linear stiffness according to the Barton-Bandis model for the discontinuities. Good agreement is obtained for the orientations and magnitudes of the displacement vectors and the HNFEMP modelling. Excellent agreement was obtained for joint shear displacements and joint conductive aperture with the MUDEC modelling.

VALIDATION DE DEUX PROGRAMMES DE CALCUL SUR LA MECANIQUE DES ROCHES PAR RAPPORT AUX DONNEES EXPERIMENTALES OBTENUES SUR UN BLOC A LA COLORADO SCHOOL OF MINES

RESUME

Les auteurs ont procédé à la validation du programme de calcul à éléments finis HNFEMP et du programme micro-universel à variables discrètes MUDEC par rapport aux données expérimentales obtenues sur un bloc à la Colorado School of Mines (CSM). Le programme HNFEMP s'appuie sur un modèle qui prend en compte les propriétés de la roche intacte et les discontinuités, sans considérer celles-ci comme des entités distinctes. Dans le programme MUDEC, la modélisation de la masse rocheuse prend en compte individuellement la réaction de chaque discontinuité et de chaque bloc déformable, avec une ridigité linéaire ou non linéaire selon le modèle Barton-Bandis applicable aux discontinuités. S'agissant de la modélisation HNFEMP, on obtient une bonne concordance pour les orientations et les magnitudes des vecteurs de déplacement. En ce qui concerne la modélisation MUDEC, on observe une excellente concordance pour les déplacements des joints par cisaillement et pour les ouvertures conductrices liées aux joints.

INTRODUCTION

In the modelling of large rock masses for the Lansjärv Project [1] the problems and features to be modelled are defined and conceptual models are presented showing the sequence of modelling, the boundary conditions and the sub-structuring of the models in global and regional scale. Before the modelling tools are applied to any problem related to final storage of high-level waste, the computer codes must be validated. The thoroughly instrumented and carefully investigated block test at Colorado School of Mines have been used in the validating process of two codes, namely, the non-linear finite element method HNFEMP and the code MUDEC for the distinct element method.

MODELS OF ROCK MASSES

Jointed rock continuum model, HNFEMP

A jointed non-linear rock continuum model has been formulated by Olofsson [2] for modelling jointed rock mechanical response in continuum-based numerical codes. The recoverable normal and shear joint deformations are assumed to depend linearly on the applied stresses so that the joint elastic compliance is regarded as a constant property. The intact rock is considered to be a linearly elastic material and for an elastically isotropic material the compliance matrix may be determined from the Young's modulus and the Poisson's ratio.

The fundamental Mohr-Coulomb failure criterion is included. Equations to relate the dilation angle and other unknowns to the well-known empirical formulations of joint shear and normal behaviour by Barton and Bandis [3] and Barton's parameters JRC, JCS and joint length, L, are used in the model.

The equivalent rock mass model for the mechanical behaviour of continuous rock joints has been implemented in a finite element code called FEMP [4] , and the special version containing the non-linear model fo the rock joints is called HNFEMP.

Distinct element program for modelling jointed rock masses, MUDEC

The Micro Universal Distinct Element Code, MUDEC, provides in one code modelling variable rock deformability, plastic behaviour and fracturing of intact rock, fluid flow and fluid pressure generation in joints and voids and linear and non-linear inelastic behaviour of joints. In this study we apply both linear and non-linear behaviour of the joint parameters. The shear-dilation-conductivity coupling of each joint set is modelled in accordance with the method suggested by Barton et al. [5] and called the Barton-Bandis joint model. MUDEC contains logics to account for generation of fluid pressure in inner domains of the model. This facility has been applied in generating boundary loadings from flat jacks.

COLORADO SCHOOL OF MINES BLOCK TEST

A series of mechanical and hydrological experiments is being conducted with an in-situ block of fractured gneissic rocks at the Colorado School of Mines (CSM) Experimental Mine at Idaho Springs, Colorado. The block and the specially-excavated drift have been the sites for several rock mechanics and hydrological studies. Results from these studies are reported by Barton et al. [6] and summarized in Stephansson and Savilahti [7].

The test block is a two-meter cube of Precambrian gneiss excavated by Terra Tec, Inc. in 1979, Figure 1. They subjected the block to bi-axial and uni-axial loading at ambient and elevated temperatures, using flat jacks and a line of borehole heaters for measuring hydro-thermal-mechanical properties of the rock mass. After Terra Tec completed their program, CSM began a second series of experiments with new hydraulic flat jacks, decoupled recording displacements with a fixed external reference frame and an updated recording system [8].

Mechanical properties

Extensive laboratory tests on intact rocks from the experimental mine have resulted in the following elastic constants for the intact rock matrix:

E_R = 58.6 GPa for all rock types, all directions; $\nu = 0.25$

Based on joint parameters recorded in the mine and the known hydro-mechanical coupled joint behaviour the parameters for the Barton system of joint characterization for the foliation set and the diagonal joint set were as follows:

JRC = 8.2, JCS = 62.2 MPa and $\Phi_r = 26.5^o$

For a representative normal stress level of 3.5 MPa the cohesion, friction angle and peak dilation angle is found to be:

c = 0.4 MPa, $\Phi = 32^o$ and d = 6.7^o

The parameters associated with the joint-shear and normal stiffness properties of the CSM block have been determined by three different methods [7].

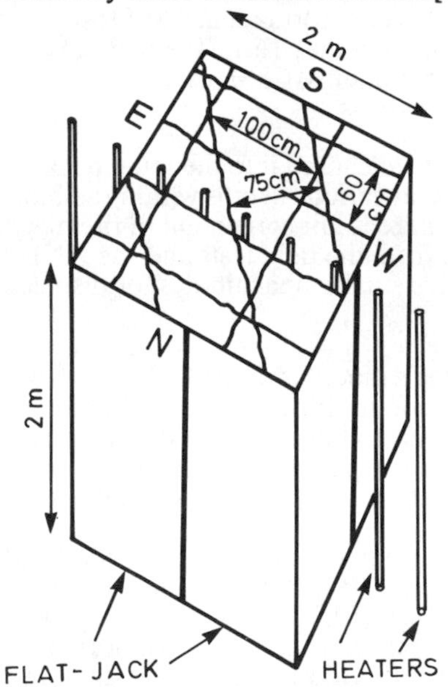

Figure 1 Schematic block diagram of the CSM block showing the relationship of the average joint structure to the flat jacks, the line of borehole heaters and the joint permeability test. After [9].

MODELLING WITH HNFEMP AND COMPARISON OF NUMERICAL AND FIELD TEST RESULTS

A continuum model of the CSM block was constructed using the HNFEMP code and six different material models were tested, Table I. Uni-axial and bi-axial loading to a peak stress of 5.14 and 5.4 MPa, respectively, were applied. A total of 18 models (6 materials and 3 loadings) were analyzed and the results in full are presented in [7].

Table I Material models tested with HNFEMP.

	Material Model					
Joint Stiffness [MPa/mm]	1 Isotropic, Linearly Elastic	2 Field Data Barton System	3 Block Test Terra Tek*	4 Block Test CSM+	5 Sensitivity Analysis	6 Sensitivity Analysis
Normal	–	67.2	117.3	35.4	117.3	17.3
Shear	–	2.7	16.7	24.0	2.7	0

* Whittmores pin [9]

+ Triangulation array [8]

The following tests are used for the validation of HNFEMP:

- Peak-load displacement vectors, CSM test
- Equivalent modulus of deformation, CSM test
- Block deformation modulus, Terra Tek test
- Rock deformation modulus, CSM tests
- Rock stresses, CSM tests

With the fixed frame anchored into the mine back in the CSM test, absolute displacements of station points at various depths within the block during loading could be measured [8]. In principle there is fair agreement in the magnitude and orientation of the displacement vector obtained from the field tests and the HNFEMP-modelling, see Figure 2. Modelling results always give smaller magnitudes for displacements than those obtained from the field data.

The effective Young's modulus of the CSM block was determined from the calculated overall strains between four block corners and discontinuous deformation analysis (DDA). The average Young's modulus of 12.4 GPa calculated from DDA is found to be in good agreement with the results (12.0 GPa) of the finite element modelling with the stiffness parameters suggested by Richardson [8], material model 4.

Stress monitoring during flat jack loading of the CSM block was conducted with two instrumentation systems - the USBM Borehole Deformation Gauge (BDG) and the Luleå University of Technology gauge (LUT). Excluding a few measuring stations at the centre of the block, the monitored directions of the principal horizontal stress were in fair agreement with the modelled results. Furthermore, the magnitude of the average maximum horizontal stress from the BDG monitoring agreed best with the results of the modelling and the applied loading from the flat jacks.

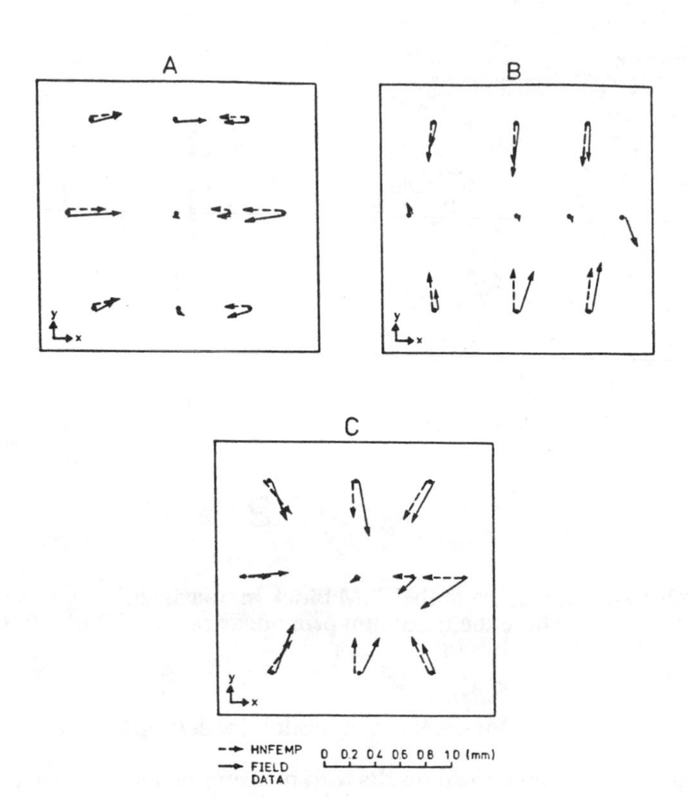

Figure 2 Displacement vectors at peak loads from field tests and modelling with material model 4, CSM test. A, Uni-axial E-W loading. B, Uni-axial N-S loading. C, Bi-axial loading.

MODELLING WITH MUDEC AND COMPARISON OF NUMERICAL AND FIELD TEST RESULTS

Two versions of the MUDEC code have been tested against the measured results from the CSM block test, namely, MUDEC-linear, where the joint properties are assumed to vary linearly with applied stress, and MUDEC-BB, where joint stiffnesses are assumed to vary non-linearly with stress in accordance with the Barton-Bandis joint model. For simplicity, the CSM block was modelled by three joints and four principal blocks, see Figure 3.

Boundary conditions were varied in an attempt to simulate flat jack loading more realistically. Block models with linear joint behaviour (Model A) and Barton-Bandis joint behaviour (Model B) were analysed with simple uniform stress boundaries [10]. More realistic boundary conditions were achieved by applying fluid pressure loading in rectangular slots simulating the flat jacks and with rigid boundaries resisting displacements behind them. These boundary conditions were applied only to the MUDEC-BB models (Model C). Typical results for the principal stresses and block rotations inside the model are shown in Figure 3 for the case of bi-axial loading of Model C (flat jack simulation and Barton-Bandis joint behaviour). Notice that stresses tend to be transmitted perpendicularly across the joints. In general, the stress magnitudes are of the order of 4 to 7 MPa compared with the maximum applied boundary stresses of 5.4 MPa. Stress concentrations associated with block rotation and corner loadings increase the stresses in the range of 9.1 - 14.3 MPa for linear joint be-

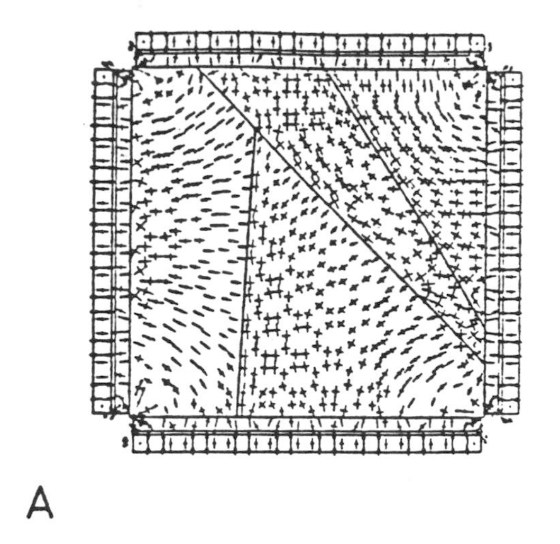

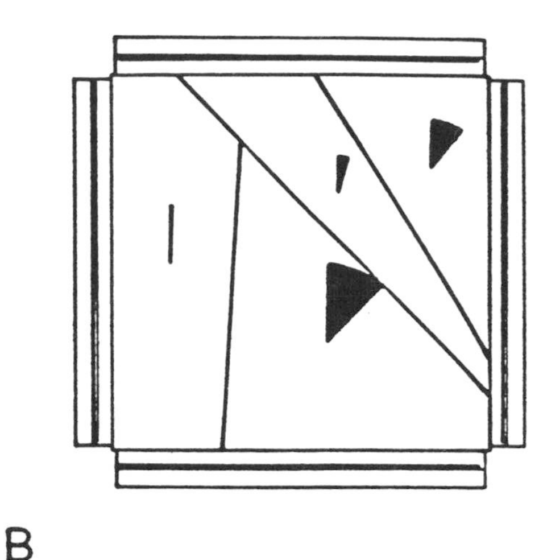

Figure 3 MUDEC-BB modelling of the CSM block in bi-axial loading and flat jack simulation. A, Principal stresses where the maximum principal stress is 17 MPa. B, Rotation of individual blocks in the model. After [6].

haviour and 7.9 - 23.2 MPa for non-linear joint models loaded without the flat jack fluid pressure simulation.

Comparison of the numerical results with measurements of displacement vectors and shear displacements [8], measurements of shear displacement, shear stiffness and conducting aperture [9], and stress measurements performed [11] show good agreement in general, cf. Table II. Excellent agreement was obtained for joint shear displacements and joint conductive apertures, cf. Table III.

Table II Summary of the numerical and experimental results. After [6].

	Maximum displacement vector [mm]	Total block displacement [mm]	Shear displacement [mm]	Maximum principal stress [MPa] DBG method	Maximum principal stress [MPa] LUT method
CSM test Richardson, 1986 [8]	0.5 Biax 0.6 N-S 0.6 E-W	0.725 Biax	- Biax 0.225 (N-S) 0.20 (E-W)		
Terra Tek test Hardin et al., 1982 [9]	* * *	~0.23 Biax ~0.23 N-S ~0.23 E-W	- Biax 0.22 (N-S) 0.13 (E-W)		
CSM test Brown et al., 1986 [11]				20.0 Biax 24.7 N-S 16.3 E-W	11.6 Biax 8.4 N-S 11.8 E-W
Numerical results from Models A B C	0.10-0.19 Biax 0.28-0.53 N-S 0.29-0.40 E-W	0.20-0.38 Biax - N-S - E-W	0.16-0.13 Biax 0.15-0.75 N-S 0.15-0.33 E-W	8.6-15.9 Biax 14.1-23.2 N-S 7.9-9.1 E-W	

Table III Summary of the numerical and experimental results from Terra Tek and Richardson tests concerning changes in mechanical and conductive joint apertures. After [6].

	Stress range [MPa]	Mechanical jonit aperture change [μm])	Conducting joint aperture change [μm]	Mechanical aperture [μm]	Conductive aperture [μm]
Terra Tek's measurements					
Diagonal joint	(0-6.9)	59 BIAX	26 BIAX	---	27-55
		94 N-S	17 N-S	---	29-46
		63 E-W	26 E-W	---	28-53
Richardson's measurements					
Diagonal joint	(0-5.2)	7-11 BIAX	---	106 BIAX	---
		---	---	---	---
		---	---	---	---
Foliation joint	(0-5.2)	2-15 BIAX	---	28 BIAX	---
		---	---	---	---
		---	---	---	---
Numerical results Models A B C	(0-7)	29 Bia (diag)	14 (Bia diag)	137-142 Bia	34.7-36.9 Bia
		27 Bia (fol.)	19 (Bia fol.)	152-162 N-S	49.6-49.9 N-S
				160-164 E-W	48.4-49.2 E-W

CONCLUSIONS

Two approaches for modelling rock masses have been applied in this study, namely the finite element equivalent continuum or smeared-joint technique (HNFEMP) and the distinct element method for modelling discretely jointed rock (MUDEC). Despite the different modelling approaches their application to the CSM block test gave similar results and fair to good agreement with data recorded in the field test. The moderate loading of the block and only minor exceeding of shear strength of existing joints favoured elastic deformation and enhanced the agreement between the modelling techniques and between the models and the block test.

ACKNOWLEDGEMENTS

U. Singh and T. Olofsson made important improvements of the code HNFEMP, and M. Christianson, Itasca, made major improvements to the MUDEC code. Parts of this study were conducted during a sabbatical leave of O. Stephansson to the Colorado School of Mines. The helpful attitude and stimulating discussions with W. Hustrulid and his students are gratefully acknowledged. This study was initiated and supported by the Swedish Nuclear Fuel and Waste Management Co. Valuable discussions were held with G. Bäckblom and we thank him.

REFERENCES

[1] Bäckblom, G. and Stanfors, R. (Eds.): "Interdisciplinary study of post-glacial faulting in the Lansjärv area, Northern Sweden". Technical Report TR 89-31, Swedish Nuclear Fuel and Waste Management Co., Stockholm, 1989.

[2] Olofsson, T.: "Mathematical modelling of jointed rock masses". Doctoral Thesis 1985:42D, Luleå University of Technology, Luleå, 1985.

[3] Barton, N.R. and Bandis, S.C.: "Effect of block size on the shear behaviour of jointed rock". Proc. 23rd U.S. Symposium on Rock Mechanics, 1982, 739-760.

[4] Nilsson, L. and Oldenburg, M.: "FEMP - An interactive, graphic finite element program for small and large computer system". Technical Report 1983:07T, Luleå University of Technology, Luleå, 1983.

[5] Barton, N.R., Bandis, S.C. and Bakhtar, K.: "Strength, deformation and conductivity coupling of rock joints". Int. J. Rock Mech. Min. Sci. & Geomech. Abstr., 22, 3, 1985, 121-140.

[6] Barton, N.R. and Chryssanthakis, P.: "Validation of MUDEC against Colorado School of Mines block test data". Technical Report TR 88-14, Swedish Nuclear Fuel and Waste Management Co., Stockholm, 1988.

[7] Stephansson, O. and Savilahti, T.: "Validation of the rock mechanics HNFEMP code against Colorado School of Mines block test data". Technical Report TR 88-13, Swedish Nuclear Fuel and Waste Management Co., Stockholm, 1988.

[8] Richardson, A. M.: "In-situ mechanical characterization of jointed crystalline rock". PhD Thesis T-3256, Colorado School of Mines, Golden, Colorado, 1986.

[9] Hardin, E.L., Barton, N.R., Lingle, R., Board, M.P. and Voegele, M.D.: "A heated flat jack test to measure the thermomechanical transport properties of rock masses". Office of Nuclear Waste Isolation, Columbus, Ohio, ONWI, 1982.

[10] Barton, N.R. and Chryssanthakis, P.: "Distinct element modelling of the influence of glaciation and deglaciation on the behaviour of a faulted rock mass". Working Report AR 89-19, Swedish Nuclear Fuel and Waste Management Co., Stockholm, 1989.

[11] Brown, S.M., Leijon, B.A. and Hustrulid, W.A.: "Stress distribution within an artificially loaded jointed block". Proc. Int. Symposium Rock Stress and Rock Stress Measurements, Centek Publishers, Luleå, 1986, 429-439.

MODELLING THE ROOM 209 EXCAVATION RESPONSE TEST IN THE CANADIAN UNDERGROUND RESEARCH LABORATORY

T. Chan, E. Kozak, B.W. Nakka
Atomic Energy of Canada Limited
Manitoba, Canada

A. Winberg
Swedish Geological Company
Göteborg, Sweden

ABSTRACT

Finite-element simulations have been performed to predict the mechanical response of the granitic rock mass to the excavation of a tunnel through a narrow water-bearing fracture zone and the near-field hydraulic response of the fracture zone. Predictions for the displacements and stress changes were reasonable, pressure drops at boreholes and inflow to the tunnel were overpredicted, and fracture permeability changes were underpredicted, often opposite to the measured direction. A revised post-excavation hydraulic model indicated the formation of a thin "skin zone" of significantly altered permeability around the tunnel after excavation. The permeability change is considered too large to be solely stress-induced.

MODELISATION DE L'ESSAI DE REACTION A L'EXCAVATION DE LA CHAMBRE 209 DU LABORATOIRE DE RECHERCHES SOUTERRAIN CANADIEN

RESUME

On a exécuté des simulations par la méthode des éléments finis pour prédire la réaction mécanique de la masse de roche granitique à l'excavation d'une galerie à travers une étroite zone de fracture aquifère ainsi que la réaction hydraulique de la zone de fissures en champ proche. Les prédictions de déplacements et variations de contraintes sont raisonnables, les chutes de pression au niveau des trous de forage et l'entrée d'eau dans la galerie sont surévaluées et les variations de perméabilité de la fracture sont sous-évaluées et se produisent souvent dans la direction opposée à la direction préférentielle mesurée. Un modèle révisé de réaction hydraulique après excavation indique la formation, après excavation, d'une mince "envelope extérieure" de perméabilité considérablement modifiée autour de la galerie. On pense que la variation de perméabilité est trop grande pour être uniquement provoquée par les contraintes.

1. INTRODUCTION

An in situ excavation response test was conducted by Atomic Energy of Canada (AECL) Research Company [1] at the Canadian Underground Research Laboratory (URL). The test was performed in conjunction with the drill-and-blast excavation of a tunnel (Room 209) at 240-m depth through a narrow, near-vertical, water-bearing fracture zone oriented almost perpendicular to the tunnel axis. Prior to excavating the tunnel through the fracture zone, an array of instruments was installed within boreholes to measure the mechanical response of the granitic rock mass and the near-field hydraulic response of the fracture zone to excavation (Figure 1). The instrument array included eight multi-anchor borehole extensometers, one sliding micrometer, four pairs of convergence pins, eight CSIRO triaxial strain cells and nine straddle packer hydraulic testing/hydraulic pressure borehole monitoring systems [1]. The test was conducted according to the following procedure (see Figure 1): a) excavate Room 209 to Face 1; b) drill boreholes, install instruments and perform pre-excavation characterization of fracture zone; and c) excavate to Face 2 while monitoring instruments with datalogger as well as conducting single-borehole hydraulic tests before and after each excavation blast sequence.

Modellers were supplied with the necessary input data and requested to predict the rock displacements, stress changes, induced fracture permeability changes, piezometric head changes, and groundwater seepages into the tunnel in response to the excavation. The experimental report of measured responses [2] was withheld from the modellers until after they had filed their prediction report [3]. This paper briefly describes the modelling approach [3] and compares the predicted and measured responses.

2. GEOMECHANICAL MODELLING

Three-dimensional (3-D) linear elastic continuum finite-element analyses (FEAs) were performed using the ABAQUS code [4]. Excavation damage effects were not accounted for in the initial predictive models. The model domain was a 200 m x 200 m x 225 m rectangular block of rock discretized by approximately 5000 eight-noded hexahedral elements (Figure 2). Three meshes were constructed, each concentrating elements in a particular area of interest: the extensometer array, the CSIRO triaxial strain cell locations, and along the fracture plane. All three meshes concentrated elements around the planned tunnel. The excavation sequence of ten pilot rounds and nine slashing rounds was simulated by sequential removal of element stiffness matrix and load vector contributions. Zero normal displacement boundary conditions were applied to all exterior boundaries.

The rock mass was assumed to be homogeneous and isotropic, with an in situ rock-mass deformation modulus (E) of 40 GPa based on displacements associated with the excavation of the shaft at depths of 15 m and 183 m [5,6]. The value of 0.257 used for Poisson's ratio was based on laboratory testing. The magnitude and direction of the in situ stresses, determined by overcoring measurements [1], are given in Table I. These values are the average stresses based on two overcoring methods: AECL-modified CSIR and USBM.

The normal and shear stresses acting on the fracture before and after excavation of the tunnel were obtained by projecting the stress tensor onto the plane of the fracture zone. Changes in fracture permeability resulting from the stress redistribution caused by tunnel excavation were calculated using Gale's empirical normal stress-permeability relationship [7]. Generally, no shear dilation was predicted, based on an analysis of shear stresses on the fracture permeability induced by tunnel excavation using Barton, Bakhtar and Bandis' shear stress-permeability relationship [8], except in a small area where the ratio of the shear stress to the normal stress exceeds 0.51. If the fracture is assumed to be well mated, in accordance with Barton's expert opinion (private communication), there is no shear contraction.

3. HYDRAULIC MODELLING

The MOTIF finite-element code developed by AECL [9,10] was used for the hydraulic modelling. The pre-excavation hydraulic model was a two-dimensional (2-D) model comprising planar elements with the same orientation in 3-D space as the natural fracture. The MOTIF formulation for this planar element correctly accounts for the contribution of elevation gradient to the total hydraulic gradient for a fracture of arbitrary orientation. The approximately rectangular mesh representing the fracture was 36 m wide by 40 m high, in accordance with field estimates [1]. It contained 170 planar quadrilateral elements, with element concentrations around the pilot drift and slash regions. It was assumed that the fracture zone was hydraulically equivalent to a single planar fracture with spatially varying aperture, $2b_{esf}$, as inferred by the experimenters [1] from the results of single- and multiple-borehole straddle-packer hydraulic tests conducted prior to tunnel excavation. The intrinsic permeability distribution was calculated from the equation $k_f = (2b_{esf})^2/12$. A porosity of unity was assumed for the fracture, with the rest of the rock assumed impermeable. A transient flow simulation indicated that steady state was reached in a matter of seconds. Accordingly, saturated steady-state flow was assumed in all subsequent runs.

No-flow boundary conditions were imposed at the sides and bottom of the model. The top of the model corresponding to the intersection of a major low-dip fracture zone (Zone 2 splay), was assigned a constant head value of 153 m. This is equivalent to assuming the major fracture zone has "infinite" storage of water compared to the small flow rates in the near-vertical fracture.

After a simulated excavation of the pilot drift and slash through the fracture, piezometric pressures were computed at the intersection of the hydrology holes with the fracture. In addition, seepage rates into the tunnel were computed from the predicted velocity field.

4. COMPARISON OF PREDICTED AND MEASURED GEOMECHANICAL RESPONSES [11]

4.1 Displacements

The predicted displacements were found to agree reasonably well with measured values. As an example, the predicted and measured displacements between anchor points for the Bof-ex extensometer (EXT-07 in Figure 1) are plotted against blast step number in Figure 3. Between the first two anchors (at 0.1- and 0.5-m depth), the predicted displacement (0.02 mm) is one-quarter of the measured displacement (0.08 mm), possibly because of excavation damage. Apart from this case, the predicted and measured responses are all within 0.03 mm, and within a factor of two, of each other. For anchor point pairs at radial distances greater than about 2 m from the tunnel wall, the model over-predicts the displacement. This indicates that the actual rock-mass deformation modulus (E) at the 240-m level is higher than the 40-GPa value used in the predictive model. Back calculation of the in situ E is discussed in Section 4.3 below.

4.2 Stress Changes

The predicted stress changes at the triaxial strain cells were found to agree reasonably well with the measured values. Figure 4 shows the measured and predicted stress changes versus blast step number for N1, the north wall borehole farthest from the tunnel. The stereonet plots confirm the good agreement between measured and predicted orientation, and the predicted stress change magnitudes follow the measured values in trend. It was noted that the absolute value of the measured minimum principal component of stress change, $|\Delta\sigma_3|$, was consistently larger than the predicted value for most of the triaxial strain cells (c.f. Figure 4). A similar trend was also observed during shaft excavation [6]. This may be related to the interpretation of the measured strains or to excavation damage.

4.3 Back Calculation of Rock-Mass Deformation Modulus

A better match between predicted and measured displacements, as illustrated by the solid points in Figure 3, was achieved by allowing the rock-mass deformation modulus (E) to vary as a function of radial distance from the tunnel wall. The back-calculated E (Figure 5) varied from 10 to 35 GPa within the first metre of the tunnel wall, and from 35 to 60 GPa over the next 2.5 m. Further into the rock, E was practically uniform with a value of about 60 GPa, comparable to the value from standard laboratory testing at similar confining stresses. The decreasing modulus near the tunnel is probably due to a combination of excavation damage and the confining stress dependence of the modulus. These results appear reasonable and are consistent with observations at the 240-m level. Apart from the narrow near-vertical fracture zone, there is no other long natural fracture in the rock mass within a few tunnel diameters of Room 209. The rock at the 240-m level, therefore, deforms like intact rock. At shallower depths, the rock mass is fractured and, hence, softer than intact rock. Clearly, well-characterized in situ properties are necessary for successful predictive modelling.

4.4 Fracture Permeability Changes

The model predicted relatively small changes in permeability of the fracture zone due to the stress redistribution induced by excavating the tunnel through the fracture. The permeability was predicted to increase slightly as a result of excavation in some areas of the fracture and to decrease slightly in other areas.

In situ test results conducted before and after tunnel excavation identified significant, permanent decreases in single-fracture hydraulic aperture at boreholes R1, R2, N1 and N2 (see Figure 6 for locations) amounting to 36, 36, 16 and 12%, respectively [6]. By contrast, the model predicted the aperture at R1 was to decrease by only 6%, and the aperture at R2, N1 and N2 to increase by 3, 1 and 6%, respectively [5].

5. COMPARISON OF MEASURED HYDRAULIC RESPONSES WITH PRE-TEST MODEL PREDICTIONS [11]

5.1 Piezometric Pressure Changes

Pressure drops at the intersection of the nine hydrology boreholes with the fracture, measured relative to the pre-excavation pressure in the fracture, were not accurately predicted. For borehole pairs N1/N2 and R1/R2, and far-field borehole OC1 (12.6 m from the north wall), the pressure drops were overpredicted by 6 to 16 times the measured pilot excavation values, and by 4 to 7 times the measured slash excavation values. Pressure drops at borehole pair S1/S2 were less consistently predicted (6 to 9 times higher for pilot, 0.8 to twice for slash). However, the comparison with S1 was questionable because instrumentation monitoring S1 was damaged and hydraulic testing could not be completed [2].

Pressure drops at the floor borehole pair F1/F2 were consistently underpredicted at 70% of the measured values for pilot and slash. Since F1 and F2 lost hydraulic pressure after intersection with the pilot round, a possible explanation for the underprediction is that pressures measured at F1 and F2 were much reduced by blast damage in the tunnel floor, which increases the permeability there [2].

5.2 Groundwater Seepage Into The Tunnel

Measured and estimated seepages into the tunnel following the advance of the pilot drift and slash through the fracture were reported to be 300 mL/min for the pilot drift and 450 mL/min for the slash, respectively [2]. Much higher values were predicted by the pre-excavation hydraulic model, viz., 2077 mL/min after the pilot drift and 3565 mL/min after the slash [11].

6. HYDRAULIC MODEL CALIBRATION [12]

In view of the large discrepancies between predicted and measured hydraulic responses and seepages into the tunnel, the conceptual hydraulic model was revised by a) adjusting mesh geometry to reflect new information collected during tunnel excavation; b) incorporating boundary conditions inferred from a far-field borehole; c) including dense discretizations around boreholes near the tunnel to more accurately simulate hydraulic testing; and d) adjusting the permeability distribution to obtain the best match between predicted and measured head drops and seepages into the tunnel. The top boundary was still assumed to have been fixed head, unaffected by the excavation, but there was a steep hydraulic gradient from NW to SE.

The resulting calibrated model indicated that, after excavation of the pilot drift, a 0.5-m to 1.0-m-thick "skin zone" of altered permeability formed around the tunnel periphery. Where this skin zone intersected the fracture, the permeability was about an order of magnitude lower in the tunnel roof and wall areas, and one to two orders of magnitude higher in the floor than it had been in the pre-excavation state.

Figure 6 compares the predicted head drops based on the calibrated model, to the in situ measured values for the nine hydrology boreholes. The figure corresponds to excavation of the pilot drift through the fracture. The intermediate models represent various stages in the adjustment of the permeability distribution before arriving at the final model. A much improved agreement between values predicted by the final model and measured values is evident, with the predicted-to-measured head drop ratios ranging from 0.28 to 2.7 for the indicated boreholes.

Predicted seepage into the tunnel based on the calibrated model for advance of the pilot drift through the fracture, was 300 mL/min [12]. While this figure compares much more favourably with the measured value, the predicted distribution of seepage around the tunnel does not compare well with the measured/inferred distribution [2]. Measured seepage from the roof amounted to 100 mL/min; estimated seepage from the north wall and floor amounted to 200 mL/min. By contrast, the calibrated model predicted virtually all the seepage entering from the floor, with the north wall and roof contribution underpredicted by 50 times.

7. CONCLUSIONS AND RECOMMENDATIONS

Predicted displacements and stress changes caused by tunnel excavation through the fracture were in reasonable agreement with measured values. The agreement in displacements is significantly improved if the back-calculated rock-mass deformation modulus E, which varies with radial distance from the tunnel wall, is used. For the rock mass more than 2.5 m from the tunnel wall, the back-calculated E was essentially the same as for intact rock. Drill and blast excavation caused softening of the rock in a 2.5 m annulus around the tunnel, with the most severe softening in the first half metre. Fracture permeability changes based on predicted excavation-induced stresses were much less than measured values and often in the wrong direction.

The pre-excavation hydraulic model grossly overpredicted seepage into the tunnel and poorly predicted the pressure drops in the fracture at monitoring boreholes. A revised hydraulic model calibrated to pre- and post-excavation hydraulic responses indicated that following the pilot tunnel excavation, a 0.5- to 1.0-m-thick "skin zone" of altered permeability developed around it. Compared to pre-excavation state, the permeability of the fracture in the skin zone was much reduced in the tunnel roof and walls areas, and was greatly increased in the tunnel floor. These permeability changes (generally an order of magnitude or more) appear too large to be stress-induced, based on predicted stresses and existing empirical stress-permeability relationships. At present, the causal mechanisms for the large permeability changes have not been identified.

For model validation, it is preferable to use mechanical excavation or to introduce pressure or stress perturbations without excavation.

8. ACKNOWLEDGMENT

The authors would like to thank D. Martin, C.C. Davison and Dr. K.W. Dormuth for reviewing the draft and making helpful suggestions. The hydrology experiment was conceived and directed by C.C. Davison.

9. REFERENCES

1. Lang, P.A., Everitt, R.A., Kozak, E.T. and Davison, C.C. (1988). "Underground Research Laboratory Room 209 Instrument Array: Pre-Excavation Information for Modellers", Atomic Energy of Canada Limited, Report, AECL-9566-1.

2. Lang, P.A., Kuzyk, G.W., Babulic, P.J., Bilinsky, D.M., Everitt, R.A., Spinney, M.H., Kozak, E.T. and Davison, C.C. (in preparation), "Room 209 Instrument Array: Measured Response to Excavation, Atomic Energy of Canada Limited Report, AECL-9566-3.

3. Chan, T., Griffiths, P. and Nakka, B. (1989). "Finite Element Modelling of Geomechanical and Hydrogeological Responses to the Room 209 Heading Extension Excavation Response Experiment: I. Pre-Test Modelling", Atomic Energy of Canada Limited Report, AECL-9566-2, Volume 1, 21-328.

4. Hibbitt, Karlson and Sorensen, Inc., (1988) "ABAQUS: Theory, Users and Example Problems Manual", Providence, Rhode Island, U.S.A.

5. Chan, T., Lang, P.A. and Thompson, P.M., "Mechanical Response of Jointed Granite During Shaft Sinking at the Canadian Underground Research Laboratory", Proc. 26th U.S. Symp. Rock Mechanics, June, 1985.

6. Lang, P.A., T. Chan, C.C. Davison, R.A. Everitt, E.T, Kozak and P.M. Thompson (1987), "Near-Field Mechanical and Hydraulic Response of a Granitic Rock Mass to Shaft Excavation", Proc. 28th U.S. Symp. Rock Mechanics, Tucson, A2, 1987 June 29-July 1, pp. 509-516.

7. Gale, J.E. (1987), "Roughness, Contact Area and Pore Structure, Under Given Normal Stress Conditions, of Selected Natural Fractures in the Lac du Bonnet Granite", Preliminary report to AECL, unpublished.

8. Barton, N., and Bakhtar, K. (1987), "Description and Modeling of Rock Joints for the Hydrothermomechanical Design of Nuclear Waste Vaults", Atomic Energy of Canada Limited Technical Record, TR-418, Vols. 1 and 2*.

9. Guvanasen, V. (1984), "Development of a Finite Element Hydrogeological Code and its Applications to Geoscience Research". Proc. 17th Information Meeting of the Nuclear Fuel Waste Management Program, 554-566, Atomic Energy of Canada Limited Technical Record TR-299*.

10. Chan, T., Guvanasen, V. and Reid, J.A.K. (1987), "Numerical Modelling of Coupled Fluid, Heat and Solute Transport in Deformable Fractured Rock". Coupled Processes Associated with Nuclear Waste Repositories, Ed., C-F, Tsang, pp. 605-625, Academic Press, Orlando, Fl.

11. Chan, T., Griffiths, P. and Nakka, B., "Finite Element Modelling of Geomechanical and Hydrogeological Responses to the Room 209 Heading of Extension Excavation Response Experiment: II. Post-excavation Analysis of Experimental Results", 1988. Unpublished interim progress report.

12. Winberg, A., Chan, T., Griffiths, P., Nakka, B., (1989) "Post-Excavation Analysis of a Revised Hydraulic Model of the Room 209 Fracture, URL, Manitoba, Canada", SKB Technical Report 89-27.

---

* Unrestricted, unpublished report, available from SDDO, Atomic Energy of Canada Limited Research Company, Chalk River, Ontario KOJ 1J0

TABLE I

IN SITU STRESSES

	σ_1			σ_2			σ_3		
	MPa	Plunge	Trend	MPa	Plunge	Trend	MPa	Plunge	Trend
CSIRO	26.5	16	231	16.4	40	127	9.1	46	338
USBM	30.0	15	213	17.1	31	113	12.9	54	325

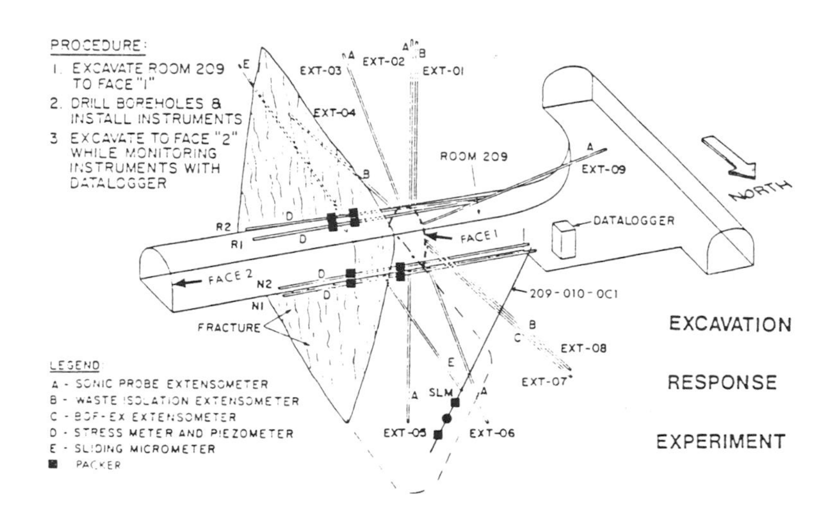

Figure 1. Isometric View of Room 209, the Room 209 Instrument Array and the Tunnel Extension [1].

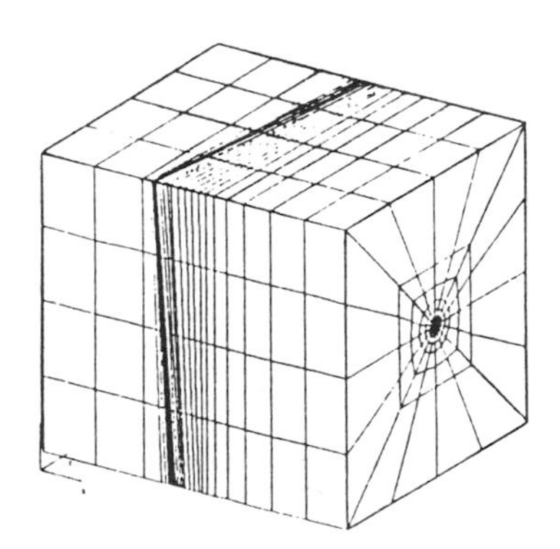

Figure 2. Global Cut Away Perspective View of the Finite-Element Mesh Used in Computing the Permeability Change [3].

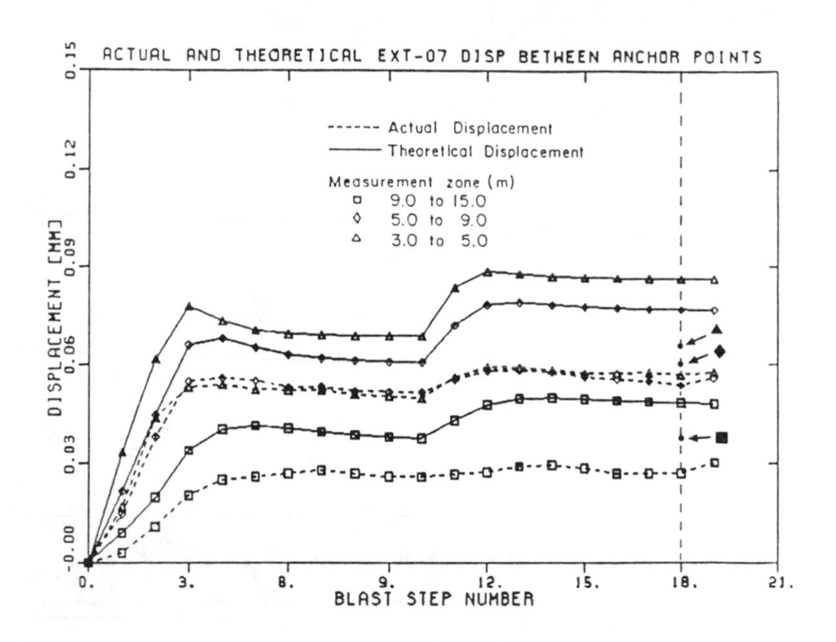

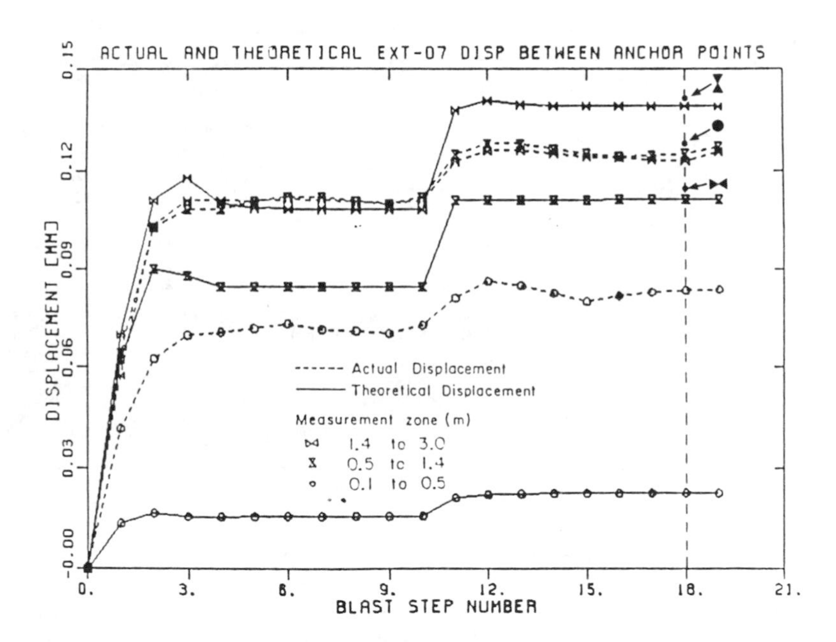

Figure 3. **Predicted and Measured Displacements versus Blast Step Number in Extensometer #7 (the Bof-ex). Blasts 1 to 10 were pilot and 11 to 19 were slash rounds. The plot is of the displacements between pairs of adjacent anchors. The anchor depths in the borehole are shown in the legend. The solid symbols at Blast Step 18 use the back-calculated modulus to compute displacements [11].**

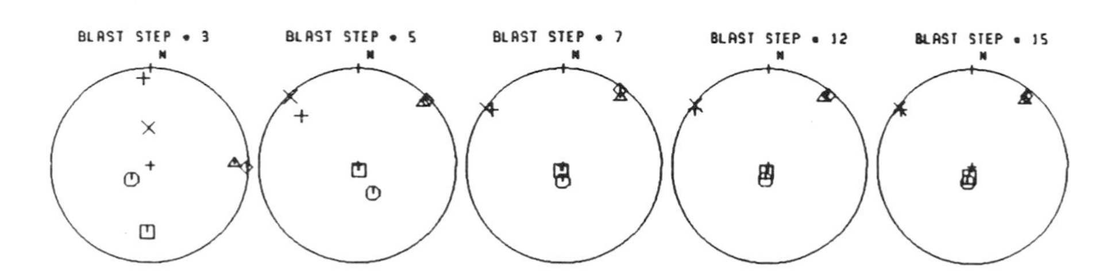

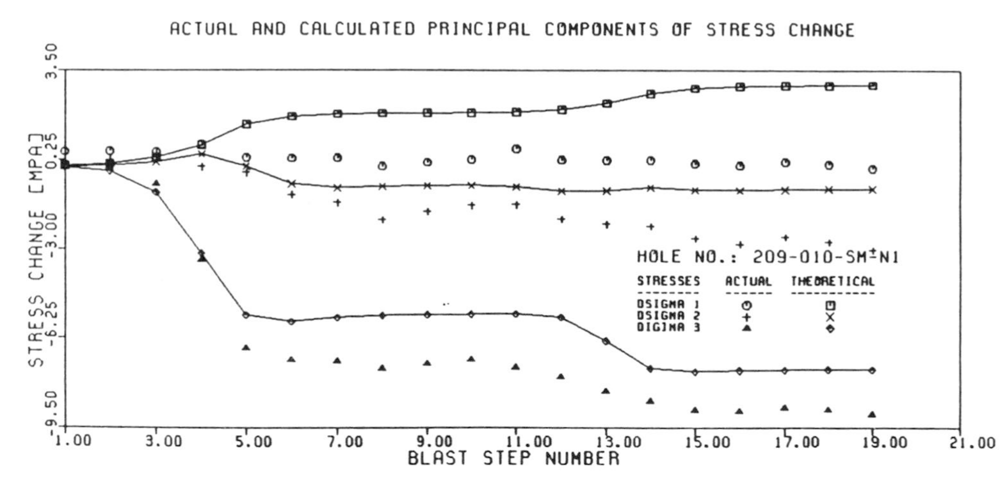

Figure 4. Comparison of Measured and Predicted Principal Components of Stress Change versus Blast Step Number at the Triaxial Strain Cell in Borehole N1 [11]. The stereonets show the orientation of the principal components of stress change and the plot shows the magnitudes of change. Positive values represent increased compression. Blast 1 to 10 were pilot rounds and 11 to 19 were slash rounds.

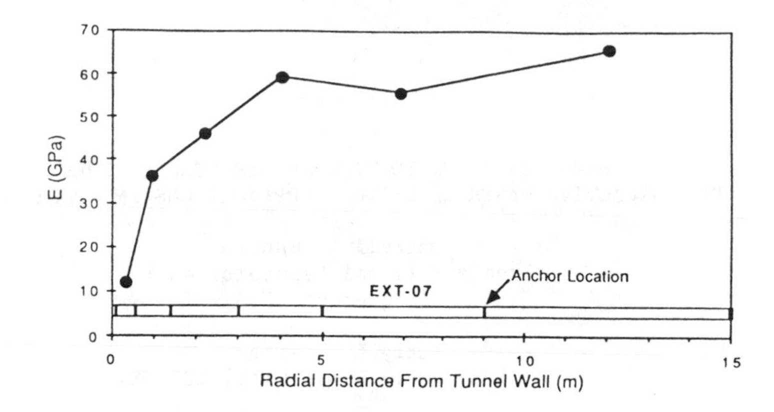

Figure 5. Rock-Mass Deformation Modulus versus Radial Distance from the Tunnel Wall. The values of modulii were back calculated [11] from displacements measured between adjacent anchor pairs in Extensometer #7. The anchor locations are indicated.

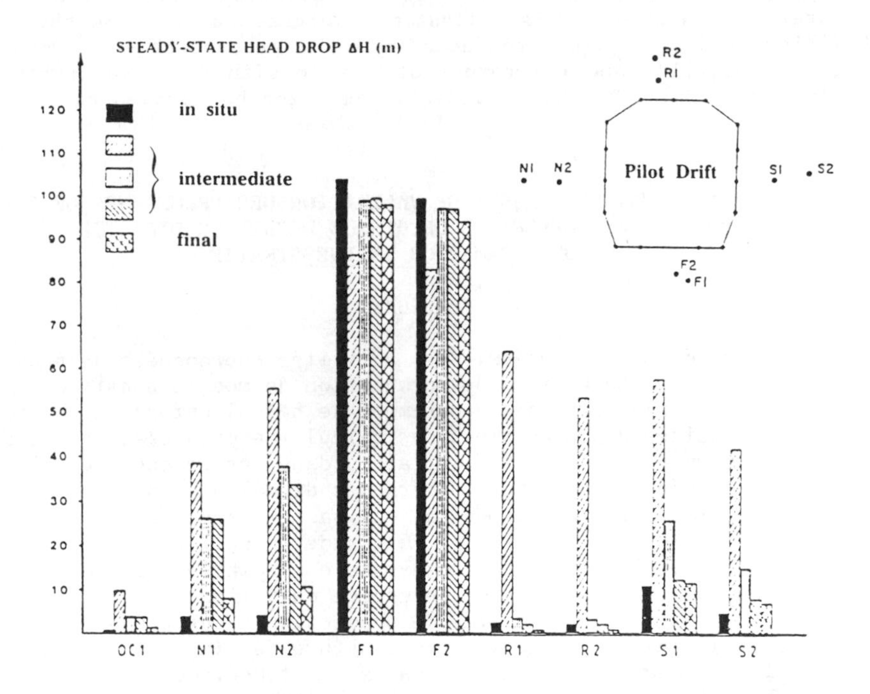

Figure 6. Graphical Representation of Calculated and Measured Head Drop in the Boreholes (see inset) Following the Intercept of the Pilot Tunnel with the Fracture [12]. Far-field borehole OC1 is located 12.6 m from north wall of tunnel.

53

PROGRESS IN VALIDATION OF STRUCTURAL CODES FOR RADIOACTIVE WASTE REPOSITORY APPLICATIONS IN BEDDED SALT[1]

Darrell E. Munson
Sandia National Laboratories[2]
Albuquerque, USA

Kerry L. DeVries
RE/SPEC Inc., Rapid City, SD, USA

ABSTRACT

Over the last nine years, coordinated activities in laboratory database generation, constitutive model formulation, and numerical code capability development have led to an improved ability of thermal/structural codes to predict the creep deformation of underground rooms in bedded salt deposits. In the last year, these codes have been undergoing preliminary validation against an extensive database collected from the large scale underground structural in situ tests at the Waste Isolation Pilot Plant (WIPP) in Southeastern New Mexico. This validation exercise has allowed the prediction capabilities to be evaluated for accuracy. We present here a summary of the predictive capability and the nature of the in situ database involved in the validation exercise. The WIPP validation exercise has proven to be especially productive.

ETAT D'AVANCEMENT DES TAVAUX DE VALIDATION DES PROGRAMMES DE CALCUL POUR L'AMENAGEMENT DE DEPOTS DE DECHETS RADIOACTIFS DANS UNE FORMATION SALINE STRATIFIEE

RESUME

Ces neuf dernières années, des activités coordonnées de production de base de données de laboratoire, de formulation de modèle constitutif et de développement des possibilités des programmes de calcul ont abouti au perfectionnement des capacités des programmes de calcul thermiques/structurels visant à prédire le fluage des chambres souterraines dans les couches de sel de noyage. Au cours de l'année passée, ces programmes de calcul ont fait l'objet d'une validation préliminaire par rapport à une base exhaustive de données rassemblées à partir d'essais souterrains à grande échelle, in situ, réalisés à l'usine pilote d'isolation des déchets (Waste Isolation Pilote Plant - WIPP), au sud-est du Nouveau Mexique. Cet exercice de validation a permis d'évaluer le degré de précision des capacités de prédiction. Nous présentons ci-après un résumé des capacités de prédiction et la nature de la base de données in situ sur lesquelles l'exercice de validation s'est fondé. Cet exercice de validation à la WIPP s'est avéré être tout particulièrement fructueux.

---

(1) Work supported by U.S. Dept. of Energy (DOE), contract DE-AC04-76DP00789.
(2) A U.S. DOE facility.

INTRODUCTION

The purpose of the Waste Isolation Pilot Plant (WIPP) Project is to develop a technology base for underground disposal of radioactive Transuranic (TRU) waste from the U.S. defense projects in a bedded salt deposit. One requirement is to assure that the repository will isolate the waste from the biosphere until it is no longer a threat to man. In salt, isolation depends in part upon the creep of the salt to aid in sealing the repository and encapsulating the waste. As a result, a predictive technology for the structural behavior of salt must be developed to determine response of repository seals and closure time of the rooms. The extremely large forward extrapolation in time of the prediction places demands on the technology that greatly exceed previous engineering practice. In general, overall technology development is the responsibility of Sandia National Laboratories, with the Thermal/Structural Interactions (TSI) program specifically organized to address the structural aspects.

As originally defined, the TSI program approach to a predictive capability was based on a mathematical model of constitutive behavior and laboratory determination of relevant material parameters. The model was to be based on first principles, where possible, or laboratory empirical inputs, as necessary. Results from in situ testing at the WIPP facility would be used as final validation of the predictive technology, not as a database for obtaining backfitted solutions [1]. Separation between the independently developed predictive capability and the means of validation is considered essential to the requirements of a radioactive waste disposal technology.

Technology demonstration implies compliance with the Environmental Protection Agency (EPA) regulations. To demonstrate the repository technology in anticipation of establishing a repository, a pilot plant facility has been constructed in salt, 659 m underground, in southeastern New Mexico. This facility also provides a setting for the large-scale, in situ tests which are supplying part of the databases for technology development and validation.

It is the intent of this paper to show our progress in the development of the structural predictive capability and the results of initial validation comparisons between calculations and in situ test data.

STRUCTURAL RESPONSE PREDICTIVE TECHNOLOGY

To permit orderly calculations for project needs, an initial reference creep law, together with reference material properties and stratigraphy, was established by 1984 [2]. These reference conditions did not preclude further development of more sophisticated creep models or updates in material creep properties and stratigraphic representations. The 1984 reference was based on a thermally activated, steady-state function (with a seldom-used, first-order kinetics, transient representation) compatible with integration schemes of existing numerical codes at that time. Material properties were determined from laboratory creep tests of core specimens obtained from deep boreholes at the proposed WIPP site, and the stratigraphy was constructed from core descriptions from the same deep boreholes at the site. While existing code capabilities were being used for repository and in situ test parametric studies, these codes were also undergoing benchmark exercises against boundary value problems intended to verify their numerical adequacy [3]. However, benchmark verification exercises are insufficient to guarantee code accuracy for the highly nonlinear constitutive models believed representative of the behavior of salt; thus, confidence in prediction technology depends in part upon validation against the extensive in situ database now becoming available.

In the first comparison of calculations using the 1984 reference conditions and actual in situ closure data from the South Drift, Morgan et al. [4] found that the calculations underpredicted the closure and closure rates by a factor of three. Further examination revealed that the known uncertainty in material properties did not permit enough variation to explain the results [5]. A fundamental resolution of this discrepancy was essential to the development of an acceptable predictive capability. The resolution effort has resulted in an extensive reevaluation of the entire predictive technology.

REEVALUATION STUDIES

In a preliminary effort to determine the relative influence of the various aspects of the predictive technology, Munson and Fossum [6] used a multimechanism, steady-state creep model with a workhardening/recovery transient response, and two different flow potentials. These results suggested the importance of the stress generalization (flow potential), the need for a precise description of the transient creep, and the significance of omissions of important strains from the creep generated databases.

Flow potential: The flow potential is the significant factor in extending laboratory creep data obtained in uniaxial stress tests to any generalized, three-dimensional calculation. Various representations of the flow potential, even though they degenerate to the same value in uniaxial stress, can vary greatly for other stress states. This difference in flow potential is exaggerated in calculations of creep closure because the creep rate is a very strong function of the stress. Investigation of the flow potential for salt creep was undertaken using laboratory tests of thin-walled cylindrical specimens. As a result of these tests, it appears that a maximum shear stress or Tresca criterion better describes the flow surface [1], in contrast to the assumption of a von Mises criterion used previously [2].

Constitutive model: Initially, a common assumption was that for long-term creep closure of the rooms, only steady-state creep behavior needed to be considered. However, this was questioned on the basis that strain, not time, governs achievement of steady state. As the stress field around the rooms expands with time (for the WIPP configuration), large volumes of previously unstressed salt progressively come under stress. As a result, there are always large volumes of previously unstrained salt undergoing the pronounced transient strains exhibited by salt. These transient strains in the new material encompassed by the expanding stress field contribute significantly to the integrated room closure displacements, even though the stresses producing the strains may be small. Thus, inclusion of the proper description of the small strains of transient creep into the constitutive model is essential.

The constitutive model currently used by the Project is based on the deformation-mechanism map specific to the expected WIPP conditions. As most recently modified [1], this constitutive model (modified M-D model) consists of multimechanism, steady-state creep with a workhardening/recovery transient creep response, treats recovery (stress unloading), and includes a quadratic approximation to the small strain portion of the transient response. The model includes three distinct mechanisms: (1) a dislocation climb mechanism at high temperatures, (2) an undefined (but empirically well specified) mechanism at low stresses and temperatures, and (3) a dislocation slip mechanism at high stresses, all acting in parallel. The modified M-D constitutive model is used in all of the preliminary validation calculations presented here.

In addition to the salt layers, bedded natural evaporite deposits also contain substantial layers of anhydrite. As initially defined by Krieg [2], the time-independent deformation of the anhydrite is described through a Drucker-Prager yield function. This description has been retained for the preliminary validation calculations.

The shear response of clay seams commonly found in bedded evaporite deposits is accommodated through incorporation of slip lines in the numerical codes. Shear along the slip lines is governed by Coulomb friction [2].

Material parameters: The emphasis on small strain response resulted in reevaluation of the methods for determining material parameters. The reevaluation centered on including previously ignored "lost" strains into the parameter determination. In the past, parameters were typically determined from laboratory creep tests, where time and strain are rezeroed when the loading to the stated stress level is complete. Consequently, creep databases exclude the significant, although small, loading strain associated with creep testing. In addition, creep specimens are obtained from either deep boreholes or from around underground excavations into the "virgin" salt beds. Under these conditions, specimens can accumulate excavation strains prior to testing which are not recaptured by the laboratory testing. In order for the material parameters to correctly reflect the excavation of the WIPP facility from virgin salt beds, these loading and excavation strains must be included in the database from which the parameters are determined [1,6].

Based on the reevaluation, two major bedded salt layer types are evident, clean salt and argillaceous salt, for which new material parameters are determined. For the anhydrite layers, the Drucker-Prager constants are determined from laboratory quasi-static compression tests. The coefficient of friction of clay seams cannot be determined in the laboratory and reasonable values must be assumed. This process leads to the reevaluated parameter set used in the calculations, as given elsewhere [1]. At this time, only the coefficient of friction parameter of the required parameters cannot be defined through laboratory tests, and remains a free parameter.

Stratigraphy: Although too complex to describe here, the reevaluation of the stratigraphy based on underground observations indicates significantly more beds consist of argillaceous salt than the 1984 reference stratigraphy [2] indicated. The current stratigraphy [1] reflects this difference.

IN SITU TESTS

A number of test rooms and room complexes have been excavated in the underground facility, as shown in Figure 1. Available structural validation tests are summarized in Table 1. The structural response of these tests is being measured to create the in situ test database required for the validation exercise. The intent is to measure the salt displacements around the test rooms in exceptional detail from the first moment of excavation. However, measurements in some of the early excavations were outside the research effort and did not meet this intent. In these cases, it is necessary to reconstruct the early, lost displacements. The test rooms or room complexes represent a number of different geometries, including a circular cross section room, with individual rooms situated in three somewhat different stratigraphic settings. Also, two of the four shafts have been well instrumented at as many as five discrete horizons in the salt.

Although data continue to be gathered from these tests, the database is already very extensive, with sufficient data to permit meaningful comparison to preliminary validation calculations. At this time, we have completed calculations of the isolated, single test rooms or drifts, which we report here. The complex multiple room configurations which represent somewhat more difficult calculations are now under study, so these calculations will be forthcoming.

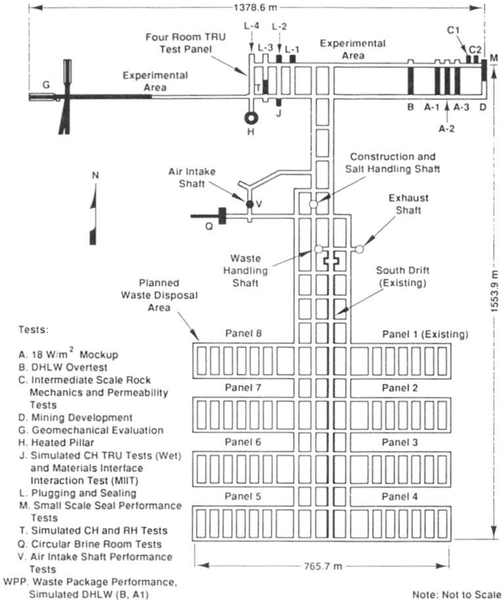

Figure 1 Plan View of the WIPP Underground Facility

PRELIMINARY VALIDATION CALCULATIONS

Although the South Drift represents the first underground excavation to be compared to calculation with the previously noted three-fold prediction discrepancy [4], it was not the first of the current preliminary validation calculations using the M-D model and reevaluated parameter set. Preliminary validation efforts initially focused on Room D, where closure measurements were taken at gage stations within an hour of excavation of the virgin salt. The attention to obtaining very early displacements for the in situ validation database is the equivalent of incorporating all specimen strains in the database for parameter determination, which assures both databases correspond to "virgin" conditions. The Room D calculation as reported by Munson et al. [1] used the reevaluated parameter set and stratigraphy (Set A), except for a few minor differences. This calculation used a 40% larger value of transient strain limit for argillaceous salt and substituted argillaceous salt layers for anhydrite layers (Set D). In this calculation, as given in Figure 2, agreement between calculated and measured vertical and horizontal closures is within 2% and 15%, respectively, at 1000 days. The results strongly suggested that the earlier prediction discrepancy has been resolved through use of a better flow potential, a more precise transient creep characterization, the reevaluated material parameters, and an updated stratigraphy. Although the agreement would be slightly modified through the use of the Set A rather than the Set D material parameters and stratigraphy, this would not actually alter the adequacy of the agreement to any significant degree.

Room B is identical to Room D in setting and configuration and gave the same measured response as Room D until onset of Room B heating on day 354, which resulted in a marked increase in observed closure rates. Reevaluated material parameters and updated stratigraphy, except for retaining the 40%

Table 1. Summary of WIPP Structural Test Rooms, Room Complexes, and Shafts

Room(s)	Stratigraphy	Width (m)	Height (m)	Length (m)	Pillar (m)	Remarks
A	salt/arg. salt	5.5	5.5	93.3	18.0	Heated 3 room complex
B	salt/arg. salt	5.5	5.5	93.3	-	Heated overtest
D	salt/arg. salt	5.5	5.5	93.3	-	
G	arg. salt/anhyd.	6.1	3.05	183.9	-	Long room
H	arg. salt/anhyd.	11.0	3.05	round	11.0	Heated round pillar
TRU Pn.	arg. salt/anhyd.	10.1	3.96	91.4	30.5	TRU 4 room test panel
So.Drft	arg. salt/anhyd.	7.6	2.44	1117.0	-	Long south drift
1st Pn.	arg. salt/anhyd.	10.1	3.96	91.4	30.5	First 7 room panel
Q	arg. salt	2.44	diameter	106.8	-	Circular room
CSH Sft	various	3.66	diameter	659.0	-	Construction shaft
AIS Sft	various	6.17	diameter	659.0	-	Air intake shaft

larger argillaceous salt transient creep limit parameter (Set B), was used in the Room B preliminary validation calculation, as given by Munson et al. [7]. This calculation, also shown in Figure 2, was complicated by the heating of the room and provision was necessary to account for heat loss by conduction and infiltration through the doors at the end of the test room. The thermal calculation, while not shown here, was brought into very good agreement by extracting heat along the room periphery. Until day 354, when heating began, the calculated results differ from those for Room D because of the use of different stratigraphies. The heating in the calculation is reflected in an abrupt increase in both vertical and horizontal closure rates. The agreement between measured and calculated vertical and horizontal closures is within 4% and 18%, respectively, at day 1045. We suspect much of the discrepancy in the vertical results is caused by the onset of separation and failures in the roof. Small roof cracks appeared as early as day 650, with definite formation of a thin roof slab starting on day 1045. The continuum creep constitutive model does not treat the localized separation and fracture processes.

Room G is a very long isolated room intended to be "two-dimensional" to exactly match the requirements of two-dimensional numerical solutions. Figure 3 compares the calculated and measured closure results using the Set A reevaluated material parameters and stratigraphy. Although the agreement appears to be very good, the order of the vertical and horizontal closure magnitudes is reversed between calculation and measurement. This result suggests that minor discrepancies can still occur, even though the major prediction discrepancy appears to be resolved. While these minor prediction discrepancies are thought to be well within acceptable uncertainty limits, it may eventually be necessary to resolve some of these minor discrepancies.

A calculation of the South Drift was made using reevaluated parameters and stratigraphy (Set A). Delay (up to 30 days) in gage installation was corrected by reconstructing the missing displacements from the calculated displacements. The "average" of the measured results for ten measurement stations (locations) along the 1117-m-long drift are represented by the measured closures at station 2950, as given in Figure 4. The data from individual stations form a series of essentially parallel curves, with an apparent scatter of ± 10 mm in displacement magnitude at day 1400 caused by an uncertainty of ± 0.5 day in the time between drift excavation and first measurement. Regardless of the measurement uncertainty, the agreement between

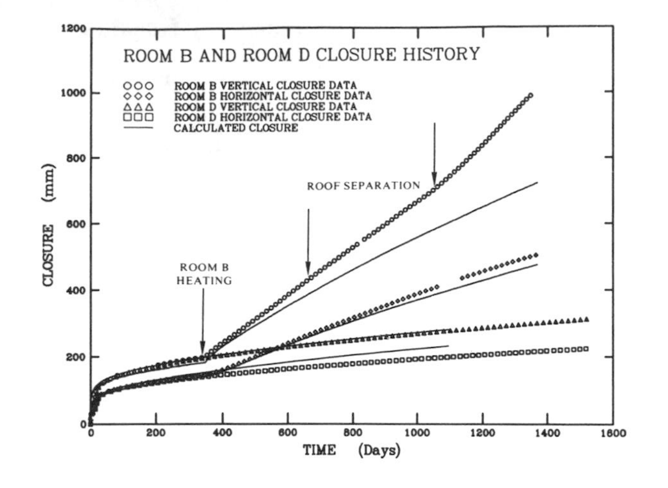

Figure 2 Measured and Calculated Room D and Room B Closures Parameter Set D and Set B, Respectively [after [7])

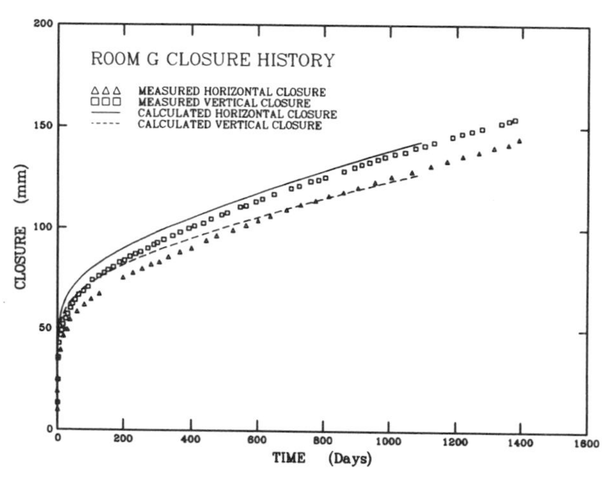

Figure 3 Measured and Calculated Room G Closures, Parameter Set A

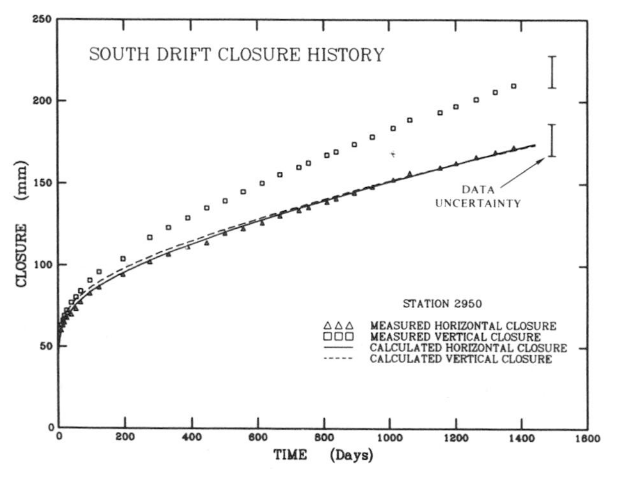

Figure 4 Measured and Calculated South Drift Closures, Parameter Set A

calculation and measurement, as shown in Figure 4, is very good, especially for the horizontal closures where it is well within the known uncertainty of the measurement. The South Drift exhibits the underprediction of vertical closure seen in the Room D, Room B, and (perhaps) Room G comparisons. Because slab separations are observed in both roof and floor of the South Drift, there is a strong possibility that the common discrepancy between calculated and vertical closure is the result of this unaccounted for progressive separation, as is clearly the case for Room B where the separations are quite pronounced.

CONCLUDING REMARKS

Although considerable work remains to be done to demonstrate code validation, preliminary calculations for three isolated rooms of different dimensions and stratigraphic setting, one of which is heated, are quite encouraging. These results suggest the use of a Tresca flow potential, a transient as well as steady-state creep response, the reevaluated material parameters, and an updated stratigraphy is warranted. It remains to be seen if closures of the remaining test rooms and shafts can be successfully calculated, and further, if slab separation effects must be included.

REFERENCES

[1] Munson, D.E., A.F. Fossum, and P.E. Senseny : "Approach to First Principles Model Prediction of Measured WIPP In Situ Room Closure in Salt", Proc. 30th U.S. Symp. on Rock Mechanics, A.A. Balkema, Brookfield, MA, 1989, 673-680.

[2] Krieg, R.D. : Reference Stratigraphy and Rock Properties for the Waste Isolation Pilot Plant (WIPP) Project, SAND83-1908, Sandia National Laboratories, Albuquerque, NM, 1984.

[3] Morgan, H.S., R.D. Krieg, and R.V. Matalucci : Comparative Analysis of Nine Structural Codes used in the Second Benchmark Problem, SAND85-0324, Sandia National Laboratories, Albuquerque, NM, 1981.

[4] Morgan, H.S., C.M. Stone, and R.D. Krieg : "The use of Field Data to Evaluate and Improve Drift Response Models for the Waste Isolation Pilot Plant (WIPP)", Proc. 26th U.S. Symp. on Rock Mechanics, A.A. Balkema, Boston, MA, 1985, 769-776.

[5] Morgan, H.S., C.M. Stone, and R.D. Krieg : Evaluation of WIPP Structural Modeling Capabilities Based on Comparisons with South Drift Data, SAND85-0323, Sandia National Laboratories, Albuquerque, NM, 1986.

[6] Munson, D.E., and A.F. Fossum : "Comparison between Predicted and Measured South Drift Closures at the WIPP using a Transient Creep Model for Salt", Proc. 27th U.S. Symp. on Rock Mechanics, Soc. of Mining Engineers, Littleton, CO, 1986, 931-939.

[7] Munson, D.E., K.L. DeVries, and G.D. Callahan : "Comparison of Calculations and In Situ Results for a Large, Heated Test Room at the Waste Isolation Pilot Plant (WIPP)", Proc. 31st U.S. Symp. on Rock Mechanics, A.A. Balkema, Brookfield, MA, 1990 (in press).

MODELLING THE EFFECT OF GLACIATION, ICER FLOW AND DEGLACIATION ON LARGE FAULTED ROCK MASSES

P. Chryssanthakis, N. Barton
Norwegian Geotechnical Institute, Oslo (Norway)
J. Lanru, O. Stephansson
Luleå University of Technology, Luleå (Sweden)

ABSTRACT

Two modelling techniques have been applied in this study of rock mass response to glaciation, deglaciation. The first approach (HNFEMP code) considers the properties of the intact rock and the faults without including the faults as separate entities. The second approach models the faulted rock masses by considering the discontinuities and the intact rock separately (MUDEC code). MUDEC has been applied in two versions with linear fault stiffness properties and non-linear stiffness according to the Barton-Bandis joint model. A two dimensional 4 x 4 km rock mass with three different geometries has been modelled. An initial ice mass of 3 km height x 4 km simulated the glaciation procedure. The ice load was reduced in steps to simulate the deglaciation. The results from the three analyses showed no dramatic differences and gave the same approximate values for the glacial rebound of the ground surface after deglaciation (4 m). The geometry with two intersecting joints sets (Model 3) was given emphasis since it represents the Lansjärv region. MUDEC modelling with the Barton-Bandis joint model permitted also the determination of mechanical and conductive fault or joint apertures.

MODELISATION DE L'EFFET DES GLACIATIONS, DE L'ECOULEMENT DES GLACES ET DE LA FONTE DES GLACES SUR DES MASSIFS ROCHEUX FAILLES DE GRANDE DIMENSION

RESUME

Deux techniques de modélisation ont été utilisées dans cette étude de la réaction des massifs rocheux à la glaciation et à la fonte des glaces. la première méthode (programme de calcul HNFEMP) prend en compte les propriétés de la roche intacte et des failles, ces dernières n'étant pas considérées comme des éléments distincts. Dans la seconde méthode de modélisation les discontinuités et la roche intacte sont traitées séparément (programme MUDEC). On a utilisé deux versions du programme MUDEC, l'une en retenant des caractéristiques de rigidité linéaire pour les failles, l'autre en retenant des caractéristiques de rigidité non linéaire suivant le modèle de joint Barton-Bandis. On a modélisé un massif rocheux bidimensionnel (4 km x 4 km), avec trois formes géométriques différentes. Pour simuler le phénomène de glaciation on est parti d'une masse de glace initiale de 3 km d'épaisseur et de 4 km de longueur. Pour simuler la fonte on a réduit par étapes la masse de glace. Les résultats de ces trois analyses n'ont pas fait ressortir de différences importantes et ont donné des valeurs approximativement égales concernant la remontée isostatique du sol à la suite de la fonte des glaces (4 m). Le modèle simulant l'intersection de deux séries de joints (modèle 3) a reçu une attention particulière dans la mesure où il représente la région de Lansjärv. Utilisé avec le modèle Barton-Bandis le programme MUDEC a aussi permis de définir les caractéristiques des ouvertures mécaniques et conductrices associées aux failles ou aux joints.

1. INTRODUCTION

The effect of glaciation and deglaciation has been studied in two-dimensional generic models of faulted rock masses. Six loading cases were modelled two-dimensionally for a 4 x 4 km rock mass, Fig. 1. After consolidation and application of boundary stresses, where the models were loaded by horizontal stresses from one side and both sides respectively (case 1), an ice sheet, three kilometres thick, was applied at the surface of the model (case 2). A special interface was introduced at the ice-rock boundary. Ice flow was then simulated either by a velocity applied at the vertical boundary or by shear stresses applied at the top surface of the rock mass (case 3). For simulation of deglaciation the ice thickness was reduced in two steps. First 2/3 of the ice sheet was removed and the response in stress and displacement was studied (case 4). Later all ice was removed from one half of the model and a triangular ice load was left (case 5). Finally, all ice was removed to represent today's situation (case 6). In modelling the effect of glaciation and deglaciation the displacements at the bottom of the model were restricted to be vertical. In nature the ice load will cause subsidence and rebound due to the viscoelastic behaviour of the earth's mantle. These effects are not considered in this study.

Three different fault set geometries were analysed, Fig. 2. Models 1 and 2 consist of a rock mass with one set of idealized faults having a spacing of 500 m and oriented horizontally and vertically, respectively. Model 3 has two sets of faults with a spacing of 200 m and 500 m, respectively. The fault geometry of Model 3 reflects the situation at Lansjärv.

Three different codes where applied for analysing the fault set geometries of Fig. 2.

a) HNFEMP code on all models.
b) MUDEC with linear fault stiffness on all models.
c) MUDEC with the Barton-Bandis non-linear joint model. (Model 3 only).

Emphasis was given to the modelling with MUDEC using the Barton-Bandis joint model (model 3) which reflects the situation at Lansjärv region. Details on the appliction of this code can be found in Barton and Chryssanthakis (1988) and Chryssanthakis (1989). Modelling with the MUDEC and Barton-Bandis joint model permits determination of mechanical and conductive (hydraulic) fault or joint apertures. These two joint properties are very important parameters in deciding the final location of a repository for radioactive waste.

2.1 Modelling with HNFEMP

Each of the Models 1-3 is subjected to an initial horizontal stress

$$\sigma_H = 5 + 32 \cdot z$$

acting at one of the vertical boundaries and a vertical stress $\sigma_V = 27 \cdot z$, where z is the depth in kilometres and the stresses are in MPa. The finite element mesh contains 289 nodes and 256 elements and plane strain conditions are assumed. Two types of material properties were considered, namely, a soft fault stiffness and a fault system 60 times stiffer, see Table I.

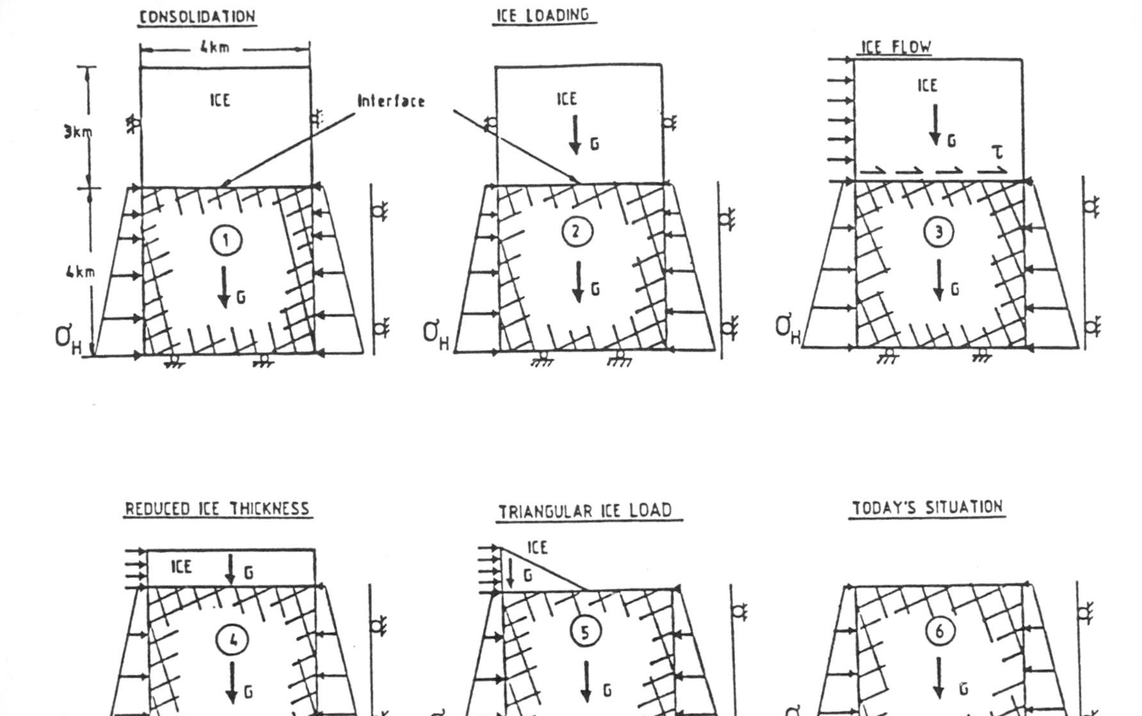

Fig. 1 Generic models for glaciation, glacial flow and deglaciation. Loading cases, boundary conditions and geometry. Two different boundary conditions were used. During consolidation, loading case 1, no ice load was applied.

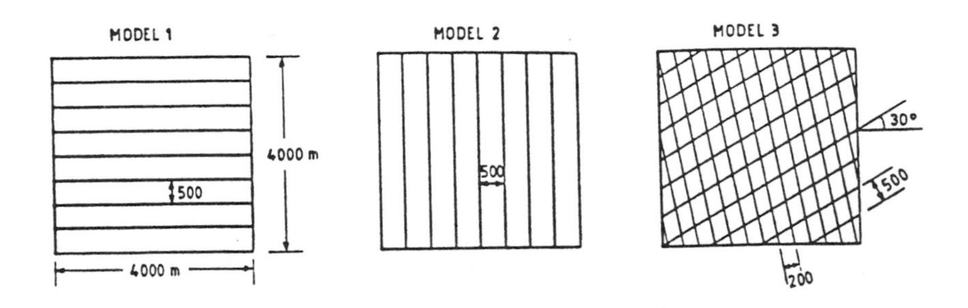

Fig. 2 Fault set geometries in generic modelling.

Table I Material properties for HNFEMP-modelling of glaciation and deglaciation.

Material	Parameter	Soft Faults	Stiff Faults
Intact rock	Density, kg/m^3	2700	2700
	Young's modulus, GPa	40	40
	Poisson's ratio	0.2	0.2
Fault	Normal stiffness, MPa/m	10.0	600.0
	Shear stiffness, MPa/m	1.0	60.0
	Cohesion, MPa	0.1	0.1
	Dilation angle, degree	5.0	5.0
	Tensile strength, MPa	0.0	0.0
	Friction angle, degree	35.0	35.0
	JRC	6	6
	JCS, MPa	50.0	50.0
	Φ_r, degree	25.0	25.0
Ice	Density, kg/m^3	900	900
	Shear stress, MPa	0.1	0.1

The modelling approach with HNFEMP uses a material constitutive model that accounts for the properties of the intact rock and the joints without including the joints as separate entities. This modelling technique results in continuous displacements and stresses throughout the model.

Models with a soft fault stiffness gave large displacements. The total vertical displacement is governed by the thickness of the model. The results obtained demonstrate the relative importance of fault stiffness and fault geometry. Ice loading and removal cause changes in the stresses as shown in Fig. 3 for a point located 500 m below the ground surface. The stress changes are $\Delta\sigma_x \sim 3$ MPa and $\Delta\sigma_y \sim 27$ MPa for Model 3 with two sets of intersecting faults.

2.2 Modelling with MUDEC, linear fault stiffness

The three generic models, Models 1-3 with different fault set geometries, were analysed with the MUDEC linear fault stiffness code (Jing and Stephansson, 1988). The 4 x 4 kilometre rock mass was treated as an assembly of elastic rock blocks separated by discontinuities with linear stiffnesses and Mohr-Coulomb failure properties. Two different joint stiffnesses were used, the ice was simulated as an elastic medium and a distributed load at the ground surface, respectively. The interface between the ice and the rock mass was simulated by means of a soft, low-strength and low-friction discontinuity. Two different tectonic, horizontal stresses were simulated and a total of seven models were analysed and presented by Jing and Stephansson (op. cit.). Results for Model 3 with identical geometry, material properties and boundary stresses to those presented in section 2.1 are shown in Figs. 4 and 5. Horizontal and vertical stresses versus loading cases for points along the central line and various depths below the ground surface are shown in Fig. 4. Since modelling is conducted in time steps, the ficticious time is also shown. The flattening of the stress curves for each loading case and depth indicates stable conditions. At a depth of 400 m, changes of horizontal stress during ice loading and unloading are of the order of $\Delta\sigma_x \approx 10$ MPa. The stress magnitudes are less

obtained from the HNFEMP-modelling, cf. Fig. 3. Slip along inclined faults causes the stress release. The change in vertical stress reflects the loading and removal of the ice sheet and the results obtained from HNFEMP and MUDEC linear modelling are in close agreement.

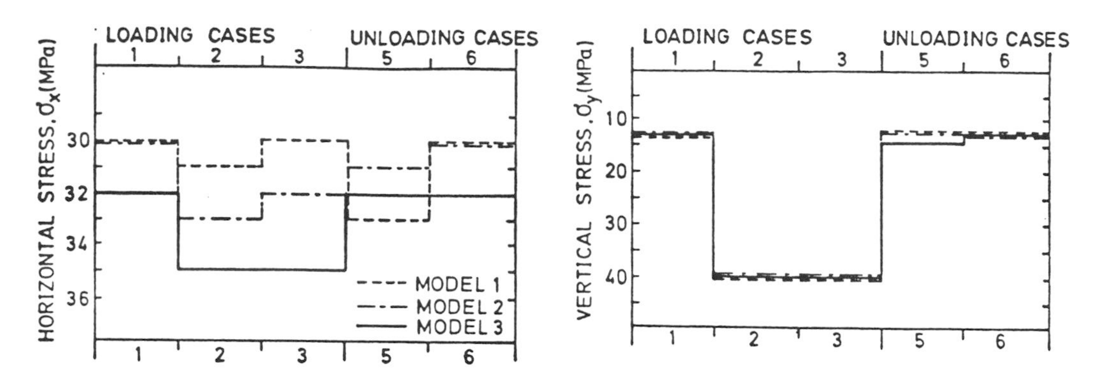

Fig. 3 Results for HNFEMP-modelling of horizontal and vertical stress at 500 m depth i the centre of Models 1-3. Stresses are similar for soft and stiff faults. Loading and unloading cases according to Fig. 1

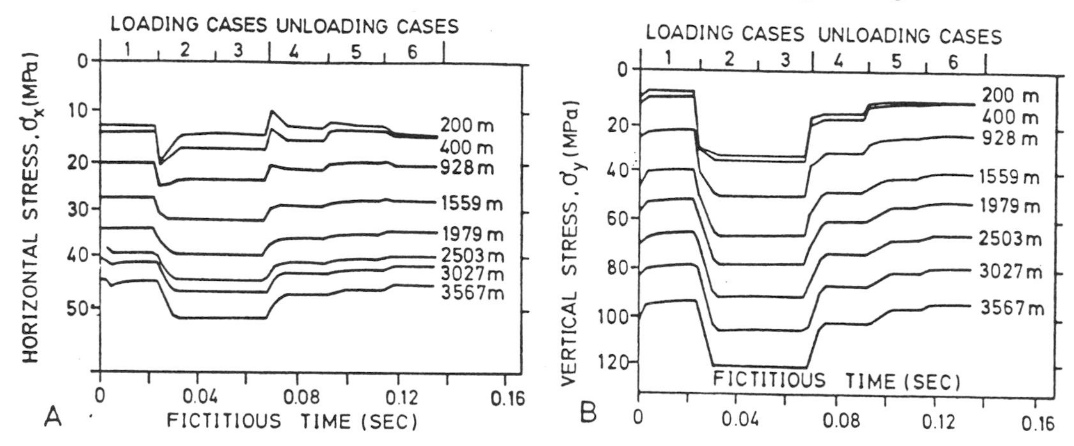

Fig. 4 Results from MUDEC linear modelling of Model 3 with linear fault stiffness. Normal stresses versus loading cases and ficticious computational time is shown for points at various depth. A, Horizontal stress. B, Vertical stress.

The individual displacement of the uppermost blocks of the model at the ground surface is shown in Fig. 6, where the relative displacement between individual blocks are of the order of 5 to 10 cm. A complete deglaciation of the model gives a rebound of 3 to 4 m as indicated by the uppermost line for the blocks. The total rebound of the ground surface is represented by the uppermost solid line. Relative displacement between individual blocks and rebound are slightly less but of the same magnitude at 500 m depth.

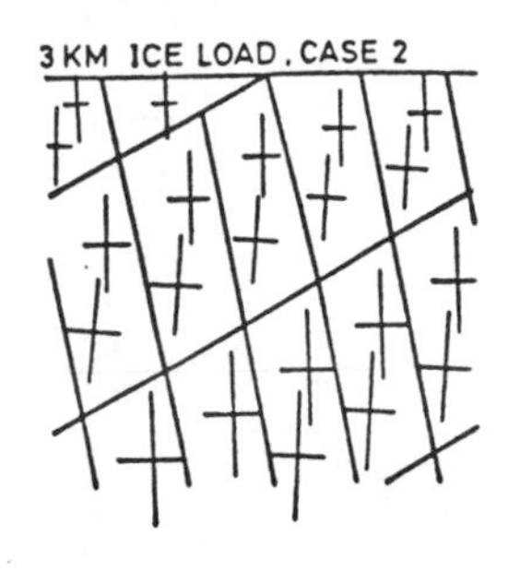

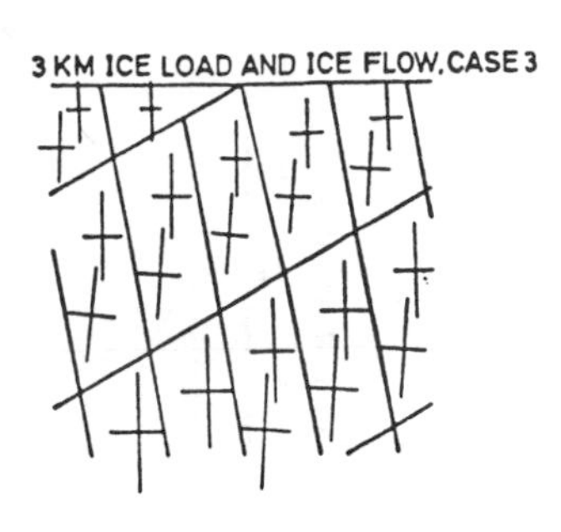

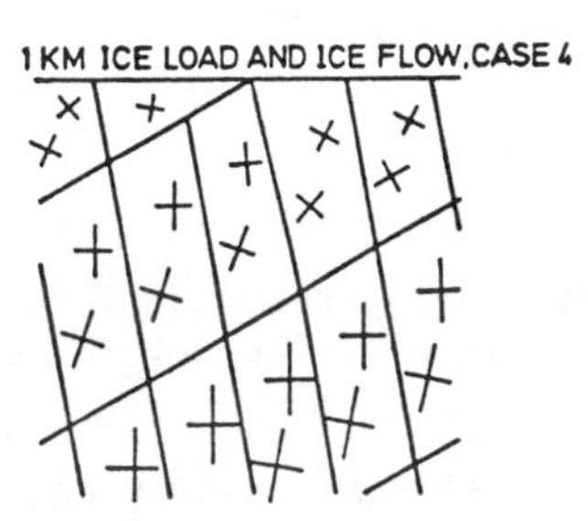

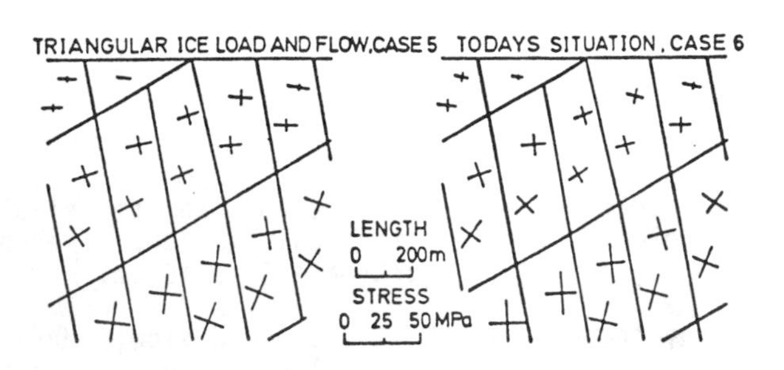

Fig. 5 MUDEC linear modelling of generic Model 3 with stiff faults. Changes in magnitude and orientation of principal stresses for the uppermost 1000 m of the rock mass.

The results obtained from the HNFEMP and MUDEC modelling are in close agreement, see Fig. 3. The effect of a glacial sheet on the remaining stress state will be almost negligible when the process of ice retreat is completed. However, the stresses do tend to rotate and become aligned with the boundary of the rock blocks after deglaciation, as indicated for case 6 in today's situation, Fig. 5. At greater depth, the reorientation is less pronounced.

2.3 Modelling with the MUDEC non-linear Barton-Bandis joint model

Three different generic models of large rock masses with a fault geometry as for Model 3 (see Fig. 2) were analysed with the non-linear MUDEC code and the Barton-Bandis joint model, Chryssanthakis (1989). Results from two of the generic models - called Model 3A and Model 3B - are presented. The geometry, boundary conditions and unloading cases are the same as for modelling with linear MUDEC, except that effective tectonic stresses were applied to Model 3A. By the term effective tectonic stresses we mean absolute stresses minus pore water pressure. Model 3A has fault stiffnesses that are two to four orders of magnitude stiffer than the faults for Model 3B. The material properties are presented in Table II.

The large problem size and the difference between the joint stiffnesses and the surrounding rock mass in Model 3B, resulted in a very long consolidation proce-

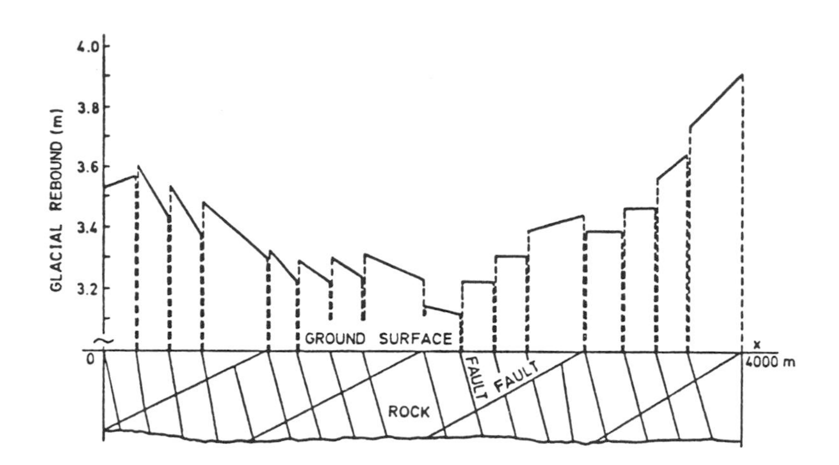

Fig. 6 MUDEC-modelling of Model 3 with stiff faults showing rebound of the ground surface after deglaciation. The relative displacement between adjacent rock blocks are 5 to 10 cm.

dure (loading case 1) which proceeded slowly during successive loading and unloading stages. The vertical displacements of the ground surface after ice loading were found to be 3.78 m and 3.90 m for Models 3A and 3B, respectively. This is in close agreement with the results obtained from the HNFEMP and linear MUDEC modelling, see Fig. 6. The application of effective tectonic stresses to the model make no major changes in magnitudes of stresses and displacements.

An important difference in the behaviour of model 3A and 3B can be seen in the last stage (case 6) today's situation where all ice has been removed from the rock mass. Model 3A exhibits an upwards movement of 0.77 m indicating a moderate hysteresis effect with no overshooting. On the contrary, model 3B (soft faults) shows no hysteresis and as much as 3.8 m of overshooting. The overshooting effect of model 3B can be understood easier by considering the forward shear and reversal shear behaviour of a rock joint. In general when shearing a joint in the reverse direction it exhibits lower dilation and shear strength values, due to damage in roughness (JRC reduction). The Barton-Bandis joint model integrated in MUDEC contains a shear reversal "loop" similar in principle to the one reported by Celestino and Goodman (1979) for a plaster replica of a joint in sandstone.

The principal stress plots for Models 3A and 3B analysed indicate no dramatic changes in terms of stress rotation under the various loading and unloading cases. Changes in horizontal and vertical stresses at 500 m depth below ground surface are in close agreement with other modelling results except for the low stresses to deglaciation in Model 3B, cf. Fig. 7.

The computational mesh used in the central and uppermost 100 m of Model 3B was made finer to study in detail the change in displacements and stresses in the vicinity of a repository. Figure 8A shows a slight re-orientation of the principal stresses due to ice loading and ice flow from left to right for Model 3B.

Table II Rock material and joint properties used for Model 3A and Model 3B. After Chryssanthakis (1989).

Properties	Intact rock	Model 3A †		Model 3B	
		Subvertical	Subhorizontal	Subvertical Fault Set	Subhorizontal Fault Set
Young's Mod. (GPA)	39.9*				
Poisson's ratio	0.19*				
Density (kg/m^3)	2700				
Bulk Mod. (MPa)	2.1EA*				
Shear Mod. (MPa)	8.2e3*				
JRC (Joint roughness coefficient)		5	10	5	10
JKN (MPa/m) Joint normal stiffness limit		2.4E8	1E9	2.4E4	1E5
JKS (MPa/m) Joint shear stiffness limit		2.9E4	7.8E4	2.0E2	6.0E2
KN (MPa/m) Point contact normal stiffness		2.4E7	1E8	2.4E3	1E4
KS (MPa/m) Point contact shear stiffness		2.9E3	7.8E3	2.0E1	6.0E2
APER (mm) Zero load aperture		0.150	0.200	0.150	0.200

NOTE:

* equivalent properties
† effective tectonic stresses applied

JCS (joint wall compressive strength) is 150 MPa for all joint sets.
ϕ (residual friction angle) is 300 for all joint sets.
L_0 (laboratory scale joint length) is 0.1 m for all joint sets.
σ_c (uniaxial compressive strength of intact rock) is 220 MPa for all joint sets.

Modelling with the MUDEC and Barton-Bandis joint models permits determination of mechanical and hydraulic joint/fault apertures. The hydraulic/conductive aperture shows direct similarities with the mechanical aperture since it is derived from the empirical formula $e = E^2/JRC^{2.5}$ where

e = conducting aperture
E = mechanical aperture
JRC = joint roughness coefficient

The mechanical apertures during the different loading cases are relatively stable for Model 3A, which exhibits a small variation in apertures and closures down to a depth of about 200-300 metres. After this depth the joints of model 3A (properties from table II) cannot close any further, they have reached maximum closure (100 µm for the sub-horizontal and 40 µm for the sub-vertical joint

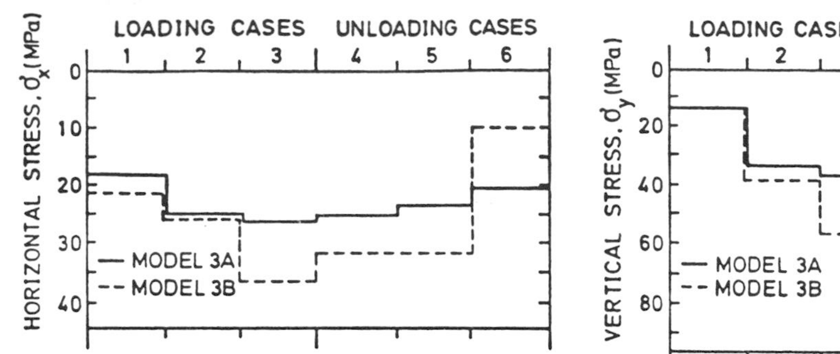

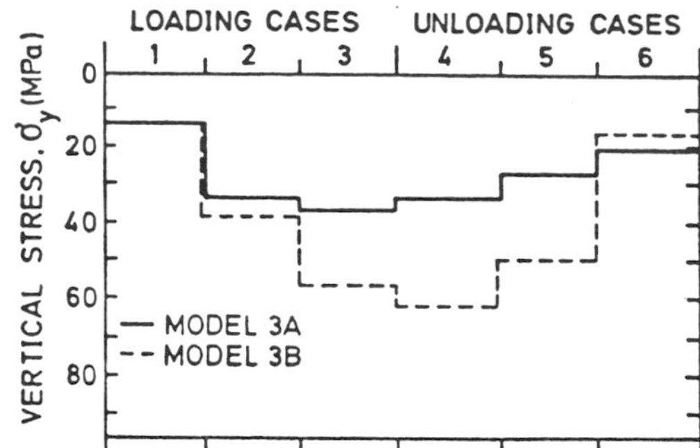

Fig. 7 Horizontal and vertical stresses at 500 m depth versus loading and unloading cases for Models 3A and 3B, MUDEC with Barton-Bandis joint model. Based on data taken from Chryssanthakis (1989).

set). In the last loading case Model 3A exhibits a slight opening of 40 µm of the sub-vertical joint set down to a depth of 800 metres. The behaviour of the "soft" faults of Model 3B differ significantly. The fault apertures change with depth during the loading cases. The apertures at the top and bottom of Model 3B vary between 90 µm and 40 µm for the sub-vertical fault set. The corresponding values for the sub-horizontal fault set are 160 µm and 110 µm respectively. During the last unloading case, Model 3B exhibited an opening for both fault sets down to a depth of about 1500 m. The maximum variations for the sub-horizontal and sub-vertical fault set at the top and bottom of the model were 90 µm and 100 µm respectively. A typical mechanical aperture plot is shown in Fig. 8B.

The shear displacements obtained varies between 1.2 mm and 80 mm for Model 3A down to a depth of 1000 m. A maximum shear displacement of 0.89 m was obtained for the "soft faults" in Model 3B. For more details about mechanical and hydraulic apertures the reader is referred to the work by Chryssanthakis (1989)

3. CONCLUSIONS

1. HNFEMP-modelling with stiff faults gave a vertical displacement (rebound) at the ground surface between 2.5 and 4.2 m. The model with two intersecting faults gave the largest displacements. Horizontal stresses at a depth of 500 m varied between 30 and 35 MPa for the loading and unloading cases. The load of a 3 km thick ice sheet superimposed on the virgin vertical rock stress gave a vertical stress of 40 MPa.

2. MUDEC-modelling with linear fault stiffnesses gave smaller horizontal stresses but similar vertical stress compared with the HNFEMP-modelling results. The loading and unloading of the models caused variation in magnitude and orientation of the principal stresses for the uppermost ~ 1000 m of the rock mass. A major part of the re-orientation of the stresses was caused by rigid body translation and rotation of the rock blocks.

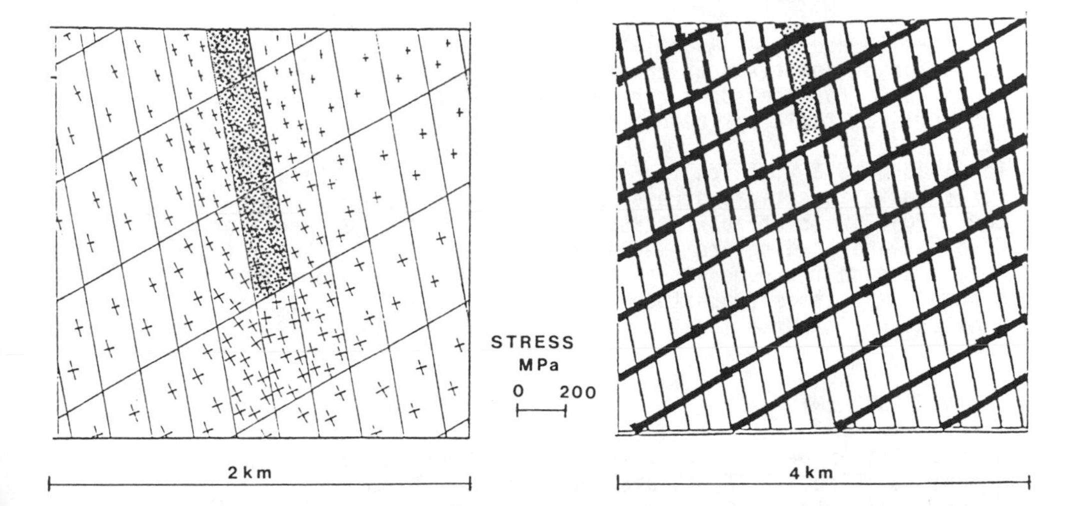

Fig. 8 Loading case 3 with 3 km of ice and ice flow analysed with the MUDEC non- linear and Barton-Bandis joint model.
A) Principal stress vector in the central portion of Model 3B. There is no significant stress rotation observed.
B) Mechanical aperture plot of the whole Model 3B. On the sub-vertical fault set the variation of apertures on the top and bottom of the model ranges between 90 µm and 40 µm. (After Chryssanthakis 1989).

3. For MUDEC-modelling with linear fault stiffness the glacial rebound of the ground surface after deglaciation was found to be 3.2 to 4.0 m and about the same as that obtained from the HNFEMP-modelling. The relative vertical displacement between individual rock blocks was 5 to 10 cm.

4. For MUDEC-modelling with the Barton-Bandis joint model, the vertical displacement of the ground surface after glaciation was found to be 3.8 and 3.9 m for Models 3A (stiff faults) and 3B (soft faults) respectively. This is in fair agreement with results obtained from the HNFEMP and linear MUDEC modelling. Modelling with stiff faults was satisfactory while modelling two sets of intersecting fault sets and soft fault stiffnesses gave very high vertical stresses when ice flow was introduced. Application of effective tectonic stresses, i.e. total stress minus pore pressure, made no major change in the magnitude of stresses and displacements.

5. On the final unloading case (today's situation) where all ice was removed, the stiff joints simulated in Model 3A (stiff joints) showed elastic response with a hystersis effect while the simulated "soft" faults of Model 3B exhibited an elastic response of 3.8 metres. This elastic response can cause changes in the mechanical and conductive aperture of the faults and consequently change the groundwater flow during deglaciation.

6. Modelling with the MUDEC and Barton-Bandis joint model permitted determination of mechanical and hydraulic fault apertures. Aperture changes of 40 and 100 µm for the sub-vertical and sub-horizontal joints sets respectively were obtained for Model 3A down to a depth of 200-300 metres. In

contrast to Model 3A, Model 3B exhibited apertures changing with depth. The variation of apertures at the top and bottom of the model ranged between 90 µm to 40 µm for the sub-horizontal and between 160 µm and 110 µm for the sub-vertical fault set respectively.

7. The load of a 3 km thick ice sheet superimposed on the virgin vertical rock stress gave a vertical stress of 40 MPa. This stress was superimposed on the virgin state of rock stress. Construction of a repository, thermal loadings and loading from swelling betonite will further influence the stress around the repository. The results of the modelling demonstrate the change in stresses (magnitudes and directions) due to glaciation and deglaciation and favour a location of a repository around 500 m depth.

4. ACKNOWLEDGEMENTS

The work of this paper was supported by the Swedish Nuclear Fuel and Waste Management Co. Special thanks are due to G. Bäckblom for his interest and support during this work.

5. REFERENCES

Bakkelid, S. 1986. The determination of rates of land uplift in Norway. Tectonophysics, Vol. 130, pp. 307-326.
Barton, N. and Bandis, S. 1982. Effect of block size on the shear behaviour of jointed rock. Proc. 23rd U.S. Symposium on Rock Mechanics, pp. 739-760.
Barton, N., Bandis, S. and Bakhtar, K. 1985. Strength, deformation and conductivity coupling of rock joints. Int. J. Rock Mech. Min. Sci. & Geomech. Abstr., Vol. 22, No. 3, pp. 121-140.
Barton, N. et al. 1986. Numerical analyses and laboratory tests to investigate the Ekofisk subsidence. Proc. 27th U.S. Symposium on Rock Mechanics, Univ. of Alabama, June 23-25, 1986.
Barton, N. and Chryssanthakis, P. 1988. Validation of MUDEC against Colorado School of Mines block test data. SKB Technical Report 88-14, Swedish Nuclear Fuel and Waste Management Co., Stockholm, 94 p.
Celestino and Goodman 1979. Path dependency of rough joints in bi-directional shearing. Proc. of the Fourth Int. Cong. on Rock Mechanics, Montreaux, Vol. 1, pp. 91-98.
Chryssanthakis, P. 1989. Distinct element modelling of the influence of glaciation and deglaciation on the behaviour of a faulted rock mass. SKB Working Report 89, Swedish Nuclear Fuel and Waste Management Co., Stockholm, 41 p. and 84 App.
Cundall, P.A. 1980. UDEC - A generalized distinct element program for modelling jointed rock. Report PCAR-1-80, Peter Cundall Associate, Contract DAJA 37-79-C-0548, European Research Office, U.S. Army.
Goodman, R.E. 1976. Methods of geological engineering in discontinuous rocks. West Publ. Co., St. Paul.
Jing, L. and Stephansson, O. 1988. Distinct element modelling of the influence of glaciation and deglaciation on the state of stress in faulted rock masses. SKB Working Report 88-23, Swedish Nuclear Fuel and Waste Management Co., Stockholm, 16 p.
Nilsson, L. and Oldenburg, M. 1983. FEMP - An interactive, graphic finite ele-

ment program for small and large computer system. Technical Report 1983:07 T, Luleå University of Technology, Luleå.
Olofsson, T. 1985. Mathematical modelling of jointed rock masses. Doctoral Thesis 1985:42 D, Luleå University of Technology, Luleå.
Sing, U., Savilahti, T and Stephansson, O. 1988. Generic model analysis of large faulted rock masses with soft faults using HNFEMP computer code. SKB Working Report 88-21, Swedish Nuclear Fuel and Waste Management Co., Stockholm, 13 p. and 13 App.
Singh, U., Savilahti, T. and Stephansson, O. 1988. Generic model analysis of large faulted rock masses with hard fault stiffness. SKB Working Report 88-22, Swedish Nuclear Fuel and Waste Management Co., Stockholm, 13 p. and 15 App.
Stephansson, O. 1983. The need of discontinuities and horizontal stresses in geological storage of radioactive waste. Proc. 4th Int. Conf. on Basement Tectonics, pp. 35-47.
Stephansson, O. 1987. Modelling of crystal rock mechanics for radioactive waste storage in Fennoscandia. SKB Technical Report 87-11, Swedish Nuclear Fuel and Waste Management Co., Stockholm, 79 p.
Stephansson, O., Bäckblom, G., Groth, T. and Jonasson, P. 1978. Deformation of a jointed rock mass. Geol. Fören. Stockholm Förh., Vol. 100, pp. 387-394.
Stephansson, O., Särkkä, P. and Myrvang, A. 1986. State of stress of Fennoscandia. Proc. Int. Symposium on Rock Stress and Rock Stress Measurements, pp. 21-32, Centek Publishers, Luleå.
Zimmerman, R.M. and Blanford, M.L. 1985. Evaluation of the accuracy of continuum-based computational models in relation to field measurements i welded tuff. Proc. Int. Symposium on Fundamentals of Rock Joints, pp. 233-245, Centek Publishers, Luleå.

STOCHASTIC THREE DIMENSIONAL JOINT GEOMETRY MODELLING INCLUDING A VERIFICATION TO AN AREA IN STRIPA MINE

P.H.S.W. Kulatilake, D.N. Wathugala
Department of Mining and Geological Engineering
University of Arizona, Tucson, USA

O. Stephansson
Division of Rock Mechanics
Luleå University of Technology, Luleå, Sweden

ABSTRACT

At present, a three dimensional joint geometry modelling scheme which investigates statistical homogeneity, incorporates corrections for sampling biases and applications of stereological principles, and, also includes a formal verification procedure is not available in the literature. This paper shows development of 3D joint geometry models with aforementioned features. Joint data from Stripa mine were used in the investigation. Verifications performed so far have indicated the need to try out different schemes in modelling joint geometry parameters and to consider a number of combinations of such schemes in order to arrive at a realistic 3D joint geometry model which provides a good comparison with field data during verification. Further verification studies are recommended to check the validity of the suggested modelling schemes.

GEOMETRIE DES JOINTS : MODELISATION TRIDIMENSIONNELLE STOCHASTIQUE ET VERIFICATION DANS UNE ZONE DE LA MINE DE STRIPA

RESUME

A ce jour, on ne trouve pas mention dans la bibliographie scientifique pertinente de projet de modélisation tridimensionnelle de la géométrie des joints permettant d'étudier l'homogénéité statistique, de corriger les erreurs systématiques d'échantillonnage et d'appliquer les principes stéréologiques et prévoyant en outre une procédure structurée de vérification. Les auteurs exposent les travaux de mise au point de modèles simulant la géométrie des joints en trois dimensions et possédant les caractéristiques mentionnées ci-dessus. Des données relatives aux joints tirées de la mine de Stripa ont été utilisées pour ces recherches. Il ressort des vérifications effectuées à ce jour qu'il faut explorer diverses possibilités dans la modélisation des paramètres relatifs à la géométrie des joints et examiner plusieurs combinaisons de ces possibilités pour parvenir à un modèle tridimensionnel réaliste de la géométrie des joints qui donne des résultats qui concordent bien avec les mesures de vérification effectuées sur le terrain. Des études de vérification supplémentaires sont recommandées pour s'assurer de la validité des projets de modélisation suggérés.

1 INTRODUCTION

Presence of discontinuities strongly affects the mechanical and hydraulic behaviour of rock masses. Therefore, joint geometry plays a vital role in civil, mining and petroleum engineering projects associated with discontinuous rock. The need for a realistic representation of joint geometry has been recognized.

Since joint geometry pattern can vary from one statistically homogeneous region to another, each statistically homogeneous region should be represented by a joint geometry model. Therefore, the first step in the procedure of joint geometry modelling in a rock mass should be the identification of statistically homogeneous regions. To model joint geometry in three dimensional (3D) space, for a statistically homogeneous region, it is necessary to know the number of joint sets, and for each joint set, the intensity, spacing, location, orientation, shape and dimension distributions. These joint parameters are inherently statistical. Sample values of joint parameters provided by the field data usually subject to errors due to sampling biases and represent only one or two dimensional properties. Therefore, before inferring these parameters from sampling values, sampling biases should be corrected on field data. In addition, principles of stereology need to be used in order to infer 3D parameters of the joint sets from 1D or 2D parameter values. At present, a 3D joint geometry modelling scheme which incorporates the aforementioned features and also includes a formal verification procedure is not available in the literature [1]. This paper shows development of 3D joint geometry models to an area in Stripa mine, including investigations for statistical homogeneity, corrections for sampling biases and applications of stereological principles. The paper also shows a verification for a built-up 3D model. Due to space limitation, only a summarized presentation is given in this paper.

2 DEVELOPMENT OF 3D JOINT GEOMETRY MODELS

2.1 Investigation of statistical homogeneity

Figure 1 shows the flow chart used to develop 3D stochastic joint geometry models. Orientation data from a thirty six meter long stretch of the ventilation drift, Stripa mine [2] were used to investigate the statistical homogeneity of the rock mass. A ten meter long stretch from this thirty six meter stretch was identified as the largest statistically homogeneous region [3].

2.2 Joint orientation modelling

Using the method given in [4], orientation data in this region were delineated into four joint sets as shown in Fig. 2. For each joint set, high dispersion of the orientation data can be seen very clearly from Fig. 2. For such joint sets orientation bias correction may be significant. Joint data for each joint set come from the east and west walls and the floor of the drift. A general procedure applicable for sampling domains of any orientation was developed to correct for orientation bias [5]. This procedure was applied to each joint set in order to correct for orientation bias.

Raw orientation data as well as orientation data corrected for sampling bias were subjected to chi-square goodness-of-fit tests to check the suitability of Bingham distribution and the hemispherical normal distribution in representing the statistical distribution of data. The maximum significance level at which the tried probability distribution is suitable to represent the statistical distribution of data was computed for both raw and corrected data of each cluster. This significance level should be at least 0.05 to accept the tried probability

Step 1: Obtain a statistically homogeneous region.

Step 2: Delineate joint sets.

Step 3: Apply correction for orientation bias for each joint set.

Step 4: Model the true orientation distribution for each joint set.

Step 5: Determine joint spacing distributions along scanlines taking into account the sampling bias on spacing.

Step 6: Infer joint spacing distribution along the mean pole direction (true spacings) for each joint set.

Step 7: Estimate the trace length distribution for each joint set taking into account the sampling biases.

Step 8: Infer joint size distribution for each joint set taking into account the sampling biases (joints are considered as circular discs).

Step 9: Estimate the mean joint center density in 3D (number/volume) for each joint set.

Step 10: Obtain a distribution for the random variable, " number of joint centers per chosen volume".

Step 11: Suggest a 3D stochastic joint geometry model by describing the joint geometry parameters:

(a) number of joint sets

(b) orientation distribution for each joint set

(c) spacing distribution for each joint set

(d) distribution for density in 3D for each joint set

(e) size (diameter when joint shape is considered as circular) distribution for each joint set.

Figure 1 Flow chart of the procedure used for development of stochastic joint geometry model

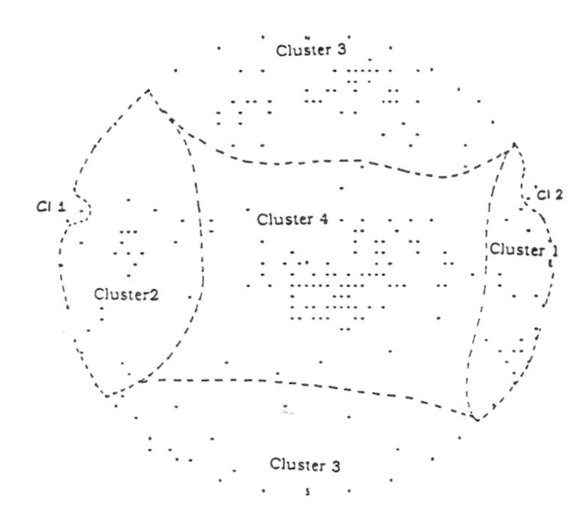

Figure 2 Polar equal area projection of orientation data in upper hemisphere

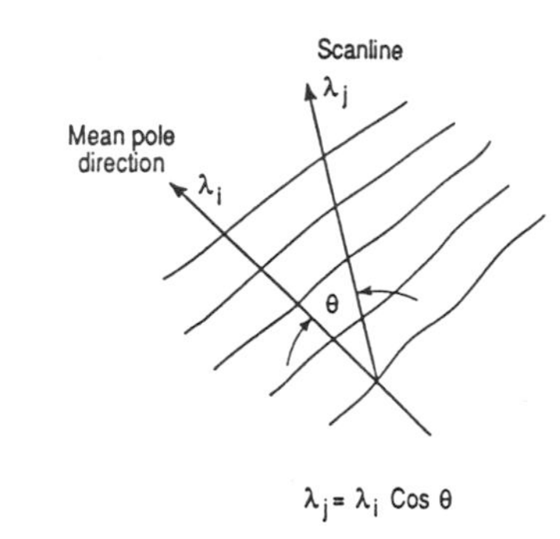

Figure 3 Influence of direction on mean linear intensity estimation

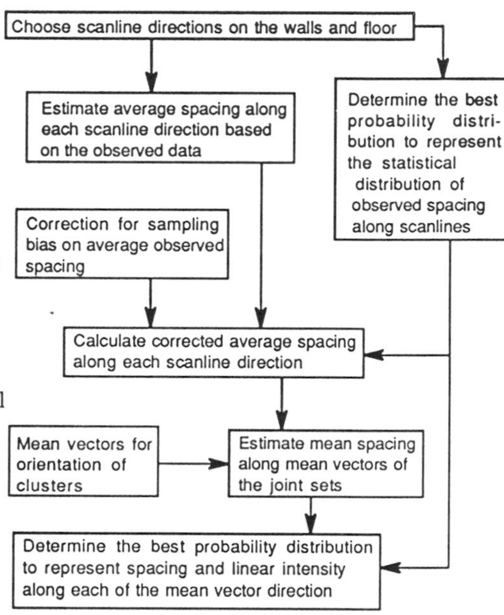

Figure 4 Flow chart used to estimate spacing and linear intensity distribution along the mean vector directions

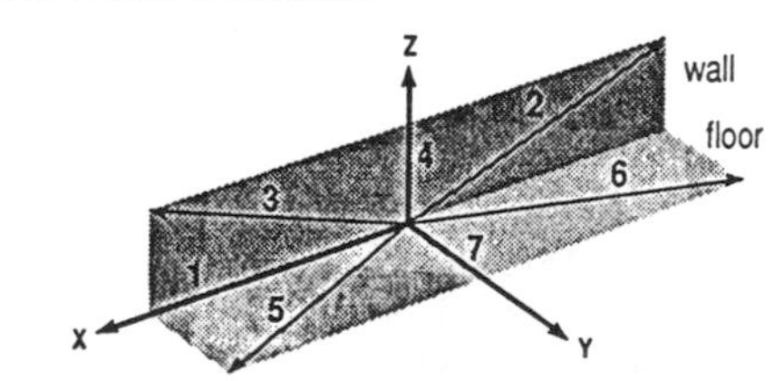

Figure 5 Chosen scanline directions for spacing and intensity analysis

distribution to represent the data. According to the results, only cluster 4 follows a Bingham distribution for raw data; neither of the two probability distributions satisfied the corrected data. Therefore, empirical distributions obtained for corrected data were chosen to represent the statistical distributions of joint set orientations. For three clusters, considerable differences were observed between the chi-square values for raw and corrected data. That reflects the significance of sampling bias correction.

2.3 Joint spacing, linear intensity (frequency) and location modelling along scanlines and mean vectors of joint sets

These tasks were performed according to two methods.

2.3.1 Method 1

For a joint set, mean spacing and mean linear intensity (average number per unit length) estimates depend on the chosen direction as shown in Fig. 3. For a fixed direction, mean linear intensity is the reciprocal of the mean spacing. The estimations along the mean vector (mean pole) directions can be considered as the true values. Figure 4 shows the flow chart used to obtain spacing and linear intensity distributions along the mean vector directions starting from the observations made on several scanlines. Seven scanline directions (Fig. 5) were chosen to analyze spacing and intensity in different directions to have a good coverage in 3D. Along each direction, several parallel scanlines were drawn either on the walls or on the floor of the drift, having joint traces coming from all four joint sets, to estimate spacing distribution as well as observed mean spacing. For each direction, observed spacing values were subjected to chi-square and Kolmogorov-Smirnov goodness-of-fit tests to check the suitability of exponential, gamma, lognormal, normal, uniform and triangular distributions in representing the observed spacings. For all seven directions the exponential distributions were found to be the best distributions to represent the distributions of observed spacing by satisfying the goodness-of-fit tests at very high significance levels. Chosen scanlines were of finite size. To obtain unbiased estimates of spacing and intensity, scanlines should be of infinite size. The equation given in [1] was used to correct this sampling bias on spacing and to obtain corrected mean spacing from the observed mean spacing. This error was found to be respectively about 0.1, 2 and 5% for length of scanline/mean spacing ratios of 9, 6, and 5.

Hudson and Priest [6] expressed mean linear intensity resulting from several joint sets in any arbitrary direction in 3D in terms of the mean linear intensities along the mean pole directions of the joint sets. Using this expression, Karzulovic and Goodman [7] suggested a procedure based on least square method to estimate the mean linear intensities along the mean vector directions of joint sets using the mean linear intensities estimated along several scanline directions. This procedure was used with aforementioned seven scanlines in order to estimate the mean linear intensities along the mean vectors of the four joint sets. The relative standard error obtained in the least square procedure was less than four percent with respect to any mean linear intensity estimated along scanline directions indicating quality estimations. These findings lead to the conclusion that joint spacing along mean vector directions follow exponential distributions with mean spacing values obtained in respective directions. Thus, according to statistical theory, linear intensity and linear location distributions along the mean vector direction for each joint set follow respectively Poisson and uniform distributions.

2.3.2 Method 2

First, joint traces on the two walls and the floor were sorted out into the four joint sets. Then for each cluster, the following analyses were conducted.

The flow chart shown in Fig. 4 was followed for each cluster until mean corrected spacings and mean linear intensities were found along the seven scanline directions. Then the relationship shown in Fig. 3 was used to estimate the corresponding mean linear intensities along the mean vector direction of the cluster. Finally, these values were averaged to obtain the final mean linear intensity estimate for each cluster. Again, the conclusions regarding the probability distribution types for spacing, intensity and location are same as for method 1.

2.4 Joint trace length and 3D joint size modelling

First, joint traces appearing on the two walls and the floor were sorted out into the four joint sets. Then for each cluster, the following modelling was performed; one on the wall data and the other based on the floor data. For example, Fig. 6 provides the subsequent modelling performed on the wall data for each joint cluster.

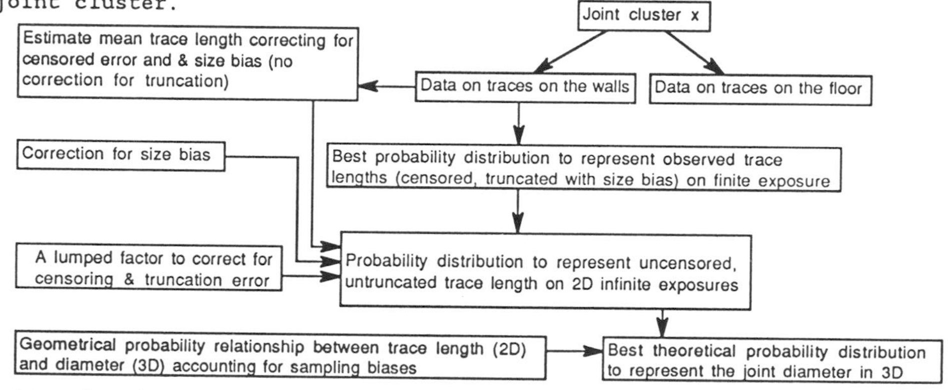

Figure 6 Flow chart for modelling joint size in 3D for each cluster

Due to finite size of the walls (10m x 4m) and the floor (10m x 4m), the observed traces are subject to censored and size biases. Since observed traces of length less than about 0.4m were neither mapped on the walls nor on the floor, it was assumed that the truncation limit for joint traces is 0.4m. Area sampling technique was used in sampling the joint traces on both the walls and on the floor. As for spacing, goodness-of-fit tests were carried out on the sampled joint traces. For all four joint clusters, gamma distribution was found to be the best distribution to represent observed trace length distribution on both the walls and on the floor.

The method given in [8] was used to correct for censoring error and size bias, and to estimate mean trace length on an infinite 2D exposure using the observed trace data from a finite 2D exposure. Differences up to sixty percent were found between the observed mean trace length and the estimated mean trace length on infinite 2D exposures. Such results imply the importance of the correction. An attempt was made to express the probability density function for the trace length distribution on 2D infinite exposure, f(l), from the probability density function obtained for observed traces on the 2D finite exposure, g(l).

Assuming joints as finite circular discs, joint diameter distribution for a joint cluster was estimated from f(l) using the procedure given in [9]. Diameter distributions for all joint clusters were found to be of Gamma type.

2.5 Three dimensional joint intensity (number per unit volume) modelling

3D joint intensity was estimated according to the following four methods.

2.5.1 Method 1

For each cluster, mean value for 3D joint intensity was estimated using the following equation:

$$(\lambda v)_j = \frac{4(\lambda_l)_j}{\pi E(D^2)} \tag{1}$$

where $(\lambda_l)_j$ and $(\lambda_v)_j$ are respectively mean linear intensity along its mean vector and mean 3D intensity of j the joint set and $E(D^2)$ is the expected or the mean value of squared joint diameter. This equation was derived based on an equation presented by Oda [10]. For each cluster, $E(D^2)$ can be estimated based on the diameter distribution obtained either from the wall data or from the floor data. Values obtained from Section 2.3.1 were used for $(\lambda_l)_j$. Since linear intensity follows a Poisson distribution, it is reasonable to conclude that 3D intensity follows a Poisson distribution with corresponding $(\lambda_v)_j$. Mean 3D intensity for the rock mass was obtained through equation (2).

$$\text{mean } 3D \text{ joint intensity} = \sum_{j=1}^{4}(\lambda_V)_j \tag{2}$$

2.5.2 Method 2

The same procedure as for Method 1 was used for this case. However, for this case, values obtained from Section 2.3.2 were used for $(\lambda_l)_j$ in computing $(\lambda_v)_j$.

2.5.3 Method 3

For each cluster, mean value for 3D joint intensity was estimated using the following derived equation:

$$(\lambda_v)_j = \frac{(\lambda_a)_j}{E(D)E(|\sin \nu_1|)} \tag{3}$$

where $(\lambda_a)_j$ and $(\lambda_v)_j$ are respectively, mean areal intensity (mean number of joint centers per unit area) and mean 3D intensity of j th joint set, and, E(D) and $E(|\sin \nu_1|)$ are the expected or mean value of the diameter and $|\sin \nu_1|$ respectively. The angle ν_1, is the angle between the joint plane and the sampling plane on which $(\lambda_a)_j$ was estimated. From a sampling domain, what we can directly estimate is the average number of joints per unit area. The procedure given in [11] was used to estimate $(\lambda_a)_j$ from the average number of joints per unit area.

2.5.4 Method 4

Values of $(\lambda_v)_i$ obtained from the previous three methods were averaged to obtain these estimations.

3 JOINT SYSTEM MODELLING

Section two dealt with modelling of joint geometry parameters for the chosen statistically homogeneous region. Results obtained in Section two were used in this section to build joint geometry networks in 3D. For each joint cluster the following statistical models were used to generate joints in 3D: (1) Number of joints per certain volume is Poisson distributed with mean value $(\lambda_v)_i$ (2) Location of joint clusters in 3D is uniformly distributed (3) Orientation is distributed according to the empirical distribution obtained for corrected data (4) Diameter is gamma distributed with the parameter values obtained.

4 JOINT GENERATION IN 3D, PREDICTION IN 2D, AND VERIFICATION

Only one of the several verification studies performed for the largest joint set (cluster #4) is discussed in this paper. The joints were generated in the volume shown in Fig. 7, according to the statistical model given in Section 3 using Monte Carlo simulation. In Fig. 7, $\bar{D}_{max}$ is the largest mean diameter of the four joint sets. In the chosen volume, the vertical plane EFGH of size (10m x 4m) (Fig. 7) was chosen to simulate the tunnel wall. For this case, $(\lambda_v)_i$) obtained from Section 2.5.4 was used. Diameter estimation obtained based on wall data was used. Joint traces appearing on EFGH were censored and truncated. A truncation limit of 0.4m was used. These joint traces were used to estimate the parameters given in Table 1. Several such simulations were repeated. Mean

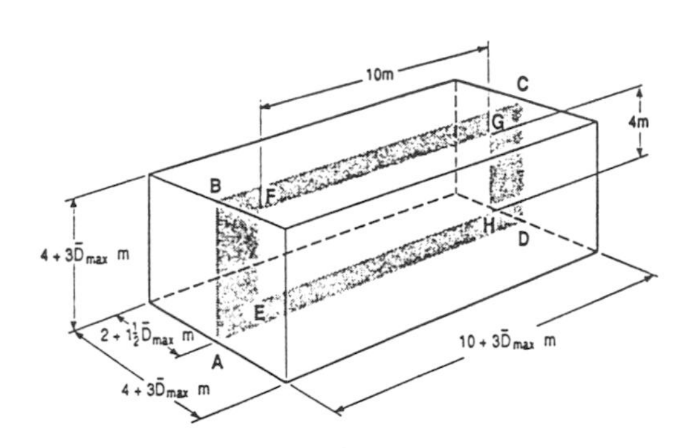

Figure 7 Volume and vertical section used in verification

Table 1. A comparison between the field data and predictions on 2D from Monte Carlo simulation of the parameters in the model for joint cluster 4.

Parameter	Field value	Predicted range	Average prediction
R_0	0.01	0-0.02	0
R_1	0.18	0.07-0.33	0.17
R_2	0.81	0.67-0.93	0.83
mean ϕ	14.9	7.09-18.6	13.2
mean θ	245.1	223.3-257.0	238.8
l_{obs} (m)	1.38	0.87-1.39	1.14
$l_{corr.}$ (m)	0.79	0.23-1.27	0.61
Number of joints	40	30-49	41

where
θ - dip direction (deg.)
ϕ - dip angle (deg.)
l_{obs} - mean trace length on the finite window
l_{corr} - corrected mean trace length

Note: $R_0 = N_0/N$, $R_2 = N_2/N$ and $R_1 = N_1/N$
where
N_0 - expected number of joints with both ends censored intersecting the window
N_1 - expected number of joints with only one end observed intersecting the window
N_2 - expected number of joints with both ends observed intersecting the window
$N = N_0 + N_1 + N_2$

predictions were computed using the results obtained through thirty simulations. Table 1 shows a comparison between predictions and actual field data. In statistical sense, the agreement is very good.

5 CONCLUSIONS

Verification studies performed so far has indicated the need to try out different schemes in modelling joint geometry parameters and to consider a number of combinations of these schemes in order to come up with a realistic 3D joint geometry model which provides a good comparison with field data during verification. The verification reported in the paper shows that the modelling scheme used for the verification has very good capability in producing 2D predictions which provide very good agreement with 2D field data for the cluster studied. Further verification studies for other clusters of the same site as well as for different sites are recommended to check the validity of the suggested modelling schemes.

ACKNOWLEDGEMENTS

The Swedish Natural Science Research Council and the Arizona Mining and Mineral Resources Research Institute under USBM Grant No. G 1194104 provided partial financial support for this study.

REFERENCES

[1] Kulatilake, P.H.S.W. 1988 & 1989. Stochastic joint geometry modelling: state-of-the-art. Proc. 29th U.S. Symp. on Rock Mech., Minneapolis, Minnesota: 215-229. Also invited papers at the conferences held in Switzerland, Sweden, Mexico, and Thailand.

[2] Rouleau, A., Gale, J.E. and Baleshta, J. 1981. Fracture mapping in the ventilation drift at Stripa: procedures and results. Research report LBL-13071, Lawrence Berkeley Laboratory, Berkeley, California.

[3] Kulatilake, P.H.S.W., Wathugala, D.N., Poulton, M. and Stephansson, O. 1990. Analysis of structural homogeneity of rock masses. Int. J. Engineering Geology (in press).

[4] Shanley, R.J. and Mahtab, M.A. 1976. Delineation and analysis of centers in orientation data. Math. Geology. 8: 9-23.

[5] Wathugala, D.N., Kulatilake, P.H.S.W., Wathugala, G.W. and Stephansson, O. 1990. A general procedure to correct sampling bias on joint orientation using a vector approach. Computers and Geotechnics (submitted for publication).

[6] Hudson, J.A. and Priest, S.D. 1983. Discontinuity frequency in rock masses. Int. J. Rock Mech. Min. Sci. 20: 73-89.

[7] Karzulovic, A. and Goodman, R.E. 1985. Determination of principal joint frequencies. Int. J. Rock Mech. Min. Sci. 22: 471-473.

[8] Kulatilake, P.H.S.W. and Wu, T.H. 1984. Estimation of mean trace length of discontinuities. Rock Mech. and Rock Engineering. 17: 215-232.

[9] Kulatilake, P.H.S.W. and Wu, T.H. 1986. Relation between discontinuity size and trace length. Proc. 27th U.S. Symp. on Rock Mech., Tuscaloosa, Alabama: 130-133.

[10] Oda, M. 1982. Fabric tensor for discontinuous geological materials. Soils and foundations. 22: 96-108.

[11] Kulatilake, P.H.S.W., and Wu, T.H. 1984. The density of discontinuity traces in sampling windows. Int. J. Rock Mech. Min. Sci. 21:345-347.

ROCK MASS RESPONSE TO GLACIATION AND THERMAL LOADING FROM NUCLEAR WASTE

B. Shen, O. Stephansson
Division of Rock mechanics
Luleå University of Technology, Luleå, Sweden

ABSTRACT

The response of the SKI reference site to glaciation and thermal loading of nuclear waste is modelled numerically by a three-dimensional distinct element code (3DEC). Four stages of a glaciation cycle are studied based on a 11x10x7 km^3 global model. After completing a glaciation cycle, stresses and displacements did not differ significantly from their initial value. This implies that the rock mass with the SKI fracture zones behaves elastically. A local model of 2x2x2 km^3 was used to model the rock mass response for thermal loading of waste repository. The heat release increases the stresses and rotates the principle stresses in the rock mass near the repository. Fracture closure also occurs in the vicinity of the repository

REACTION D'UNE MASSE ROCHEUSE A LA GLACIATION ET AU CHARGEMENT THERMIQUE INDUIT PAR DES DECHETS NUCLEAIRES

RESUME

La réaction du site de référence du Service suédois d'inspection de l'énergie nucléaire (SKI) à la glaciation et au chargement thermique provoqué par des déchets nucléaires est modélisée numériquement au moyen d'un programme de calcul tridimensionnel à éléments distincts (3DEC). Quatre étapes d'un cycle de glaciation sont étudiées sur la base d'un modèle global de 11x10x7 km^3. A l'issue d'un cycle de glaciation, les contraintes et les déplacements ne diffèrent pas sensiblement de leur valeur initiale. Il en résulte que la masse rocheuse contenant les zones de fissures qu'étudie le SKI réagit de façon élastique. On a utilisé un modèle local de 2x2x2 km^3 pour modéliser la réaction de la masse rocheuse au chargement thermique induit par un dépôt de déchets nucléaires. Le dégagement de chaleur augmente les contraintes et entraîne une rotation des principales d'entre elles dans la masse rocheuse à proximité du dépôt. Il se produit également une fermeture des fissures au voisinage du dépôt.

1. INTRODUCTION

Glaciation and waste thermal loading are two crucial factors in the stability and safety of a nuclear waste repository. Stephansson (1988) and Barton (1989) studied the two dimensional model for the glaciation effects on the behaviour of rock masses containing two groups of regular joint sets. Wang et al (1988) described the thermal impact of waste emplacement with emphasis on the thermal and thermo-mechanical response of the rock mass. Numerical thermo-mechanical analysis for repository stability are also presented by Tiktinsky (1988). However, due to the theoretical difficulties and program limitation, most of the previous works were based on the assumption of a two-dimensional geometry. The study presented in this paper is based on a large-scale, three-dimensional geometry. The three-dimensional distinct element code (3DEC), developed by Cundall(1988) and expanded by Mack & Hart(1989) to deal with thermal loading, has been used. This study is part of the work for Project-90 of the Swedish Nuclear Power Inspectorate (SKI). Data on fracture zones are taken from the SKI reference site (Lindbom et al, 1989), and the thermal density and rock thermal properties follow KBS-3 concept, SKB(1983).

2. MODELLING OF GLACIATION

A 3DEC global model is established for the mechanical analysis of the effects of glaciation, see Figure 1. Three types of fracture zones presented for the SKI reference site are included in the model, namely major, intermediate and minor fracture zones. The properties of intact rock and fracture zones are listed in Table 1. Five loading steps of a glaciation cycle are modelled, as shown in Figure 2. Previous modelling (Stephansson, 1988) indicated that the glacial flow has only a minor effect on the stress state and therefore is omitted. A maximum ice thickness of 3 km is believed to appear when glaciation occurs in the repository area.

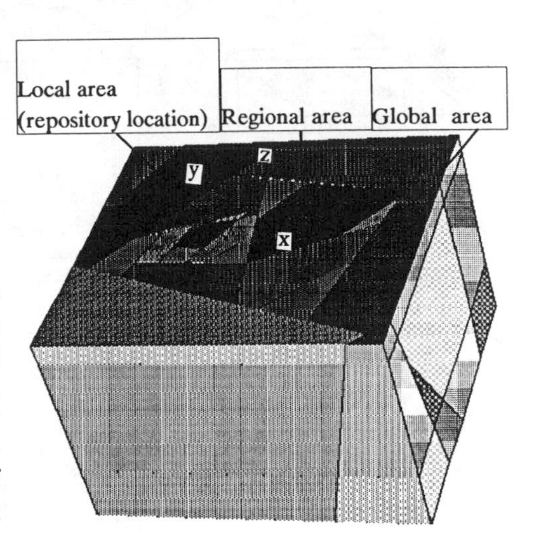

Figure 1. 3DEC global model for modelling the rock mass response to glaciation.

Stephansson et al. (1986) have presented the general in-situ stress distribution in Sweden. This is described by the following equations:

$$\sigma_x = 2 + 28|y| \text{ (MPa)} \qquad \sigma_z = 4 + 32\,|y| \text{ (MPa)} \qquad \sigma_y = 27\,|y| \text{ (MPa)}$$

Table 1. Rock and fracture properties.

Properties	Intact Rock	Ice	Major Fracture	Intermediate Fracture	Minor Fracture
Young's Modulus E (GPa)	40	9			
Poisson's ratio υ	0.2	0.33			
Density ρ (kg/m^3)	2700	900			
Normal stiffness Kn (GPa/m)			1.0	10.0	100.0
Shear stiffness Ks (GPa/m)			0.33	3.33	33.33
Friction angle ϕ (degree)			20	25	30
Cohesion c (MPa)			0	0	0

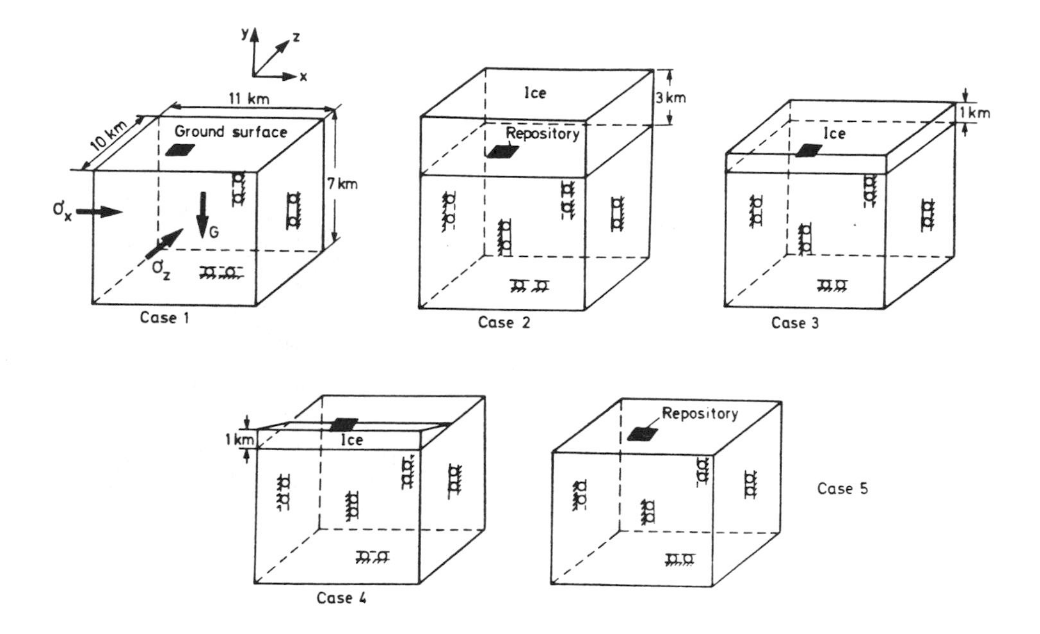

Figure 2. Ice loading and unloading sequence for modelling a glaciation cycle.

$|y|$ is the depth in kilometres. The boundary conditions during the ice loading and unloading are shown in Figure 2. Three repository locations at the depths of 500m, 1000m and 1500m are considered in the analysis. The modelling results are presented as stresses, surface displacements and fracture deformations.

2.1 Stress distribution in the rock mass.

The rather complicated fracture zone geometry results in a complicated stress distribution in the rock mass. The block corners where two or more fracture zones intersect cause large stress concentrations and stress rotations, see Figure 3. The stress concentrations and stress rotations exist during consolidation of the model and enhanced by glaciation of 3km ice on the ground surface. There are three block corners in the lower-right part of Figure 3 where stresses are distorted. These areas are potential earth quake sites.

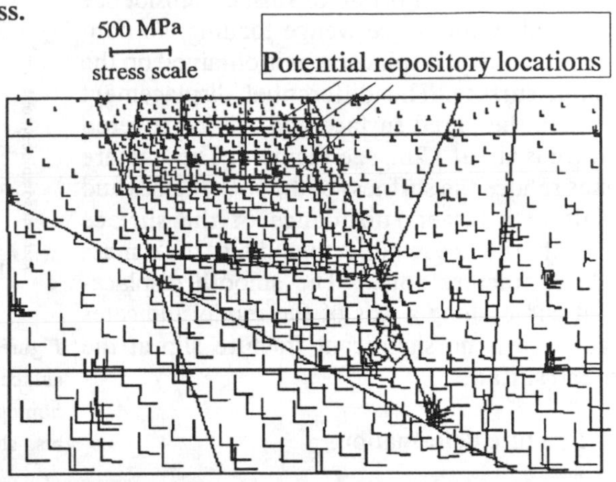

Figure 3. Principle stress distibution in plane z=0 when a 3km ice load is applied to the ground surface. The stress overlap in some corners indicates there is a fine mesh.

Since most of the fracture zones are orientated vertically and horizontally in the area where a repository may be located, the stress distribution is more uniform near the potential repository locations. Normal stresses at the centers of three potential repositories during glaciation and deglaciation are shown in Figure 4. The excess in vertical stress is about 27MPa during glaciation. The stress increase caused by ice load is 1-2MPa larger for a deep seated repository location.

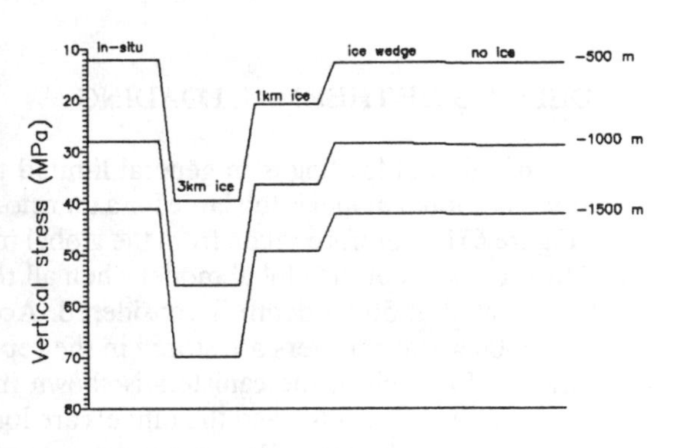

Figure 4. Vertical stress at three repository locations during glaciation and deglaciation.

When the glaciation retreated and became an ice wedge, a major reorientation of principle stresses occurs beneath the ice wedge.

2.2 Ground surface displacement

A 3km thick ice loading causes 5m of subsidence of the ground surface. Since the complicated fracture zones are usually limited in the rock mass and few penetrative slip paths can develop, the model behaves like an elastic medium during glaciation and deglaciation. After deglaciation, only less than 0.15m of surface subsidence remained. With an ice wedge loading, an non-uniform vertical displacement is obtained on the ground surface. The differential displacement between the bared surface and the ice covered surface is about 1.2m, see Figure 5. The fracture zones reduce the stiffness of the rock masses and result in a larger subsidence of the surface beneath the ice wedge than in a rock mass without fracture zones. The smooth displacement curve along z axis of the model indicates that there is no significant fracture slip at the ground surface.

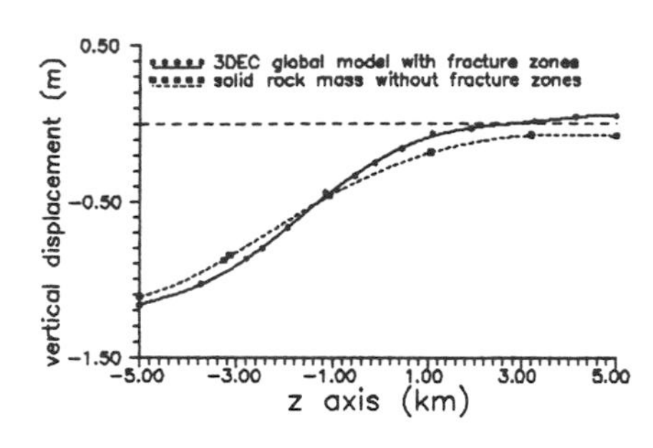

Figure 5. Vertical displacement of the ground suface caused by an ice wedge. The results for a homogenous rock mass without fracture zones are also shown.

2.3 Fracture deformation

3km of ice loading results in 1.9cm of closure along the horizontal fracture zones and almost 6cm shearing along inclined major fractures outside the local region. Most of the fracture deformations are elastic and recovered during deglaciation. There is no significant normal and shear displacement along the minor fracture zone close to the potential repository locations at any stage of the glaciation cycle.

3. MODELLING OF THERMAL LOADING

The effect of thermal loading is in general limited to the vicinity of the radioactive waste repository. The global model is too large for a thermo-mechanical analysis. A 2x2x2 km^3 local model (Figure 6) is therefore token from the global model with its stress boundary condition defined by the results of the global model when all the ice has disappeared. A double-layer repository located at 500m depth is considered. According to the KBS-3 concept, a total number of 4400 waste canisters are stored in the repository. The initial heat power and the decay function for each of the canisters is shown in Figure 7. The distance between two canister depositing holes is 6m and the tunnels are located 33m apart. The distance between the two storage layers is 100m. The upper layer is loaded with wastes 15 years after the lower layer is sealed. The thermal properties of rock mass are: thermal conductivity $= 3.0 W/(m^{\circ}C)$; specific heat $= 2.0 MJ/(m^{3}{}^{\circ}C)$ and linear expansion coefficient $= 8.5 \times 10^{-6}\ 1/{}^{\circ}C$

3.1 Temperature distribution

Heat radiation and conduction increase the rock temperature in the vicinity of the repository. As a simplification in 3DEC, the rock mass is assumed to be an infinite and thermo-isotropic medium. In our modelling, the semi-infinite problem is defined by using an isothermal boundary at the ground surface.

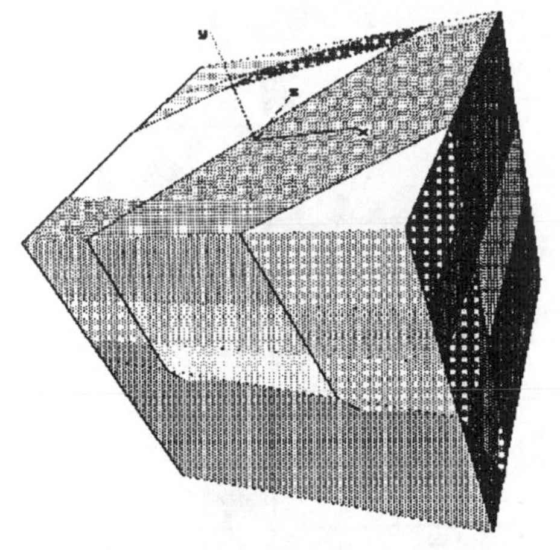

Figure 6. Local model for thermo-mechanical analysis.

Figure 7. Heat decay curve for each waste canister.

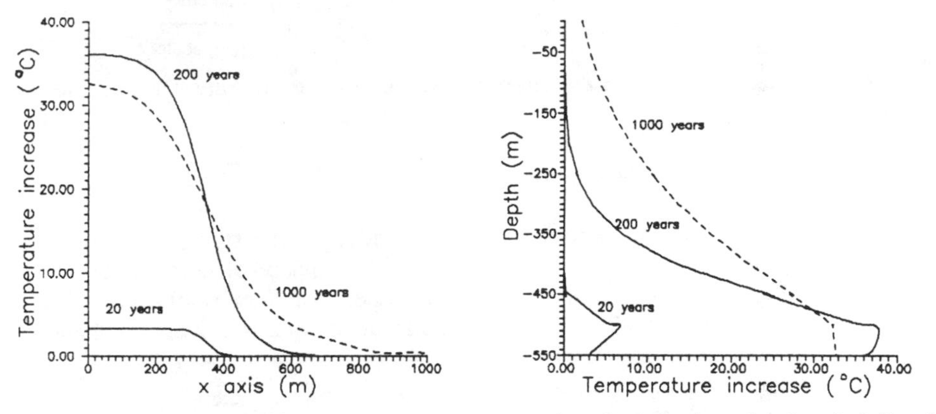

Figure 8. Temperature increase versus distance from the repository in the horizontal and vertical directions.

Figure 8 shows the temperature changes with distance from the repository center at 20, 200 and 1000 years after complete sealing of the repository. A maximum increase in temperature of 36°C is obtained after 200 years.

3.2. Thermal stresses

The rock expansion due to the temperature increase induces the stress redistribution and stress reorientation. The stress distribution after 200 years is shown in Figure 9(A). Major rotations of principle stresses in the vicinity of the repository are observed. The thermally induced stresses are mainly reflected in the direction of heat conduction. The maximum principle stress therefore tend to rotate towards the center of the repository. The maximum rotation after 200 years is about 30 degrees and occurs near the boundary of the repository.

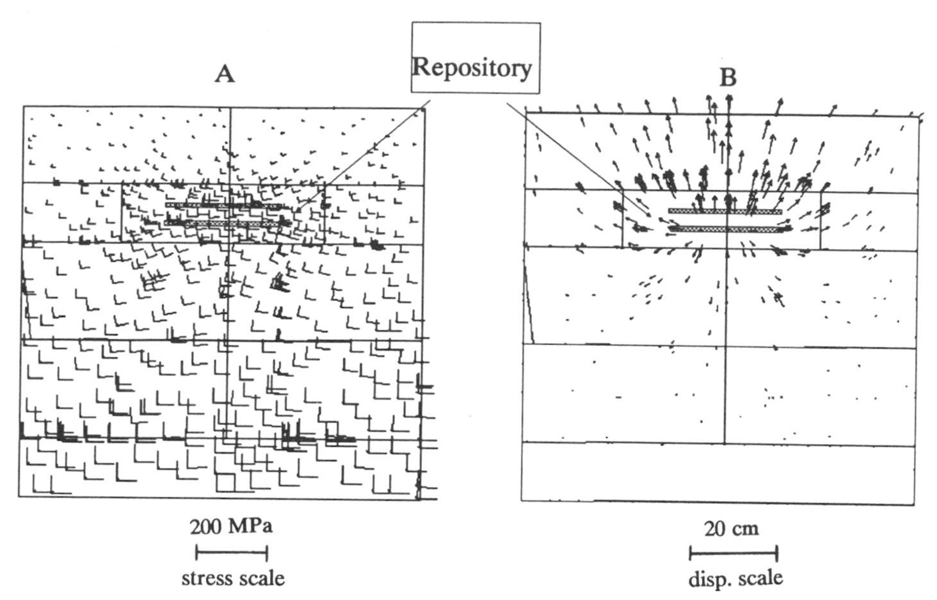

Figure 9. Rock mass response in section z=0 after 200 years. A. Principle stress distribution. B. Displacement distribution.

3.3 Thermo-induced rock displacements

Non-uniform displacement in the rock masses will endanger the stability and safety of the repository. Inevitably, thermal loading is the source of the rock displacement due to the rock expansion, see Figure 9(B). A maximum displacement of 5cm is observed above the repository after 200 years. After 20 years the rock displacement is relatively small and limited to the vicinity of the repository. After 1000 years the maximum displacement is 7.8cm and occurs closer to the ground surface. The displacement response lags behind the stress response.

The uplift of the ground surface during thermal loading is shown in Figure 10. The maximum uplift after 1000 years is 7.6cm and occurs directly above the center of the repository.

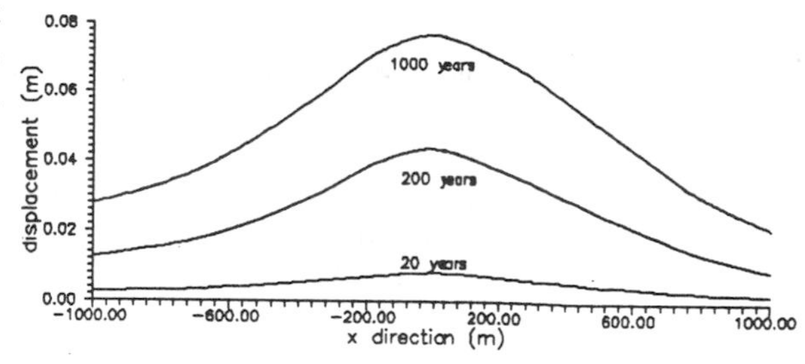

Figure 10. Vertical displacement of the ground surface during thermal loading.

3.4 Fracture deformation during thermal loading.

Thermal loading causes fracture closure near the repository. Figure 11 and 12 show the normal displacement for the horizontal fracture just beneath the repository and the vertical fracture intersecting the center of the repository. Minor fracture opening occurs outside the repository. The peak shear displacement occurs near the boundary, see Figure 13. The maximum shear displacement is 5cm and obtained along the horizontal fracture zone just beneath the repository.

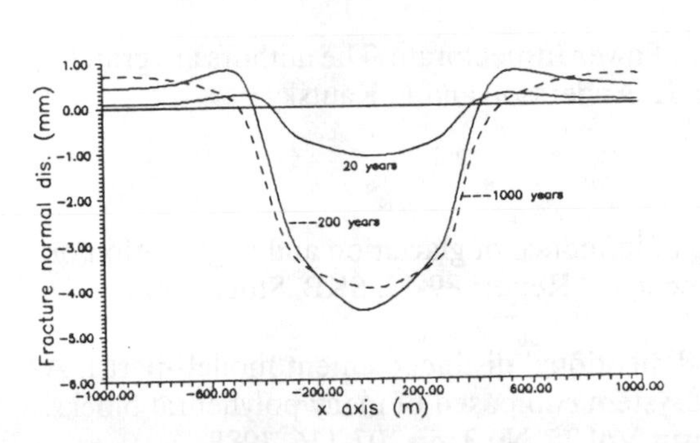

Figure 11. Normal displacement of horizontal fracture zone just beneath the repository. Positive sign indicates an opening displacement.

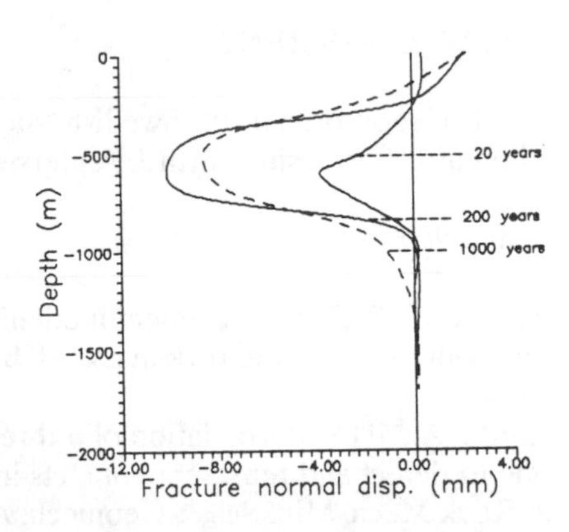

Figure 12. Normal displacement of vertical fracture zone intersecting the repository.positive sign indicates an opening displacement.

Fracture opening and shearing in the vicinity of the repository have a major effect on the safety of the repository. Fracture opening means more water flow into the repository, and large fracture shearing may cause waste canister failure. Thermal loading appears to have a minor effect since it causes only fracture closure and does not cause significant shearing close to the repository.

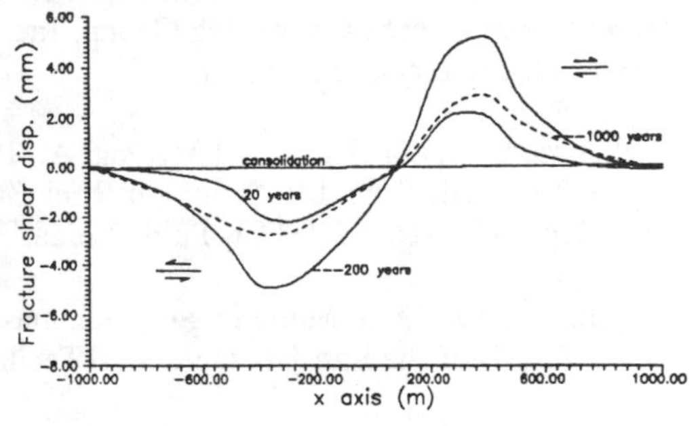

Figure 13. Shear displacement of the horizontal fracture zone just beneath the repository.

3.5 The effect of deposition depth.

A repository location at the depths of 1000m and 1500m have been modelled. The distance of the ground surface has a minor effect on the temperature distribution in the repository and it is almost the same at the three repository locations.

After 200 years, the maximum stress at the repository locations are 37.4MPa, 51.2MPa and 66.8MPa at the 500m, 1000m and 1500m depths respectively. The stress caused by thermal loading are almost same (17MPa) for all repository locations.

Thermally induced rock displacement is related to the overburden distance to the surface. As the repository depth increases, the displacement decreases. For instance, the maximum displacement in the rock masses after 1000 years is 7.8cm, 5.7cm and 5.5 cm when the repository depths are 500m, 1000m and 1500m respectively.

ACKNOWLEDGEMENT

This study is sponsored by the Swedish Nuclear Power Inspectorate. The authors are grateful for stimulating discussions with J. Andersson, K. Andersson and F. Kautsky

REFERENCES

Barton, N., 1989. Distinct element modelling of influence of glaciation and deglaciation on the behaviour of a faulted rock mass. SKB Technical Report 89-19, SKB, Stockholm.

Cundall, P.A., 1988. Formulation of a three-dimentional distinct element model--part I. A scheme to detect and represent contacts in a system composed of many polyhedral blocks. Int. J. Rock Mech. Min. Sci. & Geomech. Abstr. Vol.26. No.3. pp.107-116. 1988

Lindbom B., Lundblad,K. & Markström A. 1989. Initial groundwater flow calculations at the SKI reference site. SKI Technical Report 89:02, SKI, Stockholm, Sweden.

Mack, M. & Hart, R.D. 1989. Verification of thermal logic in the three-dimensional distinct element code. Itasca Consulting Group, Inc. Minnesota 55414, U.S.A. Prepared for U.S. Nuclear Regulatory Commission.

Stephansson, O., Särkkä, p. and Myrvag, A., 1986. State of stress in Fennoscandia. In: Ove Stephansson (ed) Proc. Int. Symp. on Rock Stress and Rock Stress Measurements, Stockholm, Sept.1-3, 1986. CENTEK Publ., Luleå.

Stephansson, O., 1990. Modelling of rock masses. In: Bäckbolm and Stanfors. An interdisciplinary study of the Lansjärv fault. SKB Technical Report 1990, SKB, Stockholm. (in press)

SKBF/KBS, 1983. Final storage of spent Nuclear fuel -- KBS-3. Swedish Nuclear Fuel Supply Co/Division KBS, Stockholm, Sweden.

Tiktinsky, D.H., 1988. Numerical parametric sensitivity study of thermal and mechanical properties for a high level nuclear waste repository. In Cundall et al (eds) Proc: Key Question in Rock Mechanics. 29th U.S. Symp. on Rock Mech.

Wang, J.S.Y., Mangold, D.C. and Tsang, F.C., 1988. Thermal impact of waste emplacement and surface cooling associated with geologic disposal of high-level nuclear waste. Environ. Geol. Water Sci. Vol.11, No.2.

THREE DIMENSIONAL COUPLED THERMO-HYDRAULIC-MECHANICAL ANALYSIS CODE WITH PCG METHOD

Y. Ohnishi, S. Akiyama
Kyoto University, Kyoto, Japan

M. Nishigaki
Okayama University, Okayama, Japan

A. Kobayaski
Hazama Corporation, Saitama, Japan

PROGRAMME D'ANALYSE TRIDIMENSIONNELLE AVEC COUPLAGE THERMO-HYDRO-MECANIQUE

RESUME

Pour évaluer l'évacuation des déchets de haute activité et comprendre les phénomènes qui se produisent dans le site, il faut étudier le plus précisément possible les comportements couplés thermiques, hydrauliques et mécaniques. A cet effet, les auteurs ont déjà élaboré le programme de calcul avec couplage bidimensionnel et ont analysé quelques expériences réalisées sur le terrain. Toutefois, le dépôt se déploie dans les trois dimensions et cela a un effet non négligeable. Aussi, les auteurs ont-ils mis au point un nouveau programme d'analyse prenant en compte dans les trois dimensions les phénomènes couplés thermiques, hydrauliques et mécaniques dans des milieux saturés et non saturés. Ce programme, appelé THAMES3D, tire parti de la méthode préconditionnée des gradients conjugués pour économiser le temps de stockage et le temps de calcul d'ordinateur.

Les auteurs présentent les principes de ce programme et les résultats de la vérification effectuée, puis ils simulent une installations représentative de dépôt profond et comparent les résultats des calculs tridimensionnels avec ceux obtenus au moyen de l'analyse bidimensionnelle.

Three dimensional coupled thermo-hydraulic-mechanical analysis code with PCG method

Y. Ohnishi, Kyoto Univ., Kyoto, JAPAN
M. Nishigaki, Okayama Univ., Okayama, JAPAN
A. Kobayashi, Hazama Corporation, Saitama, JAPAN
S. Akiyama, Kyoto Univ., Kyoto, JAPAN

ABSTRACT

For the assessment of the high level radioactive waste disposal and the understanding of the phenomena which occurred at the site, coupled thermal, hydraulic and mechanic behaviors have to be examined as exactly as possible. We have already developed the two dimensional coupling code for this purpose and have analyzed a few field experiments. The geometry of the depository is, however, three dimensional and that effect is not negligible. Therefor, we developed the new analysis code for three dimensional coupled thermal, hydraulic and mechanic phenomena in the saturated-unsaturated fields, which is called THAMES3D. This code takes advantage of preconditioned conjugate gradient method in order to save the computer storage and calculation time.

In this paper, we present the method of this code and verification results, and then simulate an example problem of a deep depository system and compare the three dimensional results with the ones obtained from the two dimensional analysis.

1.INTRODUCTION

The phenomena occurring near a high level radioactive waste repository site are very complicated. It is, however, necessary for the safety assessment to understand the complex behavior of groundwater flow in the vicinity of a repository.

Shortly after disposal, the temperature near the repository will increase and thermal stresses will occur, which have an influences on the permeability of rock mass. Furthermore the fluid flow by buoyancy induced by the increase of the temperature may occur in the rock mass around the repository. These phenomena have to be understood for the exact prediction of ground water flow.

We have already developed the two dimensional fully coupled thermal, hydraulic and mechanical analysis code called THAMES (Thermal, Hydraulic And MEchanical System analysis) [1] for the examination of such a phenomena mentioned above. The Buffer mass test and macropermeability test conducted at Stripa project have been analyzed with THAMES and the phenomena have been considered from the various points of view [2],[3]. It was found that such an analysis was effective for understanding the phenomena occurred at the experiment site.

We think that the detail analysis is essential for the assessment of deep geologic isolation of nuclear wastes. Thus we try to develop the THAMES to the three dimensional code for more exact analyses.

THAMES3D is a three dimensional finite element code developed to perform analyses of fully coupled thermal, hydraulic and mechanical behaviors of a saturated-unsaturated geologic medium. The mathematical formulation of the model utilizes the Biot's Theory with the Duhamel-Neuman form of Hooke's law and the energy balance equation. The governing equations are derived with the fully coupled thermal, hydraulic and mechanical relationships as shown in Fig.1. The three coupled equations are solved simultaneously.

Nonlinear parameter of granite rocks related to its heat conductivity, specific heat and thermal expansivity may be used in the calculation. For the analysis of fractured rock mass, the dependency of permeability in fractures on stresses proposed by Iwai [4] may be utilized. The permeability as a function of void ratio may also be utilized for the analysis of soils.

THAMES3D has a number of limitation, however, which must be recognized. Perhaps the most serious limitation is the fact that presently the code accounts for only elastic media. Considerable investigation is necessary to introduce the elasto-plastic theory under conditions which allow a change in the temperature. Other limitation of the code include its inability to simulate phase change of water induced by temperature change.

In this paper, we present the analysis method of THAMES3D, the verification of the code and then demonstrate the ability of THAMES3D to assess the phenomena occurred around a repository.

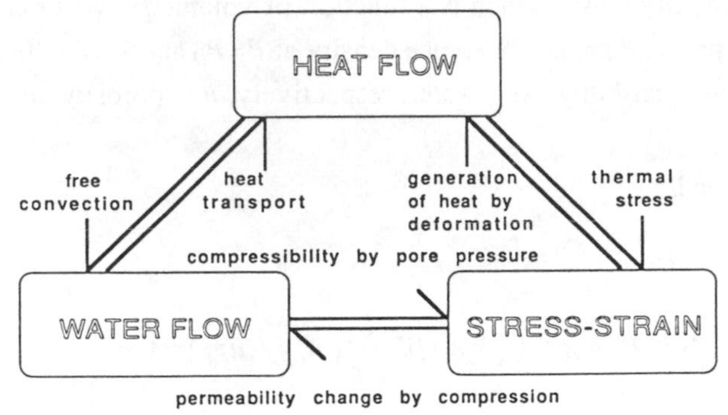

Fig.1 Iteration mechanisms of fully coupled thermal, hydraulic and mechanic analysis.

2. GOVERNING EQUATIONS

2.1 Assumptions

The governing equations are derived under the following assumptions.

1) The medium is isotropic and poro-elastic.
2) Darcy's law is valid for flow of water in the saturated-unsaturated medium.
3) Phase change of water between liquid and gas are not considered.
4) Heat transferred among three phases (solid, liquid and gas) are neglected. Energy flow occurs only in the solid and liquid phase.
5) Fourier's law holds for heat flux.
6) Water density varies depending upon the temperature and pressure of water.

2.2 Governing Equations

Using the above assumptions, we derive the governing equations like follows.

(1) Equilibrium Equation

$$\left[\frac{1}{2} C_{ijkl}(u_{k,l} + u_{l,k}) - \beta\delta_{ij} (T - T_0) + \chi\delta_{ij}\rho_f h\right]_{,j} + \rho b_i = 0 \quad (1)$$

where C_{ijkl} is an elastic matrix, ρ is the density of the medium, ρ_f is the density of water, b_i is the body force, χ is a nonlinear function of S_r (degree of saturation), T is the temperature, u_i is the deformation vector and β=(3λ +2μ)α. $(-\beta\delta_{ij}(T\text{-}T_0))_{,j}$ is the term which means that the heat transfer influences the equilibrium equation. λ and μ are Lamé's constants and α is the thermal expansivity coefficient. Subscript 0 means that the parameter are in a reference state. δ_{ij} is Kronecker's delta.

(2) Continuity Equation for Groundwater

$$\left\{ \rho_f k(\theta)_{ij} h_{,i} \right\}_{,i} - \rho_{fo} n S_r \rho_f g \beta_P \frac{\partial h}{\partial t}$$

$$- \rho_f C(\psi) \frac{\partial h}{\partial t} - \rho_f S_r \frac{\partial u_{i,i}}{\partial t} + \rho_{fo} n\, S_r \beta_T \frac{\partial T}{\partial t} = 0 \quad (2)$$

where $k(\theta)_{ij}$ is permeability tensor which is a function of volumetric water contents, θ, h is total head, P is pore water pressure, ρ_{f0} is a reference density at $P=P_0$ and $T=T_0$, β_T and β_P are thermal expansivity and compressibility of water, respectively, n is porosity and $C(\psi)$ is a specific water content .

(3) Energy Conservation Law

$$(\rho C_v)_m \frac{\partial T}{\partial t} + n s_r \rho_f C_{vf} V_{f\,i} T_{,i} - K_{Tm} T_{,ii}$$

$$+ n s_r T \frac{\beta_T}{\beta_P} k(\theta)\, h_{,ii} + \frac{1}{2}(1\text{-}n)\, \beta T \frac{\partial}{\partial T}(u_{i,j} + u_{j,i}) = 0 \quad (3)$$

$$(\rho\, C_v)_m = nS_r\rho_f C_{vf} + (1\text{-}n)\rho_s C_{vs}$$
$$K_{Tm} = nS_r\, K_{Tf} + (1\text{-}n)K_{Ts}$$

C_v is specific heat, K_T is a coefficient of heat conduction, V_f is the substantial velocity and T is temperature.

2.3 Finite Element Discretization

We discretize the above equations to finite element with the Galerkin method and then make the matrix form, $Gx = f$, where G is the global matrix, x is the unknown vector and f is the nodal force vector. This matrix is solved by the method mentioned below.

3.PRECONDITIONED CONJUGATE GRADIENT METHOD

Gauss Elimination method (GE method) is often used for solving the matrix form. GE method is very useful for solving the matrix of which band width is not large. The band width of the matrix is dependent on the number of unknown variables and the connectivity of the elements. It is generally very difficult to number nodes so as to make the band width small for three dimensional mesh. Furthermore five unknown variables have to be considered (displacements of X-Y-Z directions,total water head, temperature) in this three-dimensional coupled analysis. Thus a rather large band width of the global matrix is often produced for the analysis in this code. Large memory storage of computer and calculation time are needed if GE Method is applied for the examination with this code. From these reasons, we have to use the more effective method to solve the matrix.

THAMES3D takes advantage of the Preconditioned Conjugate Gradient Method (PCG method). In this iteration method only the relations between the neighboring nodes are considered. Therefore, the memory needed is smaller than in the case of the GE method.

The algorithms of PCG method can be explained as follows:

(1) First Step (initial state)

$$r_0 = b - Ax_0, \qquad P_0 = B\, r_0 \tag{4}$$

in which A is G^TG. G is the global matrix to be solved and superscript T means transposed matrix. b is G^Tf. f is the nodal force vector, x is the unknown variables vector, P is a correction vector, r_i is the error at the ith iteration and subscript 0 means the initial set. B is a preconditioner.

(2) Iteration Calculus

Firstly, we get the magnitude of correction,

$$\alpha_k = \frac{(P_k, r_k)}{(P_k, AP_k)} \tag{5}$$

where (P,r) is the inner product of P and r, α is the magnitude of correction, subscript k is the iteration number. Then the unknown vector is corrected with the next equation.

$$x_{k+1} = x_k + a_k P_k \tag{6}$$

Using this unknown vector, the error at the (k+1)th iteration is calculated by the following equation.

$$r_{k+1} = r_k - \alpha_k AP_k \tag{7}$$

Sequentially, the correction vector at (k+1)th iteration is obtained like the followings.

$$\beta_k = -\frac{(r_{k+1}, AP_k)}{(P_k, AP_k)}, \qquad P_{k+1} = r_{k+1} + \beta_k P_k \tag{8}$$

This iteration continues until P_k becomes zero. At this point, r_k also becomes zero and therefor x_k becomes exact . B is called a preconditioner which reduces the iteration cycles. In this case only the diagonal of global matrix A is involved in the preconditioner B.

4.VERIFICATION OF THAMES3D

In order to verify the functions of THAMES3D, theoretical solutions for the mechanical-hydraulic, mechanical-thermal and hydraulic-thermal problems are compared with the ones calculated with the THAMES3D. This code has to be verified for the fully coupled hydraulic, mechanic and thermal problem. This problem is, however, not solved analytically at the present. The validation of the THAMES3D is not carried out in this paper. This subject remains unsolved.

As to the comparison of the mechanical behavior with the hydraulic behavior, a one dimensional consolidation problem was investigated. Fig.2 illustrates the finite element mesh and the loading condition. The upper boundary is set to be a drainage condition and the other boundaries are set to be a no-flux condition. Displacement is allowed for the z-direction only. The numerical results are compared with Terzaghi's solutions as shown in Fig.3. As can be seen in this figure, the calculated results agree with the theoretical ones very well.

The mechanical-thermal analysis is applied to a two dimensional thermal stress problem as shown in Fig.4. This simulates the thermal stress induced in the thick walled cylinder with a constant temperature gradient. The horizontal and vertical boundaries are fixed and the inner and outer ones allow a displacement. The temperature distribution is given by $t = 5°C/r$ in which r is the radial distance and t is the temperature. The comparison of numerical results with analytical ones are shown in Fig.5. It is found from this figure that the calculated radial and tangential stresses agree with the analytical ones very well.

The heat transfer in moving groundwater field is the third problem to be consider for the THAMES3D verification analysis in this paper. The fluid flows to x-direction only with the constant velocity, 0.05m/sec.The temperature is fixed at 1 °C at the upstream boundary and 0 °C at the downstream boundary as shown in Fig.6. The results are shown in Fig.7. As can be seen in the figure, the theoretical results coincide with the numerical results very well. This type of advection - dispersion problem is difficult to solve for the advection dominate. The nature of the advection-dispersion equation can be conveniently characterized by the Peclet number. In this case it is 0.25. If the problem has the higher Peclet number, the THAMES3D will not solve the problem well. However, the problem treated for the high level radioactive waste disposal has the very low velocity field. Thus, this does not appear to be a problem for practical purpose.

5.EXAMPLE PROBLEM OF A DEEP NUCLEAR WASTE DEPOSITORY

High-level radioactive wastes are planed to be disposed in deep geological formations after a interim storage. In our model calculations, the region is considered having an area of 2000 x 2000 m and a height of 1500 m as shown in Fig.8. The geology is supposed as granite. The depository is assumed to be located within a 500 m times 500 m square at the depth of 1000 m. Table 1 gives the properties of the rock mass considered.

This model is analyzed by using the two-dimensional code,THAMES and the three-dimensional code, THAMES3D. Fig.9 is the schematic figure of two dimensional model and Fig.8 is that of three dimensional one. The depository is simulated by the heat sources indicated by circles in Fig.8,9. We set the heat source to be 100°C during the analyses. The temperature in the initial state is a function of the depth, D_{ep},

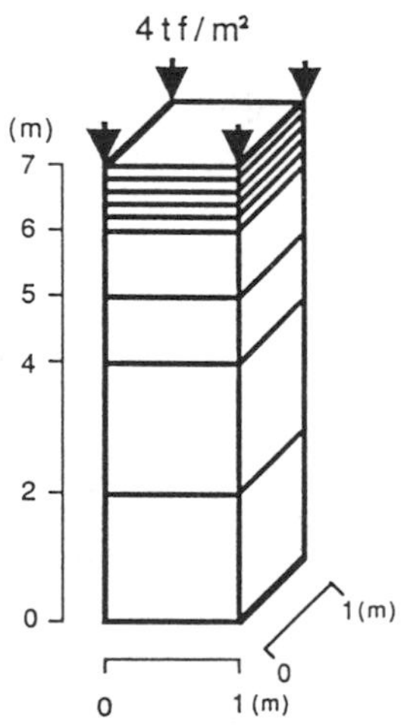

Fig.2 Finite element mesh

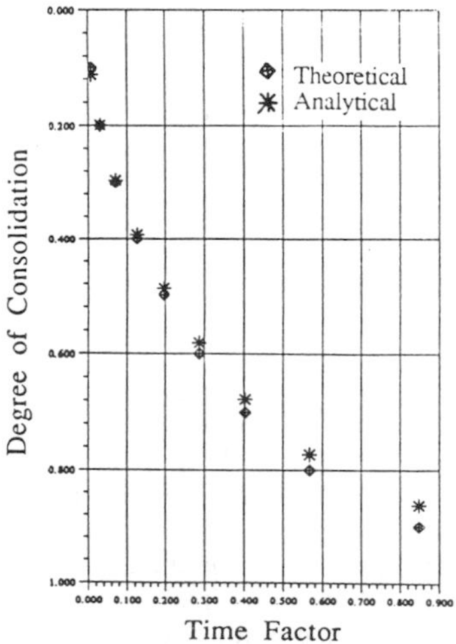

Fig.3 Results of consolidation analysis

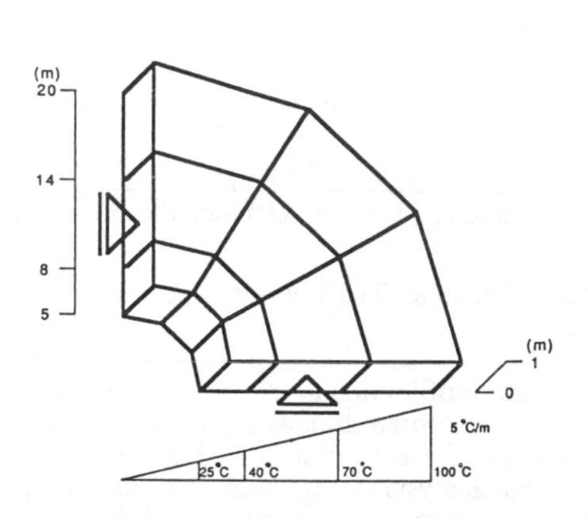

Fig.4 Finite element mesh

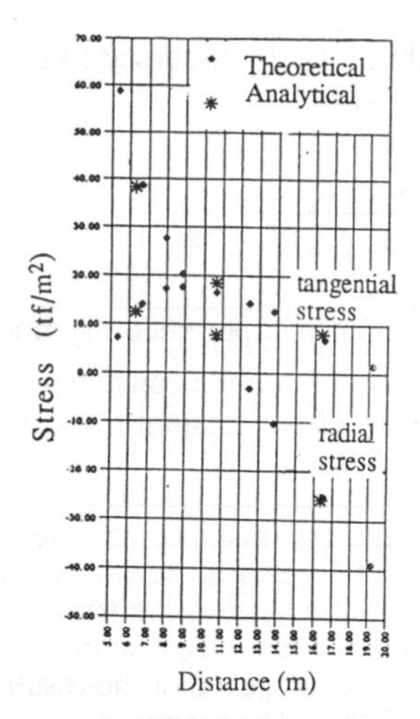

Fig.5 Results of thermal stress analysis

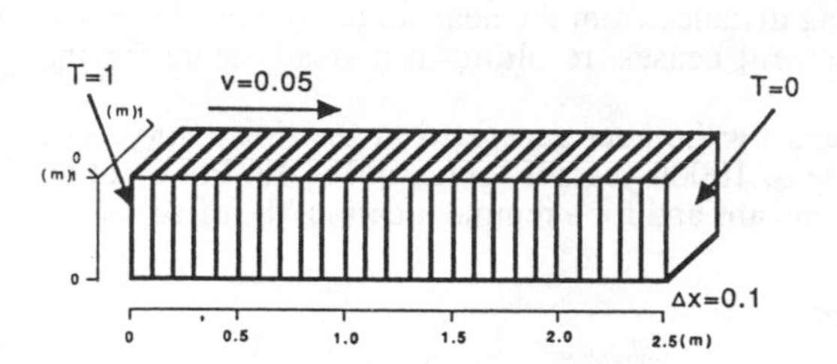

Fig. 6 Analysis model of heat transfer

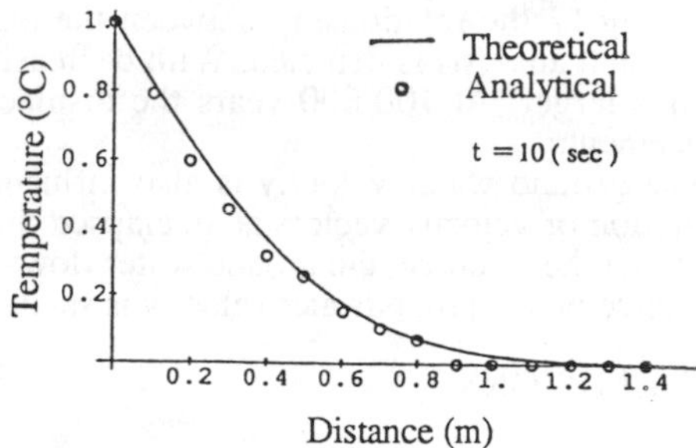

Fig.7 Results of hear transfer analysis

Table 1 Parameters used in the analysis

Parameters	Values
Young's Modulus	500000tf/m^2
Poisson's Ratio	0.3
Unit Weight	2.7tf/m^3
Initial Void Ratio	0.02
Instinct Permeability (x-direction)	10^{-12} m^2
(y-direction)	10^{-12} m^2
(z-direction)	10^{-12} m^2
Specific Heat of Soil	42.6 m/ C
Heat Conductivity of Fluid	6.12x10^{-5}tf/ °Cs
Heat Conductivity of Soil	6.12x10^{-5}tf/ °Cs
Thermal Expansion Coefficient	6.0x10^{-6} 1/ °C

$$t = 20 + 3xD_{ep}/100 \quad (°C) \tag{9}$$

Therefore, the temperature at the depository (= GL. -1000m) is 50°C.

The permeability is also assumed to be the function of the depth. We use the following equation proposed by Iwai [4].

$$\frac{k_e}{k_{od}} = \frac{\{1 + A(\sigma_{eo}/\zeta)^{t'}\}}{\{1 + A(\sigma_e/\zeta)^{t'}\}} \tag{10}$$

where k_e is the intrinsic permeability under the stress, σ_e and k_{od} is the one under the reference stress, σ_{eo}. A,ζ and t' are the constants.

Total water head is constant for all regions in the initial state. The boundary conditions are indicated in Fig.8,9.

The distribution of displacement vectors at y = 0 (frontal profile) due to thermal stress after an elapsed time of 100 years is shown in Fig.10 and Fig.11. The three dimensional analysis gives similar results as the two dimensional one at the vicinity of the heat source. However, with increasing distance from the heat source the three dimensional analysis yields smaller displacements. The reason for the divergence of the results is due to the assumption of plane strain condition in the two-dimensional case. The thermal stress are isotropic and the depository has the finite length for y-direction in Fig.8.

In Fig.12 the relationship between the elapsed time and the displacement from the three dimensional analysis is depicted. With decreasing distance from the heat source the displacement becomes larger. At 100 000 years the displacement ceases, resulting in a steady state for the displacement.

The ground water velocity is also influenced by the heat generated at the depository. The distribution of velocity vectors at an elapsed time of 10000 years is given in Fig.13. For location close to the heat source, the ground water flows upward and for a remote location, the effect of the heat source on the groundwater velocity is small.

6.CONCLUSIONS

For detailed assessment of the phenomena occurring at a high level radioactive waste repository, we developed a new analysis code for the assessment of the fully coupled thermal, hydraulic and mechanic behavior. This code used the PCG method in order to save the computer storage and calculation time. In this paper, we introduced the basic concept of this code and then presented verification of the results. The verification is carried out for the consolidation problem, thermal stress problem and heat transfer problem in the advection-dispersion field. Furthermore, the ability of this code for the assessment of disposal was demonstrated by using an ideal repository in a granite rock mass.

The followings are clear from the above examinations.

1) This code called THAMES3D has the ability to explain exactly the coupled thermal, hydraulic and mechanic behavior.

2) The two dimensional analysis estimates larger thermal expansion due to the heat generated at the depository than the three dimensional one,

3) The fluid flow due to the buoyancy is not samll at the vicinity of site.

REFERENCES

[1] Y. Ohnishi, H. shibata and A. Kobayashi: " Development of Finite Element Code for the Analysis of Coupled Thermo-hydro=mechanical Behaviors of Saturated-unsaturated medium", Proc. of Int. Symp. on Coupled Processes Affecting the Performance of a Nuclear Waste Repository, Berkely, (1985), pp.263-268.

[2] Y.Ohnishi, A. Kobayashi and M.Nishigaki: " Finite Element Coupled Process Analysis of Buffer Mass Test in the Stripa Rock Mass", GEOVAL87, Stockholm, (1987),pp.665-676.

[3] Y.Ohnishi, A. Kobayashi and M.Nishigaki: " Thermo-hydro-mechanical Behavior of Rocks around an Underground Opening", ISRM CONFERENCE, Montreal, pp.207-210.

[4] K.Iwai:" Fundamental Studies of Fluid Flow through a Single Fracture", ph.D. Dissertation, Univ. of California, Berkeley, 1976.

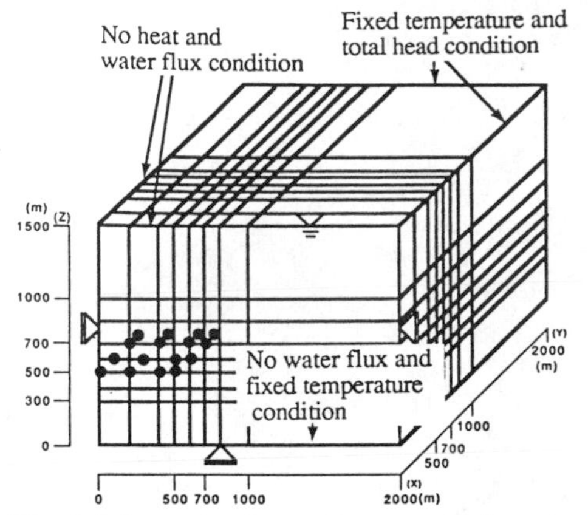

Fig. 8 Schematic view of 3D analysis model

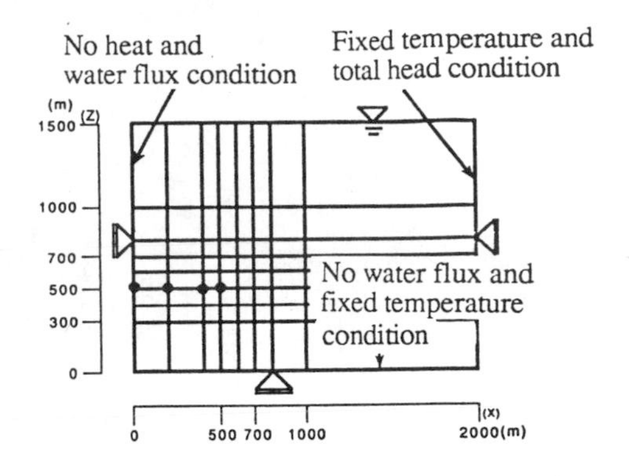

Fig.9 Analysis model of 2D analysis

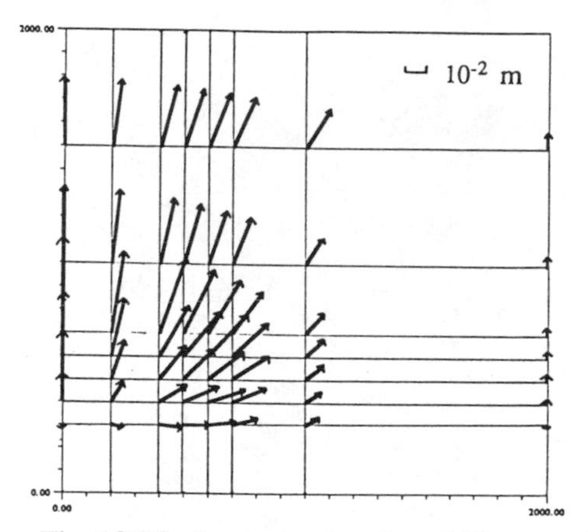

Fig. 10 Displacement vectors at 100 years (2D analysis)

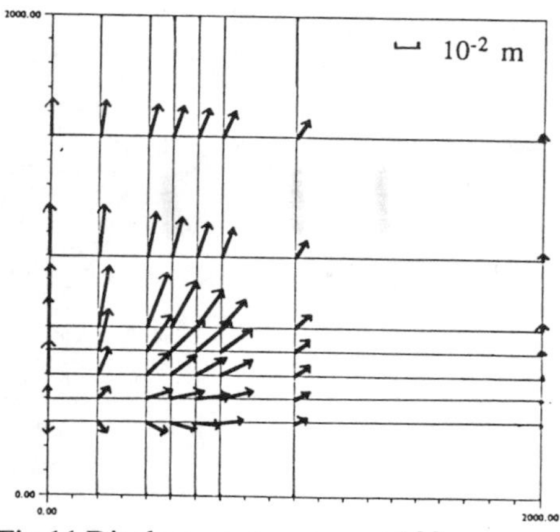

Fig.11 Displacement vectors at 100 years (3D analysis)

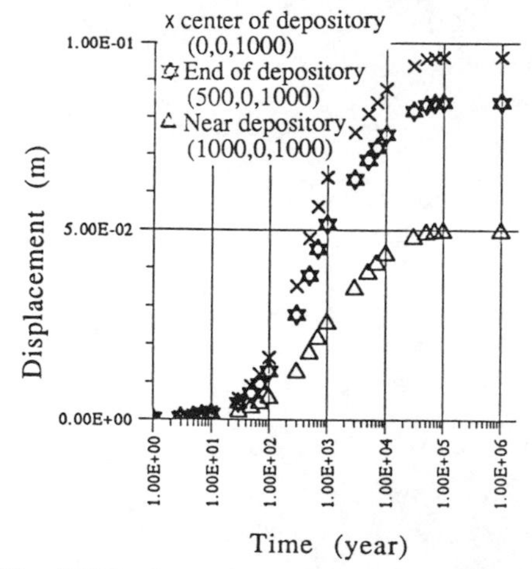

Fig. 12 Displacement as a function of time

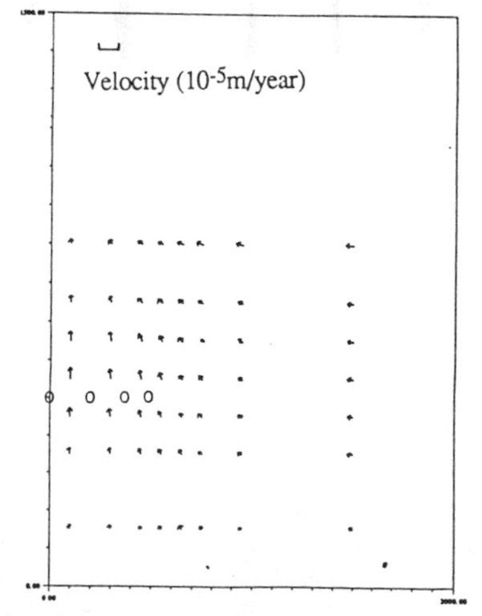

Fig.13 Velocity vectors at 10000 years

Session 5

VALIDATION STRATEGY

Séance 5

STRATEGIE DE VALIDATION

Chairman - Président

P.E. AHLSTRÖM
(Sweden)

VALIDATION OF MATHEMATICAL MODELS

Robert G. Sargent
Simulation Research Group
Syracuse University, USA

ABSTRACT

This paper discusses the validation of computer-based mathematical models as it is performed in the fields of operations research and systems engineering. The various approaches for deciding model validity are described, a description relating model verification and validation to the model development process is given, various validation techniques are defined, a recommended validation procedure is presented, and a limited bibliography is furnished. The recommended validation procedure consists of steps that as minimum should be performed in the various phases of the model development process. This procedure is commonly used in practice; in particular for the validation of simulation models.

VALIDATION DES MODELES MATHEMATIQUES

RESUME

L'auteur examine la validation des modèles mathématiques informatisés telle qu'elle est exécutée en recherche opérationnelle et en ingénierie des systèmes. Il décrit les diverses méthodes utilisées pour décider de la validité des modèles, détaille les liens entre la vérification et la validation des modèles, d'une part, et le processus de mise au point des modèles, d'autre part, définit diverses techniques de validation, expose une procédure de validation recommandée et présente une bibliographie restreinte. La procédure de validation qui préconise l'auteur consiste en une série minimale d'opérations qu'il conviendrait d'exécuter dans les diverses phases du processus de mise au point du modèle. En pratique, cette procédure est communément utilisée, notamment pour valider les modèles fondés sur la simulation.

1. INTRODUCTION

Computer-based mathematical models are increasingly being used in problem-solving and to aid in decision-making. The developers and users of these models, the decision-makers using information derived from the results of the models,and people effected by decisions based on such models are all rightly concerned with whether a model and its results are "correct". This concern is addressed through model verification and validation. Model validation is usually defined to mean "substantiation that a computerized model within its domain of applicability possesses a satisfactory range of accuracy consistent with the intended application of the model" [22] and is the definition used here. Model verification is often defined as ensuring that the computer program of the computerized model and its implementation is correct, and is the definition adopted here. A related topic is model credibility (or acceptability) which is developing in the (potential) users of information from the models (e.g., decision-makers) sufficient confidence in the information that they are willing to use it.

A model should be developed for a specific purpose or use and its validity determined with respect to that purpose. If the purpose of a model is to answer a variety of questions, the validity of the model needs to be determined with respect to each question. Sometimes, different models of the same system are developed for different purposes. Several sets of experimental conditions are usually required to define the domain of a model's intended application. A model may be valid for one set of experimental conditions and be invalid in another. A model is considered valid for a set of experimental conditions if its accuracy is within the acceptable range of accuracy which is the amount of accuracy required for the model's intended purpose. This generally requires that the variables of interest be identified and their required accuracy determined. If the variables of interest are random variables, then properties and functions of the random variables such as their means and variances are frequently what is of primary interest and are what are used in determining model validity.

The substantiation that a model is valid, i.e., model validation, is part of the *total model development process* and is itself a process. The *validation process* consists of performing tests and evaluations within the model development process to determine whether a model is valid or not. Several models or versions of a model are usually developed in the modelling process prior to obtaining a satisfactory valid model.

It is often too costly and time consuming to determine that a model is ***absolutely*** valid over the complete domain of its intended application. Instead, tests and evaluations are conducted until sufficient confidence is obtained that a model can be considered valid for its intended application [19, 24]. The relationships of the cost (and similarily for time) of performing model validation and the value of the model to the user as a function of model confidence are illustrated in Figure 1. This cost is usually significant; inparticular where extremely high confidence is required because of the consequence of an invalid model.

The major objective of this paper is to describe the basic approaches used in determining the validity of computer-based mathematical models in operations research and systems engineering. Model validation (or assessment) is receiving increased interest in these fields, in both research and practice. The validity of simulation models has received the most attention.

2. VALIDATION PROCESS

There are three basic approaches used in operations research and systems engineering for the determination of whether a computer-based mathematical model is valid. Each of these approaches require the model development team to conduct verification and validation as part of the model development process and this is discussed below in some detail. The most common approach is for the model development team to make the decision on whether their model is valid. This decision is a subjective decision based on the results of various tests and evaluations conducted as part of the model development process.

Another approach (sometimes called independent verification and validation) uses a third (independent) party to make a subjective decision on whether the model is valid. The third party is independent of both the model development team and the model sponsor/user(s). Their evaluation is inaddition to that done by the model development team and is usually done after the model has been developed. This evaluation can be as simple as reviewing what verification and validation has been performed in the model's development or can be a complete verification and validation effort. Some experiences over this range is contained in [25] based on third party evaluation of energy models. One

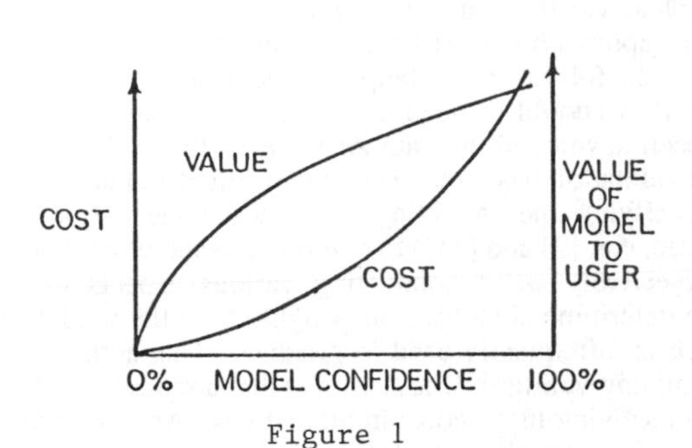

Figure 1

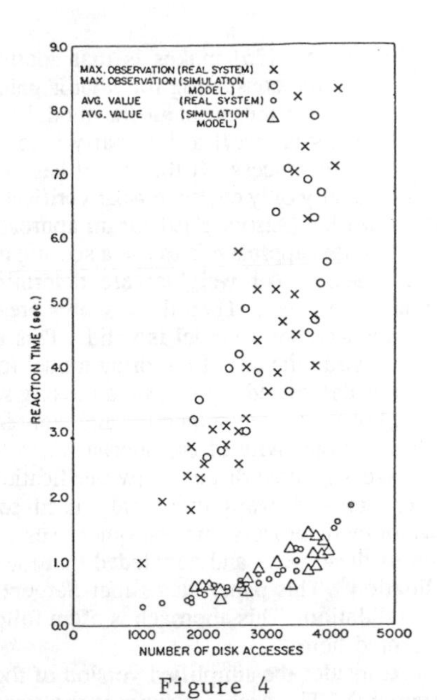

Figure 4

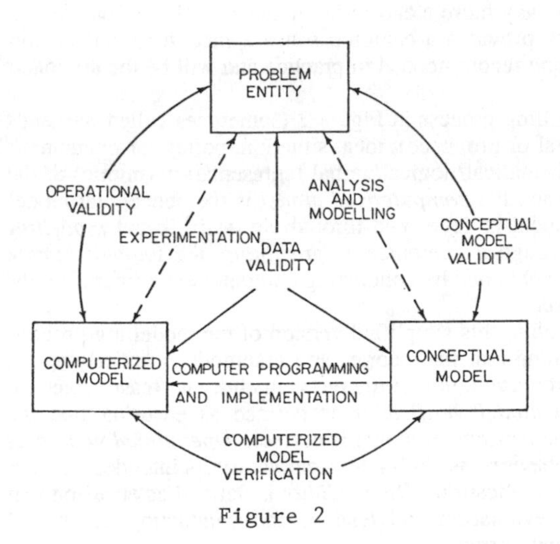

Figure 2

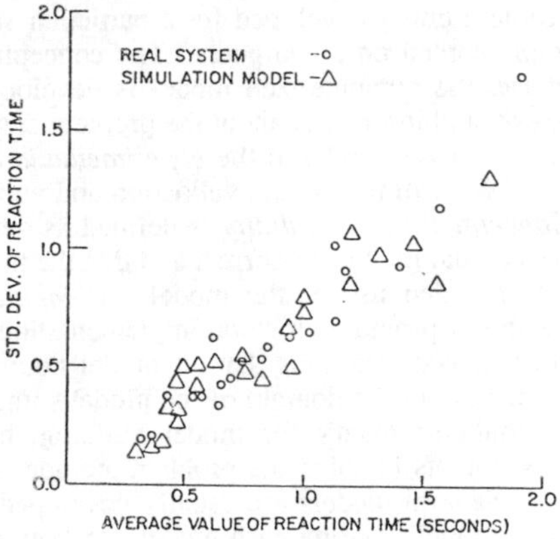

Figure 5

	OBSERVABLE SYSTEM	NON-OBSERVABLE SYSTEM
SUBJECTIVE APPROACH	• COMPARISON OF DATA USING GRAPHICAL DISPLAYS • EXPLORE MODEL BEHAVIOR	• EXPLORE MODEL BEHAVIOR • COMPARISON TO OTHER MODELS
OBJECTIVE APPROACH	• COMPARISON OF DATA USING STATISTICAL TESTS AND PROCEDURES	• COMPARISON TO OTHER MODELS USING STATISTICAL TESTS AND PROCEDURES

Figure 3

conclusion that [25] makes is that another complete verification and validation effort is extremely costly and time consuming for what is gained. This approach is usually used when there is a large cost associated with the problem the model is being used for and/or to help in model credibility. This author's view is that if a third party is to be used, they should be used and involved **during** the model development process. If the model has already been developed, this author believes that a third party should usually only evaluate what verification and validation has already been performed and not repeat earlier work. (Also see [9] for an approach to specifying and validating model simultaneously).

The last approach is to use a scoring model (see, e.g. [2] and [11]) to determine whether a model is valid. Scores (or weights) are determined subjectively when conducting various aspects of the validation process. Then these scores are used to determine some overall weight(s) for the model for deciding whether a model is valid. This approach is infrequently used in practice. This author does not believe in the use of a scoring model for determining validity. One reason is that a model can have a serious defect and yet receive a passing score by receiving high scores in other areas. Another reason is that the scores are subjective and combining them for a number which then decides validity tends to hide the subjectivity of the approach.

We will now discuss how verification and validation relates to the model development process. There are two ways commonly used to view this relationship. One way uses a detail model development process and the other uses a simple model development process. [8] reviewed work in both of these ways and concluded that the simple way more clearly illuminates model verification and validation. This paper states that Sargent's work provides a common sense approach to verification and validation. This approach is often followed and recommended in practice and will be the approach presented here.

Consider the simplified version of the modelling process in Figure 2 (sometimes called Sargent's paradigm). The *problem entity* is the system (real or proposed), idea, situation, policy, or phenomena to be modelled; the *conceptual model* is the mathematical/ logical/verbal representation (mimic) of the problem entity developed for a particular study; and the *computerized model* is the conceptual model implemented on a computer. The conceptual model is developed through an *analysis* and *modelling phase*, the computerized model is developed through a *computer programming and implementation phase*, and inferences about the problem entity are obtained by conducting computer experiments on the computerized model in the *experimentation phase*.

We will now relate validation and verification to this simplified version of the modelling process. *Conceptual model validity* is defined as determining that the theories and assumptions underlying the conceptual model are correct and that the model representation of the problem entity is "reasonable" for the intended use of the model. *Computerized model verification* is defined as ensuring that the computer programming and implementation of the conceptual model is correct. *Operational validity* is defined as determining that the model's output behavior has sufficient accuracy for its intended purpose (or use) over the domain of the model's intended application. *Data validity* is defined as ensuring that the data necessary for model building, model evaluation and testing, and conducting the model experiments to solve the problem are adequate and correct.

Several models are usually developed in the modelling process prior to obtaining a satisfactory valid model. During each model iteration, model verification and validation are performed. A variety of (validation) techniques are used, which are described below. Unfortunately, no algorithm or procedure exists to select which techniques to use. Some of their attributes are discussed in [19].

3. VALIDATION TECHNIQUES

This section describes various validation techniques (and tests) used in model verification and validation. Most of the techniques described here are found in the literature (see [5] for a detailed bibliography), although some may be described slightly different. They can be used either subjectively or objectively. By objectively, we mean using some type of statistical test or procedure, e.g., hypothesis tests and confidence intervals. A combination of techniques is usually used. These techniques are used for validating and verifying both the overall model and submodels.

Animation (Operational Graphics): The model's operational behavior is displayed graphically as the model moves through time. Examples are (i) the graphical plot of the status of a server as the model moves through time, i.e., is it busy, idle, or blocked, and (ii) the graphical display of parts moving through a factory.

Comparison to Other Models: Various results (e.g., outputs) of the simulation model being validated are compared to results of other (valid) models. For example, simple cases of a simulation model may be compared to known results of analytic models. (In some cases, the model being validated may require modification to allow comparisons to be made between it and analytic models.)

Degenerate Tests: The degeneracy of the model's behavior is tested by removing portions of the model or by appropriate selection of the values of the input parameters. For example, does the average number in the queue of a single server continue to increase with respect to time when the arrival rate is larger than the service rate.

Event Validity: The "events" of occurrences of the simulation model are compared to those of the real system to determine if they are the same. An example of events are deaths in a given fire department simulation.

Extreme-Condition Tests: The model structure and output should be plausible for any extreme and unlikely combination of levels of factors in the system, e.g., if in-process inventories are zero, production output should be zero. Also, the model should bound and restrict the behavior outside of normal operating ranges.

Face Validity: Face validity is asking people knowledgeable about the system whether the model and/or its behavior is reasonable. This technique can be used in determining if the logic in the conceptual model is correct and if a model's input-output relationships are reasonable.

Fixed Values: Fixed values are used for all model input and internal variables. This should allow checking the model results against hand calculated values.

Historical Data Validation: If historical data exist (or if data is collected on a system for building or testing the model), part of the data is used to build the model and the remaining data is used to determine (test) if the model behaves as the system does. (This testing is conducted by driving the simulation model with either Distributions or Traces [4, 6]).

Historical Methods: The three historical methods of validation are *Rationalism*, *Empiricism*, and *Positive Economics*. Rationalism assumes that everyone knows whether the underlying assumptions of a model are true. Then logic deductions are used from these assumptions to develop the correct (valid) model. Empiricism requires every assumption and outcome to be empirically validated. Positive Economics requires only that the model be able to predict the future and is not concerned with its assumptions or structure (causal relationships or mechanisms).

Internal Validity: Several replications (runs) of a stochastic model are made to determine the amount of internal stochastic variability in the model. A high amount of variability (lack of consistency) may cause the model's results to be questionable, and, if typical of the problem entity, may question the appropriateness of the policy or system being investigated.

Multistage Validation: [15] proposed combining the three historical methods of Rationalism, Empiricism, and Positive Economics into a multistage process of validation. This validation method consists of (1) developing the model's assumptions on theory, observations, general knowledge, and function, (2) validating the model's assumptions where possible by empirically testing them, and (3) comparing (testing) the input-output relationships of the model to the real system.

Parameter Variability - Sensitivity Analysis: This validation technique consists of changing the values of the input and internal parameters of a model to determine the effect upon the model and its output. The same relationships should occur in the model as in the real system. Those parameters which are sensitive, i.e., cause significant changes in the model's behavior, should be made sufficiently accurate prior to using the model. (This may require iterations in model development.)

Predictive Validation: The model is used to predict (forecast) the system behavior and comparisons are made to determine if the system behavior and the model's forecast are the same. The system data may come from an operational system or experiments may be performed, e.g., field tests.

Traces: The behavior of different types of specific entities in the model are traced (followed) through the model to determine if the model's logic is correct and if the necessary accuracy is obtained.

Turing Tests: People who are knowledgeable about the operations of a system are asked if they can discriminate between system and model outputs. ([23] contains statistical procedures).

4. DATA VALIDITY

Even though data validity is usually not considered part of model validation, we discuss it because it is usually difficult, time consuming, and costly to obtain sufficient, accurate and appropriate data, and is frequently the reason that initial attempts to validate a model fail. Basically, data is needed for three purposes: for building the conceptual model, for validating the model, and for performing experiments with the validated model. In model validation, we are concerned only with the first two types of data.

To build a conceptual model, we must have sufficient data (and understanding) about the problem entity in order to develop the mathematical and logical relationships in the model for it to adequately represent the problem entity for its intended use. It is also highly desirable to have data to develop theories that can be used in building the model and to test the model's underlying assumptions. The second type of data desired is behavior data on the problem entity to be used in the operational validity step of comparing the problem entity's behavior with the model's behavior. (Usually, these data are system input/output data.) If these data are not available, high model confidence usually cannot be obtained because sufficient operational validity cannot be achieved.

The concern with data is that appropriate, accurate and sufficient data are available, and if any data transformations are made, such as disaggregation, they are correctly performed. Unfortunately, there is not much that can be done to ensure that the data are correct. The best that one can do is to develop good procedures for collecting data; test the collected data using such techniques as internal consistency checks and screening for outliers and determine if any outliers found are correct; and develop good procedures to properly maintain the collected data. If the amount of data is large, a data base should be developed and maintained.

5. CONCEPTUAL MODEL VALIDATION

Conceptual model validity is determining that the theories and assumptions underlying the conceptual model are correct and that the model representation of the problem entity and the model's structure, logic, and mathematical and causal relationships are "reasonable" for the intended use of the model. The theories and assumptions underlying the model should be tested using mathematical analysis and statistical methods on problem entity data. Examples of theories and assumptions are linearity, independence, stationary, and Poisson arrivals. Examples of applicable statistical methods are fitting distributions to data; estimating parameter values, mean, variance, and correlations between data observations; and plotting data to see if it is stationary. In addition, all theories used should be reviewed to ensure they were applied correctly; for example, if a Markov chain is used, does the system have the Markov property and are the states and transition probabilities correct?

Next, each submodel and the overall model must be evaluated to determine if they are reasonable and correct for the intended use of the model. This should include determining if the appropriate detail and aggregate relationships have been used for the model's intended purpose, and if the appropriate structure, logic, and mathematical and causal relationships have been used. The primary validation techniques used for these evaluations are face validation and traces. Face validation is having experts on the problem entity evaluate the conceptual model to determine if they believe it is correct and reasonable for its purpose. This usually means examining the flowchart or graphical model, or the set of model equations. The use of traces is the tracking of entities through each submodel and the overall model to determine if the logic is correct and the necessary acuracy is maintained. If any errors are found in the conceptual model, it must be revised and conceptual model validation performed again.

6. COMPUTERIZED MODEL VERIFICATION

Computerized model verification is ensuring that the computer programming and implementation of the conceptual model is correct. To help ensure that a correct computer program is obtained, program design and development procedures found in the field of Software Engineering should be used in developing and implementing the computer program. These include such techniques as top-down design, structured programming, and program modularity. A separate program module should be used for each submodel, the overall model, and for each function (e.g., random variate generators and integration routines).

One should be aware that the use of different types of computer languages effects the probability of having a correct program. The use of a special purpose language such as a simulation language generally will result in having less (programming and implementation) errors than if a higher order language is used. Not only does the use of special purpose languages increase the probability of having a correct program, they usually reduce programming time.

After the computer program has been developed and implemented, the program must be tested for correctness and accuracy. There are numerous static and dynamic testing methods (see, e.g. [21] and [26]). One must continuously be aware in checking that errors may be caused by the data, the conceptual model, the computer program, or the computer implementation (see Figure 2).

7. OPERATIONAL VALIDITY

Operational validity is primarily concerned with determining that the model's output behavior has the accuracy required for the model's intended purpose over the domain of its intended application. This is where most of the validation testing and evaluation takes place. The computerized model is used in operational validity and thus any deficiencies found can be due to an inadequate conceptual model, an improperly programmed or implemented conceptual model on the computer (e.g., due to programming errors or insufficient numerical accuracy), or due to invalid data.

All of the validation techniques discussed in Section 3 are applicable to operational validity. Which techniques and whether to use them objectively or subjectively must be decided by the model developer and other interested parties (e.g. model sponsor). The major attribute effecting operational validity is whether the problem entity (or system) is observable or not, where observable means it is possible to collect data on the operational behavior of the program entity. Figure 3 gives one classification of the validation approaches for operational validity. The "explore model behavior" in this figure means to examine the behavior of the model using appropriate validation techniques for various sets of experimental conditions from the domain of the model's intended use and usually includes parameter variability-sensitivity analysis.

To obtain a *high* degree of confidence in a model and its results, comparison of the model's and system's input-output behaviors for at *least* two (and preferably several more) different sets of experimental conditions is usually required. There are three basic comparison approaches used: (i) graphs of the model and system behavior data, (ii) confidence intervals, and (iii) hypothesis tests. Graphs are the most commonly used approach and confidence intervals are next.

7.1. Graphical Comparison of Data

The model's and system's behavior data are plotted on graphs for various sets of experimental conditions to determine if the model's output behavior has sufficient accuracy for its intended purpose. See Figures 4 and 5 for examples of such graphs. A variety of graphs showing different types of measures and relationships are required. Examples of measures and relationships are (i) time series, means, variances, and maximums of each output variable, (ii) relationships between parameters of each output variable, e.g., means and standard deviations, and (iii) relationships between different output variables. It is important that appropriate measures and relationships be used in validating a model and that they be determined with respect to the model's intended purpose. For an example of a set of graphs used in the validation of a model, see [1].

These graphs can be used in model validation in three ways. First, the model development team can use the graphs in the model development process to make a subjective judgement on whether a model possess sufficient accuracy for its intended purpose. Secondly, they can be used in the face validity technique where experts are asked to make subjective judgements on whether a model possess sufficient accuracy for its intended purpose. Third, graphs can be used in Turing tests.

7.2. Confidence Intervals and Hypothesis Tests

Confidence intervals (c.i.), simultaneous confidence intervals (s.c.i.), and joint confidence regions (j.c.r.) can be obtained for the differences between the population parameters, e.g., means and variances, and distributions of the model and system output variables for each set of experimental conditions. These c.i., s.c.i., and j.c.r. can be used as the model range of accuracy for model validation. It is usually desirable to construct the model range of accuracy with the lengths of the c.i. and s.c.i. and the sizes of the j.c.r. as small as possible. The shorter the lengths or the smaller the sizes, the more meaningful the specification of the model range of accuracy will usually be. The lengths and the sizes of the joint confidence regions are affected by the values of confidence levels, variances of the model and system response variables, and sample sizes. The lengths can be shortened or sizes made smaller by decreasing the confidence levels and increasing the sample sizes. A tradeoff needs to be made among the sample sizes, confidence levels, and estimates of the length or sizes of the model range of accuracy. In those cases where the cost of data collection is significant for either the model or system the data collection cost should also be considered in the tradeoff analysis. (See [6] for details).

Hypothesis tests can be used in the comparison of parameters, distributions, and time series of the output data of a model and a system for each set of experimental conditions to determine if the model's output behavior has an acceptable range of accuracy. The first step is to state the hypotheses:

H_0: Model is valid for the acceptable range of accuracy under the set of experimental conditions.

H_1: Model is invalid for the acceptable range of accuracy under the set of experimental conditions.

Two types of errors are possible in testing this hypothesis. The first or type I error is rejecting the validity of a valid model; the second or type II error is accepting the validity of an invalid model. The probability of a type error I is called *model builder's risk* (α) and the probability of type II error is called *model user's risk* (β). In model validation, model user's risk is extremely important and must be kept small. Thus *both* type I and type II errors *must* be considered in using hypothesis testing for model validation.

The amount of agreement between a model and a system can be measured by a *validity measure*. There is a direct relationship among builder's risk, model user's risk, validity measure and sample size of observations. A tradeoff among these *must* be made in using hypothesis tests in model validation. In those cases where the data collection cost is significant for either the model or system, the data collection cost should also be considered in performing the tradeoff analysis. For details of the methodology of using Hypothesis Test see [3, 4, 7].

8. RECOMMENDED MODEL VALIDATION PROCEDURE

There are currently no algorithms or procedures available to identify specific validation techniques, statistical tests, etc. to use in the validation process. Various authors suggest (e.g., [24]) that as a minimum the three steps of (1) Face Validity, (2) Testing of the Model Assumptions, and (3) Testing of Input-Output Transformations be made. These recommendations are made in general and are not related to the steps of the modelling process discussed in Section 2.

This author recommends that, as a minimum, the following steps be performed in model validation:

(1) An agreement be made between (i) the modelling team and (ii) the model sponsors and users (if possible) on the basic validation approaches and on a minimum set of specific validation techniques to be used in the validation process ***prior*** to developing the model.

(2) The assumptions and theories underlying the model be tested, when possible.

(3) In each model iteration, at least face validity be performed on the conceptual model.

(4) In each model iteration, explore the model's behavior using the computerized model.

(5) In at least the last model iteration, comparisons be made between the model and system behavior (output) data for at ***least*** two sets of experimental conditions, when possible.

(6) Validation discussed in the model documentation.

(7) If the model is to be used over a period of time, a schedule for periodic review (and possible revalidation) of it be made and followed.

With respect to documentation, [13] contains a description of a set of documentation for computer-based models. The documentation on verification and validation should include what tests and evaluations were performed along with their justification and supporting data, and the conclusions drawn with the supporting reasons. Tables that have the following format are frequently helpful. The first column lists the issues being addressed, e.g., each underlying assumption and the accuracy of each output variable. The second column contains the techniques used for each issue along with justifications for their use. The third column contains the conclusions drawn in applying each technique and for each issue with the supporting reasons. A confidence level such as low, medium, and high, might be attached to each set of conclusions. A separate table can be used for conceptual model validity, computer model verification, operational validity, data validity, and an overall summary. These tables can be referred to in the documentation text, used for drawing general conclusions, and read as a summary.

Models occasionally are developed to be used over time. A procedure for reviewing the validity of these models over their life cycles needs to be developed. No general procedure can be specified as each situation is different. For example, if no or only a little data were available on the problem entity (e.g. system) when the model was initially developed and validated, then revalidations of it should take place prior to each time the model is used if additional data or system understanding has occurred since its last validation.

10. SUMMARY

Model validation is one of the most critical issues faced by a modeller. Unfortunately, there is no set of specific tests that can be easily applied to determine the validity of the model. It is usually much easier to show that a model is invalid than valid. Furthermore, no algorithm exists to determine what techniques or procedures to use. Every new modelling project presents a new and unique challenge. There is considerable literature on model validation [5]. Articles given in the limited bibliography can be used to further your knowledge on model validation.

LIMITED BIBLIOGRAPHY

[1] Anderson, H.A. and R.G. Sargent (1974). An Investigation into Scheduling for an Interactive Computer System, *IBM Journal of Research and Development*, *18*, 2, pp. 125-137.

[2] Balci, O. (1989). How to Assess the Acceptability and Credibility of Simulation Results, *Proceedings of the 1989 Winter Simulation Conference*, Washington, DC., pp. 62-71.

[3] Balci, O. and R.G. Sargent (1981). A Methodology for Cost-Risk Analysis in the Statistical Validation of Simulation Models, *Communications of the ACM*, *24*, 4, pp. 190-197.

[4] Balci, O. and R.G. Sargent (1982). Validation of Multivariate Response Simulation Models by Using Hotelling's Two-Sample T^2 Test, *Simulation*, Vol. 39, No. 6, pp. 185-192.

[5] Balci, O. and R.G. Sargent (1984). A Bibliography on the Credibility, Assessment and Validation of Simulation and Mathematical Models, *Simuletter*, Vol. 15, No. 3, pp. 15-27.

[6] Balci, O. and R.G. Sargent (1984). Validation of Simulation Models via Simultaneous Confidence Intervals, *Amer. J. of Math. and Management Science*, 4, 3 and 4, pp. 375-406.

[7] Banks, J. and J.S. Carson II (1984). *Discrete-Event System Simulation*, Prentice-Hall, NJ.

[8] Banks, J., D. Gerstein, and S.P. Searles (1988). Modeling Processes, Validation, and Verification of Complex Simulations: A Survey, *Methodology and Validation*, Simulation Series, Vol. 19, No. 1. The Society for Computer Simulation SCS, pp. 13-18.

[9] Davis, E.A. (1986). Use of Seminar Gaming to Specify and Validate Simulation Models, *Proceedings of the 1986 Winter Simulation Conference*, Washington, D.C., pp. 242-247.

[10] DOD Simulations: Improved Assessment Procedures Would Increase the Credibility of Results (1987). United States General Accounting Office, PEMD-88-3.

[11] Gass, S.I. (1977). A Procedure for Evaluation of Complex Models, *Proceedings of the First International Conference on Mathematical Modeling*, University of Missouri, pp. 247-257.

[12] Gass, S.I. (1983). Decision-Aiding Models: Validation, Assessment, and Related Issues for Policy Analysis, *Operations Research*, 31, 4, pp. 601-663.

[13] Gass, S.I. (1984). Documenting a Computer-Based Model, *Interfaces*, 14, 3, pp. 84-93.

[14] Guidelines for Model Evaluation (1979). U.S. States General Accounting Office, PAD- 79-17.

[15] Naylor, T.H. and J.M. Finger (1967). Verification of Computer Simulation Models, *Management Science*, 14, 2, pp. B92-B101.

[16] Oren, T. (1981). Concepts and Criteria to Assess Acceptability of Simulation Studies: A Frame of Reference, *Communications of the ACM*, *24*, 4, pp. 180-189.

[17] Rao, M.J. and R.G. Sargent (1988). An Advisory System for Operational Validity, *Artificial Intelligence and Simulation: The Diversity of Applications*, SCS, San Diego, CA, pp. 245-250.

[18] Sargent, R.G. (1981). An Assessment Procedure and a Set of Criteria for Use in the Evaluation of Computerized Models and Computer-Based Modelling Tools, RADC-TR-80-409.

[19] Sargent, R.G. (1984). Simulation Model Validation, *Simulation and Model-Based Methodologies: An Integrative View*, edited by Oren, et al., Springer-Verlag.

[20] Sargent, R.G. (1986). The Use of Graphic Models in Model Validation, *Proceedings of the 1986 Winter Simulation Conference*, Washington, D.C., pp. 237-241.

[21] Sargent, R.G. (1988). A Tutorial on Validation and Verification of Simulation Models, *Proceedings of 1988 Winter Simulation Conference*, pp. 33-39.

[22] Schlesinger, et al. (1979). Terminology for Model Credibility, *Simulation*, 32, 3, pp. 103-104.

[23] Schruben, L.W. (1980). Establishing the Credibility of Simulations, *Simulation*, 34, 3, pp. 101-105.

[24] Shannon, R.E. (1975). *Systems Simulation: The Art and the Science*, Prentice-Hall.

[25] Wood, D.O. (1986). MIT Model Analysis Program: What We Have Learned About Policy Model Review, *Proc. of the 1986 Winter Simulation Conference*, Washington, D.C., pp. 248-252.

[26] Whitner, R.B. and O. Balci (1989). Guidelines for Selecting and Using Simulation Model Verification Techniques, *Proc. of 1989 Winter Simulation Conf.* Washington, D.C., pp. 559-568.

A PROPOSED STRATEGY FOR THE VALIDATION OF GROUND-WATER FLOW AND SOLUTE TRANSPORT MODELS

P.A. Davis
Sandia National Laboratories, Albuquerque, NM, USA

M.T. Goodrich
GRAM, Inc., Albuquerque, NM, USA

ABSTRACT

Ground-water flow and transport models can be thought of as a combination of conceptual and mathematical models and the data that characterize a given system. The judgment of the validity or invalidity of a model depends both on the adequacy of the data and the model structure (i.e., the conceptual and mathematical model). This report proposes a validation strategy for testing both components independently. The strategy is based on the philosophy that a model cannot be proven valid, only invalid or not invalid. In addition, the authors believe that a model should not be judged in absence of its intended purpose. Hence, a flow and transport model may be invalid for one purpose but not invalid for another.

PROJET DE STRATEGIE POUR LA VALIDATION DE MODELES SIMULANT L'ECOULEMENT DE L'EAU SOUTERRAINE ET LE TRANSPORT DE SOLUTES

RESUME

On peut considérer les modèles de transport et d'écoulement de l'eau souterraine comme une conjugaison de modèles théoriques et mathématiques, d'une part, et de données propres à un système particulier, d'autre part. Pour juger si un modèle est valide ou non, on se fonde à la fois sur la pertinence des données et sur la structure du modèle (c'est-à-dire le modèle théorique et mathématique). Les auteurs proposent une stratégie de validation permettant de vérifier séparément ces deux critères. Elle repose sur l'axiome suivant : on ne peut pas prouver qu'un modèle est valide, on peut seulement prouver qu'il est inadéquat ou pas inadéquat. En outre, les auteurs estiment qu'il ne faut pas juger un modèle sans tenir compte de l'objectif visé. Ainsi, un modèle de l'écoulement et du transport peut être inadéquat en fonction d'un objectif donné et ne pas l'être en fonction d'un autre.

This work was supported by the U.S. Nuclear Regulatory Commission and was performed at Sandia National Laboratories which is operated for the U.S. Department of Energy under Contract No. DE-AC0476DP00789.

Ces travaux ont reçu l'appui de la Commission de la réglementation nucléaire des Etats-Unis (U.S.NRC) et ont été exécutés aux Laboratoires nationaux Sandia pour le compte du Ministère de l'énergie des Etats-Unis au titre du contrat n° DE-AC0476DP00789.

1.0 INTRODUCTION

Numerical simulations of ground-water flow and solute transport will be relied on in the assessment of radioactive waste disposal sites. Therefore, these models must provide accurate descriptions of the processes occurring at a particular site. In other words, the models must be validated. The purpose of this paper is to propose a model validation strategy that can be used to test the validity or invalidity of a given model against a given experiment or set of experiments. The question of the validity of a model as applied to an actual radioactive waste site is left to a later paper. This paper presents a strategy that is based on the opinions and philosophy of the authors and is not intended to represent an accepted model validation strategy of either the U. S. Nuclear Regulatory Commission (NRC), the U. S. Department of Energy (DOE), or Sandia National Laboratories.

2.0 WHAT IS VALIDATION?

The NRC [1] defined validation as the process of obtaining "assurance that a model, as embodied in a computer code, is a correct representation of the process or system for which it is intended". The DOE [2] feels that validation is "a process whose objective is to ascertain that the code or model indeed reflects the behavior of the real world." Others [3] follow the International Atomic Energy Agency definition that validation is confirmed when the model and computer code "provide a good representation of the actual processes occurring in the real system." All the definitions are consistent, although derived from different perspectives. That is, they all are concerned with providing assurance that a model represents reality. This is different from the classical scientific method which consists of proposing a hypothesis and then designing tests to disprove that hypothesis. Thus, science is not concerned with validating models but, rather, invalidating them. More than 30 years ago Popper [4] perceived that "...whenever we propose a solution to a problem, we ought to try as hard as we can to overthrow our solution, rather than defend it." This idea also has a basis in ordinary statistics whereby one proposes an idea, then attempts to disprove it by testing the null hypothesis. This philosophy is analogous to United States judicial process where the null hypothesis is that one is "not guilty." A jury must either fail to reject the null hypothesis (the individual is "not guilty"), or reject the null hypothesis (not "not guilty"). Clearly, just because an individual is judged not guilty does not mean that the person is innocent. It simply means that there was enough evidence for the jury to fail, beyond a reasonable doubt, to reject the null hypothesis. Hence, from a regulatory perspective, declaring a model not invalid does not guarantee validity. It only provides a means for the modeler to demonstrate reasonable assurance that the model is not incorrect. Hence, a model can be declared "invalid" or "not invalid"; however, it is the premise of this paper that one can never say for sure that a model is "valid."

The desire for validated models arises from a regulatory decision-making framework, either for designing a repository or for providing assurance that the assessment of long-term repository performance is meaningful. Therefore, the distinction between a scientific approach to developing and testing models and a regulatory approach to validating or invalidating models is critical. For one, the scientific approach generally would ask for a complete and unattainable detailed explanation for all observed phenomena, whereas the regulatory approach would ask only for an adequate description of the phenomena for a given purpose (e.g., for the licensing of a repository). Thus, a bounding or conservative model may be valid for regulatory purposes but, by definition, not provide a detailed description of all phenomena.

Geospheric flow and transport models are a combination of the site-specific data that describe a particular geologic setting or experiment and the physical process models contained in the relevant computer codes. Hence, the term model, as defined herein, includes the conceptual and mathematical models as well as the computer code and its associated input data. Therefore, validation of geospheric models is truly a site-specific issue. However, the testing of codes on various geologic media will aid in gaining (or losing) confidence that the process models contained in the codes are adequate representations of what is occurring in nature.

Both the NRC and the DOE define validation as a process and not as a fixed product. They apparently are not asking for absolute proof that models are correct representations of reality. Instead, they are asking only for assurance that the models are adequate representations (i.e., that compliance with current regulations can be shown). Thus, the goal of a model validation exercise should not be viewed as providing a set of "validated" models. Rather, the goal should be obtaining added confidence that the models are able to simulate the behavior of specific hydrogeologic systems.

The modeler and regulator alike should realize that the ultimate decision on model validity will be based on the intended use of the model. Moreover, that decision will be, to a large degree, subjective.

3.0 WHY IS THERE A NEED FOR VALIDATION?

The need for model validation in the high-level nuclear waste program of the United States arose from at least two sources. First, the NRC and the DOE have interpreted the need for validation to be implicit in the U.S. Environmental Protection Agency (EPA) and the NRC regulations (40 CFR Part 191 and 10 CFR Part 60, respectively) that require that the DOE provide assurance that the proposed site will meet the stated performance criteria. In fact, it is stated clearly in 10 CFR Part 60.21(c)(1)(ii)(F) that "Analyses and models that will be used to predict future conditions and changes in the geologic setting shall be supported by using an appropriate combination of such methods as field tests, in-situ tests, laboratory tests which are representative of field conditions, monitoring data, and natural analogue studies." Second, intervenors could use validation as an issue in litigation against either the DOE (before a license application for construction of a repository) or the NRC (after a license is granted). In fact, the issue of validation was the basis for the decision in a court case involving the State of Ohio and the EPA (23 ERC 2091 [6th Cir. 1986]). In this instance, the court ruled that the EPA had acted arbitrarily in using the CRSTER computer code [5] as a basis for establishing limitations on sulfur dioxide emissions from two electric utility plants. The Court decided that the EPA had failed to establish the accuracy or trustworthiness of the model as compared with the actual discharge from the plants. In other words, the EPA did not perform a site-specific validation of the model.

4.0 VALIDATION STRATEGY

Given enough time and resources, all of the models used to assess the performance of a HLW repository could be validated using site-specific data. However, limitations in time and resources lead to alternative approaches to validation. Foremost, they lead to the use of as many revelant experiments as possible to test the models. Thus, a validation strategy should include the use of so-called generic validation experiments as well as experiments performed at the proposed repository location. In general, both generic and site-specific experiments will include laboratory tests, field tests, and natural analogs. The strategy outlined below is to be applied in evaluating a model or models against an experiment or a set of experiments. Thus, the issue of combining all validation experiments and the site-specific experiments and data in determining the validity of a model as it is applied to a specific radioactive waste site is left for a later paper.

The following strategy is based on a clear definition of the model purpose which leads to performance measures and acceptance criteria that are based on that purpose. In addition, the strategy attempts to account for input data uncertainty as well as uncertainty in experimental results. This strategy is not presented as the only means of assessing model validity but only as one possible approach.

1. Define a validation issue. Defining validation issues should be one of the first steps taken by the modeler. Key unresolved issues should be identified and experiments sought that shed light on these issues. For example, modelers routinely use the convection-dispersion equation when simulating solute transport. In practice, this equation is often applied assuming a homogeneous and isotropic media. Hence, one validation issue is whether this approach is valid for solute transport in natural media which are always heterogeneous.

2. Develop a conceptual model or models. A conceptual model is a simplification of the processes and geometry of a real system. The conceptual model can be thought of as a summary of simplifying assumptions that may be derived from a regulatory viewpoint, a management viewpoint, an experimentalist's viewpoint, a modeler's viewpoint, or all of the above. Bear and Verruijt [6] suggest that a conceptual model describes assumptions regarding the problem geometry, material heterogeneity, mode of flow (i.e. dimensionality), the effect of dissolved solids, temperature, and viscosity, the volume or area over which parameters are averaged, and sources and sinks of water and contaminants. By changing the assumptions, one has in essence created a new conceptual model. Multiple conceptual models can be applied to a given experiment.

3. Develop a mathematical model. A mathematical model is a numerical expression of the conceptual model. It consists of a set of equations that describe the initial and boundary conditions, the governing equations, and the type of numerical solution used. The mathematical model may or may not be implemented in a computer code.

4. Identify and/or design an experiment that addresses the validation issue. All chosen experiments should be directly related to the validation issue and not necessarily related to a given models capabilities. Previously completed experiments should be avoided if possible for two reasons. First past experiments may not provide all of the information necessary to test the models validity because these experiments were generally performed for other purposes. Second, the modeler may have familiarity with the experimental results which could lead the effort to be one of calibration not validation.

5. *Define performance measures to be used for model comparison.* In order to judge the performance of the model relative to the experimental results, some representative measure(s) of the system response must be defined. A performance measure must be a quantity that is of regulatory interest or directly related to a quantity of regulatory interest. In almost all countries, the regulatory interests are in concentrations of contaminants, integrated quantities of contaminants, and/or geospheric travel time of contaminants. Performance measures that are more removed from regulatory requirements (pressure and moisture content distributions, for example) should only be used as a last resort because of the potential for misinterpretation of the model validation results. For example, a model may be judged invalid if it cannot reproduce a three-dimensional distribution of moisture content. However, the same model may provide a perfectly adequate prediction of regulatory quantities, such as ground-water flux or contaminant concentrations.

6. *Quantify the uncertainty associated with the input data and the data available for comparison with the model output.* The ability of a model to adequately predict the performance measures cannot be judged without consideration of the uncertainty of the input parameters and the accuracy of the experimental results. There are several types of error that could be associated with the data. These include data collection errors (e.g. instrument error, human error) and interpretation errors (e.g. fitting a Theis curve). Some types of errors are quantifiable while others can only be estimated subjectively by the experimentalists. In any case, the responsibility of quantifying the errors associated with input parameters and experimental results lies with the experimentalists, not with the modelers.

It is the modelers responsibility to account for the uncertainty in input parameters in simulating the experiment. This is more difficult than accounting for measurement errors in the experimental results. To illustrate this difficulty, consider the following two extremes. First, the input parameters could be known with complete certainty. Given that the model is an acceptable representation of the real system (i.e., it is not an invalid model), then the model results should agree with the experimental results within the accuracy of the experimental measurement error. However, if the model does not agree with the experimental results, the model is invalid. Second, consider the possibility that the input parameters are completely uncertain. This situation can occur when a particular parameter is not measured during an experiment. In this case, the modeler is free to use any reasonable value for this variable in an attempt to obtain agreement between model and experimental results. In the event that the modeler is unable to obtain agreement in this manner, the model is invalid. Normally, however, the modeler is able to achieve some degree of agreement between the model and the experimental results. Given the lack of knowledge of the data, a large degree of uncertainty in the model validity remains. Thus, uncertainty in the model input must be taken into account, but the judgment about the validity of the model should be tempered by this uncertainty.

7. Define the acceptance criteria or acceptable model error based on regulatory requirements and data uncertainty. This is the most difficult step in the validation process because the judgment of the adequacy of the model predictions relative to the experimental results is ultimately subjective. In judging the validity of model results, the uncertainty in experimental results should be taken into account. That is, the model can only be judged relative to the accuracy of the measurements. Accounting for errors in the experimental results can be accomplished directly by defining acceptance limits which correspond to a band defined by the experimental measurements, plus and minus their associated error.

The more difficult problem is defining quantitative acceptance criteria. Consideration of two types of uncertainty, data and model structure, leads us to define at least two acceptance criteria for a given case. The first is a measure of the accuracy of the model input parameters relative to the experimental results, while the second is a measure of the adequacy of the model structure (i.e., the underlying conceptual and mathematical model) in describing the system behavior. To illustrate the two types of criteria consider a simple tracer tests through a column with a steady-state uniform velocity field and a constant inlet concentration. In this example, the conceptual model to be tested is a homogeneous isotropic porous medium, the behavior of which is assumed to be governed by the classic convective-dispersion equation. To elucidate the criterion used to judge the adequacy of the model input data, *assume that the underlying model structure is correct*. For the example described above, this means that the model and experimental results both display the classic S-shaped contaminant breakthrough curve. Now consider the possibility that the model predicts that the contaminant arrives earlier than it really did. Depending on how close the two curves are, the model results could still be acceptable. The proposed acceptance criterion is based on a combination of the distance between the curves, the uncertainty in input values, and the uncertainty in experimental results. Given this approach, several possibilities arise.

First, there may be no overlap between the model-predicted values resulting from any combination of model input parameters and the experimental results (taking into account the uncertainty in the experimental results). In this case, the model is invalid and the invalidity arises out of errors in or incomplete knowledge of the values of the input variables. The next possible condition for this acceptance criterion is that the model results agree with the experimental results. That is, some combinations of input variables result in model predictions that lie within the band created by the plotting of the experimental results and the associated experimental error. In this case, one cannot state that the model is valid but only that the model is not invalid. The model still would need to be tested for the full range of expected environmental conditions. Our hypothetical tracer test would have to be repeated under different conditions (for instance, the experiment could be repeated under a range of imposed velocities). To illustrate this point, consider a model for tracer movement that treats retardation as a constant. Conceivably, the modeler could find a value of retardation that allowed the model to reproduce the experimental results. However, if the retardation was actually a function of velocity, then the model would be invalid for all velocities other than the one used in the experiment. Thus, without performing additional experiments very little confidence is gained in the model.

A second acceptance criterion arises out of the need to determine whether the model structure (i.e., the underlying conceptual and mathematical model) is adequate. This criterion is based on the form of the output predicted by a given conceptual model relative to the form of the experimental results. To illustrate this point, consider the classical S-shaped concentration breakthrough curve predicted by the conventional convective-dispersion equation. If the convective-dispersion equation is an adequate representation of the physics governing a particular experiment, then the experimental results should be of the same s-shaped form. Deviations from this form indicate an incorrect model structure. For example, matrix diffusion or diffusion into dead-end pores result in a concentration breakthrough curve which falls underneath the classical curve at late times. One possible approach to assessing the adequacy of the model structure is to start with the "best fit" model obtained in the data input uncertainty analysis described above. That is, start with the combination of data input that results in the closest fit to the experimental results. Use the best fit results as the representation of the behavior of the model and search for systematic differences between the model results and the experimental results. The basis for this approach is the assumption that random deviations between the model and the experiment are caused by measurement error but systematic errors result from errors inherent in the model structure. One additional assumption contained in this approach is that there are no systematic errors in the experimental results. This search could be accomplished by subtracting the conceptual model results from the experimental results and testing these residuals for white noise (i.e., no trend in the residuals). Thus, if the residuals do not reveal any detectable trend, the conceptual model adequately represents the experimental results. On the other hand, the conceptual model is in error if a trend exists in the residuals.

8. Simulate the experiment. Simulation of the experiment may be done using simple analytical methods or with a computer. Most of the ground-water flow and transport calculations faced by modelers today are so numerically complex that a computer is the only reasonable way to arrive at a solution. However, this is not always true and a simple calculation may be adequate, depending on the purpose of the model. Regardless of how the calculations are done, they must include an analysis of the uncertainty of the input data. Several uncertainty analysis techniques are available for this purpose [7, 8]. One popular method is to use a Monte Carlo approach based on the Latin Hypercube Sampling (LHS) scheme [9]. Using this approach, one can be sure that the entire range of parameters are tested.

9. Perform the experiment in the laboratory or field. The field or laboratory experiment should be performed after the model simulations to assure that the modeling effort is not simply a calibration effort based on the experimental results. This is not possible for every laboratory and field experiment, and is never possible for natural analogues. At any rate, an attempt should always be made to simulate the experiments without reliance on the experimental results.

10. Evaluate model results based on the acceptance criteria. At this step, the performance measure predicted by the model is compared to the experimental results using previously defined acceptance criteria. This allows one to conclude whether the model structure is correct and whether the model input data have been adequately estimated. In the event that the model structure and the input data are adequate, two questions remain. First, would the same model be able to adequately predict different experimental conditions and, second, are other model structures (conceptual models) also able to adequately simulate the experimental results? The first questions can only be answered by new experiments under different conditions while the second can be answered by applying the validation strategy to a different conceptual model of the experiment. If multiple conceptual models adequately simulate the experiment, then understanding the differences in conceptual models should allow the modelers to recommend experiments that can only be adequately simulated by one of the models.

Comparison of model predictions and experimental results may lead one to conclude that the input data were not adequately estimated. Again, the modeler then should be able to provide the experimentalist with a list of the desired data, including some indication as to which data are most important. This final distinction could be arrived at using a sensitivity analysis of the model results.

Finally, comparison of model predictions and experimental results may lead to a conclusion that the model structure is inadequate for describing the experiment. The modeler would then have to propose a new model structure and repeat the simulations and model comparisons. However, the modeler then has knowledge of the experimental results and, thus this effort will be just calibration unless a new experiment is performed that follows the final model simulations.

5.0 CONCLUSIONS

Ground-water flow and transport models can be thought of as a combination of conceptual and mathematical models and the data that characterize a given system. The judgment of the validity or invalidity of a model depends both on the adequacy of the data and the model structure (i.e., the conceptual and mathematical model). This report proposes a validation strategy for testing both components independently. The strategy is based on the philosophy that a model cannot be proven valid, only invalid or not invalid. In addition, the authors believe that a model should not be judged in absence of its intended purpose. Hence, a flow and transport model may be invalid for one purpose but not invalid for another.

The proposed strategy is based on the definition of a performance measure that is related to the intended use of the model (in this case, the licensing of a radioactive waste disposal site), the definition of acceptance criteria for the input data and the model structure, the explicit treatment of input data uncertainty, and a continual iteration between experimentalist and modeler.

It should be emphasized that the concept of model validation arises out of regulatory concerns, not scientific concerns. Additionally, we believe that the NRC and the DOE are not seeking to perfectly describe the geosphere but, rather, to provide the public with some reasonable assurance that the chosen site is a safe place in which to dispose of waste. Therefore, it must always be re-emphasized that the final decision regarding the acceptability of a model is subjective and should be based only on the intended use of that model.

6.0 REFERENCES

1. NRC, 1984. A Revised Modelling Strategy Document for High-Level Waste Performance Assessment, U.S. Nuclear Regulatory Commission, Office of Nuclear Regulatory Research, Washington, D.C.

2. DOE, 1986. Environmental Assessment - Yucca Mountain Site, Nevada Research and Development Area, Nevada. DOE/RW-0073, Vol. 2, U.S. Department of Energy, Office of Civilian Radioactive Waste Management, Washington, D.C.

3. Borgorinski, P., B. Baltes, J. Larue, and K.-H. Martens, 1988. The Role of Transport Code Verification and Validation Studies in Licensing Nuclear Waste Repositories in the FR of Germany, Radiochimica Acta, 44/45, pp. 367-372.

4. Popper, K., 1958. The Logic of Scientific Discovery, Harper and Row, New York, NY, 479 pp.

5. EPA, 1977. User's Manual for Single Source (CRSTER) Model, EPA 450/2-77-013, U.S. Environmental Protection Agency, Washington, D.C.

6. Bear, J. and A. Verruijt, 1987. Modelling Groundwater Flow and Pollution, D. Reidel Publishing Co., 414 pp.

7. Zimmerman, D.A., K.K. Wahi, A.L. Gutjahr, and P.A. Davis, 1989. A Review of Techniques for Propagating Data and Parameter Uncertainties in High-Level Radioactive Waste Performance Assessment Models, NUREG/CR-5393, SAND89-1432, Sandia National Laboratories, Albuquerque, NM, in press.

8. Doctor, P.G., E.A. Jacobson, and J.A. Buchanan, 1988. A Comparison of Uncertainty Analysis Methods Using a Groundwater Flow Model, PNL-5649/UC-70, Pacific Northwest Laboratory, Richland, WA.

9. Iman, R.L. and M.J. Shortencarier, 1984. A FORTRAN 77 Program and User's Guide for the Generation of Latin Hypercube and Random Samples for use with Computer Models, NUREG/CR-3624, SAND83-2365, Sandia National Laboratories, Albuquerque, NM.

NATURAL ANALOGUE STUDIES USEFUL IN VALIDATING REGULATORY COMPLIANCE ANALYSES

Donald H. Alexander
Office of Civilian Radioactive Waste Management
U.S. Department of Energy, Washington, USA

Abraham E. Van Luik
Battelle Memorial Institute
Pacific Northwest Laboratory, Washington, USA

ABSTRACT

The U.S. Department of Energy (the Department) is evaluating the suitability of the Yucca Mountain site, Nevada, as the location for a mined geologic disposal system for high-level waste and spent nuclear fuel. If the site is found suitable, performance assessment calculations will be used to provide a basis for giving reasonable assurance of regulatory compliance. These performance assessment calculations must be certified, which includes - to the extent practicable- validation using field, laboratory and natural analogue studies. In the Department's published Site Characterization Plan for the Yucca Mountain candidate site numerous proposed studies are listed, some of which include natural analogue studies to obtain information over scales of time and distance not addressable in field and laboratory work. In this paper, the Site Characterization Plan's treatment of natural analogues is examined. Natural analogues in a variety of environments may be useful in terms of identifying and understanding specific processes, but an analogue may be particularly useful if key environmental characteristics are similar to those of the geohydrological setting under consideration for a repository.

ETUDES D'ANALOGUES NATURELS UTILES POUR VALIDER LES ANALYSES DU RESPECT DES DISPOSITIONS REGLEMENTAIRES

Donald H. Alexander
Office of Civilian Radioactive Waste Management
U.S. Department of Energy, Washington, Etats-Unis

Abraham E. Van Luik
Battelle Memorial Institute
Pacific Northwest Laboratory, Washington, Etats-Unis

RESUME

Le Ministère de l'énergie des Etats-Unis (le Ministère) a entrepris d'évaluer si le site de Yucca Moutain (Nevada) se prêtait à l'implantation, dans des formations géologiques, d'une installation d'évacuation de déchets de haute activité et de combustible irradié. Si la réponse est positive, on s'appuiera sur des calculs d'évaluation du comportement pour pouvoir avancer avec une assurance raisonnable que les dispositions réglementaires seront respectées. Ces calculs d'évaluation du comportement doivent être certifiés et devront donc comprendre - dans la mesure du possible - une validation reposant sur des études menées in situ, en laboratoire, et dans des analogues naturels. Dans le document publié par le Ministère sur le site éventuel de Yucca Mountain et intitulé Site Characterization Plan (Plan de détermination des caractéristiques du site) de nombreux projets d'études sont énumérés, dont certains prévoient des études d'analogues naturels pour obtenir des informations portant sur des échelles spatio-temporelles impossibles à reproduire dans les travaux in situ et en laboratoire. Les auteurs examinent la façon de traiter les analogues naturels dans le Plan de détermination des caractères du site. Les analogues naturels peuvent être utiles dans de multiples environnements pour définir et comprendre des processus spécifiques, mais un analogue peut devenir particulièrement intéressant si ses principales caractéristiques environnementales se rapprochent des conditions hydrogéologiques anticipées pour un dépôt.

1. INTRODUCTION

The U.S. Department of Energy (the Department) is evaluating the suitability of the Yucca Mountain site, Nevada, as a final repository for spent fuel and high-level radioactive waste. If the site is found suitable, performance assessment calculations are to be used to demonstrate compliance with regulatory performance objectives. In order for these calculations to provide a basis for giving reasonable assurance of compliance with regulatory performance targets, they must be certified. This requires that expert reviewers concur with the applicability of the calculational approach, codes used have been properly verified, and -to the extent practicable- models used have been validated using field, laboratory and natural analogue studies. The need to validate models predicting long term behavior of a repository in the unsaturated tuffs of Yucca Mountain suggests a need for field, laboratory and natural analogue studies focused on processes operative in unsaturated tuffs. The focus of this paper is on natural analogues, and emphasis is placed on analogues in unsaturated tuffs. This is not to suggest, however, that processes of interest are not also operative in other geohydrologic settings.

No exact analogue to a radioactive waste disposal system exists, and the search for useful natural analogues should focus on the processes controlling the performance of repository subsystems and components. The search for potentially suitable natural analogues should, therefore, focus on specific aspects of the repository system. In terms of model validation, the purpose of the analogue study is to support the choices of processes and process-hierarchies used to conceptualize the system.

2. PLANNED USES OF NATURAL ANALOGUES

The Department's Site Characterization Plan for Yucca Mountain, Nevada, discusses a number of natural, and some anthropogenic, analogues that may be useful in evaluating postclosure processes specific to the unsaturated tuff site to be characterized [1: Section 4.3]. The anthropogenic analogues are not discussed here.

The types of natural analogues explicitly discussed in the Site Characterization Plan are 1) warm and hot springs in tuffaceous rocks and 2) uranium and thorium ore bodies. The Site Characterization Plan contains one explicit proposal for a natural analogue study [1: Section 8.3.1.3], a scoping study that includes 1) identifying a hydrothermal system in tuff that is a suitable analogue to the Yucca Mountain tuff, 2) evaluating paragenesis of the hydrothermal minerals, and 3) characterizing the groundwater. The results will be used to evaluate geochemical/reaction-path codes and to judge the usefulness of knowledge that may be gained from hydrothermal tuff systems in enhancing long term predictive capabilities for the behavior of a waste repository system at Yucca Mountainy. Existing data from a number of hydrothermal sites in tuffaceous rocks -such as the Yellowstone National Park geyser basins - are also to be used to evaluate geochemical models.

Uranium and thorium ore deposits may yield data on the stability and release characteristics of specific radionuclide solids, as well as on migration characteristics for selected radionuclides for specific ranges of geohydrochemical conditions. The suites of radionuclides that may be available for study are relatively limited, except for the case of the Oklo,

Gabon, deposit which experienced nuclear criticality for a period on the order of 10^5 years, about 2 x 10^9 years ago. Other uranium and thorium ore deposits have been and are being studied. For example, the Site Characterization Plan discusses the natural analogue in the Alligator Rivers region of northern Australia. The Alligator Rivers deposits may allow study of processes controlling radionuclide migration in relation to the unsaturated/saturated portions of the deposits and the oxidizing regional groundwater.

The Department's interest in natural analogues is shown by its active participation in the Poços de Caldos, Brazil, natural analogue project and its current evaluation of direct participation in the Alligator Rivers studies. In addition, the Department's Yucca Mountain Project is evaluating participation in the Cigar Lake, Canada, natural analogue which may allow the study of oxidative and radiolytic processes. In addition, a number of studies of natural geohydrologic features at sites near Yucca Mountain are described in the Site Characterization Plan. Examples include the investigation of rock mechanics properties at Busted Butte, adjacent to Yucca Mountain, and of mechanical and hydrologic properties at Rainier Mesa, northeast of Yucca Mountain. These are also natural analogue studies. Finally, an internally proposed initiative is being evaluated that suggests a vigorous natural analogue program be started while the Department is awaiting access to the Yucca Mountain site.

3. ANALOGUES THAT MAY BE USEFUL IN EVALUATING LONG TERM PERFORMANCE MODELING OF A YUCCA MOUNTAIN SITE REPOSITORY

Potentially useful natural analogues for specific facets of a mined geologic disposal system are discussed below in terms of previous work and present needs.

3.1 Stability of Excavations

The postclosure engineered barrier system for a repository in Yucca Mountain, Nevada, is to include an air gap between the waste package and the borehole wall. This air gap breaks hydrologic continuity between the waste package and the unsaturated host rock. If this gap is compromised by rubble, silt, or the shifting of waste packages over time, a diffusive pathway may be established. The stability of the waste package environment is to be addressed through in-situ testing of excavation and thermal loading effects on emplacement hole integrity [1: Section 8.3.4.2.4.3]. Host rock response to possible earthquake ground motion or underground faulting are also to be investigated [1: Section 6]. Prototype testing of in-situ investigative techniques is planned for nearby tuffaceous blocks such as Busted Butte or Rainier Mesa on the Nevada Test Site [1: Sections 2 and 4]. If needed, other locations could be found in similar geohydrologic settings.

3.2 Tectonic and Magmatic Processes

It has been speculated that tectonic events may reduce the fracture apertures that allow rapid water movement in the saturated zone under Yucca Mountain, thus raising the water table above current levels [1: Section 8.3.1.8]. This potential effect is to be investigated through studying the effects of past changes in groundwater conditions at Yucca Mountain. It seems that finding natural analogues of this particular mechanical/hydrologic coupled effect elsewhere in the Basin and Range Province could be helpful. Finding such examples may be difficult, however, since tectonism may also result in the formation of new flow paths rather than the loss of flow paths.

Magmatic process effects on the repository may range from creating a hydrothermal system, to destroying the integrity of one or several waste packages, to exhuming portions of the emplaced waste. The study of magmatic intrusions into tuffaceous rocks may be useful in describing credible effects. The study of regional volcanic occurrences may help bound potential effects from this disruptive scenario.

3.3 Metal Barrier Behavior

The selection of materials for primary containment barriers should take into account a complex number of factors, from failure-mode susceptibilities to fabricability. Materials selection is in progress, thus selecting appropriate natural analogues is not possible at this time. If a copper based or iron-nickel based alloy is to be used in Yucca Mountain, with its oxidizing, unsaturated environment, natural analogue choices seem limited to iron-nickel meteorites and metallic copper occurrences. A recent discussion of corrosion studies, including the potential use of metal natural analogues by Van Orden and McNeil [2] suggests that there may be support for the prediction of very long term corrosion behavior from short-term corrosion tests if the processes that locally affect the corrosion environment are taken into account.

It is not clear at this point if copper or copper alloys are to be used at all in the engineered barrier system of a Yucca Mountain repository. Copper occurs in tuffs, but typically as copper or copper-iron sulfide hydrothermal vein deposits at the edges of complex calderas, protected from oxidizing solutions by envelopes of tight silica and clay minerals. The analogue value of these types of deposits in terms of metal durability is low. However, there may be some value in studying the durability of copper and copper-oxide containing copper-oxidation haloes overlying supergene-enrichment mantos, even though such assemblages may not occur in tuff lithologies.

3.4 Waste Form Behavior

Studies of reactions at natural glass surfaces may be useful in determining the interactions between groundwater and the glass waste form. Natural glasses in proper settings may be useful as natural analogues for waste glass stability. White et al. [3] found natural and synthetic silicate glass dissolution kinetics to be similar, for example, but found calculated diffusion coefficients to be ten orders of magnitude higher than predicted from extrapolating high temperature diffusion data. The difference was thought due to surface hydration reactions. This type of work may be an example of the utility of natural analogues in identifying operative processes that should be included in a competent conceptual model.

The Oklo natural reactor site may be a suitable natural analogue for the long term behavior of spent nuclear fuel. Curtis et al. [4] report the interpretation of Oklo data as a spent fuel analogue. Their conclusions are that the retention of fission products is relatable to their partitioning into the uraninite or secondary mineral assemblages either in zones physicochemically controlled by the uranium dioxide grains or in the ambient environment. Those partitioned into the latter mineral assemblages were largely lost over time, pointing to the importance of small uranium dioxide grains in controlling the microenvironment. Technetium was partly retained, perhaps in a metallic phase similar to that found in spent nuclear fuel. A comprehensive chemical, mineralogical, and hydrological description of the microenvironments in and adjacent to the reaction zones may be useful in formulating a conceptual model for spent fuel dissolutioning or for estimating metal barrier corrosion product effects on the microenvironment of spent fuel.

3.5 Hydrologic System

The Yucca Mountain Project has studied the hydrologic system at Rainier Mesa, northeast of Yucca Mountain on the Nevada Test Site, to gain understanding of the dynamics of matrix/fracture flow in unsaturated tuffs. Rainier Mesa, as described by White et al. [5], has an elevation of about 2300 m above sea level, and consists of about 600 m of Tertiary tuffs overlying Paleozoic carbonates. The tuffs have been generally classified into three hydrologic units, from top to bottom: 1) welded and partly welded ash flow tuffs with low interstitial porosities and conductivities but extensive open joints and faults, 2) friable vitric tuffs with relatively high interstitial porosities and conductivities and few open fractures, and 3) zeolitized tuffs with restrictive interstitial porosities and conductivities and numerous open fractures. It is thought that the dominant water flow mechanism in each hydrologic unit is 1) unsaturated fracture flow, 2) interstitial matrix flow, and 3) fracture and matrix flow with saturated zones in the matrix and erratically distributed saturated and unsaturated fractures. Thus, the third hydrogeologic unit has perched water about 600 m above the water table.

White et al. [5] report the collection and analysis of interstitial waters obtained by centrifuging and squeezing 19 core samples from about the 130 m to the 530 m depths. Fracture flow waters were collected in a tunnel below from 50 to 400 m of overburden. Comparisons were made and correlated with mineralogical observations. Anion contents of interstitial and fracture flow waters from the 300-350 m depth were found to differ, with the fracture flows being bicarbonate waters and the interstitial waters being chloride and sulfate enriched in comparison. The differences were not readily explainable, but could indicate significant isolation of interstitial waters from fracture flow waters. Since previous work on tritium data has suggested that fracture flow waters 400 m below the mesa surface are only a few years old, age dating the interstitial waters seems like a logical way to evaluate differences in the ages of interstitial and fracture flow waters, if any. Perhaps natural radionuclide decay chain disequilibrium studies, such as reported by Laul and Maity [6], could be useful in terms of dating the interstitial water and investigating other differences between fracture and interstitial waters.

Work reported by Norris [7] on the hydrologic interpretation of ^{36}Cl data at Rainier Mesa and Yucca Mountain illustrates the potential for using natural analogue work to aid the interpretation of site-specific processes. It seems the application of this work to a modest number of tuffaceous rock sites in the Basin and Range Province that cover a range of altitudes (infiltration rates) could lend interpretive depth to the ongoing discussion of a conceptual model for the Yucca Mountain hydrologic flow system.

3.6 Radionuclide Transport

Specificity to an unsaturated tuffaceous rock environment is not a requirement for a competent analogue of transport processes operative in such an environment. However, the existence of tuffaceous rock locations throughout the southwestern U.S. invites the identification of suitable analogue sites for work on identifying retardation processes, perhaps including work on naturally occurring radioelements [6]. Thus, for the purposes of this paper, a number of uranium bearing near-surface, readily accessible sites in tuffaceous rock were reviewed. Of the sites considered that could allow the study of the processes affecting the migration of uranium and a limited suite of other elements, one location is described here, the McDermitt Caldera, Nevada-Oregon.

The McDermitt Caldera, according to Dayvault et al. [8], is located on the Nevada-Oregon border, with the bulk of the caldera being in Nevada, in the northern Basin and Range Province. The uranium deposits of the caldera are of five general types, schematically illustrated in Figures 1 and 2. Figure 1 shows deposits of reduced U associated with pyrite and titanium phases in basalt flow tops; U associated with hydrothermal or syngenetic cinnabar occurrences; and stratabound U in a tuffaceous sediment layer within volcanoclastic lakebed sediments. Vein deposits associated with pyrite, fluorite, and clayey zirconium, and U occurrences in hydrothermally affected silicified and argillized rhyolitic and volcanoclastic rocks are illustrated in Figure 2. The silicified rhyolitic tuff, the vein, and the lava flow top occurrences are genetically related in terms of mineralization processes, but each reflects a different mineralogical/geochemical setting and has different associated trace element assemblages.

This variety of hydrothermal mineralization environments in tuffaceous rocks suggests the McDermitt Caldera could be a promising natural analogue site. The tuffaceous rocks associated with the caldera include rhyolitic ash flows, tuffaceous sandstones, and air fall tuffs. The McDermitt Caldera seems promising as a natural analogue site because a number of ingredients are present that would appear germane to Yucca Mountain: significant U and associated heavy metal concentration gradients in hydrothermally altered and unaltered, saturated and unsaturated, tuffaceous rocks.

It is possible that a number of mineralogic, hydrologic, hydrothermal, and element migration processes may be usefully addressed through selected investigations at this site or a site like this one. Ongoing evaluations of strategies for enhancing the scientific credibility of the Yucca Mountain modeling effort may wish to give serious consideration to this or comparable potential natural analogue locations.

4. CONCLUSIONS

The U.S. Civilian Radioactive Waste Program is currently charged with determining the suitability of Yucca Mountain as a location for a high-level waste repository. The opportunities for benefitting from the exploration and modeling of natural analogues are being examined. Analogue sites to address hydrothermal system effects are being evaluated. The selection of such sites should consider multiple analogue uses that can take advantage of the

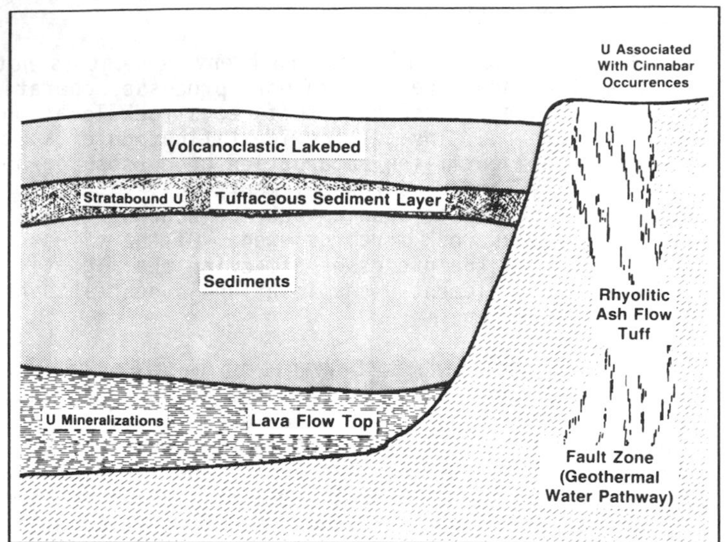

Figure 1. Schematic of lava flow top, stratabound, and cinnabar-associated U occurrences at the McDermitt Caldera (based on text and figures in Dayvault et al. [8]).

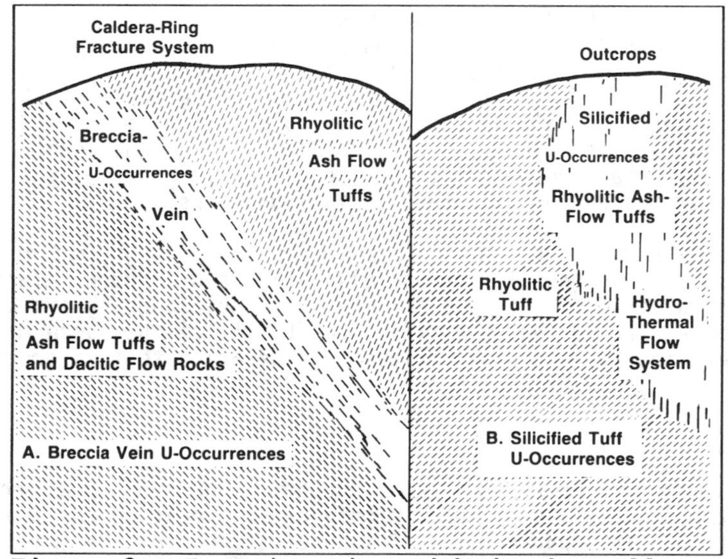

Figure 2. Breccia vein and hydrothermally affected rhyolitic tuff type deposits at the McDermitt Caldera (based on text and figures in Dayvault et al. [8])

hydrologic system descriptions created for the hydrothermal work. Natural radionuclide distributions, ^{36}Cl measurements, and hydrologic flow questions should also be addressed to help foster the synergism that is characterisitic of well planned and well executed multidisciplinary efforts. Although specificity to an unsaturated tuff environment is not an absolute requirement for an analogue allowing study of processes operative in unsaturated tuffs, such specificity may still be desirable. A search for near-surface, readily accessible uranium deposits in unsaturated tuffs was conducted for this paper, and the McDermitt Caldera, Nevada-Oregon, is described as one candidate for possible further evaluation.

5. ACKNOWLEDGEMENTS

The authors are indebted to a number of reviewers, and wish to acknowledge helpful ongoing discussions of ore occurrences in tuffs with Alex Livnat of the Roy F. Weston, Inc. Technical Support Team. This work was performed by the U.S. Department of Energy and supported under contract DE-AC06-76RLO 1830.

REFERENCES

[1] U.S. DOE. Nuclear Waste Policy Act (Section 113) Site Characterization Plan, Yucca Mountain Site, Nevada Research and Development Area, Nevada. DOE/RW-0199. 8 Vols. U.S. Department of Energy, Office of Civilian Radioactive Waste Management, Washington, D.C. (1988).

[2] Van Orden, A. C. and M. B. McNeil. "Mineralogical Issues in Long-Term Corrosion of Iron and Iron-Nickel Alloys." Proceedings of the Electrochemical Society Fall Meeting, (extended abstracts), 88-2:211. Chicago, Illinois, 1988. The Electrochemical Society, Pennington, New Jersey (1988).

[3] White, A. F., W. F. Passchier and B. F. M. Bosnjakovic. "Surface reactions of natural glasses." Nuclear waste management II, Advances in ceramics 20:713-722. Third International Symposium on Ceramics in Nuclear Waste Management, Chicago, Illinois, 1986, American Ceramic Society, Westerville, Oklahoma (1986).

[4] Curtis, D., Benjamin, T., Gancarz, A., Loss, R., Rosman, K., De Laeter, J., Delmore, J. and W. Maeck. "Fission Product Retention in the Oklo Natural Fission Reactors." Applied Geochemistry 4:49-62 (1989).

[5] White, A. F., H. C. Claessen and L. V. Benson. The Effect of Dissolution of Volcanic Glass on the Water Chemistry in a Tuffaceous Aquifer, Rainier Mesa, Nevada. U.S. Geological Survey Water Supply Paper 1535-Q, U.S. Government Printing Office, Washington, D.C. (1980).

[6] Laul, J. C. and T. C. Maiti. "Natural Radionuclides in Groundwater from J-13 Well at the Nevada Test Site." In press: proceedings of the International High Level Radioactive Waste Management Conference & Exposition, April 8-12, 1990, Las Vegas, Nevada, published by the American Society of Civil Engineers (1990).

[7] Norris, A. E. "The Use of Chlorine Isotope Measurements to Trace Water Movements at Yucca Mountain." LA-UR-89-2573, Los Alamos National Laboratory, Los Alamos, New Mexico.

[8] Dayvault, R. D., S. B. Castor and M. B. Berry. "Uranium Associated With Volcanic Rocks of the McDermitt Caldera, Nevada and Oregon." In: Uranium Deposits in Volcanic Rocks. Proceedings of a Technical Committee Meeting, El Paso, Texas, 1984, pp. 379-409. Panel Proceedings Series, International Atomic Energy Agency, Vienna, Austria (1985).

Sufficient validation: The value of robustness in performance assessment and system design

C. McCombie, I.G. McKinley, P. Zuidema
Nagra, 5401 Baden/Switzerland

ABSTRACT

Quantitative performance assessment relies on a chain of models describing the evolution of the entire repository system. Assessment models, along with their associated databases, have varying requirements for validation depending on the complexity of the system modelled, the temporal and spatial extrapolation required and the sensitivity of overall system performance to the specific processes being modelled. A model can be considered robust if we have confidence that the results are **either** correct (i.e. that they are sufficiently realistic) **or** that they over-predict detriment (i.e. err on the side of conservatism).

Validation is the process of confirming that a model is sufficiently close to "the truth"; an adequately validated model is therefore obviously robust. A model may, however, be systematically wrong but still robust if it consistantly errs on the conservative side. To derive a robust model of a real system, all potentially detrimental processes need to be explicitly considered but demonstrably favourable processes only need to be considered to the extent that they are required to provide sufficient performance. This simplification in modelling "reality" can decrease validation requirements.

A robust model with no excessive demands on validation results from a combination of a simple, well understood system with large reserves of performance and a conservative approach to choice of methods and data. On a case by case basis, the feasibility of achieving reasonable validation needs to be examined (complete validation is never possible for natural systems). If acceptable validation is not achieved, either the conceptual model framework must be changed or the system itself must be improved (designing for robustness).

CONCEPTS AND DEFINITIONS

Quantitative performance assessment for radioactive waste repositories involves a chain of mathematical models which describe the evolution of the various component barriers (engineered and natural). Detailed, process-orientated **research models** are constructed, based on a theoretical framework supported by laboratory and field studies. These research models (and their associated databases) are generally simplified and grouped together to form **assessment models**. The assessment models are then linked in a model chain in order to describe the overall performance of repository system. The numerical accuracy of individual research models can be tested (verified) by comparison with exact analytical solutions or by intercomparison with other independent models. Proof that the model provides sufficiently accurate description of reality (validation) is a much more difficult problem.

The problem of validation becomes clear from consideration of the general characteristics of a good physical system model - it should make predictions of future behaviour and such predictions should be capable of being disproven[1]. In repository performance assessment, the spatial and temporal scales for prediction can be so long that testing opportunities are very limited. In any case, models can only be disproven (invalidated) - rigorous validation is not possible.

Validation is, therefore, concerned with building sufficient confidence that the predictions of a model are correct (a close representation of the truth). Fortunately, for performance assessment, we rarely need to demonstrate that a model is correct - merely that, if incorrect, it errs on the side of conservatism (maximising consequences as assessed by a global performance measure such as hazard or radionuclide release). We call a model robust when we are confident that any errors either will have little effect on performance or will be on the conservative side. Such a robust performance assessment model will have the following characteristics:

- A description (conceptual model) of all key processes occurring
- Quantitative consideration of all potentially detrimental processes
- Quantitative evaluation of positive (impact-reducing) processes only when they are certain to operate (validated by their dependence on fundamental laws of science or proof from experimental/analogue evidence)
- Reasonable support of both the process sub-models (or their supporting research models) and the overall model chain from laboratory, field and analogue studies (i.e. validation, in a weaker sense)
- Parameter values (or ranges) which are either well justified through direct evidence or are demonstrably conservative.

Generally, a robust model would be a simulation of well-understood processes in which the required databases are well defined. Critical, however, is that the overall repository performance (as measured by a global parameter like release or dose) is relatively insensitive to, or is demonstrably conservative in the case of, variations in the conceptual model framework or individual parameters within established ranges. This second condition means that performance assessment predictions with large uncertainties can be acceptable provided that the band of results lie well below defined targets. A robust model with no excessive demands on validation results from a combination of a simple system with large reserves of performance and a conservative approach to choice of models and data. On a case by case basis, the feasibility of achieving reasonable

---

1) N.B:: We discount models which simply explain past observations - these are exercises in the statistics of curve fitting and have little to do with science.

validation needs to be examined (complete validation is never possible for natural systems). If acceptable validation cannot be achieved, either the conceptual model framework must be changed or the system itself must be improved (designing for robustness).

This definition is rather abstract on its own and is best illustrated by some examples of models of varying robustness. Thereafter, the more problematic issue of improving robustness will be considered before explicitly discussing validation requirements.

A ROBUST MODEL OF THE SWISS HLW NEAR FIELD

Although the near field is a chemically complex system, it can be very well defined in its initial state. The Swiss HLW reference disposal concept involves vitrified HLW encapsulated in a massive steel canister which is emplaced horizontally in tunnels which are backfilled with compacted bentonite (Figure 1).

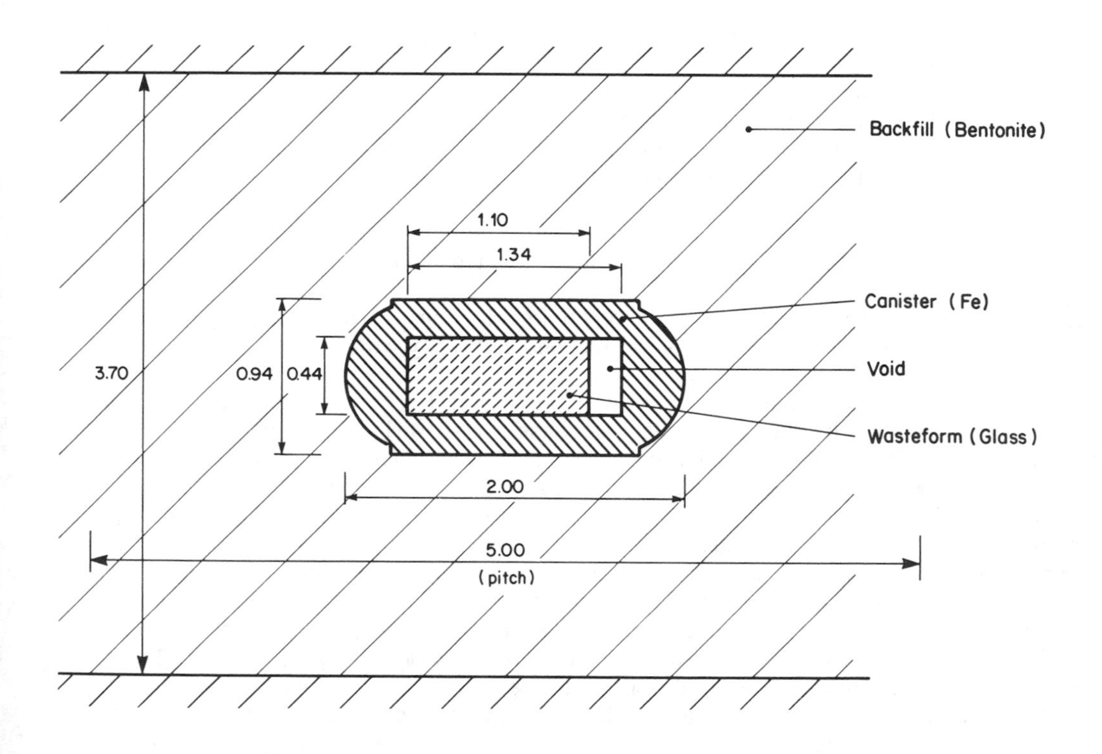

Figure 1: HLW emplacement geometry (dimensions in metres)

Derivation of the conceptual near-field model requires sufficient understanding of the following processes:

- Slow resaturation of the bentonite, swelling to seal any fabrication gaps, very slow bentonite alteration over a 10^7 year timescale
- Oxic canister corrosion until all trapped oxygen is consumed followed by an anoxic corrosion and canister failure (mechanical) after more than 10^3 years. Canister corrosion products buffer all radiolytic oxidants
- Slow glass leaching following canister failure with the release of nuclides constrained either by their rate of release from the glass matrix or by their solubilities
- Slow diffusion of dissolved radionuclides through the bentonite.

In the early stages after emplacement, the radiogenic thermal pulse is taken into account while the groundwater chemistry and advective flux through the surrounding rock sets constraints on solute exchange between near- and far-fields.

The performance assessment model utilised predicts that most of the activity inventory decays to insignificance within the near field, with only very low releases of a few radionuclides. This model conservatively ignores several favourable processes (nuclide immobilisation in glass alteration products, sorption on canister corrosion products, etc.) and, thus, large safety reserves are available, which is one of the preconditions for robust modelling. The model is, additionally, very robust in that the processes involved are well defined and can be modelled in a simple manner (e.g. mass balance) and the releases are relatively insensitive to data variations within acceptable ranges.

These properties imply that adequate validation should be feasible; this is achieved by a combination of comparisons with experiments and with natural analogue systems. Some more explicit pointers to the validation of the individual processes models required in the near field are:

1) Although modelling resaturation with a coupled thermal pulse is rather complex, the results fall within a range which little affects releases because of the very large safety reserves in the assessment of canister performance. Predicted slow bentonite mineralogical alteration is well supported by theoretical, laboratory and analogue studies. A complicating factor could be loss of swelling due to interaction with a steam phase. If current studies indicate that this could be significant, the problem could be avoided by decreasing the heat load resulting in lower temperatures (i.e. design for robustness by lowering waste loadings or increasing storage times before disposal).

2) Canister corrosion rates measured in the laboratory match well those evaluated from relevant archeological artifacts. The corrosion allowance for mechanical failure is chosen conservatively but, in any case, early canister failure has little effect on net releases.

3) Glass leaching has been extensively measured and the basic models agree with natural analogue studies. The model assumes free availability of water which is probably over-conservative. Elemental solubility limits are calculated for a well defined water chemistry (buffered by the canister corrosion products and the bentonite); here the associated databases are somewhat weak. It is possible that solubilities could be underestimated if either an important solution species is missing from the database or if the system is not in chemical equilibrium. Solution concentrations, however, are inherently limited by glass degradation rates while the solubilities predicted can be validated experimentally or by natural analogue studies.

4) Diffusive transport through the bentonite is readily calculated. The hydraulic properties of this material are well understood which, additionally, allow its rôle as a colloid barrier to be accepted. Retardation mechanisms in bentonite are less well understood, but empirical measurements agree well with analogue observations.

The robustness of the overall near-field model can be illustrated by examining the consequences of altering the boundary conditions set by the host rock. In the base case model, the bentonite is assumed to act only as a colloid barrier and to prevent release of short-lived/strongly sorbed radionuclides. A more realistic model (Figure 2a) shows that there is considerable retardation/decay of many important radionuclides, which is the case even if the effect of low water fluxes through the host rock is totally ignored by setting a zero concentration boundary outside the bentonite (Figure 2b). In a scenario where the zero concentration outer boundary may be reasonable (massive flux of groundwater, some tens to hundreds of m^3/year through the entire repository), the concentration of most nuclides would be lowered by dilution to the point where requirements for further concentration reduction in the geosphere/biosphere are minimal. For the fully realistic case, the spatial distribution of the water flux should be included but, as this will only further decrease releases, it can be conservatively ignored to give further robustness.

The stability of the bentonite clearly plays a major rôle in providing robustness. This stability is guaranteed by mass balance constraints (rate of supply of K^+) which can, if necessary, be further supported by kinetics arguments and analogue evidence of bentonite mineral (montmorillonite) persistence in relevant geochemical environments.

The general concept of geochemical immobilisation of a wide range of elements in a diffusive controlled reducing environment is, additionally, well supported by observation of natural ore bodies. More specifically, detailed studies of redox halos in Swiss rocks have shown that very reducing conditions, of the type ensured by the iron canister, can lead to enrichment of many elements from the groundwater. Such an observation is enhanced by the observed efficiency of iron minerals/corrosion products for scavenging many relevant elements (U, Th, Se, Sn, Pd, I).

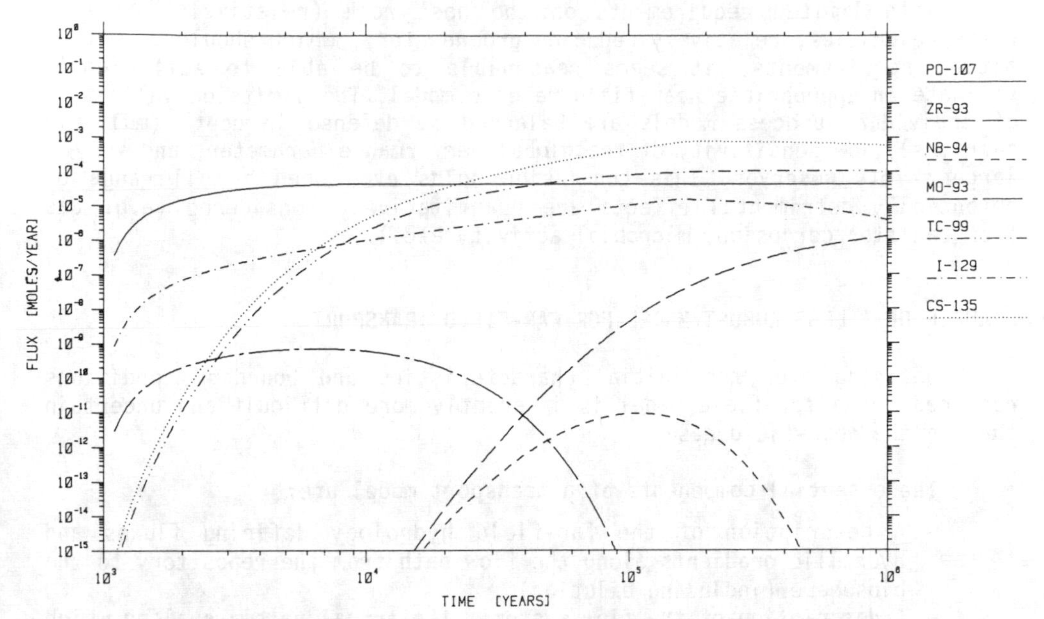

Figure 2a: Release to geosphere via backfill
Mass transfer boundary, selected fission products

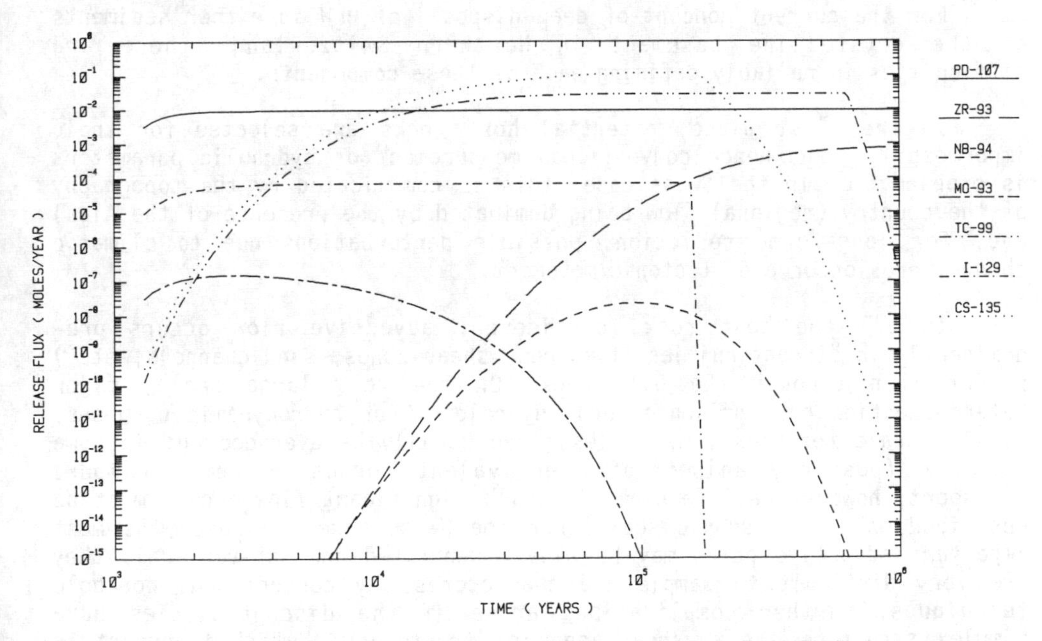

Figure 2b: Release to geosphere via backfill
Zero concentration boundary, selected fission products

With limited requirements on the host rock (relatively low flow rates/velocities, relatively reducing groundwater), which should be met by siting requirements, it seems reasonable to be able to **sufficiently** validate an appropriate near-field release model. The limits on validation of individual process models are balanced by defense in depth (multiple barriers), low sensitivity of the global performance parameter, and very large safety reserves. This conclusion holds even when a full range of potentially detrimental effects are quantitatively considered (e.g. gas from canister corrosion, microbial activity etc.).

EXAMPLE OF A LESS ROBUST MODEL FOR FAR-FIELD TRANSPORT

Defining even the initial characteristics and boundary conditions required for a far-field model is inherently more difficult and uncertain than in the near-field case.

The essential components of a transport model are:

- A description of the far-field hydrology defining fluxes and hydraulic gradients along the flow path from the repository to the biosphere (including dilution)
- A description of the flow system - the actual pathways along which transport occurs
- A representation of solute retardation during transport.

For the current concept of deep disposal of HLW in either sediments or the crystalline basement in Northern Switzerland, there are difficulties in reliably defining each of these components.

In the first place, potential host rocks are selected for their impermeability and hence conventional measurement of hydraulic parameters is problematic. In the Swiss case, this is complicated by the topography of the country (regional flow being dominated by the presence of the Alps) and, for long-term prediction, possible perturbations due to climatic change, erosion or even tectonic movement.

In all the host rocks considered, advective flow occurs predominantly in discontinuities (fissures, shear zones, sand channels, etc.) rather than through the bulk rock. On the very large scale, major heterogeneities can influence local hydrology. For hydrodynamic purposes, smaller scale features (≈m, or less) can usually be averaged out in some kind of (possibly anisotropic) equivalent porous medium. To model transport, however, all features in which significant flow occurs must be described on a microscopic scale. For the Swiss case, in which the most important advective paths may be widely separated and sub-vertical, they are very difficult to sample and characterise by conventional borehole techniques. Further complications arise if the discontinuities have complex structure - e.g. shear zones or fractures in which transport is localised in channels.

Finally, after the flow system has been defined, processes which retard the movement of particular radionuclides need to be quantified. In Swiss performance assessement, retardation is modelled in terms of a simple sorption parameter (Kd approach) and consideration of matrix diffusion into a limited depth of non-flowing, interconnected porosity surrounding the advective feature.

With considerable input of expert opinion, an integrated transport model can be constructed which shows, for most parameter ranges, that extensive retardation of most radionuclides would occur. The model has, however, very strongly non-linear response to the variation of some parameters. A particularly sensitive parameter is the sorption coefficient assumed which, in turn, is very dependent on the conceptual model of the flow system. This is illustrated by calculations for 2 different flow systems through a sedimentary host rock. In the first case (Figure 3) the very clearly non-linear response in breakthrough concentration to variation in Kd is shown. In the second, an even more dramatic dependency on assumed permeability can be seen for an alternative flow system (Figure 4).

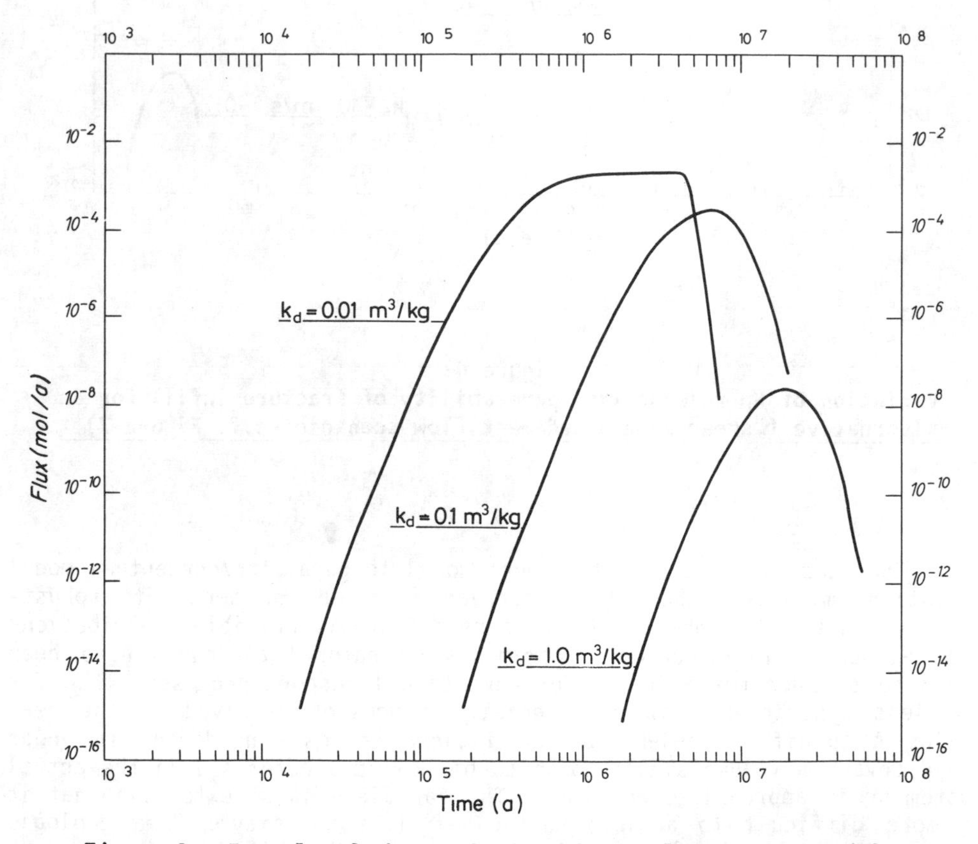

Figure 3: Example of the variation of Np release curves with assumed sorption distribution coefficient (Kd) for a particular flow scenario ("isolated sandstones") in a sedimentary host rock

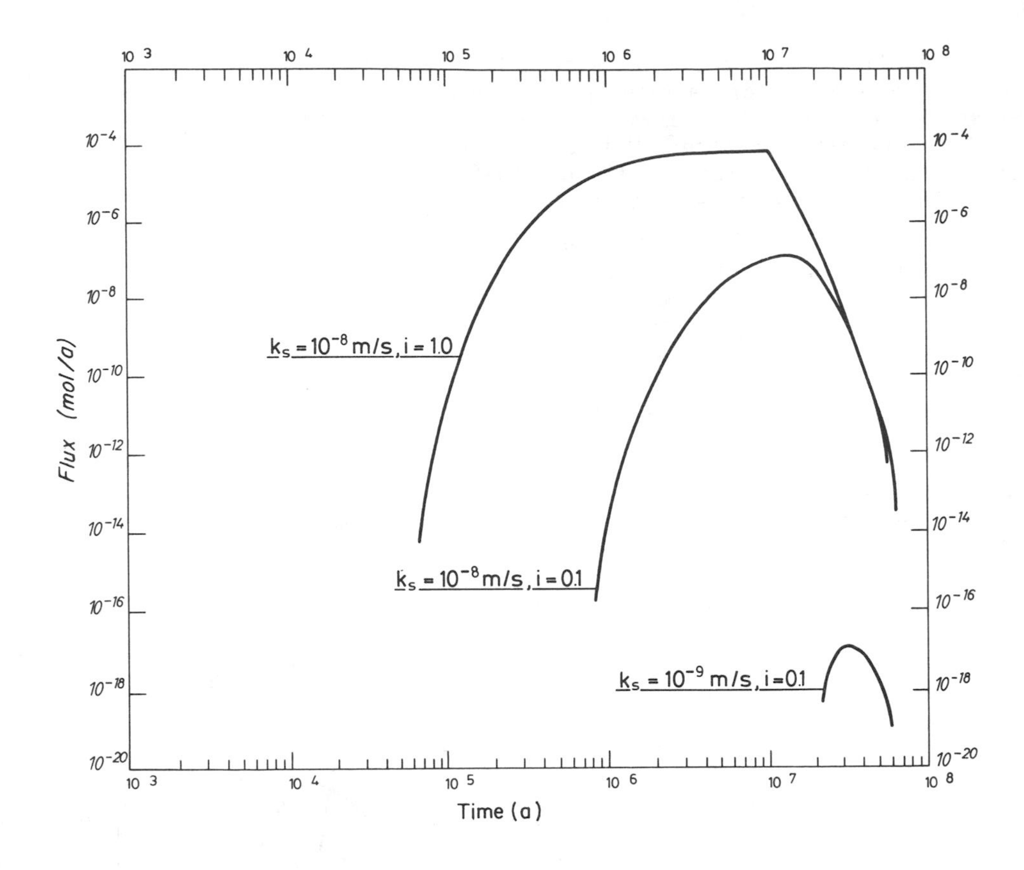

Figure 4:
Variation of Np release with permeability of fracture infill for an alternative ("shear zone") sediment flow scenario (c.f. Figure 3)

The sensitivity of the transport model to parameter/conceptual model variations minimises the safety reserves involved and hence its robustness. Validation is inherently much more difficult. Possible contributions from either field/laboratory experiments or natural analogues have been found to be much more limited for advective transport processes than for the less dynamic diffusion/geochemistry processes involved in the near field. A specific problem is the distance scales considered. The near field covers a volume small enough to be well characterised in the actual system or in appropriate analogues. The far field is so extensive that it is more difficult to achieve sufficient data "density". Some geologic media (e.g. tight clays) may be less problematic in this respect.

IMPROVING ROBUSTNESS

In general, there are two fundamental ways of improving the robustness of a performance assessment - either upgrading the models used (and their associated databases) or altering the actual system itself (designing for robustness).

Considering the latter point first, we can see that opportunities here lie in alteration of the design, layout or operational practices of the repository in order to improve the robustness of the near-field model.

A good example for HLW is increasing the intermediate storage time in order to minimise the potentially perturbing effects of the radiogenic thermal transient. This can be particularly attractive if it can be shown that a potentially detrimental process can be avoided completely - e.g. if temperatures in the near field remain below 100°C, then any perturbation caused by steam can be neglected from further consideration. Similarly, if organic materials can be excluded from the near field, speciation and solubility modelling becomes much simpler. In general, moves should be towards simplification of the overall system and not involve the addition of additional components. Although complicated backfills with exotic getters or additional canisters, liners and shields may seem very attractive when a particular process is considered in isolation, they will almost inevitably perturb other processes and demonstrating that there has been a net gain in performance is likely to be difficult.

In terms of the geosphere, the only major alteration possible involves changes in siting. For example, location of the repository in a hydraulic regime where flow paths are relatively easy to predict may be a considerable advantage. The ability to ensure adequate characterisation with minimum remaing uncertainty could be an important criteria. Design changes may also be used to increase robustness by obviating possible short-circuits via features such as shafts or tunnels. A good example of a robust design to avoid such short-circuits is provided by the access tunnel layout for a potential Swiss L/ILW repository site (Figure 5). In this case, the hydraulic gradient is defined by surface topography and the access tunnel is curved to ensure it approaches the repository area down-gradient. The requirements for performance of designed tunnel seals are thus fairly minimal.

For an established site/repository design, the robustness of the performance assessment can be increased by improving individual models and their associated databases. Often, this will involve not taking credit for potential safety features in order to ease validation requirements. An example of this is the treatment of matrix diffusion in Swiss HLW analyses. Although there is direct evidence of open porosity surrounding water-carrying features in crystalline rock, the extent to which connected porosity extends throughout an unweathered granite is very difficult to determine. The presence of unlimited matrix diffusion very greatly increases retardation in the geosphere, but is more difficult to test in any rigorous manner.

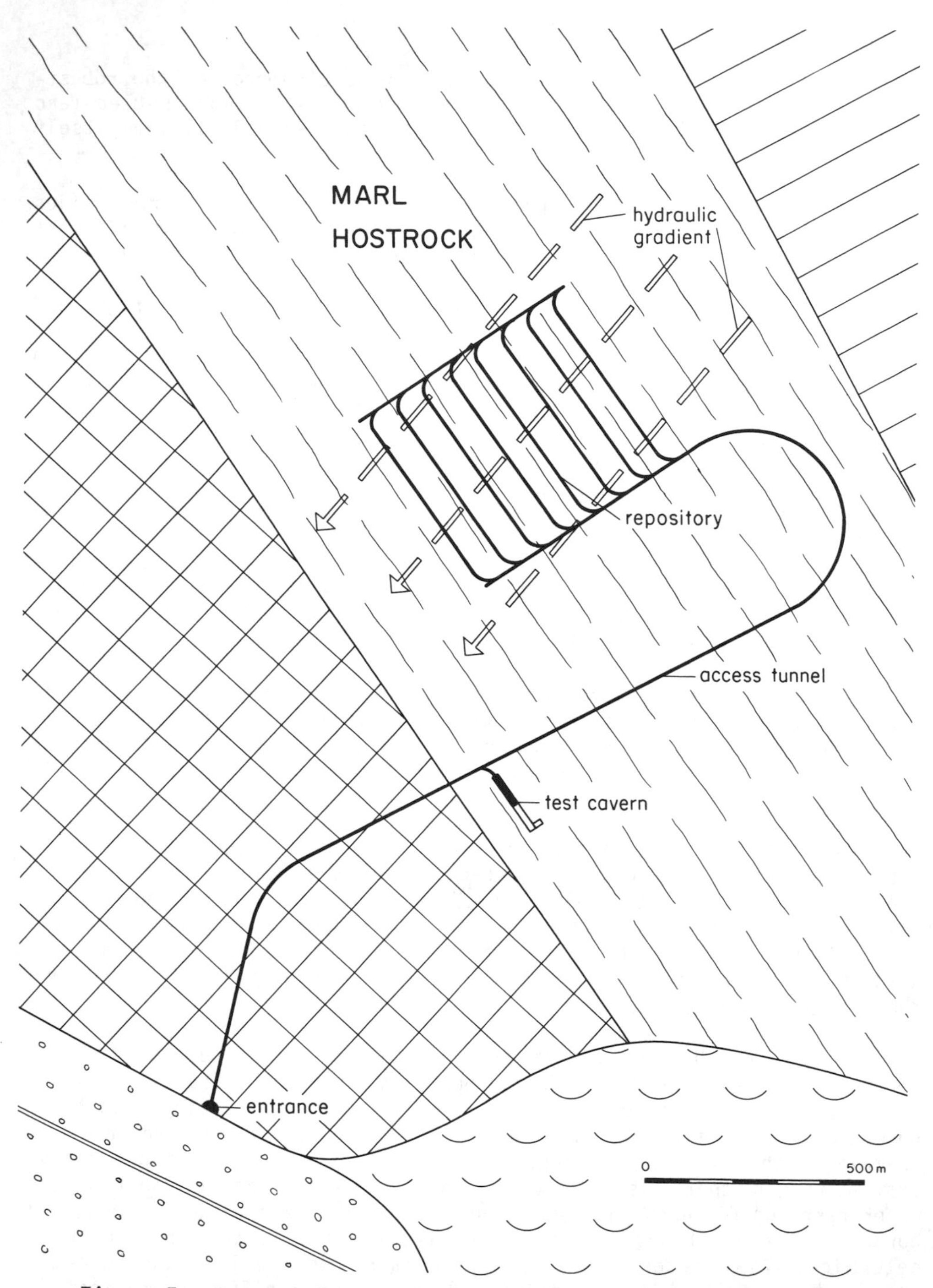

Figure 5: Potential layout for a potential repository for L/ILW

Thus, the concept of limited matrix diffusion was adopted - taking credit only for the zone in which connected porosity can be seen microscopically and matrix diffusion in-situ can be validated by natural analogues (natural series radionuclide profiles). As long as sufficient safety can be demonstrated with a chain of such robust models, credit for extra conservatisms involved can be gained in qualitative discussion of the results.

Generally, as far as the geosphere barrier is concerned, it is desirable that adequate performance can be assured on as small a spatial scale as possible. If only a limited thickness of good rock is required to provide an acceptable barrier, the needs for both exploration and characterisation are decreased and the potential safety reserves are increased. Conversely, if safety depends on the properties of a very long (many km) flow path, assuring sufficient robustness will be very difficult.

CONCLUSIONS AND THE NAGRA PERSPECTIVE

From the discussion above, it is clear that validation of performance assessment is difficult and can never be perfect. Some sub-system models can be termed robust in that we can be confident that any uncertainty involved lies in the direction of conservatism. These are generally relatively simple, contain large safety margins and hence can be sufficiently well validated. Robustness can be improved by either changing the system modelled or the model itself.

In a formal application of safety assessment (e.g. for licensing), the model chain must be sufficiently robust to demonstrate clear achievement of performance goals. Not all component models require the same degree of robustness due to the redundance involved in the multiple barrier concept. Best estimate evaluations along with sensitivity analysis, using realistic assessment models which can be less robust, will indicate minimum performance requirements for individual barriers. Sufficiently large safety margins, which are required for the overall model chain, can be demonstrably achieved by the most robust sub-model components.

The importance of the use of realistic models to analyse sensitivity and indicate performance objectives must be emphasised. The robust models may be so conservative that the true interaction of the model components is lost. This can lead to problems when attempting system optimisation; a danger of designing for robustness is the possibility of complete overdesign. Sensitivity analyses with a realistic model chain will indicate when enhancement of safety barriers (e.g. thicker canisters, backfills; longer cooling times) reaches the point of diminishing returns.

Finally, we can illustrate the issue of performance assessment robustness in a more concrete manner by examining the current situation in Switzerland. Present concepts for HLW, as indicated previously, involve massive, multiple engineered barriers which result in a robust prediction of very low releases from the near field. Because of this strong near field, requirements on the far field are decreased and models/parameters can be selected which are so conservative that they are reasonably robust.

For LLW containing primarily radionuclides of short half-life, the toxicity of the waste is rather low and hence the performance requirements are smaller for both the near field and the far field. The Swiss concept of deep disposal of such waste gives sufficient safety margins to assure robustness.

For ILW, particularly that containing significant long-lived α-emitters (TRU), the position is less clear. Both the waste itself and the other engineered barrier components are much more complex then is the case for HLW. Devising a robust model of the near field which assures long-term isolation is more difficult. The requirements on the far field may thus be greater than those for HLW. At present, effort is focussed on improvement of ILW near-field models to determine how much credit can be derived from current designs. Additionally, the option exists of strengthening the geosphere by emplacing the most toxic component of such waste in a deeper repository together with HLW, rather than in a L/ILW facility. Nevertheless, it may eventually be necessary to redesign the TRU engineered barriers for increased robustness.

In conclusion, although rigorous validation of a performance assessment model is fundamentally impossible, models of sufficient robustness to ensure meeting performance requirements can be constructed. When examined from this viewpoint, it is interesting to note that most problems arise with certain classes of ILW, while HLW is relatively unproblematic. A fundamental point is that it is **not** necessary within the scope of a performance assessment to aim for "absolute truth" or for perfect accuracy. The best possible understanding of system behaviour and a realistic modelling of important processes involved should, of course, always be aimed at. In a formal performance assessment - in particular within the scope of a licensing procedure - conservative models and bounding analyses will continue to play a key rôle. We should not allow our laudable, intensive efforts at improving model validating to blind us to this important fact or to obscure the continuing need for robust disposal systems and robust performance assessments.

A FRAMEWORK FOR VALIDATION AND ITS APPLICATION ON THE SITE CHARACTERIZATION FOR THE SWEDISH HARD ROCK LABORATORY

G Bäckblom[1], G Gustafson[2], R Stanfors[3], and P Wikberg[1]
GEOVAL - 90, Stockholm May 14-17, 1990

[1] Swedish Nuclear Fuel and Waste Management Co, Box 5864, S-102 48 Stockholm

[2] Chalmers University of Technology, Dept of Geology, S-412 96 Gothenburg

[3] IDEON, Ole Römers väg 12, S-223 70 LUND

INTRODUCTION

SKB plans to construct an underground research laboratory, the Hard Rock Laboratory close to the CLAB facility in the southeast part of Sweden. The project is a part of the comprehensive R&D programme, SKB (1989) to implement a safe final disposal of the spent fuel in Sweden.

Pre-investigation for the HRL was launched 1986. These investigations will be concluded during 1990, before construction of the Laboratory begins. The construction of the Laboratory provides an unique opportunity to evaluate the value of the pre-investigations.

During the course of the project it was found to be fruitful to structure all these possible predictions with respect to geometric scale and to major issues as related to design and safety analysis for a HLW-repository. The rationale for this approach is further discussed in this report.

Furthermore the paper outlines the adopted validation procedure and gives the rationale for:

- which predictions will/can be made before start of construction
- how the outcome during construction will be compared with earlier hypotheses
- how a judgement will be formed whether a prediction is good enough and validated or if characterization should continue into greater detail

INVESTIGATION STAGES

Site characterization for a final repository can be divided into three parts:

- Characterizations from surface and in boreholes prior to any construction activities

- Characterization investigations during access of a tunnel/shaft to a potential repository level

- Characterization investigations during construction of the repository.

These quite evident stages are also applicable to the HRL-project.

The pre-investigation period has further been divided into three phases. Every phase has and will give an integrated data set structured to different geometrical scales. This evaluated dataset has later been the basis for the appropriate numerical model, mostly on groundwater flow. The first evaluation was based on surface investigations, mostly in regional scale. The second integrated evaluation mostly on km^2 scale was concluded on supplementary surface data and three deep cored holes. This data set was as well the basis for detailed modelling of groundwater-flow, including fracture network models as well as predictive modelling of a long-time pumping test.

The last phase of investigations will utilize data from several cored holes, interference tests etc to establish the final predictions before the construction starts according to the format presented later in this paper.

PREDICTION SCALES AND THEIR RATIONALE

One of the features of the programme for the HRL is that predictions are set up in different scales for each step in the characterization. Appropriate numerical models are applied where possible. It will thus later be possible to check the possibilities to assess the rock mass properties in different scales at various steps in the site characterization programme. These assessments are of importance as a base for decisions on repository design as well as analysis of the safety. How the scales can be applied during characterization for a repository is outlined in the following.

Characterization in a regional scale >> 1000 m forms a basis for the detailed investigations. Assessment in regional scale can be used to select a suitable outcrop for the repository site. Areas of recharge and discharge can be defined. The regional assessment will also provide the basis for long-term descriptions of where the discharge area can be in the future as well as where potential zones of movement can be found.

The site scale characterization, 100-1000 m can be used to locate major fracture zones and/or major flow paths. These investigations are essential as they will provide guidance on what depth the repository should be placed at as well as to define a potential repository volume. Characterization in this scale also defines the farfield groundwater flow through the repository and to the biosphere.

Block scale assessment 10-100 m will be used to position deposition tunnels and later to position the canisters. Essential assessments are on the transport of solutes from possibly failured canisters to major flow paths.

The detailed scale 0-10 m may be the most important scale, as the properties in this scale defines the geohydrological, chemical and mechanical near-field to the canisters. By proper selection of canister positions it will be possible to influence the service life of (copper) canisters and the dissolution of the spent fuel and transport and fixation of radionuclides.

However, it should be borne in mind, that the choice of geometrical scale has been more or less subjective.

KEY ISSUES

With respect to **issues** it is appropriate to discuss - to every scale - the geological-structural setting, the groundwater flow, the (groundwater) chemistry, transport of solutes and mechanical stability during construction as well as long-term assessments.

The principle is depicted in *Figure 4.1*.

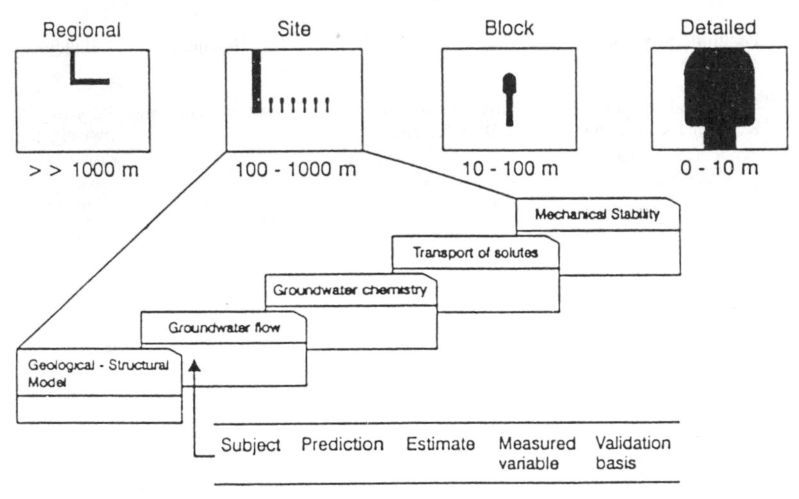

Figure 4.1 Overview of geometrical scales and key issues

PREDICTIONS AND FIELD MEASUREMENTS

An important concern in modelling of site properties and responses is that the same items that are predicted also will and can be measured in the field. To make that sure, lists of what will be predicted and measured are set up and matched for the different key issues and in the adopted prediction scales. Two such lists in site scale are shown in tables 1 and 2.

Table 1

PREDICTIONS OF GEOLOGICAL - STRUCTURAL MODEL - SITE SCALE (500 m)

SUBJECT	PREDICTION	ESTIMATE	MEASURED VARIABLE	VALIDATION BASIS
Lithological units				
Different rock types	Position and extension of lithological bodies	Maps of 100 m slabs % in tunnel	Rock contacts	Lithological mapping
Rock boundaries				
Rock mass description	Number of positions of rock boundaries	Maps of 100 m slabs	Rock contacts	Lithological mapping
Significant zones				
Different tectonic zones	Position, strike dip extension.	L/leg $\pm$%	Rock boundaries	Lithological mapping
	Width and character RQD(z) for crossings	RMR for crossings RQD(z) for crossings	RMR for crossings	Pilot tunnel investigations crossings

Table 2

PREDICTIONS OF GROUNDWATER FLOW - SITE SCALE (500 m)

SUBJECT	PREDICTIONS	ESTIMATE	MEASURED VARIABLE	VALIDATIONS BASIS
Hydraulic conductivity	Distribution of hydraulic conductivity in space	K (x,y,z)	K (x,y,z)	Pilot hole investigations
Water bearing zones	Position and transmissivity of hydraulic conductors	T (x,y,z)	T, (x,y,z)	Hydrogeological mapping Pilot hole investigations
Boundary conditions	Head or gradients at model boundaries	h (x.y,z) $\frac{dh}{dn}$ (x,y,z)	h (x,y,x) $\frac{dh}{dn}$ (x,y,z)	Groundwater level observations
Flow	Total inflow to tunnel	Q_{tot}	Q_{tot}	Total pumpage from tunnel + vapour transport
Pressures	Pressures in percussion and core boreholes under natural and disturbed conditions	p at x section for t timesteps	p	Pressures measured in boreholes during constructions
Flux distribution	Groundwater fluxes at natural and disturbed conditions	q at x sections for t timesteps	C(t)	Dilution tracer tests at different borehole sections
Inflow to tunnel	Inflow from identified zones F_x1--F_xN	Q_{FX} at x zones for t timesteps	Q_{FX}	Inflow measurements in sections
	Inflow to tunnel legs	Q_{leg} for z timesteps	Q_{leg}	Flow in sections
Salinity	Electric conductivity in boreholes and inflows to tunnels	Conductivity for t sections $Cond_{leg}$	$Cond_{BH}$ $Cond_{leg}$	Electric conductivity in borehole sections and inflow sections

AN APPROACH TO GEOLOGICAL MODELLING

In the tables above some entities are a result of numerical modelling and some are not, such as positions and extension of lithological bodies etc. In the context of modelling we prefer to use they are said to be predictions of the geological structural model, or the structure. This structure is basically a three-dimensional generalized description of the geology with properties assigned to the different sub-units.

A correct result of a numerical model provides normally a sufficiently accurate and complete model structure. In the validation procedure therefore not only the calculations must be observed but also the geological structural model that underlies the calculations.

APPROACH TO VALIDATION OF THE HRL PROJECT

According to the definitions given in Ref 2.3 the validation comprises a comparison and a judgement. To this can be added an analysis of the underlying processes. In short the three elements of the process of validations are:

- A systematic comparison of prediction and outcome
- A scrutiny of the underlying structure and process
- A judgement whether the prediction is good enough

Before the validation is performed, the accuracy of the predictions can be made better by calibration against known experiments or field measurements.

COMPARISON OF PREDICTION AND OUTCOME

As a first step of the validation of the model a systematic comparison of predictions, both of the structure and the resulting predictions and on the outcome shall be made. This comparison can always be done by the same persons that made the prediction. They are supposedly familiar with all the tools needed and have a thorough knowledge of all data.

In some stage of a project, when predictions and the underlying models are to be validated, it is suitable to produce a list of what is going to be predicted and what is going to be mapped or measured as shown in *Tables 1 and 2*. This will considerably facilitate the comparison of prediction and outcome.

SCRUTINY OF UNDERLYING STRUCTURE AND PROCESS

If prediction and outcome do not match, the reason for this may be either in structure or in the process or both, and even if the outcome matches the prediction, both may have their flaws.

In a scrutiny of the structure, the completeness is a critical issue. Based on pre-investigations, a set of features compatible with the geological structure are defined. If some important feature that affects the outcome is only encountered in the outcome, the structure is incomplete. This can be a serious factor concerning the validity of a model.

Also, the process model must be scrutinized in order to make sure that the applied processes accurately describe the significant responses to tests and experiments.

ACCEPTANCE

The final judgement wether a certain model is good enough follow normally from the scientific process; results are published and those results that last for long time are considered to be some kind of scientific truth. In the long run the results from the HRL-project must follow this process also.

In order to get an earlier judgement a peer review is required. Basic reports, predictions and experimental results are put forward to a group of peers that after a thorough analysis give a judgement if the used models are good enough.

CONCLUDING REMARKS

All predictions of the different key issues have a rationale in that they are relevant with respect to one or several aspects governing the general understanding, performance/constructions and safety of a final repository.

Consensus of process and investigations

In order to know that the right methods are used a, thorough knowledge of the properties of the crystalline rock and the processes that occur must be demonstrated. This knowledge shall not only make it possible to show how things are, but also to show how things cannot be. This insight must also be recognized by the scientific society, so that consensus of processes and investigation methods is reached.

Items governing the performance of a repository

An obvious task for a pre-investigation is to give a necessary basis for localization and design of a repository. Design means both geometry of the repository and as well design of engineered barriers. The design shall aim at an optimal performance of the system with regard to safety, acceptance and economy.

Safety analysis of a repository

The objective for the safety analysis, however, is to calculate the maximum possible impact on the environment. Also for this kind of analysis a set of relevant parameters are required, not necessarily the same as for the performance assessment.

With respect to these free aspects our conceptual and numerical models of a future repository for high level nuclear waste needs to be validated so we can rely with confidence on that the waste handling problems will be solved satisfactory.

REFERENCES

1 SKB, 1989, Handling and final disposal of nuclear waste. Programme for research, development and other measures, Stockholm.

2 IAEA-TECDOC-264, 1982, Radioactive Waste Management Glossary, International Atomic Energy, Vienna.

3 U.S. NRC, 1987, Proceedings of a Workshop of Mathematical Models for Waste Repository Performance Assessment.

SOME CONSIDERATIONS FOR VALIDATION OF REPOSITORY PERFORMANCE ASSESSMENT MODELS[1]

N. Eisenberg
U.S. Nuclear Regulatory Commission

ABSTRACT

Validation is an important aspect of the regulatory uses of performance assessment. A substantial body of literature exists indicating the manner in which validation of models is usually pursued. Because performance models for a nuclear waste repository cannot be tested over the long time periods for which the model must make predictions, the usual avenue for model validation is precluded. Further impediments to model validation include a lack of fundamental scientific theory to describe important aspects of repository performance and an inability to easily deduce the complex, intricate structures characteristic of a natural system. A successful strategy for validation must attempt to resolve these difficulties in a direct fashion. Although some procedural aspects will be important, the main reliance of validation should be on scientific substance and logical rigor. The level of validation needed will be mandated, in part, by the uses to which these models are put, rather than by the ideal of validation of a scientific theory. Because of the importance of the validation of performance assessment models, the NRC staff has engaged in a program of research and international cooperation to seek progress in this important area.

1. The views stated in this paper are those of the author and do not represent those of the U.S. Nuclear Regulatory Commission or its staff.

REFLEXIONS SUR LA VALIDATION DES MODELES D'EVALUATION DU COMPORTEMENT DES DEPOTS[1]

RESUME

La validation est un aspect important des utilisations de l'évaluation du comportement à des fins de réglementation. Il existe une bibliographie abondante décrivant comment on aborde généralement la validation des modèles. Les modèles simulant le comportement des dépôts de déchets nucléaires doivent donner des prévisions à long terme, aussi est-on privé du moyen classique de validation des modèles, à savoir l'essai en temps réel. D'autres obstacles s'opposent à la validation des modèles, notamment un manque de théories scientifiques fondamentales pour décrire des aspects importants du comportement des dépôts et une incapacité à sérier facilement les structures interdépendantes et complexes caractéristiques d'un système naturel. Pour élaborer une stratégie de validation efficace, il faut s'attacher à résoudre ces difficultés de façon directe. Bien que certains aspects de la procédure revêtent de l'importance, la validation devrait s'appuyer en premier lieu sur une base scientifique solide et une logique rigoureuse. Le niveau de validation requis dépendra, partiellement, des utilisations qui seront faites de ces modèles, plutôt que d'une norme idéale découlant d'une théorie scientifique. Compte tenu de l'importance que revêt la validation des modèles d'évaluation du comportement des dépôts, les chercheurs de la Commission de la réglementation nucléaire des Etats-Unis participent à un programme de recherches et de coopération internationale pour faire progresser les connaissances dans ce domaine important.

Les vues exprimées dans le présent document sont celles de leur auteur. Elles ne représentent pas celles de la Commission de la réglementation nucléaire des Etats-Unis ni de son personnel.

1. Importance of Validation in the NRC Program

Performance assessment, i.e. predictive modeling, is central to the evaluation of the adequate safety provided by a geologic repository (1,2)[2]. Although validation of models is not an explicit requirement of either NRC or EPA regulations, the central role of modeling in compliance assessment implies it. Furthermore, the NRC regulations do require (3): "Analyses and models that will be used to predict future conditions and changes in the geologic setting shall be supported by using an appropriate combination of such methods as field tests, in situ tests, laboratory tests which are representative of field conditions, monitoring data, and natural analog studies." The NRC regulations also require a Performance Confirmation Program, during which adequacy of modeling assumptions and performance predictions is to be checked. Another motivation for validation is that regulatory decisions of other agencies have been successfully challenged because the models used to support such decisions were not sufficiently validated (4).

2. Background discussion of validation

Because the author of this paper has found that discussions of validation often are impeded by semantic arguments, definitions of some key terms are cited. Some of the substantial literature on modeling is also cited.

Modeling. One source (5) defines a model simply as: "A mathematical model is an abstract, simplified, mathematical construct related to a part of reality and created for a particular purpose." Although this definition is informal and brief it contains several important concepts. First, the model abstracts a part of reality. A model need not describe all the relations among all parts of the universe or even part of it. Rather, only the part of the universe of interest and only the behavior of interest of that part need be of concern to the modeler. Second, the model produces quantitative information or at least gives instructions about how to obtain such information. Several sources (6, 7) distinguish between modeling and simulation. Modeling is described as the process of abstracting the real world to obtain a mathematical description of the system and behavior of interest. Simulation is described as the process of quantifying the mathematical description embodied in the model. Figure 1 (based on a diagram of Ziegler) shows the relationship between modeling, simulation, and the real world.

Ziegler and Spriet use abstract mathematical formalism to describe modeling. Ziegler develops an axiomatic framework for modeling and simulation, while Spriet develops a system modeling formalism. Both arrive at a system modeling formalism, which defines a system (or input-output system) as a set structure. Although much of this theory may be too detailed for considerations in waste management, an important concept inherent in this description is that the output of the system may depend upon the internal state of the system, which is not only a function of the system inputs, but

---

[2] Numbers refer to the reference list.

also a function of the initial internal state of the system, determined by its history. Ziegler points out that the state of the system may be unobservable, but still have a significant effect on the observed behavior of the system. This is not an unusual issue in waste management. For example, in modeling hydrologic systems, an assumption is often made that the hydrologic system is in equilibrium, when in fact, it may be in a transient state.

Modeling is often considered applied science, not pure science. In modeling the behavior, of interest of a particular system is deduced on the basis of the axiomatic acceptance of a set of "laws." Two items are important to note. First, a model is an abstraction and only the behavior of interest need be modeled. This is an important issue in model validation, because to be valid a model does not need: (1)to explain all the behavior of the system, (2)to explain different aspects of the behavior of a system with equal precision, (3)to bear a similitude to the physical system modeled except in the behavior of interest. Second, although a conclusion about system behavior obtained deductively from a set of axioms should be unique, in practice, the axioms are supplemented by a set of additional assumptions intended to simplify the analysis. Because the set of assumptions are not unique, a model with one set of assumptions can be more valid than a model with a different set. These assumptions may be generally classed in three categories: (1)what processes and phenomena must be included in the model, (2)what geometry should be assumed for the system being modeled, and (3)what environmental and boundary conditions should be assumed. Some discussions emphasize the first two of these (8, 9), while others emphasize all three (10). Thus, validation activities should be directed at determining the validity of these assumptions, not at determining the validity of the underlying physicochemical principles.

Although the distinction made above between the development of theory in pure science and the development of models in applied science is correct from a strict philosophical point of view, the difference is less distinct on a practical basis. Ziegler (6) mentions that the goal of modeling can be (1) to understand nature better (a scientific pursuit) or (2) to serve more practical goals, such as optimizing systems, providing a surrogate for expensive or impossible testing, facilitating interpretation and reduction of system behavioral data. Spriet (7) discusses this "duality in goals" at some length.

Validation. Ziegler (6) defines model validation informally as: "The modeling relation concerns the validity of the model, that is, how well the model represents the real system. In the first instance, validity is measured by the extent of agreement between real system data and model-generated data, as portrayed symbolically in the equation:

$$\text{real system data} \overset{?}{=} \text{model-generated data."}$$

Both Ziegler and Spriet define three levels of validity:
1. Replicative Validity (valid at the behavioral level). The model matches data already acquired from the real system.
2. Predictive Validity (valid at the state-structure level). The model can match data from the real system, before such data are acquired or, at least, seen by the model. Alternatively, the model can be put into a state

(synchronized or calibrated) such that the future behavior of the real system in uniquely predicted.
3. Structural Validity (valid at the composite structure level). The model not only reproduces and predicts the behavior of the real system, but uniquely represents the internal operation of the real system that produces such behavior.

Note that these authors apply these terms to modeling as they view it. Simulation is another issue. In the context of Figure 1, "the simulation relation concerns the faithfulness with which the computer carries out the instructions intended by the model" (6, p. 5). Clearly, validation always depends on comparing model predictions with the observed behavior of the real system. Such comparison is possible only when the instructions provided by the model are quantified, which these authors call simulation. Determining whether the simulation is correct is usually termed verification (19) in waste management studies. Henceforth, in this paper, we will assume that the simulation is correct and we will focus on the more difficult issue of validation, i.e. is the model "correct".

Ziegler points out that the experimental frame is an important issue in model validation:

> "The experimental frame characterizes a limited set of circumstances under which the real system is to be observed or experimented with...Since an experimental frame puts constraints on the possible observation of the real system, it is possible for a relatively simple model to produce input-output pairs that agree, within some standard of comparison, with all the input-output pairs of the frame. When this happens, we say that the model is valid for that experimental frame. It is not surprising then to learn that a model may be valid in one experimental frame but invalid in another. Thus there may be many valid models (at least as many as experimental frames). We see that the validity of a model is relative to an experimental frame and the criterion by which agreement of input-output pairs is gauged."

This is a crucial issue in validating performance assessment models for a nuclear waste repository, because the time frame of application of the model (hundreds to thousands of years) is beyond the reach of experimentation.

Spriet (7) introduces the concept that the validity of a model cannot be separated from the state of scientific knowledge upon which the model is based. Figure 2 shows schematically the model building process and the hierarchy of knowledge available, a priori. Three levels of generality (broad, limited, and narrow) are shown to correspond respectively to three levels of validity (high, medium, and low). Although three categories are shown, there is obviously a continuous spectrum of quality of knowledge. Spriet states:

> "In many sub-fields [of physical science], a mature general theory is available stating its principles and laws in mathematical and, hence, quantitative terms... As a consequence, the model-builder will draw almost all a priori information from route I [in Figure 2]. He has an adequate modeling methodology and consequently, his model will have high validity."

In addition to the three levels of validation mentioned above (replicative, predictive, and structural validity), Spriet introduces the three additional considerations of validation: (1)in deduction, (2)in induction, (3)in purpose. An important issue in the validation in deduction is the validity of the a priori knowledge used to build the model. Spriet points out that not only does this knowledge vary in generality and exactitude, but the determination of the quality of a priori information is difficult to make, because of the manner in which scientific results are usually reported in the literature. Validation in induction, according to Spriet, reduces to comparison of data obtained from the model to data obtained from the real system. Comparison of the two data sets must be over the same experimental frame. A simple approach to comparing the data sets would require equality of the two sets for a valid model. But Spriet points out:

> "In practice, one has to choose a metric, distance measure, which inevitably reflects a certain weighting of features of the behaviour. Validity can be ascertained as long as deviations of the model data and the real-world data are close, with respect to the chosen metric. A second choice must be made if the real system data is is believed to be stochastically generated and/or if the model is taken to represent a stochastic process. In this case, statistical tests must be chosen in order to assess the probability that the degree of agreement found between model and real system with the finite amount of data available is truly representative of the stochastic processes from which the data was sampled."

To ascertain the validity of a performance assessment model both the performance measure(s) of interest and the metric to compare the model output to the real system data must be specified.

Validation in purpose, according to Spriet, is achieved if the goal of the model is obtained. But he further states: "Little, however, is known as to how a limited objective allows for less stringent validity requirements." For the performance assessment modeler two sources of reduced requirements for modeling are:

1. Often the goal of modeling is to demonstrate that a given performance measure is greater than or less than a specified regulatory limit. For example, cumulative release of radionuclides, individual dose, fractional release of radionuclides from the engineered barrier system all have upper limits specified in U.S. regulations.
2. Often the performance measures of interest (like those listed above) are sums or probabilistically weighted cumulations; this implies that the details of the system (and hence the model) behavior are not as important as some average or gross behavior. This may reduce some of the stringency for requiring "structural validity."

In spite of the intuitive appeal of the idea that reduced model requirements imply reduced validation requirements, the manner in which this is to be quantified, or even specified, is not clear.

3. Impediments to validation of performance assessment models.

The impediments to validating performance assessment models of repository performance include the long time periods for model predictions, the involvement of natural, heterogeneous natural systems, and the limitations of the state-of-the-art in some scientific disciplines used in modeling. The time periods for the application of repository performance assessment models are so extremely long ($\sim 10^2$ to 10^6 years) that direct comparison of system performance with model prediction cannot be accomplished; therefore, the inductive approach to model validation is inapplicable. In virtually all repository concepts, and certainly in the U.S. repository concept, the geology of the site plays a significant role in the performance of the repository and is a major motivation for pursuing mined geological disposal instead of other disposal concepts. However, the site geology, geochemistry, hydrology, rock mechanics, etc. are aspects of a highly complex, heterogeneous natural system. Because of the nature of the geologic system, the behavior of the system can only be observed inferentially and only at a finite number of measuring points in space and time. These limitations apply to both the boundary observables and to the much more difficult-to-obtain interior structure and behavior. The inability to confirm the interior structure of the system or to confirm the processes relating such structures to each other and to system performance, severely limits the degree to which validation can be carried out. Thus, in the parlance of Spriet and Ziegler, "structural validation" would appear to be limited for geological systems. The state-of-the-art of scientific theory for various aspects of performance assessment modeling are not advanced enough to assure a basis of a priori knowledge sufficient to produce valid models by deductive reasoning. In the earth sciences, corrosion sciences, and other disciplines some fundamental questions have yet to be answered and several fundamental principles are stated qualitatively, rather than quantitatively.

4. Ideas for a Model Validation Strategy.

As the discussion above indicates, the validation of models for performance assessment of a nuclear waste repository is difficult because of the nature of the systems modeled, because of the state of knowledge in the sciences involved (primarily because the systems are so complex), and because the time period for model use is inaccessible for testing on the real system. This last difficulty must be addressed frontally and vigorously, since it provides the greatest challenge to model validation. Some practical steps to address the issue of model validation are suggested below:

Step 1. Evaluate the a priori basis for the model.
1. The a priori information should be evaluated and the level of validity of this information should be determined.
2. The importance of various components of the a priori information to the prediction of the performance measure(s) to be provided by the model should be specified, quantitatively if possible (e.g. by sensitivity analysis).
3. For situations where a priori information is very important in the determination of model predictions and where the a priori information has insufficient validity (see Figure 2), additional work to improve the fundamental science involved is mandated. When the model is not as sensitive to the information or when the information is of higher quality, the a priori information available may be considered adequate.

Step 2. Obtain data on the real system or parts of it to quantitatively evaluate the model or parts of it.
1. Data should be obtained for the purpose of model validation. Such data could describe the behavior of the real system or a part of the real system. Examples of such subsystem tests might include: (1)tests on the

corrosion of a waste package, (2)tests on radionuclide migration in rock cores, (3)tests on radionuclide migration in field situations over tens of meters. As mentioned elsewhere (8, 10) the data obtained in such tests must be sufficiently redundant to determine model parameters and to compare model predictions to real system behavior.

2. To evaluate the results of the validation tests, two items must be specified before the testing commences: (1)the performance measure of interest must be specified (the physical quantity of interest, the location where the quantity is to be observed, the means of measurement and/or calculation of the performance measure from measurements) and (2)the metric for comparing model predictions and observed behavior of the real system. The specification of the metric can incorporate, in a quantitaive fashion, the relative importance of various performance measures to the end use of the model. Specification of a metric for validation tests provides for a quantitative criterion for determining whether a model is valid. Specification of a metric also limits the extent to which certain aspects of model performance are important; these limitations on certain aspects of the modeling, are driven by the end use of the model, rather than the ideal of a complete, comprehensive, valid scientific theory.

3. To be successful this step must be accomplished with real testing and quantified metrics, which have a strong, rational basis. This step cannot be accomplished by vague procedural activities.

Step 3. The fact that the model predictions must be for time periods much longer than testing can be performed must be addressed by a strategy intended to reduce the impact of this obstacle.

1. The a priori set of knowledge used to construct the models must be exploited (and perhaps expanded) in an attempt to derive predictions on how the behavior of the system will change as the prediction time becomes very large. If a sufficient a prior basis can be compiled for estimating system performance into the far future, the need for testing the ability of the model to accurately make such predictions is lessened.

2. Data from different lines of evidence, covering different time scales, should be brought together to determine how well the model can predict over these different time scales. Also, these data should be compared to the theoretical projections made in item 1, above. As discussed elsewhere (8, 11, 12) data representing different time scales could be obtained, in ascending order of time scale, from: laboratory tests, in-situ tests, field measurements, natural analogues. One difficulty in applying such disparate data to a single model is that the data available to specify model parameters is of greatly varying quality. Thus, the boundary conditions, such as flow rate and source radionuclide concentration, that are significant in a migration model, can be well regulated in a laboratory environment, but may be impossible to deduce in any given natural analogue. Under such circumstances, agreement or lack thereof cannot be solely ascribed to the differences in the temporal scale of the different sources of data. Nevertheless, pursuit of such an approach is indicated, even though many investigators of natural analogues believe that such quantitative uses of natural analogue data are not supportable (13).

5. NRC Involvement in Validation Projects

In the late 1970's a primary focus of research at the NRC was the development of computer programs to analyze the deep geologic disposal of high-level nuclear waste, concentrating on models of radionuclide migration in ground water and techniques for sensitivity and uncertainty analysis. The current focus of research has been towards evaluating the validity of these models and gaining confidence in these techniques. Validation efforts have concentrated on validation field work and participation in international simulation projects, such as INTRACOIN (14, 15), HYDROCOIN (16), and INTRAVAL (8).

REFERENCES

1. Brooks, P. and S. Coplan. The role of verification and validation in licensing repositories for disposal of high-level waste. in GEOVAL 1987, p.41. Swedish Nuclear Power Inspectorate, Stockholm, 1988.

2. Eisenberg, N. A., D. H. Alexander, W. W.-L. Lee. The role of validation activities in the U.S. Department of Energy's Office of Civilian Radioactive Waste Management Program. in GEOVAL 1987, p.27. Swedish Nuclear Power Inspectorate, Stockholm, 1988.

3. U.S. Nuclear Regulatory Comm. Disposal of high-level radioactive waste in geologic repositories, Title 10, Code of Federal Regulations, Part 60.

4. Walsch, W. J. Legal and policy role in the use of transport and fate modeling. in Proceedings: Environmental research conference on groundwater quality and waste disposal, p. 13-1. EPRI, Palo Alto, March 1990.

5. Bender, E. A. An introduction to mathematical modeling. John Wiley & Sons, Inc., 1978, New York.

6. Zeigler, B. P. Theory of modeling and simulation. John Wiley & Sons, Inc., 1976, New York.

7. Spriet, J. A. and G. C. Vansteenkiste. Computer-aided modeling and simulation. Academic Press, 1982, New York.

8. Swedish Nuclear Power Inspectorate. INTRAVAL Project Proposal, SKI 87:3, SKI, July 1987.

9. Tsang, C. F. Tracer travel time and model validation. in Journal of Radioactive Waste Management and the Nuclear Fuel Cycle, Vol 13, pp. 311-23. 1989.

10. Eisenberg, N. A. et al. A proposed validation strategy for the U.S. DOE Office of Civilian Radioactive Waste Management Geologic Repository Program. in GEOVAL 1987, p. 341. Swedish Nuclear Power Inspectorate, Stockholm, 1988.

11. Hayden, Nancy K. Benchmarking NNWSI flow and transport codes: Cove 1 results. Sandia National Labs, June 1985, Albuquerque. SAND84-0996.

12. Flavelle, P. A. Regulatory perspectives of concept assessment. in Proceedings of the conference on geostatistical, sensitivity, and uncertainty methods for ground-water flow and radionuclide transport modeling, p. 111-122. Battelle Press, 1989, Columbus.

13. Chapman, N. A. Can natural analogues provide quantitative model validation? in GEOVAL 1987, p. 139. Swedish Nuclear Power Inspectorate,

14. Swedish Nuclear Power Inspectorate. INTRACOIN Final Report Level 1, Code verification, SKI 84:3, SKI, September 1984.

15. Swedish Nuclear Power Inspectorate. INTRACOIN Final Report Levels 2 and 3, Model Validation and Uncertainty Analysis, SKI 86:2, SKI, May 1986.

16. NEA. The International HYDROCOIN Project - Level 1 Code Verification, Organisation for Economic Cooperation and development/Nuclear Energy Agency (OECD/NEA) and Swedish Nuclear Power Inspectorate (SKI), 1988, Paris.

FIGURE 1. SCHEMATIC OF THE RELATIONSHIP BETWEEN MODELING AND SIMULATION (after Lieberman, P.4)

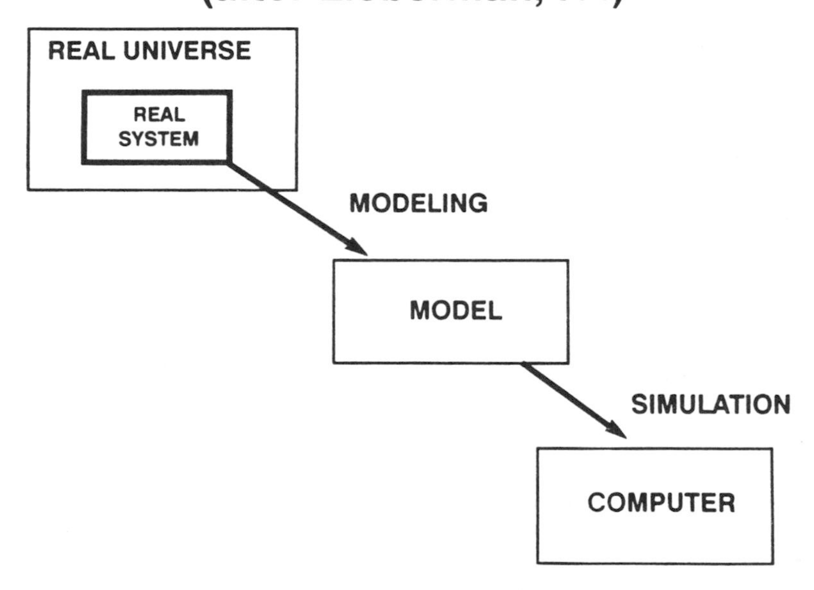

FIGURE 2. INFORMATION ENVIRONMENT FOR MODEL BUILDING (after Spriet, P.43)

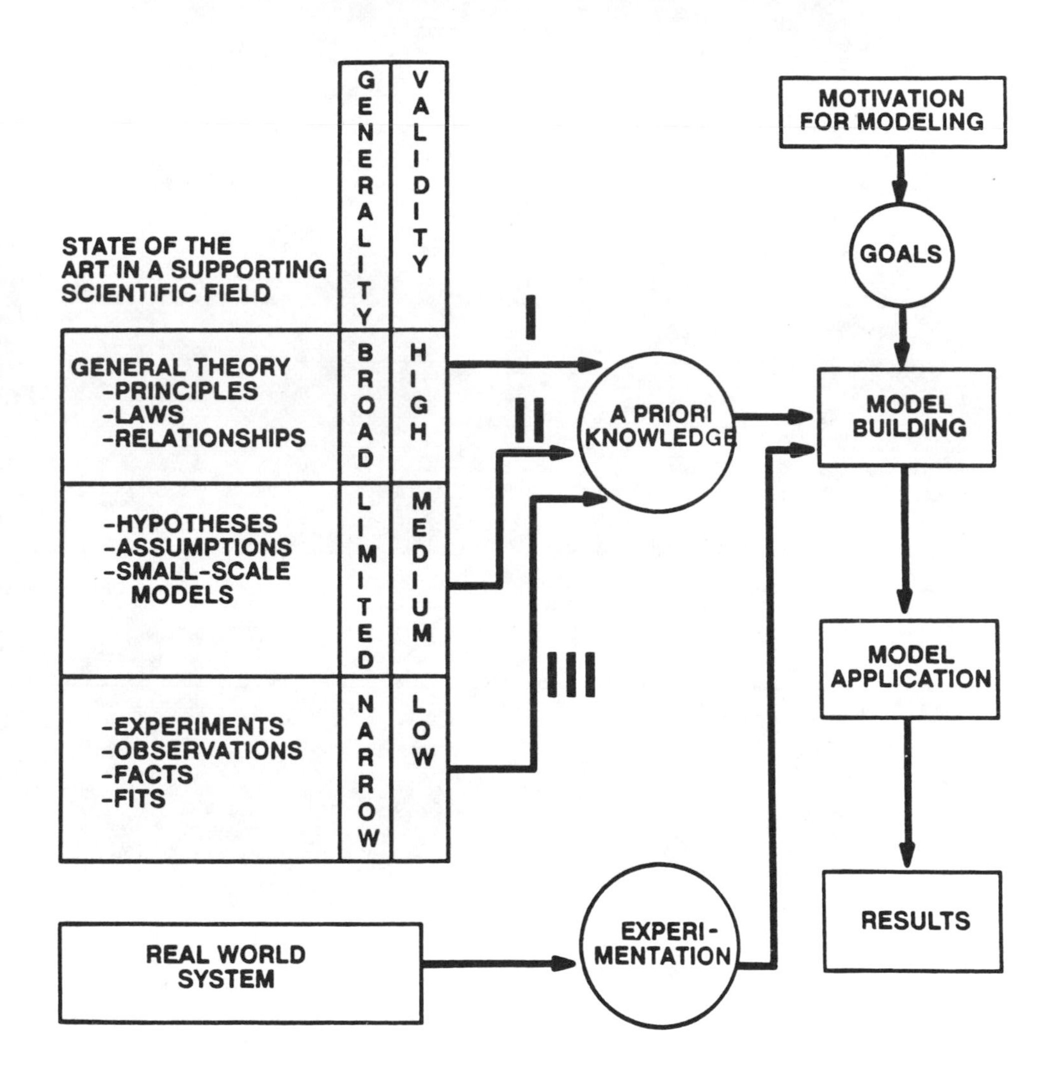

Session 6

PANEL DISCUSSION

Séance 6

TABLE RONDE

Chairman - Président

S.P. NEUMAN
(United States)

Panelists - Membres

J. ANDERSSON (SKI, Sweden)
C. McCOMBIE (NAGRA, Switzerland)
K. PRUESS (UC/LBL, United States)
P. SARGENT (AECL, Canada)
C.-F. TSANG (LBL, United States)

Panel Discussion, May 17, 1990

Chairman Shlomo P. Neuman

Presentations by Panel Members

Charles McCombie

In modelling, one should aim at a robust system which is easier to validate. I refer here to Tönis Papp who has said, "Maybe in the end one will chose a disposal system, not because it appears able to provide the most safety, but rather because it is the easiest to validate." In order to have a robust system, complicated "technical fixes" should be avoided if it is difficult to have sufficient confidence in predictions of their behaviour.

I will mention two aspects of robustness – for the engineered system and for the performance assessment. For a repository system in a geological medium there are design choices to be made. To maximise redundance, safety and robustness, one should chose elements with the simplest possible physics and chemistry. This can run counter to accepted engineering practice or intuition. For example, in our HLW disposal tunnels we request our engineers to avoid using concrete for reinforcing weak zones – the geochemistry can be negatively affected. In our L/ILW programme, on the other hand, we request the engineer to use disposal space inefficiently in order to maximise the amount of cement in the tunnels – here we deliberately want to condition the near-field chemistry. In both cases we are making design concessions which are intended to make the system as simple and as predictable as possible. High technology and highly complex technical fixes can be counter-productive if they decrease system robustness.

The choice of permissible temperatures in the repository is another example. Anyone who chooses not to wait sufficiently long enough before disposal of HLW to ensure that the residual decay heat gives only temperatures with which the chemical/geochemical modeller feels comfortable is asking for trouble. It might indeed be the case that higher temperatures could provide a real positive contribution to safety. If this is problematic to validate, however, we could be making our task more difficult.

The other aspect or robustness concerns the assessment itself. Conceptually we can separate robust engineering and robust assessment, although they are obviously interrelated. We should aim at a robust practical assessment which can be transparently demonstrated to everybody concerned. This does not, of course, exclude the need for complex research models or for good scientific research work.

The good modeller is not a brutal number-cruncher who blindly believes in all results produced by his computer. A good modeller realises the crucial importance or understanding of the system and also of quantifying this understanding; he knows that there are many sources of uncertainties in his results and tries to scope the possible consequences of these; he also realises that his work has practical applications and is prepared to seek a sufficient degree of accuracy for his model predictions and not necessarily a complete characterisation of all phenomena involved.

I should like now to make a couple of comments on the timescales for assessments. Are the long timescales involved in the safety assessment of a waste repository really so unique? They are not so long compared to geological timescales. What is unique is our explicit recognition of the long timescales. It is easy to think of other technologies for which long timescales also are important, e.g. for disposal of toxic wastes. The case is slightly more subtle for climatic effects. On what timescale can the greenhouse effect influence our environment – infinitely long?

Another example which is often overlooked is the exploitation of raw materials. We know that it took a very long time to build up the earth's reserves of raw materials, but that we could use many of them up in our own life time. The total exhaustion of a resource will have a very significant environmental influence which will affect this planet for the rest of time.

Where do we stand with validation now? Two aspects have been mentioned, the processes and the structural database. For the processes, the good news is that, generally, there are no new processes coming up that we have to consider. For ten years, I have looked around, listened and assembled a long list of processes considered in performance assessments. In recent years, the original list has narrowed in and there are not many new processes being discussed. There are exceptions, such as gas generation, which has been underestimated as an important process. But there are not many new processes appearing that would cut through the safety barriers of the whole system. Some of the old processes, like colloids, have, however, been raising questions for a long time. Maybe we are not making as fast progress there as we should do.

For the structural database we are getting more data. We now see (in INTRAVAL for example) more experiments than people to model them – which is a new, more satisfactory situation. At the same time, the awareness of how difficult it is to find good experiments, which directly help to validate models, seems to be growing, judging from the first two days of discussion here. But at least in some situations, we are aware of the range and complexity of the data we want to get, so that suitable experimental techniques and programmes are being developed. When I integrate the good and bad news, I feel that the remaining important uncertainties are not so much the conceptual uncertainties, but rather the more tractable uncertainties arising from incompleteness of databases.

What has got to be done now, and by whom, to proceed further? Formal procedures for validation are not well developed today, but such procedures are coming. We are progressing towards a consensus of what we should do, what we can expect from validation exercises – and also what we cannot expect. When we reach that consensus, it should be from the whole scientific community. I do not like the polarisation between implementors and regulators which seems to have developed in the disposal area but has not occurred, for example in the nuclear reactor area. There, the reactor scientists developed reactors, made license applications and convinced the regulators that nuclear reactors could be run safely. And then what happened? The whole community, the regulators and the implementors, went out and spread the message that nuclear reactors are okay. If you look at a report of a reactor accident on television, the people likely to be questioned and to be explaining the safety issues are regulators. Regulators are in the nuclear reactor business because they feel that nuclear reactors can be safe. That is the last message I want to bring. Technical people such as those in this meeting must try to reach a consensus on the way to proceed with implementation of safe disposal; they must do this together, do it as a scientific community!

Shlomo P. Neuman

Before we open up the floor for comments and questions, I would like to ask Johan Anderson to make some comments. Please!

Johan Andersson

We have during this symposium had extensive discussions on what validation stands for. We also know that there have been many attempts over the years to define validation, which have resulted in different but similar definitions. One suggestion, which is in fact the one that one of my colleagues at SKI has formulated, concerns predictions of flow and transport. The predictions you do for performance assessment, should have an accuracy which is acceptable within the framework of the total safety analysis. You don't need more precision in the predictions than what is needed for the safety analysis. On the other hand, you require that the predictions should include all uncertainties of all kinds, correctly extrapolated into relevant space and time.

If we are very critical, we must realise that we can never fully resolve the influence of uncertainties on the performance assessment, because we can always find instances where part of the argumentation chain will break down. Perhaps, we should not be too much concerned about this problem and instead turn to areas where we really have difficulties. There are some issues in the geosphere, that I think, we haven't resolved enough yet. I am not convinced, however, that we need a formal procedure to show that we have solved a certain problem, but we need a scientific process with continuous evaluation, with experiments and modelling to get a resolution and we need to be more specific.

At the INTRAVAL meeting in Helsinki I showed a list of some critical issues dealing with transport in crystalline rock. As to advection-dispersion we still see that we do not know everything but I want to be optimistic. We have, as David Hodgkinson expressed, developed tools on how to understand and attack such problems and I think we can formulate new experiments which will lead us further. We also know from experience that advection-dispersion is not that critical for performance assessment.

Sorption and matrix diffusion is not in the same good shape, however. We know that the processes exist. We also know from performance assessment that the amount of matrix diffusion will make enormous differences in the performance of a repository, at least for some radionuclides. The experimental and site specific data we have today will give us the range, but the range in prediction we get from site specific data today is far too big to be of use in performance assessment. If we use the lowest values we can think of, the effect of matrix diffusion is not all that large as we imagine it could have been with higher values. There is apparently much more work needed.

For thermo-hydro-mechanical effects the validation or scientific problem is of another nature. We and others have made scoping calculations with existing models that we possess today. Broadly speaking, the calculations show that if we are careful, the effects do not pose a problem for the repository. If we have the repository at a safe distance away from major fracture zones, all the movements will be taken up in the fracture zones and nothing will happen to the repository. But the scientific knowledge of the processes involved is not sufficient and the mathematical modelling of these processes is not that advanced that we really can believe that the results of the calculations are correct. In particular, we have very little knowledge of the stiffnesses and the deformabilities of faults, and that could make differences. We apparently have to increase our knowledge in this area just to know that we don't fool ourselves and that our modelling system as a whole is correct.

Scoping calculations made on colloids show that they do not present a problem. Still it is a major concern that we do not fully understand the processes involved and we need to do more work. If we resolve questions like that, I believe that a formal proof of validation of the whole system is not necessary.

When considering different validation techniques, it is often suggested to make more use of data. We have statistical analysis techniques, inverse techniques, and calculation of intervals of confidence. We should discuss how to use models to design experiments that differentiate between different assumptions. To my mind this is a very critical issue, at least if we want to maintain the given timetables of the different waste disposal programmes. Experiments take a very long time to perform and we have to be very clever in setting up new experiments.

Discussion on first two Panel Presentations

Shlomo P. Neuman

We will now open the floor to comments and questions. Let me take my prerogative as chairman of this panel and start.

What I heard from the last two speakers, was the need for consensus in the scientific community. We need to agree and achieve a warm feeling in our tommy that our repository is going to be safe. My question is: Do we have such a consensus? If not, can we achieve it within a reasonable amount of time? And if not, why not?

Before we start to discuss these and other related issues, I would like to remind ourselves of an interesting observation that professor Grenthe made yesterday or the day before. He asked the question: Why is it that in chemistry or chemical engineering models seem to be reliable as predictors, whereas in hydrology this does not seem to be the case? Our models do not seem to be doing as well, or at least we are not sure that they are doing as well, as in the chemical engineering field. He suggested in fact that hydrology or hydrogeology may lack what his own field probably possesses to a greater extent, a common paradigm.

I share his feeling. I think that we are dealing with a relatively unexplored area. We are developing a new field in the earth sciences, in geology, in geophysics. Previously the earth sciences have been interpreting the past. We have tried to reproduce or understand earth processes, but we have not tried to predict how earth-related processes will evolve with time in the future. It is this predictive aspect which is new to us and in which we are inexperienced. With this comment, I would like to open the floor for other comments and questions.

Allan Emrén

I have some objections to the statement that chemical models would always make good predictions. They do sometimes, but as you might have seen from the poster sessions, there are cases when they do not, as for example in the case of the composition of groundwater. For some species, e.g. iron, the groundwater appears to have a composition which is far from what the models predict, amounting to a difference of several orders of magnitude. If our models are not able to predict the composition of the groundwater, how can we expect anybody to believe that they will be able to predict something which has never been measured?

Shlomo P. Neuman

Would anybody like to comment on this or anything else related to the questions we are discussing?

Majid Hassanizadeh

The prediction of oil reservoirs, mentioned as an example, has been very successful. There is, however, a major difference between that kind of modelling and the one we are aiming at, which is to understand how critical the consequences would be if we are not right with our predictions. I am not sure our predictions have been successful in the same sense and with the same confidence as the prediction of oil reservoirs.

Shlomo P. Neuman

I cannot resist making a comment about oil reservoirs. In my experience with the oil industry, they never make predictions for more than a few years and every few years they recalibrate or completely change their models.

Jordi Bruno

We should not put the validity and nonvalidity of geochemical modelling or thermodynamic modelling, as opposed to hydrogeological modelling. They are two different disciplines that have evolved independently. Because of the difficulties involved in hydrology I believe chemistry is better in producing more reliable results.

I agree with Professor Neuman's comment that a paradigm is needed. Contrary to Charles McCombie's ideas, I think this paradigm requires plenty of dialectic and a lot of polemics and polarisation. That is something that I have tried to create in the waste management community for a while, but which I have missed very much in this particular meeting.

Shlomo P. Neuman

Ghislain de Marsily, did you want to make a comment?

Ghislain de Marsily

I would like to discuss the problem of predictive models. What happens on earth is essentially the consequence of large thermodynamic processes. Most of what we see in the climate (weather) is the thermodynamic process which is triggered by the sun and the radiation of the sun. We all know that this process is unstable and that at present predictive modelling for climate is only feasible on a three-day basis. That is the time scale. After that time, very small changes in the initial conditions predict large variations in the consequences, so we cannot predict further than three days. That is not a lack of data, it is just a simple instability of the system.

When we go to geologic processes, the cause of all the internal and external geodynamic processes is also thermodynamic, the thermodynamics of the interior of the earth or the thermodynamics of the climate for the exterior of the earth. Certainly, the predictive period of these mechanisms is much longer. The internal thermodynamics of the plate movements, volcanic eruptions, etc., probably have time scales of some tens of thousand of years, but definitely not very much longer than that, because of the intrinsic instability of such thermodynamic processes. What makes the earth move is actually convective instabilities and these are presently unpredictable. So there is a time limit to our predictability. I don't know the numbers, they are not three days, but they are not millions of years, either.

Coming back to what Charles McCombie said about what we should do for geologic disposal, I strongly agree that we should go to simple systems, where, no matter what happens to the environment, the validation or the proof of the validity of a concept of storage is easily feasible. If we carry that reasoning to its logical conclusion, it brings us necessarily to the seabed option, which is the most predictable solution, where the changes of the environment will have less influence on the deterioration of the quality of the repository. If we were at all logical in our decisions on disposal of waste, we would go to the seabed. For reasons that most of you know, seabed disposal is not considered politically a satisfactory solution, however, and I am sad to admit that.

What is the next best solution for disposal of waste? It's probably not going to be just any geology. Some geologies are more predictable than others and in my view the clay formations are probably best suited. Contrary to what everybody is doing here, we should, if we agree to that kind of reasoning, decide that there are a few places on the earth to put the waste, where it would be safe, predictable and validatable. If that reasoning is correct, let us select a few places, where all of our European Community countries, OECD member countries, or human beings, if I can say so, would put their waste for disposal.

Shlomo P. Neuman

Many of us would probably agree that if there is a prospect for scientific consensus about the behaviour of a geological environment as a transporting medium, it can be achieved most easily when dealing with homogenous clays. I think we know much more about these than about fractured rocks, but many of us are working with fractured rocks. The U.S. Congress has decided on Yucca Mountain as being the only option for the United States. Even though many of us may agree with you, I think we still face the problem of having to address the issue: What can we say about other media?

Peter Glasbergen

I would like to raise a more general issue concerning validation in perspective of the total safety assessment. We have made considerable progress in validation of geosphere migration modelling. But in the total safety assessment, what do we do with our results? We predict nuclear migration to the biosphere over hundred thousands or millions of years and then we give our results to the biosphere modelling people. Mr Niederer made the remark, that human intrusion cannot be modelled because we don't know anything about how to predict the human behaviour in the far future. On the other hand, we assume that the biosphere modelling people can model biosphere nuclear migration over millions of years. Should we not go back to assessing a kind of geohydrological migration criteria instead of going to biosphere modelling with all the uncertainties involved and with the inability to make real dose calculations for the far future.

Sergie Pankratov

One of the crucial questions on the methodology of validation is the predictability. As far as the predictability is concerned, I do not agree with Monsieur de Marsily. At least his example was not very successful. One should not compare the geosphere with the atmosphere. They are two entirely different dynamic systems. I will not go into technicalities, but as far as the chaotic dynamics are concerned, why we cannot predict the behaviour of the atmosphere within four days or five days, is because of a special nature of mathematical equations pertinent to the atmospheric model. It is the so called Lorentz attractor, chaotic dynamics. It is entirely different for the waste or liquid behaviour in the geosphere. Since the mathematics is different and we have a solution of a different kind, we cannot predict the behaviour of the leakage of wastes or many other things for quite a period of time. Sometimes the question arises: For which time can we not predict your system?

Being an outside observer in a way, because so far Russian scientists are more or less detached, more or less disengaged from the whole scientific community, at least in this field, I want to express certain opinions, and give certain impressions of this conference which is very good and sometimes very witty and funny, but in general good and informative. I have noticed some saddening compartmentalisation of sciences. I don't want to sound trivial, but it is one more example of scientific compartmentalisation. There are some things which have been developed in other sciences which can be very easily transferred to this field, but you just don't look at their direction. You form a very closed community. Because of this confinement to your own magic circle and the inefficiency of the process you are loosing time, maybe you are also loosing money, which you will require from some governmental or other type of organisation. I am proposing that you tend to be more versatile, more flexible and that you connect some other communities to your field and to your organisation.

Shlomo P. Neuman

This is an excellent comment, one that we should take to heart. To some extent it has been anticipated by the organisers who have invited one "outsider", Professor Robert G. Sargent, to suggest to us 'What to do with the issue of conceptual validations'.

Invited Speaker

Robert G. Sargent

You are trying to validate models for decision-making; with your decisions being where to put the nuclear waste and how to store it. There are three basic approaches to decision-making: You may use intuition, use the results from the evaluation of the performance of systems, or use models. There are empirical models which are "black boxes" developed from system data and there are causal models which need to be validated. In most application domains, validation is included as part of the (causal) model development process. A few application domains that use computer based models do separate the validation process from the model development process. It appears that your field is separating the two processes.

I have some observations to make. First, the interaction between modellers and experimentalists that is occurring, such as at this conference, is a good direction to proceed. Second, I see lots of empirical models being developed and lots of model calibration occurring. However, there appears to be a lack of causal models being developed. Third, regarding validation, I see "reinventing the wheel" occurring; i.e. you are researching how to do model validation instead of using existing model validation knowledge. Fourth, I see a lack of determining and specifying the amount of model accuracy needed in your models. Fifth, I find a lack of the use of a validity criterion to determine whether a model is valid or not, and lack of an explicit decision on whether a model is valid.

Some other observations include the following. I note that you are developing a lot of small models but not many integrated models. I believe you should start developing databases of models, model data, and system data to allow integrated models to be developed and validated. One interesting approach to developing integrated models is the use of graphical systems using databases.

I observe that most of your models are science models, i.e. models to understand the physical phenomena, instead of decision-making models. I believe more attention should be given to decision-making models because that is your basic objective. I do recognise that your field is young with respect to causal knowledge and science research is needed.

The issue of models versus modelling has come up at this conference. A large number of people are developing models but I do not know how much modelling is occurring. Modelling is abstracting, developing causal models, and modifying them until they are valid. Related to this is whether generic models with parameters should be developed or specific models developed for specific sites.

I have also prepared some recommendations. I believe you can accelerate the development of validation by using technology transfer. In other application domains, validation methodologies, validation techniques, tests, etc., have been developed that should be applicable in your application domain. You should transfer from other domains as well as develop your own.

With respect to models, I believe you need to specify over what domain they are applicable, what performance measures the models use, how much accuracy they have, and how they were validated. I find this missing from most of the models presented at this conference. More rigor is needed with respect to validation. Many of your models are calibrated but not validated. Also much curve fitting is occurring as models; and empirical models are being developed instead of validated causal models.

You should give more consideration to the use of stochastic models; for both science models and decision-making models. I was surprised to find almost no stochastic models presented here. Many things are not known with certainty, or one may not want to model at detailed level, and stochastic models should accelerate the development of your field.

You should also be concerned with performance measures beyond what comes out of deterministic models. In reality you are just going to have one realisation of the actual process over a large number of years, e.g. 10000 years. One should probably use a stochastic model to model this as a stochastic process and look at worse case behaviour, Using only the mean behaviour from stochastic models is not satisfactory as in using deterministic models.

Finally, I recommend that decision-making models be developed. Most of your work is directed towards finding science models and not decision-making models. Decision-making is what has to be done in the end.

It was earlier said that you can only use conceptual models. I do not agree with this statement. You do not need to base your decisions purely on conceptual models. Much more is known. Certain theories and assumptions can be empirically justified. Experts can be sued to do face validity. Analogues can be used. You can develop causal models for submodels of the total system. Models can be validated over short time frames. How much accuracy is needed can and should be able to be determined, Sensitivity analysis can be performed. Yes, it may be difficult to integrate submodels into an overall model and to ensure that it is valid over a long time period. Overall, however, I believe it is possible to obtain that "gut-warm feeling" of consensus that is based on more then just a conceptual model.

My last comment is that I believe different kinds of things will need to be done to reach consensus in the scientific community and with the public. I get the feeling that you are trying to do the same things simultaneously for these two audiences. I believe that you will have to use different tools to obtain credibility with the public, and the public should not be mislead.

Discussion on Robert G. Sargent's Presentation

Shlomo P. Neuman

Thank you very much. I would like to ask a specific question which is a simplification of the problem of validation, the way I see it. Suppose that you have a choice between only two models. One model views the rock as made of discrete discontinuities which we call fractures. The other model replaces all discontinuities in the rock with some fictitious continuum. As we are dealing with an extremely complex geological medium, we can examine it only with the aid of bore holes at a few selected points in space, so our information is limited. The question that we are asked, among many other questions, is: With either one of these two models, how well can we predict the transport of radionuclids from a hypothetical or real repository to what the US programme calls the accessible environment, say a certain number of kilometers from the repository within a given amount of time? Since we are dealing with a very large medium and very long time-span, no experiments exist nor will exist to compare our models against. There are small scale laboratory and field experiments. Given this paucity of experimental data to compare our models against, what specific strategy for validation do you propose?

Robert G. Sargent

I would compare the two models and do sensitivity analyses on them. I would study the submodels in each of them and how the submodels are integrated. I would try to get a consensus of the scientific community about them.

Shlomo P. Neuman

There is no consensus.

Robert G. Sargent

You have to work towards consensus. If the scientific community cannot agree, the public will never be convinced. The scientific community must be brought together and the various steps of the validation process must be worked out together. Can any agreement be made on submodels? What kind of accuracy is required? What kind of decision criteria is to be used? What domain is the submodels applicable to? What kind of integration of the submodels can be agreed upon. If necessary, go one step at a time.

Shlomo P. Neuman

So, is it your feeling that such a consensus can be built based on a logical, rational process, not necessarily supported by large-scale experimentation?

Robert G. Sargent

That is done in many application domains. You never have all the information you desire. In every application domain there are problems that cannot be solved. There is always some risk involved. To reduce risk requires time and money. When to stop testing the validity of model is often difficult, especially when the risk is not easy to accept by the parties affected. A logical process is required and that is what the scientific approach is. Of course, as much experimentation as possible should be done.

Shlomo P. Neuman

Any questions or comments from the audience?

Johan Andersson

I don't believe these two different concepts of crystalline rock that were used as examples are that different. I would think that you can make them do exactly the same predictions, if you use the same data In my opinion, they are just different ways of describing the same thing.

Shlomo P. Neuman

Just one example of the many disagreements that the community faces.

Charles McCombie

Maybe the reason why we do not get consensus is that we are not trying to reach consensus on the right questions. If we cannot get consensus on how the system actually works, why do we not tackle a less difficult question, for example, how to reach consensus on predictions of worst case behaviour. How poorly could the system function? This issue has to do with your definition of performance measures, definition of required accuracy and so on.

If we do some of the things that Professor Sargent mentioned, we will not get consensus, because we are very often not talking about the same questions. We are not aiming at the same targets. We should try to choose a target that is more amenable to consensus.

Robert G. Sargent

That was one of the things I tried to indicate that needs to be done. You have to specify what your objectives are. Why are you going through this whole process?

Paul Davis, Sandia

In working in INTRAVAL and HYDROCOIN I feel that we got consensus that we cover the issues, not consensus that anyone is exactly right. I don't view consensus in such a way that we all have to agree on one or another model, but rather that we have covered what we all understand.

Robert G. Sargent

I agree with you that more than model can often describe the same phenomena. It depends on the type of model and what is the objective of the model. There is usually only one causal model but there may be different models that can describe the behaviour of some phenomena. For example, different statistical models can often be used to describe and statistically reproduce some behaviour data from some specific system. However, there is usually only one causal model for the phenomena that contain the mechanisms of the phenomena.

Shlomo P. Neuman

Thank you very much, Professor Robert Sargent! We will now continue with Peter Sargent, who is on our panel.

Presentations by Panel Members (Continued)

Peter Sargent

Our colleague from the Soviet Union pointed out that perhaps we should learn from other disciplines. That is what we have been doing and many of us have in fact come from a number of disciplines. One of the risks that one runs is that there is a learning exercise for new people that come in.

There is some problem with reference to the worst case as well as to the stochastic treatment. Actually, when we do performance assessment those factors are taken in account. It just depends on which country you are from and which methodology you use.

I want to remind you that the whole purpose of this meeting is to validate geosphere codes used in performance assessment. One component part of that is some of the very detailed scientific codes that one uses in research. When somebody tries to bring out the actual level of validation required, that depends on the impact the code has. If it is something that addresses an issue that has no impact, I don't see any real point in validating it.

Otto Brotzen in his talk mentioned that we have to remember that we all represent different interests, different companies, and we have different concepts. The level of validation that is required for the geosphere is very dependent on the concept of the engineering that has gone into your near field. For example, if you have a container that has an infinite lifetime, you have no real interest in validating the geosphere aspect, whereas if you have a container that has a lifetime of 10 to 100 years, then obviously you have much more concern.

In the Stripa project, the consensus we are looking at is essentially that which is acceptable to our group of experts. It is the actual application of the codes that is important and it is that part that we have to be comfortable with, not just the codes themselves. You cannot generally take a code that has been validated for a given application in a given country and use it, you can only do that if it is for the same application.

The critical issues in geochemistry are very dependent on detailed site characterisation, and unfortunately there are only a few such sites. The two important geochemical issues are the solubility and speciation of the migrating species, and the retardation factors. You have heard this many times before, by Ingmar Grenthe for example. In the gathering of thermo-dynamic databases we are assuming thermodynamic equilibrium. The data are reviewed and we can generate essentially consensus on what the appropriate data is. I want to point out, however, that these reviews are based on quite old data and just a few new data points can in fact have quite a profound affect on those databases. I think there is still a lot more work to be done.

The major Achilles heal from a geochemical viewpoint is in fact quantifying sorption and then being held to defend it. All you want to know is, depending on the particular usage that you put this information to, that sorption is finite, is greater than zero. So it again depends on this application.

Let us assume you do need to know that it's greater than zero and you want to bound it. I see again some need for a peer review of the data we are all currently using. As I pointed out, I believe the NEA has taken steps to remedy that. But this is an area where we need some concerted effort, some basic fundamental understanding of the processes that are occurring, rather than just solely making empirical measurements. This doesn't mean to say that we have to wait until the information is available. I would still see the need to maintain some underlying basic research programme in geochemistry. I think we would see it in all the applied geosciences too.

Conceptually, if you take a large engineering structure, like a disposal vault which could be 2 by 2 kilometers with extensive excavations containing thousands of tons of used fuel, significant amounts of bentonite, significant amounts of concrete and in some cases significant amounts of bitumen, this structure presumably is going to have at least some effect on the geosphere. Yet, most models have ignored it.

I see no use in using generic site data other than for the development of a methodology. I would urge all of you to consider real site data, irrespective of whether it is from your own personal site or you acquire somebody else's data. Let us abandon this strange concept of going around selecting numbers from all sorts of different sites for particular geologies within different countries. This makes no sense.

Shlomo P. Neuman

Thank you. Before we open the floor to comments and questions I would like to call on Chin-Fu Tsang to make some comments.

Chin-Fu Tsang

Thank you, Mr Chairman. I have put five questions to guide my comments.

1. What is implied in "validated models"?
2. Is it possible to have validated models?
3. Who needs validation anyway?
4. What have we been doing all these years?
5. What can and what should we do in the coming years?

To answer the first question I try my best to see what is implied. An analogy I can think of is related with a medical doctor, say, John Smith, MD. If you consult him, he will let you know that he is a medical doctor right away. You know that he has been somewhat validated. What does that mean? That means that he has studied for N years and passed M tests. I think this might be a good analogy, but of course none of us ask a Doctor what grades he got, did he get A or B or do he just barely pass all the tests. It is interesting that he would sign his name as John Smith MD with MD at the end automatically. I wonder whether you wish to have the codes validated in the same manner, signed "validated" after the code name. But this in a way is the implied feeling that people may have.

Then the next question: Is it possible then to have validated models? The answer is 'no'. There is no such thing, in my opinion, as generic validated models. You cannot say if my model is validated so that anybody can use it in any situation and give you valid results. However, you could probably say the model is validated with respect to a given process or with respect to particular spacial structures. This means that we believe the model will work for those cases. But then you should also ask, which is perhaps more relevant in our field, whether our model or models, a group of them are validated with respect to a particular site. That means your group of models actually can make believable predictions for that site. I think that is what we really are up to. On the other hand I also agree with Bob Sargent that you cannot say that a model is absolutely validated, because it is really a validation process.

Now the next question. Who needs validation anyway? Suppose we do not want validation, and we do not have GEOVAL symposia, is then something missing in the world? The answer is 'there is something missing' because we do need a certain degree of confidence and assurance in what we are doing. Nature is complex, and we cannot model it easily and fully. There is also a need to collect test cases. We need to collect many test cases from around the world so models can be tested in different kinds of situations. Furthermore, we need extensive and intensive discussions. I do not quite agree that we should make decisions just by a majority opinion. In science often somebody comes with a minority opinion that was eventually recognised as correct through discussions and study. We see that happen again and again in the history of science. Then we need validation to establish credibility and to build up confidence, first among the scientific public, then the general public.

The next question is what have we been doing in the last few years? Between GEOVAL–87 and GEOVAL–90 most of the work has been devoted to looking at validation with respect to processes or some particular structures. I am happy that a large number of test cases have been collected and properly documented. Actually, that in itself would be a very valuable document which will be much used by students in the field. There are also extensive discussions among us which is also extremely important. There is a certain convergence of scientific fields in certain areas. For example, there seems to be some consensus that the matrix diffusion is probably proven; and so is the concept of channelling. David Hodgkinson actually made a very strong statement: "Unless you can prove otherwise, you need to include these effects in your models." There are many other phenomena yet to be studied in depth, such as coupled processes. Rock mechanics may become more and more important. In a paper from Switzerland on tidal effects it was suggested that these effects would cause a change of water within the fracture matrix. That might be a minority opinion right now, but who knows? As we look at it more it might be of similar magnitude as matrix diffusion.

Finally, what can and should we do with validation before GEOVAL–93 and beyond? My feeling is that we should continue with the validation of different processes, but we should turn our emphasis to discuss, practice and learn about validation with respect to a site. If you define a process or define a particular structural feature, the problem is relatively simple. But if you have to make predictions for a specific site, the problem is different and is orders of magnitude more difficult.

Let us assume that you identify a number of processes. You have done a good job, so you know you can model these processes. Then you have to put them all together. That is the start of what I call the modelling process. You first need to validate the data and the information. Next you have to construct conceptual models. Here I define a conceptual model as a quantitative structural and process model. You look at the structure of the geological formation such as the layering and the faults and then you ask: What are the processes that might be involved? You construct a model and consider potential scenarios. What might happen in the next thousands years? You also need to decide the performance criteria and what the requirements are. Finally, you get the calculation model, perform prediction calculations and sensitivity and uncertainty analyses and you evaluate the results. After you validate the results, you might find that the prediction is acceptable, because the uncertainty is acceptable within the criteria. We can then proceed and try to construct our nuclear waste repository. On the other hand, it might not be acceptable. Then you can go back to iterate with the calculation models, improve your models, reduce the uncertainty or even change the performance criteria. Perhaps the performance criteria required by the agencies are too strict. They might not be obtainable within any reasonable effort. Perhaps you can loosen them and still have reasonable safety. One needs this kind of iteration.

Furthermore, one can also go and get more data and repeat the modelling exercise with these new data. Thus this is an iterative process. You might come to a stage where any new data at reasonable cost and within reasonable time-frame do not reduce the uncertainty and you may have to say 'stop', no repository at this particular site. But every step needs to have some kind of validation.

Bob Sargent talked about face validation, rank validation, etc. You apply these different means to see whether we do every step of the modelling process correctly.

In addition to the above site-specific validation exercise, we also recognize that there are many unknowns in a geological system. You cannot possibly know everything. As we are talking about a ten thousand year time-frame, there is a need for a lot of fundamental studies, including improvement of measurement techniques and improvement of understanding of processes. Site-specific validation exercise and fundamental studies together form the direction in which we should proceed.

Discussion on last two Papers

Shlomo P. Neuman

Let me open the discussion by asking the following question. One purpose of performance assessment is to predict a possible outcome for a given scenario or class of scenarios. The other way in which performance assessment models have been used is to attempt to predict the range of the possible outcomes. We all agree that there is an uncertainty in our prediction of mean system behaviour. This is why we do sensitivity analyses and uncertainty analyses. We probably also agree that to predict mean behaviour, we need mean parameters. To predict uncertainty we require a good size statistical sample of measurements, meaning more information. If we are uncertain about our ability to predict mean behaviour into the distant future, why are we not equally concerned about our ability to predict uncertainty, the confidence band to place about this mean behaviour?

Chin-Fu Tsang

I really don't have a full answer to that. I think some people in the audience might have. We do need to know some kind of limits of uncertainty, and I believe that we can establish certain error limits even without knowing the mean values. Suppose you have a repository at 500 meters or 1000 meters depth. If you put nuclear waste down there, it is not going to come up tomorrow. That we know. It is not going to come next year, pretty reasonable. Then it's not going to come up in 10 years. You could reduce your uncertainty by backing away from the worst end. There are certain physical principles that predict that things would not move so fast. So even in a primitive way, one could establish some kind of uncertainty limits estimate. Of course, I don't think the technology is there yet to allow us to have a very narrowly bounded estimate.

Shlomo P. Neuman

If all we can do is predict worst or best conditions or upper and lower bounds on our predictions, then I think many of our sites may turn out to be unacceptable. We should try to do better than that. The question is: How should we convince ourselves and others that we indeed can do better and what is needed?

Thomas Nicholson

After listening to the discussions over the last few days, I would evaluate the problem of modelling radioactive waste disposal more from a scientific viewpoint. The reason for this view is my concern for understanding the natural system into which the radioactive waste is to be disposed. This hydrogeologic system will be subject to dynamic forces. The fractures are not static, there are tremendous stresses imposed on them which may change their characteristics. The groundwater is not static, it moves. The groundwater chemistry has evolved over thousands, if not millions of years depending upon the depth and location of the ground water. We have to understand that we are dealing with a complex system. One of the validation issues should be: Do we understand the natural system in which we are going to place the radioactive waste?

Often people, when they model geochemistry, use what I call chemical engineering models, quite similar to water quality "continuous stirred tank" models, in which they represent the groundwater flow as some large-scale beaker in which they put in the waste and simulate the distribution of the radionuclides. They give credit for a certain amount of water movement and certain interactions, but generally speaking, they forget the complex ambient hydrochemistry. As presently understood, the groundwater chemistry has evolved over thousands, if not millions of years and has taken on certain diagnostic chemical characteristics. Bill Back from the U.S. Geological Survey has introduced the concept of hydrochemical facies. We have to understand the hydrochemical system as well as the groundwater flow system in which we are placing the waste. The modelling being done today is nothing more than a caricature of that system. We may never know the system in all the details but we will know it in certain simplified aspects.

How do we find the truth in this caricature? The only way to do that is to go to actual sites and obtain data from investigations made there. The people investigating the WIPP and the Mol sites, which are represented at GEOVAL–90, appear to be making some excellent progress in this regard which hopefully will become part of INTRAVAL Phase 2. There appears to be detailed databases which warrants our attention. When we began Phase 1 of INTRAVAL, many of the test case databases were not oriented towards validation studies of groundwater flow and transport models. They included field tracer tests and a variety of laboratory studies, few of which were designed specifically for examining validation issues. My hope is that more detailed, well-designed field experiments can be developed and pursued which will provide us with the knowledge and experience for understanding natural systems similar to those in which radioactive waste may be disposed.

Shlomo P. Neuman

You may recall that Charles McCombie showed a little caricature earlier this afternoon in which we are painting ourselves into a corner. This came from a discussion he and I had, where I mused that perhaps by placing so much emphasis on validation we may perhaps be painting ourselves into a corner for the following reason. An increasing number among us are becoming convinced that we will never validate our models so as to become sufficiently convinced that we can predict what will happen, or even a confidence band on what will happen, over the long time-spans that we must deal with. By putting so much emphasis on model validation, which probably stems from a tradition where for many years we have placed more emphasis on computer models than on the science behind these computer models, by talking so much about validation, we are convincing ourselves and others that we will be able to validate our models. When it comes to a licensing test we may be surprised and stand empty-handed when people ask: Show us how you are validating your models? It seems to me that a better strategy might be to recognise that the public is not naive, that it includes scientists who will help it ask us hard and difficult questions. There is only one way that we will be able to convince them, and that is to demonstrate that we have done our best to understand the science and to develop the engineering of repositories and the surrounding systems, to admit that we cannot predict with confidence what is going to happen in the distant future, but to argue that the public must make an informed decision despite such uncertainty, because there is a problem that needs to be solved.

Sven Follin

I am geohydrologist here in Stockholm at the Royal Institute. It seems to me that the word 'consensus' has become the name of this symposium. We had a speech referring to Thomas Kuhn in the beginning. I am not quite sure that Thomas Kuhn meant that a common paradigm or consensus avoids an error of type 2 failing to reject a hypothesis although it's wrong. Rather I would like to bring forward another philosopher, Feierabend, who, among many other statements, said 'let hundred flowers bloom' meaning that we should have some kind of Brownian motion in science in order to know what we are dealing with.

We talk much about validation which is the topic of this symposium. However, I am worried about the physics and what we put into these models, especially the parameter estimations. Shlomo Neuman brought up the question concerning continuum models and discrete models for fractured media and claimed or stated that they might be quite different viewpoints on the same medium. Johan Andersson thought it might be saying the same thing in different ways. In a way I agree with you both. There are different concepts, but both concepts use the same type of measurements. So far, we have mainly used single-hole packer tests for evaluation of the hydraulic conductivity, which is an interpretive parameter and not a measured parameter. In the discrete fracture approach, we talk about fracture transmissivity, in the continuum approach we talk about hydraulic conductivity as a property of the medium. I am waiting for much more testing.

When we use interference tests, we check principally for reciprocity. Doing two-way tests, as in an interesting experiment reported by Lars Birgersson and the Neretnieks group, I was waiting for an overhead showing how these tracer tests were carried out when they reversed the direction of flow. Unfortunately I did not get any such results. We have to do much more testing to find out at what scale we can use our results in our models.

Shlomo P. Neuman

As far as hydraulic measurements in granite are concerned, are you familiar with Paul Hsieh's work on crosshole testing in the Oracle granite? There we definitely did find a very nice reciprocity when we injected into one borehole and measured the response in another and then reversed the process. However, I can show theoretically, and I am sure that others in this room, especially Dennis McLoughlin, can show that if you have a system which is heterogeneous in a statistically nonhomogeneous way then, depending on the strength of this nonhomogeneity, there is generally not going to be a reciprocity in a tracer test. If you inject a tracer and watch it transported one way, I expect on purely theoretical grounds that its mode of transport will differ in the opposite direction. Now, how are we going to test this? If we run a tracer test in the field and we obtain reciprocity, have we proven anything? Do we now know whether the system is homogenous or nonhomogeneous in the statistical sense? The only way to tell is from a very large sample or by actually observing a lack of reciprocity. These are some of the small problems we face.

Rod Ewing

This is a question for Peter Sargent. As Peter pointed out, the basis for the application of the geochemical codes is the assumption of thermodynamic equilibrium. At the same time many people refer to the valuable information that we might get from natural analogues. From my own experience with natural analogues for the corrosion of borosilicate glass using basaltic glass or corrosion of spent fuel using natural uraninites, the one consistent observation I can make is that the resulting phase assembly is a metastable phase assemblage. My question, and I think it is a question people will ask when we finally present our models for as having been validated, is: How do we rationalise the application of these geochemical codes with the observations that appear in large scale longtime experiments in natural analogues?

Peter Sargent

Yes, that is a good point. You can examine natural analogues that are in equilibria, but you are referring to ones that are not. The other assumption that is made in waste disposal, is that the temperature transient is going to be in the order of 100 degrees. For granites that is one of the lowest temperatures that minerals were laid down at approximately, so one has some justification then of using the assumption of thermodynamic equilibrium, since you can show that present groundwater is in equilibrium.

When we use some of these very large geochemical codes we would like to do just that what you suggest, in fact to determine what the metastable states are and what the implications would be, whether this would necessarily be enhanced solubility or not, for example.

As another example, you may from a thermodynamic code under certain conditions predict that you would have uranates present. You may then go to certain deposits and you cannot find them, they are just not there. So, they are eliminated from the calculation. Apparently, there are examples where there are other assemblages present.

I was trying to highlight that you must regard the application of the code. You cannot just take a canned code and database and use it everywhere. For example, I couldn't necessarily use the sort of information that we would have at Cigar Lake and apply it to some of your analogues because the situations are entirely different.

Jordi Bruno

In my opinion, nobody should think that thermo-dynamics is something that makes us understand how things work. Normally, they make us understand the things that would not work, they give you a 'positive negative' answer. There are many kinetic hinders regarding geochemical problems that we try to understand and try to model with the codes. Metastable phases will never show up there, but they are the ones that we want to understand. Thermodynamics give us a nice frame, but there is a lot more fine structuring in between.

Ghislain de Marsily

I fully agree with Tom Nicholson about the need to do validation on sites and in my first presentation I said: 'Collect data that would make it possible to validate sites.'

The second point is about averaging. For performance assessment we need to have model parameters, some kind of mean and perhaps some kind of variance. There are areas of physics where we know how to go from chaos at one scale to order at another scale. For instance, if it is a problem of permeability, we are in good shape. We have what is called emergence, we can go from Stokes' to Darcy's law and there are presently research programmes on what is called homogenisations. If we go further on another scale, we can go from local Darcy's law to average values of hydraulic conductivity and we know how to average that. We get in general the geometric mean, depending on the number of dimensions of space. We are in good shape for that kind of averaging and it gives us not only the mean but also the variance for some of the physical processes we are studying.

However, if we go to the geochemical processes and, without going into the details, we are trying to catch up on what Peter Sargent said, I am not sure that we have yet got the method or the physical reasoning or the intuition or whatever you want to call it, to do that emergence. How do we go from the very complex thermodynamic equilibrium or kinetic theory, that we know applies at a given scale, to another scale? We can do the scaling directly and measure K_d's. But we all know that these K_d's are not related to the thermodynamic data and we cannot solve that problem. Even if we have the K_d's, and we want to extrapolate those K_d's to a larger scale formation, we don't know how to do that. So, we have to develop the equivalent of the emergences, not only in space but also in time on these thermodynamic problems. I don't know how to do that but I invite your suggestions.

Let me just say some words to my good friend, Sergei Pankratov on the third point. I am quite aware that the behaviour of the interior of the earth is not equivalent to the dynamics of the atmosphere. I claim, however, that in the mantle, the conditions of thermodynamic disequilibrium, which are based on the Rayleigh numbers, create convective cells in the earth and that not only the convective cells exist but that we are in the range of Raleigh numbers where these conductive cells are unstable. They are not in steady state, they are unstable and they will change over time and time in an unstable way. So, I still believe that my comparison is valid.

Shlomo P. Neuman

We will return now to the panel and I would like to call on Karsten Pruess to present his comments.

Presentations by Panel Members (Continued)

Karsten Pruess

Nuclear waste disposal technology as all nuclear things are being developed under intense scrutiny of the public, which is not only skeptical but often worried, if not outright hostile. One reason for that is that the development of nuclear technology has been littered with exaggerated claims and counter claims, exaggerated promises and anxieties. It is rather clear, that waste disposal technology will only be successful if we gain public acceptance. The reason I am saying that is that I have a feeling that in some places to this day this is not fully appreciated. There are still people and forces at work who feel that it is enough to sort of take people for a ride in a sufficiently elegant fashion and you can then get them where you want. I believe that this is never going to work and the sooner we come to the full realisation of that the better. The key to public acceptance is to show that what we are doing has scientific credibility. From a scientific, technical viewpoint, that is the essence of model validation. I regard validation as largely a regulatory term and a regulatory need. I am very well aware of the importance of model validation, but from my viewpoint as a researcher, who is trying to contribute in this area, I am much more comfortable with the notion of scientific credibility. If I am successful with that, then validation will follow suite in due course through processes that will involve people outside science, e.g. lawyers and what have you.

In addition to performance measures and agreements, a critical issue in establishing the credibility of models is that models need to be based on sound and accepted scientific principles. In our area we don't have much of a broad track record of dealing with these kinds of natural systems that are part of nuclear waste engineering. As we go about this engineering problem, we at the same time have to pay attention to establishing this track record which will be broad enough to carry the engineering systems that we will create.

The issue of large space and time scales in attaining credibility for models is a serious one, but I don't view that as the main obstacle. There are examples in other areas of science where models have been created for very large space and time scales, for example, in describing the motion of celestial bodies, where I think we all walk on very firm ground. The crucial issue with a natural system, that we are trying to engineer in and with, is that we are dealing with messy systems, that have a high degree of internal invariability that will of necessity always remain incompletely known to us. Because the moment we would know it completely, we would have destroyed a potentially beautiful repository rock. This is a boundary condition for all of the things that we can undertake. One consequence of this is that we should not have the notion that we will ever arrive at the one approach or the one model that will really describe everything to everybody's satisfaction and become the one that is validated and will carry the licensing procedures forward. I firmly believe that we need many different approaches, conceptually and otherwise. We should have a humble attitude with regard to these natural systems in the sense that we do not mix up the models that we make of them with reality itself. I believe that our models always only capture certain facets and aspects of this reality. The much talked about warm feeling of consensus must ultimately arise from some kind of convergence among people that the various models, each within their own realm of applicability and hopefully with sufficient overlap in those realms and agreement with data, will collectively substantiate the engineering that we are proposing.

If one acknowledges that credibility and validation of necessity will remain imperfect and limited and that they are open-ended processes, I would propose a notion that confidence in models and other things is good but control is better. I would suggest that in devising nuclear disposal systems we should consider including performance monitoring and confirmation measures and give ourselves an ability to intervene in the system. I mean not necessarily intervene in the sense that we take everything out and move it some place, but to be able to take some kind of measures at some point in the future. It seems to me to be just common sense and prudence that we don't deprive ourselves of the considerable capabilities that we have in these areas.

I have discussed this notion of monitoring privately with some people in the audience, and my feeling is that some people are worried that the moment we open the door for these kinds of things, the public is really going to get worried. It might appear tantamount to admitting that the system isn't altogether that safe as maybe we want to claim and that is why we still have to keep our tabs on it. I feel quite the contrary. If we admit that nothing that we can ever engineer will be perfect, I find it quite reassuring that we don't have to sort of force ourselves to forget where the repository is and just trust that, no matter what we do, we will never be harmed by it, but that we can in fact bring other resources to bear on it.

I have one more final comment that I want to make regarding specifically the proposed US disposal site at Yucca Mountain. I wrote down a number of points on what I consider potential weak spots in Yucca Mountain that research efforts ought to be focused on. Obviously you cannot validate a model until you have one. What worries me a little bit, and I am saying this perhaps mostly to my American colleagues, is that in this very complex multiphase system that we are encountering at Yucca Mountain, where we have multi-phase behaviour, nonisothermal flow and rough walled fractures in a very complex hydrogeological setting, we have not in some important areas even begun the process of model building, much less substantiating the models and confirming and validating them. This is something that ought to be pursued with much more urgency than it has been.

Shlomo P. Neuman

Talking about the ability of scientists in other disciplines to extrapolate beyond their experimental base, I have a very warm feeling that an apple is light and if it falls on my head, nothing very bad is going to happen to me. If I were an astronaut to be placed in orbit, however, I would have a rather heavy feeling, had the engineers and designers not prior experience with gravity in space, based on the theory of relativity.

I prepared these comments based on a speech I gave in Berkeley in October. There I talked about Thomas Kuhn. I now find it interesting that the same philosopher of science was referred to in one of the introductory lectures at this meeting. Thomas Kuhn in his "The Structure of Scientific Evolutions" noted that it makes little sense to suggest that verification is establishing a correspondence between fact and theory. We talk about models, he talked about theory. In Kuhn's opinion, all historically significant theories have agreed with the facts but only more or less. Few philosophers of science still seek absolute criteria for verification or validation. No theory can ever be exposed to all relevant tests and it is only when crucial new tests appear, which cannot be explained by existing theories, that one starts questioning the latter. Of course, such experiments may never come up in the context of repository far-field transport because they require more than a life-time.

I gave in Berkeley a simple example of what I mean. Think of classical models of sub-surface transport which regard advective velocity and dispersion as two distinct phenomena, represented by two separate terms in the advection-dispersion equation. This equation has been amply "validated" in laboratory column experiments, supporting the Fickian concept that dispersivity is a constant and local medium property, and as such, appears to be small and relatively unimportant for performance assessment.

There is a large number of data to support this. Yet work in recent years by Jean Pierre Sautie in France and many others, including recent experiments at Borden, Ontario, Canada, have demonstrated that on a larger scale this process is predominantly non-Fickian and that extrapolating the laboratory-validated Fickian model to the field is not justified. Recalling Kuhn, 'Ask not whether the theory has been verified but rather ...it's probability in light of the evidence ...' But you need to have the evidence that the existing theory may not always work; without experiments to bring out such "anomalies" one will never discover that the theory is unsatisfactory. Hence validation is limited to the existing realm of man's empirical experience.

'Compare the ability of different theories to explain the evidence', says Kuhn. It makes a great deal of sense to ask which of two actual and competing theories fits the facts better. But until we had these new, "anomalous" field experiences the Fickian model did just fine and there was no need to ask how other models would do.

We now have quite a few competing theories to explain available field experiments. I mentioned ours but there are other due to Dagan, McLaughlin, etc. These explain the observed behaviour by assuming that the medium is statistically homogenous. All existing stochastic theories make this assumption and they are able to reproduce some field experiments. They explain why dispersivity in general grows with plume travel distance and show that it is not a local constant medium property. Dagan, Shapiro and Cvetkovic demonstrate that dispersion and travel-time distribution are two sides of the same coin. Hence to claim that the dispersion is not important to far-field transport, is to say that travel time distribution is equally irrelevant and I wonder what the NRC would say to that, given its travel-time performance objective.

The Borden experimental data show a measure of the spread of a tracer plume in the longitudinal and transverse directions as functions of time over approximately three years. They show that the slopes of these curves are not constant, implying that dispersivity is not constant. Various theoretical attempts to fit the data have been made and most appear quite successful. Yet later things happen which these theories cannot explain, because they deal only with statistically homogenous media. These theories predict that the longitudinal dispersivity becomes asymptotically Fickian or constant as plume travel distance increases. The latter is contradicted when one looks at a collection of many tracer tests in the field. A collection of a large number of longitudinal dispersivities from laboratory and field tracer studies in a variety of porous and fractured media in a diversity of hydrogeological settings show a systematic increase with plume travel distance, a phenomenon called the "scale effect". This is clearly a non-Fickian phenomenon which, however, the Borden validated theories, including our quasilinear theory which fitted the Borden data well, fail to predict. The implication with respect to travel times is that their variance is not fixed but increases continuously as radionuclides migrate from the repository to the accessible environment.

Are the existing theories any good? 'If any and every failure to fit theory to data', says Kuhn, 'were ground for theory rejection', as was suggested in some of the talks earlier this week, 'all theories ought to be rejected at all times'. Though existing theories cannot predict the scale effect, they nevertheless serve a more limited purpose. They apply locally to homogenous geological units. Furthermore, the quasi linear theory and other similar theories can explain the scale effect, once observed. According to my interpretation of the scale effect, and I am sure many others view it the same way, geological media consist not of one statistically homogenous log-hydraulic conductivity field or mode, with one unique correlation scale, but the superposition of many such modes, with a hierarchy of correlation scales that combine to render the media random fractal. For example, a fractured rock may be thought of as the superposition of at least two media, the fractures and porous blocks, each with at least one (but possibly many more) characteristic scale. One may conclude that statistical homogeneity is at best a local phenomenon limited to random, relatively narrow scale intervals. Hence, one must question the utility of associating medium properties with local parameters or Representative Elemental Volumes (REV's), except on such narrow scale intervals.

Fickian behaviour, which is the basis of our current transport models, constitutes at best intermittent episodes during what is otherwise an inherently non-Fickian mode of transport. Hence, one must question the utility of relying on Fickian transport models to predict dispersion and associated travel time distributions over more than a limited range of scales.

It appears that porous and fractured media scale in very similar manners. This raises a question about the validity of many of the distinctions that we routinely make between these media.

This is true when one has little information about the hydraulic properties of specific heterogeneities, such as fractures, channels and other pathways. If one has such information one can build it into the model. But usually one has only scanty information, and one can provide only a crude description of advective velocities and travel times. As one collects more information, as more details become available about the space- time distribution of velocities and travel times, the rate of dispersivity growth is seen to slow down. Therefore, separating the notion of dispersivity and travel time distribution from the available information content is inappropriate. As I said before, one cannot validate a generic model, one can only talk about the model within the context of a given database. One must ask: How does information affect the ability to model?

It seems to me that this requires a new paradigmatic approach to geosphere flow and transport modelling. Many may resist it due to what Kuhn described as the 'assurance that the older paradigm will ultimately solve all our problems, that nature can be shoved into the box the paradigm provides', and we could add the practical need for answers now.

I would like to finish with the following three questions:

1. How many of our current models are such boxes?
2. How can we tell?
3. How can we resolve this issue now given that we cannot conduct large scale experiments now and given further, the "inevitable misunderstanding between the existing and new schools of thought, considering that a generation some times is required to affect the change?"

If we admit that new experiments may reveal new and unexpected behaviour, then we must admit (given the limitation of our current experimental base) that geosphere models will never be validated over the space-time scale in which we are interested. A warm feeling in the tommy may be comforting but misleading as would Newton's theory of gravitation to areas where relativity dominates.

Discussion on the last two Presentations and General Discussion

Charles McCombie

I have one reaction to your very last statement. Newton's theory is perfectly good today for building bicycles, cars, aeroplanes and so on. If you know that your range of application is to build up to aeroplanes and not space craft, then it is a good theory and you can work with it.

Shlomo P. Neuman

You should not use it outside its proven range of validity.

Chin-Fu Tsang

The question whether we are validating ourselves into a corner, is a little bit unfair. In all the efforts we have in INTRAVAL, there is a strong emphasis in trying to understand what is going on, rather than to arrive at a model validation. That has been emphasis all the time. Actually, that is the particular reason why INTRAVAL and HYDROCOIN are so successful from our point of view.

I agree that it is extremely dangerous to extrapolate into a system which is beyond the range of experimental information. That is the reason why I have more and more come to the view that the validation process should go on all the way until you construct the complete repository. By that time you have a big hole in the ground, and you are looking at a much bigger rock mass. Furthermore, you will be looking at it for fifty years, which is the time-frame for the retrievability period. You are then making measurements at much larger scales, large in respect to both space and time. The extrapolation will then be better based than today.

Johan Andersson

We have during this week mentioned scales in order of kilometers and thousands of meters. I want, however, to recall the point Otto Brotzen made in one of the first speeches this week. If we manage to understand transport in the first hundred meters, say even fifty meters of the rock, we would be very well off. If we believe in the present performance assessment, most of the interesting things happen in this short distance. That makes me somewhat comfortable, because it is feasible to carry out experiments in that scale.

Peter Sargent

At this meeting, we have not talked much about natural analogues, because they are so well covered, perhaps even at nauseam at a number of other meetings. For the purpose of the final record, we should make some reference to the usefulness of natural analogues.

Karsten Pruess

I am very glad that the focus on validation has helped us to concentrate on experimental efforts. I got very worried for some years that computer models were just mushrooming. The threshold to get involved in computer modelling was so small, you just wheeled yourselves over to your terminal and you got going. It is a very healthy development that we now are focusing much more on laboratory and field experiments.

Shlomo P. Neuman

I would like to make one final comment which is a quotation from Kuhn. You may disagree. 'Probabilistic theories disguise the verification situation as much as they illuminate it.' We talk about probabilistic or stochastic methods of validating our models and they do apply, but they apply within the existing conceptual frame work that we work with, which is based on our existing experience. "The theories and observations at issues are always closely related to ones already in existence. Verification picks out the most viable among the actual alternatives. Whether that choice is the best that could have been made if still other alternatives, larger scale experiments had been available or if the data had been of another sort, different boundary conditions, cannot be answered by any methods." That means, we cannot validate what we haven't seen.

Majid Hassanizadeh

We have heard the stories of success with validation efforts over the past few days. Within INTRAVAL the Mol experiment was an interesting example. Yet, although we hear that it is impossible or not feasible to talk about validation, to hope to do that. Looking back at the past days and trying to organise my thoughts, I think we can differentiate between two types of validation efforts. One is research model validation and the other one is safety assessment modelling validation. The validation we have with research models is a tool, whereas the validation we are aiming at regarding safety assessment modelling is a goal. The tool is going to help us to understand the conceptual models with their uncertainties, how to learn to interpret the predictions and similar things, whereas the goal which is the validation of models for performance assessment is going to tell us to how to make decisions. In order to reach the goal, we have to prepare the tools, sharpen them and learn to use them.

Charles Fairhurst

I am here under the auspices of the US Board of Radioactive Management. This is a research conference and it is not surprising that the conclusion of research conferences is that we need to do more research and that we know very little. As Charles McCombie and Ghislain de Marsily pointed out, we actually do know a lot. We could probably give the public a strong reassurance right now that the problem is not unsolvable. The sub-seabed is not a popular opinion, but if it is one that we believe is a real one, we should say so.

About a week ago I was one of a group of four or five scientific groups in the Unites States asked to testify before Congress. We were asked a very specific question by the legislators. Can we build a safe repository? More specifically will WIPP be safe? Even though the members of the groups came from both sides of the fence, i.e. interveners and so on, they answered unanimously with 'Yes'. So I think it would be unfortunate for a conference such as this to leave with the idea that we cannot trust Newton's laws of motion, or that we must worry about relativity, etc. We must not leave the idea that there is so much uncertainty that we cannot say anything for sure.

Shlomo P. Neuman

Thank you very much. I will not answer because I don't want to spoil the positive note on which you finished.

Shaheed Hossain

The main question, as Majid Hassanizadeh said, is that we have so far discussed validation concerning research models, but our goal is validation of a concept for assessment models. That is what Mr McCombie also looked forward to.

Göran Bäckblom

Dr Tsang and others have presented the idea of validation being a process. I acknowledge that concept and additionally, that validation must be site specific. If we examine the processes and the structures, we are, as Charles McCombie told us, quite confident about the processes. We have to recognise, however, that validation of the structures is a stepwise procedure, because it deals with a real site. We know that we can get a small amount of structures from surface investigations. When we continue with drillings we get a better picture and thus arrive at a stepwise validation. But we must realise that most of the data on the structures will be obtained during repository construction phase.

A KBS-3 repository in Sweden as it is envisaged would have a minimum of 40 kilometers of tunnels. Before the start of construction we can give a mere stochastic description of the structures, but only when we have constructed the repository, we can give an adequate deterministic description of the structures in the near field.

For the repository we have two major decisions to make: One is when we should start to encapsulate the spent fuel, the other is when we should seal the repository. Let us work stepwise and try to use performance assessment to make the lay-out of a repository as flexible as possible. Hopefully, performance assessment can also give us some clues to how we should do site characterisation.

Shlomo P. Neuman

I would suggest that we do not leave this room with a defensive attitude which confuses two things. One thing, which I think many here share, is that we cannot validate long-term predictive models. A completely different thing is to decide, based on our engineering and scientific know-how, whether a given repository should be licensed or not. I do not think that these two things must go together. All I am arguing against is this continuous attempt to prove to ourselves and to others that we can do something which I have not seen anyone yet do and that should have nothing to do with the decision to license a repository.

Allan Emrén

We have been talking about heterogenous properties of the nature and of the rock in particular. Now, what scales do we have to consider? The range is from the atomic scale to several hundreds of meters. We have to find out where we in a safe way can stop the procedure of going down in scale knowing that the scaling effects do not influence the properties any more.

Shlomo P. Neuman

Any other questions or comments? If not then I would like to close this session. For me, it was an extremely stimulating discussion. The comments made by the audience and the panel members are going to be very useful to us and I would like to return the chairmanship to Johan Andersson.

Johan Andersson

It is now time to call on Kjell Andersson, who will conclude this GEOVAL–90 symposium.

Concluding Remarks

Kjell Andersson

This symposium has covered the complex area of geosphere model validation. We have addressed experimental and modelling achievements in flow and transport in various media, geochemical and thermo-hydromechanical effects as well as overall validation strategies and international programmes.

Thomas Nicholson emphasised the progress made since the GEOVAL Symposium in 1987. Certainly, this has been demonstrated during the symposium and the discussions and achievements within the INTRAVAL project also show that we now have a much more mature understanding of what validation really means, its possibilities and limitations. More development and new experiments have been made and further experimental programmes are being planned. Scientific progress has been made and more evidence has been collected for specific processes. Systematic procedures have been developed for validation.

Yet, we are not satisfied with the present state of knowledge, at least some of us are not. More experiments and analyses are needed before we have sufficient understanding of flow and transport processes like channelling and matrix diffusion. The work also needs to be extended to include thermo-mechanical effects on flow and transport properties. The questions then arise: What is sufficient knowledge? Which are the most important issues?

It is obvious that performance assessment has a key role in answering such questions. Ideally, the importance of uncertainties should be evaluated within the framework of performance assessment and the results can then direct further research. It is thus very encouraging that performance assessment methodologies have now been so well developed.

However, we have to remember the very first paper at this symposium by Ueli Niederer. He addressed the scientific aspects of validation and the differences between the two "cultures" of scientists on one hand and modellers on the other. When we use performance assessment models to evaluate the importance of uncertainties we need scientific review. There is otherwise a risk that the only uncertainties included are those that exist within the conceptual world of the modeller and the models he uses, especially because of his nature to press ahead and get results. The use of performance assessment as a driving force in the disposal programmes thus needs a critical and constructive dialogue with the basic scientific disciplines.

A second and equally important process in identifying the most important questions and determining criteria for sufficient validation is a critical and constructive dialogue between the implementors and regulators.

It is the task of the regulator to review the implementors' assessments and to set conditions for his activities including constructing and operating a repository. By performing independent assessments the regulator can also, from his own perspective, identify and maybe sometimes demonstrate important questions, gaps of knowledge, and validation needs.

Who has an interest in validation? Modellers and experimentalists are obvious interest groups. I mentioned the dialogues between assessors and scientists, implementors and regulators. However, the "validation dialogue" must also involve the general public. After all, they are the final decision makers. To realise a repository, the implementor needs confidence in his solutions and the regulator needs confidence in his evaluation methods. Maybe we have here the real challenge for the 1990's.

Now, how do we proceed? One important step is to go site-specific. The problems to do much further progress in disposal programmes which are predominantly generic have been mentioned several times during this week. Suppose we have a new GEOVAL Symposium in 1993. Then three candidate sites have been chosen by SKB, if the present programme is followed. Certainly, this will have a large impact on the validation discussion in Sweden.

There might be reasons, political or others, to delay the time schedule. If this is done too much or too often, there is a risk of dilution of efforts and decreasing motivation of involved project people and scientists as Bob Bernero of the U.S. NRC, pointed out in his keynote speech in Paris last fall: Only when we become site-specific, we will see clearly which of the conservative assumptions we have made in generic studies, that are really pessimistic, and which ones were of a more hypothetical nature.

This symposium has reached its end point. However, the validation work will go on and international cooperation will continue within the NEA, CEC, and IAEA programmes as well as in projects like INTRAVAL.

This has been a very stimulating and I think also successful meeting. I thank all those that have contributed: the authors, our chairmen and rapporteurs, the panel and finally the members of the programme committee, especially Johan Andersson who made the heavy part of the work. I especially want to thank the NEA for good cooperation in organising the meeting. The meeting is closed. Thank you very much!

Shlomo P. Neuman

Allow me to use the microphone for one last statement as chairman of this panel. I would like to express our mutual thanks to the organisers who did an exceedingly good job in bringing together an extremely strong and stimulating group. I found this week to be very interesting and useful.

LIST OF PARTICIPANTS LISTE DES PARTICIPANTS

AUSTRALIA - AUTRICHE

DUERDEN, P., Australian Nuclear Science & Tech. Or, Bag 1, Post Office, MENAI NSW 2234

BELGIUM - BELGIQUE

PUT, M., SCK/CEN, Boeretang 200, 2400 MOL

CANADA

CHAN, T., Atomic Energy of Canada, Ltd., Whiteshell Nuclear Research Establishment, Pinawa, MANITOBA ROE 1LO

FLAVELLE, P., Atomic Energy Control Board, P.O. Box 1046, Station "B", OTTAWA, K1P 5S9

SARGENT, P., Atomic Energy of Canada Limited, Whiteshell Nuclear Research Establishment, Pinawa, MANITOBA ROE 1LO

SCHEIER, N., Atomic Energy of Canada Ltd., Whiteshell Nuclear Research Establishment, Pinawa MANITOBA, ROE 1LO

FEDERAL REPUBLIC OF GERMANY - REPUBLIQUE FEDERALE D'ALLEMAGNE

ARENS, G., GSF, Theodor-Heuss-Strasse 4, 3300 BRAUNSCHWEIG

BESENECKER, H., Niedersachsisches Umweltministerium, Ärdnivstrasse 2 3000 HANNOVER 1

BOGORINSKI, P., Gesellschaft fur Reaktorsicherheit, Schwertnergasse 1, 5000 KOLN 1

BRANDES, Hauptabt. Kerntechnik und Strahlenschutz, Postfach 810551, 3000 HANNOVER 81

FEIN, E., GSF, Theodor-Heuss-Strasse 4, 3300 BRAUNSCHWEIG

ILLI, H., Bundesamt fur Strahlenschutz, Postfach 100149, 3320 SALZGITTER 1

LIEDTKE, L., Federal Institute for Geosciences, Stilleweg 2, 3000 HANNOVER 51

MARTENS, K-H., Gesellschaft fur Reaktorsicherheit, Schwertnergasse 1, 5000 KOLN 1

NEUSS, M., Nieders Landesauit f. Bodenforidr., Stilleweg 2, 3000 HANNOVER 51

RINKLEFF, L., Hauptabt Kerntechnik und Strahlenschutz, Postfach 810551, 3000 HANNOVER 81

SCHELKES, K., Bundesanstalt fur Geowissenschaften und Rohstoffe, Stilleweg 2, 3000 HANNOVER

FINLAND - FINLANDE

AHONEN, L., Geological Survey of Finland, 021 50 ESPOO

HAUTOJÄRVI, A., VTT, Technical Research Centre of Finland, Box 169, 001 81 HELSINKI

HÖLTTÄ, P., University of Helsinki, Dep of Radiochemistry, Unioninkatu 35, 001 70 HELSINKI

PELTONEN, E., Teollisuuden Vioma Oy, Nuclear Waste Office, Fredrikinkatu 51-53 B, 00100 HELSINKI

SIITARI-KAUPPI, M., University of Helsinki, Dep of Radiochemistry, Unioninkatu 35, 001 70 HELSINKI

FRANCE

BRUN-YABA, C., Commissariat à l'Energie Atomique, BP 6 IPSN/DAS, 92265 FONTENAY-AUX-ROSES CEDEX

COULON-CASALOTTI, C.N., CEA, DEMT ISNTS Cen. Saclay, 911 91 GIF-SUR-YVETTE

DE MARSILY, G., Université PEM Curie, Lab. Geologie A liguée, 4 place Jussien, 75252 PARIS

DEWIERE, L., CEA/ANDRA/DESI, Route du Panorama Robert Schumann, BP 38, 92266 FONTENAY AUX ROSES

GOBLET, P., Centre d'Informatique Geologique, 35 rue Saint Honoré, 77305 FONTAINEBLEAU

JACQUIER, P., Commissariat à l'Energie Atomique, IPSN/DAS/SAED B.P.N 6, 92265 FONTENAY-AUX-ROSES CEDEX

JAMET, P., Ecole des Mines de Paris CIG, 35, rue St Honoré, 77305 FONTAINEBLEAU CEDEX

JAUZEIN, M., French Atomic Energy Commision, CEA/CENG-SAR/SAT, B.P. 85 X, 38041 GRENOBLE CEDEX

LEWI, J., Commissariat à l'Energie Atomique, IPSN/DAS/SAED CEN/FAR B.P.N 6 92265 FONTENAY-AUX-ROSES CEDEX

MONAVON, A., CEA, CEN-SACLAY DEMT/SMTS/TTMF, 91191 GIF-SUR-YVETTE CEDEX

PRIEM, T., Commissariat à l'Energie Atomique, IPSN/DAS/SAED CEN/FAR B.P.N 6 92265 FONTENAY-AUX-ROSES CEDEX

RAIMBAULT, P., Andra, Route du Panorama Robert Schuman, B.P.N 38, 92266 FONTENAY-AUX-ROSES CEDEX

SARDIN, M., CNRS-LSGC, Ensic, 1 Rue Grandville, BP 451, 54001 NANCY CEDEX

SAUTY, J-P., BRGM-4S, BP 6009-45060, ORLEANS CEDEX 02

SCHWEICH, D., CNRS-LSGC, Ensic, 1 Rue Grandville, BP 451, 54001 NANCY CEDEX

TEVISSEN, E., CEA/ENG., BP 85X, 38041 GRENOBLE CEDEX

JAPAN - JAPON

KIMURA, H., Jaeri, Tokai-mura, Naka-gun, IBARAKI-KON 319-11

KOBAYASHI, A., Hazama corporation, 4-17-23, Yonohonmachi-nishi, YONO-SHI, SAITAMA 338

TANAKA, Y., CRIEPI, 1646 Abiko, Abiko-Shi, CHIBA-KEN 270-11

YOSHIDA, H., PNC 959-31 Sonodo, Jyorinji, 509-51 TOKISHI, GIFUKEN

KOREA

HAHN, P., Korea Atomic Energy Research Institute, P.O. Box 7, Daeduk-Danji, TAEJEON

KIM, C-S., Korea Atomic Energy Research Institute, P.O. Box 7, Daeduk-Danji, TAEJEON

NETHERLANDS - PAYS BAS

GLASBERGEN P., RIVM, Box 1, 3720 BA BILTHOVEN

HASSANIZADEH, S.M., RIVM, Box 1, 3720 BA BILTHOVEN

TAAT, J., Delft Geotechnics, Postbox 69, 2600 AB DELFT

NORWAY - NORVEGE

CHRYSSANTAKIS, P., Norwegian Geotechnical Institute, P.O. Box 40 Tåsen, 0801 OSLO 8

SPAIN - ESPAGNE

BAJOS, C., Enresa C/ Emilio Vargas 7, 28043 MADRID

DEL OLMO, C., Enresa C/ Emilio Vargas 7, 28043 MADRID

FERNANDEZ GIANOTTI, J., ITGE, C/Rios Rosas 23, 28003 MADRID

GARCIA CORTES, A., Instituto Tecnologico Geominero De Es, Rios Rosas 23, 28003 MADRID

GRIMA OLMEDO, J., ITGE, Rios Rosas 23, 28003 MADRID

MAYOR, J.C., Enresa C/ Emilio Vargas 7, 28043 MADRID

SWEDEN - SUEDE

AHLBOM, K., Conterra AB, Box 439, 751 06 UPPSALA

AHLSTRÖM, P-E., SKB, Box 5864, 102 48 STOCKHOLM

ALMEN, K-E., SKB, BOX 5864, 102 48 STOCKHOLM

ANDERSSON, K., Swedish Nuclear Power Inspectorate, Box 27106, 102 52 STOCKHOLM

ANDERSSON, J., Swedish Nuclear Power Inspectorate, Box 27106, 102 52 STOCKHOLM

ANDERSSON, P., SGAB, P.O. Box 1424, 751 44 UPPSALA

ANDERSSON, K., Lindgren & Andersson HB, Åsbacken 28, 445 00 SURETE

BENTLEY, B-M., SKB, Box 5864, 102 48 STOCKHOLM

BIRGERSSON, L., CHEMFLOW AB, C/O Kemakta Cons. Co, Pipersgatan 27, 112 28 STOCKHOLM

BJELKAS, J., Kemakta Konsult AB, Pipersgatan 27, 112 28 STOCKHOLM

BOGHAMMAR, A., Kemakta Konsult AB, Pipersgatan 27, 112 28 STOCKHOLM

BRANDBERG, F., Kemakta Konsult AB, Pipersgatan 27, 112 28 STOCKHOLM

BROTZEN, O., FBAB Yngvevägen 13, 182 64 DJURSHOLM

BRUNO, J., Kungliga Tekniska Högskolan, Oorganisk Kemi, Valhallavägen 79,
100 44 STOCKHOLM

BÄCKBLOM, G., SKB, Box 5864, 102 48 STOCKHOLM

CARLSSON, T., Swedish Nuclear Power Inspectorate, Box 27106, 102 52 STOCKHOLM

CRONHJORT, B., National Board for Spent Nuclear Fuel, Sehlstedtsgatan 9,
115 28 STOCKHOLM

CVETKOVIC, V., Kungliga Tekniska Högskolan, Vattenvårdsteknik, 100 44 STOCKHOLM

DE CAMPOS PEREIRA, A., University of Stockholm, Department of Physics,
Vanadisvägen 9, 113 46 STOCKHOLM

DVERSTORP, B., The Royal Institute of Technology, Water Resources Engineering
100 44 STOCKHOLM

ELERT, M., Kemakta Konsult AB, Pipersgatan 27, 112 28 STOCKHOLM

EMREN, A., Kärnkemi, CTH, 412 96 GÖTEBORG

ENG, T., SKB, Box 5864, 102 48 STOCKHOLM

ERIKSSON, E., University of Uppsala, v. Ågatan 24, 752 20 UPPSALA

FOLLIN, S., Kungliga Tekniska Högskolan, Mark-och Vattenresurser,
100 44 STOCKHOLM

FORSBECK, I., Swedish Nuclear Power Inspectorate, Box 27106, 102 52 STOCKHOLM

GRENTHE, I., KTH, 100 44 STOCKHOLM

GRUNDFELT, B., Kemakta Konsult AB, Pipersgatan 27, 112 28 STOCKHOLM

GUSTAFSON, G., CTH/ VIAK, Möndalsvägen 85, 412 85 GOTEBORG

GUSTAVSSON, E., SGAB P.O. Box 1424, 751 44 UPPSALA

HÖGBERG, L., Swedish Nuclear Power Inspectorate, Box 27106, 102 52 STOCKHOLM

HÖGLUND, L-O., Kemakta Konsult AB, Pipersgatan 27, 112 28 STOCKHOLM

ISMAIL, M.S., University of Stockholm, Department of Physics, 113 46 STOCKHOLM

JOHANSSON, G., National Institute of Radiation Protection, Box 602 04,
104 01 STOCKHOLM

JOHANSSON, C., SKI, Box 27106, 102 52 STOCKHOLM

KARLSSON, F., SKB, Box 5864, 102 48 STOCKHOLM

KAUTSKY, F., SKI, Box 27106, 102 52 STOCKHOLM

KUNG, KTH, 100 44 STOCKHOLM

LAAKSCHARJU, M., KTH, OOK, 100 44 STOCKHOLM

LARSSON, A., Kemakta Konsult AB, Pipersgatan 27, 112 28 STOCKHOLM

BLINDBERG, H., Kemakta Konsult AB, Pipersgatan 27, 112 28 STOCKHOLM

LINDBOM, B., Kemakta Konsult AB, Pipersgatan 27, 112 28 STOCKHOLM

LINDGREN, M., Kemakta Konsult AB, Pipersgatan 27, 112 28 STOCKHOLM

LUNDEN, I., Chalmers University fo Technology, Department of Nuclear Chemistry, 416 69 GOTEBORG

MARKSTRÖM, A., Kemakta Konsult AB, Pipersgatan 27, 112 28 STOCKHOLM

MORENO, L., Royal Institute of Technology, Department of Chemical Engeneering, 100 44 STOCKHOLM

NERETNIEKS, I., Royal Institute of Technology, Department of Chemical Engineering, 100 44 STOCKHOLM

NILSON, L., Royal Istitute of Technology, Department of Chemical Engineering KTH, 100 44 STOCKHOLM

NILSSON, L.B., SKB, Box 5864, 102 48 STOCKHOLM

NORDQUIST, R., SGAB, P.O. Box 1424, 751 44 UPPSALA

NORDQUIST, W., The Royal Institute of Technology, 100 44 STOCKHOLM

NORRBY, S., Swedish Nuclear Power Inspectorate, Box 27106, 102 52 STOCKHOLM

PAPP, T., SKB, Box 5864, 102 48 STOCKHOLM

PERS NEDBAL, K., Kemakta Konsult AB, Pipersgatan 27, 112 28 STOCKHOLM

RASMUSON, A., Kemakta Konsult AB Pipersgatan 27, 112 28 STOCKHOLM

ROMERO, L., Royal Institute of Technology, 100 44 STOCKHOLM

RUTQUIST, J., Luleå University, Division of Rock Mechanics, 951 87 LULEÅ

RYDELL, N., SKN, Sehlstedtsgatan 9, 115 28 STOCKHOLM

SCHERMAN, G., SKN Sehlstedtsgatan 9, 102 52 STOCKHOLM

SELLIN, P., SKB, Box 5864, 102 48 STOCKHOLM

SHEN, B., Luleå University of Technology, Division of Rock Mechanics, 951 00 LULEÅ

SKAGIUS, K., Kemakt Konsult AB, Pipersgatan 27, 112 28 STOCKHOLM

STANFORS, R., Roy Stanfors Consulting AB, Ole Römmersväg 12, 223 70 LUND

STEPHANSON, O., Luleå University of Technology, Division of Rock Mechanics, 951 00 LULEÅ

STRÖM, A., SKB, Box 5864, 102 48 STOCKHOLM

SUNDSTRÖM, B., Swedish Nuclear Power Inspectorate, Box 27106, 102 52 STOCKHOLM

THORSON, E., Swedish Nuclear Fuel & Waste Management Co., Box 5864, 102 48 STOCKHOLM

WERME, L., SKB, Box 5864, 102 48 STOCKHOLM

WIBORGH, M., Kemakta Konsult AB, Pipersgatan 27, 112 28 STOCKHOLM

WIKBERG, P., SKB, Box 5864, 102 48 STOCKHOLM

WINGEFORS, S., Swedish Nuclear Power Inspectorate, Box 27106, 102 52 STOCKHOLM

ÅHAGEN, H., SINTAB, Valhallavägen 78, 114 27 STOCKHOLM

SWITZERLAND - SUISSE

GILBY, D., Colenco AG, Parkstrasse 27, 5401 BADEN

HADERMANN, J., Paul Scherrer Institute, 5232 VILLIGEN PSI

KUHLMANN, U., Laboratory of Hydraulics, Vaw-Dol, ETH-Zentrum, 8092 ZURICH

MCCOMBIE, C., Nagra, Parkstrasse 23, 5401 BADEN

NIEDERER, U., Swiss Nuclear Safety Inspectorate, 5303 WUERENLINGEN

SMITH, P., Paul Scherrer Institut, 5233 VILLIGEN PSI

ZUIDEMA, P., Nagra, Parkstrasse 23, 5401 BADEN

TAIWAN

HUANG, C.T., Energy Research Laboratory, No 195, Sec 4, Chung Hsing Road, Chuntung, HSIN-CHU.

LIU, S-J., Institute of Nuclear Energy Research, Box 3-7, LUNGTAN.

SHIH, C-F., Institute of Nuclear Research, Box 3-7, LUNGTAN.

SOONG, K-L., Institute of Nuclear Energy Research, Box 3-7, LUNGTAN.

UNITED KINGDOM - ROYAUME-UNI

BOURKE, P.J., Harwell Laboratory, Waverly, Picklershill, ABINGDON, OX14 2BA

BROYD, T., WS Atkins Engineering Sciences Ltd, Woodcote Grove, Epsom, SURREY, KT18 5BW

BUTTER, K., HMIP, A 521, Romney House, 43 Marsham Street, LONDON SW1 3PY

CHAPMAN, N., Intera-ECL, Park View House, 14B Burton Street, Melton Mowbray, LEICESTERSHRE, LE13 1AE

GRINDROD, P., Intera-ECL, Chiltern House, 45 Station Road, HENLEY-ON-THAMES, OXFORDSHIRE

HODGKINSON, D., Intera-ECL, Chiltern House, 45 Station Road, HENLEY-ON-THAMES, OXFORDSHIRE

JACKSON, C., AEA Technology, Theoretical Studies Department, Harwell Laboratories, OXFORDSHIRE, OX11 ORA

KINGDON, R.D., Harwell Laboratory, Theoretical Physics Division B424.4 Harwell Laboratory, Didcot, OXFORDSHIRE, OX11 ORA

LINEHAM, T., AEA, B10-30, Harwell Laboratory, Didcot, OXFORDSHIRE, OX11 ORA

SHARLAND, S., AEA Technology, Harwell Laboratory, Didcot, OXFORDSHIRE, OX11 ORA

TEASDALE, I., Electrowatt Engineering Service (UK), Grand Ford House, 16 Carfax, Horsham, WEST SUSSEX, RH 12 IUP

UNITED STATES - ETATS-UNIS

ALEXANDER, D.H., Department of Energy, 1000 Independence Avenue, WASHINGTON DC

BEAUHEIM, R., Sandia National Laboratories, Division 63, Box 5800, ALBUQUERQUE, NEW MEXICO

DAVIES, P., Sandia National Laboratories, Division 63, Box 5800, ALBUQUERQUE, NEW MEXICO

DAVIS, P., Sandia National Laboratories, Box 5800, ALBUQUERQUE, NEW MEXICO 89185

EWING, R., University of New Mexico, Department of Geology, ALBUQUERQUE, NEW MEXICO 89131

FAIRHURST, C., National Academy of Sciences, University of Minnesota, 122 CME Building, 500 Pillsburg Drive, MINNEAPOLIS, MN 56455

GLASS, R., Sandia National Laboratories, Division 6315, P.O. Box 5800,
ALBUQUERQUE, MN 87185

HOXIE, D.T., US Geological Survey, Box 25046, MS 425, Denver Federal Center,
DENVER, Colorado 80225

HUNG, C., US Environment Protection Agency, ANR 460, 401 M Street, SW
WASHINGTON DC

KULATILAKE, P., University of Arizona, Department of Mining & Geological Eugg.
ARIZONA

LA VENUE, M., Intera, 8100 Mountain Road NE 204 D, ALBUQUERQUE NM 87110

LEHMAN, L., L. Lehman & Associates, 1103 W. Burnsville Parkway-Suite 209,
BURNSVILLE, MN 55337

LUIS, S., M.I.T., 10 Cottage Avenue, ARLINGTON MA 02174

MUNSON, D., Sandia National Laboratories, Division 6346, P.O. Box 5800,
ALBUQUERQUE, NM 87185

NEUMAN, S., University of Arizona, Department of Hydrology and Water-resources,
TUCSON AZ

NICHOLSON, T., U.S. Nuclear Regulatory Commission, Office of Research, Mail
Stop NL/S-260, WASHINGTON DC 20555

PRUESS, K., Lawrence Berkeley Laboratories, Mail Stop 50E, BERKELEY, CA 94720

SARGENT, R.G., Syracuse University, 439 Link Hall, SYRACUSE, NY 13244

TSANG, C-F., Lawrence Berkeley Laboratory, Earth Sciences Division,
1 Cyclotron Road, BERKELEY, CA 94720

USSR - URSS

PANKRATOV, S.G., Nuclear Safety Institute of the Academy of Sciences, MOSCOW

YUGOSLAVIA - YUGOSLAVIE

LUKACS, E., Republic Administration for Nuclear Safety, Kardeljeva Plascad 24,
61113 LJUBLJANA

COMMISSION OF THE EUROPEAN COMMUNITIES
COMMISSION DES COMMUNAUTES EUROPEENNES

COME, B., CEC, Rue de la Loi 200, 1049 BRUSSELS

INTERNATIONAL ATOMIC ENERGY AGENCY
AGENCE INTERNATIONALE POUR L'ENERGIE ATOMIQUE

HOSSAIN, S., IAEA, Box 100, 1400 VIENNA

OECD NUCLEAR ENERGY AGENCY
AGENCE DE L'OCDE POUR L'ENERGIE NUCLEAIRE

OLIVIER, J-P., OECD/NEA, 38 Boulevard Suchet, 75016 Paris

THEGERSTRöM, C., OECD/NEA, 38 Boulevard Suchet, 75016 Paris

WHERE TO OBTAIN OECD PUBLICATIONS – OÙ OBTENIR LES PUBLICATIONS DE L'OCDE

Argentina – Argentine
Carlos Hirsch S.R.L.
Galería Güemes, Florida 165, 4° Piso
1333 Buenos Aires Tel. 30.7122, 331.1787 y 331.2391
Telegram: Hirsch–Baires
Telex: 21112 UAPE–AR. Ref. s/2901
Telefax:(1)331–1787

Australia – Australie
D.A. Book (Aust.) Pty. Ltd.
648 Whitehorse Road, P.O.B 163
Mitcham, Victoria 3132 Tel. (03)873.4411
Telex: AA37911 DA BOOK
Telefax: (03)873.5679

Austria – Autriche
OECD Publications and Information Centre
Schedestrasse 7
5300 Bonn 1 (Germany) Tel. (0228)21.60.45
Telefax: (0228)26.11.04

Gerold & Co.
Graben 31
Wien I Tel. (0222)533.50.14

Belgium – Belgique
Jean De Lannoy
Avenue du Roi 202
B–1060 Bruxelles Tel. (02)538.51.69/538.08.41
Telex: 63220 Telefax: (02) 538.08.41

Canada
Renouf Publishing Company Ltd.
1294 Algoma Road
Ottawa, ON K1B 3W8 Tel. (613)741.4333
Telex: 053–4783 Telefax: (613)741.5439
Stores:
61 Sparks Street
Ottawa, ON K1P 5R1 Tel. (613)238.8985
211 Yonge Street
Toronto, ON M5B 1M4 Tel. (416)363.3171

Federal Publications
165 University Avenue
Toronto, ON M5H 3B8 Tel. (416)581.1552
Telefax: (416)581.1743

Les Publications Fédérales
1185 rue de l'Université
Montréal, PQ H3B 3A7 Tel.(514)954–1633

Les Éditions La Liberté Inc.
3020 Chemin Sainte–Foy
Sainte–Foy, PQ G1X 3V6 Tel. (418)658.3763
Telefax: (418)658.3763

Denmark – Danemark
Munksgaard Export and Subscription Service
35, Norre Sogade, P.O. Box 2148
DK–1016 Kobenhavn K Tel. (45 33)12.85.70
Telex: 19431 MUNKS DK Telefax: (45 33)12.93.87

Finland – Finlande
Akateeminen Kirjakauppa
Keskuskatu 1, P.O. Box 128
00100 Helsinki Tel. (358 0)12141
Telex: 125080 Telefax: (358 0)121.4441

France
OECD/OCDE
Mail Orders/Commandes par correspondance:
2 rue André–Pascal
75775 Paris Cedex 16 Tel. (1)45.24.82.00
Bookshop/Librairie:
33, rue Octave–Feuillet
75016 Paris Tel. (1)45.24.81.67
(1)45.24.81.81
Telex: 620 160 OCDE
Telefax: (33–1)45.24.85.00

Librairie de l'Université
12a, rue Nazareth
13090 Aix–en–Provence Tel. 42.26.18.08

Germany – Allemagne
OECD Publications and Information Centre
Schedestrasse 7
5300 Bonn 1 Tel. (0228)21.60.45
Telefax: (0228)26.11.04

Greece – Grèce
Librairie Kauffmann
28 rue du Stade
105 64 Athens Tel. 322.21.60
Telex: 218187 LIKA Gr

Hong Kong
Swindon Book Co. Ltd.
13 – 15 Lock Road
Kowloon, Hongkong Tel. 366 80 31
Telex: 50 441 SWIN HX
Telefax: 739 49 75

Iceland – Islande
Mál Mog Menning
Laugavegi 18, Pósthólf 392
121 Reykjavik Tel. 15199/24240

India – Inde
Oxford Book and Stationery Co.
Scindia House
New Delhi 110001 Tel. 331.5896/5308
Telex: 31 61990 AM IN
Telefax: (11)332.5993
17 Park Street
Calcutta 700016 Tel. 240832

Indonesia – Indonésie
Pdii–Lipi
P.O. Box 269/JKSMG/88
Jakarta 12790 Tel. 583467
Telex: 62 875

Ireland – Irlande
TDC Publishers – Library Suppliers
12 North Frederick Street
Dublin 1 Tel. 744835/749677
Telex: 33530 TDCP EI Telefax : 748416

Italy – Italie
Libreria Commissionaria Sansoni
Via Benedetto Fortini, 120/10
Casella Post. 552
50125 Firenze Tel. (055)645415
Telex: 570466 Telefax: (39.55)641257
Via Bartolini 29
20155 Milano Tel. 365083
La diffusione delle pubblicazioni OCSE viene assicurata dalle principali librerie ed anche da:
Editrice e Libreria Herder
Piazza Montecitorio 120
00186 Roma Tel. 679.4628
Telex: NATEL I 621427

Libreria Hoepli
Via Hoepli 5
20121 Milano Tel. 865446
Telex: 31.33.95 Telefax: (39.2)805.2886

Libreria Scientifica
Dott. Lucio de Biasio "Aeiou"
Via Meravigli 16
20123 Milano Tel. 807679
Telefax: 800175

Japan– Japon
OECD Publications and Information Centre
Landic Akasaka Building
2–3–4 Akasaka, Minato–ku
Tokyo 107 Tel. (81.3)3586.2016
Telefax: (81.3)3584.7929

Korea – Corée
Kyobo Book Centre Co. Ltd.
P.O. Box 1658, Kwang Hwa Moon
Seoul Tel. (REP)730.78.91
Telefax: 735.0030

Malaysia/Singapore – Malaisie/Singapour
Co–operative Bookshop Ltd.
University of Malaya
P.O. Box 1127, Jalan Pantai Baru
59700 Kuala Lumpur
Malaysia Tel. 756.5000/756.5425
Telefax: 757.3661

Information Publications Pte. Ltd.
Pei–Fu Industrial Building
24 New Industrial Road No. 02–06
Singapore 1953 Tel. 283.1786/283.1798
Telefax: 284.8875

Netherlands – Pays–Bas
SDU Uitgeverij
Christoffel Plantijnstraat 2
Postbus 20014
2500 EA's–Gravenhage Tel. (070 3)78.99.11
Voor bestellingen: Tel. (070 3)78.98.80
Telex: 32486 stdru Telefax: (070 3)47.63.51

New Zealand – Nouvelle–Zélande
Government Printing Office
Customer Services
33 The Esplanade – P.O. Box 38–900
Petone, Wellington
Tel. (04) 685–555 Telefax: (04)685–333

Norway – Norvège
Narvesen Info Center – NIC
Bertrand Narvesens vei 2
P.O. Box 6125 Etterstad
0602 Oslo 6 Tel. (02)57.33.00
Telex: 79668 NIC N Telefax: (02)68.19.01

Pakistan
Mirza Book Agency
65 Shahrah Quaid–E–Azam
Lahore 3 Tel. 66839
Telex: 44886 UBL PK. Attn: MIRZA BK

Portugal
Livraria Portugal
Rua do Carmo 70–74
Apart. 2681
1117 Lisboa Codex Tel. 347.49.82/3/4/5
Telefax: 37 02 64

Singapore/Malaysia – Singapour/Malaisie
See "Malaysia/Singapore – "Voir "Malaisie/Singapour"

Spain – Espagne
Mundi–Prensa Libros S.A.
Castelló 37, Apartado 1223
Madrid 28001 Tel. (91) 431.33.99
Telex: 49370 MPLI Telefax: 575 39 98
Libreria Internacional AEDOS
Consejo de Ciento 391
08009 –Barcelona Tel. (93) 301–86–15
Telefax: (93) 317–01–41

Sweden – Suède
Fritzes Fackboksföretaget
Box 16356, S 103 27 STH
Regeringsgatan 12
DS Stockholm Tel. (08)23.89.00
Telex: 12387 Telefax: (08)20.50.21
Subscription Agency/Abonnements:
Wennergren–Williams AB
Nordenflychtsvagen 74
Box 30004
104 25 Stockholm Tel. (08)13.67.00
Telex: 19937 Telefax: (08)618.62.36

Switzerland – Suisse
OECD Publications and Information Centre
Schedestrasse 7
5300 Bonn 1 (Germany) Tel. (0228)21.60.45
Telefax: (0228)26.11.04

Librairie Payot
6 rue Grenus
1211 Genève 11 Tel. (022)731.89.50
Telex: 28356
Subscription Agency – Service des Abonnements
4 place Pépinet – BP 3312
1002 Lausanne Tel. (021)341.33.31
Telefax: (021)341.33.45

Maditec S.A.
Ch. des Palettes 4
1020 Renens/Lausanne Tel. (021)635.08.65
Telefax: (021)635.07.80
United Nations Bookshop/Librairie des Nations–Unies
Palais des Nations
1211 Genève 10 Tel. (022)734.60.11 (ext. 48.72)
Telex: 289696 (Attn: Sales)
Telefax: (022)733.98.79

Taiwan – Formose
Good Faith Worldwide Int'l. Co. Ltd.
9th Floor, No. 118, Sec. 2
Chung Hsiao E. Road
Taipei Tel. 391.7396/391.7397
Telefax: (02) 394.9176

Thailand – Thaïlande
Suksit Siam Co. Ltd.
1715 Rama IV Road, Samyan
Bangkok 5 Tel. 251.1630

Turkey – Turquie
Kültur Yayinlari Is–Türk Ltd. Sti.
Atatürk Bulvari No. 191/Kat. 21
Kavaklidere/Ankara Tel. 25.07.60
Dolmabahce Cad. No. 29
Besiktas/Istanbul Tel. 160.71.88
Telex: 43482B

United Kingdom – Royaume–Uni
HMSO
Gen. enquiries Tel. (071) 873 0011
Postal orders only:
P.O. Box 276, London SW8 5DT
Personal Callers HMSO Bookshop
49 High Holborn, London WC1V 6HB
Telex: 297138 Telefax: 071 873 8463
Branches at: Belfast, Birmingham, Bristol, Edinburgh, Manchester

United States – États–Unis
OECD Publications and Information Centre
2001 L Street N.W., Suite 700
Washington, D.C. 20036–4095 Tel. (202)785.6323
Telefax: (202)785.0350

Venezuela
Libreria del Este
Avda F. Miranda 52, Aptdo. 60337
Edificio Galipán
Caracas 106 Tel. 951.1705/951.2307/951.1297
Telegram: Libreste Caracas

Yugoslavia – Yougoslavie
Jugoslovenska Knjiga
Knez Mihajlova 2, P.O. Box 36
Beograd Tel. (011)621.992
Telex: 12466 jk bgd Telefax: (011)625.970

Orders and inquiries from countries where Distributors have not yet been appointed should be sent to: OECD Publications Service, 2 rue André–Pascal, 75775 Paris Cedex 16, France.
Les commandes provenant de pays où l'OCDE n'a pas encore désigné de distributeur devraient être adressées à : OCDE, Service des Publications, 2, rue André–Pascal, 75775 Paris Cedex 16, France.

12/90

OECD PUBLICATIONS, 2 rue André-Pascal, 75775 PARIS CEDEX 16
PRINTED IN FRANCE
(66 91 01 3) ISBN 92-64-03343-2 - No. 45424 1990